Evaluation of human work:

A practical ergonomics methodology

Evaluation of human work

A practical ergonomics methodology

Edited by
John R. Wilson and E. Nigel Corlett
University of Nottingham

Taylor & Francis
London   New York   Philadelphia

UK   Taylor & Francis Ltd, 4 John St., London WC1N 2ET

USA   Taylor & Francis Inc., 1900 Frost Road, Suite 101, Bristol, PA 19007

British Library Cataloguing in Publication Data

Evaluation of human work.
  1. Ergonomics
  I. Wilson, John, *1951–*
620.8′2

    ISBN 0-85066-480-2 Pbk
           0-85066-479-9 Hbk

Library of Congress Cataloging in Publication Data is available

Cover design by Jordan & Jordan, Fareham, Hants

Typeset by Photo·Graphics, Honiton, Devon
Printed in Great Britain by Taylor & Francis (Printers) Ltd, Basingstoke, Hants.

Contents

NOTE
A classification of ergonomics methodology containing advice on which chapters of the book contain information on particular methods and techniques appears as Table 1.3, pp. 16–21.

Preface

For a long time there existed few books on ergonomics or human factors methodology; Chapanis' *Research Techniques in Human Engineering*, published in 1959, was probably the earliest, as well as the best-known. Lately there has been a slow increase in what is available. For instance one of the contributors to this volume, David Meister, has produced two books dealing with methods (Meister, 1985, 1986) and the present editors have also been involved in two collections of conference proceedings concerned with new methods and techniques (Laboratory of Industrial and Human Automatics, 1987; Wilson *et al.*, 1987).

The books by Meister, excellent in many respects, concentrate upon investigations of large-scale (military) systems design, simulation and evaluation. The two sets of conference proceedings, whilst containing a range of methodological developments and applications, represent what was selected from the papers submitted for presentation, and cannot pretend fully to represent the field. Also produced recently is the authoritative *Handbook of Human Factors*, edited by Salvendy (1987). This does have much to say about methods and techniques, both as separate chapters or as parts of other chapters; nonetheless its intention is to be a comprehensive, general text, with explanation of theories, principles, data and application, as well as of methods.

Our aim with this volume on ergonomics methodology is to produce a text on methods and techniques that is both broad and deep. We intend it to be a companion to the major general textbooks on ergonomics and human factors, particularly and most recently those of Bailey (1982), Grandjean (1988), Kantowitz and Sorkin (1983), Oborne (1987), Salvendy (1987) and Sanders and McCormick (1987). All of these are well known to students, teachers and practitioners of ergonomics, as well as to many of those from other disciplines who take a personal or professional interest in ergonomics. There is, though, little opportunity in such texts to emphasize and make explicit the major part of methodology.

Therefore we have set out to produce a general text on ergonomics methodology. As the book's title implies we are primarily concerned with people at work and with applied rather than basic research. However, the former concern has not ruled out contributions relevant to people's activities at home, leisure or on the road; nor does the latter concern invalidate descriptions of laboratory-based methods — these can have outcomes that are as practically applicable as are those from field investigations.

The contents of the book are intended to be interesting and useful for a wide range of people, including: *students*, to give them a feel for ergonomics investigation and to complement their learning of theory and principles; *industrial and business personnel at all levels*, to allow them to understand better what ergonomics can do for them, why, and how; and *ergonomics practitioners, researchers and teachers*, to give them a compendium of methods and techniques available. For all these groups the contributions here will also point to further sources for more detail on specific topics.

Our text on evaluating human work has brought together experts from many branches of ergonomics theory and practice, and has allowed them the space to introduce and give detail on those methods and techniques of value to them. Since ergonomics is both a science and a technology, these methods can of course be concerned with collecting data or with applying their own or others' data. The primary thrust of each contribution may be general method (e.g. direct observation or protocol analysis), or particular fields of application for several types of method (e.g. mental workload or the climatic environment). Whilst there will no doubt be omissions — of branches of methodology or of techniques within one area — regretted by some readers, we trust that most will find the book to be a comprehensive, readable and useful source of ergonomics knowledge and practice. Certainly we believe that for those students or readers from industry who are relatively new to ergonomics, one of the most interesting and valuable ways to learn about it is through its rich and varied methodology.

July 1989

John Wilson and Nigel Corlett
University of Nottingham

References

Bailey, R.W. (1982). *Human Performance Engineering: A Guide for System Designers*. (London: Prentice-Hall), pp. 656 + xxviii.

Chapanis, A. (1959). *Research Techniques in Human Engineering*. (Baltimore: John Hopkins Press), pp. 316 + xii.

Grandjean, E. (1988). *Fitting the Task to the Man: A Textbook of Occupational Ergonomics*, 4th Edition. (London: Taylor and Francis), pp. 363 + ix.

Kantowitz, B.H. and Sorkin, R.D. (1983). *Human Factors: Understanding People-System Relationships*. (New York: John Wiley & Sons), pp. 699 + xii.

Laboratory of Industrial and Human Automatics, 1987, *New techniques and ergonomics*: Proceedings of an International Research Symposium. (Paris: Hermes).

Meister, D. (1985). *Behavioural Analysis and Measurement Methods*. (Chichester: John Wiley & Sons), pp. 509 + ix.

Meister, D. (1986). *Human Factors Testing and Evaluation*. (Amsterdam: Elsevier Science), pp. 424 + xi.

Oborne, D.J. (1987). *Ergonomics at Work*, 2nd Edition. (Chichester: John Wiley & Sons), pp. 386 + xvii.

Salvendy, G. (editor) (1987). *Handbook of Human Factors*. (New York: John Wiley & Sons), pp. 1874 + xxiv.

Sanders, M.S. and McCormick, E.J. (1987). *Human Factors in Engineering and Design*, 6th Edition. (New York: McGraw-Hill), pp. 664 + viii.

Wilson J.R., Corlett, E.N. and Manenica, I. (1987). *New Methods in Applied Ergonomics*. (London: Taylor & Francis), pp. 283 + x.

Acknowledgements

Our first debt with this book is to our contributing authors, all of whom have responded to our various requests with great patience, and have produced chapters of high quality within, in some cases, a very limited time. Amongst these authors we must mention those who were with us in the initial discussions about the book at the 2nd International Occupational Ergonomics Symposium at Zadar, Yugoslavia; they were Lisanne Bainbridge, Colin Drury, Ted Megaw, Ken Parsons, Pat Shipley and Rob Stammers. Colin Drury in particular has contributed much in terms of individual chapters and the overall content and style of the book.

We would like to thank our colleagues at Nottingham University for contributing to a working environment in which we feel able to embark on and complete this and other publishing ventures. One of us (JW) must also thank the Department of Industrial Engineering and Operations Research, University of California, Berkeley, for allowing him time and facilities to work on this book during periods there as a visitor in 1987 and 1988.

Our editors at Taylor and Francis — David Grist, Sarah Waddell and, for most of the time, Robin Mellors — have been exceedingly supportive, even in the face of a project which seemed to grow exponentially! The style of the book has been enhanced tremendously by the artwork of Tony Aston and cartoons of Moira Tracy. Despite both editors being away from Nottingham for substantial periods of time the production of the book has rolled on relatively smoothly; our colleagues would say this was because we left this and much else in the hands of our excellent secretaries, Lynne Mills and Ilse Browne, to whom we are immensely grateful.

Chapter 1

A framework and a context for ergonomics methodology

John R. Wilson

"Methods? Pah! Its all common sense"

Introduction

When a discipline gains a sufficient level of maturity it moves on from debates about definition—remit, boundaries, *raison d'être*, even its name—to considerations of operation. Thus ergonomics (and human factors, regarded as synonymous in this chapter and book), after many years of discussion of its nomenclature, direction and so on, has matured to the extent where the methods we use are more the focus of attention. In a way the early debates have been carried on into methodology. Where methods involve data collection concerning the body's structure, functioning and behaviour they are largely derived from the fields of anatomy, physiology and psychology. Methods that are particularly the province of ergonomics are often seen as those concerned with the application of these primary human data. Ergonomics methodology is also required in the evaluation of the systems we design, consequently enlarging our performance and behaviour data base through subsequent assessment of human–machine systems.

The original source of our methods though, is something of a red herring. Ergonomics is both a science and a technology and thus has need of techniques for both data collection (basic or functional) and application. The debts we owe to other disciplines are obvious: as Singleton (1982, p. 9) puts it, the two integral parts of ergonomics are an interdisciplinary research activity based upon anatomy, physiology and psychology and an operational activity which 'usually find expression through one of the two established technologies of medicine and engineering'. However, as we gain experience and confidence and as our armoury of knowledge and methodology grows, so the debt is being repaid. Methods developed or adapted within ergonomics will be employed in turn by other disciplines.

Bearing in mind the applied nature of ergonomics, if we try to draw parallels with three basic processes of design—analysis, synthesis and evaluation—which iterate throughout the design process (e.g. Markus, 1969), then we might divide ergonomics methodology into:

1. Methods of data collection concerning people; this is the *analysis* of our subject of interest—the human race. This produces the scientific base of ergonomics and, as pointed out above, methods here often have been taken directly (or adapted somewhat) from other disciplines.

2. Application of data to design; here we must *synthesize* data into ergonomically sound design concepts, prototypes and final design outcomes. Methods then concern the translation of basic data about people into criteria and other information useful for the particular design task, and also concern the process of such design and development.

3. Subsequent *evaluation* of designs; here we evaluate a system design—and consequently in a way also evaluate how well we have applied ergonomics to design. Evaluation implies measurement first; we must both assess the extent of something (measurement) and also put it into context (evaluation).

Definitions of ergonomics

Different definitions of ergonomics exist but the differences are more to do with where one might draw the boundary of what is ergonomics than with fundamental disagreements on approach. A wider view is that ergonomics is the 'study of human abilities and characteristics which affect the design of equipment, systems and jobs . . . and its aims are to improve efficiency, safety and . . . well-being.' (Clark and Corlett, 1984, p. 2); a narrower view is perhaps presented by Wickens (1984, p. 3), that human factors is to do with designing machines that accommodate the limits of the user. Pithy definitions of ergonomics are that it concerns designing for human use, or that it is the approach which fits systems (or machines or jobs or processes) *to* people and not vice versa. More detailed definitions exist; one which, whilst lengthy, captures the essence well was suggested as part of an exercise to provide definitions for constituent parts of the discipline: 'that branch of science and technology that includes what is known and theorized about human behavioural and biological characteristics that can be validly applied to the specification, design, evaluation, operation, and maintenance of products and systems to enhance safe, effective, and satisfying use by individuals, groups, and organizations.' (Christensen *et al.*, 1988).

Possibly most important though is to see ergonomics as an approach, as a philosophy, as a way of taking account of people in the way we design and organize things—'Designing for People'. Thus ergonomics itself is primarily a process, to an extent a meta-method. This makes the clear understanding and correct utilization of individual methods and techniques even more important.

Models of ergonomics

As a consequence of the somewhat different ways in which ergonomics may be defined or interpreted, whether by teachers, researchers or practitioners, we can place ergonomics methodology in a variety of contexts. Individual ergonomists may employ different starting or focal points for their work at different times. They may concentrate upon how what they do is applied and to what factors, or they may focus upon the aims of such application and implications for non-application. Other ergonomists base their work around knowledge or models of people and still others place their activities within some overall process of design. We can term these different contexts as being application-, consequence-, human model-, and process-oriented, although there is much overlap between them. Looking at these different models within ergonomics, developed admittedly for different purposes, has the added advantage of providing a broad view of the concerns and coverage of the ergonomics profession, as well as giving us an idea of how methods might be differentiated. By understanding these contexts or models, some wholly and some partially embracing ergonomics, we can get a feel for the range of issues, processes, applications and conditions on which we may wish to bring our methods to bear.

Application-oriented models of ergonomics

A traditional view of ergonomics is that, like the epidemiological model of 'host–agent–environment' in disease control or accident prevention, it is concerned with interactions between people and the things they use and the

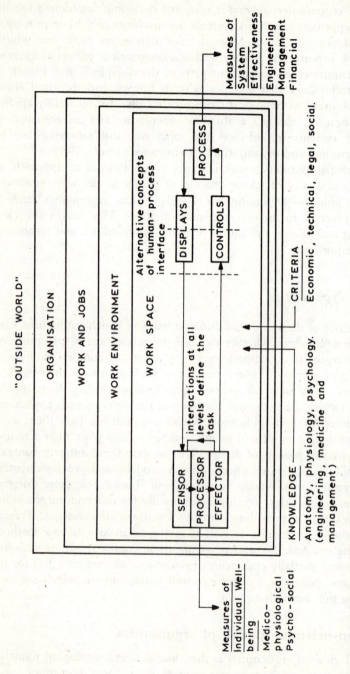

Figure 1.1. Application oriented model. Model of ergonomics from the viewpoint of to what issues (at work) it can be applied. Primarily this model stresses: the person–process interaction; that this interaction defines, and is defined by, tasks; the different concepts of the interface; the interactions with 'layers' of contextual factors; and the interaction with the outside world in terms of knowledge required, criteria to be accounted for, and measurements.

environments in which they use them. Most ergonomics and human factors texts open with a simple illustration of the interface between people and the processes with which they interact, whether this 'process' be a toothbrush, training manual, motor car, power plant control room and so on. Leamon (1980) expanded upon this and Figure 1.1 is an amended version of his model.

The person and the process form a closed loop system (but not a closed system); the output characteristics (such as the person's hands, feet, speech, or the process displays) of each must match the input characteristics (human sensory mechanisms or the process controls) of the other. Often this is denoted as achieving a person–process or, more usually, a user–system fit or match, and the human–machine interface is the subject of many ergonomics studies and focus of many methods. Generally the displays and controls themselves will be regarded as comprising the interface; however in systems which are highly automated, where the operator acts merely as a monitor, the interface may be seen as being to the human side of the displays and controls. The latter are then considered part of the process.

The human–machine interaction does not take place in a vacuum; it will be affected by the physical workplace, the physical work environment and the social environment or organization of jobs and work, and also by factors from the world outside. Within such a model we can see ergonomics methodology as comprising the techniques needed to predict, investigate or develop each of the possible interactions; person–process (hardware or software), person–environment, person–job, person–person, person–organization, and person–outside world.

Consequence models of ergonomics

In the more extensive definitions of ergonomics we find a listing of objectives or criteria that drive the application of ergonomics; for instance, jobs, systems or products that are comfortable, safe, effective and satisfying. Aims of ergonomics are often divided into those of gains for the individual (employee or user), and those for the organization (employer or producer). There is much cross-over here however, whereby we can see that improvements for the individual can well be carried over into advantages for the organization (Figure 1.2).

Taking such an outlook on consequences of good or bad work design and having also a catholic view of the legitimate concerns of ergonomics, we have developed a model for ergonomic intervention in work design (Wilson and Grey, 1986), which is shown in somewhat amended form in Figure 1.3. Again, ergonomics methodology could be seen in the light of, and classified by, how it matches the requirements for measurement, evaluation and system changes defined in this model.

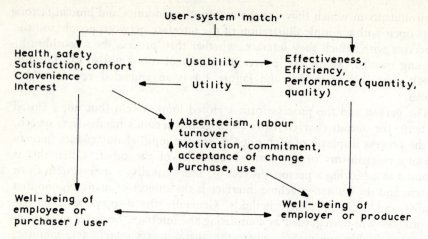

Figure 1.2. Aims in ergonomics. Improvements to the well-being of the worker or user can improve performance through making the job more possible or the product more usable, etc. Also lower absenteeism, greater commitment and raised sales or user satisfaction will themselves be of benefit to organizational objectives.

Human (performance) models of ergonomics

Another way to approach a description of the ergonomics remit is to examine what people do, how they behave, in whatever context. Then we could look at methods as providing, improving, adapting and applying information gained from such models. We will restrict consideration here to just one such model, the now ubiquitous 'human information processor' (Figure 1.4). This is widely discussed as a basis for explanation of how we behave in our environment, allowing us to test hypotheses about human performance. Thus basic human data can be seen as relevant to sensory, perceptual/cognitive and motor components of behaviour, and to associated attentional and memory processes. Application in design or evaluation can take into account the relevant stages of information processing predicted to be of importance. Methods thus could be classified according to the information processing stage implied or examined. Figure 1.5 shows an adaptation and expansion of the model, to account for individual differences of importance, for use in understanding road user behaviour as a preliminary to developing measures to improve such behaviour.

Design process models of ergonomics

If we regard ergonomics as primarily an approach or, more specifically, a process, then it is useful to conceptualize the design and development process and to place methods within this. Many such processes and complementary methods have been identified or proposed (see Jones, 1980 or Cross, 1984).

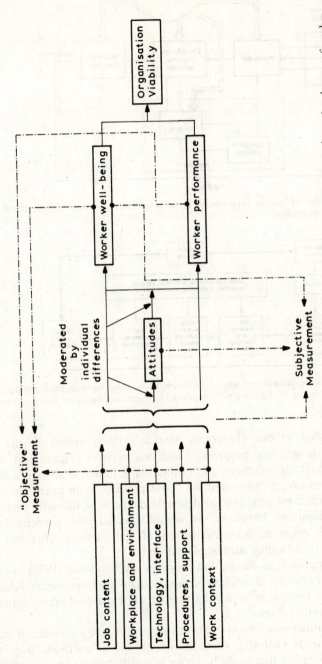

Figure 1.3. Consequence-oriented model. A model for changes in work design. Five groups of factors can determine degrees of worker well-being and performance directly or through the attitudes formed by workers' perceptions of those factors and the reasons for them. Factors can be objectively determined as being present or absent and in what degree; outcomes can be objectively or subjectively measured. Redrawn from Wilson and Grey (1986).

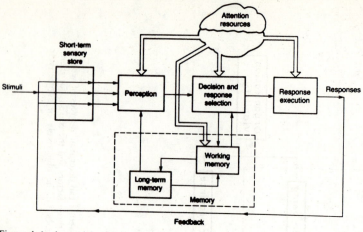

Figure 1.4. A model of human information processing. From Wickens (1984), reprinted by permission of the publisher.

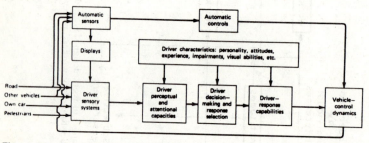

Figure 1.5. Simplified diagram of driver functions in a driver–vehicle–road system. From *Psychology on the Road* by D. Shinar (1978), reprinted by permission of John Wiley & Sons Ltd.

Already discussed in this chapter is Markus' (1969) notion of the two dimensions of design, one progressive and one iterative (Figure 1.6). Thus a sequence of analysis, synthesis and evaluation is found at all stages of the development process. At first sight we could perhaps, with profit, associate our methods with these generic stage descriptions, perhaps also with particular phases of development. However, we would soon see that, particularly for analysis (which often includes evaluation of the existing situation) and evaluation, we could utilize almost all methods.

A different approach to design was proposed by Singleton (1974) in terms of man–machine systems (nowadays termed human–machine systems) design. This is seen as a process in which the human and hardware sub-systems are developed in parallel (Figure 1.7).

Within the human–machine systems design process are a mixture of stages (e.g. training needs analysis), methods or groups of methods (e.g. task analysis), and elements that could be either (e.g. allocation of function). This last is sometimes seen as comprising only a checklist method of selecting

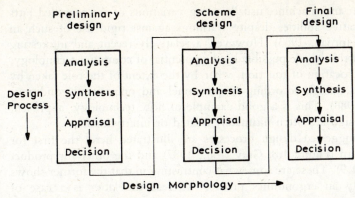

Figure 1.6. A two-dimensional model of the design process. Adapted from Markus (1969).

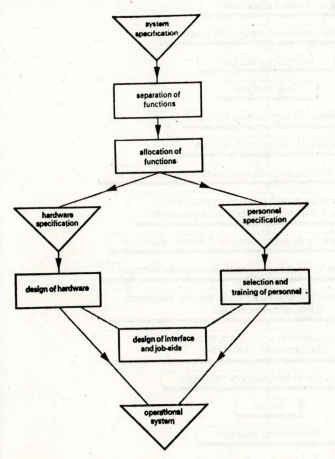

Figure 1.7. The systems design process. From Singleton (1974), reprinted by permission of the author.

between human and machine, usually using variations of the so-called Fitts List of comparative abilities despite warnings against rigid use of such an approach (e.g. Jordan, 1963). However, a relatively recent and interesting development of this, made possible by the potential of computer technology, is the flexible allocation of function, whereby the extent of the role taken by the human can change according to personnel and operational conditions (Clegg *et al.*, 1988). This is a good example of how technology, approach, method and design are often intimately related or intertwined.

Two other suggested design processes are illustrated here, the first for workspace design (Figure 1.8; Grey *et al.*, 1987) and the other for product design (Figure 1.9). These are somewhat contrasting in that the former shows what is entirely an ergonomics process whereas the latter is a case of ergonomics interventions being made in a wider development process. The

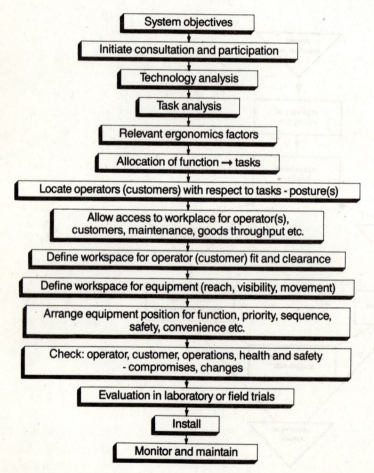

Figure 1.8. A design process specified for Electronic Point of Sale Workplaces. From Grey *et al.* (1987).

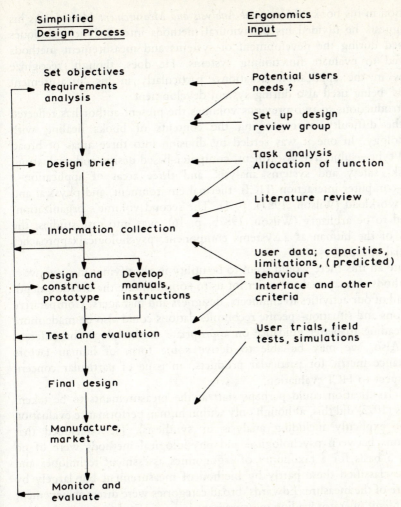

Figure 1.9. One possible simplified product design process showing types of ergonomics input.

chapters in this book by McClelland and by Christie and Gardiner also describe development processes which are similar, for products and computer interfaces, respectively.

Classification of methods

It should be apparent from the foregoing that attempts to classify ergonomics methodology in terms of parts or stages of any of the models of ergonomics will be fraught with difficulties, replete with ambiguities and probably not a very useful or fruitful exercise. Meister (1985) attempted a very gross distinction in his book, *Behavioural Analysis and Measurement Methods*. As his title suggests he divided his behavioural methods into analytic techniques employed during the development of systems and measurement methods employed to evaluate functioning systems. He does, though, recognize overlaps in the application situations, particularly in his 'measurement methods' being used also during system development.

In introductions to two previous volumes the present author has reflected upon the difficulty of organizing the contents of books dealing with methodology. In one it was settled by division into three areas of broad method type—subjective assessment, computer-based design and evaluation, and task, safety and systems analysis, and three areas of application—human–computer interaction (HCI), thermal environment, and physical and mental workload (Wilson, 1987a, p. 1). The second volume's organization, admitted to be arbitrary (Wilson, 1987b, p. 16), was into case studies, the analysis of the human as a systems component, psychological approaches and social approaches.

Despite all this there would seem to be utility in classifying, in some way, our methods. It may make it easier for us to communicate the breadth, depth and detail of our activities to engineers, designers and managers. Comparative evaluations and situation-specific recommendations could also be made more easily, leading to better guidance on appropriate methodology for different needs. Also, we may be able to derive some form of human factors performance metric for particular products, an issue of particular concern with respect to HCI evaluation.

Such classification could perhaps start at the measurements to be taken. Edwards (1973) did this, although only within human performance evaluation (i.e. not explicitly including analysis or synthesis). He concluded that distinctions between psychological and physiological methods were of no value as a basis for a taxonomy of ergonomics assessment techniques and therefore classified these partly by method of measurement but largely by the nature of the measure. Edwards' broad categories were direct achievement measurement, operator loading measurement and correlated function measurement and particular measures (or as he calls them techniques) within these are listed in Table 1.1.

Colleagues at the Ergonomics Information and Analysis Centre (EIAC) at Birmingham University have developed an ergonomics classification to keep up with the changing nature of the discipline over the years. The part of this currently dealing with methods only is shown in Table 1.2.

Starting from Edwards' taxonomy, then reviewing the EIAC classification, and with a need to structure teaching in performance measurement and evaluation a few years ago, the author has attempted an initial taxonomy of ergonomics methodology. This taxonomy recognizes differences in approach, method (or method group), technique, and measure. It may also be possible to add 'criterion' to these, the criteria underlying the required measures. The present state of such a general classification scheme or taxonomy is shown

Table 1.1. Taxonomy of assessment techniques

Assessment techniques (measures)	Examples
Direct achievement measurement	
Absolute scores	Distance travelled, numbers produced
Speed scores	Time, time per. ., reaction time
Precision scores (continuous)	Tracking performance
Error scores (discrete)	Faults, missed signals
Information scores	Rate of information handling, redundancy
Multiple scores	Combinations of the above
Operator loading measurement	
Extracted outputs	ECG, EMG, EEG, GSR, O^2 uptake
Secondary tasks	Task loading through secondary activity
Alternative tasks	Decrement assessed on new task
Time function changes	Fatigue, learning
Operator reports	Interviews on, say, fatigue
Correlated function measurement	
Variability	Consistency as measure of skill
Operator reports	Critical incidents
Observer reports	Pilot assessment

in Table 1.3. Despite some redundancy, omissions and some inconsistencies this taxonomy works to a limited extent and a special subset is being developed for human–computer interface evaluation.

Use and usefulness of methods

As far as methods are concerned, the proof of the pudding is in the eating. What to one researcher or practitioner is an invaluable aid to all their work, a crutch even, may to another be vague or insubstantial in concept, difficult to use and variable in its outcomes. More than this, the validity, reliability, sensitivity and so on of methods may well be application specific. Perhaps then, only examination of their utility and generalizability would help us select or prioritize between methods in general.

Meister (1986) reports a survey he conducted using a questionnaire, returned by 21 members of the Human Factors Society interested in testing and evaluation (pp. 387–406). (Cursory examination of the list of respondents reveals none from academia and 16 from military-based groups.) From an admittedly limited survey, it appears that those methods which rated consistently highly on all three criteria of frequency, usefulness and ease of use, are those of indirect observation—questionnaires and rating scales for instance. The exception is attitude measurement, rated low for frequency and particularly ease of use, but this could reflect the military bias in the sample! Interestingly the relative difficulty in use reported for observation and for test plan do not seem to preclude their widespread use. For dynamic mock-ups, static mock-ups, automated data recording, activity analysis and

Table 1.2. Classification (part) from Ergonomics Abstracts

Approaches and methods	Methods and techniques
63	**64 Techniques** / **65 Measures**

63 Approaches and methods

- 63.1 Modelling and simulation
  - 63.1.1 modelling human characteristics
  - 63.1.2 modelling system characteristics
  - 63.1.3 modelling environmental characteristics
- 63.2 Use of simulators
  - 63.2.1 use of test rigs
- 63.3 Mock-ups, prototypes and prototyping
- 63.4 Manikins and fitting trials
- 63.5 Systems analysis
  - 63.5.1 task analysis
  - 63.5.2 job analysis and skills analysis
- 63.6 Human reliability and system reliability
- 63.7 Physiological and psychophysiological recording
- 63.8 Work study
  - 63.8.1 method study
  - 63.8.2 work measurement
- 63.9 Data collection and recording methods
  - 63.9.1 human recording
  - 63.9.2 self recording
  - 63.9.3 instrument recording
  - 63.9.4 experimental design
  - 63.9.5 laboratory versus field
- 63.10 Data analysis and processing methods
  - 63.10.1 statistical analysis and psychometrics
  - 63.10.2 signal processing and spectral analysis
  - 63.10.3 image processing
  - 63.10.4 textual analysis and parsing
- 63.11 Psychophysics and psychological scaling
- 63.12 Use of expert opinion
- 63.13 Protocol analysis
- 63.14 Approaches to equipment testing
- 63.15 Cost benefit analysis
- 63.16 Job appraisal

64 Techniques

- 64.1 Observation techniques
  - 64.1.1 participative observation and group decision making
  - 64.1.2 visible observation
  - 64.1.3 unobtrusive observation
- 64.2 Checklists
- 64.3 Classification systems and taxonomies
- 64.4 Interviews
- 64.5 Questionnaires and surveys
- 64.6 Rating and ranking
- 64.7 Application of test batteries
- 64.8 Experimental equipment design
  - 64.8.1 hardware design for experimentation
  - 64.8.2 software design for experimentation
- 64.9 Critical incident technique

65 Measures

- 65.1 Comparison of measures
- 65.2 Time and speed
- 65.3 Error, accuracy, reliability and frequency
- 65.4 Event frequency
- 65.5 Response operating characteristics
  - 65.5.1 sensitivity
  - 65.5.2 response bias
- 65.6 Output and productivity
- 65.7 Combined measures and indices
- 65.8 Subjective measures
  - 65.8.1 ratings and preferences
  - 65.8.2 opinions
- 65.9 Usage

computerized methods their perceived difficulty of use seems to determine their frequency of use, despite them being seen generally as very useful.

Multiple methods

Given all that has been said above about methods it is not surprising that we will often look to use more than one, and often several, in any one study. This is particularly so when we are carrying out evaluations in the field. Technically this is known as triangulation (see Denzin, 1970) and, although it can encompass data or investigators, triangulation is most often used to denote methodological triangulation. This is the use of two or more methods to improve the efficiency, completeness, and insight of a study; weaknesses in one method can be balanced by strengths in another. To take one simple example, only by observing operators in complex systems *and* also recording and taking a protocol analysis of their concurrent verbal reports can we begin to understand something about their decision making activities.

A multiple-methods study may utilize a mixture of qualitative and quantitative, field and laboratory techniques. A typical battery of methods which could be employed in the evaluation of work and the production of recommendations for its redesign is:

Questionnaires.
Attitude surveys.
In-depth, informal discussions.
Group decision meetings.
Written record of activities.
Videorecording.
Photographic recording.
Physical measurmeent of workplace dimensions.
Physical measurement of environmental variables.
Computer workspace modelling.
Simulation and test trials.

Only by use of most or all of these methods may a full evaluation be possible and effective suggestions for redesigning job content, tasks, workstations and environments be arrived at.

Structure of the book

The problems of providing a complete, non-redundant and unambiguous classification of ergonomics methods have been alluded to already. One manifestation of this difficulty is knowing how to structure chapters in a book such as this. In the end a compromise has been reached; the structure comprises a mixture of classification by context-specific groups of methods, context-free and general methods of analysis and evaluation, and basic methods and techniques.

Table 1.3. A classification of methods, techniques and measures used within ergonomics

Approach	Group	Method		Technique	Measure/outcome	Chapters
		Subgroup				
Collection of information from/about people	Direct observation (laboratory or field)	Unobtrusive, participative, or visible		Human recording: checklists, rating, ranking, critical incident technique, charts (time, spatial, sequence, link)	Event frequency, sequence Times, errors, accuracy Overload, underload Descriptive, evaluative, diagnostic measures of performance	2(6, 10, 12, 20, 22, 30)
				Hardware recording: video, film, tape, event recorder, position/movement recording, computer real time recording		
	Indirect observation (laboratory or field)	Psychometrics/scaling		Surveys, questionnaires, rating, ranking, scaling, diaries, critical incidents, yes-no checklists, group discussions, interviews	Attitudes; feelings; perceived effort, difficulties, advantages, disadvantages, preferences	3(6, 10, 12, 13, 14, 15, 16, 20, 22, 24, 25, 26)
	Perceptual/cognitive performance	Ability testing		Mental or cognitive tests (e.g. general aptitude test battery); perceptual tests	Prediction of performance	24, 26
		Psychophysics		Method of limits, method of average error, method of constant stimuli, e.g. aesthesiometer, hearing loss audiometry	Thresholds and levels of perception, sensitivity	16, 17, 26
	Knowledge acquisition	Knowledge elicitation (from expert)		Interviews (structured, unstructured), protocol analysis (verbal, shadowing, behavioural), conceptual mapping, goal decomposition, automatic techniques	'Rules', reasoning, explanations	13(7)
		Other		Interpretation of records, standards, guidelines, criteria		

Method	Techniques	other descriptive statistics	(refs)
dynamic)	photography, CODA, fitting trials, computer modelling		22
Biomechanics	Dynamometer, strength gauges, goniometer	Values, descriptive statistics	21, 22, 23
Performance	Eye movements, acuity		26, 27
Physiological measurement	ECG, EEG, EMG, ERP; O² uptake, GSR, pupil diameter, etc.		9, 21, 24, 25, 27
Models	Computer, mechanical, conceptual, mathematical		8, 15, 16, 23, 29

Evaluation of human–machine system—performance or consequence (actual or potential)

Method	Techniques	other descriptive statistics	(refs)
Task analysis (TA)	Hierarchical TA, tabular TA, ability requirements analysis, TA for knowledge description, link analysis, cognitive TA, formal mappings (e.g. TAG), job analysis charts	Consequence of tasks, task sequences, times, probable error rates	6(10, 12)
Archives, databases, published information	Production, activity, quality control or personnel records; standards, guidelines, etc.	Output, times, quality, etc.; Absenteeism, labour turnover, health data	4
Interface evaluation — User trials	Techniques of direct and indirect observation, and physical performance measurement—individuals or groups	Time, reaction time; Accuracy, errors; Opinions, attitudes, responses; Physical fit; Workload, stress	10, 11, 12, 18, 20
User models, formal mapping	GOMS, CLG, TAG, etc.		6, 8, 12
Expert analysis	Walkthrough, checklists		12
Prototyping	Rapid prototyping, storyboarding		12
Introspection	Techniques for protocol analysis		7, 13
Evaluation environment	CAFE OF EVE		12
Electronic monitoring	On-line (performance) record, gripe button		

Table 1.3. *Contd.*

Approach	Method		Technique	Measure/outcome	Chapters
	Group	Subgroup			
	Work system analysis	Job Analysis, Work Analysis	All		
	Text analysis		Readability formulae (Gunning Fog, Flesch, etc); cloze procedures; judgements (rate, rank, etc); protocol analysis; scan/read tests	Normative scores, ratings	11
	Models		Task network (SAINT, Siegel–Wolf, etc); control theoretic; microprocess (HOS, etc); cognitive (GOMS, etc).	Performance predictions	6, 8, 15, 16, 23
	Simulation		Mathematical; computer, including CAD (e.g. SAMMIE) Physical mock-up, walkthrough		8, 10, 11, 19
	Human reliability analysis	Error analysis Representation Quantification	SHERPA, GEMS, PHECA, etc. Fault tree, action tree, etc. THERP, HEART, SLIM, etc.	Errors, type, causes Descriptive and predictive charts Human error probabilities	28
	Statistical analysis		Signal Detection Theory	Performance measure	
	Method study		Graphical analysis, charts, filming, micromotion, etc.		2, 6
	Accident reporting and analyses		Archive records, reporting system, in-depth follow up interviews, site analysis, statistical analyses	Incidence, severity, epidemiology and aetiology	29
	Work measurement		Time study, activity analysis, synthetic analysis, electronic monitoring	Times, standards, task sequence and simultaneity	2
	Cost–benefit analysis		Investment returns; productivity—life cost, revenue calculation; health and safety valuations	Financial return	31
	Textual analysis		Parsing etc.		
	Self recording		Gripe button, diary, event recorder	Problems, incidents	

Analysis of work activity demands		Concurrent or retrospective	Explicit, implicit content, behaviour transitions, rules, knowledge	7(13)
	Introspection (+ protocol analysis)	Hardware recording, debriefing, written record, diary, critical incidents, shadowing	Explicit content, implicit content, behaviour transitions, rules, knowledge	7(13)
	Expert analysis	Checklist; walkthrough; Delphi, etc.; method study techniques, expert systems		2, 14, 15
	Archives, data base	Medical records, accident records	Incidence, severity, risk factors	4, 29
	Physical workload	Indirect observation, e.g. Borg scale. Performance records, secondary or alternative tasks of psychomotor performance; physical changes (e.g. shrinkometer)	Subjective ratings Performance decrement, etc.	21, 22
	Posture analysis	Biomechanical (mathematical) models; optical methods (CODA, Selspot); paper and pencil (posture target, body part discomfort)	'Postures' to compare with criteria Opinions of discomfort, etc.	20, 22, 23
	Physiological	Measurement of fatigue, stress, function, etc. HR, HR variability, O^2 uptake, air analysis, GSR, ECG, EMG, EEG, ERP	Objective data, to be interpreted against norms, criteria	21, 22, 24, 25
	Mental workload measurement	Primary, secondary, alternative task, Subjective assessment (eg. SWAT), physiological response	'Performance' decrement Load (subjective or objective)	24
	Stress assessment	GSR, etc; indirect observation techniques (SACL, GWBQ, etc.)		25
	Job and work attitude measurement	Techniques of indirect observation, especially rating scales, e.g. JDS, WLAS, etc; informal group or individual interviews	Satisfaction, needs, important job characteristics	

Continued

Table 1.3. *Contd.*

| Approach | Method | | | Measure/outcome | Chapters |
	Group	Subgroup	Technique		
Physical environment assessment	Measurement by instrumentation		Light (illumination, glare, etc), climate (temperature, humidity, air space, etc), noise (sound intensity-weighted), vibration, workplace dimensions	Measurements vs. norms, Comparisons	14, 15, 16, 17, 18, 20
	Subjective assessment		Psychophysical techniques, scaling, rating, surveys, etc.	Comfort, annoyance, acceptability	14, 15, 16, 17, 20
	Performance measures		Speech intelligibility index, work-rate, standard psychomotor and mental tests, etc	'Scores'	16, 26
	Modelling and simulation		Computer (e.g. SAMMIE), mechanical (mannikins), mathematical		8, 15, 19
	Response measures		Sweat rate, body temperature, heart rate etc.; hearing loss, etc.; visual acuity, contrast sensitivity, etc.; sensation loss (vibration), shrinkometer	Measurements vs. norms, Comparisons	14, 15, 16, 17
	Archives, medical and accident records			Sickness, absence, injuries	4, 29

			References
Organizational environment assessment	Organization analysis		30
	Indirect observation	Attitudes, opinions, etc. 'Improvements'	3
		Rating, ranking, etc.; group participative methods	
Design and implementation	User tests	Direct and indirect observation	10, 11, 12
	Expert analysis	Walkthroughs, audit	12, 28
	Creative techniques	Brainstorming, decision groups, focus groups	10, 12
	Participative methods	Design and follow-up groups, user representation involvement; 'education' in ergonomics	10, 32, 33
	Evaluation environment	CAFE OF EVE	12
Process promotion and dissemination	'Literature'		1
	Participative methods	Ergonomic working groups, etc., training	32, 33
	Cost-benefit analysis	Investment returns, health and safety valuations	31, 34

We open with a number of chapters on approaches and methods which are generally used in fields other than ergonomics, as well as being used generally across all aspects of ergonomics endeavour. Therefore Part I includes contributions on direct observation techniques, indirect observation or subjective assessments, the use of archival data previously collected for another purpose, and an overview of how to design ergonomics studies. There is another group of methods and techniques, again general in that they can be applied in many situations of analysis or evaluation, but which—as they are reported here—are specifically relevant to ergonomics studies (Part II). Thus task analysis and protocol analysis, especially the former, are useful across a wide spectrum of ergonomics study. The chapter on simulation and modelling, which obviously are available as methods in many disciplines, presents those techniques useful in ergonomics; computerized data collection also has a wide relevance but is looked at here specifically from an ergonomics viewpoint.

The next four major parts cover four main areas of ergonomics activity: product or system design and evaluation (Part III); assessment and design of the physical workplace and environment (Part IV); the analysis of work activities, physical and mental (Part V); and the analysis and evaluation of work systems (Part VI). Techniques of both design and evaluation, or of analysis and evaluation, have been combined in these parts, and in the constituent chapters also, because it is often impossible to describe techniques as solely applicable to one and not the other. Analysis, for instance, takes place both prior to, and also during evaluation of, design.

Finally, we conclude the book with two chapters concerned with systems implementation, (Part VII), and a look into the future of ergonomics methodology.

Many of the methods, techniques and measures referred to in the body of the book have been placed in some sort of framework in Table 1.3. This stands as an, admittedly imperfect, frame of reference for all the methods used in the evaluation of human work. Other sources of information on methods are included in an appendix to this chapter.

Conclusions

The thin line which we must tread when we become involved in discussion of methodology is summed up by the following quotes: '[psychology] . . . should not allow itself to be driven by obsession with method to the exclusion of the human problems that are its province' (Barber, 1988, p. 7, reporting Maslow, 1946), but 'anyone who wishes to reflect on how they practice their particular art or science, and anyone who wishes to teach others to practice, must draw on methodology' (Cross, 1984, p. vii). Yes, certainly ergonomists, if anyone, must be human problem driven (or human life improvement driven); but also we are always concerned to educate others (designers, engineers, politicians, accountants, managers, public, media) in our approach and the necessity for it. If we truly believe that we are, above all, promoters of an approach, of a process, then it behoves us to pay great attention to the state of our methodology, its roots, current extent and potential use, and future developments. The remainder of this book, is but one step in doing this.

References

Barber, P.J. (1988). *Applied Cognitive Psychology* (London: Methuen).

Christensen, J.M., Topmiller, D.A. and Gill, R.T. (1988). Human factors definitions revisited. *Human Factors Society Bulletin*, **31**, 7–8.

Clark, T.S. and Corlett, E.N. (1984). *The Ergonomics of Workspaces and Machines: A Design Manual* (London: Taylor and Francis).

Clegg, C., Ravden, S., Corbett, M. and Johnson, G. (1988). *Allocating Functions in Computer Integrated Manufacturing: A Review and A New Method*. Memo No. 954 of the MRC/ESRC Social and Applied Psychology Unit, University of Sheffield.

Cross, N. (1984). *Developments in Design Methodology* (Chichester: John Wiley).

Denzin, N. (1970). *Sociological Methods* (New York: McGraw-Hill).

Edwards, E. (1973). Techniques for the evaluation of human performance. In *Measurement of Man at Work* edited by W.T. Singleton, J.G. Fox and D. Whitfield (London: Taylor and Francis), pp. 129–133.

Grey, S.M., Norris, B.J. and Wilson, J.R. (1987). *Ergonomics in the Electronic Retail Environment* (Slough, UK: ICL (UK) Ltd.).

Jones, J.C. (1980). *Design Methods*, 2nd edition (New York: John Wiley).

Jordan, N. (1963). Allocation of functions between man and machines in automated systems. *Journal of Applied Psychology*, **47**, 161–165.

Leamon, T.B. (1980). The organisation of industrial ergonomics—a human machine model. *Applied Ergonomics*, **11**, 223–226.

Markus, T.A. (1969). The role of building performance measurement and appraisal in design method. In *Design Methods in Architecture*, edited by G. Broadbent and A. Ward (London: Lund Humphries).

Maslow, A.H. (1946). Problem-centring vs. means-centring in science. *Philosophy of Science*, **13**, 326–331.

Meister, D. (1985). *Behavioural Analysis and Measurement Methods* (Chichester: John Wiley).

Meister, D. (1986). *Human Factors Testing and Evaluation* (Amsterdam: Elsevier Science).

Shinar, D. (1978). *Psychology on the Road* (New York: John Wiley).

Singleton, W.T. (1974). *Man–Machine Systems* (Harmondsworth: Penguin).

Singleton, W.T. (Ed) (1982). *The Body at Work: Biological Ergonomics* (Cambridge: Cambridge University Press).

Wickens, C.D. (1984). *Engineering Psychology and Human Performance* (Columbus, OH: Charles E. Merrill).

Wilson, J.R. (1987a). A flavour of ergonomics methodology. In *New Methods in Applied Ergonomics* edited by J.R. Wilson, E.N. Corlett and I. Manenica (London: Taylor and Francis), pp. 1–10.

Wilson, J.R. (1987b). Introduction. In *New Techniques in Ergonomics*,

Proceedings of International Research Symposium (Paris: Hermes), pp. 13–18.

Wilson, J.R. and Grey, S.M. (1986). Perceived characteristics of the work environment. In *Human Factors in Organizational Design and Management II*, edited by O. Brown and H. Hendrick (Amsterdam: Elsevier).

Further information

There are few texts in ergonomics and human factors specifically concerned with (particularly applied) methods. However, many of the best texts in the area do have varying amounts to say about method; also use of such sources is highly recommended for anyone wishing to apply the methods discussed in the present book. The list, and brief descriptions, below comprise one selection only; some good sources will have been omitted by accident or through ignorance. Also, there are many human factors texts now which deal with only a part of this rapidly expanding (in size and importance) field. Other than some in human–computer interaction, these are not included.

Bailey, R.W. (1982). *Human Performance Engineering: A Guide for System Designers* (London: Prentice-Hall).

Very readable and a good juxtaposition of basic principles and findings with human factors criteria, this book is weakest in the areas of physical ergonomics (such as environment, work physiology and biomechanics). Over half the book discusses the application of human factors to design of various sorts and there are also three substantial chapters on methodology of data collection, performance testing and comparison studies. This is a good book for the areas (largely psychologically-related) it covers.

Chapanis, A. (1959). *Research Techniques in Human Engineering* (Baltimore: John Hopkins Press).

Even 30 years on this text is still relevant and valuable. Coverage is general and largely domain-independent, thus making it usable in the context of modern systems. Specific chapters examine direct observation, the study of accidents and near accidents, psychophysical, statistical and experimental methods, and articulation testing. A good deal of information on the difficulties of using different methods, and their value is included. The book is highly recommended for anyone interested in methodology.

Clark, T.S. and Corlett, E.N. (1984). *The Ergonomics of Workspaces and Machines: A Design Manual* (London: Taylor and Francis).

A manual for students, professional designers and production engineers, covering the 'traditional' areas of ergonomics work—workspace, environment, displays and controls. Flow charts summarize design and selection criteria, and design principles are clearly explained.

Grandjean, E. (1988). *Fitting the Task to the Man: A Textbook of Occupational Ergonomics*, 4th edition (London: Taylor and Francis).

One of the staple teaching texts of ergonomics, it explains clearly with many diagrams much of the general area of traditional ergonomics. Emphasis is rather on consequences for people of work of certain types rather than upon methodology or re-design of work. A recommended basic introductory text for students or others interested in ergonomics/human factors.

Kantowitz, B.H. and Sorkin, R.D. (1983). *Human Factors: Understanding People–System Relationships* (New York: John Wiley).

Similar in some ways to the book by Bailey, this also makes little attempt to cover human physical activities in terms of energy expenditure, loading, etc. There is good coverage of other aspects of human factors; application information is in this case used throughout the text, as is information on methods.

Meister, D. (1985). *Behavioural Analysis and Measurement Methods* (Chichester: John Wiley).

This book was the first since Chapanis' in 1959 to present a useful and wide-ranging review of technique and methodology. Largely the book is split into techniques to aid in systems design, evaluations of systems effectiveness during and after development, and 'generic' methods. Intended as an introduction to behavioural measurement, the book has very full descriptions of many analysis and evaluation tools and approaches, and is—as is the author's wont—written from a pragmatic, applications-oriented standpoint. Its weakness is that the author's perspective is largely that of US armed forces needs and experiences.

Meister, D. (1986). *Human Factors Testing and Evaluation* (Amsterdam: Elsevier Science).

Not as well-written or produced as the same author's 1985 text, this book does however widen the range of methods and applications to include environmental evaluation.

Norman, D.A. and Draper, S.W. (1986). *User-Centred Systems Design: New Perspectives on Human–Computer Interaction* (Hillsdale, NJ: Lawrence Erlbaum).

This is a collection of contributions from workers and their colleagues at the University of California, San Diego. The focus is human–computer interaction; many chapters are concerned with what we should know about people to design interfaces and how we could obtain such knowledge. In particular there is a concentration upon indicating new directions for design and design-related activities, rather than upon empirical methods, quantitative rules or design processes.

Oborne, D.J. (1987). *Ergonomics at Work*, 2nd edition (Chichester: John Wiley).

Taking what its author describes as an evangelical stance to the communication of the ergonomics discipline, this book seems to an extent—and despite certain omissions and weaknesses—to bridge the gap between the human (especially physical) activity orientation of Grandjean and the systems (especially engineering) orientation of Sanders and McCormick. There is only a short chapter specifically devoted to method (interestingly making reference to two specific books dated 1959 and 1966!) although some methodology is implied in earlier chapters. Readable.

Oppenheim, A.N. (1966). *Questionnaire Design and Attitude Measurement* (London: Heinemann).

Still the first port of call for many when wishing to access or develop an indirect observation technique, this well-written book covers survey and questionnaire design, checklists, rating scales, attitude assessment, and analysis.

Pheasant, S. (1986). *Bodyspace: Anthropometry, Biomechanics and Design* (London: Taylor and Francis).

Coverage here is similar to that in Clark and Corlett and in Grandjean—physical ergonomics—with one chapter only devoted to displays and controls. The measurement of human physical form and functions and use of such data in design is, however, the author's speciality and is very well covered. As far as method is concerned the use of anthropometric and biomechanical data in design is well covered; there are short sections upon collection of basic and performance data. The author's style is to write very clearly and simply in a somewhat idiosyncratic manner.

Ronan, W.W. and Prien, E.P. (1971). *Perspectives on the Measurement of Human Performance* (New York: Appleton-Century-Crofts).

The performance referred to in the title is that of interest within occupational psychology, referring largely to mental and psycho-motor tasks. The emphasis is upon shifting research attention from individual differences (as in psychological tests) to understanding human performance in the 'real' world. In addition to long sections devoted to performance reliability, dimensions of performance and organization performance and influence upon individual performance, there is also much on the reliability of performance observation and on criteria, with explanation of several methods and techniques.

Rubinstein, R. and Hersh, H. (1984). *The Human Factor: Designing Computer Systems for People* (Burlington, MA: Digital Press).

One of the best of the books which aim to increase the awareness of systems designers of human factors issues and to provide general guidance on how to improve user interfaces. The style is extremely readable and it is a well organized text. Chapters specifically concerning methods are on task analysis and models and on testing systems. Recommended.

Salvendy, G. (Ed.) (1987). *Handbook of Human Factors* (New York: John Wiley).

A mammoth tome, this is probably the most complete collection of general human factors approaches and knowledge publically available. Specific sections devoted to methdology are ones on functional analysis (five chapters from surveys to task analysis to physical workload measurement), performance modelling (six chapters, but only some useful in terms of ergonomics methodology), and system evaluation (three chapters). Many other chapters discuss methods to varying extents. Admirable as this publishing effort has been to pull together as much as possible of current ergonomics knowledge, and despite much of the volume being surprisingly usable and readable for one with 103 authors, it is not geared primarily as a book covering the range of methods and techniques used in human factors and thus is not appropriate as such.

Sanders, M.S. and McCormick, E.J. (1987). *Human Factors in Engineering and Design*, 6th edition (New York: McGraw-Hill).

One of the most widely recommended teaching texts in ergonomics/human factors (previous editions were McCormick and Sanders), this takes an application-oriented view, much of its content being related to practical needs in engineering and design. Very thorough for its size, there are, though, omissions and/or weaknesses—in the areas of job design, computer dialogue design, and models for instance. There is one chapter specifically devoted to methodologies—looking at field and experimental research, and criteria. Definitely on any shortlist of useful texts it is nonetheless not written from the perspective of methodology.

Singleton, W.T., Fox, J.G. and Whitfield, D. (Eds) (1973). *Measurement of Man at Work* (London: Taylor and Francis).

Emanating from a conference as it does, the book suffers from the weaknesses of all collections of conference papers. Coverage is usually too full of gaps and at the same time has too much overlap. However, this collection suffers less than most and many of the contributions stand up well even two decades later. Many of the then authorities in the area have chapters in the book. Coverage is split into: 'Man'—overviews of 'descriptors' of people useful in relation to human–machine interaction; 'Techniques'—data acquisition procedures about people at work; and 'Applications'. (The care given to explanation of the grouping (pp. xi–xii) shows this to be as difficult then as it has been for this book!—see earlier in this chapter.)

Singleton, W.T. (1974). *Man–Machine Systems* (Harmondsworth: Penguin).

Although short by today's standards, this is still an important book for anyone interested in the approach and stages of human–machine systems design. The whole book defines one methodological approach, with different methods or techniques comprising the stages.

Singleton, W.T. (Ed.) (1982) *The Body at Work: Biological Ergonomics* (Cambridge: Cambridge University Press).

Its title explaining well its content, this collection of contributions from various authorities takes each aspect of physical ergonomics (energy, biomechanics, vibration, climate, illumination and noise) in detail. In general each topic is discussed in terms of effects on people, measurement and application. It is a good review of methodology within the narrow field of ergonomics specified.

Smith, S.L. and Mosier, J.N. (1988). *Guidelines for Designing User Interface Software*. Report ESD–TR–86–278 (MTR10090). (Bedford, MA: Mitre Corporation).

An approach or method for enhancing the ergonomics of designs is the checklist in reverse, the use of guidelines and recommendations to act as an *aide-memoire* for ergonomists or an aid to designers. This long report is a collection of most such recommendations and criteria relevant to HCI and is about the best of its type.

Webb, E.J., Campbell, D.T., Schwarz, R.D. and Sechrest, L. (1966). *Unobtrusive Measures: Non-Reactive Research in the Social Sciences* (Chicago: Rand McNally).

An excellent overview of indirect, direct, archival and physical measures, this still widely read book contains methods and techniques from the ordinary to the weird, solid to exciting. The title the book almost had, *The Bullfighter's Beard*—referring to the tendency for this to grow faster on the day of a bull fight, for which phenomenon there are alternative hypotheses—sums up the catholic approach of the authors.

Wickens, C.D. (1984). *Engineering Psychology and Human Performance* (Columbus, OH: Charles E. Merrill).

Not really a book on applied methodology, this does however explain very clearly and comprehensively the type of cognitive experimental psychology research and results which could be applied to systems design problems. As a basic text it fills in well the gaps left by, say, Grandjean or Sanders and McCormick. The book is structured around the model of the human information processor; some design principles are given but no real guidance on their application. Nonetheless, an excellent book within its coverage.

Wilson, J.R., Corlett, E.N. and Manenica, I. (1987). *New Methods in Applied Ergonomics* (London: Taylor and Francis).

This is a record of the proceedings of the conference at which the idea for the present text was born. Many attendees are contributors to both. The earlier volume, however, apart from a few overview papers, contains contributions on 'new' methods or techniques, adapted methods, or explanations of 'technology transfer' of methodology. Particular coverage is given to subjective assessment, computerized techniques, HCI evaluation, thermal environment, task and safety analysis, and physical and mental workload assessment.

Part I

General approaches and methods

There are certain general groups of methods, which perhaps we can see as global methods, with which we can obtain insight and information in any of the human sciences. Particularly we can use many techniques of *direct observation* to 'directly' measure and assess behavioural phenomena, and techniques of *indirect observation* to collect data from subjects about their interpretations of what they are doing, feeling or thinking. In more detail, direct observation is the collection of information on subject performance either directly by the observers themselves or from objective recordings of subject behaviour; indirect observation involves the provision of mechanisms by which the subjects themselves, or other associated individuals who are not the observer, can report behaviour, attitudes and knowledge.

The former group of methods are sometimes known as 'objective' methods, and the latter as 'subjective' methods. In this view it is assumed that direct observation produces 'true' records, with no biased interpretation by the subject whereas subjective assessment (as in the title of Sinclair's chapter 3) implies that the subject or other informant provide abstracted and interpreted information. However, this ignores the fact that data collected by direct observation will be re-analysed, summarized and abstracted by the observers, from their memory, notes, tape or video recordings. Moreover, properly constructed and carried out, subjective assessment or indirect observation should be run so as to maximize as far as possible the data reliability and validity. In any case, much of psychological measurement is possible only through indirect observation; the phenomena involved cannot be directly observed or even inferred from direct observations.

Both groups of techniques can be used in laboratory or field studies, although only direct observation in the field can be naturalistic and non-interfering with behaviour. The degree of interference will in part be determined by whether the observations are made visibly, participatively or unobtrusively. The last of these should provide a record of behaviour which is unaltered through reaction to an observer's presence, but any attempts to carry out such 'hidden' studies must beware a certain inflexibility of coverage which may result and, most importantly, must first entail careful consideration of the ethics of the particular circumstances (addressed in chapter 2 by Drury also). Of course, ethical issues will depend in part on the use to which data might be put but will also be situation specific; justification for unobtrusive observation may be easier for a study of pedestrian behaviour for instance, than for following an individual industrial worker.

Participatory observation, sometimes used in sociological studies, can allow insight into, and non-interference with, subjects' behaviour but sometimes leads to a 'worm's eye' view being taken. Also, the observer may change over time, identifying (or not) with those observed, and their presence may well be a catalyst for unusual behaviour. Generally then, in laboratory or field, the ergonomist relies upon being a visible observer, yet tries not to interfere with, or contaminate, behaviour by word, action or even mere presence.

The decision of whether to collect observable data directly in real-time, using storage in memory, on paper or event recorder and so on, or by using recording instrumentation such as video or audio tape, will depend upon the circumstances. Task type, behaviour to be observed, site conditions, study resources in terms of people, equipment and time, and the type of measurements or records needed will all be taken into account. Generally, hardware recording will give unlimited data capture rates but be more limited in terms of geographical space covered. It will provide a permanent record to be reconsulted to check accuracy but requires considerable time to analyse down to a written record (between six and ten times the 'real' time for video for instance). Also it involves no human problems of data reinterpretation or misinterpretation, attention, learning and so on in the field, but these may be problems during analysis in the laboratory.

In chapter 2, Drury concentrates upon performance observation and largely upon techniques of observation and recording derived from work study. As ergonomists we have borrowed (and improved upon) much in work measurement and method study but our philosophy or approach is somewhat different—fitting tasks to people instead of vice versa which is the case at least in work measurement. Also, we have expanded the focus and methods of observation to embrace all kinds of human behaviour and system response, not just that concerned with effective performance.

In order to truly understand much of human behaviour we must not only observe people but must question them or obtain verbal and written reports from them also. This permits greater insight into what is being done and why. It also is the method by which we assess affective responses—feelings and attitudes. Indirect observation is reported as techniques of subjective assessment by Sinclair in chapter 3; this includes rating, ranking, question-naires, interviews and checklists. Psychophysical techniques are also available for such measurement.

Ostensibly, indirect observation appears to be a simple approach to data collection. Need to ensure that subjective reports do indeed reflect accurately and completely the target behaviour, thoughts or knowledge though, means that development, administration and analysis must be treated very carefully. As well as minimizing any distortion, omissions or commissions on the part of the subjects, we must ensure as far as possible that the method and techniques used do not introduce bias themselves and do not unnecessarily constrain the focus of enquiry.

Before embarking upon any of the extensive and expensive investigations often needed for direct or indirect observation, full use should be made of data that exist already. In chapter 4, Drury identifies several sources of archival data usually available within organizations, 'fixed' records (plans, annual reports, etc.), production, industrial engineering, quality control, personnel, medical, and costing records. Other sources may be meetings minutes (e.g. worker–management meetings) or maintenance records.

The beauty of all these archives of course is that someone else has already taken the time and trouble to produce the data; all we have to do is extract and interpret what we need. Therein lies the downside, the dangers that the records themselves will be incomplete, variable over time or situation, selective or biased, or that we might misinterpret or selectively interpret them. Nonetheless, they are a very useful resource for any field study.

Included as archives might well be our store of ergonomics principles, methods, data and criteria contained in textbooks, handbooks, journals and reports. Such literature archives are a necessary resource in any study.

Archival records from the organization will be used as part of field studies. However, the other two method groups can be applied in field or laboratory situations, in the case of the latter in the context of an experimental design.

The ways of doing this, the factors to be taken into account and procedures to be followed, are laid out by Drury in chapter 6. We might prefer field studies for their greater realism and thus validity and generalizability, but the production of reliable and usable data will often necessitate use of the controlled laboratory environment. So that such studies and experiments do not end up as costly, sterile exercises, great care must be taken in their design and conduct.

A final word for this introduction, and a note for the whole book, is provided in Sinclair's conclusion where he points out that although skilful use of methods requires practice we should beware of sticking rigidly to standard methods. Imagination, as well as scientific rigour, is an ingredient of successful ergonomics.

Chapter 2

Methods for direct observation of performance

Colin G. Drury

Introduction

When we collect data from the operational system by observing the system without changing it, we have an observational study. Such studies are classified in chapter 5 as having a high degree of face validity but a low degree of experimental control. For example, Cohen and Jensen (1984) had observers in two warehouses noting the behaviour of fork-lift truck drivers. In a different context, Schiro and Drury (1981) had emergency physicians observe the behaviour and performance of ambulance attendants in bringing patients into an Emergency Department. In both cases, the system (warehouse, hospital) was functioning normally except for the presence of an observer. Hence the face validity of the result was high because actual live events were recorded. But experimental control was very low as the experimenters did not change the system under study, or at least not at the time of the first observations. Causality is impossible to establish in such circumstances, but if the goal of an observational study is not to infer causality, then observation of behaviour is an appropriate technique. Issues of data contamination and inadvertent editing by choice of observed unit will be discussed later in this chapter when common concerns of all the observation methods are raised.

Initial observation of a system

Merely walking around an operating human/machine system to 'see what goes on' is hardly an observational method worthy of the name, yet it is an indispensable first step. The ergonomist needs to combine archival data (see chapter 4 by Drury), such as plant layouts and job descriptions, with broad on-site observation to ensure that the objectives and interrelationships of the system are understood before any more formal investigation takes place.

People's jobs are never simple, despite first appearances. There are subtleties of part placement, incoming quality variations and model changes which change the operation of even the most repetitive manual task. Skills developed by operators over months and years make assembly tasks look simple when they are not. The ergonomist should talk to operators, trainers and trainees to obtain a feel for what is important. If necessary, and safe, the job should be tried out. It is only by obtaining a thorough understanding of the complexities of a job or department that the ergonomist can make the main study reveal anything beyond the obvious. Remember that ergonomic insight comes from the interpretation of the system in terms of models and concepts of human functioning, so that a rapid 'walk through' will only yield the most simplistic insights.

Observation methods are useful in collecting not only quantitative data (e.g. parts per hour, errors per part), but also in collecting qualitative data on product routings, task sequences, causes of delays and error taxonomies. In this latter context, an observational study is very close to a task analysis of an existing system, so that many of the concepts and methods of task analysis apply equally well to observation methods (see Stammers *et al.* in chapter 6).

Data collection methods classification

The individual methods will be briefly described in this chapter with reference to more complete sources, as each method can easily fill a chapter (Chapanis 1953) or a book (Hansen, 1960). To help the ergonomist choose appropriately, the methods are classified by the type of data they produce. Observation of a system is necessarily an abstraction as there is too much detail available to record everything which happens. What is abstracted depends upon how the data is to be utilized. The typical charting methods which represent one outcome of an observational study can be classified by what they abstract from the raw data set. But first the raw data set needs to be defined.

Raw observation data is either of events or states. An event is an observable occurrence at a particular point in time and is usually recorded with its associated time. Thus in an observational study of queue formation in a bank, an event would be a customer arriving, a customer leaving the queue to start service, or leaving the teller after service. Each event would usually have an associated time (see Table 2.1).

Event/time records can be analysed further to provide information on sequence of events, duration of events, spatial movements or frequency of events.

The other method of data collection is to record system states at pre-specified times. A system state is a specified set of system conditions, typically constant between events. Thus to continue the previous example, system states will specify which teller positions are occupied, which occupied

Table 2.1. Events and associated times in an observational study of queue formation at a bank

Event	Time (p.m.)
Customer arrival to queue	1:24:03
Customer leaves teller No. 3	1:24:15
Customer from queue to teller No. 3	1:25:20
Customer arrival to queue	1:25:38
Customer arrival to queue	1:26:01

positions are serving customers, and how many customers are in the queue. If observations are taken at two-minute intervals, we may have the record shown in Table 2.2.

What is shown is that positions 1 and 3 are serving customers (S = serving) while 2 and 4 have no tellers (E = empty). As the queue reduces to zero, tellers 3 and then 1 are still there (O = occupied) but are not serving customers. The example shown only has one event happening between each pair of observations, but this is not typically the case.

Note that in the event/time record, data collection is event driven so that each event is recorded when it takes place. For the time/state record, data collection is time driven, with states being recorded on a predefined schedule (which need not necessarily be an equal time interval schedule). Note also that if a starting state is defined, then an event/time record allows the deduction of system states at any time during the observation period. The reverse is not true, as between state observation times many changes could have taken place. If it is wished to deduce the system state progression from the event/time record, it would be as well to allow for multiple recordings of system state as a check on the accuracy of recorded data.

Having collected the raw data, reduction and analysis are required to meet the study objectives. Thus a time study would require means and standard deviations of element times to be calculated, while in an occurrence sampling study the relative frequencies of system states would need to be calculated.

Table 2.2. A record of system states in queue formation at a bank

Time (p.m.)	Queue Length	State			
		Teller Positions			
		1	2	3	4
3:00:00	2	S	E	S	E
3:02:00	3	S	E	S	E
3:04:00	2	S	E	S	E
3:06:00	1	S	E	S	E
3:08:00	0	S	E	O	E
3:10:00	0	O	E	O	E

Some abstraction takes place on raw data collection, but more comes from data reduction and analysis. Each of the techniques preserves certain information while ignoring other aspects. Five types of information can be recognized.

1. *Sequence of activities*: which particular activity follows another activity, by the operator or by the item processed. For a fixed-sequence operation, there will only be one answer, but in more varied situations we need ways to summarize the sequences obtained.

2. *Duration of activities*: the length of time an activity takes, with statistical summaries as appropriate if the time is indeed a random variable.

3. *Frequency of activities*: how often an event occurs, e.g. how many breakdowns per day on a wave-solder machine.

4. *Fraction of time spent in states*: the fraction of time (of a person, machine, or work unit) spent in a particular state or activity.

5. *Spatial movement*: where a person, machine or work unit moves to and from during the daily activity.

These five types of information are either preserved (P), can be calculated with outside information (C), or are lost by each of the observational techniques described, as shown in Table 2.3. Note that frequency of occurrence of an *event* is only preserved in the raw event/time records. Frequency of occurrence of a *state* in occurrence sampling should not be confused with event frequency.

Each of the methods will be presented in turn, followed by some general considerations of observational studies.

Descriptions of the methods

Note that more extensive coverages of the methods of use to engineers are available elsewhere (e.g. Kadota, 1982).

Table 2.3. Information preserved (P) for each observational method; C indicates that additional data can be incorporated to calculate the information

Information type	Time study	Process chart	Process flow chart	Gantt chart	Multiple activity chart	Link chart	Occurrence sampling
Sequence		P	P	P	P	P	
Duration	P	C_1	C_1	P	P		C_2
Frequency							P
Fraction							P
Spatial		P				P	

C_1, durations on annotated process chart.
C_2, durations calculable from production records and occurrence sampling data.

Raw event/time records

Even the raw data themselves comprise, at times, a useful final form. Detailed accident and critical incident investigations for aircraft (e.g. by the FAA) are based directly on the time history of a single sequence of events. The time history is obtained from recordings of cockpit/air traffic control communications, radar plots, the flight data recorder and eye-witness accounts. These reports are often models of careful investigation and analysis, well repaying detailed study by any ergonomists studying errors or accidents (Figure 2.1).

If time is eliminated from the event/time record, it is possible to count event frequencies. Thus we can record use/non-use of seat belts in automobiles (Matthews, 1982), non-work related behaviours (Salvendy *et al.*, 1984) or law-observance behaviour of bicyclists (Drury, 1978).

Time study

Any abstraction of a raw event/time record which produces statistical data on activity times could be called a time study, for example, a frame-by-

Time	IAS(kts)	Cockpit Recorder	Event
19:57:35	0	To ATC: Ok, is 63 clear to go?	
:42	30	ATC: Ok, cleared for takeoff,	
:47	30	good day	
:55	40	We all set?	
19:58:10	95	Eighty	
:20	110		
:22	95	Hang on guys	
:25	100	Lost all our airspeed	
:30	125	V-1, Rotate	
:31	130		Nose gear unloads
:37	140	Gonna get the damn power. . .	
:38	130		Power wire impact
:41	150	Keep it going, keep it going Tommy	
:42	145	Ok, you're clear out there just keep the airspeed	Full power on
:47	165	Ok, declare an emergency	
:48	165	To ATC: Ok, 63, we just got the wires and we're gonna be airborne	
:55		To ATC: We're gonna make it	

Figure 2.1. Extract from report of aircraft N32725, Tuscon, AR, June 3, 1977, showing the effect of severe windshear from a 'dry microburst.' Adapted from NTSB report. IAS is indicated air speed.

frame film analysis of walking or manual lifting. Traditionally, though, time study has concerned itself most with finding representative times for repetitive activities. The standard reference books (e.g. Barnes, 1980; Mundel, 1978) provide overwhelming levels of detail on a method which has a history going back to the turn of the century. More evaluative presentations are also available, both in an ergonomics context (e.g. Konz, 1983: Ch 21–25) and from an engineering viewpoint (e.g. Salvendy, 1982: Ch 4.1–4.9).

For repetitive tasks, with multiple elements following each other regularly from cycle to cycle, the observation record of events and times is usually not a simple list like the earlier bank example but a matrix with rows representing cycles and columns representing elements. Figure 2.2 shows a typical form. Activities or elements are defined to occur between events. Thus the element 'get washer' might be defined to start at the event 'touch washer' and end at the event 'washer touches bolt'. Event definitions are not recorded on the observation sheet, only activity names. Any activity which occurs out of sequence (e.g. dropping a washer or answering a supervisor's questions) is dealt with by annotation as a 'foreign' element.

Analysis is performed to obtain mean values of element times and, less often, their standard deviations. If a simple event/time record is kept, statistics for element times can be recovered by calculating means and standard deviations for every event transition possibility. A repetitive task, even with some foreign elements, will only create a small number of different event transitions. If a computer program is used to record the events and times, then a simple program can be used to find the sample statistics for each transition. Figure 2.3 shows an event recording menu, the raw event/time record, and event-transitions output for a repetitive task using the programs in Drury (1987).

As noted in chapter 5, industrial time studies use two modifications to the representative time (usually the mean time neglecting 'outliers') to arrive at a *standard time* for a task. Both modifications, rating and allowances, have no value for ergonomists, although we all need to be aware of them when using archival time study data.

Process charts

A *process chart* is nothing more than a plant (or office, etc.) layout with the materials movement for one or more processes marked on it. The materials movement can be a single line (as in a simple assembly line onto a base unit), multiple coverging lines (as in a progressive assembly line with sub-assemblies) or even a set of quite disparate flows (as in a job shop). Spatial and sequence information are both preserved, while timing and workload information is lost. At times, activity duration information can be superimposed to form an *annotated process chart*.

The main use of process charts is to provide a graphical indication of movement. They are a useful aid to visualizing the effects of changes in plant

Time Study Observation Sheet

Department   Final Assembly

Operation   Mount Norwegian Blue

Workplace   Bench 23A

Operator   A. Notlob   **Analyst**   Bolton, A.

Part Nos   221657, 273318

Start Time   11:08 a.m.   **Start Date**   4/1/88

Element	1	2	3	4	5	6	7	8	9	10	Rem
1. NB from box to rail	.17	.16	.16	.17	.15	.16	.17	.15	.16	.17	
2. P U Nails	.23	.24	.23	.23	.22	.24	.24	.24	.24	.23	
3. Nail L	.14	.14	.14	.13	.14	.13	*.78	.17	.14	.13	*Dropped
4. Nail R	.13	.14	.14	.13	.13	.14	.55	.13	.14	.28	
5. Adjust to spec	.35	.36	.41	.30	.40	.51	.58	.30	.27	.31	
6. Rail + NB to cage	.41	.38	.42	.43	.34	.41	.50	.41	.39	.40	
7.											
8.											
9.											
10.											
11.											
12.											
TOTAL	1.43	1.42	1.50	1.40	1.43	1.59	2.82	1.40	1.34	1.52	

Selected Time   1.448   **Rating**   90

Normal Time   1.303   **Allowances**   10%

Standard Time   1.537 min

Figure 2.2. Time study observation sheet example.

layout as they dramatize the often inefficient pattern of movements required
to process a part from raw material to finished product. Figure 2.4 shows
an example of such a product flow change associated with the introduction
of a 'just-in-time' manufacturing cell. If the process chart is only produced
as a visual aid in presentations, an even more dramatic alternative is to

(a) Computer on-screen menu

1 Pick up wrench	6 Nut 1 to bolt 2
2 Wrench to nut 1	7 Start nut 1
3 TIghten nut 1	
4 Wrench to bench	
5 Pick up nut 1	

TO STOP PRESS 0

PRESS ANY KEY TO GO

(b) Data set

598	1	959	2	1333	3	1583	4	1956	5
2282	6	2632	7	3041	1	3261	2	3491	3
3723	4	4268	5	4400	6	4699	7	5137	1
5564	2	5650	3	6109	4	6497	5	6740	6
7034	7	7563	1	7849	2	8076	6	8381	3
8705	4	8984	5	9399	6	9727	7	9825	1
10165	2	10440	3	10728	4	11114	5	11459	6
11724	7	11812	0						

(c) Data summarization

	TIMES BETWEEN EVENTS IN 1/100 SEC UNITS				
	N	SUM	SUM SQ	MEAN	STD.DEV
FROM 0 TO 1	1	598	357604	598	0
FROM 1 TO 2	5	1644	567086	328	81
FROM 2 TO 3	4	955	274177	238	124
FROM 2 TO 6	1	227	51529	227	0
FROM 3 TO 4	5	1553	514925	310	90
FROM 4 TO 5	5	1971	813535	394	95
FROM 5 TO 6	5	1461	473999	292	108
FROM 6 TO 3	1	305	93025	305	0
FROM 6 TO 7	5	1536	476146	307	32
FROM 7 TO 1	4	1474	648570	368	187

Figure 2.3. Computer-based data collection and summarization in time study (times in 0.01 sec. units)

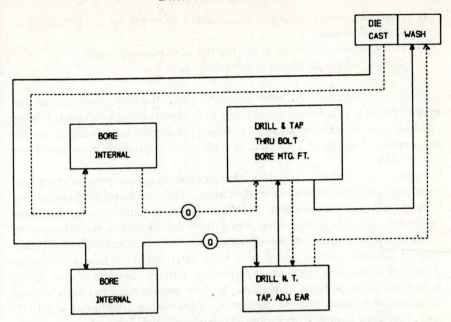

Figure 2.4. Process flow in a manufacturing cell, showing routes of two components.

videotape a part as it flows through the system. Before and after comparisons using this technique can be striking, even if the same information could be presented by simple numerical comparisons of time or distance moved.

A process chart is an ideal first technique in many studies as it allows the analyst to observe the system, ask questions, and talk to supervisors and operators in a non-threatening manner. It also serves a technical purpose in that it brings to light rework, backward flows, work in process storage areas, and multiple handling.

Note also that a process chart can be used to follow the progress of other entities apart from products. Continuous material flows (liquids, gases), movement of patients and staff in hospitals, and even information flows within a computer network are all legitimate process charting examples.

Flow process chart

A *flow chart*, or in its more restricted use, a *flow process chart*, removes the spatial information from a process chart and in its place uses conventional symbols to indicate particular aspects of the flow. The symbols are as follows:

○ *Operation*, where an object is intentionally changed.
⇐ *Transport*, where an object is moved and the movement is not part of an operation.

◊ *Inspection*, where an object is examined or tested for quality, quantity or identification.

D *Delay*, where an object waits for the next planned action.

△ *Storage*, where an object is stored for later use.

There are variations. As with a process chart, the flow process chart can refer to humans or information as easily as to objects being processed. Process symbols can be annotated with numbers to represent comments or the geographical location of activities, or with times to produce an *annotated flow process chart*.

Because they remove spatial location information, flow process charts can emphasize the logical structure of operations. They are useful for highlighting *delay and storage* elements which are expensive and in many ways the anathema of modern manufacturing. One aspect they do enhance is the structure of multi-component and branching product flows. Two branches flow together when a sub-assembly is added to a main assembly; two branches diverge whenever a choice is made at an inspection activity. Because of this ability to abstract the logic from a situation, the flow process chart has been adapted to many other areas such as computer programming (Phillips *et al.*, 1988) task analysis of complex operations (Drury *et al.*, 1987) and even signal flow graphs used in process control task analyses (Edwards and Lees, 1974). The same principles are used in charting precedence diagrams into networks. An example is the familiar PERT chart for project planning, a chart which is now available on even microcomputer project planning and control software.

In addition to their use as direct aids to visualizing a system, flow process charts for multiple product flows form the basic input data to link charts and from/to matrices (see later, p.49). Figure 2.5 shows a flow process chart for an operation in assembly of automotive thin-film ignition components.

Gantt charts

Named after its inventor in the 1920s, the *Gantt chart* is a graphical depiction of the time relationships among several activities. Figure 2.6 gives an example from a project plan. The time axis may run vertically down or horizontally to the right. There are as many bars on the graph as there are activities to be depicted, with bars starting and ending at the appropriate times. Although the chart in Figure 2.6 has a time scale covering weeks, a Gantt chart can be used for any operation with multiple activities such as maintenance to machinery or even a surgical procedure in the operating room.

The primary use of the Gantt chart is in visualizing several related activities. It is used extensively in scheduling (vehicles, orders in factories) and project planning. Microcomputer-based project planning packages use Gantt charts as a principal output. Even if the ergonomist never has occasion to plot a Gantt chart for a system under study, it is an essential tool (with a PERT network) for ensuring that ergonomic projects are completed on time.

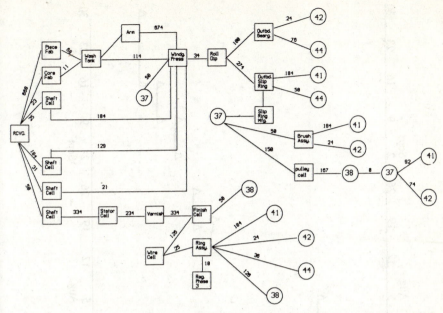

Figure 2.5. Flow Process Chart for manufacture of automotive components. Numbers on links represent flow volumes.

Multiple activity charts

The concept of the Gantt chart is modified somewhat for the *multiple activity chart*. Instead of each bar representing a single activity, the activities are grouped into continuous bars. Thus a single bar may represent the activities of a machine operator, while a second bar represents the status of the machine (e.g. off, operating, standby). Different shadings are used on the bar to represent the different activities or states. Figure 2.7 shows a multiple activity chart for the operation of a just-in-time cell. Front and rear housings are machined on six machines. One machine (column 1) operates on pairs (two fronts and two rears), two other machines operate on two rears while three deal with two fronts. The columns OP1 and OP2 show the two operators, with black representing times at which the operator is loading/unloading a machine (L) or gauging parts produced (G). Machine cycles either require the operator (O) or are unattended (CYCLE). Figure 2.7 was one of a series to determine throughput and operator workloads under different configurations of machines and operators in the cell.

A multiple activity chart is a general purpose tool. It can be annotated with exact times, comments and notes. The standard flow-charting symbols can be used in place of shading. The specific form shown in Figure 2.7 is known as the *man/machine chart* in industrial engineering texts (e.g. Salvendy, 1982). A common variant on a smaller scale is the *left hand/right hand chart*

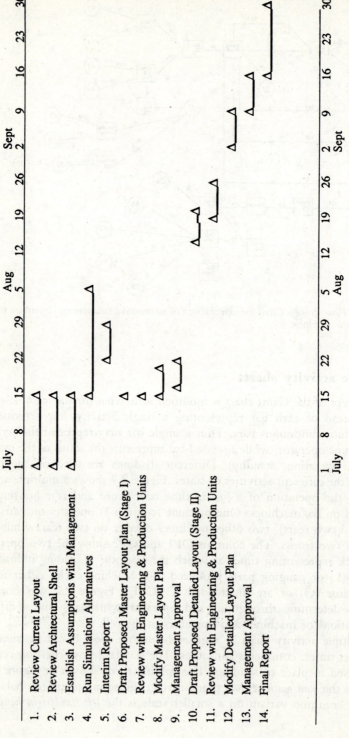

Figure 2.6. GANTT chart for design of a new facility.

REARS AND DOUBLE   FRONTS

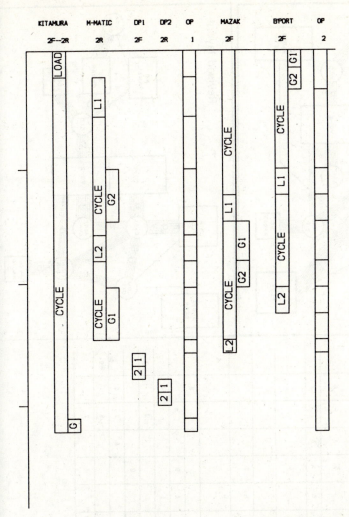

Figure 2.7. Multiple Activity Chart for a manufacturing cell. Time goes down the vertical axis.

(e.g. Konz, 1983). Here the two bars represent the two hands of an operator in an assembly task, showing which activities are being performed and the co-ordination required. In a more obviously ergonomic context, similar charts have been devised which have separate bars for cognitive and information processing activities. Crossman (1956) proposed the *sensori-motor process chart*, which has seen some use in the field of manual skills training (e.g. Seymour, 1967).

A multiple activity chart is a uniquely useful way to visualize the co-ordination of several activities. It shows conflicts (where a resource is needed

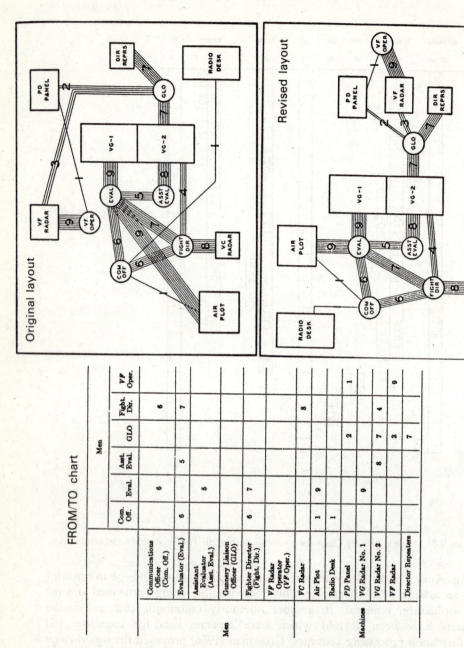

Figure 2.8. Link charts for a battle cruiser under air attack, from Chapanis (1953). The first chart represents the original layout, the FROM / TO chart extracts the movement information, and the second chart shows a revised layout with reduced congestion.

in two places at once) and idle time (where a resource is waiting for a prior operation to finish). The ergonomist should note that 'idle time' does not mean that the operator is doing nothing. Most tasks have planning and other cognitive acts which are no less required for not being visually obvious. In Figure 2.7, blocks of time are needed for preventive maintenance, machine monitoring, and communications with other parts of the plant, even if these are not activities built into every cycle of operations.

Link charts

When multiple products flow through a system using multiple paths, the resulting process or flow process charts can become so complex as to be unreadable. In these cases, the separate product identities are sacrificed to preserve only the information on how many products flow between each pair of machines. The data can be recorded as a table with identical row and column labels, one for each machine, called a *from–to chart*. In such a chart, the number in each cell represents either the number of products which flow between the machines of row and column, or a weighted number, for example, total trays of parts or total kilograms of parts. For any strictly repetitive task, only one entry will appear in each row and column, showing a single flow path. A totally random flow would have approximately equal weights or numbers in each cell.

Any from–to matrix can be equally well represented by a network, with the row and column labels as nodes and the cell entries as links between the nodes. This is a *link chart*. In most applications, the flows between nodes are represented by either numbers appended to each link or by a code of heavier link lines for larger flows. Figure 2.8 shows a classic example, from Chapanis (1953) of human and machine communications in a battle cruiser during air attack. The matrix contains exactly the same information as the first diagram, but it is much easier to visualize the relationships in the link chart. The second link chart demonstrates the reduced flow lengths and reduced path interference when the layout was revised.

As with other charting methods, there are variations. Spatial information can be removed to create a multi-product analogue of the flow process chart, even to using the standard symbols. Annotation can indicate unusual circumstances for each machine or link, e.g. different link usage under start-up and operating conditions. The same analysis and charting can be applied to sequences other than product flows between machines. Thus control use sequences on a control panel can be made into diagrams in this way. Figure 2.9 shows the information flow sequence on the panel of a metal stamping line before and after ergonomic intervention. Another classic use is to record eye movements between instruments in a complex display.

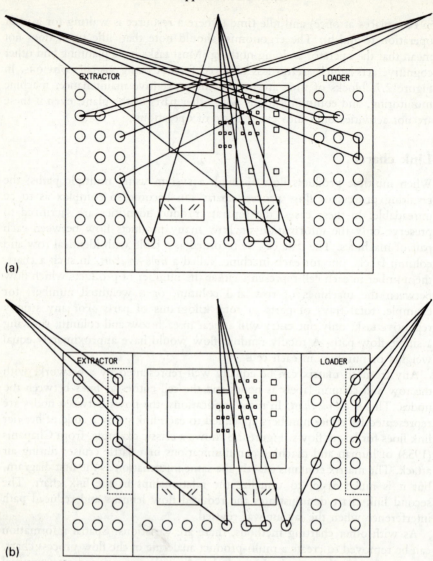

Figure 2.9. (a) Hand/eye movement chart of a control panel for a stamping operation.
(b) Chart for revised panel layout where outer panels have been laterally reversed and sub-panels installed for major sequential task.

For a from–to matrix, algorithms exist to optimize spatial layout. The science of *facilities layout* is concerned with minimizing flow costs (among other costs) by locating facilities optimally on a one, two or three-dimensional space. Simple computer programs (e.g. CRAFT, CORLAP) exist to improve plant layouts (Tomkins, 1982). Any of these can be applied to the control panel layout problem to minimize movement between controls.

Occurrence sampling

The idea of sampling system state at predetermined intervals rather than timing the instant of occurrence of events goes back to Tippett (1934), who wished to estimate the fraction of time an operator or a machine was working. Known earlier as *ratio delay* or *work sampling*, the method of *occurrence sampling* is thus qualitatively different from the others presented so far.

In occurrence sampling, the analyst observes the system at predetermined times. To perform the study we need to determine both the observation times and the system states or categories which we will record. Choosing the times means choosing the total number of observations (see later in this chapter), the time intervals between observations, and the statistical distribution of observations over time.

System states must be observable, unambiguous, mutually exclusive, and exhaustive. The analyst needs to understand the system well to develop a rational and useful set of categories. Observable states means that the analyst needs only to record what is seen, not what is inferred. Thus 'idle', 'thinking', 'planning', and 'daydreaming' all look very similar to the observer. A safer state to record would be 'no task seen' as it is both observable and unambiguous. Mutually exclusive means that on any single observations, only a single response is possible so that states must be defined so that two cannot happen simultaneously. Thus for a secretary, it is entirely possible to be on the computer while answering the telephone. The analyst must record only what is being done at the instant of recording—a key being pressed (computer use) or a word being spoken (telephone). Exhaustive means that ANY system state can be recorded. The better you know the system, the better you can anticipate all possible system states. An 'other' category is always a safe solution, although if it captures more than an isolated reading or two, more named states should be added to the study.

The choice of number of readings will be considered later but the time interval between readings is an important variable. If observations are too close to each other in time, an activity may continue from one reading to the next. This introduces sequential dependencies into the data, destroying the assumption of independent observations. Thus readings should be far enough apart in time to ensure independence, a stricture which requires at least some foreknowledge of the expected durations of activities and states. On the other hand, if observations are too widely separated, the total study duration can become so long that changes (e.g. in product mix or workload) take place during the study, making the derived probabilities meaningless.

The other choice to be made is of the observation schedule—should it be random or fixed-interval? A fixed-interval schedule is attractive in that it allows the analyst time to interleave another activity with the occurrence sampling study. However, if there is any natural periodicity in the system being observed, then there is a risk that each observation will coincide with a certain phase of the cycle, leading to highly biased data. The alternative is randomization and is always safer. If in doubt, randomize.

Having decided what to observe and when, the recording sheet needs to be designed to reduce recording errors. Such sheets should have a list of all system states at the top and leave a place for recording the code for each state against a predetermined time printed on the sheet. The alternative way of recording is to have columns representing the states and rows representing successive observation times. Here a check mark is placed in the appropriate column of each row in turn. The use of tally marks against system states is not recommended as it does not prompt the analyst with the time for each reading, leading to missing data.

A simple computer program is available for portable computers to conduct simple occurrence sampling studies (Drury, 1987). Figure 2.10 shows an example of the on-screen menu and the data file for a secretarial job. In setting up the study, the user is prompted for the total number of observations and the total study duration. The program then generates random times to give the correct number of observations in the given duration, providing an audible prompt when each observation is due.

It should be noted that inexpensive video cassette recorders (VCRs) now allow the analyst to film the system throughout the study period unattended and then to perform the occurrence sampling study on the recording. This means that the study can take up much less of the analyst's time.

An occurrence sampling study only provides information on the fraction of time the system (human or machine) spends in the various states. Time and sequence information is lost. It is thus uniquely useful in systems with no repetitive cycles (e.g. maintenance), or with too many different repetitive cycles to conveniently keep track of (e.g. an attendant serving several looms). There have been uses of occurrence sampling to determine times for repetitive

(a) Computer on-screen menu

1 Writing	7 Reference use
2 Typing	8 No task seen
3 Computer use	9 Absent
4 Telephone	
5 File cabinet	
6 Walk-in queries	

A "beep" will tell you when to enter
each observation. PRESS ANY KEY TO START

(b) Data file

1	Writing	1
2	Typing	3
3	Computer use	6
4	Telephone	5
5	File cabinet	2
6	Walk-in queries	3
7	Reference use	1
8	No task seen	3
9	Absent	1
total = 25		

Figure 2.10 Example of computer-based occurrence sampling study for a secretary.

jobs by adding more information. Thus if 74% of the time is spent operating a machine, which produced 1532 parts in a 40-h week, the operating time per part is

$$\text{operating time/part} = \frac{40 \times 0.74}{1532} = 0.019 \text{ h}$$

or 1·16 min/part. There are easier ways to perform time studies, but this way has high face validity.

General considerations

Crossing all of the methods described above are issues of who or what to observe, how often to observe and how to define the observation events.

Choice of subjects and conditions

Choosing representative subjects and time periods in an observation study is difficult. If we choose 'normal' subjects and periods (however defined), we miss many of the unusual conditions which are such a strength of methods involving direct observation of a system in its natural state. However, if we want to predict future performance, we may want to avoid New Year's day or an untrained operator or the time when the machine shop was flooded after a thunderstorm. The general considerations of sampling (such as random, stratified and clustered) which are usually discussed with reference to questionnaire surveys (e.g. Sinclair, 1975) apply equally well here. Indeed, they are explicit considerations in modern treatments of occurrence sampling (Richardson and Pape, 1982).

It may be preferable to perform different separate studies under different system conditions for many reasons. Not only would the final result be applicable more widely, but even such details as choice of events and system states may be different between conditions. The events appropriate to running a power station are quite different from those encountered in start-up or shut-down. A data recording technique to cover all conditions would be difficult to devise without it becoming unwieldy.

Ethical observation

Ethical considerations need to be explicitly addressed. We are observing an operating system and, if it is possible that the data collected may harm the participants, proper safeguards are needed. It should not be possible to tie particular data to individual subjects. This may be reasonably simple where the data recording is done by the ergonomist, but is very difficult to ensure with direct computer recording or videotaping. The workforce is now alerted to the potential misuse of keystroke recording in office tasks and timing of manual operations with VCRs, but they also need to be informed of the objectives of your study and how you intend to protect their rights and privacy. For example, the author used videotaping of traders at a major stock exchange (Drury, 1980) and then interviewed each trader while the tape was playing. Traders were assured that the tapes would be erased after use, and this assurance had to be upheld when management realized that the tapes would make a fine training tool. If a training tool was needed, new tapes should be produced with the specific permission of the subjects.

Sample size determination

The total number of observations required in a study is a trade off between accuracy (or risk of drawing the wrong conclusions) and the cost of collecting and analysing the data. Accuracy can be expressed as a confidence interval in which the true value of a measured statistic should lie with a given probability. Thus we could say that the duration of a task should be estimated so that there is a 95% chance of the true mean time lying with $\pm$ 10 s of the calculated sample mean. To calculate sample size, we need to know the statistical properties of the process generating the data, the width of the confidence interval, and the probability of the interval containing the true value.

All of the methods described produce either a time or a fraction, and hence we need distributions of times and proportions. For estimating times, we typically assume that they are normally distributed, although log–normal would often be a closer approximation for the skilled operator (Dudley, 1968). If we assume normality, then we can calculate the number of readings, N, from

$$N = \left(\frac{2\,z_{\alpha/2}\,\sigma}{A}\right)^2 \tag{1}$$

where: $z_{\alpha/2}$ is the normal deviate corresponding to a confidence interval including a fraction $(1 - \alpha)$; σ is the population standard deviation of performance times; and A is the absolute size of the confidence interval in time units.

Thus if we wish to find the time for an operation where the 95% confidence interval is 10 s and the population standard deviation is 20 s, we have

$$z_{\alpha/2} = z_{0.025} = 1.96$$

$$\sigma = 20\text{ s}$$

$$A = 10\text{ s}$$

$$\therefore N = \left(\frac{2 \times 1.96 \times 20}{10}\right)^2 = 61.5.$$

Thus 62 readings will be needed.

For occurrence sampling, a fraction or proportion is to be estimated, thus we use the binomial distribution. For large sample sizes and proportions not too close to 0·0 or 1·0, we can use a normal approximation to the binomial and thus use a similar equation to that used for times.

The standard deviation (σ) of a binomial proportion (p) is given by

$$\sigma^2 = \frac{p\ (1-p)}{N}$$

we have $N = \dfrac{2^2 p(1-p)\ z^2_{\alpha/2}}{A^2}$ (2)

where z, σ and A retain their previous definitions.

Thus to find the number of observations required to estimate a proportion to an accuracy of 0·02 with a probability 0·90, when the expected proportion is 0·75, we have

$$z_{\alpha/2} = z_{0\cdot05} = 1\cdot645$$

$$p = 0\cdot02$$

$$A = 0\cdot02$$

hence $N = (0\cdot02)(0\cdot98)\ \dfrac{(2 \times 1\cdot645)^2}{0\cdot02^2} = 530\cdot3$

or 531 observations.

Most texts will give alternate formulae for relative intervals (where A is to be a percentage of a mean), and different definitions of α, but the ones presented here cause the minimum confusion. Confidence interval length, A, is always the total length measured in the correct units (seconds or proportion), with no confusion about plus-or-minus intervals or definition of percentages.

It remains to ask where A, α, σ and p come from. The statistical parameters, A and α, are supposed to be given by the sponsors of the study but I have never found this to be the case in industry! In epidemiological work for government, one may be dealing with sponsors sufficiently versed in statistics for this to be possible, but in most industrial situations, the sponsor is more concerned with the cost, which is proportional to N, than with details of precision. The ergonomist must be prepared to advise on sensible cost/accuracy trade offs. Even a short BASIC or FORTRAN program to find N given A, α, σ (or p) can be useful in demonstrating how small increases in precision can have a large effect on study time and cost.

Estimates of σ and/or p are more difficult to obtain. The typical advice is to run a pilot test, which is always a good idea, but it does not help in costing and time estimation for studies on a new system. For research questions involving laboratory data, one can obtain gross estimates from the published literature, but in a working system consensus guesswork may be the only practical solution.

Choice of method

The variety of what to observe and how to observe it can leave the ergonomist either bewildered or practising tunnel vision to keep the number of alternatives manageable. An obvious way out of this embarrassing position is to return to the study objectives for guidance after the initial observation of the system. What are the main issues of interest—performance or well-being? If performance, how can simultaneous sense be made of speed and accuracy? Which of the methods are truly non-reactive in this system? (See chapters by Drury on studies and experiments.)

Perhaps the most telling are outcome measures, the final results of system performance. Was the mission completed? What was production/quality for the month of March? Did the plant manage to remain in business? Unfortunately, a system rarely stays still while ergonomics interventions and measurement takes place, so that such global measures are difficult to tie logically to the ergonomist's work. More often we rely on process measures, i.e., those which can be logically related both to ergonomics interventions and to final outcomes. Examples are error rates and productivity on specific jobs, improved process flow or reduced time spent waiting for work. These are certainly more comfortable to the ergonomist but need to be related directly to study goals before the study starts. It is important that the client agree to the measures and their interpretation in mission terms if later problems of meaning are to be avoided.

Meaning is the key. The purpose of any study is the interpretation rather than the data themselves. In a scientific journal, uninterpreted data are (rightly) difficult to publish. Equally, in an applied study, unless the data contribute directly to the *client's* goals, the study is sterile. As with anything an ergonomist does, it is the ergonomist's responsibility to make this interpretation. Hence there is no substitute for the ergonomist's knowledge and understanding of both the system under study and the ergonomics literature.

References

Barnes, R.M. (1980). *Motion and Time Study: Design and Measurement of Work*, 7th edition, (New York: John Wiley).

Chapanis, A. (1953). *Research Techniques in Human Engineering* (Baltimore, MD: The Johns Hopkins Press).

Cohen, H.H. and Jensen, R.C. (1984). Measuring the effectiveness of an industrial lift truck safety training program. *Journal of Safety Research*, **15**, 125–135.

Crossman, E.R.F.W. (1956). Perceptual activity in manual work. *Research,* **9**, 42–49.

Drury, C.G. (1978). The law and bicycle safety. *Traffic Quarterly,* **32**, 599–620.

Drury, C.G. (1987). Hand-held computers for ergonomics data collection. *Applied Ergonomics,* **18**, 90–94.

Drury, C.G. (1980). Task analysis methods in industry. *Applied Ergonomics,* **14**, 19–28.

Drury, C.G., Paramore, B., Van Cott, H.P., Grey, S.M. and Corlett, E.N. (1987). Task analysis. In *Handbook of Human Factors,* edited by G. Salvendy (New York: John Wiley), pp. 371–401.

Dudley, N.A. (1968). *Work Measurement: Some Research Studies* (London: Macmillan).

Edwards, E. and Lees, F.P. (1974). *The Human Operator in Process Control* (London: Taylor and Francis).

Hansen, B.H. (1960). *Work Sampling* (Englewood Cliffs, NJ: Prentice-Hall).

Kadota, T. (1982). Charting Techniques. In *Handbook of Industrial Engineering* edited by G. Salvendy (New York: John Wiley).

Konz, S. (1983). *Work Design: Industrial Ergonomics* (Columbus, OH: Grid).

Matthews, M. (1982). Seat belt use in Ontario four years after mandatory legislation, *Accident Analysis and Prevention,* **14**, 431–438.

Mundel, M.E. (1978). *Motion and Time Study,* 5th edition (Englewood Cliffs, NJ: Prentice-Hall).

Phillips, M.D., Bashinski, H.S., Ammerman, H.L. and Fligg, C.M. (1988). A task analytic approach to dialogue design. In *Handbook of Human Computer Interaction,* edited by M. Helander (Amsterdam: North-Holland).

Richardson, W.J. and Pape, E.S. (1982). Work sampling. In *Handbook of Industrial Engineering,* edited by G. Salvendy (New York: John Wiley).

Salvendy, G. (1982). *Handbook of Industrial Engineering* (New York: John Wiley).

Salvendy, G., McCabe, G.P., Souminen, S. and Basila, B. (1984). Non-work related movements in machine-paced and self-paced work: an industrial study. *Applied Ergonomics,* **15**, 21–24.

Schiro, S.G. and Drury, C.G. (1981). Emergency medical communications: an ergonomic evaluation. In *Case Studies in Ergonomics Practice,* Volume 2, edited by H. Maule (London: Taylor and Francis), pp. 65–81.

Seymour, W.D. (1967). *Industrial Skills* (London: Pitman).

Sinclair, M.A. (1975). Questionnaire design. *Applied Ergonomics,* **6**, 73–80.

Tippett, L.C.H. (1934). Statistical methods in textile research. *Shirley Institute Memoirs,* **13**, 35–93.

Tomkins, J.A. (1982). Plant layout. In *Handbook of Industrial Engineering,* edited by G. Salvendy (New York: John Wiley).

Chapter 3

Subjective assessment

Murray A. Sinclair

Characterising the ergonomist's tasks

Assume you have to carry out some design project. Considering only the information and knowledge that you need to do this, we can picture it as in Figure 3.1, as an 'egg' of knowledge, divided as shown.

The important thing about this representation is that it emphasises the need to gain knowledge as the project continues; almost never are you in a position to say, 'I know enough'. Of course, this applies not only to design projects, but to almost all projects.

The next general point to make is to characterise the activities of ergonomists. Let us take an example familiar to us all—the educational process. Imagine you have been asked to improve the efficiency with which statistics knowledge is transferred to students on a given course. A simple model of this might be Figure 3.2.

Given that this represents the process, you as ergonomist might have three roles to play:

- *Explorer:* your first task is to show that this model does in fact represent what happens. One way in which you might wish to gain relevant information is by talking to people involved in the current system
- *Optimiser:* knowing the goals of the system, your task is to optimise the flow down the various channels to ensure both more efficient learning and control of the learning process. For evaluation purposes, you might wish to talk to those involved about their perceptions of the changes you have made.
- *Innovator:* you introduce new channels of flow to improve even more the efficiency of the system—see Figure 3.3. Again you might wish to evaluate the changes by talking to people involved.

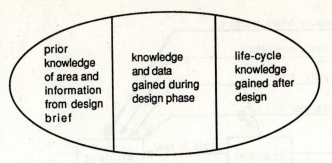

Figure 3.1. Representation of knowledge required for a design project.

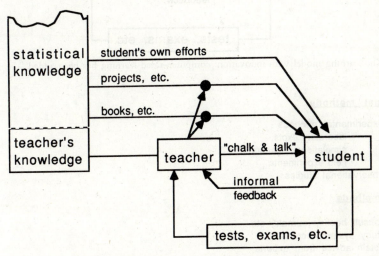

Figure 3.2. A simple model of system for transferring knowledge to student. The black blobs on the lines represent the teacher's ability to control these information flows.

For each of these activities, then, you will have to gain knowledge. This chapter is about a class of methods for accomplishing this, known as 'Subjective Methods'.

What are subjective methods, and why use them?

The first thing to say is that there are many methods for gleaning information, and there are several classifications for them. Edwards (1974) offers one based on the intrinsic nature of the measurement, another is given by Alluisi (1975) based on the purpose of the measurement exercise, and a third is by Meister and Rabideau (1965) based on the source of the data. An edited version of the last is given in Figure 3.4.

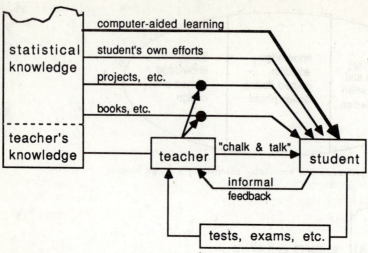

Figure 3.3. The learning model with innovation: computer-aided learning has been added.

Observational methods

- Experimental methods
    - laboratory expts.
    - simulations & games
    - field experiments
- Observational studies

Database methods

- consult books, journals, etc.
- study system records
- obtain advice from experts

Subjective methods

- questionnaires & interviews
- rankings & ratings
- critical incident techniques, etc.

Figure 3.4. Classification of data-gathering methods, edited from Meister and Rabideau (1965). It should be noted that class boundaries in this scheme are fuzzy.

The first class, 'Observational methods', includes most of the methods that are derived from those used in the so-called 'hard' sciences, such as physics. Laboratory experiments, activity sampling methods and so on are examples of these. The thing in common among these methods is that some degree of formal 'objective' measurement is involved; the idea is that you are reaching out to touch 'reality' as directly as possible. It should be noted however, that while you may measure accurately what you decide to measure, what you decide to measure may not be a good measure of reality. Discussions of this and other related points will be found in chapter 2 on direct observation by Drury in this book and in Chapanis (1967), Simon (1976) and Jung (1971). Feyerabend (1975), Ziman (1980), and Hudson (1972) give other interesting views on the role of experiments in the scientific enterprise.

The second class constitutes the historical class, where records of one sort or another or the collected, conflated and disseminated wisdom of others in books, journals, reports and so forth are used. The main problems here are firstly, that in many cases the information is not collected for the kinds of purposes that ergonomists have in mind and important types of data are either not collected, or are lost during data conflation. Secondly, where you are consulting expertise, it is often the case that because the expertise is intended to be generic for a whole class of problems, it is good on rules and standard procedures, but poor on specific facts and specific situations. Meister and Rabideau (1965) discuss these points at length and the following chapter (4) by Drury in this book is on the use of archival data.

The third class constitutes the subjective methods. The common thread here is to use the people involved in the system that you wish to study as measuring instruments. In effect, you rely on people to come to some sort of conclusion about the system, then access that conclusion as a measurement of the system. This class contains any method that draws its data from the psychological contents of people's heads.

This chapter is concerned with the last of these classes. This particular class of subjective measurements is important for a number of reasons:

(1) There are size limits to what can be measured 'objectively'. Without enormous resources, how would you measure, assemble and digest all the objective data about the activities occurring simultaneously, interdependently, in public and in private, on a ship such as the *Queen Elizabeth II* in order to assess the efficiency of the ship's crew?

(2) There are type limits as well. Many types of information can be measured objectively, as we all know, but there are some that can't be—at least, not easily. For example, how would one measure the skills involved in figure-skating without using human judges? Rather more germane to our problems, we often wish to assess human decision-making skills. For this, one needs to know what the decision-maker was thinking, not just what data were presented and what the resulting actions were. This kind of information is not accessible without subjective methods.

(3) As a sage once remarked, 'Seeing is believing, but experience gives you an original truth'. The point of this is that people who are part of a system (for example, the 'teacher' and the 'student' in Figure 3.2) will have different views of the system, both from each other and from an external observer. Furthermore, their perceptions will almost certainly be deeper, more complex and more subtle

than those of the external observer as far as their local area of the system is concerned, and therefore will constitute a valuable source of information.

(4) There are accuracy limits. In many cases, instruments can be made more sensitive, more reliable, more precise and more accurate than humans, but this is not always so, especially where qualitative judgement is concerned, or where a particular measurement is multivariate. Taste is a good example; it is not for nothing that most food and beverage companies allocate considerable resources to taste panels. Certainly, there are problems of human bias, but the subjective methods are intended to reduce, and it is hoped to nullify such sources of error.

(5) Finally, there is the validity argument. Irrespective of the methods used, we would like to know that we have valid data. The most common approach to try to ensure this is to use the notion of 'Convergent Validity'. The principle is that if you use two different, independent methods to get data about the same topic, and both produce the same (or nearly the same) results, then it is likely that the data are valid.

The arguments in favour of subjective measurement thus lie in the independence of the measures and the ability to acquire data that cannot be obtained, or cannot be obtained easily, by other methods. The two main arguments against the use of such methods are firstly the inherent biases in human judgement, and secondly the resource requirements necessary to get reliable data. These two arguments will be considered later, after a discussion of some of the basics of subjective assessment.

Typical methods used in subjective measurement

The methods available are legion. Of these, only a few will be discussed, the ones used most frequently by practising ergonomists. These are:

Ranking methods

These methods are concerned with questions of the type:
'Given four typefaces, [Courier, **Helvetica**, Times, Palatino], rank them from first to last for ease of reading.' What you obtain are data that distinguish between the examples quite well, but do not tell you for example whether the best is actually easy to read.

Rating methods

These methods deal with questions of the type:
'Given the same four typefaces, rate each one for ease of reading on the following 5-point scale:

☐ Very easy to read
☐ Easy to read
☐ Acceptable
☐ Difficult to read
☐ Very difficult to read

These methods will provide information on the perceived ease of reading of the examples, but are less sensitive to differences between them.

Questionnaire methods

Questionnaire methods make use of the two kinds of questions above, as well as several other classes of questions. Typically, questionnaires presume a fixed series of questions, often with a fixed range of alternative answers. As far as the respondent is concerned, they are typically either paper-based or interviewer-based. Questionnaires are the most common method for collecting subjective data.

Interviews

Interviews are at the opposite end of a continuum from questionnaires. They are distinguished from questionnaires by the relative lack of rigidity in the questions and acceptable answers. In other words, the interviewer has more discretion and flexibility over the questions and the course of the interview.

Checklists

Checklists are much as the name suggests; they are lists of items that you expect might occur in a given situation, and you wish to record their occurrence. In use, the observer of the situation ticks off against the list whichever item is observed to occur.

The common characteristic among the methods is the environment in which they operate. This is shown in Figure 3.5. The generic problem faced by anyone in obtaining the relevant information is that the query must be transmitted to the respondent in such a manner that the respondent can understand what is required. Then the respondent must delve into longterm memory to retrieve the information required. This may not be a textual memory; it could be a sensation of comfort or of anger, or a visual image. The respondent must then turn this memory trace into a suitable form of words. In all of these steps it is possible for errors or bias to creep in.

Discussion of the scaling methods

Theoretical details regarding the various classes of methods listed above are not discussed here; whole books are devoted to these methods, and for such details you might wish to consult some of the texts listed in the references,

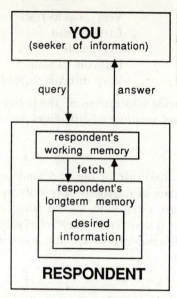

Figure 3.5. Representation of the information-seeker's basic problem. In most cases the information sought requires the respondent to access longterm memory, retrieve the information and then report it accurately to the information-seeker.

such as Coombs *et al.* (1970), Guilford (1954) and Torgerson (1958).

Instead, there is an outline of the practical principles of the method and a brief discussion of the characteristics of the techniques. In case there is some difficulty with the jargon in this section, some definitions follow:

entity   The objects that are going to be scaled. For instance, if you wished to assess cookers for ease of use, each cooker that is assessed would be an entity. Similarly, if you assess people for intelligence, each person would be an entity. If you are assessing statements typed on cards, each statement is an entity.

attribute   The property of the entity that you are scaling. For ease of use of cookers, you might be scaling the 'ease of controlling temperature'. This is the attribute that you are assessing, and it is the extent to which each cooker possesses that attribute that you are measuring with your scale.

subject   The people that you use to do the scaling. In the cooker example, you might recruit people off the street as 'lay persons' to try the cookers and register their opinions. These people are the subjects. If these people are assessing other people (for instance politicians), the latter would be entities, from the definition above.

respondent Usually used in connection with questionnaires. Again, they are the people who give their opinions which are recorded on the questionnaire, and thus are subjects, from the definition above.

judge   A subject used for special purposes, in the creation of the scales in the first place.

Ranking methods

Details of ranking methods will be found in Guilford (1954), and Nunnally (1970). The analysis of the rankings produced is considered in many texts of non-parametric statistics; examples are Meddis (1984), Mosteller and Rourke (1973), Siegel (1956) and Tukey (1977). The first is particularly appropriate.

A single question as in the example about typefaces earlier is rare; usually, there is a series of questions forming a 'battery', probing a number of different attributes of the situation in question. These batteries are usually used in experimental situations, where the subjects have a number of different alternative entities to rank. Entities are usually physical objects, but there is no reason why people, software products and suchlike should not be included. The essentials of the method involve presenting the entities to the respondent in such a way that the attribute in which you are interested can be assessed. This may involve user trials, etc. (see McClelland elsewhere in this book, as well as Kirk and Ridgeway (1970, 1971) for examples of what this entails). The objects should be presented in a random order for each subject involved. The respondent should then be presented with all the entities together, and be asked to rank them. If there is a large number of entities to be ranked, it is sometimes better to ask the respondent to rank the best ones, then the worst, and then the ones in the middle.

Characteristics

- For each attribute to be ranked in the battery, each object should possess the attribute in some degree. While this is an obvious statement, it can be difficult to establish in practice.
- It may well be the case that for several of the objects, a subject may not have any real preferences. The ranking process disguises this for an individual subject, and its occurrence can only be detected by using Coombs' Unfolding Technique (or some derivative—see Coombs *et al.*, 1970) or by using many subjects and examining the results statistically.
- It is commonly accepted that ranking methods only record real preferences for the first two or three ranks, and the last two or three ranks. In between, the rankings are possibly unreliable, hence it is unwise to give too many objects to a subject for ranking; about nine is usually taken to be the upper limit.

Rating methods

Rating methods have been studied extensively since the early 1900's. The literature is replete with texts describing exhaustively the details of the various methods for generating scales. The most commonly-used methods are:

- Simple rating scales
- Thurstone's Paired Comparisons Technique

- Thurstone's Equal-appearing Intervals Method
- Likert's Summated Ratings Method

Good introductory texts to these techniques are Oppenheim (1966) and Edwards (1957). The latter is especially good. More technical texts are Guilford (1954) and Torgerson (1958).

Simple rating scales

Simple rating scales are as shown in Figure 3.6. The 'parking bays' representation is typical.

The method requires appropriate questions to be generated about the attributes of the entities involved, typically in the form of a battery of questions, as for rankings. Scales are then created for these questions. The important points here are that the researcher must decide whether it is necessary to have a genuine 'neutral' region in the middle of the scale (the example does not have one), and how the two ends of the scale shall be 'anchored' (i.e. given unambiguous labels that will be interpreted the same way by the majority of users of the scale). The rating scale itself is usually shown as a 100-mm line; subdivisions are optional, as are labels for the subdivisions. As for the ranking methods, the respondents should be presented with the entities in a random order, and should be given an opportunity to experience the attribute of interest. The respondents should then place a mark on the scale to indicate their opinions.

CHARACTERISTICS OF RATING SCALES
- This is the most common technique used by ergonomists. This is because of its ease of use, particularly for respondents.
- You must take care about the meanings of the scale anchor points and the labels used along it. It is not easy to establish clear anchor points, nor is it easy to get good labels; if in doubt, leave them off. It appears that almost all respondents are quite good at dividing the line into equal

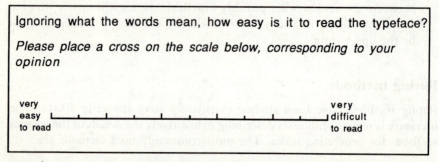

Figure 3.6. Example of a simple rating scale. It features the question to the respondent, instructions to the respondent, and the scale, as the standard 100 mm. line, in this case with tick marks. Scale 'anchors' are also given at each end, but no intermediate labels.

intervals, and in any case, if you use a 100 mm line, the usual way to get the rating converted to a numeric value is to measure to the nearest cm. This unit of measure will encompass most of the variability shown by respondents.

- There are a number of biases that can affect rating scales. A list will be found in Guilford (1954). The worst of these is the 'leniency' effect, where respondents are unwilling to be critical. A second problem is the 'halo' effect, where the respondent has already decided that one of the entities is better than the rest, and unconsciously 'adjusts' his or her ratings to demonstrate this clearly.

Paired Comparisons Technique

The Paired Comparisons technique is the 'standard' technique, against which other scaling methods are compared. In this technique, subjects are typically asked to compare two entities, A and B, and make a decision whether A is 'greater than' or 'less than' B. The entities are taken two at a time from a range of entities provided by the researcher, and the judgements are recorded for each pair. By making some statistical assumptions about the nature of human judgements, and using a pool of subjects, it is possible to derive a quantitative scale from the simple judgements, with the positions of the entities marked along it. For example, if we have compared typefaces two at a time for ease of reading, we would know where each typeface fell on an 'ease of reading' scale, and have a numerical value for that point. If instead of objects we use verbal statements, we can create a scale for measuring other objects, to be used by subjects in a field situation. For example, we may have created a scale for sitting comfort. A subject would then sit at a workstation, for example, and assess how comfortable he or she feels, then select a statement from the scale set which best represents that assessment. The numerical value for this statement is then allocated to the workstation seat.

CHARACTERISTICS OF PAIRED COMPARISIONS
- During creation of the scale, the judgement required of subjects is a relative judgement, and does not require the subject to assess 'by how much' one is greater than the other. This is a simple judgement, which most subjects find fairly easy to make about almost anything.
- Under normal circumstances, more data than strictly necessary are collected. This 'overdetermination' allows a number of internal checks for consistency to be applied to the resulting scale.
- The number of judgements required per subject rises by a factor of $[n(n-1)/2]$ as n, the number of entities, rises. This can require considerable subject time. There are penalties in this; a bored subject towards the end of the session is a very different person from the keen, perhaps slightly apprehensive, subject who started the session. There are ways of reducing the number of judgements required; see Torgerson (1958).

- The scale that results is usually taken to be unidimensional, but there is no guarantee of this (for example the comfort scale might be made up of two scales, one that measures ease and well-being, the other which measures localised pain in the rump). Small departures from unidimensionality may not be detected by the internal checks.
- Because of the resource problems it is commonly accepted that there should not be more than 9 entities to be compared.
- The success of the scale hinges on the selection of entities. If the attributes being assessed are not commensurable, or if the entities are too clearly dissimilar to each other, the scale will be invalid. In the case of scales of statements, the choice of statements is critical. One requires a set of statements that cover the range of the scale, each of which has a relatively fixed point on the scale. Most of the work should be allocated to this initial choice of statements.
- Care must be taken to ensure that the group of subjects used to create the scale is equivalent to the group of subjects who will use the scale subsequently, otherwise the scale values allocated may not be accurate.

Thurstone's Equal-appearing Interval Technique

The method of equal-appearing intervals produces the same scales as for Paired Comparisons, but is much quicker. However, there is less opportunity for internal checks. It was developed by Thurstone to provide a quick means of obtaining scales for use in the field, typically in questionnaires. The method emphasises the selection of statements to comprise the final scale; it assumes that a pool of statements can be created by the researcher, and provides a means of reducing this pool to the final selection. Once created, each scale is represented as a randomised list of statements, with no scale values. The subject using the scale is asked to select that statement which best represents his or her judgement.

CHARACTERISTICS OF THURSTONE'S TECHNIQUE

- The scales that are produced by this technique are almost comparable with the Paired Comparisons approach for validity and reliability.
- The data produced by this technique are very easy to analyse.
- Subjects can experience problems if their feelings don't quite match the statements given. This can create the impression that words are being put in their mouths, which can have fatal effects on their motivation.

Likert's Summated Ratings Method

The Likert scaling technique is based on a very different approach, but is taken to be as powerful as the Thurstone techniques. Whereas in the Thurstone techniques the subject is asked to select a statement with which he or she most agrees, in the Likert technique the subject must respond to every statement, showing his or her disagreement with it. Typically, each statement will have a simple 5-point scale associated with it as shown in Figure 3.7.

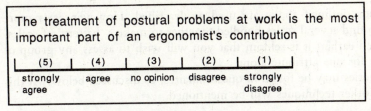

Figure 3.7. Example of a Likert scale item. It contains a statement with which the respondent can either agree or disagree, and indicate this opinion on the scale. Normally, the three scale labels in the centre are not included. The points values in brackets are never shown; they have been included here for illustration purposes only.

Each point has a scale value (for example 1 to 5), and depending on which point is selected, the points score is summated and the result represents the subject's opinion. Clearly, there is a premium in the selection of statements and in ensuring that the points allocation is matched to the statement. What one seeks is that if a subject's response to one statement obtains a score of 5, then a similar score will be obtained from a similar statement. For example, if a subject strongly agreed with the statement in Figure 3.7, scoring 5 points, one might expect the subject to disagree to some extent with the statement in Figure 3.8, again scoring high points. This enables individuals with differing viewpoints to be discriminated from each other by accumulation of these scores. Some 10 or 20 statements are used to create the scale.

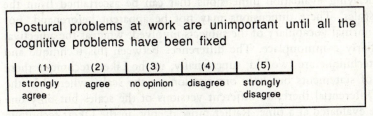

Figure 3.8. A second example of a Likert scale. Since the statement conflicts with the example in Figure 3.7, one would expect the (invisible) scoring for this item to be reversed, as shown.

CHARACTERISTICS OF LIKERT'S METHOD
 – It is said that subjects prefer the Likert scaling technique, because it is 'more natural' to fill in and because it maintains the subjects' direct involvement.
 – The Likert approach is said to require less effort to generate scales compared to the Thurstone techniques.
 – There appears to be no great difference between the Likert and Thurstone techniques in validity and reliability.

– Thurstone techniques are said to be better at measuring 'snapshot' views, whereas Likert techniques are better at measuring changes over time.

More complex methods using ratings and rankings

The methods outlined above are used as described. However, it is more common to find several scales bundled together to make a 'battery' of scales, as mentioned earlier; it is seldom that you will wish to assess any group of entities just for one attribute alone.

Such batteries may be found in questionnaires, as discussed below. At this point, two other techniques will be mentioned.

Semantic Differential Technique

This technique is described in detail by its originators, Osgood *et al.* (1975) and shorter descriptions will be found in most textbooks on attitude measurement (e.g. Moser and Kalton, 1971). Their original intention was to discover how groups interpreted the meaning of words; since then the technique has been used for all sorts of purposes, including evaluation of household goods. A series of rating scales is generated which describes the class of entities to be studied. By convention, the scales are usually seven categories long. The end points of the scales are given anchors which are single word adjectives, and are 'polar opposites' (e.g. good–bad; strong–weak; fast–slow). The subject then rates each entity according to these adjectival scales. Generally, the data resulting from a group of subjects carrying out such an exercise is subjected to a Factor Analysis, to determine what 'dimensions' are being used. The assumption in this is that the scales represent deeper underlying evaluation dimensions that can be ascertained from the adjectival scales. These dimensions may not be apparent beforehand, nor within the normal vocabulary of the subjects involved, whereas the adjectival scales are fairly commonplace. The differences between this technique and the Likert technique are twofold; superficially, in the Likert technique there is a range of statements but only one version on the scale, whereas in the Semantic Differential there are different versions of the scales but only one entity to be evaluated at a time. Rather more deeply, in the Likert technique the statements are all representations of a single underlying issue, whereas in the Semantic Differential an individual scale may be examining a unique aspect of the entity unrelated to any of the other scales.

The strength of the Semantic Differential lies in its explanatory power, in elucidating the underlying dimensions and, once these are obtained, in showing the relationships between the entities in the n-dimensional space constructed from these dimensions. These are not necessarily apparent before the exercise, though you, the investigator, may have some shrewd suspicions (and wrong ones), and it is this power to reveal what you might otherwise miss that is the appeal. Of course, the power depends critically on the quality of the data collected; statistical techniques by themselves do not provide explanatory power.

Repertory grids

This technique was developed by Kelly (1955); one of the best technical and operational descriptions will be found in Fransella and Bannister (1977). Kelly's aim was to provide a psychiatric tool to help understand an individual's representation of his or her environment. The tool has been widened in its use since the early days, and has become popular in knowledge engineering as a means of exploring experts' conceptions of their skills and knowledge; it is discussed in this context elsewhere in this book by Shadbolt and Burton.

As with the Semantic Differential, there is an assumption that there are underlying dimensions, in this case called constructs, which are to be established. Because this is a single-subject method, the typical approach used is slightly different; this is the method of triads. Once some entities have been selected (let us assume we are dealing with different telephones), they are presented three at a time to the respondent (e.g. telephones A, B, and F). The respondent is asked to arrange them so that two are similar and the third is different, and to state what the difference is (e.g. 'these two are small (B and F), and that one is bulky (A)'). The subject repeats this with the same triad (e.g. 'these two are modern (A and B) and that one (F) is old-fashioned'), until no further differences are found. Another triad is then presented (e.g. A, B, and G), and the process is repeated, until no new differences are recorded. The more common, or most interesting, of these differences are then selected, and for each of these differences (e.g. 'Small–Bulky') the respondent ranks the entities along the implied continuum. A standard statistical analysis (usually a 'Factor analysis', using correlations between the rankings) is then used to elucidate the underlying dimensions.

Questionnaires

The appeal of questionnaires lies in their ability to obtain large amounts of information from large numbers of people, at relatively low cost, and relatively quickly. Two good texts are Oppenheim (1966) and Moser and Kalton (1971). The former is easier to read, while the latter is more comprehensive and detailed.

However, there are problems with questionnaires. Unless the information required is strictly factual and fairly easily checked, their reliability and validity is not always apparent, so considerable care must be taken. Furthermore, they do make a substantial demand on resources. Designing and administering a high quality questionnaire is a skilled task; a specialist in one of the behavioural sciences, a statistician, a computer expert and a graphic designer working as a team may be needed. Then there is the fieldwork; interviewing respondents is a time-consuming and skilled task, requiring fairly large numbers of people. There are problems of getting the data into a computer system for analysis and producing reliable analyses as well. All of this costs money; as a rule of thumb, if the project is carried out commercially, the cost per completed questionnaire would be about £20 (estimated 1988 figures).

In what follows, the assumption is made that the questionnaires will be postal questionnaires, and that they are intended for groups of more than one hundred. Many questionnaires are administered to respondents by interviewers, but this is omitted from the discussion here because the topic is taken up in the next section.

Know thy respondent

Eliciting accurate, reliable information from respondents is not an easy thing. Some observations, hard-learned by experience, that indicate the necessity of careful planning of your work are given below.

- Data-gathering presupposes a fairly high level of inter-personal understanding, a common culture, and a common language. Your questionnaire must stay within the boundaries imposed by this.
- Clumsy presentation may lead respondents to 'close down', for example by giving minimal answers, devious answers, or by outright refusal.
- The opportunities for misinterpretation are much greater than you might suppose. This is considered in more detail under the heading, 'Questionnaire construction', and is illustrated in Figure 3.9 (p. 75).
- Your respondent's knowledge may not be organised usefully, it may be limited to a small range of circumstances, and it may be wrong. It is extremely difficult to guard against these problems.
- Respondents often have only a partial knowledge of the extent of their own knowledge, this is usually the case when opinion is passed off as truth. Careful question design can ameliorate this problem, typically by the use of closed questions (discussed later).
- The most common characteristic of respondents' knowledge is that they have a good grasp of generalities and rationales, but a poor grasp of particular, necessary facts. There are no solutions to this.
- Particularly for non-verbal knowledge (e.g. driving skills), a respondent may not know what the skill is, let alone be able to describe it.

Planning the questionnaire

Typically, when you administer a questionnaire to a respondent you are implicitly asking him or her to reconstruct from memory the environment or the problem area that you are interested in, and then to pluck from it the information or opinions that you require. This is not necessarily an easy task (how much can you recall of your journey to work today?) and it requires a careful, methodical approach to make your efforts worthwhile. The steps are discussed briefly below.

Definition of objectives and resources

This is the most essential and hardest step. First, you must define your objectives *in detail*. For example, it is not sufficient to have as your objective: 'Find out about accidents on turning centres.' You should be able to define precisely what you mean by 'accident' (does it include damage to clothing? Does it include near-accidents? What about minor accidents such as bruised fingers, which are almost never reported?), and what range of machines is covered by 'turning centres' (all of them, or just the 'common' ones? What does 'common' mean?). At this point, there are three important questions to be answered:

- What are the results supposed to show?
- What level of accuracy is required?
- What additional data will be required to link this survey with other people's work?

This stage must be done thoroughly, down to deciding the form of the questions you will be asking. As a rule of thumb, you will have spent enough time and effort at this stage that the questions virtually write themselves, and the format of the final report is clear. Another rule of thumb is that between a third and one-half of the total time available should be spent on this first stage.

There is no substitute for this part of the design work. No matter how sophisticated you are from here on, if you start with fuzzy thinking you will continue with fuzzy questions, and however ingenious the analyses, at best you will finish with sophisticated fuzzy answers, and a fairly clear perception of the futility of your efforts.

Sampling

This step deals with whom you will survey. First, you must define the population you wish to investigate (for example, which organisations your respondents work for, and what skills and/or grades they must possess), and then decide on the sample size (which could be 100%). This is determined by the resources available and the accuracy required—see Guilford (1954). Sample size is discussed relatively formally by Drury in chapter 5 on ergonomics studies and experiments, and pragmatically in the context of user trials by McClelland in chapter 10 of this book.

The next step is to generate a 'sampling frame', which is a list of all the members of your population (e.g. constructed from personnel records), from which you then draw your sample. You should be aware that sampling frames are not always accurate, usually because the sources are not up to date.

The aim is to eliminate any systematic bias (of course, if you have decided upon a 100% sample, this is not a problem). Bias can arise from three sources:

- Non-random sampling. For example, if the personnel records have been arranged in order of seniority, a decision to sample every tenth name in a small workforce will result in a sample with a bias towards lower levels of management.
- Bias in the sampling frame. This is usually due to the sources not being up to date.
- Non-response. In any population there are people who refuse to respond, as a matter of principle. There are also those who are never available, such as salesmen and drivers. Exclusion of these groups inevitably means introducing bias.

The chief means of overcoming bias is by random sampling. This allows you to alleviate the effects of the first two sources, but only perseverance on your part will deal with the third.

'Stratification' is often used with random sampling, to improve the representativeness of the sample. The population is arranged into strata on the basis of age groups, sex, income levels, etc. You then sample at random within the strata, but note that you need this information about each individual in your sampling frame right at the beginning, in order to sort them into strata.

Another approach used is quota sampling. This is particularly useful where you cannot generate a sampling frame. In this case each interviewer is given a quota of people to interview, subject to them falling into certain age groups, income levels, and so on. The interviewer is then left to find the people to fill the quota. This method is open to interviewer bias, but produces about as accurate results as the stratification method. There are also panels, usually groups of people to whom you make repeated reference over a period of time, to measure changes in opinions or judgements over time. The approach has its own problems of attrition or conditioning of panel members.

Having selected the sample, there remains the problem of ensuring maximum participation in the survey. This is important, because many experiments have shown that non-responders are not typical of the total population. The main causes of this together with suggested remedies are as follows:

- Units outside the sampling frame—someone who should be in the sampling frame who is not available (e.g. dead). In this case, select more people from the sampling frame.

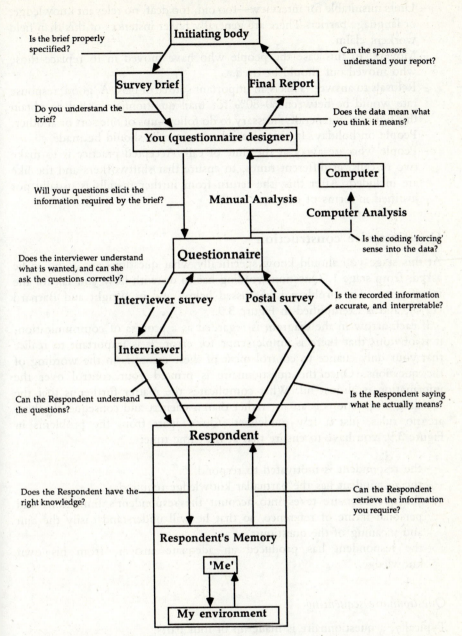

Figure 3.9. Illustration of the communication problems in questionnaire design. The arrows represent communication processes, and the questions associated with them indicate some of the problems that can occur. It should be noted that once you have designed your questionnaire, any errors are 'frozen in'. Redrawn from Sinclair (1975).

- Units unsuitable for interview—too old, too deaf, no relevant knowledge, or language barrier. There are generally fewer instances of this than field workers claim.
- Movers. In this case the people who have moved in to replace those who moved out should be used.
- Refusals to answer. This is an important source of bias. A 'good' response rate would be between 60–80% for mail questionnaires, but to obtain this figure it is generally neessary to do follow-ups of one sort or another.
- People on holiday. In this case subsequent calls should be made.
- People who are away at the time of call. Accepted practice is to make two re-calls, at different times, to ensure that shiftworkers and the like are included. After this, the return from further re-calls is usually not justified in terms of expense.

Questionnaire construction

At this stage you should know specifically what questions are going to be asked, from stage 1. Question wording has to do with how you ask for the information. The problem is discussed below and in Wright and Barnard (1975), and is exemplified in Figure 3.9.

If each arrow in the diagram is regarded as a process of communication, it is obvious that there is ample scope for error. It is important to realise that your only chance to control most of these errors is in the wording of the questions. Once the questionnaire is printed your control over the information is almost nil. What complicates the matter further is that the wording of questions is an art, rather than a science, and consequently there are no rules, just a few guidelines. Quite apart from the problems in Figure 3.9, you have to ensure that at the same time:

- the respondent is motivated to respond;
- the respondent has the particular knowledge required;
- the questionnaire takes into account the respondent's limitations and personal frame of reference, so that he will understand easily the aim and meaning of the questions;
- the respondent has produced an adequate answer, from his own knowledge.

Questionnaire sequencing

Typically, a questionnaire is made up of four parts:

(i) A prologue, which introduces the topic to the respondent, provides any information he or she will need, and tries to motivate the respondent to answer the questions. It may also include examples of difficult questions and instructions for answering them.

(ii) An information section, in which you ask your questions (*e.g.* 'How long have you worked on this machine?'. 'How many of your fingers has it chopped off?').

(iii) A classification section, in which you obtain personal data and any other background information (*e.g.* 'How old are you?'. 'How long have you been doing this job?').

(iv) An epilogue, which thanks the respondent for his efforts, and includes any further instructions, if necessary.

Within the information section, the questions should be arranged in consistent groups which follow each other in a reasonably logical way. This should be from the respondent's point of view, not from yours; it is important that the respondent should feel comfortable.

You should also consider the use of 'filter' questions. These serve two purposes: first, they determine whether your respondent has relevant knowledge or not (*e.g.* 'Have you taken a training course for . . .?'), and secondly they can guide your respondent past sections that are not applicable in his or her case (*e.g.* 'If you have had back pain, go to section . . ., if not, go to section . . .').

Degree of structure in questions

After a while it becomes apparent that questions, irrespective of content, can be classified into certain structured classes. Two of the classes have already been discussed, ratings and rankings, and some of the sub-classes within these have been outlined. Others are: factual questions, for example asking for birthdates, gender, etc. (Question 10 in Figure 3.10 is an illustration); questions with mutually exclusive answers (Question 5 in Figure 3.10); questions with multiple non-exclusive answers (Question 6 in Figure 3.10); matrix questions (Question 4 in Figure 3.10); open questions (Question 2 in Figure 3.10); and closed questions (Question 1 in Figure 3.10). There are many more. Your questionnaires will most likely be composed of a variety of these, as in the example. This structured classification is of use subsequently in organising the analytical routines.

Irrespective of the class of question, it must be worded correctly. For ratings and rankings the methods discussed above will have ensured this. It is with the other classes of questions that we are now concerned. Belson (1968) and Kalton *et al.* (1978) have a number of important, interesting and practical comments to make about the problems in these classes. The first decision that must be taken is whether the questions should be open (*i.e.* subjects compose their own answers) or closed (subjects choose an answer from a given set).

The advantages of closed questions are:

– They clarify the alternatives for the respondent and avoid snap responses being given.

PART 1 – General

1 What are the risks in your brigade area?
*Please show the distribution of risks by putting a
percentage in the box. (e.g. if your brigade is a
county with B-D risk, the B risk might represent
60% of the area, C 25% and D 20%)*

High risk	%
A risk	%
B risk	%
C risk	%
D risk	%
E risk	%
Remote rural risk	%

2 Do you consider that there is a suitable production
commercial vehicle chassis currently on the market
that will meet the demands of the Fire Service? Yes ☐
No ☐

If Yes, who manufactures it?

_____

If No, what features render them unacceptable for your use?

_____
_____
_____
_____

3 Do your drivers receive training similar to the
programmes given to police drivers? Yes ☐
No ☐

If Yes, please describe the training procedure.

_____
_____
_____
_____

4 How frequently are your vehicles involved in accidents?

	Total Calls	Total Accidents		
		On the road	Off the road	At the fire ground
1970 ...				
1971 ...				

5 In fighting fires in recent years how many times have
you used open water, with either a main or portable
pump?

More than 100 times per year	☐
50-100 times per year	☐
11-50 times per year	☐
0-10 times per year	☐

6 Have you had any vehicle failures in the last year?
Yes ☐
No ☐

If Yes, please tick the relevant boxes

Mechanical
Engine ☐
Transmission ☐
Axles ☐

Other (please specify)
_____

Electrical
Batteries
Generators ☐
Wiring

Other (please specify)
_____

Ancilliary
Pump
Compressors ☐
Tank
Ladders

Other (please specify)
_____

7 Do you have particular problems in obtaining spare
or replacement parts for your fleet? Yes ☐
No ☐

*If yes, are there any particular parts which
are always difficult to obtain?*

_____
_____

8 Are your vehicles out on the road performing fire
prevention and other duties during the day? Yes ☐
No ☐

*If Yes, can you give an approximate percentage of
those on the road against those giving cover at
Fire Stations? ...* ☐ %

9 What was the percentage of special calls against
total calls in your brigade last year? .... ☐ %

10 As brigade procedure do crew members don BA
sets on the way to a call? Yes ☐
No ☐

Figure 3.10. A page from a well-designed questionnaire, illustrating different question types, the use of different typefaces for different purposes, a layout to optimise both the use of space and the clarity of the questions, and the pleasing overall appearance of the document. (Source; Gray, 1975).

- They reduce keying errors in analysis, and eliminate the need for people to code the answers.
- They eliminate the useless answer (*e.g.* 'How long have you done this job?' 'Since I moved here').

The disadvantages are:

- It is difficult to make the alternatives mutually exclusive.
- They must cover the total response range (this presupposes that the researcher will have a good idea of what answers are likely to appear—hence the importance of pilot studies).
- They create a forced-choice situation which rules out marginal or unexpected answers.
- All the alternative answers must seem equally logical or attractive, for fear of biasing the results.
- In complex or difficult questions, subjects may dive for the safety and ease of the 'don't know' alternative.

Points to be considered in questionnaire construction

In framing the questions the following important points must be considered:

- Questions specificity. The requirement is that your questions should be precise and unambiguous. Fortunately, if you have carried out the planning stages thoroughly, this is not likely to be a problem. It implies that where you have a difficult topic, you may have to ask a number of questions to obtain the information you want, rather than try to get it in one omnibus question. As a general rule, avoid omnibus questions like the plague.
- Language. It is essential to use language relevant to the population, to make the questions easily understood by all. You should also be aware of localised interpretations. For example consider the word 'tea'—which in Britain may mean a 'pot of tea', 'high tea', or 'the main evening meal', depending on where in Britain you are. It is important to use short words rather than long ones, and to be aware of the dangers of the unconscious use of scientific or professional jargon. In this context, look at some of the editorials in the so-called 'popular press'. As a method of communication, the style is superb, whatever might be thought of the content!
- Clarity. It is a cardinal rule that questions should be short. This rule has two useful consequences; first, it ensures that you clarify your thinking and remove unnecessary words; secondly, it reduces the chance of overloading the respondent with too much information to digest. Complex questions can lead to error, as can those that contain such vague phrases as 'on the whole', 'generally', 'normally' and 'frequently'.

Double-barrelled questions should be avoided, such as 'Do you suffer from headaches or stomach pains?' (what would the answer 'yes' mean?).

– Leading questions. Clearly these must be avoided. Obvious examples such as 'Do you agree that the policies of the present government are unfair?' (which invites the answer 'yes') are quite easy to detect. More insidious examples are questions that contain such loaded words or phrases as 'get involved in', 'student' and the like. You should be aware of questions that become leading questions because of the nature of the questionnaire. As an example of this, the question 'How many cigarettes do you smoke in a day?' may be innocuous in a questionnaire about household expenditure, but may produce different answers in a questionnaire concerned with medical matters.

– Prestige bias. This is a bias that can arise in questions that involve socially desirable behaviour. Thus, a question such as 'Which magazines have you looked at recently?' is likely to reveal that such journals as *The Economist* and the *Literary Review* have a considerably larger readership than is actually the case. Great care is required to overcome this sort of bias; filter questions and careful wording should be used, so that low-prestige answers appear equally as acceptable as high-prestige ones. In the example above, you might introduce a filter question such as 'In the past seven days, have you had any time to read magazines?' to identify those who have not opened a magazine, but are not prepared to say so.

– Embarrassing questions. There is seldom an easy way to obtain information of a personal nature. If such questions must be asked, they should be placed some distance into the questionnaire, and the whole tone of the questionnaire should be personal, relaxed and permissive. This requires considerable skill in question wording: it is very easy to appeal to one segment of the population and totally offend another at the same time. Further, the use of euphemisms should be considered in the place of blunt questions.

– Hypothetical questions. These are the 'What would you do if . . .' type of question. They almost never yield reliable results, and should be avoided. There is usually a noticeable difference between people's self image in a particular set of circumstances and their actual behaviour.

– Impersonal questions. These questions tend to produce spurious answers because the respondent becomes disengaged from the subject-matter, and consequently can lose interest in the questionnaire, sometimes to the extent of refusing to answer any more questions.

Layout of the questionnaire

By this we mean the arrangement of words on the page. This aspect is almost invariably overlooked by questionnaire designers and yet it can make a difference of about 20 per cent in the response rates. Figure 3.10 illustrates a well laid-out questionnaire, and you should note the two-column layout, the use of different weights of type, the use of italics for instructions, the use of boxes to restrict the size of written answers, and the pleasing aesthetic appearance (for its date) of the page.

Piloting the questionnaire

This stage is vital. It is here that the last chance occurs to discover the fallacies and unnoticed assumptions in your thinking. It is here that the respondent's understanding of the questions and the problems of analysis are revealed. It is also the last occasion on which remedial action can be taken. It is necessary to test all aspects of the questionnaire at this stage: the introductory passage, the questions, the alternative answers (or coding frames for interpreting open questions), and the form of the analysis.

This is best done in three stages:

(i) Individual criticism: the questionnaire should be handed to a colleague or several colleagues who have some experience of questionnaires (but not of this particular one) for comment.

(ii) Depth interviewing: once the criticisms generated above have been examined and any appropriate changes made, the questionnaire should be given to a small sample of respondents (about ten) for their reaction. On completion of the questionnaire, each respondent should be questioned in detail about the answers to the questions, to ascertain what the respondent understood the question to be asking, and the exact meaning of the responses given.

(iii) Finally, the questionnaire should be given to a larger sample of respondents to investigate the implications of the desired analysis, and to detect whether any invalid or meaningless patterns of answers are occurring. This also enables estimates to be made of the reliability and validity of the questionnaire, the reliability of the sampling frame, and the likely non-response rate. This stage should be repeated until the questionnaire appears to be error-free (but you will never get rid of them all).

Fieldwork

For many surveys, where the number of respondents required exceeds 100 and the survey is to be interviewer-administered, the most appropriate course of action may be to obtain the services of one of the commercial groups. In view of the expertise and service that is provided, they are good value for money. As was stated earlier, interviewing is a job requiring training and experience: co-opting students or other 'odd-job people' does not tend to produce reliable or accurate results. Hence, if the time, trouble and cost of obtaining a field force, training and maintaining it is taken into account, there is seldom a cheaper alternative to the commercial organisations. Where the numbers are below 100, it is feasible (and highly instructive) for the designer to do the interviewing personally.

Where you intend to do a mail questionnaire, the considerations are different; the problems become those of layout and design of the questionnaire, and its retrieval. If the layout and the introduction to the questionnaire are good, the problems of retrieval are usually reduced. Nevertheless, it is usually necessary to consider some means of improving the response rate. First, there are encouraging letters, to act as reminders. It is advisable to send these within one week after the expected return date, together with a duplicate questionnaire, in case the first one was mislaid. Secondly, a shortened version of the questionnaire may be sent, to ask for the most important information. Thirdly, the use of raffles or prizes to encourage the return of questionnaires might be considered. However, the cost of these can offset the chief advantage of postal questionnaires, which is their cheapness. The final method of follow-up is to contact the non-respondents by telephone. This method has received increasing use in recent years.

You should note that for any of these follow-up methods to work, it is necessary to know who has responded and who has not, which necessitates removing the cloak of anonymity from respondents. It appears, however, that this is not generally a serious cause of non-response provided the questionnaire is well designed.

Analysis

Having completed the fieldwork and collected the completed questionnaires, there is still the problem of analysis. There are two major areas of interest here: the first is data editing, to identify any inconsistencies in the responses to questions and to take appropriate action, and the second is data tabulation, where the tables and statistics required for the report are generated.

A very brief overview of this follows. For simplicity, let us assume that the analysis will be computer based, and that the information will be transferred from the questionnaires to a computer file, for further analysis. The transfer of the data from the questionnaire into a file is normally accomplished via a human agent, and this stage is therefore prone to error. It is worth noting that computer files may also be lost, so a duplicate record of the data should always be available.

At this stage the data for a single survey unit (a data set) can be considered to be in 'raw' form; the data sets may now be edited to remove logical errors (for example a man in the household has seemingly given birth to a baby) and out-of-range errors (for example the man is 300 years old instead of 30). When these errors are checked and removed, and various new data produced (for example, volume, created from data on height, length and breadth), it may be said that one now has 'treated' data sets, which in turn should be filed.

It is the treated data sets on which the analysis is performed. If you consider the data to be in matrix form, with data items across the top and respondents (or survey units) down the side, one may then conveniently use the manipulative power of matrix algebra to simplify the analysis, and do most of the analytic sub-routines. There is a wide range of computer-based statistical packages available to accomplish this; a popular and widely-available one is SPSS (Nie *et al.*, 1975). Recently, it has appeared in a version suitable for personal computers. Whichever package is selected, it is essential to ensure that the arithmetical and statistical operations performed on the data are valid, as it is very easy to arrive at false conclusions by the use of inappropriate analyses.

Interviews

In a sense, interviews can be considered to be questionnaires carried out face to face by an interviewer, and the comments above regarding the need for planning, sampling, and so on apply here as well. However, because there is an interviewer present, the typically rigid structure of the questionnaire can be relaxed; this of course brings its own dangers unless the interviewer is skilled and experienced, and the interview is carefully planned beforehand. While interviews have always been part of the repertoire of ergonomists, the emergence of knowledge engineering has lent a new importance to this technique. An excellent, practical discussion of interviewing can be found in Macfarlane Smith (1972), written from the perspective of market research.

The big danger in deciding to carry out interviews is to use the rationale that because the interviewer is an intelligent person, he or she can adapt to the needs of the situation, and do the right thing, thereby reducing the need for careful planning at the outset. Unless you have an extremely knowledgeable and skilled interviewer, this is unlikely to happen. All too often, this argument is used as an excuse to ease the mental pain at the outset. It almost never works well; the sporting cliché, 'No pain, no gain' still applies. You will have to plan your interviews with the same dedication as for questionnaires.

However, it is because there is an intelligence in the interviewer that certain benefits do accrue, provided care is taken. In some circumstances, interviews are the best way to capture information. It is in situations where matters are very personal to the respondent, or where complex information is involved, or where you think respondents might need to have some help in giving their answers, or where different people may have markedly different views about reality, or where you genuinely don't know what is involved (as in the initial stages of piloting a questionnaire), that interviews are most helpful.

There is a continuum of interviewing styles, ranging from directed interviews to non-directed interviews. Directed interviews are those where the questions to be asked, and the order of questions, are specified beforehand. Interviewer-administered questionnaires are an example. Non-directed interviews are those where essentially the respondent controls the interview, and the interviewer is there to help the respondent express himself or herself, hopefully eliciting the important information during this process. This latter style is close to that used in psychiatry. The questions are not formulated in any detail beforehand, and the main role of the interviewer is to help the conversation along, with interjections such as 'Really?' and so on. If the respondent stops, the interviewer may start the process again by using a non-directive question, for example, by repeating the respondent's last phrase with a questioning tone of voice.

The former technique has been criticised for its rigidity and its intolerance of unanticipated individual differences, whereas the latter has been criticised for the time required, and, to quote a memorable phrase, for 'leaving behind a posse of cured souls, but not necessarily producing much worthwhile information'. Most interview techniques fall between these two extremes, where some direction of the respondent occurs, if only to direct the respondent towards the areas that you want discussed. In some cases, the initial questions might be defined, and so on.

GENERAL CHARACTERISTICS OF INTERVIEWS

- The use of an interviewer can serve to direct and accelerate the information flow
- The interviewer can explore unexpected information, or unexpected occurrences
- A well-trained interviewer will be sensitive to the individual needs of the respondents, and will adjust his or her behaviour accordingly, thereby improving the quality of the information flow.
- Interviewers can help to motivate respondents to give more information about the topic during the interview
- For the advantages above to occur, the interviewer must be well-trained in interview technique, should have at least some knowledge of the topic areas (the more the better), and must be sensitive to people. Collectively, these criteria are not easy to meet.
- It can be difficult to find and schedule people for the interview session.
- Interviewer's bias may creep in; this might be due to the interviewer's own knowledge of the topics, interpersonal relationships between the interviewer and the respondent, or to more mundane things such as fatigue, and so forth. This constitutes an extra source of error.
- Systematic recording of data is difficult, and in some cases impossible. In certain instances (an example is knowledge elicitation from experts), it may take up to three times as long to sort and assimilate the data as it took to obtain it (Shadbolt and Burton also warn us of this in chapter 13 of this book).

'Critical incident' techniques

This approach was first used by Flanagan (1954), in a study of near-accidents in aircraft, hence its title. He could not study accidents themselves, because there weren't many respondents available. The technique is basically a semi-directed interview technique in which all parties know what areas are going to be covered, and by and large the interviewer will control and direct the interview and may ask certain specific questions. However, since the antecedents of near-accidents are often unique, there has to be some room for variance. One of the main characteristics of this method is that it is not intended to be carried out with individual respondents, but within small groups. The reasoning behind this is that individuals in such situations can, usually inadvertently, encourage and assist each other to recall more information than they might as individuals. This technique, suitably adapted, is used in market research at the product definition phase, and in many other circumstances.

THE MAIN CHARACTERISTICS OF THE CRITICAL INCIDENT TECHNIQUE
- It is possible to get data on rare events, or ephemeral events, that cannot be obtained easily in other ways.
- The method can reveal much about abnormal or otherwise memorable situations, but it typically has little to say about normal conditions.
- There is a danger that because a group is involved, a single consensus view may be all that is obtained. There is a real danger of this if one of the group of respondents is a dominant character.
- For certain situations, the respondents are in effect asked to give evidence against themselves. This will only occur in the right group environment, and it requires the interviewer to have highly polished interpersonal and group skills.
- The skills required by the interviewers are not commonly distributed.

Checklists

Checklists come in a variety of forms, for a variety of purposes. There are simple lists, which may serve as procedural reminders to make sure that the right sequence of activities has occurred, there are complex hierarchical lists, using 'if . . . then branch . . .' constructions, there are activity sampling checklists, for recording behaviour at particular times during a shift, and so on. This discussion deals with the latter kind of list, for recording activities; other discussions will be found in Guilford (1954), Konz (1979), and Oppenheim (1966). The text by Konz contains many practical hints.

The first requirement for a list to record activities is that you should know comprehensively what activities you wish to record. This is a non-trivial problem. First, there may be a number of activities in which you are not interested, which may be disregarded. Second, there are the activities which are unusual, but have an important effect that you ought to record—accidents, and near-accidents are examples. These are very difficult to foresee, and therefore are difficult to record in a list a priori. While it is easy to have catch-all categories in your list such as 'Accidents', the very broadness of such categories reduces considerably their usefulness when it comes to analysis. Third, there are the activities in which you are interested, which you have on your list, and which you can record. The first difficulty here is to establish the correct level of detail for the activities. For example, is it enough to have a category of 'Monitoring behaviour'? Since this can encompass anything between alert scanning to zombie-like behaviour, you may wish to discriminate between various types. The second difficulty here lies in ensuring that you can observe and record these activities. One reason why you might not be able to do so is because the activity concerned might not be visible—perhaps the person you are studying has inadvertently turned away from you. A second reason is that other simultaneous activities might distract you. Another reason is that the preceding and following activities may be misleading.

Having generated a list of activities to record, it is essential that starting and ending cues for each activity are identified, and that these should be clearcut. As an example, consider 'Monitoring behaviour'. What establishes when it has started? Standing in front of a display does not necessarily mean that monitoring is occurring. If the activity sampling is to be undertaken by other people, these points are critical to the quality of data that will be garnered.

CHARACTERISTICS OF CHECKLISTS, ESPECIALLY IN ACTIVITY SAMPLING
- There are problems in generating unequivocal lists of behaviours and/or activities that you wish to observe.
- The problem of incomplete lists is endemic.
- They record superficial aspects of activities, not necessarily the way in which the activities are modified into sequenced skills, nor their grouping into higher level, sometimes more obscure, activities.
- The lists themselves serve efficiently as aides-memoire to the observer, reminding the observer of what should be recorded.
- The method can produce quite high correlations between observers of the same set of activities. This implies that what is recorded has some degree of validity.
- The method is relatively easy to create and administer.

Conclusion

A number of subjective assessment methods have been explored briefly in this chapter. If you wish to make use of any of the methods, the texts given in each section should be sufficient to enable you not only to get started with the method, but also to achieve reasonable results. As in most areas, using these methods skilfully requires lots of practice, not just for polish, but to make the blunders that all the experts have made in their time (including the editors and authors of this book!), from which you will learn the real lessons about the methods. As Rasmussen (1985) has said rather more cogently, it is only by making errors that you learn skills.

You should be warned: a trap into which you might fall is to stick rigidly to the methods outlined in the texts, because there is a fairly strong scientific basis to them, and a wide acceptance of them. While these are laudable attributes of the methods, what counts in the end is the quality of the data that are gathered, and what is revealed about the subject of your interest. In this regard, it is suggested that you should read the text by Webb *et al.* (1966); in its own way it is a highly entertaining book for the originality of its thoughts and the subtlety of its methods.

Finally, it should be remembered that even though the methods may seem laborious, and the statistical complexities daunting, the data that result are usually worth all the trouble that was caused, and the warm feeling engendered by this will certainly outweigh the pain you experienced at the time.

References

Alluisi, E.A. (1975). Optimum uses of psychobiological, sensorimotor, and performance measurement strategies. *Human Factors*, **17**, 309–320.

Belson, W. (1968). Respondent's understanding of survey questions. *Polls*, **3**, 52–70.

Chapanis, A. (1967). The relevance of laboratory experiments to practical situations. *Ergonomics*, **10**, 557–577.

Coombs, C.H., Dawes, R.M. and Tversky, A. (1970). *Mathematical psychology: an elementary introduction*. (New Jersey: Prentice-Hall).

Edwards, A.L. (1957). *Techniques of Attitude Scale Construction*. (New York: Appleton-Century-Crofts).

Feyerabend, P.K. (1975). *Against Method: Outline of an Anarchistic Theory of Knowledge*. (London: New Left Books).

Flanagan, C. (1954). The critical incident technique. *Psychology Bulletin*, **51**, 327–386.

Fransella, F. and Bannister, D. (1977). *A manual for Repertory Grid Technique*. (London: Academic Press).

Gray, M. (1975). Questionnaire typography and production. *Applied Ergonomics*, **6**, 81–89.

Guilford, J.P. (1954). *Psychometric Methods*, 2nd edition. (New York: McGraw-Hill).

Hudson, L. (1972). *The cult of the Fact*. (London: Cape).

Jung, J. (1971). *The Experimenter's Dilemma*. (Toronto: Harper & Row).

Kalton, G., Collins, M. and Brook, L. (1978). Experiments in wording of opinion questions. *Applied Statistics*, **27**, 149–161.

Kelly, G.A. (1955). *The Psychology of Personal Constructs*, volumes I & II. (New York: Norton Press).

Kirk, N.S. and Ridgeway, S. (1970). Ergonomics testing of consumer products 1: General considerations. *Applied Ergonomics*, **1**, 295–300.

Kirk, N.S. and Ridgeways, S. (1971). Ergonomics testing of consumer products 2: Techniques. *Applied Ergonomics*, **2**, 12–18.

Konz, S. (1979). *Work Design*. (Columbus, Ohio: Grid Press).

Macfarlane Smith, J. (1972). *Interviewing in Market and Social Research*. (London: Routledge & Kegan Paul).

Meddis, R. (1984). *Statistics using ranks*. (London: Blackwell).

Meister, D. and Rabideau, G.F. (1965). *Human Factors Evaluation in System Development*. (New York: John Wiley).

Moser, C. and Kalton, G. (1971). *Survey methods in Social Investigation*. 2nd edition. (London: Heinemann).

Mosteller, F. and Rourke, R.E.K. (1973). *Sturdy Statistics*. (New York: Addison Wesley).

Nie, N.H., Hull, C.H., Jenkins, J.G., Steinbrenner, K. and Bent, D.H. (1975). *SPSS—Statistical package for the social sciences*. (New York: McGraw-Hill).

Nunnally, J.C. (1970). *Introduction to Psychological Measurement*. (New York: McGraw-Hill).

Oppenheim, A.N. (1966). *Questionnaire Design and Attitude Measurement*. (London: Heinemann).

Osgood, C.E., Suci, G.J. and Tannenbaum, P.H. (1957). *The measurement of meaning*. (Urbana, Ill.: University of Illinois Press).

Rasmussen, J. (1985). Trends in human reliability analysis. *Ergonomics*, **28**, 1185–1196.

Siegel, S. (1956). *Non-parametric Statistics for the Social Sciences*. (New York: McGraw-Hill).

Simon, C.W., (1976). Analysis of Human Factors Engineering Experiments. NTIS Document. AD–A038–184/8GA.

Sinclair, M.A. (1975). Questionnaire design. *Applied Ergonomics*, **6**, 73–80.

Torgerson, W.S. (1958). *Theory and Methods of Scaling*. (New York: John Wiley).

Tukey, J. (1977). *Exploratory Data Analysis*. (New York: Addison Wesley).

Webb, E.J., Campbell, D.T., Schwartz, R.D. and Sechrest, L. (1966). *Unobtrusive measures*. (Skokie, IL: Rand McNally).

Wright, P. and Barnard, P. (1975). Just fill in this form—a review for designers. *Applied Ergonomics*, **6**, 213–220.

Ziman, J. (1980). *Reliable Knowledge*. (Cambridge: Cambridge University Press).

Chapter 4

The use of archival data

Colin G. Drury

Introduction

This chapter covers the collection and the use of data which are pre-existing in the sense that the ergonomist had no part in their original collection. The ergonomist's role is to access, copy, edit, compile and interpret data collected for other purposes. Obvious examples are the use of medical records to estimate crash injury severity (e.g. Baker, 1974), the use of company accident records to assess causes of accidents (e.g. Saari, 1976) or the use of quality control records to measure inspector performance (Drury and Addison, 1973).

Sources are available for most existing systems the ergonomist needs to study, such as factory, office, hospital and transportation systems. Archival data look attractive in terms of collection effort, but have their own unique pitfalls. Hence the chapter is organized by sources of data followed by general comments on suitability of consulting the archives.

Data sources

Fixed data

All organizations have sources of data which are relatively invariant over typical project time-scales and which are essential to provide proper background to an ergonomics investigation. A factory or office will usually have a layout drawing, given in more or less detail, showing the physical size of the work areas involved, sizes of storage areas and locations of fixed facilities (e.g. toilets, cafeterias). More detailed layouts will include location of services (e.g. electricity, water, drains, compressed air), positions of lighting fixtures or permissible floor loadings. It is difficult to conceive of working in a plant (or airport or hotel) without such layouts, particularly if equipment is to be replaced or relocated. The accuracy of such fixed information is rarely in doubt, although locations and names of machines or other movable equipment may show differences between drawing and reality. It pays to check, and even have your own set of updated drawings, in order to plan lighting, noise or space surveys. As organizations convert to CAD (computer-aided design) systems, changes and updates should be easier for both the factory and the ergonomist.

Another type of fixed information is that provided by company annual reports or other legal documents. These show sales, profits, production volumes, inventories, fixed assets and total employed for a company, or equivalently numbers of beds and patients for a hospital, or numbers of aircraft and passenger miles for an airline. Such data are a great help in putting the ergonomics work into context before, during and after the study. At start-up the data show where to choose the various sites for ergonomics interventions to have the most generality. During the study, the fixed data show reasonable breakdowns for data collection, e.g. by department, line or product. After the study the data can be used to obtain multiplying factors to judge the overall impact on the company as implementation proceeds.

An obvious danger with the use of fixed data is that they can easily be outdated without the ergonomist realizing it. There is no substitute for checking 'on the ground' that the machine shop really does have three Heald ICF90 grinders, and so on.

Fixed records represent the input resources an organization has with which to achieve its goals. There is no guarantee that the appropriate quantities of machines, buildings, capital and personnel will actually achieve the goals, but it is often true that insufficient input resources will of themselves prevent goals being reached. The float process in flat glass manufacture is a good example. Until it was developed, all companies competed with different processes, but once one company had developed it, the cost and quality advantages were so overwhelming that soon it was licensed by all major competitors.

A final form of useful fixed information is found in planning documents. Strategic plans and operating plans based on where the enterprise and its products are headed are found in most organizations. They represent a useful source of future information about product types, changes in technology or marketing strategy and product volumes anticipated. Hence the ergonomist will be able to find out whether markets for products are growing or shrinking, whether individual items are going to need larger or smaller castings, whether new chemicals are to be introduced and how long production runs will be between product changes. All of these can impact production process design. More rarely there will be planning documents related to such ergonomics matters as payment policy or total employment expected, but talking to key people in the personnel department and union office will often make the ergonomist aware of any obvious trends.

Production records

If the fixed records represent the 'order of battle' for an organization, the day to day records represent the communications during the battle. A company must know how many items it is ordering, receiving, producing or shipping if it is to meet delivery targets, control inventory and schedule its internal processes. In manufacturing, the record side of production is often known as *order execution*, and is in itself the object of much study. We now frequently encounter computer-based *factory information systems* (FIS) and *management information systems* (MIS). Order execution proceeds from receipt of an order through a *materials requirements planning* (MRP) phase to ensure that the correct raw materials are available at the appropriate times. As jobs are released into the manufacturing shop either a push system or pull system is used to ensure that the correct machining and assembly operations are carried out. Finally, completed goods will be accumulated until the order is complete, checked and shipped.

Records are generated at each stage. For example the original order is entered onto a form as a data base, purchase orders for raw materials are released and so on. On the manufacturing floor jobs are sequenced with routing slips or cards. Each line on the card is an operation, so that an operator signs for receiving a certain number of parts and for completing one operation on another number of parts. Incoming and outgoing numbers are different because of scrap, rework, missing parts and so on. Often a copy of the operator's part of the routing document will be retained by the operator to prove completion of a task, and hence get paid for that work. In continuous, as opposed to batch, production, each unit is covered by a separate record, rather than each batch. The record will specify, for example, a car's colour, engine type and options.

Routing slips show what work was performed at what times on what machines. Thus they give information on production levels, times, scrap rates and machine utilization. What they do not usually include is downtime, set-ups vs. running time, inventory levels or process logs, all of which are (or can be) recorded separately.

Production records in other situations can be very different, but the same principles apply. Thus a patient in a hospital has a time record by the bed to show tests and medications given and to record vital signs. Similarly, each airline flight has a manifest with passengers carried, route, weights and places to record safety-related task completions.

While records have a tremendous air of authenticity, they can conceal rather than reveal reality. On a production floor, a worker can 'bank' routing slips on a good day to have some excess production in hand for more difficult days. Recent press reports have shown that workers' time can be charged to different jobs to conceal cost overruns, and this has been found in jobs other than just the military ones.

Industrial engineering records

The industrial engineering (IE) or production engineering department of a plant or service enterprise is the source of many useful records, but with some caveats. Long associated with 'methods study and time study' (e.g. Barnes, 1980; Konz, 1983), the IE department has traditionally been concerned with the details of how operators perform their jobs. Records are kept of task and subtask breakdowns of jobs which form a remarkably good starting point for any task analysis (e.g. Drury, 1980). However, the ergonomist should be aware of the fact that most task breakdowns were made for the purpose of estimating times for task completion. Hence, there is rarely any information on errors and variability or on processes difficult to time (e.g. cognitive tasks). Additional sources of data which could contribute to task analysis are in quality control and from training and safety, both of which are typically personnel functions (see later).

There are two major uses for task time completion data: planning and control. For planning purposes (bidding, estimating delivery dates, scheduling) we need accurate forecasts of time taken on each operation. For control purposes, (e.g. production control, efficiency measurement, incentive payments) we need times which everyone concerned will accept as a reasonable yardstick against which to be measured. Thus planning and control lead to different potential biases in the times produced, and these biases can easily be the undoing of the naive ergonomist. Particularly if these times are for control purposes as they will directly affect people's measurable performance, so that pressures towards estimating larger times than actual will be beneficial to those whose performance is measured. As a 'standard time' for a job is composed of a measured time, a rating by the analyst of how effectively the operator is working and an allowance for non-productive time, there are ample opportunities for bias. Many time estimates are now made with predetermined motion time systems (PMTS), either by hand (e.g. MTM-2, WOFAC) or on a computer (e.g. MOST, ADAM), but the ergonomist should be aware that these same biases were built into the data bases on which PMTSs are based. Just because a computer gives the answer to the nearest millisecond does not mean that the answer is necessarily either valid or reliable.

Having said that, time study data have a particularly valuable part to play in increasing the precision of ergonomic studies. Any job which is variable from minute to minute or day to day is particularly difficult to measure, for example, before and after comparisons of ergonomics change. Examples of such jobs are in batch production, where the product is not always the same, and in maintenance, where jobs recur only rarely. Having a standard time for each job allows the ergonomist to measure *relative* productivity rather than *absolute* production. Productivity is defined as a ratio of output to input, and thus is often expressed as a percentage of the expected production or inversely as allowed hours divided by actual hours for a given volume produced. Drury and Wick (1984) were able to demonstrate productivity improvements due to ergonomic changes in a shoe factory, despite frequent shoe style changes at each workplace, by measuring percentage productivity (rather than number of units produced) before and after the change.

Such normalization can also help in other ways, for example by allowing us to aggregate performance across different tasks such as set-up and production. Normalization is, of course, only as trustworthy as the standards used to normalize. Existing standards should never be accepted uncritically.

Quality control records

Quality is assessed, or ensured, at many different points in an organization. Often the work of a department labelled 'Quality Control' is only a fraction of the total quality effort. Hence in using quality measures for ergonomics studies, the ergonomist needs to be aware of the danger of inferring too much from departmental or job titles. Particularly with the current push toward ensuring quality production, rather than employing an army of inspectors to check a fault-prone production (Taguchi, 1978), quality control departments are tending to take on more of an advisory role while the primary quality responsibility rests with the operator.

Overall quality is measured by fulfilment of customer requirements and so logically includes design aspects as well as manufacturing aspects. Within an organization, however, a more restricted definition, such as 'quality of conformance', is accepted (e.g. Drury, 1982). This means that specifications exist against which quality can be measured to ensure conformance. With this definition, quality is measured in two ways: by attributes and by variables (e.g. Schilling, 1982). Attributes inspection refers to whether or not a particular attribute is present and typically leads to a dichotomous measurement scale of conforming/non-conforming, acceptable/rejectable, good/defective, etc. Measures include fraction defective, process yield (= $1 \cdot 0$ − fraction defective), defects per 100 units, as well as more detailed data on frequencies of particular types of defects. Variables data is where a continuous variable is measured (e.g. length, temperature, viscosity, resistance) and sample statistics (mean, range, standard deviation) are used to infer population characteristics from sample measurements. Both attributes and variables can form the basis for two distinct control systems. Acceptance sampling answers the questions 'Should this batch be released to the next stage or customer?' while control charts answer 'Has the process producing these items changed sufficiently to require adjustment?'. The former is an after-the-fact judgement, whether applied to incoming parts, work in-process, or finished goods. Hence it represents a relatively slow-acting control loop. Control charting is a form of in-process quality control, which is aimed at ensuring that defective items are never produced. Such a fast-acting control loop is obviously preferable in a competitive manufacturing environment, although it requires both training and acceptance of responsibility by the operator.

Given the measures which exist, how can the ergonomist use them in studies? The first way requires no counting, just classification. In any process control study the first essential is to document all of the variables involved. Defect lists from operators, quality control and customer returns make an excellent starting point. As has been noted before (e.g. Drury and Sinclair, 1983) the discipline of collecting, and agreeing upon, defect names and standards is an arduous process, but a necessary precursor to successful studies in process control inspection. The variables documented can be treated as head events in a fault tree analysis (Brown, 1976) or output variables in a signal flow graph of the process (Edwards and Lees, 1974). Again, do not accept data at their face value as defect names are often surprisingly local, and causes of defects can be shrouded in process mythology.

To use quality information in a quantitative way means using the fault rates, yield values and variable means described above. Most processes have multiple testing or inspecting functions, each of which acts as a data capture point for the ergonomist (Sinclair, 1984). Existing data should be usable in tracing error rates, and because re-inspection is not uncommon, revealing rates of inspection errors. Drury and Addison (1973) show how re-inspection data can be used to reconstruct initial yield and inspection error rates.

Having said that error applies to the inspection process itself, it should come as no surprise to be warned that inspection data are at least as imperfect as other existing records. A major caveat is that departments and individuals will do what they must to avoid ownership of specific faults. If there is doubt over the origin of an error, the blame may be freely passed from person to person before coming to rest, often on a person absent at the time. Less blatant biases occur when previous processes have to share the responsibility for the faults, and time-accumulation scales differ between departments. The author has seen examples where the defect rate goes negative every Friday, because that is when a previous process sends a representative to take responsibility for certain defects. All previous defects collected over a *week* are subtracted from one *days* total, providing data which can easily confuse the unwary.

Personnel data records

Three types of data typically come from the personnel or human resources function: fixed data on jobs, data on individual employees and safety data. All have their corporate uses, and cross-referencing data between compartments can often yield useful insights.

Job data

In order to hire workers, place and promote them, a company needs descriptions of the jobs to be performed. Such job descriptions should (theoretically) exist for every job in the company; they should be detailed and regularly updated. Such utopia is rare, but job descriptions do exist in sufficient numbers to make them a useful initial tool in systems and task analysis. Although typically lacking in ergonomic detail, some schemes for job descriptions can tell much about the job. Job evaluation schemes of the 1970s had operator and specialists rate jobs on many dimensions to determine parity of job difficulty and hence of pay scales. Even more detailed are the schemes based on one of the ergonomic systems developed in the USA (e.g. Position Analysis Questionnaire or PAQ; McCormick *et al.*, 1977) or in Germany (the Arbeitswissenschaftlichen Erhebungsuerfahoens zur Tatigkeit-sanalyse or AET; Landau and Rohmert, 1981). Unfortunately such systems are not yet in widespread use so that the ergonomist may have to install one rather than find data in the archives.

Individual data

Personnel departments keep records of each individual employed, from which can come useful data. However these data on individuals are particularly sensitive and all of the ethical considerations of using such data must be fully realized before the study starts. Discriminations in hiring, placement and advancement are all too easy: the ergonomist should be seen to be sensitive to such issues. Data on age, gender, race, work history, medical history and company evaluations may provide useful co-variates or matching variables in ergonomic studies, but think ahead to the outcome. If you help to show that males or older workers or immigrants are unsuited to certain jobs, even as a by-product of your main hypothesis, you may put yourself and the company into an embarrassing position, as well as deny advancement to individuals. One must also be aware of subjectivity and bias in interview records, evaluation documents, citations for awards and even medical absence data.

In many companies and countries, research access to data as sensitive as medical records is restricted without appropriate safeguards. Even so, such data on pre-conditions, smoking and injury history, can be used to select homogeneous groups of subjects to reduce inter-subject variability and hence keep down sample sizes.

Accident and injury data

Human resources departments are often the home of safety and/or medical functions, both of which provide useful archival data. Medical histories have already been discussed under Individual Data, so that accident and injury data, whether in medical or safety departments, will be considered here.

In most endeavours, injuries rather than accidents are the events which precipitate a record in the archives. Despite pious articles on 'Total Loss Control', the typical enterprise will only record those events which it is legally required to report. In the USA, there are specific criteria which define what must be reported in industry (via OSHA), in aviation (via FAA), in mining (Bureau of Mines) and so on. As many writers (e.g. Hale and Hale, 1972) have pointed out, accidents are not synonymous with injuries, but merely a necessary condition. Indeed, the injury is seen as one event in an accident sequence by Monteau (1977), by Haddon (1973) and by the US Consumer Product Safety Commission (CPSC, 1975).

Recorded injuries are thus the typical archival substitute for accident data, whether accessed by medical logbook, accident report forms or in-depth investigations. Because of this, there are many potential biases or artifacts in such data. First, non-injury producing accidents do not get reported, or go by another route such as scrap report. Second, not all injuries are reported. The '2000 Accidents' study (Powell *et al.*, 1971) clearly showed that low-severity injuries are remarkably under-reported. So are the less visible injuries, such as back strains and mental health problems. One suspects that *all* amputations are reported!

Given that an injury is reported, it may not be recorded, particularly if it is of low severity. OSHA in the USA requires all visits to a medical facility to be recorded in a medical log, but only injuries meeting certain criteria stimulate an accident or injury report. Beyond this only very serious injuries cause an in-depth investigation to be undertaken. Thus the ergonomist has a progressively filtered system, with broad but shallow coverage in the medical log, more detailed coverage of fewer events in the accident/injury report and deep coverage of a minimal number of accidents via in-depth investigations.

From a reasonable accident/injury report comes certain standard information on the victim (age, gender, height, weight, job category), the task (job at time of accident, accident precipitating event, injury event), the equipment used (agent of accident, agent of injury), and environment (time of day, unusual conditions). The care with which forms are devised and filled out are both highly variable. Most forms are strong on medical facts and individual differences and blame-pinning and weak in details of task, equipment and environment. The archetypal form reports that the victim sustained an (alleged) back sprain due to improper lifting technique and that the remedial action taken was to instruct the victim not to lift that way again. Here we have examples of blaming the victim (who was after all the only person involved), scepticism (strains are always alleged, amputations never are), circular arguments (for a back injury to occur, the lift *must* have used the wrong technique) and inappropriate intervention (which assumes that the employee did not have the knowledge or motivation to prevent the injury). One can go into any plant and find at least one such report within the thirty most recent reports.

If we are sceptical of the archival accident/injury data, why do we use it at all? First, it is of immediate economic importance. Each report represents a loss of revenues as well as large direct costs to the company and the employee. Second, accident/injury reports have a high level of face validity.

After the events at Three-Mile Island everybody believed that there were dangers in current nuclear power stations. Third, archival reports can be used in a boot-strap procedure to devise better classifications (Drury and Wick, 1984), to see which departments are in most urgent need of ergonomic help (Drury and Wick, 1984), and to derive new and more detailed accident investigation procedures (Monteau, 1977, Drury and Wick, 1984). If the ergonomist looks at the accident/injury reporting system with detachment and foreknowledge, useful data will be found. (See also the chapter by Brown in this book).

Costing data

It is often important to make a direct financial case for ergonomic change (see Simpson and Mason, chapter 31), although we all recognize that the most visible cost elements are rarely the whole story. Organizations keep detailed financial records from the annual balance sheet down, but ergonomists are rarely equipped to understand the concepts and terminology of the cost accounting systems employed. Excellent standard texts are available (e.g. Horngren, 1967) and most introductions to business (e.g. McGarrah, 1963) and engineering management (e.g. Garrett and Silver, 1973) provide useful, if shallow, primers.

Unfortunately, the questions an ergonomist wishes to ask are rarely the ones the costing system is designed to answer. We typically want to know how much it costs now to perform a certain operation, and how much less it will cost if specificied changes are made. Direct costs are relatively easy to measure. Labour costs per hour, scrap costs per unit and even energy costs per cycle, should present no difficulties. But the operation we are studying does not exist in isolation; indirect costs are involved. A lathe operation will use a certain amount of space, environmental energy (heat, light), inventory, quality control, maintenance, personnel, washrooms, pension plans, sales executives and even ergonomists. These indirect costs, or overhead burden, are notoriously hard to allocate to direct operations, but ultimately all must be paid for from product sales and all need to be accounted for. In one company, where workers were paid at $15 per hour, the time cost of an hour's work was pegged at $60 for any labour-saving calculations as a more reasonable value for the true cost of employing an operator.

Overheads may be assigned by labour hour, by square feet of space occupied, by product being produced or by any other formula the company desires. The intent is to stress the factor which is critical to operations, whether it be direct labour or space. With the advent of plentiful and powerful computers, it is now possible to allocate overheads in many different ways to more properly reflect the true costs of doing business. Some functions which were previously financed from overheads are now being asked to operate as cost centres. Thus the ergonomics department, personnel offices or medical centre must hire out its services to the operating departments, and at least break even each year. Such systems focus attention onto the business essentials, but do have the obvious side-effect of discouraging a long-term view.

How does all this help the ergonomist to find the cost savings potential of an intervention? In truth, it does not; but largely because the ergonomist is not skilled in formulating the questions which still need to be asked and answered. Perhaps the best advice is to read up on this subject, to understand at least the rudiments, and then develop a close relationship with the costing functions in any organizations you enter.

The joys and perils of archival data

It should be obvious by now that archival data represent a valuable resource for the ergonomist, but are full of hidden traps for the unwary.

Measurements from archival data complement the more usual ergonomic surveys and experiments in many ways. Primarily they represent a long-term view of the effects of change over months or years rather than a snapshot of one particular process over a few days. This long-term view enhances the face validity, because the real system is operating, with all of its day to day idiosyncrasies as opposed to a 'best behaviour' controlled experiment. Additionally, the raw data are already being collected, and so the additional data collection costs look minimal. Inexpensive, face-valid data which reach back into the past and will reach forward into the future are certainly worth considering, but this is not the whole story.

Existing data were *never* collected for ergonomic purposes and so will never answer the ergonomist's particular questions except by fortunate coincidence. We must accept a high level of aggregation, with, for example, accident rate being aggregated across many individuals whose jobs vary to a greater or lesser extent. Similarly, productivity or quality data are accumulated over weeks or months of highly variable conditions, about which the ergonomist is likely to remain forever ignorant. Thus cause and effect are difficult to establish. The author has participated in studies in which scrap rates and accident rates halved from the year before the intervention to the year after, but it is naive to think that the ergonomic intervention was the sole innovation by the organization in two years. All of the workers responsible for poor quality or high accident rates (assuming such people existed) could have been fired in the second year, but without a very close relationship on an on-going basis it would not be obvious to the ergonomist that such a change had taken place.

Additionally, there is much scope for bias and distortion in organizational records. Thus a particular customer (e.g. the Government) may require particular cost reporting procedures which differ from the regular ones. Alternatively, accident causation can be classified by a supervisor in a way different from that used by an ergonomist investigating the same accident.

People play games with quality and productivity records, legal and ethical considerations prevent access to certain data sources. As the world moves towards computer integration in manufacturing and other enterprises, disparate data sources can now be integrated and correlated. While there is much potential for increased depth and breadth of understanding of as complex an organism as the human at work, there is also the potential for increased control and abuse. Ergonomists must know where they stand, and be prepared to back up their decisions with appropriate action.

References

Baker, S.P., O'Neill, B., Haddon, W. and Long, W.B. (1974). The injury severity score: a method for describing patients with multiple injuries and evaluating emergency care. *Journal of Trauma,* **14**, 187–196.

Barnes, R.M. (1980). *Motion and Time Study: Design and Measurement of Work,* 7th edition (New York: John Wiley).

Brown, D.B. (1976). *Systems Analysis and Design for Safety* (Englewood Cliffs, NJ: Prentice-Hall).

CPSC (1975). *In-Depth-Investigations* (Washington, DC: Consumer Product Safety Commission).

Drury, C.G. (1980). Task analysis methods in industry. *Applied Ergonomics,* **14**, 19–28.

Drury, C.G. (1982). Improving inspection performance. In *Handbook of Industrial Engineering,* edited by G. Salvendy (New York: John Wiley).

Drury, C.G. and Addison, J.L. (1973). An industrial study of the effects of feedback and fault density on inspection performance. *Ergonomics,* **16**, 159–169.

Drury, C.G. and Sinclair, M.A. (1983). Human and machine performance in an inspection task. *Human Factors,* **25**, 391–400.

Drury, C.G. and Wick, J. (1984). Ergonomic applications in the shoe industry. *Proceedings of 1984 International Conference on Occupational Ergonomics,* pp. 149–493.

Edwards, E. and Lees, F.P. (1974). *The Human Operator in Process Control* (London: Taylor and Francis).

Garrett, L.J. and Silver, M. (1973). *Production Management Analysis* (New York: Harcourt, Brace, Jovanovich).

Haddon, W. (1973). Energy damage and the ten countermeasure strategies. *Human Factors,* **15**, 355–366.

Hale, A.R. and Hale, M. (1972). *A Review of the Industrial Accident Research Literature* (London: HMSO).

Horngren, C.T. (1967). *Cost Accounting—A Managerial Emphasis* (Englewood Cliffs, NJ: Prentice-Hall).

Konz, S. (1983). *Work Design: Industrial Ergonomics* (Columbus, OH: Grid).

Landau, K. and Rohmert, W. (1981). *Fallbeispiele zur Arbeitsanalysi* (Bern: Hans Huber).

McCormick, E.J., Mecham, R.C. and Jeanneret, P.R. (1977). *PAQ Technical Manual* (West Lafayette, IN: PAQ Services).

McGarrah, R.E. (1963). *Production and Logistics Management: Text and Cases* (New York: John Wiley).

Monteau, M. (1977). *A Practical Method for Investigating Accident Factors* (Luxembourg: Commission of the European Communities).

Powell, P.I., Hale, M., Martin, J. and Simon, M. (1971). *Two Thousand Accidents* (London: National Institute of Industrial Psychology).

Saari, J. (1976). Typical features of tasks in which accidents occur. In *Proceedings of 6th Congress of IEA, Human Factors Society*, Santa Monica, pp. 11–16.

Schilling, E.G. (1982). *Acceptance Sampling in Quality Control* (New York: Marcel Dekker).

Sinclair, M.A. (1984). Ergonomics of quality control. *Workshop for the International Conference on Occupational Ergonomics*, Toronto.

Taguchi, G. (1978). Off-line and on-line quality control systems. *Proceedings of International Conference on Quality Control*, Tokyo, Japan.

Chapter 5

Designing ergonomics studies and experiments

Colin G. Drury

Introduction

This chapter will address the broad issues of study design to reduce the chances of the ergonomist leaping to obvious solutions without fully considering the alternatives. Thus the design and conduct of experiments *per se* are treated as a special case of ergonomic studies, to be used where a conscious decision has been made that they are appropriate.

There are two major choices to be made in study design: what to measure and how to measure it, known respectively as *measurements* and *methods*. Unfortunately, they are not entirely independent, so that the process of choosing alternatives is often iterative. Starting with the methods section is most logical, as major decisions must first be made on the overall technique (observation of natural behaviour vs. designed experiment) before decisions on independent and dependent variables are needed.

As with any goal-oriented activity, designing studies is largely a matter of accurate goal specification, followed by logical (and/or economic) choice of steps to meet that goal. What is the goal of a study? Typically, this would be set by those employing the ergonomist, e.g. to test different computer keyboard features for user acceptance. In many circumstances the goals will be defined in a formal request for proposal (RFP), but in other cases the ergonomist will have some ability to question the study goals. Perhaps the greatest freedom of goal-choice is in academia, where a student or faculty member can run studies with self-generated goals, subject only to the ergonomists' ability to find funding and satisfy human subjects' protection rules. Any good textbook must advise the ergonomist to question the goals of the study, but must also caution that much effort has been wasted on this activity over the years. We will assume that you have a set of goals and proceed as if these were now engraved on stone tablets. The object of this chapter is to provide an orderly way in which these goals can be turned into a designed study, in much the same way as a statistical textbook will demonstrate how to turn a research hypothesis into a statistical hypothesis.

If a goal is the input to a study, then the output is a study design. The specificity of the design is particularly important as minor details can have major consequences. One only has to look at critical evaluations of the research literature on vigilance (e.g., the special issue of *Human Factors, 24.6,* 1988) to see that after 40 years of work, differences between major theories still hinge on the minutiae of study design. A good test of the completeness of the design is that it can be given to a competent technician to carry out, and it will always be done in the way the ergonomist intended. In statistical design of studies, the hypotheses to be tested are defined in terms of the measurements to be taken and the tests to be performed on the data *before* any data are collected. A well-designed ergonomic study should meet the same criteria—you must know what to do with the data you have collected if you are to avoid the statistical (and economic) error of collecting data *and then* looking for interesting interrelationships. This does not mean that all studies require a formal research hypothesis to justify every measurement. Data collection is expensive, so that it may be possible to augment the study by including some measurements which can be conveniently collected but for which no formal hypotheses are made. In that case, these augmented variables cannot be used to test hypotheses after the fact, but they can be used to formulate new hypotheses for testing in subsequent, independent experiments. While the major goals of any study must always be tested hypotheses, it would be foolish to overlook the opportunities presented by major studies to learn more about human interaction with the environment.

Choice of technique

Chapanis' (1953) original methodology text classified the techniques available to the ergonomist into observation and experiment, the distinction being whether the world was observed by the observer or manipulated by the experimenter. This distinction is still important, but we should now recognize more than just two levels. The major difference between the levels is in their reactivity—how much or how little they change the system under study.

At the extreme low end of reactivity are *task analytic methods*. At the earliest levels of system design, they do not even rely on a system to study. Tasks are described and analysed for their ergonomic impact by logically deducing the task description from system goals and functions and by deriving task demands and human capabilities from the logic and the literature. Later stages of task analysis may well observe the real system, or even call for experiments (e.g. Drury *et al.*, 1987) but by then we are using the techniques which follow (see Stammers *et al.*, chapter 6 in this book).

At the next higher level of system manipulation are the methods of *direct observation*. Here a functioning system is studied by reading its records or observing its behaviour (see chapters by Drury on archival data and direct observation). We often think that such techniques are non-reactive, but they do involve actively interfering with the system, i.e. no observations would be taken were the study not taking place. The reactivity may only be potential, as in studying ambulance effectiveness from patient records (Baum and Drury, 1976) or it may be very real, as in a stop-watch time study (Konz, 1983).

Questionnaires and *rating scales* (see Sinclair in this book), the next higher level, obviously affect the system but are often used as if they did not. For example, the study of emergency telephone responses (Baum and Drury, 1976) probably caused many subjects to think through their emergency response behaviour for the first time as less than 15% had ever used the telephone to call for fire, police or ambulance.

Finally, any *direct manipulation* of the system must be classed as an experiment, whether it is a multifactorial design used in a research laboratory or a case-study (or other pseudo experiment; Kerlinger, 1969) run in a factory. The system is deliberately changed so that the results of the changes can be observed. This can never be done in a non-reactive manner.

Given that manipulating and changing systems must be costly and potentially dangerous, what does this increase in reactivity buy for the ergonomist? The two major advantages of a more highly reactive design are:

(1) The ability to be in the right place at the right time to observe (Chapanis, 1953). This is particularly important in ergonomics studies where the system behaviour it is desired to observe is rare and unexpected, e.g. accidents or breakdowns.

(2) The ability to use more obviously invasive, but information-rich, measurement techniques. For example, in inspection research, the response to each individual item inspected can be of considerable detail in an experiment (e.g. Drury and Sinclair, 1983), whereas in the real situation only a simple accept/reject response is often given.

If the ergonomist gains in experimental control and measurement detail by using highly reactive designs, what else is lost? The major loss is in face validity. If we observe a system in its natural state, those associated with the system, and possibly those who commissioned the study, can be convinced that the study is realistic, however that is defined by those to whom it is important. An experiment, particularly one performed in a laboratory with artificial stimuli and non-representative subjects, requires much more persuasion on the part of the ergonomist to gain accceptance. The author was once involved in two studies of fork-lift truck control. One (Drury and Dawson, 1974) involved real drivers using real fork-lift trucks in a real warehouse to study lateral control behaviour. The other (Drury *et al.*, 1974) involved real drivers controlling a toy train in a laboratory to study longitudinal control behaviour similar to Fitts law tasks. It is obviously much easier to quote the former study to convince warehouse managers of its design implications.

There are many degrees of realism within the experimental paradigm. Experiments can be conducted using the operational system (e.g. real aircraft), a faithful simulation (e.g. flight simulator) or a task only logically related to the operational system (e.g. tracking or dichotic listening). The issues of fidelity in experimentation are essentially the same as those in simulator design (see Meister, chapter 8 in this book). Only a thorough knowledge of the ergonomics literature can guide the experimenter on what it is safe to leave out of an experiment. As an example, there are thousands of references on the component processes of an industrial inspection task (visual search, signal detection, vigilance) but only a few dozen which meet the stringent requirements of experimental techniques used by real inspectors with real products.

These issues of reactivity, control, measurement detail, and validity are summarized in Table 5.1.

Choice of independent variables

Independent variables are what you change, while dependent variables are how you measure the effects of the change. Choice of independent variables comes primarily from the hypotheses: if you hypothesize that control–display compatibility affects reaction time, you have chosen compatibility as your independent variable. If life were that simple, this section would not be needed. In order to measure reaction time under different values of the independent variable, you must get down to specifics on other matters: who will be the subjects? What will be the instructions? Will the experiment be run sitting or standing, morning or evening, and so on? Thus choice of independent variables is not merely a matter of making obvious deductions from the hypotheses.

It will be simpler to talk of independent variables as if they were manifested in an experiment, implying that the ergonomist actively implements the choices of values of each variable. However, the same principles apply if the technique is observation or questionnaire although choice is more a matter of selecting, from among already existing alternatives, those to be included in the study. In a similar manner, some of the language of experimental

Table 5.1. Comparison of techniques by different criteria

Technique	Reactivity	Face validity	Control	Measurement detail
Task analysis	zero	high	—	—
Observation/records	low	high	zero	low
Questionnaires/ratings	medium	medium	low	medium
Experiments	high	low	high	high

design can be applied to observation and questionnaires. Thus an independent variable is a factor and the value which the independent variable takes is the level of that factor. Both naming conventions will be used throughout this section.

The design of any study first consists of choosing levels of all of the independent variables of relevance. Because we must be specific in our study design, then we must specify not only what we will vary, i.e. what levels of which factors, but what we will not vary, i.e. what we will keep constant. One has only to look at the *post hoc* rationalizations of unexpected study results in our journals to see that ergonomists do not always recognize what must be kept constant. Taken with the frequent finding of insignificant effects of a major factor, it recalls the scientist's version of Murphy's law: 'variables won't and constants aren't'. The least that an uncontrolled variable can do is to contribute to random experimental errors: typically it does more and introduces real biases into a study.

Systematic techniques are required if we are to bring our ergonomic knowledge to bear upon the design of effective and efficient studies. We need a technique for listing all possible factors which can affect the outcome of the experiment, and a technique for deciding what to do with each factor.

To generate a list of all possible factors which could affect the dependent variables, it is simplest to use the categories which we, as ergonomists, regularly use: human, machine and environment. To these should rightly be added task, to cover the instructions, restrictions and goals under which the system operates. Thus the four categories are *task, operator, machine* and *environment*. Under each of these is listed the factors likely to affect performance. For example, in a study of compatibility and reaction time say, obvious factors under operator are: age, experience of similar systems, training on system under test, and national stereotypes for control/display relationships. Less obvious factors, but ones which can certainly affect reaction time, are: intelligence, visual capabilities, risk acceptance/aversion, and motor co-ordination. Some factors *may* affect performance, but our ergonomics insight tell us that the probability is small: body size, gender, and time since last meal.

This list could obviously continue well into the trivial, as could similar lists for task, machine and environment. Note that the lists will be different for different hypotheses; in manual materials handling systems, body size and gender would be of primary importance.

We now have four somewhat orderly lists of possible factors and face the question of what to do with each factor on each list. There are only five alternatives:

1. *Build the factor into the experiment at multiple levels.* This ensures that our hypotheses will be general across the whole range of this factor used in the experiment. Thus if we have five age groups, covering the decades of the working population (e.g. 15–25, 25–35, 35–45, 45–55, 55–65), we can be sure that our compatibility conclusions are valid for all working ages, or we

will know that they only apply to older workers, depending upon the specific outcome of the study. This is the preferred treatment of each factor, as it maximizes the impact of our studies. It is also by far the most expensive option, especially as each factor tends to multiply the size of the experiment by the number of its levels. Typically, only factors specifically named in our hypotheses will be built in at multiple levels.

2. *Treat the factor as a co-variate.* Often it is not feasible to choose levels of a major variable in advance, as we may need expensive measurements to determine what the level actually is. For example, although age and body size are simple enough to measure or even to ask for by telephone when scheduling subjects, more subtle factors such as perceptual style or visual reaction time will need the subject to undergo tests before being allowed into the experiment. In such cases, the factor of interest can be measured, typically before the main experiment, and used as a co-variate in an analysis of covariance design. This means that any correlation between the dependent variable and the co-variate is taken out as a specific term in the analysis, typically before the other factors are analysed (Nie *et al.*, 1975). The co-variate option is usually used for operator variables, although it is not limited to them. In observational and questionnaire studies, many of the variables are treated as co-variates, such as number of miles ridden per year in studies of motorcycle accident frequency. At times co-variates are the major focus of a design, as for example the classic Fleishman and Rich (1963) study of factors affecting task performance at different stages of learning. At the very least, carefully chosen co-variates reduce the waste associated with any study which finds that individual differences account for a large portion of the overall study variance. To give a recent example, significant correlations between perceptual style and inspection performance (Gallwey, 1982) have enhanced our understanding of the inspection task. Obviously, if multiple subjects are to be a part of the design for other reasons, then treating a factor as a co-variate is much less expensive than building that factor in at multiple levels.

3. *Fix the factor at a single level.* Here a factor is treated as a constant. We could run our study with a single, narrow, age group; we could choose only third-generation Americans; we could run all experiments in a well-lighted room at constant temperature. By only having a single level of a variable, we do not increase the size of our study but the price we pay is that the results will not generalize beyond the constant conditions we specify. We may wish to extrapolate the study results, based on other data and other models, but false extrapolation has plagued many disciplines including ergonomics. Human performance has a way of constantly surprising us. Typically, the fixing of a factor at a single level is used for those factors we know are important, but about which we do not need to generalize.

4. *Randomize the effects of a factor.* If a factor may be important, but we cannot control it in one of the ways above, then randomization will prevent the factor from biasing the study results. By randomly assigning subjects to

factor level combinations or stimulus presentation order to subjects, any uncontrolled variability is given an equal chance of affecting all levels of the particular factor. Thus the random variability may be increased (resulting in a weaker study) but we will avoid reaching biased conclusions. A well-known example of the use of randomization are the random assignment of subjects to treatment groups, rather than putting the first volunteers in group A, the rest in group B, and so on. People who volunteer early may be different from those who volunteer late and randomization ensures that this volunteer bias does not result in a biased study.

5. *Ignore the factor*. This final strategy is only included for the sake of completeness. It is never safe to ignore a factor in studies involving entities as complex as human beings. The old adage, 'If in doubt, randomize,' applies here. Ignore factors at your peril.

At this point, we have an assignment problem—how to assign each factor to each alternative. In practice, the assignment is usually quite straightforward, because the real work was performed in listing the potential factors in an organized and ordered manner. What has been achieved is that a systematic technique has been used to reduce the chance of the study finding unexpected conclusions for the wrong reasons. The whole procedure may seem laborious, but it forces the ergonomist to think through the issues *before* the study starts and to use ergonomic insights to advantage. A useful exercise to practice these skills is to take a paper from the literature and to use these techniques to see how you would have designed the study. Here you have the advantage of hindsight—you know what the results were. In ergonomics practice you need to develop foresight if you are ever to get a second chance at designing a study.

Choice of dependent variables

Traditionally, ergonomists have measured the effects of their factors on a single variable (e.g. reaction time, error percentage, heart rate), but advances in the ability to record and analyse multivariate data have to some extent stopped this trend. Now we tend to think in terms of sets of dependent variables, each illuminating a different aspect of man/machine fit. Despite this trend, the current section will concentrate on single measures, as the choice of sets of measures depends to some extent on the branch of ergonomics involved. For example, it is pointless to run a signal detection experiment without measuring (or controlling) both type 1 and type 2 errors.

Before classifying and describing possible measures, it is necessary to have available some criteria by which to judge the adequacy of measures. Social science texts (e.g. Kerlinger, 1969) see three main criteria:

1. *Validity*. Does the measure have a direct relationship to the phenomena described in the hypothesis? Is heart rate a valid measure of metabolic cost? Only under particular circumstances such as lack of heat stress, static muscular

tension and emotional stress. Is reaction time a valid predictor of the relative performance of three computer keyboards? Only when certain major variables such as information content per stimulus are kept constant and only when minimum response time has some relevance to the final criteria of the system evaluation. Technically, validity is the correlation between the variable measured and the phenomenon represented in the hypothesis. A measure may be valid *a priori* (face validity) because it patently measures the phenomenon. Thus accident frequency is a face-valid measure of plant safety. A measure may be valid because it can be logically related to the phenomenon (construct validity). Thus reaction time is a construct-valid measure of control–display compatibility because it is the reciprocal of information processing rate (for constant information per stimulus) and thus is logically related to compatibility. Finally, a measure may be proven valid experimentally. Thus Borg (1982) has validated two scales of rated perceived exertion against heart rate by experimentally correlating the two. Clearly, choosing an invalid measure will ensure that we answer the wrong question.

2. *Reliability*. Does the measure give consistent results when used repeatedly? Is heart rate too variable pulse to pulse or minute to minute to be a reliable measure of metabolic load? Do we get the same reaction time to the three computer keyboards if we measure them today, tomorrow and next week? If validity asks how well a measure correlates with an external phenomenon, reliability asks how well the measure correlates with itself. While there are many reliability measures, the two most common are test/retest reliability and split-half reliability. Test/retest reliability correlates the measure obtained on a subject in the first (test) period, with that obtained during a subsequent (retest) period. Split-half reliability correlates two halves of the measure obtained during a single time period. Thus in an intelligence test, the score on even-numbered questions should correlate well with that on odd-numbered questions. Similarly, the reaction times measured in the first and second halves of a trial with computer keyboards should correlate highly. Reliability coefficients close to 1·0 are desirable, with values of 0·8 and upwards being used regularly in social science work. Reliability sets a limit to the internal consistency of our studies. Unreliable measures, even if they are valid, cannot prove a hypothesis because we cannot have faith in them to tell how the phenomenon of the hypothesis changes.

3. *Sensitivity*. Does the measure react sufficiently well to changes in the independent variable? It is quite possible that the measure chosen may be valid and reliable, but will not show a large enough effect to be measured easily. Thus, reaction time may not differ between different keyboards because it is not sufficiently sensitive to record the subtleties of keystroke force/distance characteristics. Similarly, heart rate may not react by more than a few beats per minute to differences in the design of handles in manual lifting tasks (Deeb and Drury, 1986) requiring a relatively large design to obtain statistical significance.

Any measure must pass all three tests (validity, reliability and sensitivity) before it can be used, although not every measure will need to perform equally on all three criteria. Measures themselves fall into two broad classes: (1) performance—measures of the effect of the human on the system; and (2) stress—measures of the effect of the system on the human, or cost of the performance to the human.

In general, it is difficult to conceive of measuring one without measuring (or at least controlling) the other; at times this control is implicit. Thus, when we measure reaction times in a laboratory, it is usually implied that the subject performs at the maximum level, i.e. at a constant level of stress or cost. Similarly, in a manual lifting task, the task is often fixed (box size, lift distance, speed) so that the physiological cost can be estimated using oxygen consumption, heart rate or even subjective ratings. At other times, the joint measurement of performance and stress will be explicit, as in multivariate assessment of jobs where both performance and stress need to be measured separately in a realistic setting to determine how subjects choose to allocate their processing resources to tasks of variable demand.

Performance measures

Performance itself can be subdivided into two aspects; speed and accuracy. Speed refers to the time for task (or subtask) completion, to the amount produced in a given time period (e.g. output per shift) or to the speed of movement, (e.g. driving). At times it can be a straightforward measure such as task completion time, or it can be normalized with respect to some external criterion, for example speed expressed as a percentage of maximum speed. Normalization can be a source of increased accuracy, as in measurement of accident rates rather than accident frequencies, or it can be a source of potential confusion, as in measuring industrial efficiency with respect to time standards. Productivity is another normalized speed measure, typically defined as output from a system for a given level of input resources.

The other face of speed is accuracy, if only because a speed/accuracy trade off can be seen in most human tasks (Wickens, 1984). Because of this trade off, accuracy and speed are typically co-measured or one of the two is controlled. Thus reaction time trials are repeated if the wrong response is made. Accuracy can be measured by errors or by their consequences, such as injuries or system failures.

An error rate or error probability is defined as

$$p(\text{error}) = \frac{\text{number of errors}}{\text{number of opportunities}}.$$

The measurement of both numbers can be an error-prone activity itself. Thus errors have reporting biases and classification problems. Number of opportunities can be easy to calculate in some circumstances but difficult in others. For repetitive tasks, error rate is typically defined per cycle, so that we have sewing errors per 100 pair of jeans, near misses per flight, or polarity errors per component mounted on a circuit board. For some repetitive tasks, such as inspection, there is more than one input and thus more than one error rate. An inspector receives both good components and faulty components and thus has two error rates with different denominators as well as numerators:

$$\text{Type 1 error} = p(\text{reject/good}) = \frac{\text{number of good items rejected}}{\text{number of good items}}.$$

$$\text{Type 2 error} = p(\text{accept/faulty}) = \frac{\text{number of faulty items accepted}}{\text{number of faulty items}}.$$

In such tasks, an overall error rate, obtained by dividing all errors by the total items inspected, is a meaningless quantity, although techniques do exist for combing both errors into other measures with more meaning (e.g. Drury, 1982).

For highly variable tasks, choosing the error rate denominator becomes more difficult. Thus in driving, process control, or logging we can define two denominators with different meanings:

Output based: errors per mile driven
                   errors per barrel of oil processed
                   errors per ton of logs harvested

Time based: errors per hour driven
                   errors per month of oil processing
                   errors per day of logging.

There are interesting moral questions in the choice of denominator; an output base has meaning for the company while a time base is more important to the individual at risk.

To some extent each task has its own unique error characteristics, as can be seen by examining defect reporting sheets an industrial processes. However, there are some useful, more general error taxonomies in use in ergonomics. Perhaps the simplest was the early splitting of errors into omission and commission errors (Miller, 1963).

Since then, more detail has been added. Meister (1971) recognized that errors could be classified by their causes, consequences or where in the system development process they occurred. He broke error causes into:

1. System-induced error,
2. Design-induced error, and
3. Operator-induced error.

It should be noted that bias by the experimenter can lead to different classifications. Thus most accident reports cite operator error, whereas an ergonomist prefers to look behind the operator to see what design or systems' problems caused the operator error.

Error causation has been classified by Swain and Guttman (1980) in a rather different way, by defining the operator errors more precisely:

1. Omission.
2. Commission.
3. Extraneous act.
4. Sequential error.
5. Time error.

Note that 3, 4 and 5 are really all varieties of commission error.

Error consequences are classified in computer terminology into fatal and non-fatal errors, but a more detailed breakdown is provided by Van Cott and Kinkade (1972):

1. Hazard to personnel or equipment.
2. Degradation of system performance.
3. Degradation of subsystem performance.
4. Degradation of component performance.
5. Little effect on system performance.

System development stage provides a useful classification basis where the whole system's development process is under review by the ergonomist. Table 5.2 from Meister (1971) presents a possible classification. A more useful classification for ergonomists may be by the stage of human information processing which fails, particularly the model by Rasmussen (1982) which builds stages of processing into a well-developed error taxonomic scheme (Figure 5.1).

Finally, error probability is a useful classification scheme if we wish to concentrate on likely errors. Van Cott and Kinkade (1972) use a three-level breakdown for the error probability p:

1. Highly likely $0.5 \leqslant p \leqslant 1.0$,
2. Likely $\Delta \leqslant p < 0.5$,
3. Unlikely $0 \leqslant p < \Delta$,

where Δ is a small probability, presumably defined by the investigating team.

All of these taxonomies provide a framework for studying systems errors, but in most experiments and observational studies much finer classification is needed. For example, Kasprysk *et al.* (1980), in their study of calculator notation, broke errors into sequence errors, where the key being pressed was logically wrong, and keyboard errors, where a closely adjacent key was

Table 5.2. Classification of system errors, after Meister (1971). Upper table classifies errors by system development stage and lower by human function performed.

	Stage of System Development			
	Design	Production	Test	Operation
Type of Error	Design Error	Fabrication error Inspection error	Operating error Installation error Maintenance error	Operating error Installation error Maintenance error
Causal Factors	Inappropriate function allocation Failure to implement requirements Poor human engineering design	Incorrect blueprints Incorrect instructions Inadequate tools Inadequate environment Inadequate training/skill Poor human engineering design of equipment Poor workplace layout	Inadequate/incomplete technical data Inadequate logistics Poor human engineering design Poor workplace layout	Inadequate/Incomplete technical data Inadequate logistics Inadequate training/skill Inadequate motivation Poor human engineering design Poor workplace layout Inadequate environment Overload conditions Task complexity Poor personnel selection
Error Consequences	Inadequately designed equipment	Scrapped/reworked equipment Production delays High cost Malfunctioning equipment classified as functioning Good equipment rejected	Delay in systems operations Human-initiated malfunctions System breakdown Failure to accomplish test Degradation in system performance Possible danger and loss of life	Delay in systems operations Human-initiated malfunctions System breakdown Failure to accomplish mission Degradation in system performance Possible danger and loss of life

Functions	Potential Errors
Sensing, detecting, identifying, coding, and classifying	Failure to monitor Failure to record or report a signal change Recording or reporting a signal change when none has occurred. Recording or reporting a signal change in the wrong direction. Failure to record or report the appearance of a target. Recording or reporting a target when none is in the field. Assignment of a target to the wrong class.
Sequencing	Making a below-standard response. Omiting a procedural step. Inserting an unnecessary procedural step. Mis-ordering procedural steps.
Estimating and tracking	Failure to respond to an obvious target change. Premature response to a target change. Late response to a target change. Inadequate magnitude of control action. Excessive magnitude of control action. Inadequate continuance of control action. Excessive continuance of control action. Wrong direction of control action.
Decision making	Incorrect weighting of responses to a contingency. Failure to apply an available rule. Application of a correct, but inappropriate rule. Application of a fallacious rule. Failure to obtain or apply all relevant decision information. Failure to identify all reasonable alternatives. Making an unnecessary or premature decision. Delaying a decision beyond the time it is required.
Problem solving	Formulating erroneous rules or guiding principles. Failure to use available information to derive needed solution. Acceptance of inadequate solution as final.

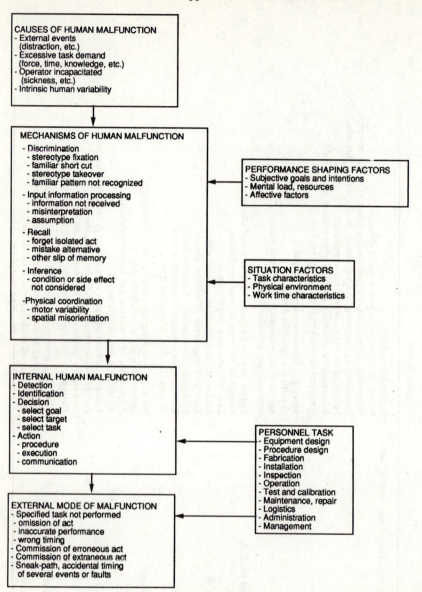

Figure 5.1. Taxonomy of events involving human malfunction, after Rasmussen (1982).

pressed. This enabled the authors to make a much more detailed comparison between algebraic and reverse polish calculator operating systems.

Error measurement of a different kind is required for a continuous task such as tracking. Here, man (aided or unaided) attempts to follow an input, time-varying, course, $x_i(t)$, with a vehicle or follower tracing an output track, $x_o(t)$. At any instant, the error is the difference between the output and the input:

$$x_e(t) = x_o(t) - x_i(t) \, .$$

In order to quantify this error, it is necessary to combine values of $x_e(t)$ over some finite time period, $O-T$. The error function $x_e(t)$ has amplitude, frequency and phase information, although all are rarely preserved in any omnibus measure. Amplitude and frequency information are preserved in the power spectral density (PSD) function which is a transform of the autocorrelation function of $x_e(t)$ into the frequency domain. The PSD will show which frequencies are present in the error function, and how rapidly error power falls with increasing frequency. Both are useful characteristics diagnostic of servomechanism functioning, but are often too complex as a single error measure.

The area under the PSD curve represents the total error power, essentially preserving amplitude information while relegating frequency information to a weighting role.

Other less sophisticated continuous error measures are possible, although with inexpensive and rapid computers with Fast Fourier Transform programs, they are less required than in the past. Time-on-target measures, which define an arbitrary target width and count the percentage of time spent within that width, were discredited many years ago, although they retain some traditional uses such as in pursuit rotor tests. Any more direct measure of $x_e(t)$ needs to distinguish between the mean error value, or bias, and the amount of error variability around that mean value. Bias is typically defined as:

$$\text{Bias} = \frac{1}{T} \int_o^T x_e(t) \mathrm{d}t$$

or, if x_e is measured at N discrete instants,

$$\text{Bias} = \frac{1}{N} \sum_{j=1}^{N} x_{e_j}$$

Obviously, positive and negative errors tend to cancel out so that the bias is small in most tracking tasks. Of more interest is a measure related to the variability about the mean. This is measured by squaring the error to remove negative signs, averaging the squared error, and then taking the square root to preserve the original measurement units. As such, it is known as the root mean square (RMS) error where:

$$\text{RMS error} = \left(\frac{1}{T} \int_o^T x_e^2(t) \, \mathrm{d}t \right)^{\frac{1}{2}}$$

or for discrete instants

$$\text{RMS error} = \left(\frac{1}{N} \sum_{j=1}^{N} x_{e_j}^2\right)^{\frac{1}{2}}$$

defining both for zero bias.

In general RMS error is an easily understood and easily calculated measure of tracking performance, and as such it has gained widespread acceptance (e.g. Sheridan and Ferrell, 1974).

Stress measures

Measures of the cost to the operator of achieving the desired measure of performance are referred to here as stress measures. Technically, stress may be defined (Cox, 1978) as the difference between the perceived demands of the task and man's perceived capacity to cope when coping is important. This is a narrower definition of stress than 'cost to the operator', and is covered in more detail elsewhere in chapter 25 by Cox. Here only a briefer, broader survey is attempted.

Stress can be measured only by its effects. Thus the heart rate required to perform a continuous lifting activity and the muscle tension required to swing a cricket bat are both effects on the body of the stresses demanded by their respective tasks. The concept of validity is especially important in stress measures. If the measure reflects changes in the bodily subsystem stressed, then we have face or construct validity.

For example, heart rate measures cardiovascular demand (at least above a level at which stroke volume is constant) while EMG in the flexor muscles of the fingers reflects the force demands of a gripping task. Such direct physiological measures are easy to validate, but others are not. For example, both heart rate variability and eye pupil response have been correlated with information-processing load ('workload'). They have a certain degree of construct validity but rely in the main on experimental validations, many of which have been equivocal (e.g. Wierwille and Casali, 1983).

In addition to physiological measures of stress, there are a number of well-validated rating scales for different aspects of stress. Examples include *rated perceived exertion* (Borg, 1982), *body part discomfort* (Corlett and Bishop, 1976), and the modified Cooper–Harter scale of workload (Wierwille and Casali, 1983), all shown in Figures 5.2 to 5.4. (See also chapters 25 by Cox and 24 by Meshkati *et al.* in this book.)

Finally, there are outcome measures related to stress. In a factory or office, increases in stress can be logically (and often experimentally) related to sickness, absence, tardiness and accidents. Such measures are only available in large-scale studies but have great face validity and what the Americans call bottom-line impact.

0	Nothing at all	
0.5	Extremely weak	(just noticeable)
1	Very weak	
2	Weak	(light)
3	Moderate	
4	Somewhat strong	
5	Strong	(heavy)
6		
7	Very strong	
8		
9		
10	Extremely strong	(almost max)
*	Maximal	

Figure 5.2. Rated Perceived Exertion (RPE) scale of Borg (1982).

Study design

While choice of technique, independent and dependent measures have to a large extent fixed the design of the study, there are still more decisions to be taken before the study can be finalized. We must know how many subjects are to be used, how many data points per subject are to be collected, and how the experimental conditions and subjects are to be assigned to each other. These next steps are the province of statistical design of experiments (e.g. Winer, 1972), but this section helps bridge the gap between the ergonomic and statistical concepts.

The main issue in study design is human variability and how to obtain reliable results despite this variability. People differ from each other in anthropometry, in physiological performance, in information processing ability, and in their reaction to external stressors. This is obvious and any study design must take these facts into account. Equally obvious, but less often considered by the laity, is the fact that any individual differs on the same measures from year to year, from day to day, and from minute to minute. Thus we have two sources of variability which any design must take into account, known as: (1) between-subject, or inter-subject variability; and (2) within-subject, or intra-subject variability.

As discussed previously, both variability sources can be minimized by correct choice of independent variables. Thus choosing all females for a backpacking task would reduce inter-subject variability somewhat in the energy costs measured. Similarly, choosing trained users of spreadsheets in a computer-aided calculation task would minimize trial-to-trial variability in

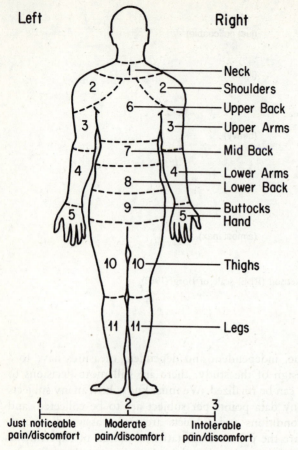

Left

Right

Neck
Shoulders
Upper Back
Upper Arms
Mid Back
Lower Arms
Lower Back
Buttocks
Hand
Thighs
Legs

1	2	3
Just noticeable pain/discomfort	Moderate pain/discomfort	Intolerable pain/discomfort

Figure 5.3. Body Part Discomfort scale (BPD) adapted from Corlett and Bishop (1976).

performance and so reduce intra-subject variability. However, within any single, even narrowly-defined, population, both sources of variance will be too large to ignore. In addition, the price of a restricted sample is lack of generality in the study findings.

Human variability is explicitly recognized in how we interpret our results. Statistical tests are used to establish the statistical significance (or otherwise) of a result by estimating the probability of so extreme a result having been obtained through pure chance. The pure chance here is the inter- or intra-subject variability. For example, if use of backpack A gives a mean oxygen consumption 0·1 l/min higher than the use of backpack B, we would interpret the fact differently if we have four subjects with an inter-subject standard deviation of 0·15 than we would if 40 subjects with an 0·05 standard deviation had been tested.

A test statistic (such as *t* or Mann–Whitney's '*U*') is typically calculated as the ratio of the size of an effect to the appropriate variability. The test statistic increases in absolute magnitude as the size of the effect increases and as the variability decreases. Size of an effect is represented by the difference between two means (or the variance between several means for an *F* test) and is thus a physical fact, although subject to sampling error. For example, given enough subjects and measurements, the true difference between backpacks A and B may be 0·12 l/min. We will only conclude that such a difference is significant, however, if the variability is low enough for our test statistic to be in the region of rejection.

The appropriate variability for a test statistic is the standard error of the difference between two means, which in simple cases is:

$$SE = \frac{\text{standard deviation}}{(\text{number of datapoints})^{\frac{1}{2}}} = \frac{SD}{N^{\frac{1}{2}}}$$

where SD is the standard deviation of a set of *N* data points. Thus the test statistic is:

$$\text{Test statistic} = \frac{M_1 - M_2}{SE} = \frac{(M_1 - M_2)\,N^{\frac{1}{2}}}{SD}$$

where M_1 and M_2 are the means of the conditions being compared.

It is now obvious that the three ways to increase the size of the test statistic are:

1. Increase $M_1 - M_2$.
2. Increase N.
3. Decrease SD.

Each represents a valid option in study design and will be considered in turn.

Increasing difference between means

A large effect will of course be easier to detect than a small effect. But ergonomics is a 'real-world' discipline and the size of effect may be fixed. However, there are a number of steps that can be taken to make the effects larger.

1. *Use a more sensitive measure*, as previously outlined. Oxygen consumption may be less sensitive than, for example, ratings of perceived body part discomfort in comparing two backpack designs. Part of the ergonomic insight comes from a careful reading of the literature to be able to predict which measures will be sensitive in a new study.

2. *Make the two conditions more extreme*. This implies that each mean is physically as different as possible from the other. One tried and tested method

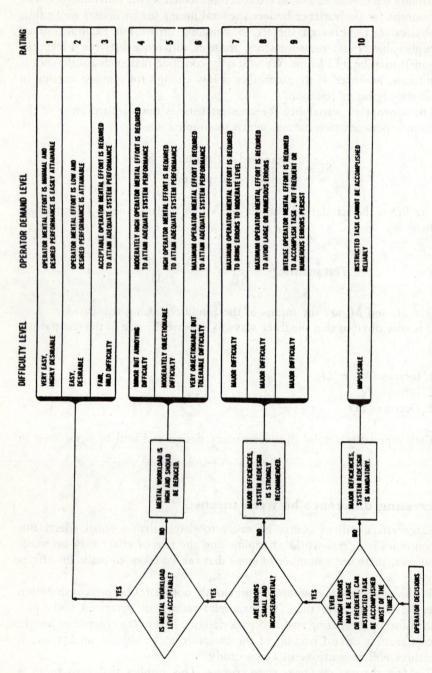

Figure 5.4. Modified Cooper Harter (MCH) scale of Task Difficulty.

is to choose subjects extreme on a measure likely to be related to the task if the difference in means is between subject characteristics. Thus aging research often compares 20–30 year olds with 60–70 year olds, rather than comparing the 30–40 with the 40–50 age group. Similarly, rather than split subjects at the median in perceptual style in an inspection task (Schwabish and Drury, 1984), it would have been possible to choose only subjects scoring in the upper and lower quartiles on the impulsive/reflexive scale. With task variables, an example would be choosing two weights in a box handling task which would be different enough to show a difference in heart rates. Deeb and Drury (1986) used 7 kg and 13 kg to represent average and difficult tasks based on a large survey of box weights handled in industry. More extreme weights have often been used, e.g. by Fish (1975), who had subjects lift barbells weighing either 0·2 kg or 20 kg to measure disc compressive forces. The danger is that one or the other extreme is unrealistic or dangerous. Physicists often refer to this technique as increasing the magnitude of the forcing function.

3. *Increase task difficulty*. This is not just making M_1 more different from M_2 but increasing the difficulty of both tasks so that differences are more likely to show up. Thus if backpacks A and B differ in the amount of padding, then testing both at heavy pack loads will show up the differences more clearly. Another example is in road tests of automobiles where the test drivers negotiate a tight slalom course and measure the performance time. Pushing cars to their limits can reveal differences that may be unimportant in ordinary driving but which become suddenly critical in emergency situations.

Variations on this theme are legion. Instead of changing the task difficulty itself, environmental stressors can be added to push subjects nearer to the limit and hence observe subtle performance changes. High noise levels, poor lighting, external pacing, or even competition have all been used in this way.

The dangers, as with making the conditions more extreme, are that the task becomes unrealistic or dangerous. Both apply directly to ergonomics practice and may contravene human subjects' protection laws. A more insidious danger is that of hitting a 'ceiling effect', which occurs when both conditions are so difficult that performance is equally impossible in both. Backpacks weighing 100 kg would represent an obvious ceiling effect. It should be noted that the opposite of a ceiling effect is a 'floor effect' where both conditions are so easy that either 100% performance is achieved whatever the condition or that some other factor limits performance. In a series of studies of self-paced tracking reviewed by Drury (1985), a floor effect is seen whenever the width of the track is so great that speed and errors are limited by vehicle maximum speed or human willingness to go faster.

Increasing number of data points

It is a truism in statistical testing that any difference, no matter how small, could be found significant given a large enough sample size. Ergonomists pride themselves on practicality and hence the need to ask how large a difference must be in order to achieve practical significance. We should thus aim to make the difference which achieves practical significance the same as that which meets the criteria for statistical significance. Doing this we can solve the equation defining the test-statistic for the sample size N and hence have a rational way of limiting our sample. In order to solve for N, we need to know:

1. The practical difference we wish to detect.
2. The critical value of the test statistic for a specific level of type 1 (alpha) error.
3. The appropriate standard deviation.

Unfortunately, while 1 and 2 may be relatively straightforward to specify, no data may exist on the variability to be expected. One can at times obtain it from the literature, but rarely has the same set of conditions been used as you want to test. The recommended method is a pre-test to estimate the variability, but such a recommendation is cold comfort when a proposal is being costed and the prototype is as yet unbuilt. In such a case, many ergonomists guess, although they will rarely admit it in print. There are obviously some broad guidelines. Upper limits on N are provided by cost and lower limits by the face validity. If you showed by calculation that only two subjects are needed to give the appropriate level of significance, your sponsor may not feel that such a low number is representative. Such misuse of statistics causes despair among statisticians who rightly point out that if the calculations were correct and the two subjects were indeed chosen randomly, there is no need to run more subjects. Life is not always kinder to statisticians than it is to ergonomists. Journals, for example in the behavioural sciences, may need much convincing that small numbers of subjects are adequate, although in anatomical and physiological studies sample size seems to be less of a controversial issue.

Note that the test statistic formula includes N as a square root term. Thus to double the test statistic means to increase sample size, for instance, from 25 to 100, as $100 = 2^2 \times 25$. Increasing sample size as a strategy is likely to meet with rapidly diminishing returns. However, the cost of a study is usually composed of a fixed cost and a varibale cost linear in the number of data points. If the fixed cost is relatively low, then four times as many subjects will mean almost four times the cost for the study. However, if the fixed cost is large, a very large increase in the number of data points will only have a small effect on total study costs. For example, the study already quoted which used 30 subjects in a manual materials handling task (Deeb and Drury, 1986) had most of its time (and hence cost) spent on the experimental set-up, pre-testing and analysis. Decreasing or increasing the number of subjects by 25% would have had less than a 10% effect on total time or total cost. In a more recent extension of that study, each subject took two half days to test and about two weeks to fully reduce and analyse the data. Total study time and cost was closely related to sample size in this case.

Decreasing appropriate standard deviation

In the previous discussion of sample size, the meaning of N was left rather ambiguous. Did it mean the number of subjects, the number of replications of a data point on one subject, or both? This ambiguity was deliberate as the arguments used apply to both meanings of sample size, as do arguments concerning the 'appropriate standard deviation'. Not only are there general ways of decreasing variability, which will be considered first, but also ways of altering which is the 'appropriate' standard deviation by varying the experimental design.

First we consider variance reduction techniques. Careful attention to ergonomic detail is the only way to minimize variability. All of the arguments which apply to the choice of independent variables need to be reviewed to ensure that there are no unexpected sources of variability. All variables which can affect the dependent variable(s) under study should now have been fixed or otherwise controlled. There are, however, some obvious precautions which apply to all studies. All involve standardizing the experimental procedure so that it does not differ between or within subjects. This means consistent, and consistently presented, instructions to subjects. General admonitions to 'do your best' rarely produce the consistency that specific instructions on speeds and errors can provide. Subjects must choose some speed–accuracy trade off, but if left to their own devices will choose trade offs which are different from subject to subject and even from trial to trial for the same subject. Specifying a payoff-matrix would be ideal, but many experiments cannot be reduced to dollars or pounds or yen. As important as the specification of instructions is their consistency of presentation. Written or tape-recorded instructions ensure that each subject receives the same set, without unwanted variables such as the inflexions of the experimenter's voice. When the experimenter is conducting studies with speakers of another language, he/she must be particularly aware of consistent and clear instructions.

Following consistent instructions are written procedures for running trials on subjects. Experimenters learn during an experiment so that the first subject is unlikely to be treated in the same way as the last unless written procedures are adhered to. Obviously these procedures must be perfected on pilot subjects. Such pre-testing of subjects (whose results are not included in the final data) not only removes bugs from the experiment but gives the experimenter enough practice to prevent adding to the experimental variability. The same concept applies to analysis. When any data reduction must be performed, again the experimenter learns. Analysis of pilot subjects' data allows the experimenter to learn analysis without adding to final bias or variance.

Types of design

The choice of appropriate standard deviation is a major factor in experimental design. Between-subjects variability almost always exceeds within-subjects (or trial to trial) variability and thus any test in which the appropriate standard deviation includes only within-subject variability will be more likely to detect significant differences than one which includes between-subject variability. This section is not meant to be a treatise on experimental design, for comprehensive treatments have wide circulation (e.g. Winer, 1972), but the aim is to point out the major choices available to the ergonomist, with their advantages and pitfalls.

Broadly, the choice is between a within-subjects design in which each subject performs in a number of experimental conditions and a between-subjects design in which different subjects are chosen for each condition, although a third design route will also be mentioned.

Between-subjects designs

Here each subject is only tested in a single condition. For example, Laughery and Drury (1979) used a between-subjects design in a study of optimization skills because it was suspected that techniques learned during the solution of one type of optimization problem might transfer in an inconsistent manner to other problems, with an adverse effect on bias and variability. Thus five subjects were used in each condition, which meant that any comparison between conditions had to be made against between-subject variability. The groups were kept reasonably homogeneous (engineering students) but this in turn limits the generalizability of the results. Because between-subjects variability was large, only large effects could be found with the given sample size.

Within-subjects designs

Here each subject receives many experimental conditions or, to put it another way, each subject is their own control. Thus a subject with a slow reaction time is likely to be slow under all conditions when reaction time is measured. Techniques such as analysis of variance or regression can estimate the size of the between conditions effect by using only the deviations from the subject's own mean performance. Hence the slow subject will not add to the standard deviation used to test differences between conditions. As an example, a study of the biomechanics and physiology of handle positions on boxes used ten subjects, each performing a box holding task using ten handle positions. The within-subject's design allowed small differences to be detected despite the limited sample size.

An obvious question is why were the experimenters concerned about inconsistent transfer in optimization but not in box holding? The answer is equally obvious: no changes to the subject were expected during the box holding experiment, but changes were expected in optimization. Change occurs in humans in the short-term as they fatigue and in the long-term as they adapt or learn. With appopriate rest periods, no fatigue was expected (or found) in the box holding task and certainly an hour or two of experimentation on a well-practiced task is unlikely to change either a subject's body strength (adaptation) or box holding technique (learning). Here a biomechanical and physiologically limited task is unlikely to exhibit what Poulton's famous (1974) paper called asymmetrical transfer effects.

The same cannot be said for most intellectual skills. What you learn in first solving one calculus or chess problem is quite likely to affect your performance in solving the next. The transfer can be positive, if the same solution techniques are useful in both problems, or negative if the solution to the first problem is inappropriate in solving the second. An optimization task is *a priori* likely to be closer to an intellectual task than to a biomechanical one, hence the choice of a between-subjects design.

Any human functions, even anatomical ones, will adapt or change given time, but the key question is not whether or not change will occur but whether enough will occur to bias or desensitize the experimental comparison. This depends upon both the length of the experiment and the resistance to change of a function. We can run short studies to minimize the change, but we are limited by other experimental constraints of how much data needs to be collected. We can increase resistance to change by either choosing to experiment on systems which are inherently resistant to change (e.g. bone length) or by deliberately ensuring that a performance plateau has been reached. Thus athletes make good subjects for physiological tests as their aerobic capacity, for example, is well trained and unlikely to change during even relatively long experiments. Similarly, experienced bus drivers are unlikely to develop new bus driving skills during a few hours of tests on buses they have already driven. Finally, for 'new' skills, we can give subjects sufficient practice in all conditions to ensure that plateaus have been reached in each condition. Often the learning/training process required to achieve this level can become of interest in itself (e.g. Bishu and Drury, 1985) so that the time spent achieving stable performance is not time wasted.

Much of the above discussion has centred around creating conditions under which a within-subjects design can be used. Such a design is clearly more efficient, but the fact that we need to take special precautions means that this efficiency is bought at the price of potential danger. Transfer between conditions is always possible; what we do is reduce its probability and magnitude. A between-subject's design is always safer, and sometimes no other is possible. For example, an experiment comparing training techniques in inspection (e.g. Czaja and Drury, 1981) must always use different subjects in different conditions because a fact or skill can only be learnt once. Conversely, some experiments must always use a within-subject's design. If it is desired to derive a functional relationship for each individual subject, then clearly multiple conditions must be given to each subject. Examples are the measurement of the speed-accuracy trade off or the utility function for each individual subject.

Matched-subjects designs

There is a third option for experimental design. If we could somehow clone a subject we could present different conditions to each clone, preserving both the elimination of between-subject variance and the lack of unwanted transfer effects. Obviously this is currently impossible, as well as being morally and legally dubious. We can do the next best thing however, and carefully match subjects between conditions. Thus two conditions could be compared using pairs of identical twins, although it would be rather difficult to find sufficient subjects who were both, say, airline pilots. Comparing three conditions would lead us to triplets and so on.

A less perfect form of matching is to use unrelated (non-family) subjects who are likely to respond in the same way; the art lies in discovering who will be likely to react similarly. If the skill being tested is one of intellectual activity, then one would expect IQ or educational level to determine similarity of response rather than shoe size. On the other hand, shoe size is likely to be correlated with body size and hence strength, making it a likely (although unusual) matching variable for experiments where strength was the limiting human subsystem. Technically, the benefits of matching are determined by the correlation between the matching measure and the experimental dependent variable. If this is denoted by r, the reduction in between-subjects variance is directly proportional to r^2. Hence we need to choose matching variables correlating highly with the dependent variable. Again the only guide is the literature, often embodied in models of the human appropriate to the current experiment. Matched subjects are used extensively in medical and epidemiological studies. For example, Whitfield (1954) studied accidents in coal miners using subjects carefully matched in job type and experience, one of whom was involved in an accident and one who was not. Similarly Saari (1976) matched not subjects but situations to study task and environmental factors in accident causation.

It should be noted that instead of matching individual subjects, groups of subjects can be matched. Thus in many studies parameters such as the mean age, height and weight of each group are kept constant to reduce group differences. Group matching is not as powerful statistically as individual matching. Finally, a caution should be raised that the tests used to establish matching criteria should not in themselves produce inadvertent transfer to the experiment proper. Thus a pre-test of a tracking task on a flight simulator to determine RMS tracking error would almost certainly transfer to future simulator tasks for novices.

Conclusion on types of design

What then is the conclusion? Should we go with the costly between-subjects design or the more efficient within-subjects design, or match our subjects? There are no universal answers. If experiments are very costly and subjects are limited in number, we may have no alternative to the within-subject design. For example, Drury (1973) measured the speed/accuracy trade off for inspection using four subjects inspecting four batches under four speed conditions with a Graeco-Latin square design. There were only four skilled inspectors in the plant so that this was the only design possible. The complex design used only four trials per subject to measure the effects of subjects, trials, batches and speeds. Clearly obtaining four main effects in 16 readings forced the author to make many untested assumptions which he would be unwilling to make in the less demanding environment of a research laboratory. But answers were needed and without the experiment there would be none, so that design decisions would be made without ergonomic input. In that case, it was better to make the assumptions than to abdicate responsibility.

Conversely, if transfer effects are known to exist in a situation, there will be no alternative to the between-subjects design; mention has already been made of training experiments in this context. For any between-groups experiment, matching is always good practice, i.e. it never hurts. Whether it is worth the time and effort depends upon the correlation, if known to the experimenter in advance, between the matching variable and the experimental measure. This in turn depends, like so many things in designing ergonomics studies, upon the breadth and depth of the ergonomist's knowledge.

References

Baum, S. and Drury, C.G. (1976). Modelling the human process controller, *International Journal of Man–Machine Studies, 8*, 1–11.

Bishu, R.R. and Drury, C.G. (1985). A study of a location task. *Proceedings of the Human Factors Society 19th Annual Meeting*, Santa Monica, CA.

Borg, G.A.V. (1982). Psychological bases of perceived exertion. *Medicine and Science in Sports and Exercise, 4*, 377–381.

Caplan, R.D., Cobb, S., French, J.R.P., Van Harrison, R. and Pinneau, S.R. (1975). *Job Demands and Worker Health* (Washington, DC: US DHEW/NIOSH, Superintendent of Documents).

Chapanis, A. (1953). *Research Techniques in Human Engineering* (Baltimore: The Johns Hopkins University Press).

Corlett, E.N. and Bishop, R.P. (1976). A technique for assessing postural discomfort. *Ergonomics, 19*, 175–182.

Cox, T. (1978). *Stress* (London: Macmillans).

Czaja, S.J. and Drury, C.G. (1981). Training programs for inspection. *Human Factors, 23*, 473–484.

Deeb, J.M. and Drury, C.G. (1986). Hand positions and angles in a dynamic lifting task. Part 2. Psychophysical measures and heart rate. *Ergonomics, 29*, 769–778.

Drury, C.G. (1973). The effect of speed of working on industrial inspection accuracy. *Applied Ergonomics, 4*, 2–7.

Drury, C.G. (1982). Improving inspection performance. In *Handbook of*

Industrial Engineering, edited by G. Salvendy (New York: John Wiley).

Drury, C.G. (1985). The influence of restricted space on manual materials handling. *Ergonomics,* **28**, 167–175.

Drury, C.G. and Dawson, P. (1974). Human factors limitations in fork-lift truck performance. *Ergonomics,* **17**, 447–456.

Drury, C.G. and Sinclair, M.A. (1983). Human and machine performance in an inspection task. *Human Factors,* **25**, 391–400.

Drury, C.G., Cardwell, M.C. and Easterby, R.S. (1974). Effects of depth perception on performance of simulated materials handling task. *Ergonomics,* **17**, 677–690.

Drury, C.G., Paramore, B., VanCott, H.P., Grey, S.M. and Corlett, E.N. (1987). Task analysis. In *Handbook of Human Factors*, edited by G. Salvendy (New York: John Wiley). pp. 370–401.

Fish, D.R. (1978). Practical models of human postures and forces in lifting. In *Safety in Manual Materials Handling*, edited by C.G. Drury, DHEW (NIOSH) Publication No. 78–185, pp. 72–77.

Fleishman, E.A. and Rich, S. (1963). Role of kineasthetic and spatial visual abilities in perceptual-motor learning. *Journal of Experimental Psychology,* **66**, 6–11.

Gallwey, T.G. (1982). Selection tests for visual inspection on a multiple fault type task. *Ergonomics,* **25**, 1077–1092.

Kasprysk, D.M., Drury, C.G. and Bialas, W.F. (1979). Human behavior and performance in calculator use. *Ergonomics,* **22**, 1004–1019.

Kerlinger, F.N. (1969). *Foundations of Behavioural Research* (London: Holt, Rinehart and Winston).

Konz, S. (1983). *Work Design: Industrial Ergonomics* (Columbus, OH: Grid).

Laughery, K.R. and Drury, C.G. (1979). Human performance and strategy in a two-variable optimisation task. *Ergonomics,* **22**, 1325–1336.

Meister, D. (1971). *Human Factors: Theory and Practice* (New York: John Wiley).

Miller, R.B. (1963). Task description and analysis. In *Psychological Principles in System Development*, edited by R.M. Gagre (New York: Holt, Rinehart and Winston).

Nie, N.H., Hull, C.H., Jenkins, J.G., Steinbrenner, K. and Bent, D.H. (1975). *Statistical Package for the Social Sciences* (New York: McGraw-Hill).

Poulton, E.C. (1974). *Tracking Skill and Manual Control* (New York: Academic Press).

Rasmussen, J. (1982). Human errors. A taxonomy for describing human malfunction in industrial installations. *Journal of Occupational Accidents,* **4**, 311–333.

Saari, J. (1976). Typical features of tasks in which accidents occur. *Proceedings of 6th Congress of IEA, Human Factors Society*, Santa Monica, CA, pp. 11–16.

Schwabish, S.L. and Drury, C.G. (1984). The influence of the reflective-impulsive cognitive style on visual inspection. *Human Factors,* **26**, 641–647.

Sheridan, T.B. and Ferrell, W.F. (1974). *Man–Machine Systems* (Cambridge, MA: MIT Press).

Swain, A.D. and Guttman, H.E. (1980). *Handbook of Human Reliability Analysis with Emphasis on Nuclear Power Plant Applications* (Washington, DC: US Nuclear Regulatory Commission).

VanCott, H.P. and Kincade, R.G. (1972). *Human Engineering Guide to Equipment Design* (Washington, DC: US Superintendent of Documents).

Whitfield, J.W. (1954). Individual differences in accident susceptibility among coalminers. *British Journal Industrial Medicine*, **1**, 126–130.

Wickens, C.D. (1984). *Engineering Psychology and Human Performance* (Columbus, OH: Charles Merrill).

Wierwille, W.W. and Casali (1983). A validated rating scale for global mental workload measurement applications. *Proceedings of the Human Factors Society 27th Annual Meeting*, pp. 129–133.

Winer, B.J. (1972). *Statistical Principles in Experimental Design* (New York: McGraw-Hill).

Part II

Basic ergonomics methods and techniques

There are some groups of methods and associated techniques which are particularly a part of ergonomics methodology, or are widely used throughout ergonomics investigation. As an early and vital part of ergonomics studies we have *task analysis* (chapter 6, Stammers *et al.*), the term denoting both a stage in such studies and also the set of recording and reporting techniques available. General opinion, followed in the chapter here, is that task analysis comprises both the description and analysis of tasks (task descriptions being the document(s) used); within task analysis, data are collected, represented in an appropriate descriptive form, and are analysed to assess task requirements, environment and behaviour.

Task analysis is widely used since it can be applied during the analysis of existing systems, the design of new ones, and the evaluation carried out subsequently. Information gained is useful both in development and as criteria against which to assess what is developed. What is represented and analysed are the actions or behaviour that individuals must perform or exhibit in order to fulfill the requirements of the task within constraints from the task environment and from individual or general human limitations. Task analysis and its techniques were originally distinguished from method study both by underlying purpose but also by the concentration upon operator decisions as much as upon actions. Lately this has been extended and we have much interest in cognitive task analysis, which obviously demands different methods of data collection, reporting and interpretation. We must find out about the information the subjects attend to, how they interpret it and make decisions to act upon it; multiple methods of direct and indirect observation, expert analysis and so on are required. What then is particular to task analysis is the description format selected to best allow requirements analysis; many such formats are described by Stammers *et al*.

One method which can be used within task analysis, but which has a wider utility, is *protocol analysis*. As described by Bainbridge in chapter 7 the concentration is upon verbal protocols, reports made of people 'thinking aloud' whilst carrying out a task; she explicitly states that they are concurrent rather than retrospective reports. There is a relationship here with cognitive task analysis in that we are seeking insight into non-observable processes, the thinking which may or may not lead to subsequent action. As with task analysis also, stages of data collection, representation and analysis are implied. Data are collected in the form of verbal—usually tape recorded—reports; reports will be transcribed and placed in the context of a simultaneous record of behaviour to produce the actual protocols divided into phrases or phrase groups; subsequent analysis will be made of explicit and implicit content and of inferred connecting material.

Bainbridge quite clearly points out many of the problems of protocol analysis, both in its operation but also in the very concept. However, verbal protocols offer one of the few ways, if not the only way, presently practicable for making inferences in the field about cognitive processes and for understanding complex behaviour in relation to similar past events, present circumstances or predictions of the future.

Both task and protocol analysis techniques will be applied when human performance can be observed or reported or at least, in the case of task synthesis, easily predicted. When this is not the case, or when the range of potential behaviour is large, then we must turn to other means. Behaviours, or task performance, and also related equipment and events, must be modelled or simulated, differentiated by Meister in chapter 8 as involving physical and symbolic representation respectively. There must be some degree of overlap between the two methods though in that models may be used in the simulation of systems and simulators may be built around models. For instance, computer workspace modelling, considered in chapter 19, is a modelling technique which can also produce simulations of activities in the workspace.

Simulation can involve game or role playing but is described by Meister as especially the physical representation of reality. Hardware plus software simulations with a good deal of fidelity to the actual system are used in system design and evaluation and in training. Expert systems also are seen by Meister as a form of simulator. At a lower level of sophistication simulations can involve expert analysis, including 'walkthrough', of two-dimensional (drawings) or three-dimensional 'mock-up' models of equipment or work places.

Models of various types will be employed usually when an actual system or even a physical simulation is not available, and thus models are used generally in new system development or in needs analysis. As seen by Meister, models of most use to ergonomists are of behaviour in human–machine systems, and are symbolic representations of human performance. The mathematical, usually computer-based, nature of the model allows manipulation of variables to determine system outcomes; they may be used within task analysis (or synthesis) in assessing task requirements or consequences.

The last chapter in this section, chapter 9 by Drury, also covers computer-based methodology. A wider look is taken at computerized data collection and analysis in ergonomics. Many of the techniques described earlier as a part of direct or indirect observation methodology entail the collation of information about a number of variables from a large number of data points or over a long time period. Automated or instrumented observation techniques can be controlled by computer in terms of what data are collected and when, and the measurements taken fed directly into the computer for subsequent reduction and analysis.

Just as developments in computer technology have enabled great improvements in model and simulator power and utility, so can general data collection and analysis be much more efficient and comprehensive. However, within Drury's chapter lie the seeds of the need for caution. In contrast to the large multi-measure evaluations possible now, the more limited investigations previously possible did have the advantage of producing relatively simple statements on performance or stress which colleagues in design, implementation and management might need. For all the methods described in this section there is a danger of the methodology acquiring a momentum of its own, of analysis being performed way beyond the needs of the situation and of the ergonomist losing sight of the original problem. Methods and techniques are only tools, to be selected and applied with a clear understanding of the design or evaluation objectives.

Chapter 6

Task analysis

Rob B. Stammers, Michael S. Carey and Jane A. Astley

Introduction

History of task analysis

The observation of others at work is a perennial human activity. In order to provide training programmes, and to design equipment, people must always have observed others at work. Very few exact notation or recording systems have been developed; examples can be found in the arts, say, for recording music. Alternatively depictions of human activities can be found in such things as military drill books. Systematic observation of tasks, with some specific purpose in mind, has more recent origins. It has become a central part of the human resource disciplines which have developed this century. The development of task analysis methods is inextricably tied up with the development of ergonomics.

Some of the earliest approaches to systematic recording of human activity are to be found in the work-study approaches developed by Gilbreth and Taylor in the early part of the century. In the main, they were used for observable actions, and attempted to codify them on paper in an easily learned scheme. The techniques were mainly restricted to perceptual motor tasks, of a highly proceduralized and repetitive nature. Gilbreth's system used symbols to codify actions. These symbols were termed therbligs (the source for this term being 'Gilbreth' spelt backwards). The symbols, rather than a longhand description, were then used to represent the actions. They could also be placed in temporal order to show the sequence of actions in a task.

The overall objective of the work-study approach was 'efficiency'. It was used in order to determine an optimum sequence of actions and to reduce inefficient or wasteful activity. This whole approach of 'scientific management' came in for criticism for the way it viewed human activity. For our purposes the most important criticisms were that it represented human skill as a very fixed thing and made little reference to the central processing demands of real life speed skills (see Drury on direct observation in this text for further discussion of the work study approach).

These aspects began to be recognized in the 1950s (Conrad, 1951) at a time when theories of skill were being far more clearly articulated. The bringing together of elements of work-study with a deeper understanding of the reality of perceptual–motor skills was achieved by Crossman (1956). He saw the need for 'mental therbligs' as the elements to be recorded. His technique took into account such activities as planning and controlling actions. He produced the 'sensori-motor' process chart. This attempts to record skill cycles in a time-based way, showing the important links between planning, initiating and executing actions. Although this technique does not appear to have been widely used in industry, developments of it can be seen to have influenced the method of Seymour (1966). His technique, which has become more commonly known as 'skills analysis', requires the analyst to record elements of a task and to draw attention to the senses and motor effectors being used to carry out the task. This technique was specifically directed at training, and in the 1960s and early 1970s was probably one of the more widely known techniques in manufacturing industry.

Alternative approaches came from post-war developments in the military human factors field, particularly in the United States, which led to a recognition of the need for more systematic approaches through task analysis. The work of Miller (1962) is probably the best known in this area. This work had a psychological basis and attempted to represent tasks in charts using various codifying techniques which also took into account temporal patterns of the tasks. Some used psychological categories as an underlying taxonomic approach to classifying performance. These techniques have had a continued development and the taxonomic approach has more recently been revived (Fleishman and Quaintance, 1984). Two camps seem to emerge, one concerned with theoretical robustness and elegance of such schemes, and another more concerned with the utility of schemes to the applied worker (Miller, 1967). Developments in the last three decades have led to a multitude of techniques being developed in a variety of contexts.

Developments in the UK have been more closely allied to the development of techniques for training applications. Hierarchical task analysis (HTA), first proposed by Annett and Duncan (1967), draws on a number of ideas such as the hierarchical representation of human activity suggested by Miller *et al.* (1960).

This overview has so far concentrated on techniques for representing information. Parallel with this have been developments in techniques for collecting task information. We have moved through a period of focusing mainly on observational techniques, followed by a refinement of those techniques, through to current developments using such recording techniques as video tape (see Drury on direct observation methods). In addition to this, questioning techniques have been refined. These range from those based on informal discussions through to structured interviews and are discussed in the chapter by Sinclair.

In more recent years attempts to get at the cognitive processes of individuals, through the use of such techniques as verbal protocols (see chapter 7 by Bainbridge), have extended the role of the observer's comments as sources of information. These techniques have recently received increased interest with the concern to elicit job incumbents' knowledge to be utilized in expert systems (see development of this by Shadbolt and Burton in chapter 13 on knowledge elicitation).

Task analysis, therefore, has a fairly short history, closely tied in with the development of ergonomics disciplines and their concern with 'data' on which to base design decisions. The developments in this area are marked by the wide range of techniques that have been produced and a lack of agreement on any single approach being more appropriate to one context than another. In the following sections attempts are made to clarify some of the underlying concepts in the field and later in the chapter a representative range of techniques will be discussed in more detail.

The definition and process of task analysis

Despite the widespread application of task analysis techniques throughout the discipline of ergonomics, there is surprisingly little agreement over which techniques can be classified as task analysis techniques, what the component stages of the task analysis process are and even over the precise definition of what constitutes a task. To some extent the resolution of such issues can be viewed as academic and unrelated to the main concerns of the practitioner, such as the suitability of each method to a specific application, its ease of use and its efficiency. However, in order to be able to make such decisions, a basic understanding is required of what is analysed and how the analysis is performed.

Task analysis as a tool for system design and evaluation

All task analysis techniques aim to produce information that is relevant either to the design of a new human/machine system or to the evaluation of an existing system design. This is achieved through the systematic analysis of human task requirements and/or task behaviour. In design situations, the predominant approach involves the analysis of user/operator tasks in existing system contexts, in order to apply the results in the design of a new system. Task analyses may then be applied during the design process to evaluate the future task demands that will be imposed by the emerging system. Finally, the process of evaluation through task analysis may be applied at points during the operational life of the system.

When used in a design context, the value and nature of the task analysis process depends directly upon the relevance and accessibility of the existing system or systems that are available for analysis. A basic premise of the task analysis process in design is that the tasks studied in an existing system context can provide information which will be directly relevant to human behaviour in the proposed system. This only holds true whilst a direct correspondence exists between user/operator tasks in existing and future systems. Changes in the user/operator role and in interface and machine functions may each contribute to the reorganization of user/operator tasks in the new system. It follows that, unless user/operator tasks are highly proceduralized or constrained, or changes in task behaviour can be reliably predicted, there will always be an element of uncertainty in applying task data from one system to a new system.

The most relevant and reliable application of task analysis methods will be in the cases where task analysis is used in the updating of an existing system (No. 1 in Figure 6.1). Unless substantial changes are made to those elements of the system that directly impact upon the user/operator, most of the task information gathered will be of direct relevance. It is quite probable that users or operators of the existing system will be involved with the new system, carrying over with them the existing patterns of behaviour. In addition, there will probably only be minor changes in the organizational context, training, manuals and operational procedures. However, some caution still has to be applied, as it is possible for quite minor changes in the task context to generate large unforeseen changes in the strategies employed by users/operators.

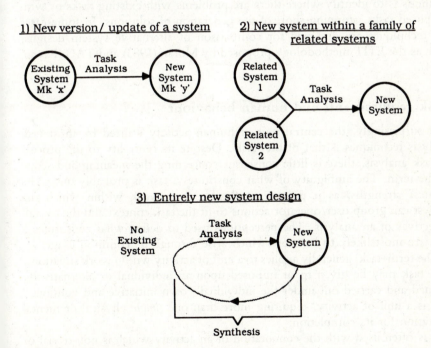

Figure 6.1. Applications of task analysis within system design.

New system designs pose problems for the use of task analysis. If systems already exist which bear some relationship to the new system under design, it is possible to analyse these systems in order to derive information that is relevant to the target system (No. 2 in Figure 6.1). Detailed information on the use of displayed information, input mechanisms, methods and approaches are only likely to be relevant at a generalized level of detail and the task description therefore can only be generated at a generalized, generic level. If more detail is required for system design, then it becomes necessary to generate the lower level detail of tasks by a process of task synthesis.

Task synthesis is essential where a system is entirely new and there are no other systems that bear any relationship, or where it is not possible to gain access to appropriate systems (No. 3 in Figure 6.1). In this case, the initial designs for a new system are used to synthesize a predicted description of human tasks. This is most appropriate in contexts where the synthesized task description can be used as the basis for mandatory training and operational procedures. Unfortunately, task synthesis becomes less and less accurate as user/operator behaviour becomes more discretionary, involving higher levels of cognitive content, and is less routine in nature. It is better in these circumstances to view task synthesis as a task design process (e.g. Moran, 1981) rather than as a task description process.

Using task analysis for system evaluation purposes does not face the same validity problems as when it is used for design. The objective with evaluation methods is to identify where there are problems with existing tasks or with proposed tasks, stopping short of the generation of solutions. In some cases the techniques are designed for comparison of alternative system designs, such as the ETIT methodology proposed by Moran (1983) and TAG (Payne, 1984).

Task requirements and human behaviour

Not surprisingly, the central unit of human activity utilized by most task analysis techniques is that of the 'task'. Despite its centrality to the process of task analysis, there is little consensus concerning the meaning and scope of the term. The ambiguity of what constitutes a task is probably one of its greatest strengths, as it provides a flexible framework within which the analyst can group user/operator actions to fit the task context and the overall objectives of an analysis. The perceptions used to define what constitutes a task are most likely to be those which arise through its common usage:

The term 'task' generally applies to a unit of activity within work situations.

A task may be given to or imposed upon an individual or alternatively, defined and carried out under the individual's own initiative and volition.

It is a unit of activity, requiring more than one *simple* physical or mental operation for its completion.

It is often used with the connotation of an activity which is non-trivial or even in some cases, onerous in nature.

The concept of a task does not restrict its use to large, global activities or to low-level operations within a system. In a number of task analysis methods, the overall activity of a user/operator is defined in terms of a task, which is then broken down into a number of subsidiary component tasks. Each of these lower level tasks may be further subdivided to give further levels of subtasks. Each of these tasks at each level fits into the general conception of a task. If the process is allowed to continue until it is impossible to subdivide the tasks any further, the bottom level of the hierarchy would consist of simple cognitive or physical actions, frequently labelled 'operations'.

Tasks can be conceived of consisting of an initial entry state which defines the conditions for the initiation of the task, a goal state or goal conditions which the task is required to achieve and the set of constraints and aids which is formed by the organizational, technological and environmental factors under which the task is performed. It should be noted that this definition does not include any mention of the actual behaviour employed to carry out the task. It is important to see the distinction between those elements of user/operator behaviour which are defined and fixed by the system context and those elements which are under the discretion of the individual concerned. It is possible in most cases to define three interacting aspects of a task (refer to Table 6.1).

The distinction between the three task aspects is vitally important for the process of task analysis. The first two aspects are determined by the system context, which incorporates the organizational context, operating requirements and limitations of the technology involved, the prescriptive elements of

Table 6.1. Components of a task

Task requirements. The goal state or goal conditions defined by the system context, given a particular initiating state or set of conditions. For example, the user of a word processing system having reached the state of the completion of a document, may be *required* to save the document on to a permanent storage device. In the same manner, the operator in a power station dealing with a sudden loss of power in the unit under his/her control is *required* to take appropriate actions to minimize loss, maximize safety and avoid plant damage.

Task environment. The factors in the work situation that constrain and direct the actions of an individual, either through restricting the types of actions that can be taken and their sequencing or by providing aids or assistance that channel user/operator actions in a particular way. Written emergency procedures, for example, would probably constrain the actions of the operator mentioned above to follow certain courses of actions, whilst should an appropriate diagnostic information display be available to the operator, he/she is most likely to make use of it.

Task behaviour. The actual actions that are performed by an individual within the constraints of the task environment in order to fulfil the task requirements. Some choices may be forced upon the user/operator by inherent psychological or physiological limitations, or a lack of appropriate knowledge or skill. The method employed may also have been developed through experience to optimize efficiency and to minimize effort (Card *et al.*, 1983 gives some examples of alternative methods selected for the same tasks).

---

training, the structure of the interface with the user/operator, operating procedures, environmental conditions and the influence of other external connected events. Almost all these elements can be observed, recorded or predicted accurately, allowing the basic framework of task activity to be accurately predicted in so far as it is determined by its context. Task behaviour, however, can vary greatly between individuals and with experience. It is difficult to predict behaviour accurately due to the influence of cognitive factors which cannot be easily observed or modelled. This is particularly the case when a task is largely cognitive in nature.

The process of task analysis

The uncertainty regarding terminology also applies to the definition of the task analysis process itself. Miller, for example, distinguishes between two processes; task description, which contains just task requirements and a separate process of task analysis, which is an analysis of the behavioural implications of the tasks identified in the description (Miller, 1962). More recently, the terms have come to be confused, with the term task analysis applied to cover the combined description and analysis processes and task description applying just to the task description document. In some cases, task analysis has even been used to refer to just the task description process. Unfortunately, the latter definition of terms is now too well established and so the following model is based upon this logically less correct usage of terms.

The process of task analysis can be broken down into three main stages of activity, varying in importance and complexity according to the method utilized, the types of tasks under examination and the final use to which the data are to be put. The first stage of the analysis process is data collection, which involves the collection and documentation of various sources of information from the source system(s). Then the task description document is formed, either from the task data or by the process of task synthesis. The task description forms the raw data for the final analysis stage generating the task specific information required by the analyst. This is represented in Figure 6.2.

The actual process of task analysis may vary from this idealized model in a number of ways. For example, in practice the process can involve some elements of iteration. Vital pieces of information may be found to be missing when generating the task description, such that it is necessary to carry out further data collection. Iterations of this kind are clearly inefficient and wasteful, though they cannot always be avoided. Another difference that often exists between the model and actual practice is that some stages are circumvented or concatenated together. It is possible, for example, to miss out the formal description phase, generating the analysis data from the task data utilizing a subjective understanding of the task. This is possible with analyses of very simple tasks or where only partial task information is

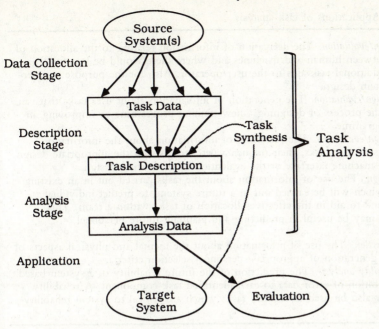

Figure 6.2. The processes employed in task analysis.

required (e.g. error data). In general though it is desirable to generate a task description document, ensuring completeness and providing a means of retracing the basis upon which recommendations and design decisions were made. In a similar manner, the task description is often generated without any formal description of the data collected, such that the data collection and description processes are carried out at the same time.

Figure 6.2 is a representation of the formal stages involved in task analysis. An equally important but informal aspect of task analysis is the depth of understanding of the task gained by the analysts. The task description can be designed to provide both a formal description of the task and a means of passing the analysts' conceptions of the task onto the recipients of the analysis output. This is particularly important when the recipients are involved in system or interface design and need to possess an accurate mental model of the future users or operators and their task needs.

The final stage in Figure 6.2 represents the application of the output from the task analysis process, either to a target system or in the form of a system evaluation. The types of uses for task information are shown in Table 6.2. Within each of the main categories, different information will be required for different system contexts.

It is an obvious but often overlooked point that in order for a task analysis to be useful, it must produce findings which can be applied. In fact, the question of how the data produced by a task analysis are to be applied is

Table 6.2. Applications of task analysis

System design/evaluation. The derivation of information relating to the allocation of functions between human and machines and where there would be benefits from providing additional task aids to the user/operator. Also for the purpose of 'user-centred' system design.

Training design/evaluation. The generation of information about user tasks that can be used in the process of designing a new training programme or improving an existing programme.

Interface design/evaluation. The provision of information about the information requirements of user tasks, their frequency, etc. to assist in the appropriate design of a human–machine interface or the evaluation of an existing design.

Job/team design. The use of information about the tasks carried out in an existing system or which will be carried out in a future system, to predict individual workloads and to aid in the effective allocation of tasks within a team. This information may be useful in predicting and planning future personnel requirements.

Personnel selection. The use of information about the mental and physical aspects of tasks in the generation of appropriate personnel selection criteria.

System reliability analysis. The prediction of the future reliability of a system based on the application of error data to each identified task component. A reliability analysis may also be used to identify tasks which are critical to system reliability and safety.

vital in the selection and use of a particular technique. It is rarely necessary to collect exhaustive information about a task, so the identification of exact task information requirements is an important component in ensuring the efficiency of the task analysis process. Table 6.3 covers just some of the types of information that might be required.

The categories of task information required determine the types of data collection techniques that can be employed. Observation or activity sampling, for example, could be used to obtain information about the identities of tasks, their frequencies, sequencing and success/error rates. (This is described as occurrence sampling in Drury's chapter on direct observation.) Interview techniques can provide a wide range of information, though such techniques cannot be relied upon to give accurate or complete information on the more cognitive aspects of tasks. In fact, the cognitive elements of task information are particularly difficult to identify. The most comprehensive technique that can be employed is verbal protocol analysis, though it does require a high level of access to the task situation or an off-site simulation. A large amount of information regarding task requirements and the task environment may be available away from the system itself in the form of manuals, training personnel, policy documents, supervisory staff and operational procedures.

The task description must also be able to represent all the categories of information required from a task analysis. In selecting the method or methods of representation to employ, four aspects of use of the task description must be considered (see Table 6.4).

Table 6.3. Types of task information

Identity of component subtasks. A listing of the sub-activities involved in a task.
Grouping of component subtasks. An organized, possibly hierarchical, listing of the sub-activities involved in a task, showing how subtasks cluster according to functional or temporal considerations.
Commonalities and interrelationships between subtasks. An indication of the extent to which component subtasks are employed, and are different aspects of the overall task, or of the way that subtasks cluster in terms of the information they require or the methods that are used in their execution.
Importance or priorities of component subtasks.
Frequency of component subtasks.
Success/failure rates of component subtasks.
Sequencing of component subtasks.
Decisions made in the execution of component subtasks.
'Trigger' conditions for task execution. Many tasks begin execution simply as a result of the completion of a previous task or following a decision, though it is possible for the execution of a task to depend upon a particular stimulus or command originating within the task environment.
Objectives or goals of each task.
Information required by each task.
Information generated by each task.
Knowledge employed in making decisions.
Knowledge of system employed in performing task.

Table 6.4. Main requirements for task description

It is required to be complete, so the representation chosen must be amenable to checking by the analyst.
As a form of validation, the task description may be shown to current users/operators for comment. The method of representation must, therefore, be relatively simple to understand without extensive training and must also be readable.
The task information must be presented in a clear and concise manner.
The task description should be designed to communicate the analyst's understanding of the task domain to the reader.

The most commonly used forms of task representations are diagrammatic and tabular methods, but there are other alternatives including computer-based methods which are only just becoming feasible (see Table 6.5). Of course, there is no reason why only one form of representation should be employed. In order to meet the multiple objectives outlined above for a task description, a set of interrelated representations may be optimal. The production problems involved in their generation are rapidly being reduced by the application of increasingly sophisticated computer packages.

The final step of extracting the required information for use may just involve the generation of summary tables or a set of conclusions. In many

Table 6.5. Methods of representing task data

Diagrammatic. The most common form of representational technique utilized in task analysis. Includes the charts employed in hierarchical task analysis, flowcharts, finite state transition charts, critical path networks and link charts.

Prose. A simple, textural description of task behaviour.

Tabular. A tabulated presentation of task information, such as the schedule tables provided in hierarchical task analysis.

Structured text. A textual representation partly conveying task information through its format or highlighting, such as the GOMS method of task description (Card *et al.*, 1983).

Code. Either actual program text or in the form of pseudocode as employed in program specification. This includes both procedural or declarative types of programming languages.

Grammar. A textual representation obeying strict grammatical rules. For example, CLG (Moran, 1981), TAG (Payne, 1984) and the grammar proposed by Reisner (1982).

Algebraic. A formula employing some form of mathematical notation such as simple algebra, set notation or calculus.

Animation. Animated presentation of some aspect of task activity.

Simulation. A representation of one or more aspects of task activity that may be manipulated by the end-user or analyst and then 'run' to give an animated or static output.

Interactive. A computer-based technique that can integrate any of the methods described above, providing rapid switching between multiple representations of a single task element, multiple methods of navigation and tools for integrating and extracting information.

---

cases, it will involve further analysis such as the derivation of estimates of physical or mental workload. Unfortunately, an account of this step of the task analysis process is frequently missed out of descriptions of particular techniques. In many cases, the analysis process appears to be based upon the analyst interpreting the task description using his/her intuition and experience to draw out conclusions.

Some analysis techniques

There is a vast range of task analysis methods available for analysing all varieties of tasks and applications. Some aim to be general in their approach, but most methods and techniques are directed towards specific applications such as training, workplace layout and human–machine interface design. Many fit into a specific phase of the design cycle of a system, and can be used for the design of new systems or the evaluation of existing systems.

The application for which an analysis is to be used is one of the important features in the selection of a method. A method may be selected, developed or adapted to suit a particular purpose within a system. In large scale systems the general purpose of the analysis may not be covered adequately by a single method of analysis and so a battery of methods may need to be applied to obtain the required task information. The method used is also dependent on the task information that is available or can be collected by the analyst prior to analysis. The effectiveness of an analysis is dependent not only on the technique but also on the quality of the information input. This information needs to be in a form acceptable for use in the analysis and of a quality and in sufficient detail to allow the analysis to be a full representation of the task.

Analysts have at their disposal a wide variety of methods; however some methods are better documented than others. Also some have been widely applied and are proven to be effective in use and application. A selection of techniques will now be outlined, some are in common use, others less so. The aim is to indicate the range of techniques available, to describe the methods, and to discuss the advantages and disadvantages of using the different techniques. A matrix illustrated in Figure 6.3 gives examples of the types of application that task analysis can be used for, in both evaluation and design. Not all the techniques outlined in the table are detailed in the text. For example, the very recent FAST method is not covered. Also some techniques are similar in principle to others and are only mentioned in the introduction to a more detailed description of one technique, e.g. PAQ is similar to the technique described under 'Abilities Requirements Analysis'.

Hierarchical task analysis

Introduction and background

Many task analysis methods adopt a hierarchical approach to analysing a task. In a hierarchical approach the task is analysed by breaking it into task elements or goals which become increasingly detailed as the hierarchy progresses. There are many possible application areas for hierarchical methods and often other task analysis methodologies incorporate a hierarchy at various points in analysis as a means of gaining more detailed task information.

Hierarchies can allow a 'quick and dirty' approach to be taken to task analysis, where just the task areas of interest can be studied in detail; alternatively the hierarchical framework can provide a systematic and thorough means of analysing a task completely. Some techniques restrict the number of levels in the hierarchy, for example GOMS (Card *et al.*, 1983). Others are more flexible in the number of levels allowed and incorporate 'stopping rules' to help the analyst decide when analysis is complete.

As mentioned in the introduction, one of the most commonly used and applied hierarchical techniques is hierarchical task analysis (Annett *et al.*, 1971). The technique has been widely applied in the process and other industries, for both training and human–machine interface design. It has evolved and been adapted to meet the needs of the changing roles of the user in line with increased automation and advances in process technology (Shepherd, 1976; Piso, 1981). The technique analyses a task by a process of

APPLICATION / METHOD	Allocation of Function	HMI: Design/ Evaluation	Personnel Selection	Workspace Design	Communications Links	Knowledge Elicitation	Task Performance Evaluation	Complex Cognitive Control Tasks	Training	Operating Procedure Design	Workload Assessment	KEY REFERENCE
HTA		✓							✓			Annett *et al.*, 1971.
GOMS		✓						✓				Card *et al.*, 1983.
FAST	✓									✓		Williams, 1988.
Cognitive Task Analysis		✓						✓				Hollnagel *et al.*, 1981.
Link Analysis		✓		✓							✓	Chapanis, 1959
TAKD		✓							✓			Johnson *et al.*, 1984.
Abilities Requirements			✓									Fleishman and Quaintance, 1984.
TAG		✓				✓						Payne, 1984.
JPC		✓			✓							Tainsh, 1985.
Miller, R.B.									✓			Miller, 1962.
Berliner *et al.*							✓					Berliner *et al.*, 1964.
PAQ			✓				✓					McCormick, 1976.

Figure 6.3. Task analysis methods and their applications.

progressive redescription, where each task element or *operation* is redescribed into a series of more detailed task elements. The output of the analysis is a diagram showing the task elements and their interrelationships (see Figure 6.4) and a table giving detailed training or human machine interface information.

Method

The analysis begins with an overall definition of the task goal in a verb–noun type statement, for example:

Run computer programme,
Operate plant,
Start pump.

This task goal or *operation* is then redescribed into a series of more detailed operations, which together constitute the superordinate task operation. Each of these operations is then redescribed in turn. This process is known as *progressive redescription*. Redescription stops when one or more of the criteria (stopping rules) are met. The primary stopping criteria is the 'probability $\times$ cost' rule ($p \times c$ rule). This is applied by taking a task element and estimating

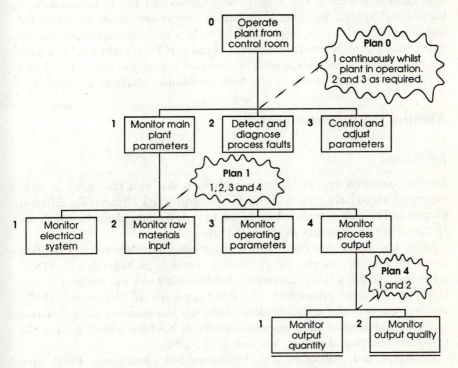

Figure 6.4. Example from a hierarchical task analysis.

the cost to the system if it is not performed successully. This is then multiplied by the probability that the element will not be successfully performed. If the resultant value is acceptable in terms of overall system functioning, then analysis is complete for that part of the analysis. In practice the $p \times c$ rule is difficult and time consuming to apply as expert judgement and a thorough knowledge of the task in question is needed. More practical are the rules outlined by Shepherd (1981), for training applications.

Within the analysis the relationships between the task elements under a single superordinate need to be specified, including the timing and sequencing of the operations. This is achieved by means of *plans*. There is a plan for each superordinate operation and they are annotated on to the task diagram. Another feature of the technique is a table where more detailed information is collated. The information used in the analysis will typically come from a number of sources. Additionally, it is unlikely that a complete analysis will be produced in a single 'pass'; a number of iterations are likely to be needed to produce the final version.

Discussion

The method of analysis provides a systematic and thorough framework for analysing a task, and it is flexible enough to be used for many applications. The method is clearly and widely documented and the technique can be learned and applied by other than human factors specialists. However the hierarchical view is often misunderstood to be a prescription of task activity rather than a convenient representation of tasks. HTA is often used as a basis for more detailed analysis specific to areas of application not covered in a HTA. One example of this is as a basis for human reliability assessment.

Abilities requirements analysis

Introduction

Another common approach to task analysis is that of a taxonomy, a wide variety of which have been developed to categorize task elements for different purposes. The task is analysed by breaking it down into elements which are allocated to the categories of the taxonomy.

Many taxonomies are aimed at describing behaviour, others break the task down in a more analytic way. Familiar methods include Miller's (1967) taxonomy, Gagné's (1977) taxonomy for learning tasks and Berliner *et al.*'s (1964) performance taxonomy. The PAQ approach of McCormick (1976) involves the use of a questionnaire, where job elements are rated in terms of their psychological and contextual factors. It has been widely used in the personnel selection area (e.g. Sparrow *et al.*, 1982).

Fleishman and colleagues (e.g. Fleishman and Quaintance, 1984) have carried out extensive research into the development of a taxonomy to identify abilities requirements for given tasks. Abilities are seen as enduring individual traits which are carried from one task to another. The aim is to optimize performance by identifying the abilities required to perform certain tasks effectively. The method has been extensively researched and tested both on laboratory and real world tasks. Although the work is currently incomplete for certain areas of human performance, the taxonomy is sufficiently developed to give an indication of the range of human abilities required to perform most tasks.

The method outlines 48 abilities, each of which has an associated detailed definition and rating scale. Each task is rated for each of the abilities, using a variety of means, such as expert subjective judgment or laboratory experimentation. Each ability is explicitly defined and descriptive anchor points and examples are given for each rating scale. The resultant task profile indicates which are the abilities featuring most prominently for the task in question.

The approach has been applied successfully to a range of occupations such as firefighters, the military and warehouse clerks. It has also been used as a basis for assessing job performance and so would be useful for both personnel selection and training applications.

Method

Two approaches can be used to identify the ability requirements of the task or job in question. Experimental studies can be used to determine the ability ratings, or subjective expert judgments can be utilized. The greater the number of experts and the wider their experience, the more effective the ratings will be. Similarly the greater the number of subjects in an experiment the more accurate and representative the task information will be. A manual is available for guidance on the use of the ability requirements scales (e.g. MARS or the *Manual for Abilities Requirements Scales*; Fleishman and Quaintance, 1984).

For each of the 48 abilities defined a task or job must be rated on a scale used to estimate the level of that ability needed for that particular task. To facilitate judgment, each ability is very carefully defined and examples are given at the bottom or top of the scale of tasks that would be rated at that point on the scale. Each scale usually has 7 points. To give an overall view of the rating of a particular task or job the abilities and rating scales can be combined into a table (see Figure 6.5) which also helps the comparative assessment of tasks.

Discussion

Taxonomies do provide a useful way of breaking down tasks, especially in applications where specific information needs to be included in the taxonomy. Categories can be incorporated to make this task information explicit; for

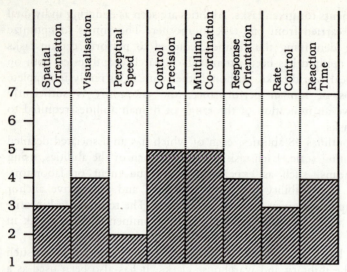

Figure 6.5. Extract of a table showing an overview of the abilities requirements for a task.

example, task information that would be particularly useful for part of the design process of a particular system could be included by adapting a taxonomy. Other types of task analysis method may not be so flexible.

However, it is difficult to ensure that a taxonomy is comprehensive in its coverage of a task. Ideally it should aim to be exhaustive and have mutually exclusive categories. If the definitions of categories are inadequate then confusion can occur when allocating task items. Many taxonomic methods have been developed in the context of specific applications; however they can often be used for analysing other tasks and may be amenable to adaptation.

Task analysis for knowledge description

Introduction

As systems have become increasingly complex and more highly automated, the role of the human has moved away from the physical 'hands on' approach, involving more and more complex cognitive processing. Many of the older more traditional task analysis methods do not adequately document these cognitive components of tasks. Furthermore, with the development of expert systems, the knowledge a human uses to perform a task needs to be elicited and represented before such a system can be fully developed. In order to ensure that the analysis is thorough, it is important for the knowledge to be elicited and systematically represented within an effective framework. All this has given rise to an approach in task analysis which aims to analyse and document some or all of the knowledge needed to perform a task. These task analytic methods focus on the cognitive rather than the physical aspects of task performance, an example being the cognitive task analysis approach of Rasmussen (1986). This aims to address the problems of analysing tasks in large scale systems, where the operators typically interact with a plant via an interface and the task places a high cognitive demand on the operator.

One of the more recently developed methods is *Task Analysis for Knowledge Description* (TAKD; Johnson *et al.*, 1984). This method was initially developed for syllabus design for training in computing related skills. It has a wider application, by providing an analysis of users' tasks on which to assess usability in interface design. The approach is more formal than many task analytic methods and provides both an abstract description of the task in question and a knowledge representation grammar. The aim is to produce a task analysis that is not task specific and which can therefore be easily used in design.

Method

Firstly the data for the task description is collected by identifying a range of exemplar tasks for the system in question and eliciting task information by observation, protocols and interviews. From this information a task description is completed, which can be in the form of a list of actions in sequence, or of the procedures involved in the task.

From this description, the objects and actions involved in each step of the task are compiled into two separate lists. These lists are then abstracted into lists of generic actions and objects which are identified in a form not specific to any particular task environment. The analyst then returns to the original task descriptions and expresses each task step again, in terms of generic actions and objects. A knowledge representation grammar sentence is assigned to each of the task steps. The grammar comprises one generic action and up to three generic objects upon which this action can act. These collected sentences can then be used as a basis for design. The main problems for the users of techniques such as these is to become familiar with the conceptual schemes and theories underlying them and to be consistent in applying the scheme through the different stages of document preparation.

Discussion

Difficulties can be encountered in the elicitation of task information from any task which has few covert actions and a high cognitive content. Firstly this is because the task elements are not immediately observable and secondly because it can be difficult to check that the analysis has been thorough. To ensure that the task is documented as thoroughly as possible the analysis should provide a systematic approach and framework for describing the task.

If the analysis concentrates on low-level tasks it may be more difficult to ensure the thoroughness of the analysis compared with a hierarchical approach.

A hierarchy can help to highlight areas where there may be omissions from the analysis or where the current analysis is inadequate. The sources of information for deriving knowledge about systems may be limited. Documentation on a system can provide useful information on system functioning, and 'job aids' can indicate some of the task areas where help is needed to perform a particular task.

In eliciting knowledge, a subject matter expert is often a prime source of information. However, for any system there may be only a few experts available and a wide sample may not be available from which to derive a variety of viewpoints. Such experts often have much knowlege which they find it difficult to verbalize, for example with a much used skill which is performed without a great deal of conscious effort. In such a situation the operator or user have to think very carefully in order to explain what the different task elements are and how they are performed. The analyst therefore needs to be skilled in eliciting the knowledge of the user. (See Shadbolt and Burton for a discussion of many different techniques available to elicit knowledge from experts.)

Link analysis

Introduction

The task analysis method selected will reflect the system and tasks to which the analysis is to be applied. In large scale systems where a relatively small number of personnel are required to perform a highly complex set of tasks, it is likely that a variety of task analytic methods will be used. Only then can all the task information be provided. However, in such systems there may be a need for some very specific information on a particular dimension of task data which does not warrant full scale analysis. Conversely a simple task may only require a simple level of analysis to provide adequate task information for evaluation or design.

Some methods of task analysis therefore only analyse a single dimension of a task such as timing, sequencing or spatial layout. These can be used for applications such as the assessment of workload and workspace layout.

Such methods include time-line analysis and operating sequence diagrams where a plan or map is taken of the workplace or interface and the movement of an individual within an environment recorded. Also for the design of layout of workplace or interfaces a technique known as link analysis (Chapanis, 1959) may be used. The technique aims to optimize the layout of a workspace by analysing the relationships or links between elements in the workspace in the performance of a task.

Method

Data for the analysis are collected by observation in the workplace of a representative series of tasks. The analyst notes all the possible links between items in a workplace and then notes the frequency with which they occur during task performance. In addition to frequency data each link may be given a weighting to indicate its importance in the task. The weightings and frequency information can be combined to give an overall value which indicates the importance of the relationships between elements and can be used as a basis for workspace layout. Chapanis (1959) shows how the technique was used to study the communication patterns of a large military team. A redesigned control room led to more effective communication.

Discussion

Although such analyses can be useful, they only make one kind of task information explicit. To be effective the technique may need to be reviewed in the light of other analyses that are being carried out. Link analyses can often be carried out quickly, although some task information will often have to be collected specifically for the analysis. Such methods do provide a necessary means of supplementing the task information of other analyses. However, use in large scale systems is only effective when the information is not explicitly provided by other analyses.

Formal and semiformal methods

Introduction

Recently emphasis has been placed on the development of more formal, qualitative and logical methods for task analysis. Formal methods have been extensively used in software engineering and development and this trend has carried over to the development and use of task analysis methods in user–interface design and other areas.

Some of the methods developed are used for evaluation, for example block interaction models, others are orientated toward design. Language grammars are one of the more common formats used for task analysis, for example. Command Language Grammar (CLG, Moran, 1981). Formal methods allow the task to be expressed in a concise and unambiguous way. This may give an advantage over many of the subjective methods which are dependent on natural language to express the content of the task. Natural language allows for individuality of expression and so may result in slight variations in an analysis carried out by a series of analysts.

Task Action Grammar (TAG) developed by Payne (1984) provides a psychological notation for a rigorous approach to task definition. The analysis defines the mapping from tasks to actions. The primary application of TAG is to provide a model of user knowledge which helps to predict the learnability of command languages and helps to provide a description of an interface.

Method

Firstly a task (usually computer-based) is decomposed into simple tasks. A simple task is defined as one which can be performed by an operator/user at a very low level, perhaps as part of a procedure, and which involves no tasks such as problem solving or decision making. Each simple task is then recorded in terms of its components or features. These features are compiled into a dictionary of simple tasks.

The TAG interface description consists of this dictionary of simple tasks, with a list of their features and finally a set of *rewriting rules* which map the tasks onto actions. TAG contains features which reflect some of the generalities found in command languages. The analysis also allows information to be included which is not specified in the language itself. However the focus of TAG is very specific, dealing only with the domain of command languages and user dialogue.

Discussion

Formal methods provide a very rigorous approach to defining a task and this can lead to advantages over more subjective methods; for instance, there may be greater reliability in the analysis produced. However the methods are very inflexible and not easily amenable to adaptation. This means that task information that is relevant to task performance may be omitted as it does not fit into the defined structure.

Formal methods often focus on tasks at a simple level to ensure that the task description on which the analysis is based is systematic. Another method of analysis may need to be used prior to applying the formal method. In other cases the analysis is based on tasks that exist within the functionality of, for example, a computer system. There the tasks may already be closely defined. A further drawback with some formal methods is that it is difficult to make explicit the interrelationships between task elements, they are often assumed to be sequential.

Job process charts

Introduction

Task analysis methods are often based on techniques drawn from other disciplines, especially those related to computer science or systems analysis. Flow charts are a standard means of describing systems and can similarly be applied to the development and analysis of tasks. Symbols are used to represent different kinds of task elements such as decisions and actions, and the sequence of the task is indicated by the flow lines. Sequential tasks and those involving decision making between several alternatives can be effectively represented in this way.

Job process charts (Tainsh, 1985) are based on the flow chart principle. The method was originally developed for analysing naval command tasks and has many similarities to both traditional flow charts and operation sequence diagrams (Brooks, 1960; Drury, 1983). The method aims to aid human–computer interface design with the emphasis on the communication flows and dialogue between the human and the machine.

Method

The technique uses a flow chart based method divided into a three level hierarchy. At the top level, the overall system tasks are specified including the identification of the main workstations and lines of communication. At the second level, these are broken down into their component tasks and the sequence of tasks is shown with transactions between the human and the computer made explicit. At the lowest level, each of the components is described in detail in terms of the human–computer interaction.

At each level the task is analysed and represented using job process charts. Flow chart type symbols are used to represent the different task elements and the flow chart is constrained within a table which allocates the task elements to the human or computer and which identifies the interface information and knowledge flows across the interface. Tainsh (1985) illustrates the application of the technique to a human–computer task showing how useful information can be collected and how it can be incorporated into the system design process.

Discussion

The job process chart helps to emphasize the parallel nature of the tasks of the system and the user, and makes explicit the information flows between them. Other flow chart methods can be used to show the demands of a task in terms of the different activities that the user is expected to perform and their frequency. Such methods are limited in their effectiveness and are most useful in analysing tasks that are sequential and have simple decision making paths. It may be more difficult and inappropriate to use such methods for analysing complex cognitive tasks. The methods assume the external environment to be constant and also make assumptions about the knowledge and information the user has. This is not always made explicit in the analysis.

Such methods also incorporate a taxonomy of task types, in that a flow chart symbol must be allocated to each task element to indicate type. Difficulties that could arise from this include problems of allocating elements to task types and ambiguities in the definition of categories of tasks and categories that are not mutually exclusive and exhaustive.

Discussion

Given the wide range of techniques that exist, the question arises as to how these may be selected. Astley and Stammers (1987) have suggested three core criteria for doing this, namely generality, utility and validity. To these can be added the more common psychological criterion of reliability.

Generality

For a technique to have wide acceptability across the ergonomics disciplines, it must be possible to use it in different contexts, for different tasks and for different applications. However, very few techniques have achieved this level of acceptance, and it is still very much a case of different approaches being used in different discipline or application areas. Hierarchical task analysis, which deliberately set out to be generalizable across task areas, has now been used in both the training and the interface design areas. Despite some of its shortcomings the underlying model for representing human activities is a general one.

Utility

Utility, taken to be the general effectiveness and efficiency of a technique, is the most likely factor to influence a technique's general use. In some ways this will be dependent upon such factors as generality, but will also be determined by the extent to which the technique has been clearly explained in publications, and tried out in practice by a range of users. Again, drawing on the example of hierarchical task analysis, Shepherd (1976) suggested a number of alterations to the basic HTA technique as a result of experience of using it in the chemical industry. This made it more easily understood by potential users.

Validity

Determining the validity of any task analysis is an extremely difficult thing. Pragmatically oriented task analysis techniques are unlikely to be fully accepted as theoretically valid by mainstream psychologists. This is likely to be so until there is a general agreement on a taxonomy of human behaviour and some acceptable unit of behaviour. However, applied problem solving cannot wait for the perfect theory to be developed. Validity, in terms of presenting the 'true' state of the world to a practitioner who has a problem to be solved, is a difficulty. Producing *an* acceptable set of data to work on is however what the practitioner is likely to be most concerned with. Whether this set of data is more valid than a set of data derived from another technique is the problematical issue.

Attempts could be made to assess the different outcomes of analyses through expert judgment or through some measure of the quality of the product. One way in which the validity of a technique might be established is in terms of the number of design iterations that have to be gone through in order to establish an acceptable design of equipment and/or training course. In describing HTA, Annett *et al.* (1971) and Shepherd (1976) discuss the problem of depth of analysis, that is how much information has to be collected, and in what detail. Rules for redescription are provided; these attempt to put the analyst in the situation of deciding whether enough information has been collected for adequate performance to be achieved after training. The suggestion is made that the analyst should not go into more detail if he or she is uncertain about a particular task element. The reasoning behind this is that any shortcoming in the data will be quickly revealed during the try-out of the training course material, when adequate performance will not be achieved. If, however, too much information has been collected this will not show up in a training programme. The course will contain unnecessary, and perhaps boring, detail and that will not be immediately apparent.

This thinking could be extended to interface design; if the analysis technique did not provide the information required by operators then this should show up in the early try-outs of design activities. The inherent dangers of this approach should be apparent. It assumes a systematic approach to design whereby proper evaluation of different design phases is carried out. There is always the scope, in an inadequately resourced project, for this detail not to be provided; this would be particularly so with rare events.

Reliability

Closely related to the above topic is the concern about reliability, the question here being, will the application of the same technique, to the same task, at different times by the same analyst, or at the same time by different analysts, yield the same results? Very little attention has been paid to this topic, as the pressures usually are to complete a particular project in the shortest period of time. It is more likely that a single analysis will be carried out, and the reliability is more appropriate to the particular techniques of data collection. With the main concern being *completeness* of data collection, the derivation of different amounts of data by different observers would not be of so much concern. In terms of overall efficiency of techniques, this is though an approach worth pursuing further in research. This problem also points out the need for a range of data gathering techniques to be used when assessing the same task area. For example, an observational technique, which focuses on more mundane tasks may leave out rare critical events. Therefore, supplementation of the observational approach by use of critical incident technique (Chapanis, 1959) would be one way of assuming that the full information is available.

Future approaches and possible developments

This review of task analysis has discussed the current state of the art, but it has also revealed the need for more consideration to be given to future activities. The search for a comprehensive, generally accepted taxonomy of behaviour will continue. It is also hoped that more people will report their use of techniques and show the advantages and disadvantages of such uses in particular applications. It would be helpful if existing techniques were more widely tried out by practitioners, rather than the continual development of new techniques to tackle new problems. It is also to be hoped that new technology will make an impact on the area.

Elsewhere in this book, Drury discusses ways in which computer techniques are being more widely applied in ergonomics data collection. Three particular roles emerge for computer technology in the area of task analysis. One is in the recording of information, either analyzing data or collecting data on-line. In data analysis, computer techniques can be used to quickly codify information; e.g. the coding of questionnaire replies into pre-arranged computer formats or of video tape material into a number of categories of behaviour. On-line recording has been tried in a number of situations where keyboard inputs can be used to log events. The timing of these events can then be carried out automatically.

In terms of representation, the very flexible application packages that are now available for information handling will be utilized more in the future. For example, information entered as a list can be presented subsequently as a hierarchical structure. The possibility of codifying human activity in computer programs has been recognized for some time. For example, Rigney and Towne (1969) outlined a system that enabled hierarchies of subtasks to be very quickly produced from task analysis material. Such a process was enhanced, as they envisaged, by alerting the analyst to information that has been missed and enabling additional information to be easily added. Current developments in object oriented programming offer very great potential, and it is interesting that this early work of Rigney and Towne was programmed in LISP.

The final area to be considered is the extent to which computer based modelling techniques can be used as an adjunct to task analysis activities. For example, data can be codified about a task and then various versions of it run in fast time in order to test hypotheses about performance. One such technique, the SAINT program (Chubb, 1981; Sticha, 1987; and described in some detail in Meister's chapter of this book), enables various configurations of the task to be represented in a computer program and for this program to be run several times. Data about task activity can be collected in this way. This is clearly not an analysis technique in itself, but is one way in which task *synthesis* for future system design can be carried out. Developments of these techniques, together with data bases of information on times for task actions and error probabilities are a way in which the whole process of task synthesis and modelling of behaviour in situations will be more feasible in the very near future.

References

Annett, J. and Duncan, K.D. (1967). Task analysis and training design. *Occupational Psychology,* **41**, 211–221.

Annett, J., Duncan, K.D., Stammers, R.B. and Gray, M.J. (1971). *Task Analysis* (London: HMSO), Training Information Paper No. 6.

Astley, J.A. and Stammers, R.B. (1987). Adapting hierarchical task analysis for user–system interface design. In *New Methods in Applied Ergonomics*, edited by J.R. Wilson, E.N. Corlett and I. Manenica (London: Taylor and Francis), pp. 175–184.

Berliner, D.C., Angell, D. and Shearer, J. (1964). Behaviors, measures and instruments for performance evaluation in simulated environments. In *Proceedings of the Symposium and Workshop on the Quantification of Human Performance*, Albuquerque, University of New Mexico, pp. 277–296.

Brooks, F.A. (1960). Operational sequence diagrams. *IRE Transactions on Human Factors in Electronics,* **1**, 33–34.

Card, S.K., Moran, T.P. and Newell, A. (1983). *The Psychology of Human–Computer Interaction* (Hillsdale, NJ: Erlbaum).

Chapanis, A. (1959). *Research Techniques in Human Engineering* (Baltimore: Johns Hopkins University Press).

Chubb, G.P. (1981). SAINT, a digital simulation language for the study of manned systems. In *Manned System Design*, edited by J. Moraal and K.-F. Kraiss (New York: Plenum), pp. 153–179.

Conrad, R. (1951). Study of skill by motion and time study and by psychological experimentation. *Research,* **4**, 353–358.

Crossman, E.R.F.W. (1956). Perceptual activities in manual work. *Research,* **9**, 42–49.

Drury, G.C. (1983). Task analysis methods in industry. *Applied Ergonomics,* **14**, 19–28.

Fleishman, E.A. and Quaintance, M.K. (1984). *Taxonomies of Human Performance* (New York: Academic Press).

Gagné, R.M. (1977). *The Conditions of Learning*, 3rd edition (New York: Holt Rinehart and Winston).

Hollnagel, E., Pederson, O. and Rasmussen, J. (1981). *Notes on Human Performance Analysis*. Report No. Risø-M-2285 (Roskilde, Denmark: Risø National Laboratory).

Johnson, P., Diaper, D. and Long, J. (1984). Tasks, skills and knowledge: task analysis for knowledge based descriptions. In *Interact'84*, sponsored by IFIP, London, Volume 1, pp. 1.23–1.27.

McCormick, E.J. (1976). Job and task analysis. In *Handbook of Industrial and Organizational Psychology*, edited by M.D. Dunnette (Chicago: Rand McNally), pp. 651–696.

Miller, G.A., Galanter, E. and Pribram, K.H. (1960). *Plans and the Structure of Behavior* (New York: Holt, Rinehart and Winston).

Miller, R.B. (1962). Task description and analysis. In *Psychological Principles in System Design*, edited by R.M. Gagné (New York: Holt, Rinehart and Winston), pp. 187–228.

Miller, R.B. (1967). Task taxonomy: science or technology? In *The Human Operator in Complex Systems*, edited by W.T Singleton, R.S. Easterby and D.J. Whitfield (London: Taylor and Francis), pp. 67–76. (Also *Ergonomics, 10*, 167–176.)

Moran, T.P. (1981). The command language grammar: a representation for the user interface of interactive computer systems. *International Journal of Man–Machine Studies, 15*, 3–50.

Moran, T.P. (1983). Getting into a system: external–internal task mapping analysis. In *Proceedings of CHI'83 Human Factors in Computing Systems*, Boston, pp. 45–49.

Payne, S.J. (1984). Task-action grammars. In *Interact'84*, sponsored by IFIP, London, Volume 1, pp. 1.139–1.145.

Piso, E. (1981). Task Analysis for process control tasks: the method of Annett *et al.* applied. *Occupational Psychology, 54*, 247–254.

Rasmussen, J. (1986). *Information Processing and Human–Machine Interaction* (New York: Elsevier).

Reisner, P. (1982). Further developments toward using formal grammar as a design tool. In *Proceedings of Human Factors in Computer Systems*, sponsored by Washington Chapter, ACM. Gaithersburg, MD, pp. 304–308.

Rigney, J.W. and Towne, D.M. (1969). Computer techniques for analysing the microstructure of serial-action work in industry. *Human Factors, 11*, 113–122.

Seymour, W.D. (1966). *Industrial Skills* (London: Pitman).

Shepherd, A. (1976). An improved tabular format for task analysis. *Occupational Psychology, 49*, 93–104.

Shepherd, A. (1981). *Carrying Out Hierarchical Task Analysis—A Course Manual* (Staines, Middlesex: Chemical and Allied Products Industry Training Board).

Sparrow, J., Patrick, J., Spurgeon, P. and Barwell, F. (1982). The use of job component analysis and related aptitudes in personnel selection. *Journal of Occupational Psychology, 55*, 157–164.

Sticha, P.J. (1987). Models of procedural control for human performance simulation. *Human Factors, 29*, 421–432.

Tainsh, M.A. (1985). Job process charts and man–computer interaction within naval command systems. *Ergonomics, 28*, 555–565.

Williams, J.F. (1988). Human factors analysis of automation requirements—a methodology for allocating functions. In *Proceedings of the 10th Advances in Reliability Symposium*, pp. 103–113 (Amsterdam: Elsevier).

Chapter 7

Verbal protocol analysis

Lisanne Bainbridge

Introduction

There are many complex jobs in which the outcome of thinking does not
emerge in observable action. For example, one can think out a plan of action,
assess it, and decide it is inadequate for the purpose, or one can work out
the implications of a situation, and memorize the decision for use later. If
we want to train and support these types of work, then we need information
about these processes. One apparently obvious way of getting this information
is to ask people to 'think aloud' while they are doing the task. These verbal
reports are called 'verbal protocols'. To distinguish them from other
knowledge elicitation techniques (discussed by Shadbolt and Burton in this
book), such protocols essentially report mental processes used during a task,
rather than being the answers to questions about the task which are given
while the person is not actually doing it.

This review of verbal protocol methodology will be in three sections:

1. *The status of verbal protocols as evidence*. Their validity as data, and what
types of data can be obtained.

2. *Technique*. Practical aspects of collecting verbal protocols, and devising
task situations.

3. *Analysis*. How to extract information from the verbal protocols, the
validity of the analysis.

From the outset it should be emphasized that there is as yet no simple, brief
method of collecting and analysing protocols, and it is likely to remain a
research technique.

Verbal protocols as evidence

As there is no way of observing someone's mental behaviour directly, it is
not possible to test whether there is a correlation between what someone
thinks and what they say they think. Consequently, when verbal reports do

not fit a theory of mental behaviour one does not have to reject this theory. As the main test for the scientific value of data is that they can be used to reject theories, many academic psychologists feel strongly that verbal data are useless.

In more practical contexts, there are two main uses for verbal protocol data. First, they can be a source of hypotheses about cognitive processes, and so of predictions about non-verbal behaviour. The predictions can be tested without involving verbal reports. Suppose one notes from verbal reports that someone doing a task uses prediction and complex working memory. One can then compare performance using equipment designed on the assumption that prediction is used, and equipment designed assuming that prediction is not used, and test which is best.

Secondly, if someone says something, evidently they do have this knowledge somewhere in their heads. The question then becomes: Is this reported activity and knowledge actually what is being done and used at the time that it is reported? We need to ask about the circumstances in which verbal reports are likely to correlate sufficiently highly with cognitive processes to be useful for practical purposes. Several studies give indirect evidence of the circumstances under which people do give valid reports of their thought processes.

Nisbett and Wilson (1977) report studies which need to be taken seriously by anyone involved with knowledge elicitation, and which have developed into a major research area in social psychology. Their studies are of peoples' 'self-perception' of the influences on their attitudes and judgements. Nisbett and Wilson did objective experiments in which people were observed acting in various circumstances. They also asked the people, in questionnaires, what they would do if faced with these situations, and what they thought was influencing their behaviour. Nisbett and Wilson concluded that people can give the same verbal report in two situations in which they actually behave differently. People can also claim that factors are important which did not actually influence their behaviour, and can deny that factors influence their behaviour which actually did. This strongly suggests that we are not very good at reporting the factors which influence our decisions.

We might add, though this is a hypothesis based on experience rather than being based on the same type of careful experiments, that people in working situations may have more accurate information about their behaviour, both because more detailed and accurate information is available to them about the state of the working environment and the results of behaviour, and because people are explicitly, and usually verbally, trained to respond to particular aspects of the environment.

Nisbett and Wilson (1977) concluded that people are conscious of the results of their thinking (in this case their attitudes or judgements) but have no special access to the process of thinking (how they made those judgements). They concede that an individual does know more than an observer about private facts, not only the results of their thinking but also their focus of attention at a given time, current sensations, emotions, evaluations, plans, intentions and personal history.

This suggests that knowledge elicitation techniques which concentrate on the content of thinking are more likely to be valid than reports which claim to be observations of the processes underlying thinking. (See Shadbolt and Burton in this volume for discussion of appropriate techniques under different circumstances.) Verbal protocols mainly contain factual statements, especially when people are under time pressure in a real task. The contents of verbal protocols and interviews differ in syntax and type. This is illustrated in these report fragments:

1. 'I shall have to cut (furnace) E off, it was the last to come on, what is it making by the way? E make stainless, oh that's a bit dicey, I shall not have to interfere with E then.'

2. 'If a furnace is making stainless, it's in the reducing period, obviously it's silly, when the metal temperature and the furnace itself is at peak temperature, it's silly to cut that furnace off.'

These two examples come from the same furnace operator, the first while he was making decisions about the task, the second during a lull in activity a few minutes later. The verbal protocol contains mainly observed facts and results of decisions. These are the data from which an analyst infers the underlying cognitive processes and knowledge used, rather than getting direct information about them. Cuny (1979), who explicitly compared verbal protocols and interview data from the same process operators, came to the conclusion that interviews give more general information about strategies, but omit the details of a particular working context which can mean that behaviour differs.

Types of distortion, and types of information obtainable

In practice, we can consider the ways in which a verbal report is likely to be distorted, with the aim of minimizing these effects. Ericsson and Simon (1980) have discussed the different types of data that may be available for verbal report. They suggest a general model for the way in which verbal reports may be generated, and review the available data.

Some types of distortion could occur in all verbal reporting situations. Having to give a report changes the task and may change the way the task is done. When there are alternative methods of doing the primary task, the need to give a verbal report may influence the person to use a method which is more easily described. Someone could do a task in a physical rather than a mental way, as this is better fitted to the pace of speech, or do things in sequence rather than doing several things at the same time. They may use a version of the task which is more verbal in form, for example a beginner's method which has been explained in words, or follow the official regulations more closely than usual as these are in verbal form.

There are time limitations to the technique. If people are giving verbal reports about problem solving situations, many things may quickly pass through their minds, and be forgotten before there is time to report them. If people are working under time pressures, there may not be time to mention everything that is relevant.

Giving a report is a social situation involving self-presentation, and so can be influenced by social biases. People can select what they think it is appropriate to say. People in experiments are often very co-operative, and try to say what they think the experimenter wants to hear. If the person reporting thinks that the listener is superior and perhaps powerful then there may be pressures to appear rational, knowledgeable and correct, or inversely to be uncooperative or to present a particular attitude to the work. An experienced worker will talk much more freely and fully if they think that the listener will understand what they are talking about. On the other hand, they may not mention points which they think are obvious.

Other distortions arise when the person usually does the job in a non-verbal way, but is asked to give a verbal report. People are not consciously aware of how they carry out perceptual–motor skills, such as skating or swimming. Parts of cognitive skills are also automatic. Words, pictures and the feel of movements are each dealt with by different parts of the brain. Images and movements may not be accurately represented when they have to be reported in words. If the person's vocabulary is limited, they may not be able to find an expression for all that they know. The knowledge expressed may therefore appear limited, while actually it is the language which is limited, not the skill.

If the verbal report is a distorted representation due to difficulties of translating between how the task is done and a verbal representation of it, this suggests that verbal protocols should be combined with other more conventional behaviour observation techniques, to check on the distortions and amplify the data. If the protocols can be distorted because of the task or social situation, this has implications for the technique used in collecting the protocols, which will be mentioned in the next section.

Verbal protocols are more suited to obtaining some types of information about cognitive processes than others. We have already seen that they may give data on the outcomes but not the processes, of skill. As the protocols are made while doing a task, they can give explicit information about the sequence of items considered. From this the strategy being used may be inferred, and also the working memory. As we will see, if there is enough data it is possible to identify the choice points, and what determines how the decisions are made; that is we can identify the criteria used in decision making.

Knowledge about categories, about the components and mechanisms, functions and causal relations in a machine, or memories of specific past events, will only be mentioned explicitly if the task involves some problem solving which requires the person to review this sort of evidence. Otherwise it will only be possible to infer this sort of knowledge indirectly from the fact that a person would not be able to act in a certain way if they did not have particular knowledge.

This underlines the general point that verbal protocol evidence may give a limited sample of the total knowledge available to the person being studied. Even in a controlled task situation, with any complex task it will not be possible to test a large enough number of task situations to explore all of the person's knowledge. When collecting protocols in a real task situation, then we must be aware that the evidence collected will just depend on what happened at the time.

Techniques of collecting the protocols

In outline, in order to collect protocols, the person is asked to 'think aloud' whilst performing the relevant task, and the comments are tape-recorded and transcribed. Where possible it is useful to make short recordings which can then be typed out during the total session. After the session the speaker can read through the transcripts to correct mistakes and clarify vague passages (though this can involve distortions of the interview situation).

It is possible to make a much more detailed analysis of the data if a simultaneous record is made of the task situation, the information the person receives and the actions they make. This is useful in at least two ways:

1. People talking while doing a job frequently use pronouns, e.g. 'it's at 55'. The objective record shows what is 'at 55', and therefore what it is that is being referred to.

2. To collect evidence about the total task situation, which may influence the person's behaviour although it is not mentioned in the protocol (see above on the completeness of protocol data, and below on analysis).

This observation record can be made in several ways, but does require considerably more work. An observer can log the events (for example, see Beishon, 1967, Fig. 5) (Figure 7.1). A specially instrumented task can make the log, for example with chart recorders of events. It is easiest to obtain this log when a computer-driven simulated task is used for the investigations. Video recordings, filming the workplace and the operators' movements, are a useful supplement to the audio recordings, but greatly increase the amount of data to be analysed.

The task and social situation

Asking people to 'talk to themselves' is unnatural, and some people find it a very uncomfortable requirement. More natural protocols can be collected by asking people to work together. Typical situations are for two or more people to be working together, particularly in a problem-solving situation, or for one person to be guiding another in how to do the task. Maximum

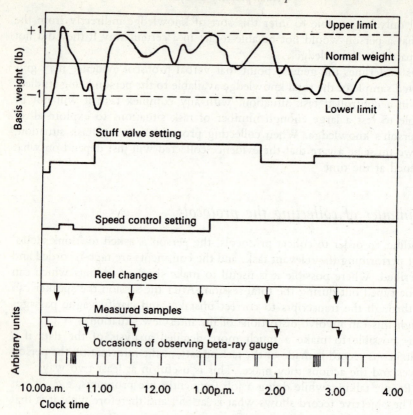

Figure 7.1. Activity graph for paper machineman (from Beishon, 1967).

communication of thoughts and knowledge, or admission of lack of knowledge, can then be natural aims rather than a source of embarrassment.

The test situation may be the real task. Otherwise it will be some sort of simulated task, which raises questions about the time scale used, and about the accuracy of the simulation.

In static simulation, the environmental changes are 'frozen', so the person has time to think things out fully and to describe all their thoughts and inferences without being under time pressure. The simulations are usually relatively simple 'paper and pencil' versions such as drawings of the workplace, interface and job aids. For a sophisticated but simple to use example, see Duncan and Shepherd (1975, p. 634) (Figure 7.2). This method is an extension of the structured interview, but reduces the imagination required, as representations of some of the actual task conditions are given. One specific problem with static simulation is that a sequence is necessarily imposed on the person's behaviour. Another is that working methods may change when people are not under the stresses and time pressures of the real situation.

Changes in the task environment over time, and feedback of the results of actions, are intrinsic aspects of some tasks which cannot be conveyed by static simulation. Dynamic simulations are therefore used in 'on-line' techniques so that, as far as possible, the operator is performing his real task.

Nevertheless, the static simulation method has many advantages. Leplat and Bisseret (1965) point out that one can isolate apsects of the work in which one is interested. People can be placed in strictly repeatable task situations. This compares with the shop-floor, or with dynamic simulation, in which different people can make different decisions and actions in the same task context. After such a decision point, the rest of the task will be different for each person, so the final outcomes differ and can be difficult to compare. Also in the real situation the person experiences an uncontrolled sample of events, and may be too busy to explain fully what they are doing and why.

Actions and reports which are not made at the actual workplace, or on a spatially and dynamically accurate (hi-fi) simulator, may represent general strategies, but are not likely to give accurate information about pattern recognition skills, or the size and timing of actions in perceptual-motor skills.

Analysis of the protocol material

Unfortunately, as well as the general problems with verbal data discussed earlier, several other problems are emphasized by on-line protocols. People may not report what is obvious to them. They may not mention information which they collect while reporting other activities, which may lead to unexplained behaviour later. Most people can think more quickly than they can talk, so only a sample of all their cognitive activity can be reported. The investigator therefore has the same problems, of inferring what lies between the items mentioned in the protocol, as there are when interpreting non-verbal behaviour. Last, but not least, analysing protocol data is both difficult to do and time consuming.

We have discussed limitations to the validity of verbal protocols as evidence of thought processes. There are also problems with the validity of the analysis. There are no objective independent techniques for doing this analysis. This means that investigators have to use both their own natural language understanding processes, and their knowledge of the task, in making sense of what is going on and inferring missing passages. This worries many investigators, who wonder whether all they are doing is rediscovering their own knowledge. There are several replies to this. One is that one does find task activities which had not been anticipated. Secondly, the analysis can give a description of the behaviour which is consistent with other data or theories, so providing converging evidence. Ideally, the analysis should be made by several investigators who work independently and then come together to agree on a final version.

As the analytic categories are developed in the process of analysing the data, neither the data nor the categories can be justified in terms of each

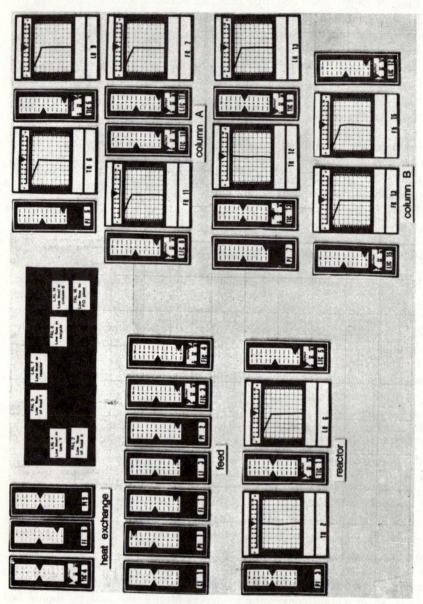

Figure 7.2. A simulation instrument panel during plant failure (from Duncan and Shepherd, 1975).

other, and some external test is necessary. One test is that the categories can be applied to the data, and with the same results, by judges who were not involved in developing the categories. The usual scientific test is that the categories should be applicable to other data. This cannot usually be done, at this stage in the development of protocol analysing techniques, because there are too few case studies, all made for different purposes. One solution to this problem is to use a split-half technique, developing the analysis using one half of the data, and testing it on the other half. It will be very interesting if or when investigators in different studies do produce similar sets of categories.

There are three main analytic problems, which require different techniques: preparing the material, analysing the explicit content, and inferring the implicit content, which includes identifying the reasons for the sequence of behaviour.

Preparing the material

The first stage is to segment the verbal report into basic units. This can be done at two levels: dividing the material into a sequence of separate phrases, or into groups of phrases.

Separating material into phrases

The following piece of report will be used in the examples (the dots indicate pauses):

> C is on oxidation now that's something you can make an estimate for it's a quality so I must leave it alone . . . oxidation average length is one hour 30 minutes for C and started at time zero no it didn't it started at time 33 minutes how confusing of it so it's got nearly one and a half hours to run . . . I'd better check that oxidation for C one hour 30 minutes started 50 minutes ago so it's got 37 minutes to go . . .

Separated into phrases this becomes:

1. C is on oxidation
2. now that's something you can make an estimate for
3. it's a quality
4. so I must leave it alone
5. oxidation average length is one hour 30 minutes on C
6. and started at time zero
7. no it didn't
8. it started at time 33 minutes
9. how confusing of it
10. so it's got nearly one and a half hours to run

11. I'd better check that
12. oxidation for C one hour 30 minutes
13. started 50 minutes ago
14. so it's got 37 minutes to go

In this section of protocol a steel-works operator is talking about a furnace called 'C'. Phrases 1–4 are concerned with general properties of the 'oxidation' stage through which the furnace is going, and phrases 5–14 make an estimate of the time at which C will finish oxidizing. The average stage length is given in a job aid booklet, and the time the stage started is on the operator's display panel.

The division of the text into phrases (which might loosely be described as minimum grammatical units, though the language in such reports is often not at all well formed) is done by natural language understanding. This can often be done by people who have no knowledge of the specialist content of the material.

Combining phrases into groups

There are two methods for combining phrases into groups, both of which make use of the semantic content. The first approach is to identify the pronominal referents, as these indicate cross-references between phrases. There are three ways of identifying the pronoun referents. The most reliable requires an independent record of the states of the environment during the task, from which what 'started at time 33 minutes' can be identified. Another method involves going through the report immediately afterwards with the speaker to check on the meaning. The third method is to use judges' semantic knowledge of the task.

After the links between phrases made by the pronouns have been identified, further groupings can be made on the basis of judges' knowledge of what items go together in the task. The result of doing this for the material above is given in Figure 7.3. One can also take advantage of the necessity to use judges' knowledge, by making an explicit record of one's own knowledge which was used in the analysis, and using this as a record of the knowledge one is attributing to the speaker. For example, lines 1–4 list properties of C furnace, and lines 5–14 recount a method of calculation. These points are also what one needs to mention to a person unfamiliar with the task before they can understand the report, as in the task explanation given earlier.

The semantics of an individual phrase may imply that the speaker is referring to wider background knowledge. Consider the phrases (from a process operator): 'I'll try to run the temperature down to about 400 degrees . . . no trouble, it's far away now.' The first phrase implies the operator knows the required future process state (400 degrees). The second phrase appears to assess the ease of making the change, which implies that he knows how to make both this change and this assessment. A full protocol analysis

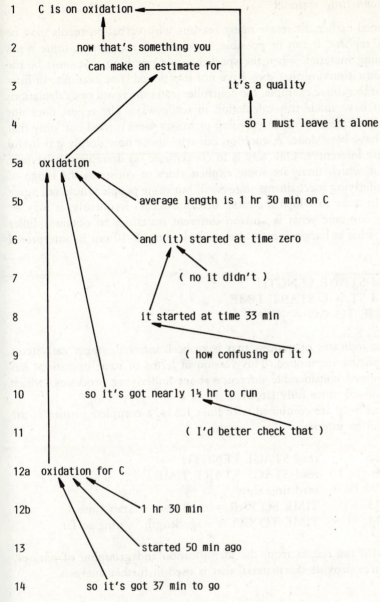

1   C is on oxidation

2   now that's something you
          can make an estimate for

3   it's a quality

4   so I must leave it alone

5a   oxidation

5b   average length is 1 hr 30 min on C

6   and (it) started at time zero

7   (no it didn't)

8   it started at time 33 min

9   (how confusing of it)

10   so it's got nearly 1½ hr to run

11   (I'd better check that)

12a   oxidation for C

12b   1 hr 30 min

13   started 50 min ago

14   so it's got 37 min to go

Figure 7.3. Cross-references between phrases in protocol sample.

could detail the types of more general knowledge which can be inferred in this way. Unfortunately, although one can infer from the protocol that such knowledge exists, there may be very little direct evidence in the protocol about its structure and use. Information about this might be obtained by following up the protocol study with interviews on these points.

Inferring connecting material

As mentioned earlier, there are many reasons why verbal protocols may be incomplete reports. It can be possible, however, to reconstruct some types of intervening material, when the speaker says something that must be the result of some thinking that they have not mentioned. For example, in lines 10 and 14 of the above example, the controller states the result of a calculation, so he must have made this calculation in some way. The report does not necessarily indicate how the intervening processes were carried out, only that they must have been done. A question can arise about how deeply it is useful to take these inferences. One way is to concentrate on describing behaviour at a level at which there are some explicit clues to constrain the range of possible underlying mechanisms suggested, but some people would not agree with this. In tasks in which the same situation recurs frequently, it may be possible to combine what is said on different occasions to obtain a fuller account of what is happening. In the example, lines 5–10 can be interpreted as:

    5.   read STAGE LENGTH
    6–8. read STAGE START TIME
    10.   TIME TO GO = X.

(Lower case indicates operations that have been inferred, upper case items that are explicitly mentioned. This version in terms of basic operations has already involved considerable inference about underlying processes, which will be discussed more fully later.)

When lines 5–10 are combined with lines 12–14, a complete picture of the processes can be inferred:

(5)	12.	read STAGE LENGTH
(6–8)		read STAGE START TIME
		read time now
	13.	TIME SO FAR = time now − start time
(10)	14.	TIME TO GO = stage length − time so far.

Together with the results from the identification and grouping of phrases, such inferences provide the material that is used in further analyses.

Explicit content: content analysis

There are many styles of analysis called content analysis. These typically involve counting frequencies of occurrence for categories of material. There are standard computer programmes that count word frequencies. This section will give brief examples of analyses using categories that are more complex than counting words. One can analyse the frequency of occurrence of particular categories of phrase types, or types of groups of phrase, or within phrases. These categories, their instances, and the frequency counts can be used as the basis for further analyses of the categories or the contexts in which they occur (see later examples).

The categories developed, to be useful, must be ones that encourage further inferences about the material, for example, the frequencies may suggest emphases in the way the speaker thinks about the topic. Alternatively the categories can be used simply as a preliminary sort before further analyses, for example, the analyst could look further at all instances of the phrases that have been categorized as 'comment on own behaviour' to see if the phrases have any common properties. If the categories are based on semantic aspects of the reports, one might want to count only the occurrence of overall concepts (e.g. birds) or to count the frequency of individual instances (e.g. robins vs. blackbirds). Different aspects of syntactic structure could be differentiated. For example, one might wish to distinguish between active and passive voice as reflecting different emphases by the speaker. Or one could differentiate between different types of conditional statement, perhaps comparing the incidence of 'A therefore B' with 'B because A'. Whether one does this depends not only on whether one is interested in this level of analysis but also on whether the concepts occur sufficiently often to make a frequency count something more informative than a simple listing. The categories may also imply inferences about the types of cognitive process underlying them, for example, 'statement of fact' and 'comment on own behaviour' may imply different types of underlying cognitive activity.

The categories are therefore always a function of particular empirical questions. If there is a set of categories that can be applied in many different circumstances, they are likely to be so general that the results of using them will not be very rich. Indeed, one of the main analytic problems is the development of categories which are both relevant to the investigation and reliably usable by the judges who assign the material to them. Rasmussen and Jensen (1974) describe the iterative method that must be used. First, several independent judges each develop a set of categories. Then they attempt to use each other's categorization schemes. This both combines the inferences the judges have made about the important distinctions to be made, and also tests whether different people can make the same allocation of material to the categories. If not, then analysis using these categories cannot give reliable data, and the categories must be revised, again with the judges working independently during development, and coming together for assessment. The categories are repeatedly defined and revised by trying to use them to analyse the protocol, until the category definitions and phrase assignments stabilize. This procedure is arduous and repetitive but must be done with precision.

To give further examples of categorizing, examples of identifying characteristic structures within phrases, at two levels will be examined: the types of referent words that it is useful to identify within the phrases, and

the characteristic patterns in which these occur. For example, the phrase:
'average length is 1 hour 30 minutes'
can be interpreted as a statement of the general form:
'VARIABLE has VALUE'.
The phrase:
'steam pressure is 101'
can then be categorized as a statement of the same type.

Having identified all statements of this type, one could then look at the category members further, e.g. identifying how many different VARIABLEs occur, and with what frequency, or concentrating on the VALUEs and seeing how accurately they are specified. One would have to consider whether to categorize:
'there's steam temperature—it's rising'
as a statement of the same type, or whether this would lose some of the information in the report, so this phrase should be identified as:
'VARIABLE has CATEGORIZED VALUE'.
This also shows how categorizing can retain information considered important for one type of analysis, but lose other aspects. This phrase is syntactically different from the previous instances, but it has been assumed that this is not relevant to the question of how VARIABLE VALUEs are thought about by the speaker. The syntactic change rather than the semantic one might, of course, be the emphasis of an analysis being made for other purposes.

Simple phrases also occur as a component in more complex ones, for example:
'I'll try to run the temperature down to about 400 degrees'
can be interpreted as:
'ACTION causes VARIABLE has VALUE'.
Statements of this type show the speaker's knowledge about how changes in the outside world can be made. Speakers may also give information about their knowledge of the conditions in which certain effects occur, for instance,
'We have to have 50 degrees superheat before we can run it up'
could be described as:
'when VARIABLE has VALUE then ACTION causes VARIABLE has VALUE'.
Or more simply, if one is not interested in distinguishing between different ways of expressing conditional knowledge (e.g. to test whether different expressions occur in different task contexts) the same phrase can be described as:
'ACTION given VARIABLE VALUE'.

These examples have categorized material within phrases. Similar methods can be used to study categories of phrase type, or of groups of phrases. As an example, the phrases in Figure 7.3 might be categorized as follows:

1. Fact.   8. Fact.
2. Strategy.   9. Comment.

3. Fact. 10. Prediction.
4. Strategy. 11. Strategy.
5. Fact. 12. Fact.
6. Fact. 13. Fact.
7. Comment. 14. Prediction.

These categories are necessarily more general, and so there may be an even more distant relation between the actual material in the report and the way in which it is described. Therefore more care may be needed to ensure that the categories are unambiguous to judges. Again, it may be useful or interesting to distinguish subcategories, for example between statements of fact about past and present, or between comments expressing general strategy compared with rules for behaviour at this particular time.

In some tasks, it is not very useful to categorize individual phrases, because similar task situations do not recur at this level of detail. In this case, analysis is done at the level of groups of phrases. As an example, the explanatory notes given for Figure 7.3 are equivalent to a categorization of types of activity in that piece of report. Rasmussen and Jensen (1974) and Umbers (1981) give examples of this type of analysis in maintenance and process control tasks, respectively.

Implicit analysis: sequences in the content

Looking for sequences in the material can also be done either at the level of individual phrases or of groups of phrases. The level of categorization at which to work depends on the material. For example, in the fairly constrained 'world' of controlling a simple industrial process, similar sequences of phrases may occur sufficiently frequently to allow one to identify a standard sequence phrase by phrase. Rasmussen and Jensen (1974) studied maintenance technicians. In each report, the speaker was working on a different piece of equipment with a different fault. Consequently, the fault finding behaviour was not repeated at the level of individual phrases. To search for the common properties of the behaviour in these different situations, Rasmussen and Jensen had to look at the data at a more global level.

The most frequently used and fully specified rigorous procedure for analysing sequences is to make a Markovian analysis, i.e. to find the probability of transition from one item to another. Unfortunately, this method gives a very limited description of the properties of a sequence. For example, one can make a Markovian analysis of the tune of 'Three Blind Mice', obtaining a table of the probability of a note at one pitch being followed by a note at each of the other pitches in the tune. This is not, however, a very helpful description of the tune because 'Three Blind Mice' is not a probabilistic sequence. It is exactly the same every time, and this feature has completely disappeared in the analysis. Markovian analysis can be a useful preliminary technique to determine which transitions are most frequent and therefore are most likely to be rewarding to study further. However, if one prefers to assume that people producing verbal reports are not acting in a random way, and that their reports reflect activity that is at least to some extent structured and repeatable, then one would like to use techniques that increase the possibility of finding more determinate sequences in the behaviour. (The techniques described can also be applied to analysis of non-verbal behaviour.)

Sequences of phrases

The sequences in which individual items are mentioned in the report can indicate the standard 'routines' or 'programmes' with which the speaker thinks about a particular topic. This is the main approach of workers who use verbal protocols as the basis for simulations of cognitive processes. There is extensive literature on this type of theory development, of which Newell and Simon (1972) is a classic example.

One can only reach strong conclusions about these activities however, if one has several examples of each type of behaviour. The reports are always incomplete at some level, and one may have several hypotheses about the processes intervening between two phrases. Unless one has other examples of the same behaviour, the analysis remains very speculative and unwieldy. For example, in Figure 7.3 it would be possible for phrases 8 and 10 to be linked by the following serial calculations:

Stage end time   = stage start time + stage length;
Time to go   = stage end time − time now.

However, we have another instance of the same behaviour, in which phrase 13 shows which of the two possible linking calculations is being used.

Sequences of groups of phrases

Identifying the sequence of groups of phrases is more difficult, because one wants to infer what influences the speaker to move from one topic to another. One cannot take the speaker's word for it, as Nisbett and Wilson (1977) have shown that speakers do not necessarily have good access to this type of information about their behaviour.

To do this analysis one needs to take into account earlier behaviour by the speaker, because previous items can affect the choice of later behaviour. Also a record of the environment is almost essential in studying dynamic tasks (ones which change over time), because it is often these changes which influence what is the most appropriate thing for the speaker to think about next.

To do this sequential analysis, one first identifies the instances of transitions from one type of behaviour to others. For example, behaviour A may be followed by behaviour B or by behaviour C. This can be identified from the Markov analysis. One then looks at the whole context of the speaker's behaviour and the environment to see whether any dimension of the task or previous thinking consistently has one value when A is followed by B and another when A is followed by C. If so, then one infers that the value of this dimension determines the choice of behaviour at this point.

An example comes from a process control task. The operator frequently made remarks such as 'it's above now', 'it's below', and the problem was to identify what dimension of the process he was using to make this judgement. There were two main candidates, the total power being used at the time, or the discrepancy between present power usage and target power usage. Table 7.1 shows the distribution of judgements at different readings of the total power display. Table 7.2 shows the distribution of judgements at different readings of the discrepancy meter. It is clear that the judgements correlate with the discrepancy meter reading and not with the total power display, so the speaker is assumed to be using the discrepancy meter when making these judgements. (Note that this example is not drawn from an analysis of phrase sequences, as no compact example is available. Here the

Table 7.1. Frequency of assessments at different total power values

Total Power	Above	Alright	Below
31–40		1	3
41–50	2	2	2
51–60	2	3	1
61–70	2		
71–80			1

Table 7.2. Frequency of assessments of discrepancy meter readings

Discrepancy	Above	Alright	Below
−41 to −50			1
−31 to −40			
−21 to −30			
−11 to −20			3
−6 to −10		3	2
−1 to −5		2	
1 to 5	1	1	
6 to 10	1		
11 to 20	2		
21 to 30			
31 to 40	1		
41 to 50	1		

technique has been used to identify pronominal referents. This illustrates that most of the techniques used in analysing verbal reports can be used to study any level of organization in the material.)

Conclusion

This methodological review has taken an optimistic approach to the difficulties of collecting and analysing verbal protocols, even though analysing verbal reports is not easy, nor are there many time-saving techniques that can be applied.

A person's choice of complex behaviour is influenced not only by immediate circumstances but also by planning in relation to the predicted future or by reference to similar past events. It is difficult or impossible to get sufficient evidence from observed non-verbal behaviour to suggest or constrain hypotheses about such cognitive activities. As a consequence, we know very little about the processes underlying complex behaviour. Verbal protocol analysis is currently one of the richest ways of investigating the nature of behaviour that is a function of either past or future. The methods described here illustrate the flexible analytic techniques that can be used.

Note

This contribution has been compiled from sections of the following papers:

Bainbridge, L. (1979). Verbal reports as evidence of the process operator's knowledge. *International Journal of Man–Machine Studies*, **11**, 411–436.

Bainbridge, L. (1985). Inferring from verbal reports to cognitive processes. In *The Research Interview*, edited by M. Brenner, J. Brown and D. Canter (London: Academic Press), pp. 201–215.

Bainbridge, L. (1986). Asking questions and accessing knowledge. *Future Computing Systems*, **1**, 143–149.

References

Beishon, R.J. (1967). Problems of task description in process control. *Ergonomics*, **10**, 177.

Cuny, X. (1979). Different levels of analysing process control tasks. *Ergonomics*, **22**, 415–425.

Duncan, K.D. and Shepherd, A. (1975). A simulator and training technique for diagnosing plant failures from control panels. *Ergonomics*, **18**, 627–641.

Ericsson, K.A. and Simon, H.A. (1980). Verbal reports as data. *Psychological Review*, **87**, 215–251.

Leplat, J. and Bisseret, A. (1965). Analyse des processus de traitement de l'information chez le controleur de la navigation aerienne. *Bulletin du CERP*, **1–2**, 51.

Newell, A. and Simon, H.A. (1972). *Human Problem Solving* (Englewood Cliffs, NJ: Prentice-Hall).

Nisbett, R.E. and Wilson, T.D. (1977). Telling more than we can know: verbal reports on mental processes. *Psychological Review,* **84**, 231–259.

Rasmussen, J. and Jensen, Aa. (1974). Mental procedures in real-life tasks: a case study of electronic trouble shooting. *Ergonomics,* **17**, 293–307.

Umbers, I.G. (1981). A study of control skills in an industrial task, and in a simulation, using the verbal protocol technique. *Ergonomics,* **24**, 275–293.

Chapter 8

Simulation and modelling

David Meister

Introduction

Both simulation and modelling are forms of representation. Simulation is physical, modelling is symbolic, representation of equipment, events and task performances. Both are most often implemented by computer, although some models are purely conceptual and some physical simulations—primarily static mock-ups—do not involve computers. Although not every simulation and model involves human performance, and most simulations in fact describe physical processes only (e.g. manufacturing, Mellichamp and Wahab, 1987; and marine operations, Park and Noh, 1987), the simulations and models ergonomists are interested in, do involve human performance. Simulators are operated by and train students; and functions in human–machine system models are performed by symbolic operators and technicians.

Although most simulators and many models replicate already existent systems (e.g. a particular type of aircraft, a particular type of ship), research simulators and models can be programmed to represent systems in general or ones not yet designed.

Simulation

When ergonomists talk about simulation they usually refer to physical simulation of human–machine systems and this is what will be described in this chapter. If, however, one defines simulation in general as an attempt to represent reality in various forms, many simulation-like activities should also be mentioned. These include what are generally termed 'games', e.g. experimental games, in-basket simulations and role playing (Cunningham, 1984).

Simulator history and uses

The very first simulators were the primitive devices used for flight training before and during World War I. The Link trainer was developed before World War II for instrument flight training. The first specific aircraft simulator was that of the Lockheed Hudson bomber developed in 1942. Electronic simulators were first used in the early 1950s by commercial airlines. The military services followed closely on.

Training simulators for systems other than aircraft have proliferated, including (the following is only a partial listing): automobiles, trucks, railroads, ship propulsion and collision avoidance systems, submarine and surface warfare systems, air traffic control, tanks, artillery, missiles, military command control, nuclear power, mining and fire fighting.

In addition to their use as trainers, simulators have been used extensively for systems research (see Parsons, 1972) and for research on workload, decision making, stress, and performance assessment (see Jones *et al.*, 1985, for a more complete list). Simulators are also used for equipment and system design, development, testing and evaluation (to 'try out' alternative design configurations), and for licensing of personnel. Training simulators have many advantages over operational equipment: *cost* (less in a simulator); *availability* (the operational equipment may not be available); reduced *hazard* (less dangerous to practise emergency procedures in a simulator); training *enhancement* (one can stop events in a simulated mission to emphasize a point, back up and repeat an activity, change conditions rapidly). Because simulators are so expensive, few are built solely for research and considerable research is performed in simulators when students are not being trained (Beare and Dorris, 1983).

Simulation characteristics

Simulators vary along two continua; their *realism* or fidelity to the operational systems they represent, and their *comprehensiveness*, which is the extent to which operational functions and environmental characteristics, etc. are reproduced in a simulator.

Fidelity, which presents the greater problem, will be discussed later. If, however, simulators are properly designed for their training function, their comprehensiveness should vary simply as a function of what is to be learned. For example in the early stages of procedural training, when students learn sequences of discrete procedural steps, it is hardly necessary to provide full scale simulation; alternatively, a total mission can be practised only in a full scale simulator.

At one extreme the simulator may be quite abstract, for example a computerized display of a ship propulsion system (Hollan *et al.*, 1984). At the other extreme the simulator may consist of the actual operational equipment to which only special control hardware and software have been added. Most simulators fall between these extremes.

A special form of simulator is the 'expert' system (Hayes-Roth *et al.*, 1983). This attempts to replicate the thinking or judgmental processes that underlie an individual's expertise (a step up from merely replicating physical events and task performance). Most expert systems today are still research tools; when fully developed they will be aids used in a great variety of activities in which judgment and decision making are required.

One need not think of a simulator as consisting solely of hardware and software. The mock-up commonly used in equipment design consists of plywood or styrofoam and may even make use of paper drawings of controls and displays (Buchaca, 1979). A form of simulation that makes use of engineering drawings or static mock-ups as simulators is the *walkthrough*, which is performed to verify the adequacy of procedures and human–machine interface design. The walkthrough is a quasi-symbolic rehearsal of the way in which an equipment under design will ultimately be operated; an engineer points to or touches a mocked up control or display and describes the actions to be taken with these.

A special form of simulation is the development of computer-assisted behavioural design tools, such as CAFES (Computer-Assisted Function Allocation Evaluation System; Parks and Springer, 1975). These automate the behavioural analyses involved in predicting workload, determining anthropometric requirements, allocating functions, and laying out control panels and work stations.

More complex simulators generally have mission scenarios, a schedule or script that drives the sequence of events with which the simulator is exercised. Part of the simulator's software generates the scenario events; another part keeps track of these events and the subject's responses. These events may have a predetermined sequence unaffected by operator actions or they may vary in accordance with algorithms in response to what the operator does.

The models to be discussed in the next major section of this chapter are often used in human-in-the-loop simulations to represent exogenous and endogenous variables and to drive the simulator's processes. A simulator also includes subsystems, for control, for monitoring progress and measuring subject performance.

Simulation fidelity

The evolution of simulation has been primarily a matter of technological advances to make simulators more accurate and hence more realistic representations of a specific system. It seems logical to developers that the more the simulator is like its real world prototype, the more confidence one can have that personnel performance, including the learning of skills, will be equivalent to personnel performance in the actual system.

Pragmatically, the more realistic the simulation, the fewer problems one should have in designing it, because one need only reproduce existing design. Complete simulation fidelity *apparently* eliminates the need to worry about the relationship between operator characteristics and simulation features; one trains or measures the same way one would if an actual system were being used. ('The term 'apparently' suggests that this is only an hypothesis.) As soon as a simulation is less than completely faithful to its original, questions arise concerning which system characteristics to modify and in what way, and what the effect of these modifications will be on the student's skill acquisition and performance.

Complete duplication of a system is costly, much of the cost arising from the need to simulate the characteristics of the system's operational environment and system-related events, such as the interaction of other systems with one's own. This is why the visual subsystem and the presentation of other aircraft are so difficult and expensive in military aircraft simulations. Because of the difficulty and expense there is strong motivation to determine how far one can deviate from complete realism without unduly reducing simulator training value and performance.

Not least of the problems presented by this is the question of how one defines *simulation fidelity*. Kinkade and Wheaton (1972) suggest that there are three types of fidelity: (1) *equipment fidelity*, or the degree to which the simulator duplicates the appearance and 'feel' of equipment; (2) *environmental fidelity*, or the degree to which the simulator duplicates the sensory stimulation from the task situation; and (3) *psychological fidelity*, or the degree to which the simulation task is perceived as being a duplicate of the operational task. Major attention has been paid to the first two, it being assumed implicitly that if the first two are provided, the third will follow automatically.

One might expect from the extensive research on simulation fidelity (see Table 8.1 for a list of the major studies performed in relation to the factors affecting fidelity) that guidelines would now be available relating simulation characteristics to performance outcomes. Whatever guidelines are available are impossibly general, as can be seen later. Perhaps the relationships we are seeking are highly specific, meaning that one must investigate the effect of each simulator's characteristics on performance in advance of simulator design or opt for as much fidelity as one can afford.

Since the purpose of the training simulator is to provide the conditions necessary for learning or for eliciting human performance very similar to what would be elicited operationally, two principles can be derived: (1) characteristics and methods of using simulators should be based on behavioural objectives; (2) physical realism is not the only, or even the optimal, means of achieving these objectives.

Adams (1979) discusses psychological principles which should underlie the design and use of any simulator. The usefulness of these principles can be illustrated by the following examples: since human learning depends on knowledge of results (KOR), every simulator should incorporate KOR; if a

Table 8.1. Summary of research on fundamental problems in human–simulator interaction (taken from Jones *et al.*, 1985)

1.	*Fundamental Behavioural Processes*	
	Structure and acquisition of skills	Gagne (1954)
		Muckler *et al.* (1959)
	Cognitive skills	Prophet *et al.* (1981)
	Motivation and learning	Gagne (1954)
		Muckler *et al.* (1959)
	Behavioural mechanisms of transfer	Gagne (1954)
	Perceptual learning	Hennessy *et al.* (1980)
	Visual perception	National Research Council (1982)
2.	*Fidelity of Simulation*	
	Important and unimportant simulator characteristics	Miller (1954)
	Effects of fidelity on transfer	Muckler *et al.* (1959)
	Interaction of fidelity with	
	(a) Instructional variables	
	(b) Experience level	Smode *et al.* (1966)
	Fidelity Requirements	McCluskey (1972)
		Huff and Nagel (1975)
		Hays (1981)
		Gaffney (1981)
	Departures from fidelity of dynamics to compensate for simulator deficiencies	NATO-AGARD (1980)
		Adams (1973)
3.	*Visual Simulation*	
	Visual display characteristics	Muckler *et al.* (1959)
		Smode *et al.* (1966)
		Huff and Nagel (1975)
		National Research Council (1975)
		Hennessey *et al.* (1980)
		NATO-AGARD (1980, 1981)
		Kraft *et al.* (1980)
	Scene content and visual cues	Matheny (1975)
		National Research Council (1975, 1982)
		Thorpe *et al.* (1978)
		Hennessy *et al.* (1980)
		NATO-AGARD (1980, 1981)
		Prophet *et al.* (1981)
4.	*Vehicle Motion*	
	Motion cues	Muckler *et al.* (1959)
		Huff and Nagel (1975)
		NATO-AGARD (1980)
		Prophet *et al.* (1981)
	Interaction of motion and vision	Smode *et al.* (1966)
		Matheny (1975)
		National Research Council (1975)
	Interaction of motion skill level	Smode *et al.* (1966)
	Effects of motion on transfer	Smode *et al.* (1966)
		Matheny (1975)
5.	*Performance Assessment*	
	Criteria for performance	Gagne (1954)
		Center for Nuclear Studies (1980)
	Performance measurement	Gagne (1954)
		Muckler *et al.* (1959)
		National Research Council (1975)

Table 8.1. Continued

	United States Air Force (1978)
	Center for Nuclear Studies (1980)
	Gaffney (1981)
	Prophet *et al.* (1981)
Automated performance monitoring	United States Air Force (1978)
	Center for Nuclear Studies (1980)
Measurement of team performance	Parsons (1972)
	Gaffney (1981)
	Prophet *et al.* (1981)
6. *Modelling*	
Models of visual and motion simulation	Waag (1981)
Sensory system modelling	United States Air Force (1978)
Model of multisensory spatial orientation	NATO-AGARD (1980)
Models of visual environment to identify variables relevant to training	NATO-AGARD (1980)
Models to predict training effectiveness	Prophet *et al.* (1981)
7. *Training*	
Critical characteristics for transfer	Miller (1954)
	Caro (1977)
Effects of change in task characteristics on transfer	Muckler *et al.* (1959)
Measurement of transfer effectiveness	Smode *et al.* (1966)
	Williges *et al.* (1973)
	NATO-AGARD (1980)
Systematic method for developing training requirements that provide guidance for simulator design	NATO-AGARD (1980)
8. *Training Methods*	Gagne (1954)
	Human Factors Operations Research Laboratories (1953)
	United States Air Force (1978)
	Center for Nuclear Studies (1980)
Sequence of training	McCluskey (1972)
Use of feedback and guidance	Prophet *et al.* (1981)
Instructional features	Prophet *et al.* (1981)
Instructional training	Prophet *et al.* (1981)
Simple (part task) versus complex (whole task) simulators	Muckler *et al.* (1959)
	Smode *et al.* (1966)
	National Research Council (1975)
Generic versus specific simulation	Center for Nuclear Studies (1980)
9. *Other*	
Documentation and use of lessons learned from past simulators	Smode *et al.* (1966)
	Caro (1977)
Experimental design	Muckler *et al.* (1959)
	Williges *et al.* (1973)

response is to be made to a stimulus, then the stimulus and the control for the response to it must be in the simulator; transfer is greatest when the similarity between the simulator and the actual equipment is high, although transfer is possible with low similarity.

Micheli (1972) makes the point that it may be more important how a device is used than how it is designed. The training effectiveness of a simulator is a function of the total training environment and not merely the characteristics of a particular piece of equipment. Wheaton *et al.* (1976) concluded after an extensive review of the fidelity literature that fidelity *per se* is not sufficient to predict device effectiveness and that the effect of fidelity varies as a function of the type of task to be trained.

These principles have limited usefulness. A possible reason for this is that no one has managed to define objectively in physical terms and to quantify what simulation fidelity actually consists of in a simulator, for example, whether it is number of identical components, number of displays in common with the operational equipment, or common units of information. What is needed are quantitative equations relating amount of fidelity (however defined) to personnel performance. Ideally, one would develop a number of simulators differing in amount of fidelity and then test personnel training and performance. However, this is impractical because of the costs involved, but if one could operationalize the amount of fidelity in the form of a metric, one could differentiate the available simulators in terms of that metric and correlate the metric with the resultant personnel performance.

Simulator development and validation

The development of a simulator, particularly a training simulator, is not or should not, be simply a matter of requiring engineers to reproduce an operational system. The proper development of a simulator begins with an analysis of what one wishes the simulator to do. This includes an analysis of the system mission, its tasks, and most particularly, what is supposed to be trained. If, for example, the simulator is designed to teach nomenclature and basic principles of equipment operation, it need not replicate the complete operational equipment. Unfortunately, many trainers are not used for the purposes for which they were ostensibly built, and many simulators satisfy their training objectives only in part.

The logical endpoint of simulator development is test and evaluation. Since the simulator is only a tool, its validation is merely verification that it accomplishes the purpose for which it was developed. Adams (1979) has pointed out that one cannot utilize high physical fidelity as a measure of simulator effectiveness. Validation is especially important for trainers because what goes on in the simulator, e.g. the number of hours spent in practice, is not necessarily the same as skill acquisition. Validation of a research simulator requires that operational personnel perform in the simulator as they do in the operational system.

The most meaningful method of evaluating the effectiveness of a trainer is by means of transfer of training studies, because the objective of training is the transfer from training to the operational job of the skills required to perform the latter. The transfer of training paradigm requires two groups of trainees: one (experimental) that receives training on the device before proceeding to perform on the operational equipment, and another (control) that receives an equivalent amount of training but only on the operational equipment. Both groups are tested on the operational system and the effectiveness criterion is performance on that system. If the experimental group performs as well as, or preferably, better than the control, the simulator is presumed to be effective. Equal performance for the two groups may represent success for the simulator, since the device is to be preferred because the cost of training with it is usually much less than that involved in using the operational equipment. The two groups must be comparable in terms of relevant prior training, experience and skill, and neither group must be permitted to engage in activities likely to influence their performance unequally on criterion tasks.

In order to employ the transfer of training paradigm effectively, measurements should be made to determine the extent to which training objectives have been met. With measures such as training time or number of training trials, it is necessary to determine (1) the training effort (TE) required to learn the job on the operational equipment without the aid of the simulator (TE − SIM), and (2) the training effort needed to learn the job when some of the training is undertaken using a simulator (TE + SIM). The difference between (1) and (2) is a measure of the training resources saved by the use of the simulator. However, the value of any savings must be considered in relation to (3) the amount of training effort required to learn the task in the simulator (TE in SIM). Roscoe (1971) has derived a transfer effectiveness ratio (TER) with the following equation:

$$TER = \frac{(TE-SIM)-(TE + SIM)}{(TE \text{ in } SIM)}$$

The transfer design is particularly advantageous because it is sensitive to both positive and negative transfer effects. Several variations of the transfer design are possible for these (see Rolfe and Caro, 1982; Meister, 1985).

Modelling

A behavioural model of a human–machine system is a symbolic representation of the actions performed by personnel in the operation and maintenance of that system. The representation must allow manipulation of extrinsic and intrinsic variables to permit the determination of the effect of model variables. The model is almost always mathematical and is exercised with a computer.

A model is not a theory. Although models often make use of theories, they are not *per se* theories of behaviour. The purpose of a theory is to describe functional relationships. A model obviously incorporates such relationships, but the model goal is to predict and control the effects resulting from the *manipulation* of model variables. Nor is the model any of the varieties of computer-assisted tools, such as computer aided design (CAD) or computer aided manufacturing (CAM). These latter are not true models because they do not describe systems, nor perform the missions these systems perform. Although CAD, for example, utilizes a computer and a software program, CAD is involved with only a small part of a total system, like a control panel, nor does it exercise that part in performing a system task.

The criteria that an effective model must satisfy are listed in Table 8.2.

Model uses

The reasons why models are developed are much the same as those for developing a simulator. Models become necessary when the actual system or a physical simulation are not available, which is usually the case when a new system is being developed and there is need to explore system parameters whose values are unknown. For example, the parameters of a new system under development can be exercised on a computer to suggest how well the future system will perform under specified conditions. In this respect the model is used the same way in which a wind tunnel is used for examining wind stresses on aircraft.

One also uses a model when the phenomena and conditions being studied are too complex to be studied operationally, e.g. predictions of available manpower in the future. Modelling has certain advantages over the real world; for example, it can reduce the latter's complexity by holding certain variables constant and varying others systematically, in much the same way in which an experiment is designed. In fact, most uses of a model are experimental. Although their functions overlap, most physical simulations are for training and most models are for research, although the research is usually highly applied, supporting concrete systems development and operations.

Table 8.2. Criteria of model effectiveness

Criterion	Definition
Validity	Agreement of model outputs with actual system performance
Utility	The model's ability to accomplish the objectives for which it was developed
Reliability	The ability of various users to apply the model with reasonable consistency and to achieve comparable results when applied to similar systems
Comprehensiveness	Applicability to various types of systems, to various kinds of system devices and to various stages of design
Objectivity	Requires as few subjective judgments as possible
Structure	Explicitly defined and described in detail
Ease of Use	Ease with which analyst can readily prepare data for, apply and extract understandable results
Cost of development/use	Includes both time and money
Richness of output	Number and type of output variables and forms of presentation

An effective model is developed for a specific purpose and has certain goals that it is intended to achieve. The assumptions that underlie model operations, for instance, that workload is some specified function of time required and time available to perform a job (see Siegel and Wolf, 1969), must be clearly stated so that they can be examined critically. The model also has to have clear-cut rules for, among other things, specifying the kind of data it requires, as well as for exercising system functions and for synthesizing the effects of system operations.

When used for support of new systems under development or system modification, behavioural models enable one to answer the following questions: (1) Will personnel be able to complete all required tasks within the time allotted? (2) Where during system operations will personnel be most over- or underloaded, and where are they most likely to fail? (3) How will task restructuring or reallocation of functions affect system functioning? (4) How much will performance degrade when personnel are fatigued or stressed? (5) How will the system's extrinsic and intrinsic variables affect operator and system performance?

Types of models

Models are of various types, the most important of which are task–network or event-driven models, control-theoretic or manual control models, and microprocess or deterministic models. Cognitive models are presently largely experimental.

Task-network models

Task–network or event-driven simulations sequentially simulate the performance of the subtasks of a given task or the events to be performed during a specified segment of activity. The goal of network techniques is to go from a statement of the functional relationships among individual task elements to an estimate of the performance of some defined aggregate of those elements (the task, the job or the system as a whole). The performance parameters of interest are typically one or more of the following: (1) the time required to complete the task aggregate; (2) the probability that the task aggregate will be completed in a given time; (3) the probability that the aggregate will be completed correctly.

All network techniques require a thorough description (frequently in diagrammatic form) of actual or intended tasks and of the interrelationships among those tasks. This means that one must decompose the total system into its component tasks and subtasks (the so-called 'top-down' approach). All models are data intensive, requiring measures of performance of each task included in the aggregate, either in the form of measures of central tendency or in the form of various distributions, e.g. normal, Poisson. The lack of data creates serious difficulties for the model developer. There is also a problem of how one combines point estimates or data distributions of individual tasks to derive data describing total system performance.

· The network approach has not been suitable for modelling of continuous activities (this has been left to the control-theoretical model). There appears however to be no inherent limitation to continuous task modelling via network formulations, since SAINT (Pritsker *et al.*, 1974) has been successfully extended to include certain families of continuous variables. The networks are not models of human performance, though they describe system structures in which performance is embedded.

Two major network models are Siegel–Wolf and SAINT. The Siegel–Wolf (Siegel and Wolf, 1969) technique, which was the first to systematically exploit computer simulation of human performance as a systems analysis tool, drew on PERT (*Program Evaluation Review Technique*) concepts, in which a project is conceived to be made up of a network of tasks and subtasks, each of which has estimated completion times—or time distributions—and probabilities of successful completion. To this Siegel–Wolf added more sophisticated, psychologically-oriented concepts to examine the impact of such factors as task-induced stress. This technology proved sufficiently attractive that the US Air Force sponsored the development of a special purpose simulation language, SAINT, specifically for the purpose of implementing network-based human performance models. SAINT has had wide acceptance and usage.

The basic approach to modelling represented in SAINT and other event-oriented modelling languages is to analyse the system flow chart into a series of processes related to each other through a contingently branching structure. A specified distribution of completion times and a probability of successful completion is assigned to each process. These values may be modified while the model is being run. Each time the model is exercised, a sample of each random variable is chosen and the contingent effects are assessed so that a unique trace through the multi-path flow chart is completed.

After a task has been completed, a decision is made as to which task should next be selected. Five decision rules are implemented in SAINT: (1) *deterministic*, in which all *n* branches are selected; (2) *probabilistic*, in which one branch is selected on the basis of a random number drawn from a uniform distribution; (3) *conditional, take first*, in which the first branch satisfying a specified condition is selected; (4) *conditional, take all*, similar to the preceding rule except that all branches satisfying a stated condition are taken; (5) *modified probabilistic*, similar to (2) above, except that branch probabilities are qualified by the number of previous completions of the task from which the branches emanate. Aggregated values for completion time and probability of completion are then derived from repeated runs of the simulation.

Control-theoretic models

The use of control-theoretic or manual control models is restricted to systems in which tracking plays a major role. In control theory the interactions between the operator and the system are represented by servo-control models. Unlike the task-network models, control theory models have a limited concept of human performance—an operator behaves in such a way that errors are minimized within fixed performance constraints. In this concept the operator is an information processing and control-decision element who relates to the system in closed-loop fashion. Feedback is central, comparing actual response with predicted or desired response. Control-theoretic models are more quantitative than other model types. Because of the explicit nature of their assumptions, inputs and outputs, they have been more thoroughly and carefully validated. The models do not, however, attempt to deal with discrete operator inputs, with monitoring or decision making, nor with the procedural aspects of tasks, all of which makes them somewhat unsuitable for describing total job performance.

Microprocess models

The outstanding example of a microprocess or deterministic model is the *Human Operator Simulator* (HOS; Strieb and Wherry, 1979). Microprocess models are very detailed representations of the operator in terms of the physical and psychological processes that are involved in carrying out a task. These are 'bottom-up' models, which is to say, they synthesize larger segments of human performance from a sequence of molecular fundamental activities such as bodily movements, reaction times, recall of events, and so on. Consequently they encounter the difficulty that the molecular phenomena they model must be combined in some manner to represent the higher order task.

In contrast to task-network and control-theoretical models microprocess models assume an operator's behaviour is explainable and not random and that an operator's actions and the times those actions will take are determined fully by the state of the system and the operator's goal at any particular point in time. Microprocess models are basically deterministic (although individual microprocesses may contain random components). In another striking contrast to the other models, HOS assumes that trained operators rarely forget procedures or make procedural errors. This means that HOS avoids the problems of dealing with error processes. HOS internally constructs an activity by using micromodels of behaviour describing information absorption, recall of information, mental calculation, decision making, anatomical movements, control manipulation and relaxation.

Cognitive models are, properly speaking, problem solving or diagnostic models in which the behaviours being modelled are those involved in, for instance, determining the cause of an equipment malfunction or diagnosing the cause of an illness. What is modelled is: (1) the task environment—that is, the objective problem to be solved—the rules to be applied in the solution, the information representing the status of the system; and (2) the program developed to solve the problem. Material for development of the model is often derived from verbal protocols of personnel performing diagnostic tasks.

Stages of model development

These include the following steps, with the understanding of course that the process of model development, like system development itself, is iterative, possessing multiple feedback loops. The following section is taken largely from Chubb *et al.* (1987).

(1) *Problem formulation*: the developer must ask, what questions must the model answer? (2) *Model building*: system operations must be decomposed into tasks, subtasks, events, etc. An appropriate simulation language must be selected. (3) *Data acquisition*: data to apply to algorithms, e.g. estimates of task duration and probabilities of task success completion, must be gathered. (4) *Model translation*: the model must be prepared for computer processing in accordance with whatever simulation language is selected. (5) *Verification*: the developer must demonstrate that model operations and outputs correspond to actual system operations and outputs. (6) *Exercise planning*: the developer must establish the experimental (design) conditions for running the model. (7) *Exercise implementation*: the developer must run the model on a computer. (8) *Analysis* of run results. (9) *Utilization*: the developer or someone else implements the decisions resulting from the model exercise. (10) *Documentation*: the developer describes the model and the results of the exercise in written form.

Model validation

Since the model is only a representation of reality (not the reality itself), the effectiveness of that representation must be demonstrated. Tests of model validity may range from informal demonstrations that it produces reasonable results to formal tests of how well model predictions fit independently derived data.

The problem of model validation is a difficult one. In one sense a model is valid if it can be used to arrive at reasonable decisions; accuracy may not be very important. More often, one looks for experimental, quantitative validation or tests of model accuracy. In this case it is necessary to compare experimental results with model predictions and to apply both engineering and formal statistical tests of the null hypothesis to determine whether or not the model should be considered valid.

If the model is to be useful as a design tool, it must be validated prior to, or at least concurrently with, the development of a system simulation. Under these conditions, system performance data against which to validate the model may not be available, in which case one must seek less rigorous means of examining model validity. If, in the course of building the model, the developer generates hypotheses about how behaviour will change with changes in critical system parameters, at a minimum he or she should be able to predict the *direction* of changes in system performance measures, if not their magnitude (Pew *et al.*, 1977).

Simulation approaches

To select a model one must consider whether the activities being modelled are discrete, continuous or combined.

(1) *Discrete* event simulation occurs when the dependent system variables change by fixed amounts at specified points in simulated time, referred to as 'event' times. The system is modified and time updated only by the occurrence of an event (normally the start or completion of an operator task in human operator modelling). To build an event model one defines the variables that portray the system's status and then identifies those events that can change system status; this is done by defining system status. The state of a system is defined by the values assigned to the attributes of system entities; these last can be, for example, physical objectives, personnel, and so on.

(2) In a *continuous* simulation model system state is represented by continuously changing dependent variables (commonly referred to as 'state variables'). This model describes only tasks involving continuous monitoring or control (e.g. tracking or maintaining process variables with narrow tolerances over substantial time periods). However, many task-network models involve continuous simulation components which lead to combined models.

(3) In *combined* (discrete-continuous) models the variables in the model may change both discretely and continuously. The behaviour of the system model is simulated by recomputing the values of the continuous variables into small time steps and recomputing the values of discrete event variables.

Selection of model variables

Since there are usually more variables than it is feasible to include in any individual model, a choice must be made. Criteria of variable selection (modified from Siegel and Wolf, 1981) suggest that a preferred variable is one which (a) is backed by available empirical data; (b) has high data reliability; (c) is sensitive to changes in system dynamics; (d) is capable of empirical (objective) measurement; (e) is free of unwarranted assumptions, excessive processing time or requirements for large amounts of memory storage; (f) is generalizable to a range of modelled situations; (g) is easily understood by model users; and (h) is most useful for answering the questions the user of the model wishes to ask. Of course, few variables ever satisfy all of the above criteria to the extent desired.

Simulation languages

This section also leans heavily on Chubb *et al.* (1987). The selection of a simulation language is frequently based on convenience (i.e. familiarity with the language and its availability). However, the important factors to consider in comparing simulation languages are the training required to use the language, its ease of coding and debugging, its portability from one model type to another, its flexibility, statistical capability and report production capabilities, its reliability of software, and speed of execution.

Special languages have been developed to fit the various types of models. Nevertheless, a simulation can sometimes be implemented by more than one language.

Discrete languages

The general purpose simulation system (GPSS) exists in a number of variations, GPSS/360 and GPSS/H being most widely used. The principal appeal of GPSS is its modelling simplicity. A GPSS model is constructed by combining sets of standard blocks into a block diagram that defines the logical structure of the system. Entities are represented in GPSS as transactions that move sequentially from block to block as the simulation proceeds. GPSS provides almost all basic simulation functions and has extensive data collection and summarization capabilities. On the other hand, GPSS works more slowly (and is consequently more expensive to run) and has a limited capability for generating random variates.

Q-GERT, developed by Pritsker (1979) has a network consisting of nodes and branches. A branch represents an activity that models a processing time or delay. Nodes are used to separate branches and to model decision points milestones and queues. Flowing through the network are entities called 'translations'. The procedures for constructing a model in Q-GERT are much the same as those for GPSS.

SIMSCRIPT developed by Kiviat *et al.* (1969) has five levels, ranging from a simple teaching language to introduce programming concepts, to statement types that are comparable in power to FORTRAN, ALGOL or PL/1, to levels that utilize entity, attribute and set concepts and provide, for example, for time advance and event processing. SIMSCRIPT is primarily event oriented, system state being defined by entities, their associated attributes and groupings of entities known as 'sets'. System structure is described by defining the changes that occur at event times. SIMSCRIPT is particularly attractive because of its free-form and English-like syntax. Like GPSS, SIMSCRIPT requires its own compiler and is therefore not always available at every computer facility.

Continuous languages

A wide variety of continuous system simulation languages (CSSL) has been developed. Most recent CSSLs are equation-oriented and have FORTRAN-like syntax (see Graybeal and Pooch, 1980).

Combined languages

Only the general activity simulation program (GASP IV) is widely used. The more recently developed Simulation Language for Alternative Modelling (SLAM) is based on GASP IV. GASP IV allows system descriptions to be written either as discrete event models, continuous models, or a combination of the two (Pritsker, 1974). It specifies procedures for writing differential or difference equations as well as methods for defining the logical conditions that affect system status variables. SLAM II (Pritsker, 1984) incorporates the process-oriented features of Q-GERT and the combined discrete-continuous features of GASP IV. SLAM employs a network structure composed of nodes and branches, which model elements in process, such as queues, servers and decision points. Their symbols are combined into a network model that pictorially represents the system with system entities flowing through the network model.

In concluding this section on simulation languages, special attention must be drawn to Micro-SAINT, a microcomputer version of SAINT (Laughery, 1985), which captures many of the features of SAINT but is designed to represent operator task networks in a more straightforward and simple manner.

Further reading

It is impossible within one short chapter to describe fully all the variables that enter into two such complex technologies as simulation and modelling. To assist readers in further explorations of these topics it is suggested that they consult the following: Naylor (1969), Emshoff and Sisson (1970), Van Horn (1971), Reitman (1971), Shannon (1975), Wilson and Pritsker (1982), and Card *et al.* (1983).

References

Adams, J. A. (1979). On the evaluation of training devices. *Human Factors*, **21**, 711–720.

Adams, J.A. (1973). Preface. Special issue on flight simulation. *Human Factors*, **15**, 501.

Beare, A.N. and Dorris, R.E. (1983). A simulator-based study of human errors in nuclear power plant control room tasks. *Proceedings of the Human Factors Society Annual Meeting*, pp.170–174.

Buchaca, N.J. (1979). *Models and Mockups as Design Aids*. Technical Document 266/Revision A (San Diego, CA: Naval Ocean Systems Center).

Card, S.K., Moran, T.P. and Newell, A. (1983). *The Psychology of Human–Computer Interaction* (Hilldale, NJ: Lawrence Erlbaum).

Caro, P.W. (1977). *Some Factors Influencing Transfer of Simulator Training*. HUMRRO Technical Report TR-77-2 (Alexandria, VA: Human Resources Research Organization).

Center for Nuclear Studies (1980). *Nuclear Power Simulators: Their use in Operator Training and Requalification*. Report B0421-8 (Oak Ridge, TN: US Department of Energy).

Chubb, G.P., Laughery, Jr., K.R. and Pritsker, A.A.B. (1987). Simulating manned systems. In *Handbook of Human Factors*, edited by G.S. Salvendy (New York: John Wiley), pp. 1298–1327.

Cunningham, J.B. (1984). Assumptions underlying the use of different types of simulation. *Simulation and Games*, **15**, 213–234.

Emshoff, J.P. and Sisson, R.L. (1970). *Design and Use of Computer Simulation Models* (London: MacMillan).

Gaffney, M.E. (1981). Bridge simulation: trends and comparisons. In *Proceedings of the U.S. Institute of Navigation Annual Meeting*, Washington, D.C.

Gagne, R.M. (1954). Training devices and simulators: some research issues. *American Psychologist*, **9**, 95–107.

Graybeal, W. and Pooch, U.W. (1980). *Simulation: Principles and Methods* (Cambridge, Massachusetts: Winthrop).

Hayes-Roth, F., Waterman, D.A., and Lenat, D.B., 1983, *Building Expert Systems* (Reading, MA: Addison-Wesley).

Hays, R.T. (Ed.) (1981). *Research Issues in the Determination of Simulator Fidelity*. Technical Report 547 (Alexandria, VA: US Army Research Institute).

Hennessey, R.T., Sullivan, D.J. and Cooles, H.D. (1980). *Critical Research Issues and Visual System Requirements for a V/STOL Training Research Simulator*. Report NAVTRAEQUIPCEN 78-C-0076-1 (Orlando, FL: Naval Training Equipment Center).

Hollan, J.D., Hutchins, E.L. and Weitzman, L. (1984). STEAMER: an interactive inspectable simulation-based training system. *The AI Magazine*, **5**, 15–27.

Huff, E.M. and Nagel, D.C. (1975). Psychological aspects of aeronautical flight simulation. *American Psychologist*, **30**, 426–439.

Human Factors Operations Research Laboratories (1953). *Flight Simulation Utilization Handbook*. Report 43 (Bolling Air Force Base, Washington, DC: Human Factors Operations Research Laboratories).

Jones, E.R., Hennessey, R.T. and Deutsch, S. (Eds) (1985). Committee on Human Factors, National Research Council. *Human Factors Aspects of Simulation* (Washington, DC: National Academy Press).

Kinkade, R.G. and Wheaton, G.R. (1972). Training device design. In *Human Engineering Guide to Equipment Design*, edited by H.P. Van Cott and

R.G. Kinkade (Washington, DC: U.S. Government Printing Office), pp. 667–699.

Kiviat, P.J., Villaneuva, R. and Markowitz, H. (1969). *The SIMSCRIPT II Programming Language* (Englewood Cliffs, NJ: Prentice-Hall).

Kraft, C.L., Anderson, C.D. and Elworth, C.L. (1980). *Psychophysical Criteria for Visual Simulation Systems.* Report AFHRL-TR-79-30 (Williams Air Force Base, AZ: Air Force Human Resources Laboratory).

Laughery, K.R. (1985). Network modeling on microcomputers. *Simulation,* **38**, 10–16.

Matheny, W.G. (1975). Investigation of the performance equivalence method for determining training simulator and training methods requirements, AIAA paper 75–108. *AIAA 13th Aerospace Science Meeting,* Pasadena, California, American Institute of Aeronautics and Astronautics.

McCluskey, M.R. (1972). *Perspectives on Simulation and Miniaturization.* CONARC Training Workshop, Fort Gordon, Georgia.

Meister, D. (1985). *Behavioral Analysis and Measurement Methods* (New York: John Wiley).

Mellichamp, J.M. and Wahab, A.F.A. (1987). Process planning simulation: an FMS modelling tool for engineers. *Simulation,* **48**, 186–192.

Micheli, G.S. (1972). *Analysis of the Transfer of Training, Substitution and Fidelity of Simulation of Transfer Equipment.* Report 2 (Orlando, FL: Training Analysis and Evaluation Group, Naval Training Equipment Center).

Miller, R.B. (1954). *Psychological Considerations in the Design of Training Equipment.* Technical Report 54–563 (Wright-Patterson Air Force Base, OH: Wright Air Development Center).

Muckler, F.A., Nygaard, J.E., O'Kelly, L.L. and Williams, A.C. (1959). *Psychological Variables in the Design of Flight Simulators for Training.* Technical Report 56–369 (Wright Patterson Air Force Base, OH: Wright Air Development Center).

National Research Council (1975). *Visual Elements in Flight Simulation.* Assembly of Behavioral and Social Sciences (Washington, DC: National Academy of Sciences).

National Research Council (1982). *Automation in Combat Aircraft.* Air Force Studies Board, Commission on Engineering and Technical Systems (Washington, DC: National Academy Press).

NATO-AGARD (1980). *Fidelity of Simulation for Pilot Training.* Advisory Group for Aerospace Research and Development, Advisory Report 159 (Neuilly sur Seine, France).

NATO-AGARD (1981). *Characteristics of Flight Simulator Visual Systems.* Advisory Group for Aerospace Research and Development, Advisory Report 164 (Neuilly sur Seine, France).

Naylor, T.H. (1969). *The Design of Computer Simulation Experiments* (Durham, NC: Duke University Press).

Park, C.S. and Noh, Y.D. (1987). A port simulation model for bulk cargo operations. *Simulation,* **48**, 236–246.

Parks, D.L. and Springer, W.E. (1975). *Human Factors Engineering Analytic Process Definition and Criterion Development for CAFES.* Report D-180-

18750-1 (Seattle, WA: Boeing Aerospace Company).

Parsons, H.M. (1972). *Man Machine System Experiments*. (Baltimore, MD: The Johns Hopkins University Press).

Pew, R.W., Baron, S., Feerher, C.E. and Miller, D.C. (1977). *Critical Review and Analysis of Performance Models Applicable to Man–machine Systems Evaluation*. BBN Report 3446 (Cambridge, MA: Bolt, Beranek and Newman).

Pritsker, A.A.B. (1974). *The GASP-IV Simulation Language* (New York: John Wiley).

Pritsker, A.A.B. (1979). *Modeling and Analysis of Q-GERT Networks* (New York: Halstead).

Pritsker, A.A.B. (1984). *Introduction to Simulation and SLAM II* (New York: John Wiley).

Pritsker, A.A.B., Wortman, D.B., Seum, C.S., Chubb, G.P. and Seifert, D.J. (1974). *Systems Analysis of Integrated Networks of Tasks, SAINT*, Volume I. Report AMRL-TR-78-126 (Wright-Patterson Air Force Base, OH: Aerospace Medical Research Laboratory).

Prophet, W.W., Shelnutt, J.B. and Spears, W.D. (1981). *Simulator Training Requirements and Effectiveness Study (STRES): Future Research Needs*. Report AFHRL-TR-80-37 (Brooks Air Force Base, TX: Air Force Human Resources Laboratory).

Reitman, J. (1971). *Computer Simulation Applications* (New York: John Wiley).

Rolfe, J.M. and Caro, P.W. (1982). Determining the training effectiveness of flight simulators: some basic issues and practical developments. *Applied Ergonomics*, **13**, 243–250.

Roscoe, S.N. (1971). Incremental transfer effectiveness. *Human Factors*, **13**, 561–567.

Shannon, R.E. (1975). *System Simulation: The Art and Science* (Englewood Cliffs, NJ: Prentice-Hall).

Siegel, A.I. and Wolf, J.J. (1969). *Man–Machine Simulation Models: Psychosocial and Performance Interactions* (New York: John Wiley).

Siegel, A.I. and Wolf, J.J. (1981). *Digital behavioral simulation—state-of-the-art and implications*. (Wayne, PA: Applied Psychological Services).

Smode, A.F., Hall, E.R. and Meyer, D.E. (1966). *An Assessment of Research Relevant to Pilot Training*. Report AMRL-TR-66-196 (Wright-Patterson Air Force Base, OH: Aeromedical Research Laboratories).

Streib, M.I. and Wherry, R.J (1979). *An Introduction to the Human Operator Simulator*. Technical Report 1400.02-D (Willow Grove, PA: Analytics).

Thorpe, J.A., Varney, N.C., McFadden, R.W., LeMaster, W.D. and Short, L.H. (1978). *Training Effectiveness of Three Types of Visual Systems for KC-135 Flight Simulators*. Report AFHRL-TR-78-16 (Williams Air Force Base, AZ: Human Resources Laboratory).

United States Air Force (1978). Scientific Advisory Board. *Ad Hoc Committee on Simulation Technology* (Washington, DC: US Department of Defense).

Van Horn, R.L (1971). Validation of simulation results. *Management Science*, **17**, 247–258.

Waag, W.L. (1981). *Training Effectiveness of Visual and Motion Simulation*. Report AFHRL-TR-79-7 (Brooks Air Force Base, TX: Human Resources Laboratory).

Wheaton, G.R., Rose, A.M., Fingerman, P.W., Korotkin, A.L. and Holding, D.H. (1976). *Evaluation of the Effectiveness of Training Devices*. Research Memorandum 76-6 (Alexandria, VA: US Army Research Institute).

Williges, B.H., Roscoe, S.N. and Williges, R.C. (1973). Synthetic flight training revisited. *Human Factors*, **15**, 543–560.

Wilson, J.R. and Pritsker, A.A.B. (1982). Computer simulation. In *Handbook of Industrial Engineering*, edited by G.S. Salvendy (New York: John Wiley).

Chapter 9

Computerized data collection in ergonomics

Colin G. Drury

Introduction

We live in an era where serious data collection *without* a computer sounds slightly perverse, yet even 10 years ago the computer was considered novel in ergonomics data collection, and 20 years ago only well-equipped laboratories and military test and evaluation sites had on-line systems. Three legitimate questions can be asked to help us understand this phenomenon, which has the proportions of a revolution:
1. How did we collect data in the past?
2. Where is computer-aided data collection now?
3. What new techniques and capabilities await the future ergonomist?

Data analysis is almost never performed without a computer and many of the more complex and precise analysis methods available would be impossible, costly and time-consuming by any other method. There is no need to review past data analysis methods, but the second and third questions above apply equally to data analysis.

Obviously, to provide comprehensive answers to these questions would require a distinguished panel of science historians, computer scientists and futurists. The aim of this chapter is somewhat more modest—to provide a review from the viewpoint of the working ergonomist who is a consumer of computer hardware and software rather than a computer specialist. Data collection will be considered first, followed by data analysis.

Data collection

Historical data collection techniques

Ergonomics has its roots in psychology, human physiology and industrial engineering, so that our primary instruments have been (like ergonomics itself) imported from other disciplines. All data collection in experiments involves providing the subject with an input (stimulus, challenge) and observing the results (response, output). Our primary measuring tools were thus time, force and distance measuring devices and chemical analysis systems, for example the stop-watch, metre rule, dynamometer and oxygen analyser. Unfortunately, many phenomena of interest such as reaction times, tracking errors and force transients happen too quickly or over too small a physical range to be measured conveniently in this way. Two solutions are possible: measure many examples of the phenomenon (repeated trials) or find more precise measuring equipment.

A good example of timing multiple trials is the use of reciprocal tapping by Fitts (and many others since) to study the speed/accuracy trade off in target aiming tasks. Another example is the use of the pursuit rotor which counts the number of times (or total time) a small target is lost rather than measuring the more complex position/time history of target and follower. The 'multiple trials' strategy is still used in class demonstration experiments and in measuring industrial performance from production records.

More precise measuring equipment for small times and distances typically involved mechanical or electrical amplification and a time-record imprint of the measure onto a continuously moving chart. Times in the millisecond range could be obtained from paper chart recorders operating sufficiently rapidly, as could transients in physiological phenomena. Strip-chart recorders were produced in increasing complexity, with less intrusive instrument dynamics and multiple channels. However, the data transcription process can be lengthy, tedious and error-prone. The advent of high quality tape recorders (including multi-channel FM recorders) did not solve the problem and only allowed the user to extend the time scale for recording very rapid events (e.g. fine-structure of EMGs) or very slow events (e.g. diurnal variations in physiological output and performance).

As will be discussed later, the purpose of data collection is ultimately data interpretation, so that the analysis stage cannot be ignored. The rise of statistical design and analysis methods for experimentation during the 1930s to 50s meant an increased demand for data collection in sufficient volume to provide statistical significance. Typically, this required larger sample sizes and analysis of individual data points rather than mean values. Happily, digital computers made their debut during this time, allowing the experimenter to attempt complex analyses of variance and regression in a reasonable time frame. However, the burden of transcription between the data collection device and the computer input device still fell onto the experimenter or onto technicians. The dedication of an expensive computer to control an experiment and collect the data was only possible with considerable money and technical help, usually only available to answer military questions.

For the laboratory experimenter in general, the rise of transistors enabled logical elements to be used in relatively inexpensive, general purpose control systems. Input devices, such as response keys, photocells and voice-sensing units, could be interfaced with logical elements such as steppers, relays and

AND/OR/NOT gates and with recording devices such as timers and counters. These same logic systems could control more complex stimulus presentation devices such as coloured light displays, numerals and even slide projectors, but the outputs were still not readable by computers.

One exception was the series of event timers and recorders which produced punched paper tape output. SETAR, and later METRA, recorded input events and the times at which they occurred, with a speed of up to 10–30 events/s. Although of low reliability when these data rates were temporarily exceeded, these machines enabled experimenters to collect complex event timing data over long periods and removed the chore of data transcription. At this point the experimenter needed computer programming skills to interpret the paper tape, particularly when dealing with the inevitable errors. Experimental data collection and data interpretation were at last coming together.

Current data collection techniques

Perhaps the real revolutions in data collection over the last 20 years have been caused by dedication of the computer itself to laboratory experimentation and, more recently, by the ability of different computers to exchange data.

To earlier experimenters 'computer dedication' meant receiving a dispensation from the operating staff of a mainframe computer to take over sole running of the machine during particularly inconvenient hours. With the advent of minicomputers it became feasible for a department, or even a laboratory, to have its own computer and hence to dedicate it to real-time data acquisition and experimental control. All that has changed in the past few years is that the price of computing power has decreased by about two orders of magnitude, and it has been recognized that the software for computer use has to be used by people other than computer enthusiasts. Microcomputers are now available in large numbers at prices even universities can consider. High-level languages have evolved which can access real-world data as well as provide display functions and calculation abilities. At one end of the scale languages such as 'BASIC' make relatively crude programmers out of experimenters, while at the other end of the scale highly structured languages such as 'C', 'PASCAL' and 'MODULA2' enable complex programs to be written by teams rather than individuals. Even older languages such as 'FORTRAN' now have the ability to display graphics (e.g. DI-3000) and communicate with external devices. Programming environments now include editors, debuggers and compilers with relatively 'friendly' user interfaces.

Networking of computers is another possibility opened up by cheap computing power and the advent of various character- and file-transmission standards. Most major computing centres have data networks which allow the small, dedicated laboratory computers to exchange files with large multiuser mainframes. Hence data collected on a machine of appropriate size for collection (mini- or microcomputer) can be analysed using a machine of appropriate size to handle massive data handling packages (mainframe or super-computer).

We now can run an experiment in a totally computer-controlled environment. Instructions, stimulus timing and feedback can all be handled by the same machine as that which records the data. In this way, complex tasks can be undertaken, bringing a sense of realism into the laboratory where appropriate. Recent examples include complex process control tasks and microscope-based inspection tasks, e.g. Synfelt and Brunskill (1986). Certain data collection methods are still relatively expensive, however. Data rates appropriate to multiple-channel physiological recording can be in the tens of thousands of samples per second. These can be difficult to handle on microcomputers, both at data input and at data storage, if collection goes beyond a few seconds. Another expensive example is the recording of three-dimensional position data over a reasonable volume of space at reasonable precision and input rates. Typical systems still cost an order of magnitude more than a microcomputer workstation in an ergonomics laboratory.

The prospect of generating data at rates of thousand of samples per second is both daunting and exciting for the ergonomist. Large, multi-measure experiments are enjoying something of a vogue (Drury and Deeb, 1986a, b) but they give complex and often messy results. In the past, a choice of a single performance or stress measure meant that system evaluation and ergonomics research had a certain happy simplicity. Thus system A performed better than system B at high speeds but not at low speeds. Ergonomists could make the relatively simple statement often demanded of them by others (usually hardware engineers) on the design team. Now with multiple measures, the picture is rarely so clear, even though techniques such as factor analysis can be used to organize data interpretation. A recent experiment on human and algorithm interaction in job-shop scheduling showed overall that humans were a necessary part of any effective system, but the various measures of effectiveness often gave directly conflicting results (Kondakci, 1985). Ergonomists and their clients are having to adapt to this new reality.

The future of data collection

While we continue to bask in our good fortune in working at a time when technological marvels aid us in the laboratory, it is necessary to examine how our tools are likely to evolve and how they will influence the data we collect and the tasks chosen for study. Outside the laboratory, computers are revolutionizing our home and professional lives while word processors have largely replaced typewriters, and spreadsheets have at least complemented calculators. Communications programs allow us to obtain on-line weather, news and shopping services as well as access to vast data bases. Even *Ergonomics Abstracts*, the bibliographic journal published by Taylor and Francis, is moving towards computer access. Today we can sort our electronic mail and fill our friends' mailboxes with electronic memos. For the future, some clear trends can be seen in the short-term.

Portability

As power requirements for integrated circuit chips decrease and the energy densities of batteries increase, portable computing becomes possible. We have used simple portable computers based on 8-bit microprocessors to enhance data collection in the field (Drury, 1986). It is now possible to design a questionnaire on a portable computer, run subjects and edit the questionnaires based on an analysis of the responses. Data collected can be passed to a micro-computer or a mainframe computer for analysis using standard packages. Similarly, a portable computer the size of a small telephone directory and weighing less than 2 kg can provide sophisticated event timing functions and even run an occurrence sampling scheme. We need no longer talk about ergonomists as if they were shackled to a laboratory. The range and complexity of portable computers is increasing. Discounting the 10 kg machines, which need mains power, the portable 16-bit and 32-bit microprocessors are highly compatible with desk-top microcomputers. Should you ever need to recalculate a large spreadsheet on the beach, this is now possible on a machine weighing less than 5 kg.

A major advantage of the simpler portable computers is their ease of programming and interfacing. Analogue and digital interfaces, similarly battery powered, are inexpensive; data transfer to a desk-top or mainframe computer is largely menu-driven. At this level of simplicity, even occasional computer users can write programs (in high-level languages) carefully tailored to the user's needs. And ergonomists can now take to the road without abandoning modern data collection techniques.

Better displays

As microprocessor power increases into the millions of floating-point operations-per-second range, so more computing power is available for the computation-intensive needs of display generation. Coupled with decreasing memory costs, for both RAM and CD-ROM, ergonomists can look forward to being able to use more complex graphics displays. Already the 1000×1000 pixel display is an inexpensive reality. With further advances in computer architecture (e.g. parallel processing) and software (e.g. graphics algorithms) colour and animation techniques, presently only available to Hollywood and the military, will be affordable in the ergonomist's laboratory. Instead of the computer controlling the presentation of complex images to the subject by manipulating slide projectors and video tape, the regular display screens will provide the same resolution with vastly increased flexibility. Incidentally, increased computer power and graphics knowledge are already merging different media. We can capture and digitize actual scenes as well as produce computer-animated scenes on video tape. Computers are already a standard method of display prototyping in the military; using computers and displays only twice to three times the price of desk-top microcomputers, rapid changes can be made in display panels and their performance assessed before the hardware is built.

For the future high resolution colour and animation will allow us to move back from what Crossman (1960) called 'symbolic' displays. Perhaps we can create a new reality for controllers of complex processes, mixing real scenes with colour-coded analogue representation of important parameters, allowing rapid changes to be displayed and even producing predictor displays. All of these remarks apply equally well to the auditory sense. Modern musicians frequently have computers between their instrument and the speakers; perhaps this technology will be found useful in studying human auditory responses and producing complex auditory displays.

Better input devices

The much-loved and much-hated typewriter keyboard is merely the traditional computer input device for human commands and can hardly be hailed as a triumph of ergonomic design. Just as the computer display is moving away from alpha-numerics into the more analogue representations of animated graphics and complex sounds, so inexpensive computing power is allowing us to utilize alternative human input devices. No longer is the required analogue-to-digital conversion and interfacing the province of computer hackers. Currently available are graphics tablets, joysticks, light-pens, trackballs, mice (for foot as well as hand use), and voice entry. The data capture and collection possibilities are seemingly endless. Subjects can make responses directly onto a display screen (light-pen), perform tracking tasks (e.g. joysticks) and give verbal responses. With more specialist input devices such as frame-grabbers and bar-code readers, we can collect data on location much more easily. Keyboards may be retained for many years, but ergonomists are no longer limited by their capabilities.

Intelligent data collection

Just because we *can* collect data does not mean that it *should* be collected. As more complex and realistic ergonomics models are developed, data collection can be guided using these models. In a strictly scientific sense data collection without an underlying model is unproductive, but in a rapidly growing field such as ergonomics how can an experimentalist keep up with the latest models from the underlying disciplines? One solution is the artificial-intelligence (AI) inspired creation of expert systems to advise the practitioner, crude forms of which already exist. Mir (1980) devised an ergonomics audit program to evaluate the level of ergonomics found in the workplaces at a plant. Many of the mathematical models and logical constructs used regularly (e.g. biomechanics) are now having their computer programs repackaged as 'Expert Systems'. Even the humble checklist appears to be headed in this direction.

A better focusing of data collection efforts can be expected as AI concepts become increasingly common. Error taxonomics can be devised with the help of an expert system. Models of human response (e.g. to thermal loads) can be used to predict what measures to take and when to take them. Standard symptoms of industrial malaise (accidents, quality, labour turnover) can be used as input to indicate probable causes and hence guide data collection efforts. Perhaps all ergonomists will carry an annually reprogrammable 'automated aide'.

Intelligent tools

As microcomputers invade more and more products (e.g. telephones, cameras, TV sets), they can easily provide the hardware for automated data capture by the ergonomist. If your automobile has an interface to the repair shop's diagnostic computer, this same interface can be used to selectively monitor the same functions during actual use. There are obvious moral questions concerning privacy, but if the subjects' informed consent is obtained then there is no reason why these interfaces should not become a relatively standard data collection tool. Already a running shoe is available whose on-board chip interfaces with your desk-top computer to analyse your daily run. Office chairs have chips to remember adjustment preferences for several individuals. How long will it be before your hammer indicates whether you are striking the nails correctly, or your pen organizes your work/rest schedule to avoid writer's cramp? More seriously, we are seeing the start of a product design revolution at least equal to the revolution caused by computers in research and industry. The possibilities for ergonomics data collection can only increase.

Data reduction and analysis

We have already alluded to the revolution in data handling caused by computers. If one re-reads experiments in the literature of the 1930s and 40s, the absence of what we now regard as commonplace statistical analyses is striking. Where massive data gathering, reduction, and analysis efforts are reported, for example Blackwell's (1946) studies of vision, one can only marvel at the patience and dedication of the experimenters.

The trend, also noted earlier, is for the integration of data collection, data reduction and data analysis. We have now reached the point where this cycle can be performed rapidly enough to provide meaningful feedback to the experimental subject and even control of the subsequent stimulus presentation. Even in the 1970s, minicomputers made possible the study of adaptive training systems where task difficulty is continuously adjusted to current performance level. The studies by Wiener (1973) in vigilance are good examples.

More traditionally though, the three functions are separated in time so that it is still useful to consider data reduction and data analysis separately.

Data reduction

There are three aspects of data reduction which need to be considered. For data which are not collected directly by computer there must be a data entry stage, followed by data verification and finally, post-processing.

Data entry ease and accuracy should be considered at the experimental design stage. An easily read source document for the subject may not be an easily read source document for the data entry person. For observational and questionnaire data it is important to minimize transcription errors, so a well-designed data entry program is essential. Three alternatives are available (all are assumed to be used on a microcomputer dedicated to the data entry task; mini and mainframe machines can be used, but connect time is usually chargeable; simple file transfer programs are available to allow use of mainframes for later analysis). First, a word processor or text editor can be used to create a data file, typing in the numerical and text entries in order with suitable delimiters. Although error correction will be easy, there is a considerable chance of forgetting the location of each item in the data file. A second, more compatible method, is to use a data base progam where each unit of data (a completed questionnaire, a set of workplace dimensions or observation records) is entered onto a custom-designed data input screen. Figure 9.1 shows such a screen for data entry in a study of postal operations. Some data bases also allow the user to enter data in a tabular format. Figure 9.2 shows further data from a postal study where body angles were entered for later calculation of biomechanical body stresses. Both methods constantly remind the entry person of the layout of the data. If it makes sense, hard copies of the format shown in Figure 9.1 can be used as field data collection forms, giving a one-to-one position correspondence between hard copy and entry screen. Third and finally, data entry programs can be written in a high-level computer language (e.g FORTRAN, C, BASIC) to accept data at particular prompts. Most sophisticated data base progams now have input screen manipulations exceeding the abilities of non-expert programmers. For example, writing a program in FORTRAN to allow full-screen editing is no simple task unless a good library of sub-routines is available.

If this represents the present situation for data entry, what of the future? The aim should be to eliminate the data entry process entirely by computer collection of data as noted previously. However, not all data are as easy to capture as switch closure, analogue signal or key press. For example, although direct data collection in a questionnaire format is possible using a keyboard or menu-and-pointer system, there will always be some subjects who find

```
WORKPLACE EVALUATION. PHASE II
  SUBJECT: 1                    HEIGHT: 61
  SEX: 0                        WEIGHT: 130
  AGE: 39                       MONTHS: 168
  EYES: 1                       SITE: 1
                                INT. MOD: 0


ENVIRONMENT
   1. SUFF SPACE: 0
   2. CROWD: 0
   3. WK ITEMS: 1
   4. STORE: 1
   5. COUNTER: 1
   6. ACCESS: 0
   7. LIGHT: 1
   8. GLARE: 0
   9. NOISE: 1
  10. DISTRACT: 0
  11. TEMP SUM: 9
  12. TEMP WINT: 1
  13. APP WORKPLACE: 1
  14. APP STATION: 0
  15. EQUIPMENT: 0

WORKPLACE/EQUIPMENT
   1. REACH KEYS: 0
   2. READ LABELS: 0
   3. READ SCREEN: 0
  3A. READ SCALE: 0
   4. READ ZIP: 0
   5. CHARACTERS: 0
   6. COUNTER HT: 0
   7. DIFF DRAWERS: 0
   8. WP ADEQUATE: 1
   9. DIFF PACKAGES: 0
  10. REACH SCALE: 0
  11. USE SCALE: 1
  12. EXCESS WALK: 0
  13. EQUIP PROBS: 0
  14. SPACE FORMS: 1
  15. SPACE PARCELS: 0
  16. STORE PARCELS: 2
  17. SLOT LAYOUT: 0
  18. ACCESS FORMS: 0
  19. EXCESS BENDING: 0
  20. EXCESS REACH: 0
```

Figure 9.1. Example of customized input screen for data base entry

that the computer interface is unsuitable. Non-readers and physically handicapped individuals will have problems unless special solutions are available. The computer can be unnecessarily obtrusive in the work situation; even recording with a lap-top portable would be difficult in a trading crowd at a stock exchange, or in a military aircraft. Pencil and paper or voice transcription are likely to remain with us despite our best efforts.

Given that data entry is a necessary evil, what aids can we expect? Computers will certainly have better interfaces to minimize transcription errors and will be faster and better interconnected to speed data on its way

SITE	BEFAFT	SUBJECT	TASK	LR	BACK	UPPER	LOWER
4	1	15	1	1	22	159	69
4	1	15	1	2	−8	156	99
4	1	15	2	1	9	165	105
4	1	15	2	2	−11	137	100
4	1	15	3	1	15	127	105
4	1	15	3	2	−10	152	112
4	1	15	4	1	2	127	99
4	1	15	4	2	3	151	88
4	1	15	5	1	10	140	77
4	1	15	5	2	−18	174	87
4	1	15	6	1	5	162	124
4	1	15	6	2	−9	168	126
4	1	15	7	1	2	166	122
4	1	15	7	2	−6	169	127
4	1	15	8	1	−4	165	107
4	1	15	8	2	−18	165	112
4	1	15	10	1	40	188	150
4	1	15	10	2	−2	159	103

Figure 9.2. Example of tabular data base entry format

to the analysis programs, but these are only small evolutionary changes. The larger changes will come from improved optical character recognition and speech recognition algorithms which allow direct entry of data which has been collected by writing on forms or recording on tape. With such devices, we can become at last independent of one of the least rewarding chores associated with research.

The second aspect of data reduction is verification. In the days of card input, this often referred to having two clerks enter data independently and then comparing cards for errors. Transcription errors were caught unless the two clerks happened to make identical errors. With full-screen editing and instant print-outs, this is now an easier problem to tackle, but it is at times neglected, resulting in unreliable data and hence conclusions.

A broader problem in data verification is the logically impossible or implausible data point. In a student sample, any age having other than two digits is suspect. Similarly, sibling birth dates closer together than the gestation period raise questions of data validity. Heart rates above 200 beats/min and negative blood pressures are other obvious examples, but there are less glaring data errors which must be detected. There is always some question of how far the ergonomist should go in editing data. Unusual values will occur legitimately from time to time, indeed there is a whole branch of statistics concerned with extreme values sampled from populations (Gumbel, 1958). Some students are over 60; premature births do occur; very rapid typists exist to cause outlying data points in word processing studies.

The ideal place for data verification is at the time of recording, when the subject is still present and the data point can be repeated if necessary. All subsequent data verification which involves data editing can be called into question. Only logically impossible data can be removed from the data base and still withstand serious questioning. Reversed dates for starting and ending employment, mechanical impossibilities, or off-scale responses are all 'safe' to remove from the data base, but how can they be replaced? If the error is a transcription error, we can re-enter the data, but if it is a recording error, generally it will not be possible to replace the reading at a later time. A subject's response cannot always be re-collected, as the subject may now have changed, leaving the problem of missing data, which is discussed later.

In the future, data verification has the potential to become more consistent as well as less time-consuming. Current data base programs already have built-in error checking routines, so that data can be tested as they are input to determine whether they meet certain criteria. In this way logical impossibilities are automatically flagged so that a decision can be taken on their inclusion. The same filtering of data can also be applied on-line to computer-recorded data. For example, heart rates based on timing the EKG 'R' wave occasionally count a spurious beat or miss a beat, leading to heart rates double or half the correct value. Data filtering routines can be built into the recording software to detect such anomalies and remove them from the final record. Ease and consistency in such processes can be expected to increase with the more widespread adoption of artificial intelligence techniques for data verification. It should eventually be possible to mimic the decisions of an intelligent and well-informed experimenter to ensure data veracity.

The final step in data reduction is post-processing—the stage between data entry and statistical data analysis. Although many hypotheses, and hence statistical analyses, are based directly on the measured data, others require transformed data. This may be as simple as subtracting the starting date from the leaving date to determine a respondent's period of employment or it may be as complex as calculating the error power spectral density function in a tracking task. Other examples are calculating signal detection theory parameters (A', d', x_c, etc.) from hit and false alarm rates, finding disc compressive forces from a biomechanical model of a manual lifting task, finding envelope measures of EMG, or even calculating heart rates from R–R interval times.

In all cases, data from various data points are combined to give a measure of greater meaning. The main post-processing decision is whether to perform this post-processing on-line as data is being collected or off-line when data is retrieved from a storage medium. On-line processing obviously requires powerful computational resources, but, as the processed data are usually less extensive than the raw data, storage requirements are minimized. As an example, if four muscles are recorded using EMG techniques at 200 Hz, we need to take $4 \times 200 = 800$ data points each second. If, however, we wish to obtain only fatigue data, then the ratio of high-to-low frequency components (e.g. Kroemer, 1984) can be calculated and recorded much less frequently, perhaps twice per second for each muscle. Our data storage requirements are reduced 100-fold, but high-capacity on-line processing is needed to filter the data by frequency and calculate ratios.

The main advantage of on-line post-processing is that it can provide useful feedback to either subject, experimenter or both. The main danger is that the data cannot be used to answer any questions which require different post-processing. To continue our example, level of effort, represented by average or envelope EMG, would not be available if only frequency ratios were recorded. For this reason, recording of the original raw data is always good practice: only record the subject's actual response, particularly in observation or questionnaire data; do not attempt to convert weeks to days just because the questionnaire requires an answer in days and the subject gives the answer in weeks; record original data long-hand and process it later without time pressure.

The future can only bring advantages for post-processing. Higher computing power, data handling rates and storage capacities in computers mean that both original data storage and on-line processing will be available simultaneously.

Data analysis

At this stage, our data are collected, entered into the computer, verified, and the calculations performed ready for analysis. A number of choices are now open for data analysis and all of the following discussion is based on computer analysis, although it is rare that no hand-work is used. Specific analyses may well be beyond the capabilities of the software package: graphs with meaningful but unusual scales may need to be drawn; a rapid calculation of chi-square for a contingency table may be preferable to re-running a batch job on a mainframe. Most ergonomists still own a pocket calculator and a box or two of graph paper.

The typical course of analysis is statistical analysis of data followed by comparison between some non-statistical model and the summarized data. Thus Chapman and Sinclair (1975) ran an experiment on inspection of food products, varying both speed of the conveyor and inspector's time on task. An ANOVA established significant differences between speeds and times on task, so that graphs were plotted to show both the speed/accuracy trade off and the vigilance decrement. Any computer package for data analysis should support both of the activities and most do.

Additionally, the handling of missing data is an important topic in its own right and may be different for different packages. In many packages, missing data within one cell of an ANOVA (e.g. four readings instead of five) can be handled easily, while the simpler packages insist on equal numbers in each cell. In other cases there may be different types of missing data. People may have no answer to a question on their age because they do not remember their age, they refuse to provide this information, the experimeter fails to record it, or an obviously wrong value is entered. A good package will allow the analyst to treat these differently, while still denoting all of them as 'missing'. Some missing data may cause other relevant data to be eliminated; e.g. a missing age in a correlation between job satisfaction score and age would cause that subject's job satisfaction score to be omitted from the correlation calculation. If there are more than two attributes and only one is missing the experimenter must choose between omitting *all* data from the subject with the attribute missing or just omitting data from calculations impossible to perform without that data.

Before considering the statistical packages, it is worth noting that some spreadsheets (e.g. Lotus 1-2-3) and many data base programs (e.g. REFLEX) for microcomputers perform rapid data summarization (means, standard deviations), some statistical functions (regression), and simple graphical plots (line, bar, scatter) which may be all that are required in simple experiments. Given means and standard deviations of two samples, a *t*-test can be performed on a hand calculator without recourse to a statistical package.

The more explicit statistical procedures are usually performed by packages on mainframes, although versions for microcomputers do exist (e.g. SPSS-PC, STAT-PAC). Unless the ergonomist is cut off from mainframe support, current microcomputers are too slow for the majority of complex analyses. Statistical packages try to balance generality against ease of use. The more statistical procedures possible, and the more choices which must be specified before analysis, the more complex the ergonomist's task in performing the analysis.

Major packages in general use include MINITAB, SPSS-X, SAS and BMDP. All will perform *t*-tests, ANOVAs, contingency table analysis and regression. All will provide summary statistics, tabular output, histograms and graphs. Their detailed capabilities will not be compared as new versions are brought out frequently, rendering such comparisons instantly dated. However, each package has its own flavour and uses. MINITAB is a rapid interactive system which is easy to learn and remember between uses. It is ideal, for example, for running regressions on means and plotting the results instantly. However, it is limited in the types of ANOVA which it can tackle and is unsuited for large data bases. SPSS-X was originally developed for social science research so that its forte is in handling multivariate data, for example by cross-tabulation or factor analysis. The days when its ANOVA capabilities were limited are long past. For graphical output, the conventional wisdom points to SAS as the package of choice. Finally, almost any model of ANOVA can be analysed by BMDP, originally a biomedical package.

There is still no universal package which meets all needs, but the ergonomist should not need to learn more than two to handle almost all needs. On-line help facilities have expanded, and interactive processing has supplemented batch processing, but getting the most out of a statistical package is still a process with a lengthy learning period. One final caution is that not all analyses are performed by a package. If the package does not meet your needs and assumptions, analyse by hand. A few hours with a calculator can produce even complex ANOVAs on a few hundred data points, whereas locating the correct package, reading the manuals, and consulting a resident expert can take much longer. Never let the package constraints dictate your experimental design or data analysis.

References

Blackwell, H.R. (1946). Contrast thresholds of the human eye. *Journal of the Optical Society of America*, **36**, 624–643.

Chapman, D.E. and Sinclair, M.A. (1975). Ergonomics of inspection tasks in the food industry. In *Human Reliability in Quality Control*, edited by C.G. Drury, and J.G. Fox (London: Taylor and Francis).

Crossman, E.R.F.W. (1960). *Automation and Skill* (London: HMSO).

Drury, C.G. (1986). Hand held computers to collect field data. *Engineering Progress*, **6**, 10–12.

Drury, C.G. and Deeb, J.M. (1986a). Hand positions and angles in a dynamic lifting task. Part 1, Biomechanical considerations. *Ergonomics*, **29**, 743–768.

Drury, C.G. and Deeb, J.M. (1986b). Hand positions and angles in a dynamic lifting task. Part 2, Psychophysical measures and heart rate. *Ergonomics*, **29**, 769–778.

Fitts, P.M. (1954). The information capacity of the human motor system in controlling the amplitude of movement. *Journal of Experimental Psychology*, **47**, 381–391.

Gumbel, E.J. (1958). *Statistics of Extremes* (New York: Columbia University Press).

Kondakci, S. (1985). *An interactive approach to scheduling of job shops with dual constraints*. Unpublished Ph.D. Dissertation, State University of New York at Buffalo.

Kroemer, K.H.E. (1984). Engineering anthropometry. In *Handbook of Human Factors*, edited by G. Salvendy (New York: John Wiley).

Mir, A. (1980). An ergonomics audit program. Unpublished M.Sc. Thesis, State University of New York at Buffalo.

Synfelt, D.L. and Brunskill, C.T. (1986). Multi-purpose inspection task simulator. *Computers and Industrial Engineering*, **10**, 273–282.

Wiener, E.L. (1973). Adaptive measurement of vigilance performance. *Ergonomics*, **16**, 353–363.

Part III

Techniques in product or system design and evaluation

The first two parts of this book contained chapters describing fundamental groups of methods and techniques, and also implicitly proposed certain approaches to ergonomics study. This third part has within it descriptions of how the basic methods and approaches may be modified and combined in four particular examples of application in design and evaluation—products, text, human–computer interfaces (HCI), and expert systems. In some cases the techniques normally available are not sufficient and so special ones have been developed.

It will be apparent that there is overlap between the four chapters in terms of the type and application of appropriate method, and that much of the methodology delineated has potential value in many other domains. However these four have been selected so as to provide an interrelated set of application domains, and as likely areas of importance into the foreseeable future. Within this whole part we will find all the methods of chapters 2–9, direct and indirect observation, archives, experiments, task analysis, protocol analysis, modelling, simulation and computerized data collection—with explanation of their strengths and weaknesses in the particular circumstances.

We start in chapter 10 with McClelland on an application area that has been of interest to ergonomists for 20 or 30 years or more—the use of user trials. Such an approach has wide applicability but has been particularly valuable in the development and assessment of products and equipment. Linkage with HCI evaluation is obvious and McClelland makes the point that current concern with usability engineering in the whole area of information technology implies amongst other things that user trials must be an integral part of an iterative design process. User trials are then seen as a way of bringing basic measurement principles into the product development process, and that their use provides both quantitative and qualitative information on product effectiveness.

In the product development process described in Figure 1.9 of chapter 1 the parallel development of product and manuals and instructions was proposed. This might seem a little optimistic or excessive but the usability and even market success of a product can be impaired by inadequate instructions for use and maintenance. Nowhere is this more so than in the world of personal computers and other computer technology, where manuals have historically been the subject of mirth or hysteria or both. Hartley then, in chapter 11, summarizes the evaluation of text, describing the sometimes specialized, sometimes general, methods to do this.

At this time much ergonomics effort is being put into the development of human–computer interfaces that match the needs and limitations of users and tasks. After a period of producing guidelines and criteria, the HCI community is now much taxed by the problems of evaluation (related of course to criteria, etc.) and of what should be evaluated. In doing so they are confronting problems, compromises and gaps in knowledge that have faced ergonomists in other domains for years. What is interesting is that the enormous concentration of ergonomics resources in HCI has allowed exploration of the use of virtually any method or technique—from psychological through to physiological measurement—and the combination of some of these into broad-based approaches to HCI evaluation. Out of much of this work are emanating results of interest and significance for all ergonomists.

Christie and Gardiner in chapter 12 give a comprehensive account of the area.

Within the sphere of computer technology one particular advance of great interest to ergonomists is that of expert systems. In particular we have a role to play in attacking what Shadbolt and Burton in chapter 13 call the bottleneck in expert systems—acquiring the knowledge necessary to build any knowledge-based system. A major, and generally *the* major source of such knowledge, will be the experts themselves, and knowledge elicitation is the name for gathering the relevant information. In some ways parallels can be drawn between this and the three preceding chapters in that all the techniques described are to elicit information, opinion and insight from people, in the first three cases from users or user 'representatives', and in the last from those whose skills the system is intended to duplicate, enhance or even replace.

Common to all four methodological approaches, to user trials, text evaluation, HCI evaluation and knowledge elicitation, is the need to understand and allow for who the users will be and what type of tasks they will carry out. McClelland and also Christie and Gardiner address explicitly the question of 'what users?'. Between them they define attributes that need to be identified to specify and characterize the target user group as: personal characteristics; physical, cognitive and attitudinal characteristics; specific skills, training, previous experience; the characteristics of their jobs, and roles; and any personal preferences. Perhaps most critical is to know whether the users will be beginners, novices, regular users or experts, whether use will be frequent or not, and the degree of discretion they have in using the product or system. Shadbolt and Burton coin three descriptors for classes of experts that might be useful in the wider context of user characterization: 'academics', 'practitioners', and 'samurai', the last being pure performance experts with little generalized theory and few heuristics.

The second main underlying need for evaluation or design is to understand what it is the users will have to do, the tasks they must undertake. In most cases task analysis will have allowed such identification, and the user characterization in terms of experience, frequency and discretion of use will also be relevant. The number complexity, sequence, simultaneity, loading and criticality of the tasks will, *inter alia*, be important to the making of design and development decisions, in determining how each of the products or systems covered in this section must be evaluated, and in selecting the appropriate methods and techniques.

Chapter 10

Product assessment and user trials

Ian McClelland

Introduction

This chapter opens by providing an overview of user trials. The second, major, section deals with the basic principles of such exercises whilst the third section considers the increasing role of user trials in product development, with reference to a specific case study. Finally some pointers are given as to future directions for this method of evalution.

What is a user trial?

The fact that it is possible to measure the interaction between people and the products, equipment, environments and services they use is the fundamental principle that underpins all ergonomics. A close second is the principle that these interactions can be measured in ways which can guide the specification, design and evaluation of the artifacts we all use. These principles have given rise to both a considerable body of data and to a range of investigative techniques which are central to the effective implementation of ergonomics. The term user trials started to become common currency in the 1970s as part of a vocabulary more suited to the environment of industrial product development. The challenge was, and still is, to market the idea that 'experiments' with people could be helpful in the design and development of products. But the term 'user trials' is not just a new label for 'experiments'. The continuing use of the term also reflects real changes in the practice of ergonomics; changes aimed at accommodating the pressures and priorities that come from working within a product development environment.

The most significant development in recent years has been the emergence of the concept of 'usability engineering'. This concept has been developed primarily as a vehicle for the application of human factors to products within the 'information technology' area (Gould, 1988; Shackel, 1986, 1987; Whiteside et al., 1988). The central purpose behind 'usability engineering' is to make the process of designing user requirements into products more systematic.

The concept involves firstly, incorporating 'usability goals' into product specifications together with associated user performance requirements that the design must meet. Secondly, user trials should be an integral part of an iterative design process so that a product can be checked against the usability goals and the performance requirements as it is developed. Bewley *et al* (1983), and Boies *et al* (1987), provide good examples of how different forms of user trials contributed to the development of specific products in this way. In general though, products are not developed on this basis at present, but the development of 'usability engineering' principles provides clear guidance on how the concept of user testing can be incorporated into the development of products in a wide range of industries as well as the information technology area.

It can thus be said that user trials have become a way of bringing the basic principles of measurement into the product development process and making them usable within industry and commerce. The central idea is that the application of these principles is product focused; i.e. that we can measure the effectiveness of products both from quantitative and qualitative points of view. User trials can provide information which is at least useful and, at times, essential if products are successfully to accommodate users' requirements. This involves using methods often associated with 'experiments', and indeed, the same basic principles of experimental design, measurement techniques and data analysis that apply to formal experimental work also apply to user trials. Chapanis (1959) describes the experimental method as follows:

> 'It is a series of controlled observations undertaken in an artificial situation with the deliberate manipulation of some variables in order to answer one or more specific hypotheses.'

In summary, user trials may be described in the same way with the one modification; . . . . . to answer one or more specific questions about the effectiveness of the product.

Many of the principles involved will be discussed later in this chapter, but for a more extensive discussion of the principles and practice of test and measurement in human factors work than covered within this chapter the reader is referred to Meister (1986). Bailey (1982), Chapanis (1959), Gould (1988), Kirk and Ridgeway (1970, 1971), Rennie (1981), Rubinstein and Hersh (1984) and Whiteside *et al*. (1988) also provide discussions on the application of ergonomics to product testing.

Setting up a user trial is primarily about creating an environment that enables the interaction between product and user to be systematically examined and measured. In creating this environment it is important to recognise that a product is usually a 'tool' of some description which helps the user to achieve some goal or objective. Using the product involves the user in carrying out some tasks. It is the combination of these tasks and the related product attributes that together make up the process of interaction between product and user. The tasks involved can generally be described as a cycle of user action, product response, user decision, user action and so on. In design terms this process of interaction is central to what is now often referred to as the 'user-interface' aspect of the product. So evaluating the effectiveness of a product from a user's point of view means evaluating the design of the user-interface.

The principal components in a user trial are:

- A group of users;
  members of the current or target user population.
- A product;
  the device(s) to be used in the trial.
- A set of tasks;
  what the users must do to achieve their goals that involves the use of the product in some way.
- User performance criteria;
  aspects of the users or their behaviour which are used to judge the effectiveness of the product.
- Measurement techniques;
  the techniques that are used to measure the user-interface against the criteria selected.
- The site for the user trial;
  the location where a product is to be evaluated.

Figure 10.1. The general scenario of a user trial.

- An investigator;
  the person who is responsible for the design and conduct of the user trial.

How simple or complex a user trial becomes will be determined by the circumstances surrounding the demand for the investigation. Many user trials can be successfully carried out based on simple pragmatic approaches, but even when running the simplest of trials all these basic aspects need to be considered.

Circumstances will also dictate which aspects present most problems and the order of priority in which the problems must be overcome. From both a technical and practical point of view all these aspects must be regarded as interrelated and must be incorporated into a comprehensive test plan.

Questions a user trial can answer

User trials can answer many different types of questions about products. The character and form a particular user trial takes will depend on many considerations, not least the technical ones; the product to be tested, the measurement techniques and so on, but what is very important to be clear about is the purpose of the user trial and the type of question that needs to be answered. For example:

Who asked for a user trial to be carried out?

- Marketing;
  particular product comparison data may be required.
- Design engineers;
  the layout of a particular workstation needs to be specified.
- Software engineers;
  a particular user group is accustomed to a particular style of interaction; will a new style be acceptable and are there any negative transfer effects?
- Training specialists;
  how long should customer training programmes take and where might particular difficulties with the product occur?
- Maintenance engineers;
  a specific time limit has to be met for the exchange of particular machine components; will the design meet this requirement?

Does a performance standard have to be met?

- An international standard, national standard, industry standard, or company standard.
- Does user experience with existing products imply certain standards of performance which new products should meet?
- Should a performance standard be developed if one does not exist?

How does one product compare with another?

- How does the product of one manufacturer compare with the products from the competition?
- How does an existing range of products compare with a new range of prototypes?

Is there a need to improve the design of a product?

- Have accidents or critical incidents occurred with users that may be attributable to the design?
- Have users reported particular problems with a product?

User trials in more detail

Although the components of a user trial are discussed separately it is worth emphasising that all these aspects are closely interrelated and should be regarded as the basic ingredients that go to make the complete recipe.

Users

Selection of users

The fact that users vary in their characteristics and requirements will come as no great surprise, but what is important to understand is that a user trial requires a sample of users which in some way reflects the product user population as a whole. This means selecting a group of users who do not just have the same characteristics as the user population but who reflect the extent to which these characteristics vary. Here it is important to remember that many of the characteristics that need to be considered are not just basic physical or psychological aspects but also variations in the patterns of product use. For example, in a study on the design of powered lawnmowers, Fulton and Feeney (1983) examined typical patterns of cleaning procedures as part of setting up a user trial.

There are two major aspects to obtaining a sample. The first is to develop a profile of the type of people who are in your user population. A very important and fundamental issue is to identify clearly what user characteristics are actually relevant to the product under investigation. Where possible data should also be identified which describe how the population varies with respect to these characteristics. An obvious example of where this can be done is with respect to certain anthropometric characteristics. If experience in the use of the product is an important consideration then some data might be available that can be used, but there are many characteristics where data simply will not be available. Part of the problem is that people can be described in relation to a wide variety of characteristics, and for the investigator the challenge is to make the right choice from the many options that could be considered. In general terms user characteristics are usually grouped under the following headings:

- General descriptive characteristics;
  age, gender and other demographic data.
- Basic physical characteristics;
  sensory processes such as visual and auditory capabilities, tactile and vestibular sensitivity.
  body dimensions such as height and weight.
  psychomotor capabilities such as strength, balance, coordination and reaction time.
- Cognitive characteristics;
  general intellectual capabilities such as problem solving and decision making.
- Attitudinal characteristics;
  motivation, personality and initiative;
- Specific skills and training;
  educational level.
  experience in specific types of work and responsibilities undertaken.
- Experience with a particular product or product type.
  novice, intermediate or experienced users.

The second main aspect concerns the actual process of selection. From the user profile a set of descriptors must be identified which can be used to describe the type of participants required for the user trial. A very important consideration is that the descriptors must be understandable to the user population and presented in terms which make it easy for users to know whether or not they comply. Although even with such a characteristic as height one might expect that people would generally know how tall they are, in practice people are quite inaccurate about their height. Consequently selection criteria tend to be described in relatively loose and general terms rather than run the risk of losing potential participants. If necessary users can be screened on a post-hoc basis to check for consistency with the user profile. In contrast, if the user population can be defined in very specific terms, such as a particular group of workers, and the investigator has the possibility of screening prior to conducting a user trial then, correspondingly, the selection criteria can be quite precise.

A major problem in the development of selection criteria is the lack of specific data that can be used to describe the user population. Investigators are often compelled to use descriptors which may correlate only loosely with the characteristics of real interest, and this can present some risks. In some cases the investigator will be on reasonably safe ground, for instance height and weight correlate with certain other anthropometric characteristics; however in other areas the assumptions may be rather more tenuous.

Comparing users with the user population

Whenever possible or practicable it is always good advice to collect descriptive data from the sample of users which enable comparisons to be made with the user population as a whole, but comparisons can only be made if the corresponding data which describe the user population are available in an appropriate form. An obvious but important point is to check what data might be available and, if they are, in what form they exist.

Screening users

If the validity of the investigation relies on users complying with specific requirements, then participants in a trial should be appropriately screened. Just because users say they comply it would be unwise to assume this to be the case if it is possible to check. In many cases the user will simply not know. For example if there is a requirement that an auditory display be tested with users who have normal hearing ability, then the hearing of users should be examined using standard test procedures.

User numbers

Providing general advice on the number of users required is always a difficult issue. To make close estimates requires specific statistical information and the use of statistical procedures, associated with formal experimental design, for making estimates of sample sizes (Wiener, 1971; Collins, 1986). Making reliable estimates depends on the answers to the following interrelated questions:

- How many variables need to be considered in relation to user characteristics, product characteristics, and environmental characteristics?
- Can the variation of the population within each characteristic be statistically described?
- What level of accuracy or precision is required of the results?

What we usually find is that the number of variables that could be included in a user trial is large. It is also often the case that accurate statistical information about how the characteristics of users within a population vary is not available and estimates based on experience have to be made. In short, this usually results in estimates of user numbers being very large, sometimes alarmingly so, even at moderate levels of accuracy.

Against this we have to balance the practical questions of costs and timescales. A product development programme can seldom tolerate elaborate and extensive user trials unless the need is critical, for instance if the consequences of making incorrect design choices could be very expensive in terms of potential accidents, lost sales etc. As a result ergonomics practitioners often have to resort to simplifying the design of user trials in order to work within practical constraints. Usually this means examining products with relatively small numbers of users and a restricted set of variables. Results from such trials have to be seen at the level of providing estimates, directions for development, detecting major design weaknesses etc. This is also in part the reason why ergonomists often advocate the stategy of iterative testing of products throughout the development cycle. Repeated testing enables designs to be evaluated in a way which fits more easily into a product development context.

In answering the question 'how many users do I need?' one must be guided then by practical experience and the constraints imposed by local circumstances. The basic objective of a user trial is to go one better than estimates based on opinions and to obtain more factual information. On that basis experience does show that dependable results can be obtained from as few as five users in a single trial (Rubinstein and Hersh, 1984). More than five users will generally always be beneficial but how many more is an open question. There are many examples where individual trials have involved no more than 30 users yet in some cases samples in excess of 100 are quoted. Drury, in his chapter on designing ergonomics studies and experiments, considers sample selection and size elsewhere in this text.

The product

The products that people use come in many shapes and sizes all of which in principle could be tested with users. In this chapter the term 'product' is meant to embrace:

- The physical attributes associated with traditional three dimensional products such as household appliances, vehicles, office equipment, medical apparatus etc.;
- Product control/display systems from simple 'knobs and dials' concepts to full software-based control systems found in information technology products;
- The material required to support the use of a product such as product instructions or help facilities.

We tend to think of products as they are sold, a complete working design as we might see it in the High Street shop window, and indeed many products are tested in this form. However, products can be successfully tested at much earlier stages in their development and the form in which a product is presented to a user can vary considerably. The form depends upon the development stage, the product type and the aspect which requires

Figure 10.2. A user's interaction with a product is made up of many attributes; the physical activities, the perceptual and cognitive processes in understanding the control of the product, and the manuals, instructions etc., which support its use.

investigation. (Christie and Gardiner in chapter 12 of this book discuss test methods relevant to stages in development of human computer interfaces.) The forms commonly tested include:

– 'Paper-based' descriptions of product concepts;
These can enable initial explorations of ideas on product functionality to be made, important usability characteristics to be identified, or 'walk through' studies of protocols for product control systems to be conducted.
– Part prototypes;
Part prototypes are used to simulate specific functional attributes of a design. The prototype may look nothing like the final product but it is essential that those aspects which are investigated are accurately represented. Part prototyping of software for product control systems through the use of rapid prototyping tools is one area where this type of testing is becoming more common.
– Full prototypes;
Full prototypes simulate the complete functionality and appearance of the product, sometimes requiring backup to support the simulation which would not be required in the final product.
– Complete products;
Complete products enable the complete user-interface to be examined. This opens up the possibility of field investigations, comparative studies with other products, in-service studies etc. to be carried out.

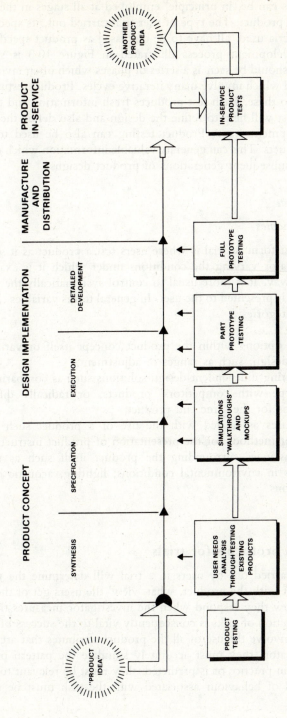

Figure 10.3. The contribution of product testing to the product development process.

User trials can be, in principle, employed at all stages in the development process of a product. The type of user trial carried out, its specific objectives and the criteria used, all have to be discussed as product specific issues. The product development process illustrated in Figure 10.3 is simplified and idealised; it should be seen as a series of phases which often involve undefined borders, and which involve many iterative cycles. Product testing can directly contribute to these. Each test produces fresh information and the experience from one test will help to refine the design and also define the needs for the next development stage. Product testing can also be used to evaluate in-service products. This can generate valuable information which can contribute directly to subsequent generations of product designs.

Product variables

The simplest form of trial is where users test a product as it stands without the investigator varying the conditions under which it is examined in any systematic way. It is more usual to control systematically the way in which the product is presented to the user. In general terms variables can be grouped into four categories:

- Design options within the product concept itself or variables inherent to the design, such as ranges of adjustment;
- Comparing independent design solutions such as comparing a complete prototype with competitors' products, or radically different design solutions for the same end product;
- Procedures associated with the use of a product such as evaluating different methods for the presentation of product instructions;
- Circumstances surrounding the product itself such as the effects of changes in environmental conditions; lighting, acoustic or mechanical conditions.

Tasks and protocols for trials

The tasks carried out by users in a trial will determine the way in which they interact with the product, what 'view' the users get of the product and therefore how they respond when the investigator measures the interaction. Correct selection of tasks is consequently vital to the success of the user trial; they must involve the use of all the product attributes that are of interest to the investigator, they must accurately simulate the pattern of product use that occurs in practice or is predicted, they must be relevant to the user, and the patterns of behaviour associated with the task must be measurable in some way.

Task sequence

The tasks performed in a user trial generally should be carried out in a sequence which is logical and relevant to the way in which a user would expect to use the product. This does not necessarily mean that all users carry out all tasks in the same sequence. The sequence would normally be varied in some systematic way in order to control for 'order effects' such as transfer of learning from task to task. This is particularly important where a user would normally expect to choose between tasks in which case a randomised or systematically varied order of task sequence is necessary.

It is important that the procedure for the trial should be consistent for each user. This includes all user instructions, measurement procedures, any response forms used, and any interviews carried out during the trial. Maintaining consistency is often easier if the investigator works from detailed notes or even reads instructions.

User trials are usually organised in the following way:

- An introduction to the trial; the purpose of the trial, the product to be used including a demonstration if necessary, any associated equipment that may be involved, the tasks to be undertaken, the measurement procedures to be followed; carrying out screening procedures;
- A 'walkthrough' of the essential aspects of the trial with the user; this usually clarifies whether or not the user has understood what is required of him.
- Data collection; the user carries out the sequence of tasks including a pilot run if necessary, and the investigator takes measurements as appropriate;
- Debriefing the user; it is always useful to enquire about the users' views of the trial even if this is not part of the data that are being collected; users also appreciate some feedback on 'how well they did' and how the results are likely to influence the product they have just used.

The set–up for a user trial should also take account of the physical or mental effort involved. If the effort is significant, rest periods should be incorporated into the procedure, unless of course fatigue is what the trial is set up to examine. Another issue is that of boredom. In long trials of 90 minutes or more users can find it difficult to maintain concentration. It is therefore worthwhile considering building some variety into the procedure in order to maintain the interest of users.

User performance criteria

Selecting user performance criteria

Measuring how users respond when they use a product is the essence of a user trial. This response is a consequence of the tasks the user has carried

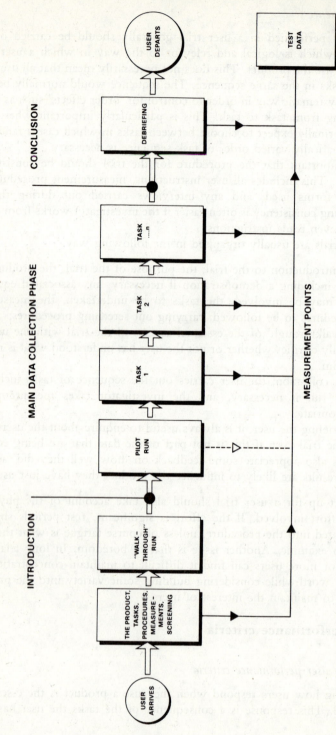

Figure 10.4. The main phases in a typical user trial. Data is usually collected during or on completion of each task, but may also be collected from a pilot run (although a pilot run is generally only used as a practice phase for the user).

out and the specific set of circumstances under which the trial was undertaken. Hence users perform in a certain way and so the emergence of the term 'user performance'. The word criteria is added to indicate the basis on which this performance is to be measured. It is, of course, essential that the criteria chosen provide a realistic indication of the effectiveness of the product. Deciding on which criteria to use to evaluate a product can be one of the most difficult aspects of setting up a user trial. In some cases the appropriate criteria may be self-evident but often the choice of the best criteria to use is an issue which in itself must be evaluated.

If it is not clear which criteria should be used then the possibilities should be considered in general terms first. The options will then require interpretation to obtain a set of criteria which are appropriate to the particular investigation. The aim is to express each criterion as a specific question that can be answered in a user trial. Each question should incorporate, implicitly or explicitly, the dimension on which the performance of the user is to be measured. In the design of formal experiments this dimension would be referred to as the 'dependent variable'.

There are always reasons for running a user trial, questions to be answered and priorities to be addressed. These should be discussed with those who require the trial to be carried out in order that the criteria chosen are appropriate. Which criteria are appropriate will also depend on the particular aspect of the user-interface which is of interest. The reason for the trial may be to check whether the basic physical, perceptual and cognitive characteristics of users are satisfactorily accommodated. There may be conflicts between what the user can satisfactorily cope with and what the product demands, that make the product unacceptable. Another reason may be to examine the use of the product in relation to the users' goals; can users achieve their goals easily or with difficulty? Is a product enabling the users to be effective in their jobs? It may be that weaknesses in the design of the product have been identified already and the purpose of the trial is to bring about improved user performance. In addition constraints derived from the context of use may make demands on the design of the product which influence the behaviour and perception of users. A user trial could be set up to check that the design satisfactorily takes such constraints into account.

There are a number of general criteria which are frequently used to evaluate user-interfaces. Below are listed some of the more common ones which are frequently used in user trials. They are not listed in any particular order of priority:

- time; speed of response, activity rate, etc.
- accuracy and errors
- convenience and ease of use
- comfort
- satisfaction
- physical health and safety

- physical fit
- physical effort and workload
- stress and mental workload

How such criteria are translated into more specific questions that can be answered in a user trial must be considered. It is usual practice to include several measures rather than to rely on one. It is also common practice to include both objective and subjective measures. For example objective measures may be used for criteria such as time, accuracy, physical fit and physical workload whereas subjective measures may be used for other criteria such as convenience and ease of use, comfort and satisfaction.

Sometimes it is not possible to measure the criteria directly and several measures have to be combined in order to arrive at an assessment. The measurement of stress or mental workload (see Cox and Meshkati *et al.* in this book) are examples where this approach is often taken; objective measures of work rate and error rates may be combined with subjective reports to assess whether the level of stress is acceptable or not.

The identification of criteria can be considered in very general terms. For example, in the area of information technology products and the design of computing systems, Bailey (1982), Gould (1988), Rubinstein and Hersh (1984), Shackel (1986, 1987) and Whiteside *et al.* (1988) discuss a variety of criteria and related methods that can be used in evaluation.

Where a product or product family is more clearly defined, more specific criteria and their related measures can be considered. For example, Spradlin (1987) discusses a range of general and specific criteria that went into the development of flight deck instrumentation for Boeing's 757/767 aircraft. This article also shows the value of considering data on accidents, critical incidents and perceived problems in determining the appropriate criteria. Fulton and Feeney (1983) provide an additional example where background data were used in this way; here accident data and interviews with accident victims were used to help develop evaluation criteria in an investigation into powered domestic lawnmowers.

For the evaluation of a specific product the criteria and the corresponding measures have to be quite specific. Here are two examples, taken from recently published work, of criteria used in relation to the evaluation of specific products.

The first was the development of the Digital Equipment Corporation (DEC) mouse. The programme of work involved comparative testing of proprietary products and prototypes. The final design emerged after several cycles of user testing with successive generations of prototypes (Hodes and Akagi, 1986). Subjective criteria used were ease of grasp, degree of hand comfort and degree of hand, wrist, and forearm strain. Objective criteria were fit in relation to hand size including hand length, palm length, palm width, and angle of palm at rest, rate of movement of the cursor over the screen (pixels/sec) and number of button pushes.

In the second case four designs of magnetic stripe readers were evaluated;

two were 'insertion' designs and two were 'slot' designs (Lewis, 1987). The results showed that the 'slot' designs resulted in superior throughput times but that there was little difference in other respects. The subjective criterion was user preference; the objective criteria were reading rate (reads/min) and percentage of good reads.

Evaluation criteria can also be developed as part of a consistent programme of product testing that extends over several generations of a product, enabling criteria to be standardised and refined to high levels of sensitivity. Repeated use of standardised criteria in controlled tests and the data that result allow benchmark performance standards to be set up as part of product usability specification procedures. This also enables improvements in the products to be monitored over time. For example, Xerox Corporation carry out routine 'operability testing' on their photocopier products (Hartman, 1987). These tests incorporate standardised procedures for machine families which test prototypes in terms of operator preferences, error rates and task times. Similarly Philips also incorporate regular user testing into the development of *Philishave* electric razors; criteria used have been refined over several years and very specific measuring techniques have been developed particular to the product family.

Collecting data

The methods used to collect data from a user trial fall into three basic categories: interview and questionnaire methods to collect data on attitudes and subjective assessments from users, direct observation of user actions, and the collection of objective data. The extent to which a user trial makes use of these data collection methods will vary although in practice a blend of all three categories is often used. The blend that is most appropriate has to be judged in relation to the particular investigation.

Useful data can be collected effectively without the need for developing complicated interview schedules, questionnaires, etc. Many user trials are successfully undertaken on the basis of quite pragmatic and simple approaches. For example, questionnaires may be no more than a few questions answered using basic rating scales; observations of user actions may consist only of counting the number of times a particular event occurs. On the other hand an investigation might require a much more elaborate approach where a full questionnaire must be developed, observation metrics compiled and interviewers trained. In this form a user trial can become a major exercise. Even within each category there are many choices to be made and some general remarks will follow about the wide variety of techniques which can be employed.

User interviews

Even at a basic level most user trials involve some form of interview, such as asking users their general opinions of the trial and the product they used.

In practice the emphasis given to an interview may be high, it might be the focal point of the whole investigation, whilst on other occasions it may only play an incidental part, but it is usually advisable to incorporate some form of interview. Approached in the right way an interview can be an extremely productive way of collecting information.

The relationship between user and product is a dynamic process which has many facets, concerning the perceptual and cognitive aspects of product use; how information is absorbed from the product, how it is interpreted, and what user actions follow as a result. Such information can usually only be obtained directly from the users themselves and is most effectively recorded through some form of structured interview. A full interview usually involves users in:

- Evaluating the product in general terms;
- Explaining difficulties they may have experienced during the course of the trial;
- Explaining why they took certain actions;
- Assessing their own physical or psychological condition;
- Commenting on how the product might be improved.

An important consideration when conducting interviews is to ensure that the user is put at ease. A user trial is an unfamiliar environment for many and, however well they are briefed, they seldom know quite what to expect. The way the interview is handled is important if the co-operation and interest of the user is to be obtained and maintained. To achieve this is not difficult providing some basic guidelines are borne in mind, chiefly to

- Receive the user with courtesy;
- Keep the user informed of what is to happen;
- Take a positive lead in directing events;
- Ensure that the user feels that his contribution is valuable;
- Aim at giving the interview the flavour of a conversation rather than an inquisition.

An interview often requires users to make judgements about product characteristics which they seldom, if ever, have considered. In everyday life people use one product or another more or less continuously without giving them a thought. Consequently a challenge for the investigator is to stimulate the critical awareness of the user and to get him or her to think positively and creatively about the product. Achieving this is largely a question of professional expertise and experience in the use of interview techniques. An important aspect is to carry out the interview in the immediate vicinity of the product under investigation. This is particularly important where the product itself is discussed and the user is required to explain actions taken or make comparisons between products or design options and so on. The presence of the product will be an invaluable aid to explaining points and prompting remarks which might otherwise be overlooked. Stimulating critical awareness applies to what the user does as well as the type of questions asked. Here are some approaches that can be taken to achieve this:

- Ensure that judgements are based on the experience of actually using the product;
- The interview should have clear structure and direction and not be simply open ended;
- Be prepared to prompt and question the user in depth;
- Encourage thoughts to be put into words as the user carries out the tasks;
- Ask the user to make choices through rating scales, ranking scales, etc;
- Use contrasting design solutions to highlight differences where choices have to be made;
- Require users to write down their views as a trial progresses.

Interview formats

The traditional format for an interview is one to one but group interviews are being used more frequently in product evaluation work. One of the main stimulants for the evolution of these approaches has been the desire to explore more productive formats for obtaining information from users.

The format of one investigator and two users has been employed, sometimes referred to as 'co-discovery'. In one example it was used to investigate the problems of users installing a small computer (Comstock, 1983). More generally this format has been advocated as an aid to protocol analysis (see Bainbridge in this book) where users are required to verbalise their interpretation of a product as a trial progresses (Rubinstein and Hersh, 1984). Lewis (1982), Olson *et al.* (1984) and Ericsson and Simon (1980) also discuss the use of this approach as part of using verbal reports in user interface evaluation.

One or two investigators and several users is an approach that is also commonly used, for instance in a study on solid fuel space heaters for domestic use (Spicer, 1987). Users individually carried out a set of tasks with the products under investigation and gave their own assessments. Subsequently they were brought together in small groups to compare their experiences. Conventional methods of subjective assessment techniques were used to record users' attitudes.

The concept of group interviews has been extended to include 'user participative' approaches to design. O'Brien (1982) has used this approach in the identification of the main user requirements and subsequent evaluations using product simulations for the design of control rooms; the work reported here has since been extended (O'Brien, 1987). Similar approaches to the design of workstations involving groups have been adopted by Davies and

Figure 10.5. More than one user can be interviewed at a time. Bringing users together in this way can be a great aid to stimulating discussion.

Phillips (1986), Murphy *et al.* (1986), Pikaar *et al.* (1985) and Stubler and Bernard (1986). Although these have involved well defined and small user groups the principles could be equally well applied to user groups representing larger populations. (See also Eason and Shipley in this book.)

Unfortunately there has been little discussion in the literature on the strengths and weaknesses of interview techniques in product development work, and more particularly any systematic evaluation of the approaches (McClelland, 1984; Meister 1986). However, most experience with interview techniques shows them to be extremely important in the evaluation of user interface design, but that there is still considerable scope for improvement.

Questionnaires

Questionnaires are often an integral part of a structured interview and the basis for documenting the results of a user trial. Typically they are paper-based, incorporating a series of questions designed to be answered in a predefined order; in their simplest form they may be only a short list of questions requiring Yes/No responses. Questionnaires can be completed in two basic ways, by the user or by the investigator; often a combination is used. The types of questions used are either open ended or closed. Open ended questions are usually used to obtain free responses from users without pre-defining the answer; they may receive a wide variety of responses which then require post-hoc coding during analysis. For large surveys this type of question can be very time-consuming to analyse. In closed questions the user responds to a set of predefined categories. A variety of these may be used, for example:

- Factual statements;
  where simple yes/no responses or specific items of information are required.
- Multiple category questions;
  where categories are specified and the user chooses which of the categories apply. They can be answered on a yes/no basis or using some form of rating scale.
- Rating scales;
  used to assess the attitudes of users to specified product attributes. Commonly used to assess comfort, degrees of convenience, ease of use, perceived degrees of difficulty and so on.
- Ranking scales;
  used to indicate the relative order of a set of conditions or products according to a specified attribute.

There is a considerable variety of methods for the presentation, wording and analysis of questionnaires. The classic texts on their design and use of Moser and Kalton (1971) and Oppenheim (1966) still provide valuable advice. More recent overviews are provided by Morton-Williams (1986) and Sinclair in this book.

Observations of user actions

The effective use of observation techniques, discussd elsewhere in this book by Drury, relies on a clear understanding of the purpose of the observation. Meister (1986) identifies three types. Modified somewhat he presents descriptive techniques, where the observer simply records events as they take place (e.g. time based, frequency based, event sequence, postures adopted, controls used, etc.); evaluative techniques, where the observer evaluates the outcome or consequence of events that have taken place (e.g. degrees of difficulty, incidence of hazardous events, errors of judgement, etc.), and diagnostic techniques, where the observer identifies the causes that give rise to the observed events (e.g. positioning of controls, inadequacy of displays, poor user instructions, etc.).

Descriptive observations are generally the easiest to set up and conduct. The degree of difficulty increases with evaluative observations and is greatest with diagnostic observations. This is primarily because of the type of on-the-spot judgements the investigator is required to make during the observation. Ability to conduct evaluative or diagnostic observations depends greatly on the background knowledge of the investigator. Detailed knowledge of the tasks in hand and the objectives behind using the product, is usually required before evaluative or diagnostic observations can be successfully carried out. It is essential that the events to be observed, what data must be recorded and how, are clearly identified beforehand. There is usually a great deal of data that could be collected but much will be irrelevant. The greatest problems come from the speed of events and the number which occur in parallel.

The simplest form of observation to make is where events occur in strict sequence and at a rate which allows them to be recorded accurately. People have limited capacity to record events that occur in parallel or where the intervals between events are very short; it often surprises people who are new to observing user behaviour just how quickly the accuracy and reliability of an investigator can break down when there are too many items to monitor. Even moderately experienced investigators, when planning investigations, often find themselves being far too ambitious in the variety of events they wish to record. Consequently it is most important that the method of observation is carefully piloted whenever possible.

Still the most common method for recording data is paper based, the effectiveness of which can be greatly influenced by the precise way in which the data are to be recorded. Wherever possible graphic methods should be used; ticking boxes, marking diagrams, use of graphic codes and so on. Spicer *et al.* (1984) provide an example of this approach where observations were made of disabled and elderly people entering and leaving cars.

Apart from paper based methods video recording is used frequently and is now regarded as standard equipment in any product usability laboratory (*e.g.* Benel and Pain, 1985; Cantwell and Stajano, 1985; Neal and Simons, 1985). Video is easy to set up and to use; it can provide a comprehensive record of events and of course it enables one to replay events. The great advantage of video is that a particular sequence of events can be analysed several times from different viewpoints, enabling an investigator to overcome many of the problems associated with analysing events which occur in parallel or very quickly. An example of this as a major consideration was in a study into warning systems for railway workers working alongside the track (McClelland *et al.*, 1983). However, a few cautionary words are in order. Firstly, detailed analysis of video recordings can be extremely time–consuming, ratios of between 1:5 and 1:10, recorded time:analysis time, are commonly quoted (Mackay *et al.*, 1988). Secondly, the need for detailed preparatory work is not reduced because a comprehensive record of events is available. The use of video requires at least as much care as paper based approaches in deciding beforehand what events are to be observed and the form in which the data are to be analysed.

Objective measures

It is always wise to obtain objective data to support data from other sources whenever it is possible and practicable, notably the subjective data from interviews and questionnaires. Sources of objective data that are frequently used can be divided into three categories. Firstly, direct objective measurements of the user; these would include body dimensions, physiological measurements such as heart rate, oxygen intake, body temperature, and so on. Secondly, directly recorded data resulting from user actions, registered by the investigator or by some remote means such as video or automatic event recording. Data generated by user actions can come in several forms. They can be time based (e.g. task duration, event duration, response time, reaction time) or based on scores of errors/accuracy (e.g. mistakes in procedures, incorrect responses to stimuli, error rates in relation to time or events). They may also be based on the frequency of events such as the rate of specific responses or task completion rates. The third type of objective data is that measured directly from the product on completion of or during the trial; for example the positon of seats, shelves, or controls, the accuracy of cursor control devices in an interactive computer system, numbers of keystrokes made, lighting levels, colour quality, brightness, contrast settings on visual displays or sound levels.

Location of a user trial

The appropriate location for a user trial will to a large extent be determined by the purpose of the user trial, but what is feasible must also be taken into account. For example it is typically much more difficult to control the variables with user trials in the field than trials carried out in the laboratory. The importance of eliminating the influence of external factors on the conduct of the trial is a significant issue in deciding where the trial is to be held. Data collection methods must also be considered. Portability of data collection methods through the use of video, telemetering, and portable computers (see Drury in this book) have all led to possibilities which in the past would have been rejected as impractical. Very important is the degree of realism required. In the final analysis the success of the product in the context for which it was designed is the acid test. Access to the location of the trial by the sample of users can be a critical issue. Related to this is whether the product is easily transportable. Finally there is the important consideration of security, both in relation to the confidentiality of the product and to the sensitivity of the data that will be collected.

The investigator

User trials should be set up and designed by personnel qualified in the design of experiments and who have knowledge of the ergonomics issues appropriate to the product under investigation. Traditionally such personnel have included people qualified in applied physiology and applied psychology as well as ergonomics or human factors. However it is clear from the wide variety of work reported that many other professional groups can successfully carry out user trials, including, for example, anthropologists, social psychologists, industrial engineers and marketing personnel. Many of the tasks involved in the day–to–day conduct of user trials can, on occasions, be undertaken by other personnel but specific training will often be required beforehand to ensure results of a satisfactory quality. Where an organisation does not have the necessary in-house expertise, appropriate staff should be recruited, or alternatively, an outside agency contracted which has experience in this type of work.

A case study

This case study is taken from the 'Study, development, and design of a mouse' by Hodes and Akagi (1986) and extends the brief description given earlier in this chapter. The paper describes the development of a new corporate mouse for Digital Equipment Corporation by the in-house Human Factors Department, together with industrial designers and design engineers.

The reasons for choosing this example in particular were several: it describes work carried out within a product development environment, the result of the work is a product in use, the strategy adopted illustrates well the importance of testing on an iterative basis and shows how the cycle of testing contributed to the successive refinement of the design. The methods used in this series of user trials were quite simple but they provided the quality of information appropriate to achieving a successful design of product. Stages 1 and 2 involved the development of design criteria and an understanding of what tasks to use as a basis for the user trials. Stages 3, 4 and 5 focused on the development of the design, with Stage 6 providing a last check. Throughout the programme of work a structured interview incorporating a questionnaire was used to assess the opinions of users. Both rating scales and ranking techniques were used to measure opinions and preferences. At one stage video recording was also used, although it is not clear whether specific data were obtained from the recordings. In the final two stages, objective performance measurements were also used to assess the designs. In all cases relatively small numbers of users took part; the most in any one stage was 24 (Stage 5). The combined results of these small studies, conducted relatively quickly and inexpensively, provided accurate predictions about the preferences of users in the design of a mouse. The method adopted here provides us with a clear example of what could be done within many other product areas.

This summary focuses only on the role of user trials in the development cycle. For a more complete picture the reader is referred to the original paper.

The development of the DEC mouse

First, what is a mouse? A mouse is a 'pointing device' for controlling the screen position of the cursor in an interactive computer system. A mouse is a hand–held desk–top device which can be moved. Movement of the mouse over the desk causes a corresponding movement of the cursor on the screen. A mouse usually has one, two or three finger–operated buttons set in its case. These buttons are used to execute various types of commands depending on the cursor position and the software.

The Human Factors Department was presented with the challenge to develop a new mouse which was a satisfactory combination of ergonomics requirements, aesthetics and production costs. The goal was to optimise the physical features of the mouse from both anthropometric and biomechanical perspectives as well as taking account of the effect that software had on use patterns; i.e. holding, moving and operating the mouse. The development occurred in several stages.

Stage 1 – Task analysis

Six users were informally observed working at their own computer terminals each with his or her normal mouse. The users worked on their own tasks, mainly word processing and file editing. The mice used were similar but not identical.

Distinct patterns of hand movement and finger positioning were identified which could cause joint strain and discomfort over time. Also users were frequently shifting visual attention from the screen to their mouse to be sure that the correct mouse buttons were being used. It was concluded that a new shape had to be created which allowed the same relaxed hand position for movement and activation of the buttons, and include strong tactile feedback to aid correct positioning of the hand and fingers.

Stage 2 – Competitive analysis

Seven mice were compared; 4 new design concepts, a current DEC product and 2 other proprietary products. Six users were involved in a user trial, all of whom were regular users of mice. The task was to move each mouse horizontally and vertically as would normally occur in practice, lift the mouse, and hold it steady to operate the buttons.

A structured interview with a questionnaire was used to collect data. Each user was asked to select their four preferred designs and to rank them in order of preference. Each design was rated for; ease of grasp; degree of hand comfort; degree of hand, wrist, or forearm strain. Users' opinions on the dimensions of the mice; height, width, and button position were also obtained.

Three mice were clearly preferred over the other four. Two of the three were existing products and the other was one of the new concepts. Within these three, differences were observed in length and width of the top surface, surface texture, and the height of the highest point. However it was unclear which precise characteristics were best. But the interview data showed that these features influenced user attitudes to the designs.

Stage 3 – Systematic design study

In stage three nine mice were examined to see what the effect was on preference of systematically varying the length of the top surface and the location of the highest point. The designs took account of known anthropometric data on hand dimensions. Eight users evaluated the nine new mouse models. The methods used in this trial were the same as in the previous stage.

It was concluded that there was a clear preference for mice which featured the highest point closest to the centre of the case.

Stage 4 – Competitive analysis

In this stage the shape of the mouse case was further refined based on testing seven models, three current products and four new designs developed as a result of the previous trial. Thirteen experienced users used each mouse for several hours. Each user used the mice in a randomised order. Otherwise the procedure for the trial was the same as in the previous stages.

Three mice were rated significantly higher than the rest; two were new designs and one was an existing product. The interview data showed clearly the importance of tactile feedback so that users could position their hands correctly without having to glance away from the screen. Also important was the effect of the shape and the rolling mechanism on the steadiness of the mouse. Positioning of the buttons on the front side of the mouse, providing them with good tactile and audible feedback were also identified as highly desirable features. From this stage one of the new mouse designs was selected to be developed into a working model for comparison with current products.

Stage 5 – Testing the new design

Twenty-four users, fourteen experienced and ten inexperienced mouse users, took part in this user trial using the new mouse in the form of a full working prototype and two current products. The test consisted of two types of task, selecting—involving coarse mouse movements— and editing—involving fine mouse movements. Both subjective ratings and objective performance measures (rate of movement in pixels/second and number of button presses) were collected. The sessions were also videotaped. The new design was consistently and significantly faster than the other two mice. Thirteen of the users chose the new mouse as their first choice and it was rated better overall than the other two mice. Also, for each attribute users selected the 'best' mouse; a clear preference for the new mouse was the result.

Stage 6 – Testing the final design

At this stage the new mouse was in production form. The final round of testing involved two parts. In the first part the new mouse was compared with two commercially available mice which were recent introductions onto the market and had not been tested previously. Twenty experienced users took part in the user trial which followed the same pattern as used in stage two.

In the second part ten users, four experienced users and six novices, used only the new mouse. The task was to manipulate a computer display; this took approximately one hour to complete. A structured interview was carried out which was based on the interview and questionnaire used previously.

In the first part the results showed that the new design was clearly preferred by users over and above the two current products. In the second part the new mouse received positive ratings on all the scales and the positive comments received in the previous stages were confirmed.

Following this final user trial the design of the new mouse was finalised and handed over for production. Figure 10.6 illustrates the mouse that was finally produced.

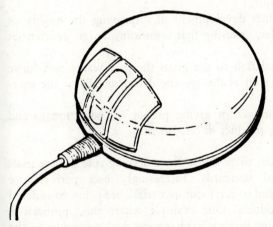

Figure 10.6. The new corporate mouse for Digital Equipment Corporation.

Future directions

Interest in the incorporation of user testing into product development has been increasing in recent years. As products become more complex, and commercial pressures become more severe, the need to examine the usability aspects of products is growing. There is increasing interest in methods which enable products to be assessed systematically and judgements made about their qualities which are based more on 'facts' and less on 'opinions'. This is precisely where the application of user trials can play a vital part in product development, but their relevance to mainstream product development will depend in large measure on the development of testing and evaluation methods that suit such an environment.

From a technical standpoint the most notable contribution in recent years has been from Meister (1986). As already noted he provides an extensive discussion of the methods available. Meister also notes the lack of attention given to evaluating the methods used and calls for further efforts from the ergonomics research community to improve our understanding of the methods that are used (see also Anderson and Olson, 1985). However many of the technical issues discussed are rather long term and not within the scope of most practitioners within industry to resolve. Practitioners within a product development environment must, of necessity, take a more pragmatic view over the short to medium term. In this context the successful implementation of user trials is likely to depend more on demonstrating their effectiveness in terms of:

- The extent to which the results from user trials with prototypes can predict how a product will perform in practice;
- How the performance of products in practice is improved as a result of user trials being carried out;
- The cost-benefit to product development of improving the quality of products and, in particular, ensuring that significant design deficiencies do not get to market;
- Evaluating products in relation to the goals the user wants to achieve by using the product as well as the specific design details of the user-interface itself;
- Demonstrating the importance of giving proper consideration to end user requirements during product development.

Outside product development there is also the emerging role of user trials in international and national standards. Increasingly user performance standards, and therefore the need to carry out user trials, are being considered as a basis for approving products. One example where this approach is already part of a standard is for reclosable pharmaceutical containers in the UK (BS 5321: 1975). Currently the approach is being considered by ISO/TC159 and TC97 for standards on keyboard design for computer terminals and dialogue design (Brigham, 1989).

References

Anderson, N.S. and Olson, J.R., (1985), Methods for designing software to fit human needs and capabilities. In *Proceedings of the workshop on software human factors*, (Washington, DC: National Academic Press).

Bailey R.W., (1982), *Human Performance Engineering*. (Englewood Cliffs, NJ: Prentice Hall Inc.).

Benel, D.C.R. and Pain, R.F., (1985), The human factors usability laboratory in product evaluation, *Proceedings of the Human Factors Society, 29th Annual Meeting*, (Santa Monica, CA: Human Factors Society).

Bewley, W.L., Roberts, T.L., Schroit, D., and Verplank, W.L., (1983), Human Factors testing in the design of Xerox's 8010 "Star" office workstation, *CHI '83 Proceedings*, Association of Computer Manufacturers.

Boies, S.J., Gould, J.D., Levy, S., Richards, J. and Schoonard, J., (1987), The 1984 Olympic messaging system, *Communications of the ACM*, September 1987.

O'Brien, D.D., (1987), Personal communication from the UK Government Home Office, London.

O'Brien, D.D., (editor) (1982), Design Methods; Seminar on Control Room Design, Lancashire Constabulary HQ.

Brigham, F., (1989), NL representative ISO/TC159/SC4, Personal Communication

BS 5321:1975 Reclosable Pharmaceutical Containers Resistant to opening by children, (London: British Standards Institution).

Cantwell, D. and Stajano, A., (1985), Certification of software usability in IBM Europe, *Ergonomics International 85 Proceedings of the Ninth Congress of the IEA*, edited by I.D. Brown, R. Goldsmith, K. Coombes and M.A. Sinclair, (London: Taylor & Francis).

Chapanis, A., (1959), *Research Techniques in Human Engineering*, (Baltimore, MD: The John Hopkins Press).

Collins, M., (1986), Sampling, In *Consumer Market Research Handbook*, edited by R.M. Worcester and J. Downham, (Amsterdam: North-Holland).

Comstock, E., (1983), Customer installability of computer systems, *Proceedings of the Human Factors Society, 27th Annual Meeting*, (Santa Monica, CA: Human Factors Society).

Davies, D.K. and Phillips, M.D., (1986), Assessing user acceptance of next generation air traffic controller workstations. *Proceedings of the Human Factors Society, 30th Annual Meeting*, (Santa Monica, CA: Human Factors Society).

Ericsson, K.A. and Simon, H.A., (1980), Verbal reports as data, *Psychological Review*, **3**,

Fulton, E.J. and Feeney, R.J., (1983), Powered domestic lawnmowers: design for safety, *Applied Ergonomics*, **14**, 91–95.

Gould, J.D., (1988), How to design usable systems, In *Handbook of Human-Computer Interaction*, edited by M. Helander, (Amsterdam: North Holland).

Hartman, W., (1987), Head Industrial Design/Operability department, Xerox Corporation, Personal Communication.

Hodes, D. and Akagi, K., (1986), Study, development and design of a mouse, *Proceedings of the Human Factors Society, 30th Annual Meeting*, (Santa Monica, CA: Human Factors Society).

Kirk, N.S. and Ridgeway, S., (1970), Ergonomics testing of consumer products, 1:General considerations, *Applied Ergonomics*, **1**, 295–300.

Kirk, N.S. and Ridgeway, S., (1971), Ergonomics testing of consumer products, 2:Techniques, *Applied Ergonomics*, **2**, 12–18.

Lewis, C., (1982), Using the 'thinking allowed' method in cognitive interface design. IBM Research Report RC 9265.

Lewis, J.R., (1987), Slot versus insertion magnetic stripe readers; user performance and preference, *Human Factors*, **29**, 465–476.

Mackay, W.E., Guindon, R., Mantel, M.M., Suchman, L., and Tatar, D.G. (1988). Panel session: Video: Data for studying human-computer interaction, *CHI'88 Conference proceedings*, (Reading, MA: Addison-Wesley).

McClelland, I.L., (1984), Evaluation trials and the use of subjects, *Contemporary ergonomics 1984 Proceedings of the Ergonomics Society Conference*, edited by E.D. Megaw, (London: Taylor & Francis).

McClelland, I.L., Simpson, C.T., and Starbuck, A., (1983), An audible train warning for track maintenance personnnel, *Applied Ergonomics*, **14**, 2–10.

Meister, D., (1986), *Human Factors Testing and Evaluation*, (Amsterdam:North-Holland).

Morton-Williams, J., (1986), Questionnaire design, In *Consumer Market Research Handbook*, edited by R.M. Worcester and J. Downham, (Amsterdam:North-Holland).

Moser, C.A. and Kalton, G., (1971), *Survey methods in social investigation*, (London:Heinemann Educational Books).

Murphy, E.D., Coleman, W.D., Stewart, L.J. and Sheppard, S.B., (1986), Case study: Developing an operations concept for future air traffic control. *Proceedings of the Human Factors Society, 30th Annual Meeting*, (Santa Monica, CA: Human Factors Society).

Neal, A.S. and Simons, R.M., (1985), Evaluating software and documentation usability, *Ergonomics International 85 Proceedings of the Ninth Congress of the IEA*, edited by I.D. Brown, R. Goldsmith, K. Coombes and M.A. Sinclair, (London: Taylor & Francis).

Olson, G.M., Duffy, S.A. and Mack, R.L., (1984), Thinking-out-loud as a method of studying real-time comprehension processes. In *New Methods in Reading Comprehension*, edited by D.E. Keiras and M.A. Just, (Hillsdale, NJ: Lawrence Erlbaum).

Oppenheim. (1966), *Questionnaire design and attitude measurement*, (London: Heinemann Educational Books).

Pikaar, R.N., Lenior, T.M.J. and Rijnsdorp, J.E., (1985), Control room design from situational analysis to final layout; operator contributions and the role of ergonomists. *2nd IFAC/IFIP/IFORS/IEA Conference; Analysis, design and evaluation of man-machine systems*, edited by G. Mancini, G. Johannsen and L. Martensson, (Oxford: Pergamon Press).

Rennie, A.M., (1981), The application of ergonomics to consumer product evaluation, *Applied Ergonomics*, **12**, 163–168.

Ross, E.H., (1983), Human factors in system development: project team approaches and recommendations, *Proceedings of the Human Factors Society, 27th Annual Meeting*, (Santa Monica, CA: Human Factors Society).

Rubinstein, R. and Hersh, H.M., (1984), *The Human Factor*, (Digital Equipment Corporation: Digital Press).

Shackel, B., (1986), Usability—context, framework, definition, design and evaluation, *Proceedings of the SERC CREST course; Human factors for informatics usability*, HUSAT Research Centre, University of Technology, Loughborough.

Shackel, B., (1987), Human Factors for Usability Engineering, ESPRIT '87; Achievements and impact, *Proceedings of the 4th annual ESPRIT conference*, Brussels, September 1987, edited by Commission of the European Communities, (Amsterdam: North-Holland).

Spicer, J., (1987), Personal communication from the Institute for Consumer Ergonomics, Loughborough.

Spicer, J., Wilkinson, S. and McClelland, I.L., (1984), Access to cars by disabled and elderly people; Report No. 3, Car trials. Working Paper WP/VED/84/10, Transport and Road Research Laboratory, Department of Transport.

Spradlin, R.E., (1987), Modern air transport flight deck design, *Displays*, October 1987, 171–182.

Stubler, W.F. and Bernard, T.E., (1986), Office ergonomics: design methodology and evaluation, *Proceedings of the Human Factors Society, 30th Annual Meeting*, (Santa Monica, CA: Human Factors Society).

Whiteside, J., Bennett, J. and Holtzblatt, K., (1988), Usability engineering: our experience and evolution. In *Handbook of Human-Computer Interaction*, edited by M. Helander, (Amsterdam: North Holland).

Wiener, B.J., (1971), *Statistical Principles in Experimental Design*, (New York: McGraw-Hill).

Woodson, W.E., (1981), *Human Factors Design Handbook: Information and guidelines for the design of systems, facilities, equipment and products for human use.* (New York: McGraw-Hill).

Chapter 11

Is this chapter any use? Methods for evaluating text

James Hartley

Introduction

How can we evaluate a piece of text? What questions must we ask, and how can we answer them? The literature in this area reveals a concern with at least four issues. We can ask questions about a text's content; about the way it is presented (in terms of its typography and layout); about the way it uses devices (such as headings) and illustrative materials (such as diagrams); and about its suitability for its intended audience. These four issues form the basis of this chapter.

Evaluating content

We can approach the task of assessing the content of textual materials from many different points of view. When the material is a procedural document, a form to be completed, or an explanatory note to accompany some other material, then the procedures used to evaluate its content will be rather different from those used to evaluate a chapter in a textbook. Our main concern will be with whether the text is fit for the job. One common way of assessing this (in addition to making one's own judgement) is to ask other people, experts and users, for their opinions. However, if the text we are assessing is, say, a textbook or a chapter in a textbook such as this one, then our main concern will be to decide whether or not the content is sufficient for our purposes. To assess this we typically skim through the material, study the headings and subheadings, examine the illustrative materials, and perhaps read the introduction and/or concluding summary. We then decide whether the coverage is sufficiently detailed and sufficiently interesting for us to pursue it in more detail. Since we may be evaluating a text for others to use, rather than ourselves, we will look to see if there are outdated materials, errors of facts or principles, unjustified inferences, important omissions, and biases of any kind—national, racial and sexual (Zimet, 1976). Textbook material, to be as useful as possible, must contain content that is accurate, unbiased, up to date and sufficient for the purpose at hand.

This may seem obvious, but what is not obvious is how we can be sure that the content meets such qualifications. Not only are the above descriptions hard to specify but also at times they may be misleading. Users of technical documents sometimes complain, for instance, that such documents contain *too much* information—that is, that they contain more than is needed for the task in hand. The same kind of problem can also arise with textbooks. How much background information, of interest historically but now out of date or even inaccurate by today's standards, should go into, say, an ergonomics textbook? And how can we be assured that selective bias is not present, except through our own rather subjective attempts to interpret today's shifting nationalistic, racial and sexual standards?

Evaluating content is at best a difficult activity, and usually a subjective one. One way to increase its objectivity is to increase the number of judges, and to provide some sort of check-list to help ensure that the judges all evaluate the same issues. Such an approach is commonly used for evaluating textbooks in countries which have state-controlled school systems. An unpublished report to Unesco from the Educational Products Information Exchange Institute entitled *Selecting Among Textbooks* contains, and critically comments on, a dozen such check-lists which teachers and subject matter experts have used when they have been assessing or making comparisons between texts. Figure 11.1 provides an illustration.

Such check-lists are usually completed *before* recommending a particular textbook for use, but there is, of course, no reason why such information could not be collected *after* the texts have been used by readers and by teachers. Information gained in this way would be useful in deciding whether or not to use the book again, and it would be helpful to authors who are planning subsequent editions. Such feedback sheets can occasionally be found in scholastic textbooks and academic journals.

The layout of text

Figure 11.1 shows that readers are often asked to judge the technical quality of text as well as the content. Generally speaking it is difficult without expert knowledge to evaluate the minutiae of typographic practice. However, it is possible to look out for some problems. Questions might be asked, for example, concerning the density of the type (are the lines too close?) and the excessive use of typographical cueing (does it look too messy?). Sless (1984) points out that the cover page shown in Figure 11.2 uses five different unrelated typefaces in the space of seven lines. Other examples of multiple

A. Format of book
1. General appearance
2. Practicality of size and colour for classroom use
3. Readability of type
4. Durability and flexibility of binding
5. Appeal of page layouts
6. Appropriateness of the illustrations
7. Usefulness of chapter headings
8. Usability of index
9. Quality of the paper

B. Organization and content
10. Consistency of the organization and emphases with the teaching and learning standards of the school
11. Consistency of the point of view of the book with the basic principles of the subject area for which the book is being considered
12. Usefulness in stimulating critical thinking
13. Aid in stimulating students forming their own goals and towards self-evaluation
14. Usefulness in providing situations for problem solving
15. Usefulness in furthering the systematic and sequential program of the course of study
16. Clarity and succinctness of the explanations
17. Interest appeal
18. Provision for measuring student achievement
19. Adequacy of the chapter organization
20. Adaptability of content to classroom situations and to varying abilities of individual students
21. Degree of challenge for the reasonably well-prepared students
22. Usefulness for the more able students
23. Usefulness for the slow learners
24. Adequacy of the quality and quantity of skills assignments
25. Provision for review and maintenance of skills previously taught

Figure 11.1. A checklist for assessing textbooks. (In the original version each item is followed by a five-point rating scale from very good to very bad.)

cueing (and its adverse effects upon the reader) can be seen in publications by Shebilske and Rotondo (1981) and Tukey (1977).

When considering the density of text it is appropriate to bear in mind that readers like text to be spacious and well structured. How can typography help one to achieve these aims? I have written frequently and at length elsewhere on how the layout of complex text can be so configured that it

the visual literacy center

presents the

PROVOCATIVE PAPER SERIES
#1

Visual Literacy, Languaging, and Learning

by

JOHN L. DEBES and CLARENCE M. WILLIAMS

Figure 11.2. An example of multiple typographic cueing (Reproduced with permission)

conveys at a glance its underlying structure. Here I shall only repeat the basic arguments. (Readers are referred to Hartley, 1985 and 1987a for more detailed discussions.) These arguments are as follows:

1. Both the horizontal and the vertical spacing of the text needs to be regular and consistent;

2. This means that interword spacing should be regular (as in typescript); and

3. That the interline, inter-paragraph and inter-section spacing should each be consistent throughout the text;

4. To achieve this the text should be set with a ragged right hand edge (technically called unjustified composition), and

5. With a variable baseline; i.e. the text should not necessarily have the same number of lines on each page;

6. The stopping point for text should be determined (horizontally) by syntactic considerations and (vertically) by sense. (For example, one should not start a page with the last line of a paragraph.)

Figures 11.3 and 11.4 provide before and after illustrations of this approach in practice.

There are several ways of assessing the effects of such typographical changes to text. In fact a whole battery of procedures may be used as appropriate. I will discuss the measures used in such comparisons later but, for the present, I shall note that they cover search and retrieval tasks, and measures of reading speed, comprehension, performance and reader preferences.

INSULATING GLOVES

1. GENERAL

1.01 This section covers the description, care and maintenance of insulating gloves provided for the protection of workmen against electric shock, and the precautions to be followed in their use.

1.02 This section has been reissued to include the D and E Insulating Gloves.

2. TYPES OF INSULATING GLOVES

2.01 All types of insulating gloves are of the gauntlet type and are made in four sizes: 9-1/2, 10, 11 and 12. The size indicates the approximate number of inches around the glove, measured midway between the thumb and finger crotches. The length of each glove, measured from the tip of the second finger to the outer edge of the gauntlet, is approximately 14 inches.

2.02 There are various kinds of insulating gloves. The original ones were just called Insulating Gloves. After that B, C, D and E Insulating Gloves were developed. As described below, the D Glove replaced the original Insulating Gloves and the E glove replaced the B and C Gloves.

2.03 **Insulating Gloves** are thick enough to eliminate the need for protector gloves and are intended for use without them. These gloves have been superseded by the D Insulating Gloves.

Figure 11.3. The first page of a piece of technical text in its original format (Reproduced with permission)

Structural devices and illustrative materials

Access structures

Readers come to text with many different purposes: they need to be able to skim, to search, to look back, to look ahead, and to read in detail. They may want to read the text themselves or to choose it for another person. Such readers need to be able to find their way around a text easily. This textual 'navigation' can be aided by good typograhic practice (as described above), and by the effective use of what Waller (1979) called *access structures*— devices which facilitate the readers' access to the text. Such access structures may be found at the beginning and at the end of texts (e.g. contents pages,

INSULATING GLOVES

1.0 General

1.1 This section covers the description,
care and maintenance of insulating gloves
provided for the protection of workmen
against electric shock, and the precautions
to be followed in their use.

1.2 This section has been reissued to include
the D and E Insulating Gloves.

2.0 Types of Insulating Gloves

2.1 All types of insulating gloves are of the
gauntlet type and are made in four sizes:
9-1/2, 10, 11 and 12.
The size indicates the approximate number of
inches around the glove, measured midway
between the thumb and finger crotches.
The length of each glove, measured from
the tip of the second finger to the outer edge
of the gauntlet, is approximately 14 inches.

2.2 There are various kinds of insulating gloves.
The original ones are just called Insulating
Gloves.
After that B, C, D and E Insulating Gloves
were developed.
As described below, the D glove replaced
the original Insulating Gloves and
the E glove replaced the B and C gloves.

2.3 **Insulating Gloves** are thick enough to eliminate
the need for protector gloves and are intended
for use without them.
These Gloves have been superseded by
the D Insulating Gloves.

Figure 11.4. The same page with a revised typographical layout

glossaries, indexes, summaries) and embedded in the texts themselves (e.g.
chapter indicators and titles at the tops of pages, page numbers, text headings,
figure and table numbers and captions).

In evaluating a piece of text, one looks for the presence of devices such
as these. It is assumed that it is easier to use a piece of text that has access
structures than it is to use one that has not. Indeed, the little research that
has been done does suggest that access structures are useful aids in text. It
appears, for instance, that reading glossaries before a piece of technical text
can help people's understanding of it (Black, 1987); that devices such as
advance organizers and summaries can aid people's comprehension (Ausubel,

1960; Hartley and Davies, 1976; Hartley and Trueman, 1982; Jonassen, 1982b); and that headings aid search and retrieval and, to a lesser extent, recall (Hartley and Jonassen, 1985). Of course, judges evaluating text have to bear in mind that there is no guarantee that because a device is provided it will always prove effective. Poor headings can mislead the reader (Swarts *et al.*, 1980), some devices (such as boxed asides) can be distracting (Hartley, 1987b), and some texts may use these devices excessively (e.g. by providing a heading for every paragraph).

Illustrative materials

In this chapter I shall for convenience refer to tables, graphs, diagrams, examples and illustrations as 'illustrative materials'. I have listed in Table 11.1 the authors of more detailed summaries of research on each of these separate textual features. The points I wish to make here, however, are basically the same, whatever the particular feature being discussed. These points are as follows:

1. To judge the effectiveness of illustrative materials one needs to know about good practice.

2. Two main features to look for in all illustrative materials are *simplicity* and *clarity*.

3. The design and presentation of the repeated illustrative materials should be *consistent*.

4. The positioning of illustrative materials is important: such materials should appear as soon as possible after their first textual reference and they should not get misplaced, or out of sequence.

Table 11.1. Useful research reviews on how to present illustrative materials in printed and electronic text

	Books	Articles
Graphs	Cleveland (1985)	Macdonald-Ross (1977a)
	Tufte (1983)	Macdonald-Ross (1977b)
Tables	Reynolds and Simmonds (1983)	Ehrenberg (1977)
		Wright (1982)
		Norrish (1984)
Diagrams		Winn and Holliday (1982)
		Winn (1987)
Illustrations	Willows and Houghton (1987)	Levie and Lentz (1982)
	Houghton and Willows (1987)	Levie (1984)
Cartoons	Harrison (1981)	Sewell and Moore (1980)
Examples		Mandl *et al.* (1984)
Flow-charts	Wheatley and Unwin (1972)	Wright (1982)

Useful general texts which cover most of these features are Easterby and Zwaga (1974), Hartley (1985), Jonassen (1982a, 1985), and Reynolds and Simmons (1983). A major review of the relevant literature has been provided by Michael (1988).

Figure 11.5 illustrates a contravention of point 3. Here, within the space of five pages, the authors used four different ways of presenting a graph. Such an approach can only be confusing for the reader. A good example of point 4 occurs in Cleveland's otherwise excellent book on graphs (Cleveland, 1985); at the beginning of Chapter One, each figure occurs not in its own text section but in the following one, thus all of the figures are out of step. Admittedly it is not always possible to follow my counsel of perfection—especially when there are several illustrations and little text, or when the illustrations are exceptionally large—but designers could try harder to minimize mismatches between text and illustration.

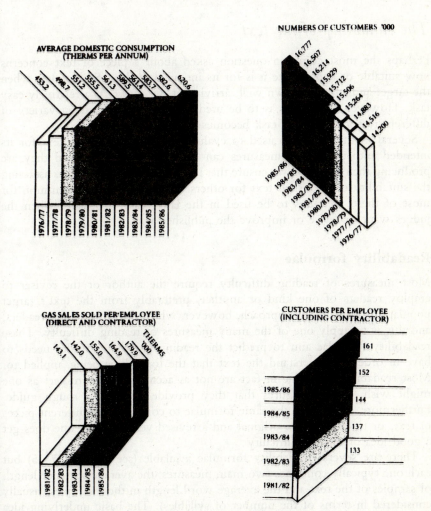

Figure 11.5. A series of bar charts from a promotional pamphlet. (The original charts were colour coded in shades of blue) (Figure reproduced with permission)

Again there are several ways of assessing the effectiveness of both access structures and illustrative materials and perhaps the most common of these is to carry out actual experiments. In my own research I have found this to be the most useful approach, although it is not without its limitations. (Usually one cannot test a sufficient number of variants of the device being tested and, although it is relatively easy to show that revised materials may be easier to use than their original versions, it is harder to compare effectively subtle variations in the revisions.) Preference measures may also prove useful in this context. I once found, for example, significantly greater preferences for a piece of text whose illustrations had been placed more on line with their textual references.

The suitability of the text

Perhaps the most common question asked about a piece of text concerns how suitable or appropriate it is for its intended readers. Sometimes, when the target audience is known well, arriving at the answer is a relatively easy task. However, if the text is to be used by multiple users for a variety of different reasons, then the task becomes more difficult.

Several methods can be used to evaluate the suitability of a text for its intended audience. Most measures can be used by *authors* when they are producing their own text (to ensure that it is effective) and by *judges* assessing the suitability of published text for others. It is perhaps more common for most of these measures to be used in the latter case, especially when the judges want to revise or improve the published text.

Readability formulae

Most measures of reading difficulty require the author or the reviser to employ readers of one kind or another, preferably from the text's target population. There is one approach, however, which does not require readers, and this is to apply one of the many measures of reading difficulty. These readability formulae aim to predict the reading age that a reader needs to have in order to understand the text that the formula is being applied to. Most readability formulae in fact are not as accurate in this respect as one might wish, but the figures that they provide do give a rough guide. Furthermore, if one uses the same formulae to compare two different pieces of text, or to compare an original and a revised version, then one does get a good idea of relative difficulty.

There are several readability formulae available (see Klare, 1974–5) but each one typically combines two main measures: the average sentence length of samples of the text and the average word length in these samples (usually considered in terms of the number of syllables). The basic underlying idea is that the longer the sentences and the more complex the vocabulary, then the more difficult the text will be. Clearly this notion, whilst generally sensible, has its limitations. Some technical abbreviations are short (e.g DNA) but difficult for people who have not heard of them. Some words are long but, because of their frequent use, become quite familiar (e.g. ergonomics in this context). Word, sentence and paragraph order is not taken into account by the formulae and nor are the effects of other devices used to aid comprehension such as typographical layout, access structures and illustrative materials. Motivation and prior knowledge, too, are ignored. Davison and Green (1987) provide other, more technical criticisms.

A readability measure then provides a *rough* guide to text difficulty. If the score goes off the scale (as is often the case with government documents) then it is clear that the text is too difficult for most readers. Several studies have been carried out which demonstrate the advantages of producing more readable text (see Hartley, 1981a; US Department of Commerce, 1984; Cutts and Maher, 1986); but it is true to say that in the majority of these studies more than just the readability of the text has been manipulated. (Revised layouts, new illustrative materials and wholesale deletions are fairly common elements in such reports.)

Some readability formulae are quite complex to calculate by hand (e.g. the Flesch) but there are tables available to ease the calculations (see Hartley, 1985, p.54). Other formulae are simpler to calculate. One of the simplest, the Gunning Fog Index, works as follows:

(a) take a sample of 100 words;
(b) calculate the average number of words per sentence in the sample;
(c) count the number of words with three or more syllables in the sample;
(d) add the average number of words per sentence to the total number of words with three or more syllables; and
(e) multiply the result by 0·4 to obtain the (American) reading grade level;
(f) add five to obtain an equivalent British reading age.

Today, with the advent of word processing systems, it is easier to carry out readability tests. For example, the Style and Diction programs of the IBM Xenith text formatting system can be applied to text to provide sets of readability data, and various other measures such as the number of sentences used and the average sentence and word lengths. Table 11.2 shows the output obtained when this program was run on the first 20 sentences of this chapter. It can be seen that the four readability formulae are roughly in agreement, and that the general picture suggests that the text is suitable for 16–19 year olds. Other kinds of computer programs are probably more useful for writers and revisers. Table 11.3 lists a sample of the programs available in *The Writer's Workbench* system developed at Bell Telephone Laboratories (Macdonald, 1983). Studies of writers, using such programs have indicated their value (e.g. see Coke, 1982; Hartley, 1984; Sterkel *et al.* 1986).

Cloze procedures

Another measure, similar in some respects to a readability formula, but this time requiring readers, is the cloze procedure (Taylor, 1953). With this technique samples of passages are presented with every *n*th word missing, and readers are required to fill in the missing gaps. Technically speaking, if say every 6th word is deleted, then six versions should be prepared with the gaps each starting from a different point. However, it is more common ————— prepare one version, and perhaps ————— to focus the gaps on ————— words. Whatever the procedure, the ————— are scored either by marking ————— correct those responses which directly ————— what the original author actually —————, or by accepting these and ————— synonyms. Since the two scoring methods correlate highly, it is more objective to use the tougher measure of matching words. (In the case above these are: to, even, important, passages, as, match, said, acceptable.) Scores can be increased slightly by varying the length of the lines to match the lengths of the missing words, providing dashes to match the number of letters missing in each word, or by providing the first of the missing letters (Hartley and Trueman, 1986). These minor variations, however, do not affect the main purpose of this measure which is to assess the reader's comprehension of the text, and, by inference, its difficulty. The cloze procedure has a strong advantage over readability formulae in that it can be used to assess the effects of the presence of other features (such as illustrations or underlining) on the comprehension of text (e.g. see Hartley *et al.* 1980a; Newton, 1983; Reid *et al.* 1983).

Table 11.2. Output from the Style program of the IBM Xenith Text Formatting System applied to the first 20 sentences of this chapter

Number of words	471
Average sentence length	23·5
Average word length	4·75
Percentage of short sentences (<19 words)	40%
Percentage of long sentences (>34 words)	20%
Longest sentences (No. 4) 44 words	
Sentence types:	
simple	40%
complex	30%
compound	15%
compound–complex	15%
Readability (US grade levels):	
Coleman–Liau	10·9
Kincaid	12·5
Auto	12·7
Flesch	13·4

Table 11.3. A sample of computer programs available in 'The Writer's Workbench' system developed at Bell Laboratories

Programs which indicate:
  spelling errors
  punctuation errors
  word repetitions
  split infinitives
  use of passives
  use of nominalizations
  use of abstract words
  acronyms
  long sentences
  sexist phrases
  awkward choices of words/phrases (with suggested improvements)

Programs which give:
  parts of speech for each word
  readability scores
  average length of sentences
  number of sentence types (simple, complex, compound, compound-complex)
  comparison figures on details such as these for other 'model' technical texts

Programs which:
  summarize the content—by listing headings
  summarize the content—by giving the first and last sentence of each paragraph
  segment the lines of the text according to line-length requirements and syntactic rules

Readers' judgements

A rather different, but useful measure of text difficulty is to ask readers to judge it for themselves. One simple procedure is to ask readers to circle on the text those areas, sentences, or words that they think *readers less able than themselves* will find difficult. In my experience if you ask readers to point out difficulties for *others*, respondents are much more forthcoming than if you ask them to point out their own difficulties. An elaboration of this technique, of course, is to use *protocol analysis* (see the chapter by Bainbridge in this book). Here readers are asked to verbalize what they are thinking about as they are reading or using text. This technique has proved extremely useful in evaluating instructional manuals (see e.g. Swaney *et al.*, 1981; Sullivan and Chapanis, 1983; Komoski and Woodward, 1985). Some critics have suggested that talking about the task whilst trying to do it can cause difficulties: such problems can be partly overcome by getting readers to work in pairs and discussing together the problems they are facing (Miyake, 1986) or by videotaping the procedure and then getting respondents to talk through the resulting tape (Schumacher *et al.*, 1984).

In this section of the chapter we may also consider *reader preferences*. When I discussed preference measures in the first edition of *Designing Instructional Text* (Hartley, 1985) I dismissed their value because I considered preferences to be 'untutored judgements'. I argued that people's preferences can be based on inappropriate considerations: for example, a person might prefer a car on the basis of its colour rather than its technical quality. Today I am not so dismissive of preference measures. My own studies have shown that such measures are sensitive to differences between novices and experts, and that they can be used to assess the effects of instruction (Hartley, 1981b; Hartley and Trueman, 1981). Furthermore, people have clear views about what they like in texts, and how they expect texts to perform (Wright and Threlfall, 1980). So, first impressions might colour one's attitudes to texts. A text which looks dull, dense and turgid is not going to encourage readers no matter how important the content. So I now use preference judgements as one of a battery of measures. Preferences can provide additional information— they can tell you whether a revised text is preferred to the original, whether people see no difference, or whether people prefer the original even though (in your eyes) it is not as effective. Such information needs to be considered, but along with other measures. (See also McClelland in this book for a view on incorporating preference measures into a series of evaluations.)

One useful tool to use if you require preference rankings for a number of say typographic designs which vary in different ways is the method of *paired comparisons*. Suppose, for example, you have 15 designs. You could ask participants to judge them (overall, or on a particular aspect) and to make paired comparisons. Essentially, this involves each judge comparing design 1 with 2 and recording the preference, then 1 with 3, 1 with 4, 1 with 5 and so on, until 1 with 15 is reached. Then the judge starts again, this time comparing 2 with 3, 2 with 4, 2 with 5 and so on until 2 with 15. This procedure is replicated in terms of 3 with 4, 5, 6, etc., 4 with 5, 6, 7, etc., until all the designs have been systematically compared. Finally, the number of recorded preferences for each design is totalized to see which one has been preferred the most often. Examples illustrating this approach can be found in Hartley *et al.* (1979) and Hartley (1979-80).

Writing and revising text

Before moving on to consider experimental methods of evaluating text, it is perhaps appropriate at this stage to say something about techniques for producing readable writing. Hayes and Flower (1986) argue that writing can be considered as a complex activity or skill, and that like all skills, writing is made up of subcomponents which have to be put together in such a way to guarantee smooth performance. Many of these subcomponents are organized hierarchically and some have priority over others, but even during the writing of a simple sentence, the writer shifts attention from one subcomponent to another. Thus, for example, the following issues have competed for my attention whilst I was writing this particular paragraph.

Who is this chapter for? Will they be involved in writing or revising?

Where should I put this bit? Should it come before or after the section on readability?

Will readers understand what I mean by subcomponents and such components being hierarchically organized? Should I explain that one has *long-term* goals—deciding what to put in the chapter; *medium-term* goals—deciding where to put it; *ongoing* goals—deciding how to say it; *specific* problems—deciding how to spell hierarchically; and *immediate* goals—writing clearly so my secretary can read what I have written.

When writing is considered in this way it is possible to suggest that in order to improve writing, writers need (a) to practise, and (b) to focus on different goals at different times. Thus, for example, one can first concentrate on getting the material down in a rough and ready fashion and then one can work on polishing it. As Wason (1983) put it, "First say it, and *then* try to say it well".

There are numerous guidelines on how to write clear text (see e.g. Klare, 1979; Hartley *et al.*, 1980b; Armbruster and Anderson, 1985) and also on how to revise published text. In my own work with secondary school children I use the guidelines for revision listed in Figure 11.6. These guidelines are based upon recent work on the revising process (see Hayes and Flower, 1986). Noticeable changes occur when procedures such as these are applied to published text. Figure 11.7 provides an illustration. Here the text presented in Figure 11.3 and 11.4 has been revised using guidelines similar to those provided in Figure 11.6.

Experimental comparisons

The methods described above can be used relatively informally: some researchers, however, may wish to carry out more formal experimental comparisons, and to use more precise measuring instruments. Some people, for instance, might be interested in measuring reading speed, ease of location and retrieval, ease of use, or ease of comprehension and recall. Some methods might be quite precise involving, for example, the use of eye-movement recording devices, or video tapes of readers using instructional texts.

Over the last ten years or so my colleagues and I have used a variety of measures (and combinations of them) to evaluate instructional text. Table 11.4 tries to encapsulate some of the strengths and some of the limitations of these measures which I have divided into five main groups. Opposite these five blocks are my estimates of the reliability of these measures and some comments about them. Table 11.4 shows that some measures are more reliable than others, and common sense indicates that some measures are more suitable than others for different purposes. Thus oral reading measures give detailed information about specific reading difficulties; search and retrieval tasks are appropriate for evaluating the layout of highly structured

1. *Read the text through.*
2. *Read the text through again but this time ask yourself:*
   - *What is the writer trying to do?*
   - *Who is the text for?*
3. *Read the text through again, but this time ask yourself:*
   - *What changes do I need to make to help the writer? How can I make the text clearer?*
   - *What changes do I need to make to help the reader? How can I make the text easier to follow?*
4. *To make these changes you may need:*
   - *to make big or global changes (e.g. re-write sections yourself);*
   or
   - *to make small or minor text changes (e.g. change slightly the original text).*
   You will need to decide whether you are going to focus first on global changes or first on text changes.
5. *Global changes you might like to consider in turn are:*
   - *re-sequencing parts of the text;*
   - *re-writing sections in your own words;*
   - *adding in examples;*
   - *changing the writer's examples for better ones;*
   - *deleting parts that seem confusing.*
6. *Text changes you might like to consider in turn are:*
   - *using simpler wording;*
   - *using shorter sentences;*
   - *using shorter paragraphs;*
   - *using active rather than passive tenses;*
   - *substituting positives for negatives;*
   - *writing sequences in order;*
   - *spacing numbered sequences or lists down the page (as here).*
7. *Keep reading your revised text through from start to finish to see if you want to make any more global changes.*
8. *Finally repeat this whole procedure some time after making your initial revisions (say 24 hours) and do it without looking back at the original text.*

Figure 11.6. The author's guidelines for revising text

text; comprehension measures are more appropriate for evaluating the effectiveness of continuous prose; and readability and preference measures are useful as additional sources of information.

Experimental evaluations of text require one to start off with a specific problem, to prepare a variety of solutions, to select the ones that seem most plausible, and then to evaluate their effectiveness. The evaluation will always be limited because (1) it is not possible to evaluate and compare every possible solution to a problem, (2) the methods which we have at our disposal for evaluation in this area tend to be somewhat limited, and (3) different measures have their own in-built assumptions. (For example most of the techniques listed in Table 11.4 seem to assume that the reader starts at the beginning of a text and reads through it steadily to the end.) None the less, these experimental comparisons do provide evidence which is of value. And, this is especially the case when several measures are combined (Wright, 1987).

The following case history demonstrates that combined measures are more informative than single ones. In this study comparisons were made between full length (three page) versions of Figure 11.3 and Figure 11.7. The results were as follows:

INSULATING GLOVES

1.0 General

1.1 This section describes how to care for
and maintain the insulating gloves
that will protect you from electric shocks.

1.2 The section has been revised to include
the D and E Insulating Gloves.

2.0 Types of Insulating Gloves

2.1 All insulating gloves are made
in the gauntlet style.
There are four sizes: 9½, 10, 11, 12.
The size indicates the approximate number
of inches around the glove across the palm.
Each glove is about 14 inches long
from the bottom of the gauntlet to the top
of the second finger.

2.2 There are various kinds of insulating gloves.
The first kind were originally just called
Insulating Gloves.
After that the B, C, D and E Insulating Gloves
were developed.
As described below, the D Glove replaced the
original insulating gloves, and the E glove
replaced the B and C gloves.

2.3 So **Insulating Gloves** have now been replaced
by D Insulating Gloves.
(Insulating Gloves could be worn without
protector gloves.)

Figure 11.7. The same material as that shown in Figure 11.3, only this time both the typography
and the text have been revised

1. In terms of *readability* the first 100 words of the original document
(excluding headings) had a reading age level of 19·5 years, whereas the first
100 words of the revised version had a reading age level of 15 years (Gunning
FOG index).

2. In terms of *reading speed* there was no significant difference between the
average times taken by two groups of ten university students to read the
three pages set in either version.

3. In terms of *factual recall* however, these same groups of students recalled
an average of 5·4 out of 10 for the original and 7·9 out of 10 for the revised
version. (This difference was statistically significant.)

4. In terms of *preferences*, seven out of ten university colleagues chose the
revised version in preference to the original when asked to judge which

Table 11.4. Methods of evaluation used in assessing text, and estimates of their reliability

Method	Estimated reliability	Test–retest correlations obtained in our studies	Comments
Reading aloud	High	0·78 to 0·95	Insensitive to differences in layouts
Reading aloud (text upside down)	Very high	0·87 to 0·99	Slows reading right down; larger spreads of scores with males than females
Scanning technical material*	High	0·75 to 0·91	Good for technical materials: sensitive
Scanning prose* (short intervals)	Moderate	0·49 to 0·68	Moderately useful and sensitive
Scanning prose* (wide intervals)	Low	0·36 to 0·83	Poor: searchers 'get lost'
Silent reading speed (without test)	Fairly high	0·53 to 0·96	The researcher does not know what has been read
Silent reading speed (with test to follow)	Fairly high	0·70 to 0·82	Knowledge of forthcoming test slows readers down markedly
Comprehension† (cloze test)	Fairly high	0·55 to 0·95	Useful for assessing relative difficulty
Comprehension (recall questions)	Low	0·46 to 0·73	Too specific to make comparisons with, if different materials are used
Preferences	Fairly high	0·72	Useful as an additional measure
Readability formulae	Very high	(not applicable)	Useful as an additional measure

*Scanning involves providing the reader with a list of phrases drawn from the text, each with a word missing. The reader has to scan (or skim) the text, find the phrase, and write in the missing word.

†The cloze test involves providing the reader with a text with, say, every sixth word missing. The reader has to supply the missing words.

figure they found 'the clearer'. (This difference, whilst pleasing, was not statistically significant.)

5. When the reading speed and factual recall measures were used again in a *replication* study with a further 20 students, the results were repeated almost exactly.

This composite picture allows me to suggest that the revised version was easier to read, that the students extracted more information per unit time, and that judges were more likely to prefer the revised version.

We may finally note in this section that a different kind of measure, not listed in Table 11.4, is that of the *cost* of production. In our experiments we have sometimes shown reductions in the cost of production without any loss in comprehension and sometimes great improvements in comprehension for a slight increase in cost (see Hartley, 1985). Results such as these point to the hidden costs of badly designed materials and to the fact that *cost effectiveness* is an important consideration (expanded on by Simpson and Mason in this book).

Cyclical testing and revising

Finally I want to point out that as far as an ergonomic approach is concerned, we may not be interested so much in making *comparisons* between texts, as in using the measures available to help us to *improve* a text. One of the most useful approaches here is that of cyclical testing and revising. This approach requires us to test the text with appropriate readers, revise it on the basis of the results obtained, test it again with another set of readers, revise it again and so on, until the text achieves its objectives. This iterative approach was used extensively in a study reported by Waller (1984) that described how he and his colleagues at the Open University set about improving a form that was to be used by unemployed people claiming supplementary benefit in the UK.

A prototype form was developed and piloted by the Department of Health and Social Security. The form was small in format (165 × 204 mm) with eight pages organized as a folded concertina (see Waller, 1984). Although the respondents found the form attractive to look at, they found it difficult to use. About 75% of the forms were completed unsatisfactorily in one way or another. This meant that the forms had to be returned and/or respondents followed up in some way before an assessment of benefit could be made and that this was a very expensive procedure.

The main sources of error in the prototype form appeared to be:

1. Problems of relevance and contextual interpretation: the form did not elicit enough information for an assessment to be made, and appeared irrelevant to a large number of claimants.

2. Problems of reading sequence: many sections did not apply to many claimants, but the form gave inadequate directions concerning which parts were to be completed.

3. Problems in graphic design: poor design practice also contributed to the problems of sequencing.

The problems of the form ranged from the fairly obvious to the subtle and the debatable. In order to redesign the form the prototype was first tested with small groups of appropriate respondents. (Interestingly, part of this assessment included the use of an eye movement recorder to assess which pieces of text were read and in what order.) The aim of this first assessment was to isolate the main causes of difficulty, and to collect data against which the redesigned forms could be compared. It appeared from this first testing that the form was not asking the right questions to gather the information that the civil servants needed. In addition, many questions were ambiguous.

Thus a redesigned version was prepared with emphasis on revising the language and the sequencing of the form: thus the content was well spaced and simply designed so that the confusions brought about by poor design practice (noted earlier) could be avoided at this stage. The revisions focused on making the branching instructions more explicit, and on giving users clearer instructions when they first encountered such a branching instruction.

This redesigned version was tested (again with small groups of appropriate respondents). It was clear that improvements had been made, but that more could be done. So a third version was then prepared. Headings for the different sections were added, and the routing instructions were further improved. The testing of this third version showed that this had solved most of the problems. Thus a fourth and final version was prepared. This version used colour coding for the main headings (earlier versions had been in black and white), a larger page-size (200 × 330 mm) and yet another re-sequenced order. Once the logical and linguistic problems were sorted out, the typography could be enhanced.

This final version was tested with larger groups of appropriate respondents. The results now indicated that about 75% of the forms were completed satisfactorily (as opposed to the original 25%). Today, after further revisions, the successful completion rate is estimated to be over 80%. These impressive results have led to massive cost benefits for the Department of Health and Social Security.

Summary

This chapter has discussed the evaluation of texts from four different viewpoints:
1. Evaluating content.
2. Evaluating typographical layout.
3. Evaluating structural devices and illustrative materials.
4. Evaluating the suitability of the text.

In each section a variety of methods for assessing text has been mentioned. Content can best be evaluated by what might be called survey techniques. Typographical layout, structural devices and illustrative materials can be evaluated both by survey techniques and by laboratory and field experiments. The suitability of text can be assessed by applying readability formulae, but such an approach would seem unduly restrictive when there are other more useful survey, field and experimental approaches. In many cases the aim will be to improve the text. Thus those approaches that emphasize combining methods and those approaches that use a test–revise–retest model are likely to be the most helpful for ergonomists.

Acknowledgments

I am grateful to colleagues who commented on earlier drafts of this chapter, and to Margaret Woodward and Dorothy Masters for their repeated processing of the text.

References

Armbruster, B.B. and Anderson, T.H. (1985). Producing 'considerate' expository text: or easy reading is damned hard writing. *Journal of Curriculum Studies*, **17**, 247–274.

Ausubel, D.P. (1960). The use of advance organisers in the learning and retention of meaningful verbal material. *Journal of Educational Psychology*, **51**, 267–272.

Black, A. (1987). Lexical support in discourse comprehension. Unpublished Ph.D. Thesis, University of Cambridge.

Cleveland, W.S. (1985). *The Elements of Graphing Data* (Monterey, CA: Wadsworth).

Coke, E.U. (1982). Computer aids for writing text. In *The Technology of Text*, edited by D.H. Jonassen (Englewood Cliffs, NJ: Educational Technology Publications).

Cutts, M. and Maher, C. (1986). *The Plain English Story* (Whaley Bridge, Stockport: Plain English Campaign).

Davison, A. and Green, G. (Eds) (1987). *Linguistic Complexity and Text Comprehension: A Re-examination of Readability with Alternative Views* (Hillsdale, NJ: Erlbaum).

Easterby, R. and Zwaga, H. (Eds) (1984) *Information Design* (Chichester: John Wiley).

Ehrenberg, A.S.C. (1977). Rudiments of numeracy. *Journal of the Royal Statistical Society A*, **140**, 227–297.

Gunning, R. (1968). *The Technique of Clear Writing* (New York: McGraw Hill).

Harrison, R.P. (1981). *The Cartoon: Communication to the Quick* (Beverly Hills: Sage).

Hartley, J. (1979–80). Designing journal content pages: the role of spatial and typographic cues. *Journal of Research Communication Studies*, **2**, 83–98.

Hartley, J. (1981a). Eighty ways of improving instructional text. *IEEE Transactions on Professional Communication*, **PC-24**, 17–27.

Hartley, J. (1981b). Sequencing the elements in references. *Applied Ergonomics*, **12**, 7–12.

Hartley, J. (1984). The role of colleagues and text-editing programs in improving text. *IEEE Transactions on Professional Communication*, **PC-27**, 42–44.

Hartley, J. (1985). *Designing Instructional Text* (London: Kogan Page).

Hartley, J. (1987a). Designing electronic text: the role of print based research. *Educational Communication and Technology Journal*, **35**, 3–17.

Hartley, J. (1987b). Typography and executive control processes in reading. In *Executive Control Processes in Reading*, edited by B.K. Britton and S.M. Glynn (Hillsdale, NJ: Erlbaum).

Hartley, J. and Davies, I.K. (1976). Pre-instructional strategies: the role of pretests, behavioral objectives, overviews and advance organisers. *Review of Educational Research*, **46**, 239–265.

Hartley, J. and Jonassen, D.H. (1985). The role of headings in printed and electronic text. In *The Technology of Text*, Volume 2, edited by D.H. Jonassen (Englewood Cliffs, NJ: Educational Technology Publications).

Hartley, J. and Trueman, M. (1981). The effects of changes in layout and changes in wording on preferences for instructional text. *Visible Language*. **XV**, 13–31.

Hartley, J. and Trueman, M. (1982). The effects of summaries on the recall of information from prose: five experimental studies. *Human Learning*, **1**, 63–82.

Hartley, J. and Trueman, M. (1986). The effects of the typographic layout of cloze-type tests on reading comprehension scores. *Journal of Research in Reading*, **9**, 116–124.

Hartley, J., Bartlett, S. and Branthwaite, J.A. (1980a). Underlining can make a difference—sometimes. *Journal of Educational Research*, **73**, 218–224.

Hartley, J., Trueman, M. and Burnhill, P. (1979). The role of spatial and typographic cues in the layout of journal references. *Applied Ergonomics*, **10**, 165–169.

Hartley, J., Trueman, M. and Burnhill, P. (1980b). Some observations on producing and measuring readable writing. *Programmed Learning and Educational Technology*, **17**, 164–174.

Hayes, J.R. and Flower, L.S. (1986). Writing research and the writer. *American Psychologist*, **41**, 1106–1113.

Houghton, H.A. and Willows, D.M. (Eds) (1987). *The Psychology of Illustration*, Volume 2. *Instructional Issues* (New York: Springer).

Jonassen, D.H. (Ed.) (1982a). *The Technology of Text* (Englewood Cliffs, NJ: Educational Technology Publications).

Jonassen, D.H. (1982b). Advance organisers in text. In *The Technology of Text* edited by D.H. Jonassen (Englewood Cliffs, NJ: Educational Technology Publications).

Klare, G.R. (1974–5). Assessing readability. *Reading Research Quarterly*, **X**, 62–102.

Klare, G.R. (1979). Writing to inform: making it readable. *Information Design Journal*, **1**, 98–105.

Komoski, P.K. and Woodward, A. (1985). The continuing need for learner verification and revision of textual material. In *The Technology of Text*, Volume 2, edited by D.H. Jonassen (Englewood Cliffs, NJ: Educational Technology Publications).

Levie, W.H. (1984). Research and theory on pictures and imaginal processes: a taxonomy and selected bibliography. *Journal of Visual/Verbal Language*, **4**, 7–41.

Levie, W.H. and Lentz, R. (1982). Effects of text illustrations: a review of research. *Educational Communication and Technology Journal*, **30**, 195–232.

Macdonald, N.H. (1983). The Unix writer's workbench software: rationale and design. *Bell System Technical Journal*, **62**, 1891–1908.

Macdonald-Ross, M. (1977a). How numbers are shown: a review of research on the presentation of quantitative data in texts. *Audiovisual Communication Review*, **25**, 359–409.

Macdonald-Ross, M. (1977b). Graphics in text. In *Review of Research in Education*, Volume 5, edited by L.S. Shulman (Itasca, IL: Peacock).

Mandl, H., Schnotz, W. and Tergan, S.O. (1984). On the function of examples in unstructured texts. Paper available from the authors at Deutches Institut fur Fernstudien an der Universitat Tubingen, Hauptbereich Forschung, Bei der Fruchtschranne 6, 7400 Tubingen 1.

Michael, D.E. (1988). User differences in graphic design: some studies with flow charts. Unpublished Ph.D. Thesis, University of Keele.

Miyake, N. (1986). Constructive interaction and the iterative process of understanding. *Cognitive Science*, **10**, 151–177.

Newton, L.D. (1983). The effect of illustrations on the readability of some junior school textbooks. *Reading*, **17**, 43–54.

Norrish, P. (1984). Moving tables from paper to screen. *Visible Language*, **XVIII**, 154–170.

Reid, D.J., Briggs, N. and Beveridge, M. (1983). The effect of picture upon the readability of a school science topic. *British Journal of Educational Psychology*, **53**, 327–335.

Reynolds, L. and Simmonds, D. (1983). *Presentation of Data in Science* (The Hague: Martinus Nijhoff).

Schumacher, G.M., Klare, G.R., Cronin, F.C. and Moses, J.D. (1984). Cognitive activities of beginning and advanced college writers: a pausal analysis. *Research in the Teaching of English*, **18**, 169–187.

Sewell, E.H. and Moore, R.L. (1980). Cartoon embellishments in informative presentations. *Educational Communication and Technology Journal*, **28**, 39–46.

Shebilske, W.L. and Rotondo, J.A. (1981). Typographical and spatial cues that facilitate learning from textbooks. *Visible Language*, **15**, 45–54.

Sless, D. (1984). Visual literacy: a failed opportunity. *Educational Communication and Technology*, **32**, 224–228.

Sterkel, K.S., Johnson, M.I. and Sjogren, D.D. (1986). Textual analysis with computers to improve the writing skills of Business Communication Students. *Journal of Business Communication*, **23**, 43–61.

Sullivan, M.A. and Chapanis, A. (1983). Human factoring a text editor manual. *Behavior and Information Technology*, **2**, 113–125.

Swaney, J.H., Janik, C.J., Bond, S.J. and Hayes, R. (1981). *Editing for comprehension: Improving the Process Through Reading Protocols*. Technical Report No. 14. Document Design Center, Carnegie-Mellon University, Pittsburgh PA.

Swarts, H., Flower, L.S. and Hayes, J.R. (1980). How headings in documents can mislead readers. Paper available from the authors, Department of Psychology, Carnegie-Mellon university, Pittsburgh, P.A., 15213, U.S.A.

Taylor, W.L. (1953). Cloze procedure: a new tool for measuring readability. *Journalism Quarterly*, **30**, 415–433.

Tufte, E.R. (1983). *The Visual Display of Quantitative Information* (Cheshire, CT: Graphics Press, Box 430).

Tukey, J. (1977). *Exploratory Data Analysis* (New York: Addison Wesley).

U.S. Department of Commerce (1984). *How Plain English Works for Business: 12 Case Studies*, (Washington: Office of Consumer Affairs, US Department of Commerce).

Waller, R.H.W. (1979). Typographic access structures for instructional text. In *Processing of Visible Language*, edited by P.A. Kolers, M.E. Wrolstad, and H. Bouma (New York: Plenum).

Waller, R. (1984). Designing a government form: a case study. *Information Design Journal*, **4**, 36–57.

Wason, P.C. (1983). Trust in writing. Paper available from the author. Department of Phonetics and Linguistics, University College, London.

Wheatley, D.M. and Unwin, A.W. (1972). *The Algorithm Writer's Guide* (London: Longmans).

Willows, D.M. and Houghton, H.A. (Eds) (1987). *The Psychology of Illustration*, Volume 1, *Basic Research* (New York: Springer).

Winn, W. (1987). Using charts, graphs and diagrams in educational materials. In *The Psychology of Illustration*, Volume 1, *Basic Research*, edited by D.M. Willows and H.A. Houghton (New York: Springer).

Winn, W.D. and Holliday, W.G. (1982). Design principles for diagrams and charts. In *The Technology of Text* edited by D.H. Jonassen (Engelwood Cliffs, NJ: Educational Technology Publications).

Wright, P. (1982). A user-oriented approach to the design of tables and flow charts. In *The Technology of Text*, edited by D.H. Jonassen (Engelwood Cliffs, NJ: Educational Technology Publications).

Wright, P. (1987). Issues of content and presentation in document design. In *Handbook of Human–Computer Interaction*, edited by M. Helander (Amsterdam: North-Holland).

Wright, P. and Threlfall, M.S. (1980). Readers' expectation about format influence the usability of an index. *Journal of Research Communication Studies*, **2**, 99–106.

Zimet, S.G. (1976). *Print and Prejudice* (London: Hodder and Stoughton).

Chapter 12

Evaluation of the human–computer interface

Bruce Christie and Margaret M. Gardiner

Introduction

Evaluation of the human–computer interface is a vast and complex subject that could easily fill a book by itself. Part of the complexity comes from the very specific requirements of particular domains within the general topic. For example, whilst measures of segmental intelligibility are essential for evaluation of synthetic speech or digitally coded human speech (Bennett and Greenspan, 1987), they are not at all appropriate for evaluating the physical ergonomics of one mouse compared with another. Emerging technologies such as expert systems (see Meister, 1987) and intelligent interfaces (see Sewell *et al.*, 1987) require their own particular evaluation procedures. Added to this, specific evaluation criteria may need to be developed for different cultures. Fang and Tzeng (1987), for example, discuss the specific problems involved in evaluating Chinese graphemic input systems.

In this extended chapter we concentrate therefore on broad principles that apply to evaluation of human–computer interfaces in general, but especially in the general office environment.

Theoretical perspectives

The modern computer is an electronic machine designed to do calculations, process information and use formal representations of knowledge. The human is a complex, highly adaptable biological system, also capable of processing information and using knowledge, as part of a broader integrated set of biological functions which include emotional and other elements. When the machine and the human come into contact they do so, by definition, at the human–computer interface, and when this happens there is normally communication across the interface. The electronic and biological systems temporarily form a new, hybrid system which has both electronic and biological components. This new system is greater than the sum of its parts. It has emergent properties that cannot be predicted very simply from knowledge of the electronic and biological components taken separately; it has its own, emergent psychology. There is as yet no complete, coherent theory of the human–computer interface, but several different theoretical perspectives are relevant and can be applied to the task of evaluation.

The cognitive perspective

The currently dominant view of human–computer interaction is that it is essentially an information processing activity, and needs to be evaluated in these terms. The human and the computer engage in an active dialogue in which the two parties exchange and interpret information in a systematic way. According to this view, the ease with which users can achieve their intended tasks depends primarily upon the degree of 'cognitive compatibility' between the human and the way the computer is presented to the human (e.g. Gardiner and Christie, 1987a; Streitz, 1987).

Such interaction does not occur in isolation of the rest of the person's life. Users of office systems or other computer systems bring with them to any given interaction knowledge they have acquired in other situations, as well as particular characteristics and limitations which determine how they use such knowledge. It is the brief of cognitive psychology to study those aspects of human information processing which determine how knowledge is acquired, how it is used actively in interacting with the environment (primarily computers, in the present context), and how knowledge is updated and modified as a result of such interactions.

Some attempts have been made to develop a comprehensive theory of all aspects of human cognitive functioning (e.g. see Anderson, 1976), but they have not met with unqualified success. One of the key lessons which has been learned from such attempts is that models of human information processing cannot be built by simply stringing together models of the several cognitive processes involved in achieving any given task.

Until relatively recently, research on human cognition focused on bottom-up (data-driven) models, in which the human is seen as passive and system characteristics are reset by successive information inputs. Rabbitt (1979) acknowledges that such models have been 'of enormous service' in human experimental psychology, but he identifies a number of limitations to them. In particular, such models assume that particular subsystems or stages in information-processing can be identified which can yield independent indices of performance, implying an additive rather than interactive set of processes. A problem with this type of model is that it draws attention away from the human's capacity for flexible, continuous control of perceptual processes and response production.

Rabbitt (1979) cites studies of reaction time which show that after extensive practice subjects can increase their speed of reaction and reduce their error rate to such an extent that there may be no overlap of response distributions. A key conclusion he draws from such studies is that the processes involved in earlier performances may not be the same as those involved in the later ones: with extensive practice subjects can discover and use increasingly efficient new ways in which to process information.

This has very significant practical implications for the evaluation of human–computer interfaces. It means that evaluations based on the performance (and presumably attitudes) of subjects with limited experience using a system may not provide a very sound basis for drawing conclusions about the performance (or attitudes) of people who have become experienced users. The two situations may not differ simply in degree; they may be qualitatively different situations, requiring separate evaluation.

An alternative to bottom-up (data-driven) models are top-down (resource-driven) control process models. These typically postulate a central decision mechanism which is in control of system input characteristics and which can autonomously reset these from time to time. Hockey (1979) proposes an example of such a model which distinguishes between storage and throughput. Under certain conditions, the human (e.g. the user of an electronic office system) must balance storage (which takes up capacity) against throughput (which enables rapid clearing of internal registers). Studies of speed–accuracy trade offs suggest there is considerable flexibility in the allocation of resources, and illustrate how different strategies may be operated at different levels of information processing, depending upon the demands of different tasks. Providing an interesting link with the psychophysiological perspective discussed below, Hockey suggests that storage mode is facilitated by low psychophysiological arousal, and throughput mode by high arousal.

This model provides further theoretical complications in regard to evaluating the human–computer interface: it is necessary in the evaluation to take account of the demands of the task, and of the psychophysiological level of arousal of the human, as well as other factors that may affect the allocation of cognitive resources.

Any comprehensive model of information processing that is eventually developed is likely to be highly complex, involving dependencies amongst cognitive processes which vary with the requirements of the task and of context, are affected by previous knowledge, and may even differ from person to person depending on the style of information processing.

No such comprehensive model can be envisaged at present in any great detail. Even so, research on specific aspects of cognitive functioning has led to the development of models of specific aspects of cognitive functioning and the identification of cognitive principles that can usefully be applied to the evaluations of human–computer interfaces. Probably the most extensive review of cognitive psychology for this purpose can be found in Gardiner and Christie (1987a), and briefer reviews are provided by Gardiner (1987) and Loftus and Schooler (1985).

An ambitious attempt to translate the most pertinent findings from the research literature into practical guidelines, that can be used in the design and evaluation of human–computer interfaces, is provided by Marshall *et al.* (1987) utilizing a set of principles from mainstream cognitive psychology. They present a methodology for interpreting and illustrating the principles in the context of systems design, and show how specific practical guidelines can be developed from them. The guidelines are organized in terms of a series of 'sensitive dimensions' relating to the design of human–computer systems and can be used both to guide design and as a basis for evaluation of different design solutions.

A somewhat different approach within the broad cognitive perspective comes from the 'engineering' approach of Card *et al.* (1983). Strongly influenced by work on artificial intelligence, whilst still trying to take account of research on human cognition, they propose a simplified model of the human cognitive system which they call the Model Human Processor. This does not attempt to be a comprehensive model of human cognitive functioning, nor a detailed model of any one aspect of human cognition. Its more modest aim is to incorporate some of the key principles of human information processing which the authors feel to be especially pertinent in the context of human–computer interaction, and to do this in a simplified way which is appropriate for the needs of the engineer.

With the Model Human Processor as background, they propose a family of models—generically called the GOMS model—for analysing human–computer interaction. The GOMS model stresses the importance of Goals in guiding human behaviour. These are organized hierarchically. For example, in the context of word processing, a high-level goal might be to type a memorandum, a low-level goal might be to delete a particular character from a word in the text. Various Operators are available to the human to help work towards the attainment of goals. In the context of word processing, an operator might be pressing the delete key or pressing a particular function key or mouse button. These operators are normally organized together into Methods. A particular way of using 'cut and paste' might be a method. The human uses Selection rules to select which method to use in which circumstances, for example, in searching for a particular word in a piece of text, the user might search visually if the text is very short, but might use the automatic 'find' function if the text is quite long.

The GOMS model can be used as a basis for more specific models. Card *et al.* (1983) present two such more specific, predictive models, called the GOMS Model UT (the GOMS model at the level of the 'unit-task') and the Keystroke-Level Model. The GOMS Model UT provides a way of analysing any given task into a hierarchy of subtasks and predicting the time taken to complete the overall task from a knowledge of the times taken to complete the subtasks. The Keystroke-Level Model provides a way of analysing human–computer interaction at the level of sequences of keystrokes and using the analysis to predict performance times.

The extent to which the simplifications involved in the Model Human Processor and in the GOMS model, and the emphasis on the criterion of performance times compared with other possible evaluation criteria, restrict the value of this approach in practical situations remains to be fully researched and the answer must depend to some large extent on the purposes of the evaluation. Card *et al.* (1983) illustrate how the approach can be used in the evaluation of alternative word-processors used for simple text editing and indicate that it has also been applied with some success to human–computer interaction in more creative tasks.

The importance of the concept of goals in the GOMS model provides an interesting link between the cognitive and the social psychological perspectives on human–computer interaction.

The social psychological perspective

The human and the computer do not interact 'in a vacuum' but in the context of a total situation, and evaluation of the interface needs to take account of this. Argyle *et al.* (1981) propose that situations emerge and have the properties they do because they enable people to attain goals, which in turn are linked to needs and other drives. Sometimes those present in a situation share the same goals, sometimes they do not.

The concept of human behaviour as being goal-driven is reflected in several different models. We have seen it, for example, in the previous GOMS model developed by Card *et al.* (1983) from a cognitive perspective, and it is recognized in the work of Cranach *et al.* (1982) as well as Argyle *et al.* (1981) from a social psychological perspective. The Cranach *et al.* (1982) work in particular draws on the earlier models of Miller *et al.* (1960) which were further developed by Hacker (1978).

The concept of the computer as being goal-driven has been reflected implicitly in many science fiction stories almost since the first computer was invented. In recent years, with developments in artificial intelligence, the possibility of actual computers being goal-driven has come several steps nearer to reality. For example, Barber (1983) has described a language called OMEGA which can be used to build office systems that support office work as goal-directed problem-solving. OMEGA recognizes that goals in the real organizational world are often ill-defined, that what information is relevant to achieving a goal is not always clear, and that organizations involve many people whose activities may interact in various ways that are difficult to predict. It is not intended to be a system to replace the human; rather, it is a language which allows tools to be built which support human problem-solving, based on the concept of the computer working towards goals agreed with the human.

A consideration in evaluating the interface between the goal-driven human and the goal-driven computer must be the extent to which the interface allows comparison of goals, adjustments to goals, and goal-sharing. Social goals as well as strictly task-oriented goals may need to be considered in this regard. Murray and Bevan (1984) have suggested that human–computer dialogue design needs to take account of this and mimic the 'social aspect' of conversation to an appropriate degree. Consistent with this view, Richards and Underwood (1984) have shown that the social style of communication adopted by the computer can influence the social style adopted by the human, and they have discussed how this could be used to advantage in facilitating speech communication between human and computer. With advances in the technology available, multimedia communication between human and computer is ever more feasible, bringing the concept of a HAL-like computer (from the science fiction film, *2001: A Space Odyssey*) nearer to reality. Gardiner and Christie (1987b) have discussed how possibilities for multimedia communication of various sorts might aid in avoiding breakdowns in the dialogue between human and computer, and Sheehy and colleagues (e.g. Sheehy, 1987; Sheehy and Chapman, 1987) have examined the non-verbal communication aspects of this in some detail.

Gardiner (1986) and others have pointed out that communication across the human–computer interface is nowadays not necessarily only between human and computer, but can be between human and other humans on the network. Increasingly computers are used to support co-operative work and, to this extent, we are beginning to see the emergence of 'human–human–machine interfaces' (Ridgway, 1987), with concomitant social psychological implications. For example, Chesebro (1985) indicates that as much as 30% of messages on computerized bulletin boards are interpersonal in nature. Similarly, Brotz (1983) has observed that the interpersonal aspects of electronic mail can be crucial to user acceptance.

Two areas are of key importance for evaluation of the social psychological aspects of human–computer interfaces to these types of systems: the type of communication channel used, and the structure of the network.

The majority of electronic mail systems are text-based. However, it is technically possible to exchange voice messages through computer networks by digitizing the spoken message, switching it through the network, storing it, and replaying it for the recipient(s). Several products (e.g. the Sydis VoiceStation, the ITL IMP, the Wang Alliance) provide this kind of facility. Chapanis *et al.* (1972) in a series of experiments demonstrated that the type of communications channel used for problem-solving can affect both task performance and the communication process. Typically in their experiments, interactive voice communication resulted in more messages being exchanged, and fewer of the sentences being complete or grammatically correct, compared with exchange of text messages; despite this, tasks were generally completed faster when voice communication was used. The voice communication in their experiments was typically interactive rather than the storing-and-forwarding of voice messages, but their research suggests that the type of channel used may be a relevant consideration in evaluating the human–computer interface to networks.

More recent, naturalistic, evidence in support of this comes from case studies of the use of electronic mail. Brotz (1983), after observing the use made of an electronic mail system (called Laurel) within the Xerox Corporation, has commented on the social psychological aspects of the communication behaviour of the users. Even though the communications were similar to conventional internal mail in that both were text-based, a number of problems were observed with the computer-based system. For example, Brotz comments that users had to learn new norms for communication and that when previously separate communities of users were brought together, they often seemed to offend each other for some extended period until they adjusted to one another and came to a common understanding of the norms for their intercommunication. Some of the problems he observed were continuing, though at a reduced level, two years after the system had been set up. He makes a number of suggestions concerning proper etiquette in regard to the use of such systems, as well as indicating types of features that can be designed into the interface in order to minimize some of the problems that were observed. In evaluating the human–computer interface to such systems, features such as those he describes need to be examined in terms of their suitability for moderating the communications behaviour of the communities using the system.

The structure of the network as presented through the human–computer interface can be expected to influence the performance and attitudes of those concerned, if social psychological research on the effects of communications networks is applicable in this context. The early experimental work on communication networks was largely laboratory based and normally used a restricted range of communications media (often text messages), not based on computer networks at all (for a review see Baird, 1977). One of the criteria in evaluating a modern human–computer interface from the social psychological perspective must be the extent to which the network structure presented by the interface is optimal for the kinds of users and tasks involved. A relevant issue in this connection is the extent to which the interface allows the users to remodel the logical structure of the network. There is some evidence (e.g. Kano, 1977) that when such flexibility is permitted, the logical structure of the network tends to move towards a particular structure that depends upon the task involved. Research on computer conferencing (e.g. Vallee *et al.*, 1975) also suggests that users often remodel logical networks based on computers according to the type of work that is being done, and that patterns of communication may evolve as communities of users become more experienced.

Further consideration of the network aspects of evaluating the human–computer interface provides a natural bridge between the social psychological perspective and the organizational perspective, considered next.

The organizational perspective

The importance of evaluating the human–computer interface as an interface to the organization has been emphasized by Gardiner (1986; see also Malone, 1985), who has argued that technology is not just supporting the organization, it is gradually becoming the organization. Two main technical developments underlie this trend.

First, the personal computer or other office workstation is no longer always seen as simply one tool in the office, used perhaps for word processing or for spreadsheet work. Increasingly, the 'electronic desk-top' is seen as being the main work area, with the office worker doing more and more different aspects of his or her work there. As office workers have become more familiar with the personal computer and the range of software has increased, a trend has been developing towards multi-tasking operating systems, allowing several different tasks to be done on screen, the user switching between them at will and the computer carrying on with some tasks behind the scenes as the user deals with others.

A second trend appears to be increasing emphasis on networking. In addition to the considerable challenge this provides in terms of the technical aspects of getting systems to work effectively, this trend poses new challenges in terms of increased presentation of the user's work environment through an electronic medium. An obvious example is the use of electronic mail for organizational communication, but this is just an early example of further developments. With networking, the organization becomes increasingly an electronic entity.

Networking provides a means of completely restructuring the organization 'at the touch of a button'. Increasingly, the possibility arises for groups, departments and divisions within the organization to be disbanded and brought together in new combinations without any physical movement of people or equipment. One aspect of the overall picture is reflected in the increasing awareness in some circles of the possibilities offered for computer support of co-operative work.

The user's view of the organization is becoming increasingly an electronic view, and how the organization presents itself through the networks becomes increasingly important in terms of all aspects of the organization's relationship with its members. The human–computer interface comes to be all or the major part of what the office workers know about the organization in which they work. The 'telecommuter' (Kinsman, 1987) is a special case of this, but it applies to varying extents to all office workers who work in an electronic office.

In the light of these trends, Gardiner (1986) has argued that, in addition to other more traditional criteria, the human–computer interface should be evaluated in terms of how well it meets organizational criteria. In particular, it is important to consider how well it adapts to the needs of the newcomer to the organization, as well as the newcomer to the computer system. There is also the question of to what extent human–computer interfaces in electronic offices need to go beyond providing passive access to powerful telecommunications and computing resources and to what extent they need to take a more active and more adaptive role in facilitating good organizational practice.

Products already differ significantly in the extent to which they present a view of the organizational network as a relatively passive entity or as a more active participant in organizational communication, acting as an electronic 'agent' for the office worker. Tapestry and Progress are two products which illustrate this point. They both provide an interface to electronic office networks but differ markedly in the philosophy which has guided their design as interfaces between the human and the electronic network.

Tapestry presents the network as a passive entity. What the office worker sees is a traditional electronic mail service, standard communication facilities for exchanging files with other computers, access to whatever standard application packages are installed on the network, and use of shared and private 'cabinets' for storing information. In short, the human–computer interface provides a set of relatively passive facilities and leaves the user to make the decisions about what to do and when.

In contrast, Progress presents the network as a much more active participant in organizational communication. Progress, from Phoebe Software Limited in Ireland, is a typical example of a 'procedure processor'. As in Tapestry and a number of other systems, the user is presented with an 'electronic workdesk', showing a range of office equipment such as in-tray, pending-tray, out-tray, forms, diary, waste bin. The user can access traditional, 'stand-alone' support functions such as word-processing, database and diary in the usual way, and has access to a traditional electronic mail system. In addition, however, the interface presents another dimension of the network, that of the network as an active participant in organizational communication, acting as an 'agent' of the office workers to help process the procedures which have been defined as making up a significant part of the work that needs to be done.

As in an earlier experimental system developed within IBM and described by Lum *et al.* (1982), the concept of a 'form' is central to the procedure processing dimension of Progress. There are several types of forms: plan, memo, schedule, information request and sign in/out. In order to use Progress to support a particular procedure, the user completes a plan form for the procedure concerned. This names each activity involved in the procedure, the conditions under which each activity occurs, and what the activity consists of. This plan is then stored in the system. In order to initiate any given procedure once it has been defined, the user completes a schedule form. This tells Progress the plan's name and exactly when it should commence. The user then puts this schedule form into the out-tray, and Progress will then act on it at the appropriate time.

Progress manages each of the activities involved, sequencing them according to the information provided by the user in the plan. At appropriate stages during the procedure, Progress will automatically deliver a document for completion to the office worker(s) concerned. These people complete the document and put it in their out-trays, at which stage Progress moves on to the next part of the plan for the procedure being processed. If there is undue delay in completing a document, Progress will send a reminder to those involved.

Further discussion of the issues involved in evaluating human–computer interfaces to electronic systems supporting communities of users working with a wide range of applications software can be found in Marshall *et al.* (1987). Discussion of the broader social–organizational aspects of modern human–computer interfaces, including evaluation in terms of stress levels and other considerations, can be found in Furnham (1987) and Sloan and Cooper (1987). Recent in-depth research on human, organizational and economic aspects of information technology in complex work environments has been conducted as part of the ESPRIT Programme (Project 1030, see Ryan *et al.*, 1987).

The psychophysiological perspective

The sections above support the view that the human interacts with the computer at several levels, all of which need to be considered when evaluating the interface between them. The psychophysiological perspective attempts to integrate these different levels and to understand the complex interactions between them. It acts as a bridge between several psychological and other disciplines, providing a co-ordinating framework (Gale, 1979). In terms of evaluation criteria, the psychophysiological perspective (Gale, 1985) suggests that the human–computer interface should be evaluated in terms of its effects on:

1. Behaviour, including specific task performance and broader aspects such as absenteeism.

2. Subjective experiences, relating to interaction with the particular interface concerned as well as the work situation more generally.

3. Physiological responses, including specific stress responses which may or may not result in physical symptoms.

Psychophysiology has developed a long way from the simple, though historically important, experiments of Pavlov. Nowadays, it can boast sophisticated methods of data collection and analysis. Elaborate physiological monitoring devices supported by powerful computers form an important element in much current psychophysiological research. These developments have helped contribute to and support an increasingly complex view of the human as a multicomponent, multimodal system made up of interacting behavioural, physiological and experiential subsystems.

This view of the human is in contrast to many older, simpler models of human functioning that focused on one or other aspect of the human in isolation, yet it is better suited to facilitating an understanding of complex human–computer systems, and helps to provide a framework for understanding some of the surprises that can sometimes be encountered. For example, the cordless telephone switchboard was seen as an important step forward in the design of exchanges and, based on a relatively traditional logical analysis of the key task elements involved, a number of benefits were anticipated in terms of efficiency of performance and operator attitudes. Contrary to expectations, however, operator performance deteriorated, with an increase in the time taken to respond to calls and increases in the number of calls lost and failures to connect. In additon, job satisfaction declined. The analysis that had contributed to the design of the switchboard had neglected aspects of the overall human–computer situation that in the event proved critical.

Wastell *et al.* (1981) showed how it was possible to build up a more complete picture of what was going on in the situation by using a variety of measures that could be made without interrupting the flow of normal work. These measures included brain-evoked responses (ERPs) and heart rate, as well as continuous measures of performance. Without intruding into the normal pattern of work, the combination of physiological and performance measures helped to provide a more detailed picture of the events in the operator's working environment as a whole which demanded attention, created stress, and led to improvements or reductions in efficiency and satisfaction.

Psychophysiology is not yet a coherent or integrated theoretical body of knowledge. However, it does provide a perspective on human–computer interaction which emphasizes a view of the overall system; a perspective in which the behaviour of the computer combines with elements in the environment and interacts with the physiological responses, subjective experiences and overt behaviour of the human. According to this view, these various aspects of the overall system being evaluated interact with one another so intimately that to separate out any one from the others in order to try and study it in isolation is to limit severely the theoretical and practical value of any explanatory models that are developed.

Consistent with the emphasis on looking at the whole system rather than one aspect in artificial isolation, contemporary psychophysiology has found ways of freeing itself somewhat from the apron strings of the traditional laboratory. Techniques have been developed which allow monitoring of physiological responses, subjective experiences and overt behaviour in real-life situations. Physiological telemetric instruments allow physiological changes to be recorded as they occur in their normal environment. Some of the distortions often associated with traditional laboratory evaluations can be reduced or avoided and data can be collected in ecologically valid situations.

Further discussion of the psychophysiological perspective on evaluating human–computer interfaces can be found in Gale (1985) and Gale and Christie (1987a).

Evaluation and the design process

The principal purpose of evaluation in the applied context is to contribute to the design of improved human–computer interfaces. There has been a marked evolution in such interfaces over recent years with much of the contributory work being done within the context of competitive production, in terms of ease of use and other usability criteria. It is useful against this background to look more closely at the relationship between evaluation and design and, especially, the role of evaluation at different stages of the design process (see also McClelland in the context of product user trials).

There are many possible models of the design process (e.g. Roach *et al.*, 1982; Foley, 1983; Novara *et al.*, 1986; Faehnrich *et al.*, 1987; Furner, 1987; Meister, 1987; Williges, 1987; Mantei and Teorey, 1988). Each company has its own way of doing things, and to some extent each product development project has its own unique requirements. Any general model will almost certainly fail to match exactly what can actually be observed on any given project. However, it is useful to have a set of conceptual pegs on which to hang discussion. The following is a convenient framework which combines key elements from earlier models (especially that of Foley, 1983) with the practical experience of the present authors.

Five main stages or aspects to the design process can be identified, as follows:

1. Pre-design information gathering.
2. Design.
3. Design review.
4. Implementation.
5. Fine tuning.

Different kinds of techniques and tools can be brought to bear at each of the above stages in order to support different data collection and evaluation tasks. Although not specifically pointed out at each stage it will be recognized that there is iteration within and between the stages.

Pre-design information gathering

This covers five main areas of work:

1. Task analysis.
2. User characterization.
3. Situational analysis.
4. Design objectives.
5. Design constraints.

Task analysis

This is treated in depth elsewhere in this volume by Stammers *et al.* Here, we focus on its relevance to the design and evaluation of human–computer interfaces. In this context, task analysis is an assessment of the work to be performed by the human–computer system (see also Drury *et al.*, 1987). It involves the identification of both the overall goals associated with the work, and the pattern of tasks and subtasks involved in achieving those goals. This kind of information provides a direct input to a number of aspects of design, namely:

1. The determination of the required *functionality* of the product as a whole, i.e. the features that need to be included. (See also Gutierrez, 1987, for a discussion of a simulation methodology for eliciting end-users' information requirements and facilitating communication between designers and end-users.)

2. General guidelines for the sequencing of operations (e.g. sequencing and content of menus; frequent or important tasks should appear earlier in the menu sequence than infrequent or peripheral tasks).

3. Indications of the optimal allocation of functions between human and computer (see also Kantowitz and Sorkin, 1987).

4. General guidelines for the speeds and interdependencies required for optimal performance using the product (e.g. tasks that need to be performed in background mode; tasks which need to have access to other tasks or to inputs from other systems).

Hammond *et al.* (1987) make a distinction between task analysis tools which have been developed largely for the purpose of representing the task and tools which have been developed largely in order to help predict performance and help choose between alternative design solutions. The distinction is not absolute but rather a matter of emphasis. Representational tools provide part of the background (a description of the task) against which later evaluation can be done. Predictive tools set some of the standards, or goals, which the interface should meet, i.e. the pitfalls and problem areas which are best avoided in design.

Conceptual analysis of the task domain (the province of representational tools) forms a normal part of the overall design process in most cases. Even when there is little or no explicit consideration of the cognitive or other psychological aspects of what is involved it can still be a useful influence (cf. Tsichritzis, 1985). Some methods such as TAKD (Task Analysis for Knowledge Descriptions; see Johnson, 1985) do, however, attempt to incorporate explicit consideration of the user in the description of the task. TAKD identifies and classifies constituent task actions and task objects, and attempts to specify the knowledge required to perform the task (see also Bösser, 1987). Some methods more clearly acknowledge the intimate relationship between the task and how the task is actually performed through the human–computer interface. One of the more ambitious examples of this is the notation developed by Kieras and Polson (1985). This represents the user's knowledge of the task in the form of a production system, and it represents the 'notional device' underlying the human–computer interface in terms of a generalized transition network. A good mapping between the two representations is seen as reflecting good design of the interface in terms of functionality.

These kinds of task analysis tools can be useful to the design team at the pre-design stage in helping to develop functional specifications—specifications of the requirements made of the system—which can be used later as a basis for evaluation of specific interface designs. Dzida and Valder (1984) have emphasized the value of formalized task analysis in terms of facilitating user involvement in the early stages of the design process. In particular, they describe how knowledge engineering techniques can be used in a way which allows the design team to enter into a relatively formal dialogue with prospective users in order to come to a common understanding of the tasks that need to be supported.

User characterization

The intention here is to identify the target user groups for the product: those user groups which are most likely to be affected by the product and which, therefore, are most likely to affect its market success. Different types of users doing different types of work will place different requirements on the user interface. Malt (1987) suggests therefore that information from task analysis needs to be combined with information about the users. User characterization involves looking at, amongst other attributes (see also McClelland in this book):

1. User personal characteristics (ages, proportion of males to females, cultural characteristics, etc.). These are important determinants of, for example, the optimal weight of hardware products, their required ease of manipulation, and so on.

2. User skills (e.g. typing ability, authoring skills).

3. Job characteristics (e.g. status, role and seniority in the organization, turnover rates, scheduling constraints). These can constrain, for example, the processing speeds required, the allocation of tasks to foreground or background activity, and the type of dialogue used.

4. Previous experience (e.g. education, job experience, training). These can affect the users' expectations of what the product should do (features) and what it should look like.

5. Usage constraints (e.g. whether usage is mandatory or voluntary). This can have effects on the trade offs that are allowable in the design process. Voluntary users are free to go elsewhere if they dislike a product or its feature list; mandatory usage pretty much forces the user to put up with what circumstances dictate, or is most economical to provide.

6. Personal preferences (e.g. different types of learning or interaction style). These can affect the choice of types of ancillary materials provided with the product (e.g. on-line tutorials or prompt cards; command-language shorthands or exclusive menu-selection).

Information about user characteristics is used again at the prototyping stage, to provide inputs to the selection of suitable participants in beta-site or usability testing.

Situational analysis

This is an assessment of the likely environment in which the intended product will be used. It covers aspects of the physical, social and organizational user environment, including such diverse factors as environmental temperature, staff availability, likely loads, company policy, and so on. Together with the task analysis and the user characterization process, it provides a complete picture of the role the product is likely to play, the changes (both positive and negative) the product is likely to make to the user environment and task performance, the flexibility it needs to have, and so on.

Design objectives

Having mapped out the domain that the product is addressing, the design team now needs to identify the specific objectives of the product, against which it can be evaluated later. (This can also provide useful marketing input.)

With regard to the design of the human–computer interface, the design objectives are often *usability goals* which the design team hopes to achieve. They form a very specific basis for the evaluation work which takes place later on in the design process. In this connection, Brooke (1986) has argued that usability should be 'modified from a vague wish to a clear statement of measurable goals in product requirements' (see also Butler, 1985). The idea is that a usability goal will provide clear guidance to the design team, and at the prototyping stage evaluation activities will have a concrete focus.

Usability goals can span many different aspects of the human–computer interface. They can address very specific aspects of system design (e.g. 'Users need to be able to install and configure this product on a PC fitted with a hard-disk in less than two hours, using the manual but with no additional assistance'), or more general features, such as the transfer of skills from one system to another (e.g. 'Users with more than three months' experience of an Apple Macintosh should be able to draw an A4-size diagram of their choice in less than an hour, with no help.'). Brooke reviews a number of different metrics that can be used as the basis for defining different usability goals, covering both performance and attitude measures. Similarly, Kirakowski and Dillon (1987) propose a computer-user satisfaction inventory as a first step towards a long-term goal of developing what they term 'System

Independent Evaluation Metrics' (SIEMs) which can provide a benchmark for commercial systems.

Design constraints

At this stage, one compares what can realistically be achieved within the time and resources available, against the design objectives and other information from the stages above. The outcome may be the identification of very real hardware, software, cost or development time constraints, which mean that maybe the 'ideal interface' cannot be built. Constraints of this sort therefore need to be identified and understood from the outset. Their documentation and analysis provides a justification for the particular design philosophy eventually adopted. The justification, in turn, provides the boundaries within which evaluation can meaningfully take place (see also Norman, 1987).

Key outputs from pre-design information gathering

Taken together, these five areas of work provide information that can be used to evaluate the potential usefulness of the product (i.e. whether it is relevant to the work of its intended users and if it takes into account the many factors impinging on the user), as well as providing the boundaries within which evaluation can meaningfully be made. The information provides a justification for design decisions.

Another output, from the task and situational analyses and the user characterization in particular, is practical 'task scenarios' for the product being developed. These provide concrete and realistic guides to the design team at the design stage, and a framework for evaluation at the prototyping and subsequent stages. A task scenario is a description of a typical task that a particular target user group will normally want to achieve using the product. It can be written at different levels of detail and in qualitative or quantitative terms. For example, it can contain indications of the time currently taken to achieve the task, and thus provide a yardstick against which to evaluate product performance; or it can simply provide a qualitative, sequential description of a task and its dependencies. Usability goals are often tied to particular task scenarios.

General techniques and tools of pre-design information gathering

A number of techniques are available for obtaining information relevant to the five main areas of work in the pre-design stage. These include:

1. Questionnaire surveys of target user groups, aimed at identifying their current work patterns, current product usage, and current 'gripes', or complaints, with the products they use.

2. Interviews with users and/or clients, aimed at exploring in detail a given domain of interest, a client's opinion about the proposed product development, etc. (see Sinclair in this book for a discussion of questionnaires and interviews).

3. Participant observation (usually at a fairly low level of participation) at the place of work of the intended users. This can provide valuable inputs to the development of a comprehensive 'user view' of the domain to be addressed, against which to evaluate alternative product designs, as well as providing specific information about such aspects as workloads and interdependencies (see Drury on direct observation in this book).

4. Functional analysis techniques (e.g. flow analysis, time line analysis; see Laughery and Laughery, 1987). These typically provide information on the sequencing, timing and other interdependencies among the subtasks involved in achieving a given work goal.

The techniques can be used on their own, but they are more commonly used as part of integrated packages, following an overall plan that guides the collection of information on the optimal functionality and desirable attributes of the intended product. Detailed comparisons of the techniques are provided by Bouchard (1976), Laughery and Laughery (1987) and Novara *et al.* (1986) amongst others.

Evaluation criteria for pre-design information gathering

We can identify the following general 'pre-design stage' criteria for evaluating any given human–computer interface. Each of these general criteria can be analysed into more specific criteria, and it is in terms of these more specific criteria that a practical evaluation would normally be done. In terms of the general criteria, a human–computer interface should:

1. Reflect an appropriate allocation of tasks between the human and the computer, taking account of the kind of human and the kind of computer involved.
2. Reflect specific design objectives that have been made explicit and agreed by the design team and the client.
3. Take account of the characteristics of the intended target user groups.
4. Take account of specific situational constraints.
5. Take account of design constraints that have been made explicit and agreed by the design team and the client.

The more specific criteria which can be derived from these general criteria will depend upon the particular development project concerned, and can be formulated to address specific issues raised by the theoretical perspectives discussed earlier—cognitive, social, organizational and/or psychophysiological. Taken together, the evaluation criteria provide a framework within which a relatively high-level specification of the human–computer interface being developed can be evaluated, even before time and money are spent on more detailed design work.

The design stage

The design stage or phase as treated here considers the human–computer interface from the user's point of view and adopts the widely-accepted four-level approach (e.g. Foley, 1983; Faehnrich and Ziegler, 1984; Christie and Gardiner, 1987). Later in this chapter, in discussing prototyping and implementation, we consider the interface from the point of view of its links into the underlying system. The four-level approach considers the user's view of the interface in terms of the following levels:

1. Conceptual level.
2. Semantic level.
3. Syntactic level.
4. Lexical level.

Conceptual level

This involves design of the basic philosophy for the product, as well as the concepts with which the user will need to be familiar in order to utilize the product to its full potential, with an acceptable degree of effort. It is especially important in shaping the overall image or model of the product or system that is presented to the user. For example, at the highest level, is the product a high-powered typewriter or a basic word-processor? Is it a sophisticated programmable calculator or a pocket computer? The answer to these questions should affect how the product is presented to the user both in terms of hardware and software, as well as in the documentation (and marketing). From the point of view of evaluation at this level the key question is: Is the basic philosophy of the product an appropriate one, and is it communicated effectively to the user, using appropriate concepts and relationships between concepts? Houston (1983) humorously discusses this question at length with reference to the relative merits of the 'digital kitchen' and 'giant squid' metaphors for electronic office systems compared with the more conventional 'desk-top' metaphor (as used in the Macintosh, GEM, and a number of other products).

Semantic level

Evaluation at this level involves assessing the appropriateness of the commands that the user is required to give the system, but only in terms of what they are and what they mean, not the particular form they take. It also covers evaluation of the kinds of information that are presented to the user, including the kinds of error feedback and user help.

Syntactic level

Evaluation here is concerned with the appropriateness of the form of the commands, including such aspects as the parameters that need to be provided by the user and the sequencing of commands that the user needs to give.

Lexical level

This involves more detailed evaluation of the commands that the user provides and the information that is presented in return. It covers such things as the design of particular symbols and the size, shape and position of buttons on the mouse. There will normally be differences between different countries at this level if not at the higher levels.

Key outputs of the design stage

Taken together, the design work done at the four levels provides the basis for the specification to which the implementation team will work, and in terms of which the user-interface to the product will be evaluated. Evaluation done at this stage provides inputs for the refinement of the specification. It may also lead to revisions of the original design objectives.

As by-products of the evaluation work, prototypes and storyboards may be developed which help to provide a vision of the product to help in the design review (considered later) and to help the implementation team interpret the specification.

Tools and techniques of the design stage

A wide variety of tools and techniques are available for evaluation of the human–computer interface during the design stage, and a number of different conceptual frameworks have been suggested for organizing use and discussion of these (e.g. Howard and Murray, 1987; Bullinger *et al.*, 1987).

Howard and Murray (1987), on the basis of some 80 references involving various aspects of interface evaluation, were able to identify five main types of formal evaluation: subject-based, expert-based, theory-based, user-based, and market-based. They then tabulated these against the techniques most appropriate for each type. Taking a rather different approach, Bullinger *et al.* (1987) distinguish between 'hard' methods (e.g. controlled laboratory experiments) and 'soft' methods (e.g. interviews with actual users) of interface evaluation. They propose that both of these types of methods must be used, in an essentially complementary fashion.

Whilst it is useful to have a broad range of tools and techniques available, it is not the techniques themselves that are to be valued but the information that can be obtained if they are used appropriately. In deciding on how to proceed with an evaluation the focus needs to be primarily on the kind of information that is most needed in terms of the decisions that have to be made. Calder (1977) makes this point clearly in regard to the 'soft' method of focus group interviewing, where he discusses the different kinds of knowledge these groups can provide and how that knowledge can be used in developing scientific theory or for other purposes. Patrick (1987) makes a similar point in relation to the 'hard' technique of error analysis, where he discusses the different kinds of errors that can be observed and their theoretical and practical significance.

Against this background, the decision of which technique(s) to use needs to take account of the practical constraints that apply (e.g. timescales and resources), so that evaluation does not add unacceptable costs to the design process either in terms of money or of product development time.

Prototypes and evaluation

In the present context, it is useful to consider the techniques available in terms of their applicability within the product design process. It is helpful to consider what can serve as a suitable object of evaluation. At the design stage (as well as at later stages in the design process), this is likely to be some form of 'specification' or 'prototype', where we use these terms broadly to cover four main levels.

Level 1. At this level, the product is in the very early stages of definition. No prototype as such exists, but various indications as to what the product will look like do, including: early written specifications, storyboards, 'slide shows' (e.g. showing key screen layouts), and 'interactive slide shows' (where it is possible to branch from 'screen' to 'screen' in different ways in order to give a feel for the dialogue envisaged).

Level 2. At this level, some partial prototypes exist in which design solutions to limited aspects of the product have been worked out in depth. For example, mock-ups of alternative keyboard designs have been produced, or perhaps a mock-up in software of a database retrieval function has been written using dummy data in order to illustrate some of the principles envisaged for the style of human–computer interaction, but there are as yet no links to a real database.

Level 3. Here, working prototypes exist of specific aspects of the product. For example the database aspect works, but the other aspects (e.g. word-processing, calendar management) do not.

Level 4. At this level, the whole product exists as a working prototype, but is not yet ready for release. It is known that various bugs and design faults exist.

Traditionally, evaluation has focused on the last two levels, with the concomitant feeling amongst many practitioners that key design decisions have already been made and that all that remains for the human factors or similar specialist to do is to 'put out fires'. In recent years, however, evaluation has begun to move to earlier stages in the design process, and this can be helpful in various ways, not all necessarily to do with the development of actual market products.

Evaluations based on prototypes can be more or less formal. At one extreme they can be used by the software engineer in much the same way as an artist may use preliminary sketches before embarking on a final drawing or painting. The increasing availability of software engineering tools which provide methods for rapid prototyping facilitates this style of working (see, e.g. Chao, 1987; Harker, 1987; these tools are discussed in more detail in the section dealing with the Implementation Stage). At the other extreme, they can be used to test aspects of the proposed design in controlled laboratory experiments (see Hoyos *et al.*, 1987, for a review of the literature on experimental prototyping). They may also be used at the design review stage to choose between different design solutions on a purely subjective basis, informed by whatever knowledge and experience is available to the reviewers at the time. Gruenenfelder and Whitten (1984) advocate the use of prototypes in augmenting generic research on the human–computer interface and in helping to apply general guidelines to specific design problems. They argue that, despite the value of generic research, it is relatively insensitive to the context of a full design. Prototyping provides a practical basis for evaluating specific approaches to design.

The four levels of specifications/prototypes described allow for evaluations of different kinds and for different purposes. For example, Levels 1 and 2 are most appropriate where the evaluation is aimed at narrowing down the possible range of design choices, or when consulting users, designers and/ or software engineers concerning a particularly novel concept that is to be implemented. They have the advantage that they are relatively quick and cheap to implement, while providing valuable information which may avoid costly mistakes later. Prototypes at Levels 3 or 4 allow for more sophisticated types of early evaluation. For example, the overall logic of the dialogue can be prototyped on-screen and some early evaluations of this can be conducted before any screen designs are written.

The following techniques apply primarily at the level indicated. Techniques which apply at the numerically lower levels also apply at the higher levels, but the reverse is usually not true. For example, a technique applicable at Level 1 can probably also be used at Levels 2, 3 and 4, but a technique applicable at Level 4 may not be applicable at Level 1.

Level 1 techniques

Three main categories of techniques are available at this level:

1. Focus groups.
2. Walk-through techniques.
3 Application of predictive models.

FOCUS GROUPS

This is a group interview technique widely-used in qualitative market research (47% of companies in a survey cited by Fern, 1982, claimed to have used focus groups for one purpose or another, often for idea generation). The technique usually involves about eight to ten individuals who are led through a more or less open-ended discussion about a product or product concept by a moderator who ensures that topics of significance to those sponsoring the group are covered.

In the present context, once some tentative ideas about the functionality of the product and the form of the interface have been developed, but still at a very early stage in the design process, it is possible to check these ideas against what typically prospective user groups have to say. This is essentially a qualitative kind of evaluation but can be very helpful in identifying problems early on and perhaps in helping to avoid going down the wrong track.

Focus groups work best when the concepts to be explored can be explained using 'slide shows', storyboards or other vehicles for embodying aspects of what it is intended to design. It is often helpful to have several alternative possibilities demonstrated, to emphasize the point that there is more than one possible solution and to stimulate discussion about common themes, gaps and problems.

It is critical to be clear about the purpose of the evaluation. Calder (1977) provides a useful discussion of what he sees as the three main approaches (phenomenological, exploratory and clinical), reflecting different purposes, and their practical and theoretical implications.

WALK-THROUGHS

Evaluation may be based on 'walk-throughs', using the prototype (or other specification materials, such as storyboards, state-transition diagrams). In a walk-through, a human–computer interface specialist (or even, in some cases, the engineers in the design team) plays the role of a typical user and 'experiences' the product as specified.

The evaluation can take several forms, for example, the specialist can simply 'audit' the interface, by checking the proposed design against known standards and guidelines (for hardware or software), and record its compliance or lack of compliance; or a more predictive approach can be adopted, loosely based on the 'critical incident' or 'critical event' analysis approach (e.g. Meister, 1985; Del Galdo *et al.*, 1986), and try to identify possible problems or bottlenecks, as well as their likely severity. Severity, in this case, is measured in terms of the likely frequency, probability of occurrence and impact on users' performance. Both the 'audit' and the more predictive types of inspection of the interface are useful—and easily carried out—at most stages of design, from the early storyboards and specifications, to the prototyping, beta-testing and fine-tuning stages.

In order to carry out a good 'audit', it is usually necessary to derive appropriate checklists, based on relevant design guidelines and standards, which cover many aspects of hardware and software but often need to be interpreted in the particular design context. They can conflict with one another, are of widely varying status in terms of their research foundations, and need to be applied intelligently and with caution. Typical examples of guidelines applying to hardware can be found in the classic volume by Cakir *et al.* (1980), and in a more recent paper by Oborne (1987). Examples applying to software can be found in Cole *et al.* (1985), Shneiderman (1986) and more recently in Marshall *et al.* (1987).

A problem for the human–computer interface specialist arises when standards that have been agreed by national or international standards bodies or which have become *de facto* industry standards conflict with guidelines based on relevant research. A classic example of this problem applies to the design of keyboards, where the QWERTY layout of keys is the accepted industry standard but was arrived at for reasons to do with the mechanics of manual typewriters rather than the ergonomics of word-processors or other electronic office equipment. Software standards are similarly beginning to become established based largely on common practice rather than through systematic research of optimal design solutions.

Part of the problem arises from the fact that researchers, either at the stage of planning their research or at the stage of reporting their results, are often not concerned with translating their findings into a form that designers can use directly. 'Pure' researchers typically do not see it as their business to do this, even where they are working in areas where their research is potentially of relevance.

Marshall *et al.* (1987) have considered the problem of capitalizing on the 'pure' research base and applying it to practical design issues, and have presented a methodology for translating pure research findings (specifically in the area of cognitive psychology) into practical design guidelines. The guidelines they present are themselves based on this methodology and have the advantage of clearly indicating specific research foundations.

APPLICATION OF PREDICTIVE MODELS

Written specifications can be used as a basis for applying some predictive models (see Meister in this book). For example, Card *et al.* (1983) illustrate how it is possible to apply their GOMS UT model (described briefly earlier) to written specifications concerning the tasks a system is being designed to support. In essence, the task is broken down into subtasks. The frequency with which each subtask needs to be done in performing the overall task is estimated, and the time required to do the subtask is also estimated. These estimates are combined (across all the subtasks involved) in order to estimate the time required to perform the overall task. They discuss the kinds of assumptions that typically need to be made in applying the model.

The more formal the specification, usually the easier it is to use it as a basis for evaluation. One type of formalization favoured by many human–computer interface practitioners (e.g. Gardiner and Christie, 1985;

Wasserman, 1985; Graesser *et al.*, 1987) is the state-transition (or augmented state-transition) network description. There are a number of variants of this but essentially it involves describing the system as seen by the user in terms of a number of states and transitions between states. The states correspond roughly to screens, but this will not always be exclusively the case if, for example, voice output is used in the system. The transitions are roughly the actions that the user needs to take in order to move from one state to the next (e.g. pressing function key 1, or clicking on the mouse button). Again, this will not always be exclusively the case if, for example, during some parts of the dialogue (e.g. during the beginning sequence) the system moves automatically from one screen to another. State-transition descriptions can be used in several ways for evaluation purposes. Graesser *et al.* (1987) discuss how they can be used to predict trouble spots if human response times are known or can be estimated for the transitions. ESPRIT Project 234 (Gardiner and Christie, 1985) demonstrated how they could be used as a basis for assessing the consistency of the dialogue, and showed how it was feasible to develop a number of predictive indices relating to different aspects of the dialogue based on general principles of cognitive psychology.

Where the application of predictive models such as the GOMS UT model require estimates of performance times or other such estimates, estimates based on experience with similar systems are often used. However, the predictions made on the basis of the models can normally be expected to be more precise and/or more accurate if at least some of the estimates can be obtained empirically from laboratory tests or in other ways. The importance of precision and accuracy will depend upon the purposes of the evaluation (which might be simply to help in choosing between markedly different approaches) and the costs associated with obtaining empirical estimates.

Level 2 techniques

Two further main categories of techniques become available at this level:

1. Informal user tests.
2. Controlled laboratory tests.

INFORMAL USER TESTS

This technique is widely-used in order to set reasonable boundaries on hardware and software parameters at an early stage in the design process. For example, major problems with the layout of a workstation may become apparent as soon as a 'space model' is available. Such a model often need only be a simple physical mock-up of the main features, perhaps in wood or cardboard. Informal tests in which potential users are asked to sit at the workstation and perform a few simple tasks such as reaching for a particular drawer or moving the mouse can often reveal unforeseen difficulties. The drawer may be in an inconvenient position, or there may not be sufficient space for the mouse on the desk-top. Such problems seem obvious once detected, but there are so many different ways in which problems of this sort can arise that it is very difficult to predict all of them beforehand and often far easier to produce a simple model and have a look. The technique is similar to the 'walk-through' using early prototypes but differs in that a few 'real users' are used instead of relying on the judgements of an expert ergonomist or other specialist. One advantage of this is that real users often do not behave as one might expect. A second advantage is that by using a few people who differ quite noticeably in critical ways (either physical ways, such as height, or in other ways, such as their style of working) it is possible to get a feel for how the proposed design is likely to cope with such variations. Computer simulations may be possible in some cases (e.g. Porter in this book; Pulat and Grice, 1985).

CONTROLLED LABORATORY TESTS

Controlled tests may be conducted for two main purposes in this context:

1. To compare different possible design solutions (e.g. different keyboard layouts, or different screen designs).
2. To obtain empirical estimates of particular performance or attitudinal parameters (e.g. estimates of transition times or subtask performance times, for use in conjunction with predictive models).

Drury discusses experimental design elsewhere in this book, and illustrations of the application of experiments in the context of human–computer interaction can be found in Monk (1984), Shneiderman (1986) and Lewis and Downton (1987). Comparisons with field studies are drawn below.

Level 3 techniques

These depend upon working prototypes being available, but the prototypes may only be partial prototypes covering some but not all aspects of the product. The key difference compared with Level 2 techniques is that the prototypes have to work, they are not simply 'space models' or mock-ups of the screens with no real functionality. Three main categories of techniques are available at this level:

1. Informal user tests.
2. Controlled laboratory tests.
3. Field tests.

The first two are essentially the same as Level 2 but differ significantly in their level of sophistication. Given working prototypes, it is possible to explore attitudes and performance during the conduct of a range of specific task scenarios. These task scenarios can be designed to test certain aspects of system performance to the limits, or they can be designed to be representative of the range of tasks that real users would normally be expected to do. (See McClelland in this book for a discussion of user trials.)

It may also be possible in some cases to conduct limited field trials. For example, if office workers are already using a network for a variety of purposes it may be possible to introduce working prototypes of specific, well-defined aspects of a new software package onto the network for certain people to use on a test basis. There are potential hazards associated with this, however, and it is necessary to consider these carefully alongside the potential benefits in the context of the implementation programme as a whole. Implementation of new systems is discussed elsewhere in this book by Eason.

The main limitation of both laboratory and field work at this level is that, by definition, the tests are conducted on particular aspects of the product or system in isolation from the product or system as a whole.

Level 4 techniques

Once complete working prototypes or early product releases are available, it is possible (in addition to the lower-level techniques) to conduct in-depth field testing. Different specific techniques can be used in this type of evaluation, many of which are discussed in detail in other chapters of this book. Especially relevant are: direct observation of behaviour (Drury), indirect observation of performance (Sinclair), verbal protocol analysis (Bainbridge), and assessment of stress (Cox), as well as sections of other chapters.

Many specific techniques derived from more general methods can be adopted according to specific circumstances and the purposes of evaluation. The following are just a few examples of techniques that have been found useful in the field testing of human–computer interfaces.

1. *Videorecording* of behaviour in the work situation (e.g. Youmans, 1987). This can be done relatively unobtrusively and can be helpful in four main ways:

(a) evaluating macro aspects of equipment layout, such as the physical relationship between peripherals such as printers, scanners, plotters;
(b) evaluating micro aspects of equipment layout, such as the positioning of disc drives, mouse pad, function keys, the accessibility or otherwise of dip switches;
(c) evaluating some aspects of the dialogue, for example, identifying areas of the dialogue where the user seems to slow down because of difficulties, has to refer to manuals or a colleague;
(d) evaluating social and organizational factors, such as the amount of time spent keyboarding compared with talking with colleagues.

A major advantage of videorecording is that the behaviour is captured and the tapes can be reviewed over and over again at the convenience of the researcher. It is also possible to ask those recorded to comment on parts of the tapes in order to clarify points. The tapes can also be edited in order to provide 'case study' illustrations when explaining points to hardware or software engineers or others involved in the design.

Tapes are, however, notorious for the amount of time that can be consumed analysing them. It is certainly important to be clear about the purposes of the analysis in order to avoid unnecessary work. Significant savings can also be achieved by using a computer to assist in the analysis. Many departments have developed their own specific analysis systems, optimized for the kinds of analyses they typically conduct, and some general-purpose systems (such as The Event Analyser from HeptaCon Ltd.) are now commercially available at low cost and can be used for a wide range of analyses.

2. *Use of a 'gripe command'.* This is a command (e.g. pressing a particular function key) which the user can give the system to call up a 'notepad' on which the user can record whatever comments he or she wishes to make. These can be descriptions of software bugs the user has identified, suggestions for improvements to the dialogue, difficulties encountered, and so on. Although normally called a 'gripe command', it can be used in any number of different ways, and can be used to obtain favourable as well as unfavourable feedback from users. The main advantage is that it is on-line. It is relatively convenient for the user, it is immediate and does not rely on the user remembering points (e.g. in an interview later), and it is relatively safe as a method of data collection (especially if the user is on-line to a network and the 'gripes' can be sent automatically through the network to the person who is responsible for collecting them).

3. *Checklists and rating scales.* These can be used to record users' assessment of particular aspects of the product or system in a standard format. In the case of human–computer interfaces it may be possible, as for the 'gripe command', to use these on-line, perhaps with the system being programmed to present the checklists or rating scales on-screen automatically at particular times or under particular circumstances.

4. *Diaries.* These can be used to monitor many different aspects of the user's behaviour, along with comments about the usefulness (for example) of the system, or problems experienced, as the person concerned uses the system for normal work over an extended period of time. It may be possible for the diary to be on-line. It may also be possible to facilitate the keeping of the diary by using the computer in various ways. For example, if the system incorporates voice store-and-forward technology it may be convenient to have the option of the user speaking his or her comments into the diary instead of typing or writing as would be done conventionally.

Many other specific techniques can be invented or adapted for use in field trials of human–computer systems. In all cases, it is necessary to balance a number of factors, primarily the costs of developing the techniques, the costs of using them, the benefits they can provide given the particular purposes of the evaluation, and the impacts (positive and negative) they might have on other aspects of the situation.

Evaluation criteria for the design stage

As for the criteria in the pre-design stage, each design stage criterion can be analysed into more specific criteria, addressing issues raised by the different perspectives on the human–computer interface. It is in terms of these more specific criteria that a practical evaluation would normally be done. In terms of the general 'design stage' criteria, a human–computer interface should:

1. Present an overall coherent environment and specific concepts which are appropriate for the kinds of tasks involved and familiar to the intended target user groups or can be expected to be learned at an acceptable cost.

2. Provide an adequate level and type of user support.

3. Provide a suitably error-cushioned environment for its users (whatever the target user group).

4. Use a dialogue whose semantic, syntactic and lexical aspects are appropriate for the kind of task and the kind of user concerned.

The design review stage

The purpose of the design review is to agree on a design before committing the necessary resources to actually coding it. If there is only one design, the decision must be whether to accept it, reject it or send it back for modification. If there are several alternative designs available, then it is necessary to select one. Both the 'pre-design stage' and the 'design stage' criteria are relevant to evaluation of the human–computer interface at the design review. The 'pre-design stage' criteria will have been used already to evaluate the high-level specification which formed the basis for the design work, but they should be reviewed now that a specific design(s) has been produced.

At this stage, if the necessary evaluations and documentation of such evaluations have been carried out properly, the process is one of 'signing off' the design. The documentation accumulated up to this point should provide the justification for the decisions made, and should include the recommendations that were made based on the results of the evaluations.

The implementation stage

Traditionally, the implementation phase is concerned with taking the agreed design and coding it. Evaluation of the human–computer interface is not really relevant to this stage, being done both before and after but not during it. This traditional view of implementation has begun to give way to one combined stage in which the design and implementation stages come together and the design of the product is iteratively modified and enhanced as the coding is done.

One of the key developments which has facilitated this approach has been the development of 'application generators' and 'computer-aided software engineering tools' (CASE tools) of one sort or another. A major stimulus or the development of these has been the need for increased productivity in he writing of applications software. The use of CASE tools can mean that applications can be written in less than a quarter of the time it would take using more conventional programming techniques.

Although productivity has been the key driving force, a number of other benefits are associated with these tools. These include some degree of standardization of approach, which can facilitate consistency in the design of he dialogue, and the potential for rapid prototyping of applications. The ease of producing prototypes as an integral part of writing the application in turn facilitates the use of prototyping in the evaluation of design solutions, smoothing the way for truly iterative design life cycles.

In some cases, such as the SET system from PA, these prototypes can be used as the 'specification' around which the actual code is written semi-automatically by the tools themselves. In this way, prototyping becomes an integral part of the implementation process rather than a separate, parallel activity. One important benefit of this is that the marginal costs associated with prototyping (and the possibility of evaluating alternative design solutions) are reduced compared with traditional methods.

The SET system is interesting in the present context as an example of the type of development system in which applications are written by designing the human–computer interface first, from which the rest of the application is then derived, rather than designing the underlying application and then adding on the interface. This puts the emphasis in design very strongly on the interface to the human. The formalism used in SET for designing the interface is based on augmented state-transition networks, discussed previously. The 'application code' itself need not be written until the application from the user's point of view is virtually complete. The application code that ensures the application actually does some useful work is then generated semi-automatically, using the human–computer interface as the specification'.

This logical separation of the human–computer interface from the underlying application has become more common in recent years. An important stimulus has once again been the need for productivity. Separation of the two means that in principle it can be easier to modify aspects of either without having to rewrite both, so long as the necessary linkages between the two are maintained. This facilitates both maintenance and portability of software, and has been a major factor in the development of the concept of UIMS (User-Interface Management Systems).

Several attempts have been made to develop a standard scheme for conceptualizing the structure of the hardware and software involved in the human–computer interface (or 'user-interface'). Gregory (1987), for example, proposes that it can be thought of in terms of seven layers which the designer (and, by implication, the evaluator) needs to address: conceptualization, hardware for interaction, window manager, presentation manager, state manager (Gregory adopts a state-transition model of the interface), application, and database manager. Watanabe (1987) proposes a three-layer human interface architecture for office systems, subdivided into six separate levels elementary information, enhancement control, media control, format control dynamic control, and physical control.

There is no standard scheme at present, different schemes being used by different software development systems. (For a further discussion see, for example, Olsen and Dance, 1988.) What is most important to note in the present context is that there is a definite trend away from the human–computer interface being inextricably bound up in the 'application code' itself towards the concept of the interface as a separable entity with a definite structure of its own. Evaluation of the human–computer interface will increasingly need to be tied to this concept.

Automated evaluation

The trend towards formalization and standardization of the interface opens up possibilities for automating the evaluation of some aspects of the interface during the design process. ESPRIT project 234 (Gardiner and Christie, 1985) demonstrated the feasibility in principle of successfully predicting certain aspects of user performance and attitudes on the basis of indices (e.g. relating to various aspects of dialogue consistency) derived from formal descriptions of the interface (based on state-transition descriptions). A further logical step would be to link an automatic evaluation system to a prototyping and software development system, in order that evaluations could be carried out routinely as design solutions were developed, without interfering with the design process.

This kind of evaluation would form only one part of the overall evaluation of the product, and design teams would need training in how to use the automatic evaluation appropriately. An obvious danger would be that, without adequate training, those using such systems could be tempted to use them 'mechanically' rather than 'interpretatively' or else to dismiss them as of no use at all because they do not address all critical aspects of evaluation. If used appropriately, however, such systems could routinely provide feedback on some key design parameters during the early stages of design without any extra cost in terms of time taken or use of personnel or equipment. In this way, automated evaluation could form a useful additional element in the evaluation repertoire, without prejudicing more in-depth evaluation by conventional methods at suitable stages in the design process.

Automated evaluation depends upon there being suitable metrics or formalized evaluation procedures available that can be incorporated into such systems. Such metrics have been proposed in some areas. Tullis (1984), for example, proposes four metrics for evaluating alphanumeric screens: overall density (in terms of the proportion of character spaces filled), local density (density in a small area around each character), grouping (specific measures based on the clustering of characters), and layout complexity (based on the distribution of horizontal and vertical distances of each label and data item from a standard point on the display). Streveler and Wasserman (1984) propose various quantitative measures of the spatial properties of screen designs based on what they call 'boxing analysis', 'hot-spot analysis' and 'alignment analysis'.

Whilst metrics can in principle be proposed for almost any aspect of the human–computer interface, well-researched areas are the most likely to have predictive validity in practical situations. Such areas may be relatively 'pure' areas of research, such as on memory or other aspects of cognitive psychology (Gardiner and Christie, 1987a) or more applied, such as the use of colour in visual displays (see Davidoff, 1987), cognitive aspects of windowing (Norman *et al.*, 1986), or hardware ergonomics (Oborne, 1987). It is likely that evaluation systems would need to take account of task and other context variables in applying many of the possible metrics, although some may have very general applicability (for example, Davidoff's (1987) review suggests that a large number of colours in a display usually increases search times, and one might hypothesize that a metric based on the number of colours in the display might have predictive validity in a variety of different types of dialogue).

The possibility of using a computer to aid in the design of a human–computer interface is not restricted to software aspects of the interface or to cognitive aspects. Computer-based design aids have also been developed for use in the older field of workstation ergonomics (see Porter in this book).

Trade offs

The huge variety of possible metrics that can in principle be developed itself suggests a need to consider trade offs in design. Improving the interface in regard to one set of metrics may well reduce its rating in some other respect. As Norman (1987) puts it, 'in design, there are no correct answers, only trade offs'.

The need to formalize trade offs becomes more pressing as evaluation becomes more automated; what might have been done intuitively before now needs to be made explicit. Norman (1987) presents some illustrations of how formal trade off analysis can be applied to human–computer interfaces.

The fine tuning stage

Fine tuning is possible once an early version of the product is available. The techniques discussed above for evaluation of prototypes apply equally well here. The important point to remember is that fine tuning involves very specific recommendations which, normally, given the time and cost constraints under which the design team must work, must not involve extensive re-coding. It is assumed that larger-scale modifications are normally carried out earlier as a result of the evaluation activities involved in the design stage. At

the fine tuning stage, the overall philosophy (conceptual model) for th
product is mainly fixed and cannot be changed.

Post-release evaluation

Evaluation need not, normally does not, and certainly should not, stop onc
the product is released. It is quite normal for software and software/hardwar
products to be revised on a more or less continuous basis. Many versions o
an applications package may be released during the commercial life of th
product. These versions normally represent improvements not just becaus
various bugs have been removed but also because of enhancements in term
of usability or functionality.

The improvements built in to a new version may result from informal o
formal feedback from various sources, including (although not as a matte
of course) formal user tests. One source of feedback that may become o
increasing importance in regard to software packages, as it has already becom
in regard to some other 'products' such as books, films and theatre plays, i
that of the professional reviewer. It is not yet the case that 'the critics' ca
kill a software package, but they can certainly have a strong influence on th
image that package has in the computing and other press, and they ca
influence fashion in this area as in any other.

Feedback from reviews in the press can be used more or less formally fo
evaluation purposes and it can have a bearing on all aspects of the product
from the overall concept down to detailed feedback on very specific aspect
of the dialogue (e.g. 'I found the printer settings menu confusing') or ver
specific difficulties with particular aspects of the hardware (e.g. 'I found th
mouse a little sticky').

Whilst such reviews represent the experiences and views of only a few
reviewers, and may not accurately represent the views of everyone wh
might use the product, they do have several positive features as an input t
evaluation: the feedback they provide is normally articulate and to the point
written by motivated people who (by virtue of their occupation) have th
time to devote to the task; it can take account of what else is on the market
and (unlike, say, responses to a questionnaire, or the findings from
controlled experiment) it is not constrained by the preconceptions of th
evaluation team—the reviewer is free to conceive of the review in any wa
he or she chooses, has a free choice of what to test, and can comment o
any aspect of the product from any point of view.

In the same spirit, evaluation of a product can often benefit from th
feedback (which can include both favourable and unfavourable comments
which users sometimes spontaneously offer in letters they write to th
software house or manufacturer concerned. Much can often be learned fron
a qualitative analysis of the 'bouquets and brickbats' offered by reviewer
and users.

Evaluation of advanced concepts

The discussion so far has focused on the role of evaluation in the design of actual products. There is also value in looking beyond current and even next-generation products to advanced concepts that cannot presently be realised as products, but which are none the less of interest as concepts for the future and which may help to guide research and development activities. Evaluation of advanced concepts can often benefit from demonstrations or simulations that capture key features of the concepts without the kind of underlying hardware or software that would be needed in order to develop real products or systems. The following examples illustrate some of the possibilities.

Before WIMP interfaces (using Windows, Icons, Menus and Pointing) became popular in commercial products, many of the essential concepts involved were demonstrated in a special 'Media Room' set up at the Massachussetts Institute of Technology (MIT) by Bolt and colleagues (see for example Bolt, 1984). The 'Media Room' used expensive technology in a special demonstration environment to simulate advanced information systems. It showed, for example, how one could conceive of a 'dataland' in which items of information (such as memos, reports and even video tapes) as well as objects (such as calculators or telephones) could be represented as icons which were laid out in a two-dimensional array taking up the whole of one wall of the room. A joystick could be used to 'helicopter' over the array to items of interest, which could then be shown in close up and read (e.g. a memo), viewed (e.g. a video tape), or used (e.g. a calculator or telephone). Spence and Apperley (1982) took some of the key concepts demonstrated at MIT and interpreted them into a concept for an 'electronic office', producing a demonstration and video tape which showed how some of the key concepts could in principle be implemented within an otherwise relatively ordinary office environment.

Evaluation of these simulations and demonstrations was generally informal, but the concepts involved had an important influence on the research and development activities of many organizations. Concepts used in commercial products such as the Xerox Star, Apple Lisa and, later, the Apple Macintosh can be seen to have their roots at least partly in the early simulations at MIT.

Evaluation of advanced concepts demonstrated by means of simulations can be formal. An example of this is the work on the 'listening typewriter' by Gould and colleagues (e.g. Gould *et al.*, 1981). Although the concept of dictating to a typewriter has been around in the science fiction literature for decades, such a product has not yet been produced. The speech recognition technology required for it does not yet exist, although there are continuous advances in this area. This does not mean that it is impossible to evaluate various human factor aspects of the interface to such a product, as an experiment by Gould *et al.* (1981) illustrates. Acknowledging that speech recognition technology was advancing but that it would be many years before it would match the performance of a human listener, Gould *et al.*

(1981) addressed the question of whether an imperfect listening typewriter would be useful. Two major factors were considered, reflecting key parameters of the technology: (a) recognition of words spoken separately ('isolated word recognition') versus recognition of words spoken in continuous speech ('continuous speech recognition', a much more difficult problem in terms of the technology); and (b) the size of the vocabulary that could be recognized (speech recognizers at that time could handle only very small vocabulary sizes, and this is still a significant limitation of the technology today).

Subjects in their experiment dictated either in isolated words or in continuous speech. A typist located in another room heard what the subject dictated and typed it, the text appearing on both the typist's own screen and the subject's screen. Before displaying the text, however, the computer checked whether the words the typist was typing were in the vocabulary allowed for or not; if a word was not in the vocabulary, the computer would insert XXXXX's into the text instead of the word. In this way it was possible to vary the size of the vocabulary available. Two sizes were used in the experiment: 1000 words, and unlimited.

The experiment suggested that for composing letters it might not be essential to have perfect speech recognition, and that some versions of an imperfect listening typewriter could produce results at least as good as writing.

Speech may in the future be useful not simply for dictating to machines but as a means of communicating commands, intentions, explanations, and so forth, in some cases as a major part of the human–computer dialogue. This possibility has been foreseen in many science fiction stories, and, as previously mentioned, a particularly memorable example being the HAL computer in the film *2001: A Space Odyssey*. Chapanis and colleagues in a series of experiments stimulated in part by the film (e.g. Chapanis *et al.*, 1972) showed how it is possible to evaluate speech as a means of communication with 'intelligent' computers without actually having to have such a computer available. In these experiments, subjects worked in pairs on various problem-solving tasks, one of the subjects taking the role of the computer and the other the role of the human. The human was typically given a problem to solve and the 'computer' was given relevant information. By restricting communication between the two (e.g. to written notes, or to speech) it was possible to evaluate different forms of communication for this type of dialogue. One of the key findings from the simulations was that problems typically were solved faster when speech was used, even though speech involved exchanging more words and the structure of the communication was less complete (only partial sentences or odd phrases) and less grammatically correct.

Towards an evaluation environment

The discussion so far suggests that evaluation of the human–computer interface is multifaceted and needs to be linked into various different stages or aspects of the design process, each with its own purposes of evaluation. In this section, we look at possibilities for bringing at least some of the different elements together in an environment especially designed to serve the purposes of evaluation of human–computer systems.

The CAFE OF EVE

The environment for office systems and other human–computer systems is continually evolving as organizations adapt to new conditions and office workers become more proficient in their use of electronic systems. Office systems manufacturers need to be able to monitor user needs, develop new concepts and evaluate products on a continuing basis. Gale and Christie (1987b) have outlined a model for a research and development environment that addresses this need, and issues relating to the concept have been examined by Gale and Scane in conjunction with a major office systems manufacturer in research at Southampton University. Such an environment would be a CAFE OF EVE, a 'Controlled Adaptive Flexible Experimental Office of the Future in an Ecologically Valid Environment'.

The essence of a CAFE OF EVE is a real working office in which the office workers are employed to do two jobs. Their job description includes both aspects, and the salaries they are paid cover both aspects. One job is to conduct the work they would normally have to do in an ordinary equivalent office in a normal organization. For example, in the Personnel Department the office workers would perform normal personnel work. The other aspects of their work in a CAFE OF EVE is to participate as subjects in research activities aimed at evaluating the electronic systems they use in their work. This aspect of their work would involve using prototype systems as well as finished products, and would also involve participating in controlled laboratory experiments and other research aimed at developing and refining product concepts, human–computer interface designs, and other aspects of product development.

The workplace would be a living research and development laboratory. Video cameras, on-line collection of performance data, presentation of rating scales on the workstation screen, physiological measurement (where appropriate) and other aspects of the ongoing research would be as accepted as the ordinary office furniture. A normal part of the day would be a debriefing session at its end. Group discussions about the nature of the normal office work done and the suitability of the various prototypes and other systems used to handle it would be normal activities, as would individual experimental sessions in controlled experimental suites that would form part of the office/laboratory complex.

The typical status differential between researcher and subject would be changed. The office workers involved would be as much researchers as subjects. They would have the status of co-researchers, although they would be trained in office work rather than research. The researchers would be participant observers, sharing in the normal office tasks, developing normal peer-group relationships, rather than being seen as outside agents. The researchers, by being co-workers in a real office, would be privy to confidential information that would be kept from outside researchers.

Both laboratory experiments and field tests have a part to play in a CAFE OF EVE, where they can be used to complement one another in the evaluation of human–computer interfaces. Considering the relationship between the two in the context of human–computer interaction, Gale and Christie (1987b) have suggested that laboratory studies allow independent variables to be manipulated systematically, the number of such variables to be limited to a manageable number, environmental influences to be controlled, events to be sampled according to an experimental plan, complete experimental designs to be constructed, different subjects to be allocated randomly to different experimental conditions, experimental replication, and a number of other positive features. Field studies, in contrast, are characterized by multivariate influences on the subjects, uncontrolled and disruptive events, restrictions on when events can be sampled, incomplete designs often with unequal numbers in different cells, non-random selection of subjects who may be obviously unrepresentative of larger populations, difficulty of replicating studies, and a number of other disadvantages.

On the other hand, the apparent advantages of laboratory studies have to be considered against their disadvantages. They do not truly allow differential partitioning of all the causal factors that operate in real-life situations, they are typically oriented towards demonstrating statistically significant effects rather than effects that are important in practical situations, often use subjects with inappropriate or minimal training and who may differ in key respects, such as educational or occupational background, from the population to which the results of the research are intended to apply. Field studies, whatever their disadvantages, do capture the richness and complexity of real-life situations. The two need to be seen as complementary, and both have a place in the evolution of human–computer interfaces. The concept of a CAFE OF EVE shows one way of bringing the two together in an integrated applied research environment.

A CAFE OF EVE offers many benefits:

1. *Ecological validity.* It offers the opportunity to research office work conducted using electronic systems in a real office environment rather than solely in simulations or in artificial experiments. The complex interplay of the many factors that affect real systems can be available for study.

2. *Salience.* The office workers would be treated as co-researchers, authorities on their own needs, their own work, their own preferences. The research could be guided by what was truly salient to the users concerned, rather than predominantly by a priori concerns of external research communities.

3. *Symmetrical status.* The emphasis on the researcher and the office workers as partners in a joint endeavour would facilitate openness and frankness. There would be less tendency for the users to say things and behave in ways

designed either to conform to what is believed to be the researcher's expectations or to 'get the researcher off the subject's back' so that the office worker concerned could get on with the 'real work'. In a CAFE OF EVE, the research would also be 'real work'.

4. *Sampling at choice.* Sampling could be done, with agreement, at any time and using any technique. The office workers concerned would help to choose the variables to measure and so problems of face validity, threat and intrusion would be greatly reduced.

5. *Strategy assessment.* Traditional laboratory studies are artificial in various ways, including the fact that they are typically brief and the subjects do not get an opportunity to develop their own, long-term strategies or methods of working. Individual, longitudinal research is possible in a CAFE OF EVE.

6. *Multiple measurement.* Human-computer systems are very complex and research could benefit from being able to measure a number of different variables at once, observing how they relate to one another. Co-operation in this regard is likely to be higher when the subjects and researchers see themselves as peers working in the same environment towards mutually agreed goals.

7. *Lowered reactivity.* Hawthorne and related reactivity effects (where any change in the environment may result in improved performance, improved job satisfaction, and other positive effects) can be expected to be reduced in an environment in which change is seen as a normal, ordinary part of the research and development context rather than as something special.

8. *Lower subject attrition.* A problem with traditional laboratory research is that it can be very difficult to get subjects to come back reliably on repeated occasions in order to examine effects over a period of time. This would be much less of a problem in a context in which the subjects are doing the research as part of their normal work. Subject drop-out would correspond to subjects leaving in order to take a job elsewhere.

9. *Valid testbed conditions.* Continuous, iterative prototyping and testing could be used in the development of new products and systems at much reduced marginal costs since the environment for doing so, including the people involved as well as the equipment and buildings and so forth, would already be there as a permanent facility.

10. *Rapid bench to field transfer.* There could be a closer relationship between laboratory work and field testing of products and prototypes. It would be easier to schedule testing so as to fit in with normal work patterns.

11. *Interdisciplinary discussion.* The personal assistants, secretaries, managers, researchers and others working in the office could develop a common frame of reference and a common language to guide discussion of user needs and other aspects of the research.

12. *Public relations, sales and training.* The office workers in a CAFE OF EVE could be expected to be very familiar with and enthusiastic about the products to which they contributed. Together with the electronic environment itself, this would provide an excellent context in which the company

concerned could display its products, train sales teams and other organizations' staff in the use of the products, and create confidence in the products and sytems being marketed.

Along with these benefits, a number of problems with the CAFE OF EVE can be identified, including:

1. Very high set up costs. Practical experience has shown that even large office equipment manufacturers, individually or even in co-operation with others, are not easily convinced that the benefits would justify the cost. Many prefer to rely on traditional product development methods and feedback from the market once the products are launched. (Simpson and Mason provide ideas on cost–benefit justifications for ergonomics in chapter 31.)

2. It is conceivable that such an environment, by introducing more elements of prototyping and evaluation, could slow down the development process. In the kind of environment in which companies are currently operating it may be more important to meet a market window than to achieve the best possible product.

3. A key aim of the CAFE OF EVE is to achieve ecological validity. At least three factors could operate against this. First, it requires recruiting office workers who are willing or positively enthusiastic about working in such an environment. Such people may not be typical of other office workers and results achieved with them may not reflect what is true of office workers elsewhere. Second, after working in the environment for a while, being exposed to many different prototypes, being involved in discussions of the nature of office work and the appropriateness of the equipment, and after being sensitized in various ways to questions of usefulness, functionality and usability, as well as to opportunities for improvements in all these areas, the office workers concerned may be very different in terms of their skills, understandings and expectations compared with office workers elsewhere. Third, some aspects of the research, such as debriefings at the end of each day, combined with the equal priorities given to the research and other aspects of the work done, could encourage patterns of working that are different from those that would otherwise be observed. For example, it is not uncommon in offices for people to stay late into the evening in order to finish important or urgent work. Giving equal priority to a research debriefing could interfere with this.

What is the future for evaluation?

In the applied context, evaluation is not an end in itself but a means to an end, the end being better design—but not better design at any cost. The key need in many cases is to get a reasonable product on to the market in time to meet the market window. Within this constraint, the product should be designed as well as possible but it is no use having an excellently-designed product that is launched too late or with too high a price tag. Evaluation needs to be an integrated part of the design process, contributing to better design at little or no marginal cost in terms of time to project completion or overall project costs. Attention needs to be given to the concept of value of information. The value of the information obtained from any proposed evaluation needs to be weighed against the costs involved (see Mantei and Teorey, 1988, for a discussion and practical illustrations of this point). This concept is common in market research, but could be developed further in regard to evaluating human–computer interfaces.

Specialists in human–computer interaction can in principle make both direct and indirect contributions to the design and evaluation process: directly, by working with or as part of the design team, and indirectly, by transferring knowledge, skills or design aids to the design teams concerned.

Direct contributions

Office systems and other human–computer systems are normally designed by teams rather than individuals working alone. The teams are multidisciplinary, and human–computer interface specialists are playing an increasingly important role in them. In many companies this type of specialist is now quite likely to be invited to join the design team as a full member, rather than being brought in as a consultant to work on specific problems as and when the team feels this to be appropriate.

When working as a member of the design team, the human–computer interface specialist normally works alongside and in close collaboration with the systems analyst, programmers and other software and hardware engineers. The precise boundaries of his or her role are often worked out pragmatically in negotiation with the other members of the design team as the project develops. It is possible, however, to define the boundaries in accordance with a more formal model. One such example of this is provided by Roach *et al.* (1982). They propose a fairly equal relationship between what they call the 'dialogue author' (concerned with designing the dialogue between the human and the machine) and the 'application programmer' (who is concerned with constructing the software behind the dialogue). Whilst distinguishing between the two roles, they see a need for communication between them and have attempted to formalize the boundaries between the two in a system called DMS (Dialogue Management System). The system has been successful to a large extent in maintaining the distinction between the two roles but it has also been observed that in practice the 'dialogue author' often feels a need to become directly involved in what formally is the work of the 'application programmer', and vice versa, perhaps suggesting that further research on appropriate role differentiation with regard to the human–computer interface aspects of design, within an 'equal partners' framework, could still be of some value.

It is also possible to develop models of the design process based on alternatives to the 'equal partners' approach. Foley (1983) argues that the human–computer interface is so important in some products and has such a wide-ranging influence on the overall design, that the design team should include a human–computer interface 'architect' (he uses the term 'user-interface architect') who, like a building architect, would be an intermediary between the users (clients) and the implementors (construction teams). He seems to suggest that this person should have the central role in the design process. For example, at the design review it should be the user-interface architect who decides whether the proposed design meets the design objectives whilst satisfying the design constraints.

Whilst human–computer interface specialists may well be full members of design teams on some key product developments, it will often be the case that they are in addition called upon to act as consultants to other projects. This style of working remains important for two main reasons. Acting as a consultant, the specialist concerned can make contributions to a number of different projects, which can be a cost-effective way of utilizing what is still in most companies a scarce resource. Second, and related to this, it means that a number of different projects can gain access to the facilities (such as prototype testing facilities) that the human–computer interface specialist may need to use. In order for this style of working to be effective, it is important to identify relatively well-defined problems in areas where relatively specific inputs from the human–computer interface specialist can have a useful influence on the design. This approach works best when there already is a human–computer interface specialist on the design team and the consultant is being brought in to provide additional support in specific areas, or when there is a good understanding between the consultant and the design team built up through previous projects.

However they operate, the human–computer interface specialist is increasingly expected to be able to draw on a substantial body of relevant theory as well as practical know how. He or she is increasingly expected to be familiar with a broad area of relevant cognitive psychology and cognitive science, and to be able to apply this to the practical problems which the team faces. Inputs from the human–computer interface specialist are increasingly likely to be sought at early stages of the design process, not just at the end, after all the key decisions have been made and when perhaps only fine tuning is possible.

Despite the increasing expectations which project managers and co-workers often have of what the human–computer interface specialist can deliver, a number of factors in practice constrain the effectiveness of direct contributions from such specialists. In many cases, those who have drawn up and agreed the budget for the project have not made explicit allocations for a human–computer interface specialist or the various associated costs. This often reflects a lack of awareness, or a belief that within the overall system of priorities this aspect of the design is one where costs can be trimmed in order to keep the overall budget down to an acceptable level.

The time-scale of the project may effectively preclude some kinds of contributions that a human–computer interface specialist might otherwise be able to make. There may be insufficient time for extensive user testing, for example, or any experimental comparisons of different design solutions. With the pace of developments in this field as high as it is, products can be effectively obsolete almost as soon as they are launched. Very often there may be a very limited 'market window', perhaps only a few months. A product which is 70% right and is launched within that window may well make money, whereas a product that is 80% right but is launched late will probably lose money.

Being too late is perhaps the most usual hazard, but it is also possible to be too early, to launch the product before the beginning of the market window for it. This can happen for a number of reasons. An existing generation of equipment may not have been depreciated. The product might be seen as being too sophisticated. This relates to the concept of 'user migration'. Users of electronic office systems and other computer systems generally become more demanding as their experience increases—they are said to 'migrate' from less sophisticated to more sophisticated products—but if a product is more sophisticated than its users are ready for it may be 'ahead of its time' and fail for that reason. This can be a problem for the human–computer interface specialist. Evaluated in terms of relevant theory, a particular interface may have a lot to commend it, but if it is 'ahead of its time' it may make better sense commercially to shelve it in favour of something nearer to what users currently expect and the hardware (and operating systems, and other aspects of the product's environment) that are currently the standard.

A further factor operating against direct contributions from human–computer interface specialists in many organizations is the relative lack of awareness of the interface as a specialist area. Although the situation is changing, it may often still be assumed that specialist software engineers who are expert in the design of computer systems are aware of what is involved in designing a successful interface to, say, an ordinary office worker and have the necessary knowledge and skills to be able to cope with that effectively. Against this background of relatively low awareness, a human–computer interface specialist normally works in an organizational context in which others are competing for the available funds. It helps if the specialists concerned have some 'success stories' to their credit, in terms of product developments which can clearly be seen to have benefited from their inputs. Building on this base, they then need to advertise the human–computer interface resource within the organization, and what the resource has to offer.

The image of the human–computer interface specialist in any given organization will depend to a significant extent on the particular contributions made to product developments. This may itself be a constraining influence. Managers and co-workers may expect the specialists concerned to be able to operate only on the kinds of problems tackled before, and may fail to appreciate that they could make equally useful contributions beyond those particular boundaries. There are other problems as well. It seems to be an unfortunate fact that bad design is more obvious than good design. Gross errors are often easily spotted and it is often relatively easy to make a plausible suggestion concerning how the design could be improved. This helps to reinforce the feeling which is still prevalent that good design is common sense. It may also reinforce any feeling there may be that the human–computer interface specialist is too 'academic' and that real design work should be left to those who have the necessary practical experience.

All these factors taken as a whole suggest that it may often be useful to complement the direct contributions that may be possible with indirect contributions aimed at transferring specialist knowledge, skills and tools to a wider population of software engineers and others who are concerned with the human–computer interface.

Indirect contributions

The human–computer interface specialist can make a number of different kinds of indirect contribution to the design or uptake of human–computer systems. Primary amongst these are training and techniques. Whether in the context of designing systems or helping user-organizations to evaluate systems, training programmes can play a key role in helping to raise awareness of what is involved in evaluating the human–computer interface. These training programmes can include a range of different elements, including lectures, workshops, tutorials and other elements involving direct contact between trainers and trainees, as well as elements based on video tapes, video discs, computer-aided learning, and other elements based on using information technology as the vehicle to deliver the training.

Practical experience suggests that training is perceived to be of much greater value when it is supported by the provision of tools which the trainees can use in their work to help them apply their new knowledge and skills to practical problems, whether in product design or within an information technology uptake programme. These tools and techniques can cover a broad range, including such things as: checklists, guidelines, generic designs, design and evaluation aids, and other tools and techniques discussed earlier.

The future

Evaluation of human–computer interfaces has traditionally been the responsibility of specialists in that area, and has often been done as an add-on to the 'main' design process. The extent to which human–computer interface specialists can afford to keep their knowledge and skills to themselves and still contribute significantly to the needs of a broad range of practical projects, and the extent to which it is necessary or desirable to take a more educational role, transferring knowledge and skills to other professionals (such as programmers and system analysts) is perhaps the single most important issue facing the development of the profession during the 1990s.

Notes

Software mentioned in the text is available from the following companies:

Progress. Phoebe Software Limited, Chambers House, Ellison Street, Castlebar, Co. Mayo, Ireland. Tel: (094) 23397.

SET. PA Technology, Computer Aided Engineering, Cambridge Laboratory, Melbourn, Royston, Hertfordshire, SG8 6DP. Tel: (0763) 61222.

Tapestry II. Torus Systems Limited, Science Park, Milton Road, Cambridge CB4 4GZ. Tel: (0223) 862131.

TEA: The Event Analyser. HeptaCon Limited, Suite 500, Chesham House, 150 Regent Street, London W1R 5FA. Tel: (01) 734 5351.

References

Anderson, J.R. (1976). *Language, Memory and Thought* (Hillsdale, NJ: Lawrence Erlbaum).

Argyle, M., Furnham, A. and Graham, J.A. (1981). *Social Situations* (Cambridge: Cambridge University Press).

Baird, J.E., Jr. (1977). *The Dynamics of Organizational Communication* (London: Harper and Row).

Barber, G. (1983). Supporting organizational problem-solving with a work station. *ACM Transactions on Office Information Systems*, **1**, 45–67.

Bennett, R.W. and Greenspan, S.L. (1987). Evaluating synthetic speech devices. In *Cognitive Engineering in the Design of Human–Computer Interaction and Expert Systems: Proceedings of the Second International Conference on Human–Computer Interaction, Honolulu, Hawaii, 10–14 August 1987*, Volume II, edited by G. Salvendy (Amsterdam: Elsevier Science), pp. 391–398.

Bolt, R.A. (1984). *The Human Interface: Where People and Computers Meet* (Belmont, CA: Lifelong Learning).

Bösser, T. (1987). The evaluation of learning requirements of IT systems. In *Cognitive Engineering in the Design of Human–Computer Interaction and Expert Systems: Proceedings of the Second International Conference on Human–Computer Interaction, Honolulu, Hawaii, 10–14 August 1987*, Volume II, edited by G. Salvendy (Amsterdam: Elsevier Science), pp. 45–52.

Bouchard, T.J., Jr. (1976). Field research methods: interviewing, questionnaires, participant observation, systematic observation, unobtrusive measures. In *Handbook of Industrial and Organizational Psychology*, edited by M.D. Dunnette (Chicago: Rand McNally), pp. 363–413.

Brooke, J.B. (1986). Usability engineering in office product development. In *People and Computers: Designing for Usability*, edited by M.D. Harrison and A.F. Monk (Cambridge: Cambridge University Press).

Brotz, D.K. (1983). Message system mores: etiquette in Laurel. *ACM Transactions on Office Information Systems*, **1**, 179–192.

Bullinger, H-J., Faehnrich, K-P. and Ziegler, J. (1987). Software-ergonomics: history, state-of-the-art and important trends. In *Cognitive Engineering in the Design of Human–Computer Interaction and Expert Systems: Proceedings of the Second International Conference on Human–Computer Interaction, Honolulu, Hawaii, 10–14 August 1987*, Volume II, edited by G. Salvendy (Amsterdam: Elsevier Science), pp. 307–316.

Butler, K. (1985). Connecting theory and practice: a case study of achieving usability goals. *Proceedings of Computer–Human Interaction (CHI) Conference, 1985: SIGCHI Bulletin, Special Issue*, April 1985.

Cakir, A., Hart, D.J. and Stewart, T.F.M. (1980). *Visual Display Terminals*. (Chichester and New York: John Wiley).

Calder, B.J. (1977). Focus groups and the nature of qualitative marketing research. *Journal of Marketing Research*, **XIV**, 353–364.

Card, S.K., Moran, T.P. and Newell, A. (1983). *The Psychology of Human–Computer Interaction* (London and New Jersey: Lawrence Erlbaum).

Chao, B.P. (1987). Prototyping a dialogue interface: a case study. In *Cognitive Engineering in the Design of Human–Computer Interaction and Expert Systems: Proceedings of the Second International Conference on Human–Computer Interaction, Honolulu, Hawaii, 10–14 August 1987*, Volume II, edited by G. Salvendy (Amsterdam: Elsevier Science), pp. 357–363.

Chapanis, A., Ochsman, R., Parrish, R. and Weeks, G. (1972). Studies in interactive communication: the effects of four communication modes on the behavior of teams during cooperative problem solving. *Human Factors*, **14**, 487–509.

Chesebro, J.W. (1985). Computer-mediated interpersonal communication. In *Information and Behavior*, Volume I, edited by B.D. Ruben (New Brunswick and Oxford: Transaction Books), pp. 202–222.

Christie, B. and Gardiner, M.M. (1987). Office systems. In *Designing for the Future: Information Technology and People*, edited by F. Blackler and D. Oborne (Leicester: The British Psychological Society), pp. 85–102.

Cole, I., Lansdale, M. and Christie, B. (1985). Dialogue design guidelines. In *Human Factors of Information Technology in the Office*, edited by B. Christie (Chichester and New York: John Wiley), pp. 212–241.

Cranach, M. von, Kalbermatten, U., Indermuhle, K. and Gugler, B. (1982). *Goal-directed Action* (London: Academic Press).

Davidoff, J. (1987). The role of colour in visual displays. In *International Reviews of Ergonomics*, Volume I, edited by D.J. Oborne (London: Taylor and Francis), pp. 21–42.

Del Galdo, E.M., Williges, R.C., Williges, B.H. and Wixon, D.R. (1986). An evaluation of critical incidents for software documentation design. *Proceedings of the Human Factors Society 30th Annual Meeting*, pp. 19–23.

Drury, C.G., Paramore, B., Van Cott, H.P., Grey, S.M. and Corlett, E.N. (1987). Task analysis. In *Handbook of Human Factors*, edited by G.

Salvendy (Chichester and New York: John Wiley), pp. 370–401.

Dzida, W. and Valder, W. (1984). Application domain modelling by knowledge engineering techniques. In *Interact '84: First IFIP Conference on Human Computer Interaction*, edited by B. Shackel (Amsterdam: IFIP, North-Holland), pp. 320–327.

Faehnrich, K.P. and Ziegler, J. (1984). Workstations using direct manipulation as interaction mode—aspects of design, application and evaluation. In *Interact '84: First IFIP Conference on Human Computer Interaction*, edited by B. Shackel (Amsterdam: IFIP, North-Holland).

Faehnrich, K.P., Ziegler, J. and Davies, D. (1987). HUFIT (Human Factors in Information Technology). In *Cognitive Engineering in the Design of Human–Computer Interaction and Expert Systems: Proceedings of the Second International Conference on Human–Computer Interaction, Honolulu, Hawaii, 10–14 August 1987*, Volume II, edited by G. Salvendy (Amsterdam: Elsevier Science), pp. 37–43.

Fang, S-P. and Tzeng, O.J.L. (1987). An evaluation model of Chinese graphemic input systems. In *Applications of Cognitive Psychology: Problem Solving, Education, and Computing*, edited by D.E. Berger, K. Pezdek and W.P. Banks (Hillsdale, NJ: Lawrence Erlbaum), pp. 201–217.

Fern, E.F. (1982). The use of focus groups for idea generation: the effects of group size, acquaintanceship, and moderator on response quantity and quality. *Journal of Marketing Research*, **XIX**, 1–13.

Foley, J.D. (1983). Managing the design of user–computer interfaces. *Computer Graphics World*, December, 47–56.

Furner, S.M. (1987). Practical information about interface operating characteristics for engineering design. *Proceedings of an IEE colloquium on Evaluation Techniques for Interactive Systems Design: I organized by Professional Group C5 (Man–machine interaction), 27 March 1987, Savoy Place, London. IEE Digest No. 1987/38*, pp. 4/1–4/6.

Furnham, A. (1987). The social psychology of working situations. In *Psychophysiology and the Electronic Workplace*, edited by A. Gale and B. Christie (Chichester and New York: John Wiley), pp. 89–112.

Gale, A. (1979). Psychophysiology: a bridge between disciplines. Inaugural Lecture, University of Southampton.

Gale, A. (1985). Assessing product usability: a psychophysiological approach. In *Human Factors of Information Technology in the Office*, edited by B. Christie (Chichester and New York: John Wiley), pp. 189–211.

Gale, A. and Christie, B. (Eds) (1987a). *Psychophysiology and the Electronic Workplace* (Chichester and New York: John Wiley).

Gale, A. and Christie B. (1987b). Psychophysiology and the electronic workplace: the future. In *Psychophysiology and the Electronic Workplace*, edited by A. Gale and B. Christie (Chichester and New York: John Wiley), pp. 315–333.

Gardiner, M.M. (1986). Psychological issues in adaptive interface design. *Proceedings of an IEE colloquium on Adaptive Interface Design organized by Professional Group C5 (Man–machine interaction), 4 November 1986, Savoy Place, London. IEE Digest No. 1986/110*, pp. 6/1–6/3.

Gardiner, M.M. (1987). Contemporary models of information processing. In *Psychophysiology and the Electronic Workplace*, edited by A. Gale and

B. Christie (Chichester and New York: John Wiley), pp. 65–88.

Gardiner, M.M. and Christie, B. (1985). Packaging cognitive psychology for user-interface design. *Ninth Congress of the International Ergonomics Association, Bournemouth, England, September, 1985.*

Gardiner, M.M. and Christie, B. (Eds) (1987a). *Applying Cognitive Psychology to User-interface Design* (Chichester and New York: John Wiley).

Gardiner, M.M. and Christie B. (1987b). Communication failure at the person–machine interface: the human factors aspects. In: *Communication Failure in Dialogue and Discourse: Detection and Repair Processes*, edited by R.G. Reilly (Amsterdam: Elsevier Science/North-Holland), pp. 309–323.

Gould, J.D., Conti, J. and Hovanyecz, T. (1981). Composing letters with a simulated listening typewriter. *Proceedings of the Human Factors Society 25th Annual Meeting 1981*, pp. 505–508.

Graesser, A.C., Lang, K.L. and Elofson, C.S. (1987). Some tools for redesigning system–operator interfaces. In *Applications of Cognitive Psychology: Problem Solving, Education, and Computing*, edited by D.E. Berger, K. Pezdek and W.P. Banks (Hillsdale, NJ: Lawrence Erlbaum), pp. 163–181.

Gregory, K. (1987). Methodology for designing a normalized user interface. In *Cognitive Engineering in the Design of Human–Computer Interaction and Expert Systems: Proceedings of the Second International Conference on Human–Computer Interaction, Honolulu, Hawaii, 10–14 August 1987*, Volume II, edited by G. Salvendy (Amsterdam: Elsevier Science), pp. 139–146.

Gruenenfelder, T.M. and Whitten, W.B., II. (1984). Augmenting generic research with prototype evaluation: experience in applying generic research to specific products. In *Interact '84: First IFIP Conference on Human Computer Interaction*, edited by B. Shackel (Amsterdam: IFIP, North-Holland), pp. 315–319.

Gutierrez, O. (1987). A system simulation system to support the elicitation of information requirements by end-users. In *Cognitive Engineering in the Design of Human–Computer Interaction and Expert Systems: Proceedings of the Second International Conference on Human–Computer Interaction, Honolulu, Hawaii, 10–14 August 1987*, Volume II, edited by G. Salvendy (Amsterdam: Elsevier Science), pp. 529–536.

Hacker, W. (1978). *Allgemeine Arbeits- und Ingenieurpsychologie* (Bern: Huber).

Hammond, N., Gardiner, M.M., Christie, B. and Marshall, C. (1987). The role of cognitive psychology in user-interface design. In *Applying Cognitive Psychology to User-interface Design*, edited by M.M. Gardiner and B. Christie (Chichester and New York: John Wiley), pp. 13–53.

Harker, S. (1987). Rapid prototyping as a tool for user centered design. In *Cognitive Engineering in the Design of Human–Computer Interaction and Expert Systems: Proceedings of the Second International Conference on Human–Computer Interaction, Honolulu, Hawaii, 10–14 August 1987*, Volume II, edited by G. Salvendy (Amsterdam: Elsevier Science), pp. 365–372.

Hockey, R. (1979). Stress and the cognitive components of skilled performance. In *Human Stress and Cognition: An Information Processing Approach*, edited by V. Hamilton and D.M. Warburton (Chichester and New

York: John Wiley), pp. 141–178.

Houston, T. (1983). The allegory of software: beyond, behind and beneath the electronic desk. *Byte*, December, 210–214.

Howard, S. and Murray, D.M. (1987). Experiences in evaluating the human–computer interface. *Proceedings of an IEE colloquium on Evaluation Techniques for Interactive Systems Design: I organized by Professional Group C5 (Man–machine interaction), 27 March 1987, Savoy Place, London. IEE Digest No. 1987/38*, pp. 2/1–2/3.

Hoyos, C.G., Gstalter, H., Strube, V. and Zang, B. (1987). Software-design with the rapid prototyping approach: a survey and some empirical results. In *Cognitive Engineering in the Design of Human–Computer Interaction and Expert Systems: Proceedings of the Second International Conference on Human–Computer Interaction, Honolulu, Hawaii, 10–14 August 1987*, Volume II, edited by G. Salvendy (Amsterdam: Elsevier Science), pp. 329–340.

Johnson, P. (1985). Towards a task model of messaging: an example of the application of TAKD to user interface design. In *People and Computers: Designing the Interface*, edited by P. Johnson and S. Cook (Cambridge: Cambridge University Press).

Kano, S. (1977). A change of effectiveness of communication networks under different amounts of information. *Japanese Journal of Experimental Social Psychology*, **17**, 50–59.

Kantowitz, B.H. and Sorkin, R.D.(1987). Allocation of functions. In *Handbook of Human Factors*, edited by G. Salvendy (Chichester and New York: John Wiley), pp. 355–369.

Kieras, D.E. and Polson, P.G. (1985). An approach to the formal analysis of user complexity. *International Journal of Man–Machine Studies*, **22**, 365–394.

Kinsman, F. (1987). *The Telecommuters* (Chichester and New York: John Wiley).

Kirakowski, J. and Dillon, A. (1987). The computer–user satisfaction inventory. *Proceedings of an IEE colloquium on Evaluation Techniques for Interactive Systems Design: II organized by Professional Group C5 (Man–machine interaction), 2 October 1987, Savoy Place, London. IEE Digest No. 1987/78*, pp. 3/1–3/3.

Laughery, K.R., Sr. and Laughery, K.R., Jr. (1987). Analytic techniques for function analysis. In *Handbook of Human Factors*, edited by G. Salvendy (Chichester and New York: John Wiley).

Lewis, S.M. and Downton, A.C. (1987). Statistical design as a vehicle for minimising experimental evaluation. *Proceedings of an IEE colloquium on Evaluation Techniques for Interactive Systems Design: II organized by Professional Group C5 (Man–machine interaction), 2 October 1987, Savoy Place, London. IEE Digest No. 1987/78*, pp. 4/1–4/6.

Loftus, E.F. and Schooler, J.W. (1985). Information-processing conceptualizations of human cognition: past, present, and future. In *Information and Behavior*, Volume 1, edited by B.D. Ruben (New Brunswick and Oxford: Transaction Books), pp. 225–250.

Lum, V.Y., Choy, D.M. and Shu, N.C. (1982). OPAS: an office procedure and automation system. *IBM System Journal*, **21**, 327–350.

Malone, T.W. (1985). Designing organisational interfaces. In *Proceedings of CHI '85: Human Factors in Computing Systems, San Francisco, April*, edited by L. Borman and B. Curtis (New York: ACM).

Malt, L.G. (1987). Skills analysis in human–computer interface: a holistic approach. In *Social, Ergonomic and Stress Aspects of Work with Computers: Proceedings of the Second International Conference on Human–Computer Interaction, Honolulu, Hawaii, 10–14 August 1987*, Volume I, edited by G. Salvendy, S.L. Sauter and J.J. Hurrell, Jr. (Amsterdam: Elsevier Science), pp. 319–326.

Mantei, M.M. and Teorey, T.J. (1988). Cost/benefit analysis for incorporating human factors in the software lifecycle. *Communications of the ACM*, **31**, 428–439.

Marshall, C., Christie, B. and Gardiner, M.M. (1987). Assessment of trends in the technology and techniques of human–computer interaction. In *Applying Cognitive Psychology to User-interface Design*, edited by M.M. Gardiner and B. Christie (Chichester and New York: John Wiley), pp. 279–312.

Marshall, C., Nelson, C. and Gardiner, M.M. (1987). Design guidelines. In *Applying Cognitive Psychology to User-interface Design*, edited by M.M. Gardiner and B. Christie (Chichester and New York: John Wiley), pp. 221–278.

Meister, D. (1985). *Behavioural Analysis and Measurement Methods*. (Chichester and New York: John Wiley).

Meister, D. (1987). Behavioral test and evaluation of expert systems. In *Cognitive Engineering in the Design of Human–Computer Interaction and Expert Systems: Proceedings of the Second International Conference on Human–Computer Interaction, Honolulu, Hawaii, 10–14 August 1987*, Volume II, edited by G. Salvendy (Amsterdam: Elsevier Science), pp. 539–549.

Miller, G.A., Galanter, E. and Pribram, K.H. (1960). *Plans and the Structure of Behavior* (New York: Holt).

Monk, A. (1984). *Fundamentals of Human-computer Interaction* (London and New York: Academic Press).

Murray, D. and Beban, N. (1984). The social psychology of computer conversations. In *Interact '84: First IFIP Conference on Human Computer Interaction*, edited by B. Shackel (Amsterdam: IFIP, North-Holland), pp. 268–273.

Norman, D.A. (1987). Design principles for human–computer interfaces. In *Applications of Cognitive Psychology: Problem Solving, Education, and Computing*, edited by D.E. Berger, K. Pezdek, and W.P. Banks (Hillsdale, NJ: Lawrence Erlbaum), pp. 141–162.

Norman, K.L., Weldon, L.J. and Shneiderman, B. (1986). Cognitive layouts of windows and multiple screens for user interfaces. *International Journal of Man–Machine Studies*, **25**, 229–248.

Novara, F., Bertaggia, N. and Allamanno, N. (1986). Designing adequate human–computer interfaces: methodological issues and considerations on usability. *Paper presented at an ESPRIT Technical Week '86 HUFIT Seminar on Designing Usable IT Products, Brussels, 1 October 1986*.

Oborne, D.J. (1987). Psychophysiology and the ergonomic demands of visual display terminals. In *Psychophysiology and the Electronic Workplace*, edited by A. Gale and B. Christie (Chichester and New York: John Wiley), pp. 139–162.

Olsen, D.R. and Dance, J.R. (1988). Macros by example in a graphical UIMS. *IEEE Computer Graphics and Applications*, January 1988, 68–78.

Patrick, J. (1987). Methodological issues. In *New Technology and Human Error*, edited by J. Rasmussen, K. Duncan and J. Leplat (Chichester and New York: John Wiley), pp. 327–336.

Pulat, B.M. and Grice, A.E. (1985). Computer aided techniques for crew station design: Work-space Organizer—WORG; Workstation Layout Generator—WOLAG. *International Journal of Man–Machine Studies*, **23**, 443–457.

Rabbitt, P. (1979). Current paradigms and models in human information processing. In *Human Stress and Cognition: An Information Processing Approach*, edited by V. Hamilton and D.M. Warburton (Chichester and New York: John Wiley), pp. 115–140.

Richards, M.A. and Underwood, K.M. (1984). How should people and computers speak to each other? In *Interact '84: First IFIP Conference on Human Computer Interaction*, edited by B. Shackel (Amsterdam: IFIP, North-Holland), pp. 33–36.

Ridgway, J. (1987). Human–human–machine interactions. *Proceedings of an IEE colloquium on Evaluation Techniques for Interactive Systems Design: I organized by Professional Group C5 (Man–machine interaction), 27 March 1987, Savoy Place, London. IEE Digest No. 1987/38*, pp. 8/1–8/3.

Roach, J., Hartson, H.R., Ehrich, R.W., Yunten, T. and Johnson, D.H. (1982). DMS: a comprehensive system for managing human–computer dialogue. *Proceedings of the Conference on Human Factors in Computer Systems, Gaithersburg, Maryland, 15–17 March 1982.*

Ryan, G., Cullen, K., Wynne, R., Ronayne, T., Cullen, J., Dolphin, C., Korte, W.B., Robinson, S., Ennis, B. and Hopkins, M. (1987). Human, organisational and economic (HOE) factors in IT uptake processes in complex work environments (IT-Uptake). In *ESPRIT '87: Achievements and Impact—Proceedings of the 4th Annual ESPRIT Conference, Brussels, 28–29 September 1987*, edited by Commission of the European Communities Directorate-General Telecommunications, Information Industries and Innovation (Amsterdam: Elsevier Science/North-Holland), pp. 1053–1065.

Sewell, D.R., Geddes, N.D. and Rouse, W.B. (1987). Initial evaluation of an intelligent interface for operators of complex systems. In *Cognitive Engineering in the Design of Human–Computer Interaction and Expert Systems: Proceedings of the Second International Conference on Human–Computer Interaction, Honolulu, Hawaii, 10–14 August 1987*, Volume II, edited by G. Salvendy (Amsterdam: Elsevier Science), pp. 551–558.

Sheehy, N.P. (1987). Nonverbal behaviour in dialogue. In *Communication Failure in Dialogue and Discourse: Detection and Repair Processes*, edited by R.G. Reilly (Amsterdam: Elsevier Science/North-Holland), pp. 325–337.

Sheehy, N.P. and Chapman, A.J. (1987). Nonverbal behaviour at the human–computer interface. In *International Reviews of Ergonomics*, edited by D.J. Oborne (London and New York: Taylor and Francis), pp. 159–172.

Shneiderman, B. (1986). *Designing the user interface: strategies for effective human–computer interaction*, (Reading MA: Addison-Wesley).

Sloan, S.J. and Cooper, C.L. (1987). Sources of stress in the modern office. In *Psychophysiology and the Electronic Workplace*, edited by A. Gale and B. Christie (Chichester and New York: John Wiley), pp. 113–135.

Spence, R. and Apperley, M. (1982). Data base navigation: an office environment for the professional. *Behaviour and Information Technology*, **1**, 43–54.

Streitz, N.A. (1987). Cognitive compatibility as a central issue in human–computer interaction: theoretical framework and empirical findings. In *Cognitive Engineering in the Design of Human-Computer Interaction and Expert Systems: Proceedings of the Second International Conference on Human-Computer Interaction, Honolulu, Hawaii, 10–14 August 1987*, Volume II, edited by G. Salvendy (Amsterdam: Elsevier Science), pp. 75–82.

Streveler, D.J. and Wasserman, A.I. (1984). How should people and computers speak to each other? In *Interact '84: First IFIP Conference on Human Computer Interaction*, edited by B. Shackel (Amsterdam: IFIP, North-Holland), pp. 125–133.

Tsichritzis, D. (Ed.) (1985). *Office Automation* (Berlin: Springer).

Tullis, T.S. (1984). Predicting the usability of alphanumeric displays. Ph.D. dissertation. Available from the Report Store, Lawrence, K.S.

Vallee, J., Johansen, R., Lipinski, H., Spangler, K., Wilson, T. and Hardy, A. (1975). *Group Communication Through Computers*, Volume 3: *Pragmatics and Dynamics*. (Menlo Park, CA: Institute for the Future), Report R-35.

Wasserman, A.I. (1985). Extending state transition diagrams for the specification of human–computer interaction. *IEEE Transactions on Software Engineering*, **SE-11**, 699–713.

Wastell, D.G., Brown, I.D. and Copeman, A.K. (1981). Evoked potential amplitude as measure of attention in working environments: a comparative study of telephone switchboard design. *Human Factors*, **23**, 117–121.

Watanabe, H. (1987). Human–interface architecture: its significance for office automation systems. In *Social, Ergonomic and Stress Aspects of Work with Computers: Proceedings of the Second International Conference on Human–Computer Interaction, Honolulu, Hawaii, 10–14 August 1987*, Volume I, edited by G. Salvendy, S.L. Sauter and J.J. Hurrell, Jr. (Amsterdam: Elsevier Science), pp. 279–294.

Williges, R.C. (1987). Adapting human–computer interfaces for inexperienced users. In *Cognitive Engineering in the Design of Human–Computer Interaction and Expert Systems: Proceedings of the Second International Conference on Human–Computer Interaction, Honolulu, Hawaii, 10–14 August 1987*, Volume II, edited by G. Salvendy (Amsterdam: Elsevier Science), pp. 21–28.

Youmans, D.M. (1987). Using video in the design process. *Proceedings of an IEE colloquium on Evaluation Techniques for Interactive Systems Design: II organized by Professional Group C5 (Man–machine interaction), 2 October 1987, Savoy Place, London. IEE Digest No. 1987/78*, pp. 2/1–2/3.

Chapter 13

Knowledge elicitation

Nigel Shadbolt and Mike Burton

Introduction

Expert systems are computer programs which are intended to solve real-world problems, achieving the same level of accuracy as human experts. There are many obstacles in such an endeavour. One of the greatest is the acquisition of the knowledge which human experts use in their problem solving. The issue is so important to the development of knowledge-based systems that it has been described as the 'bottle-neck in Expert Systems construction' (Hayes-Roth *et al.*, 1983).

Despite its central role there is no comprehensive theory of knowledge acquisition available. Many regard the area as an art rather than a science. It is not the purpose of this chapter to investigate the theoretical shortcomings of knowledge acquisition but to deliver practical advice and guidance on performing the process.

Expert systems

In the early days of Artificial Intelligence much effort went into attempts to discover general principles of intelligent behaviour. Newell and Simon's (1963) General Problem Solver exemplifies this approach. They were interested in uncovering a general problem solving strategy which could be used for any human task. In the early 1970s this position came to be challenged. A new slogan came to prominence—'in the knowledge lies the power'. A leading exponent of this view was Edward Feigenbaum of SRI. He observed that experts are experts by virtue of domain specific problem solving strategies together with a great deal of domain specific knowledge. It was the attempt to incorporate these various sorts of domain knowledge which resulted in the class of programs called Expert Systems.

Throughout this chapter we will be assuming that current commercially available expert system software will be the implementation vehicle for the programs. Thus the form in which the knowledge will be implemented is likely to be standard *production rules* with perhaps a *structured object* facility such as *frames*. For a review of the major types of expert system architecture see Jackson (1985) and of different knowledge representation formalisms see Shadbolt (1989).

The problem of acquisition

The people who build expert systems, the knowledge engineers, are typically not people with a deep knowledge of the application domain. However, it is the knowledge engineers who must gather the domain knowledge and then implement it in a form that the machine can use. In the simplest case, the knowledge engineer may be able to gather information from a variety of non-human resources: e.g. text books, technical manuals. However, in most cases one needs actually to consult a practising expert. This may be because there isn't the documentation available, or because real expertise in the problem solving derives from practical experience in the domain, rather than from a reading of standard texts. The task of gathering information generally, from whatever source, is called *knowledge acquisition*. The subtask of gathering information from the expert is called *knowledge elicitation* (KE).

Many problems arise before an elicitation of the detailed domain knowledge is ever conducted. There are possible failures in the understanding of what it is realistic to build. Sometimes the failure occurs when formulating the role of the system. On other occasions there is an inadequate understanding of the task environment. Very often the effort and resources required to build systems are underestimated: this occurs in both the development and maintenance of systems. A particularly nasty situation arises when the knowledge engineer is expected to conjure up knowledge for areas in which no evidence of systematic practice exists at all. Knowledge engineers seem to be expected to provide theories for domains where there is no theory. Providing we can avoid all of these obstacles then we get down to detailed issues of KE.

The problem of elicitation

The question in KE is this: how do we get experts to tell us exactly what they do? The task is enormous, particularly in the context of large expert systems. There are a variety of circumstances which contrive to make the problem even harder. Much of the power of human expertise lies in laid-down experience, gathered over a number of years, and represented as heuristics. Often the expertise has become so routinized that experts no longer know what it is that they do or why.

There are also commercial reasons to try to make KE more effective. We would like to be able to use techniques which will minimize the effort spent in gathering, transcribing and analysing the knowledge. We would like to minimize the time spent with expensive and scarce experts. And, of course, we would like to maximize the yield of usable knowledge.

This chapter will continue by describing, in sufficient detail for the reader to apply them, examples of major KE methods. We will then mention other techniques and where the reader can find out more about them. In later sections we will review aspects of expertise and cognition that are likely to directly affect the KE process. Finally, we describe the construction of programmes of acquisition.

Methods of knowledge elicitation

The structured interview

Almost everyone starts in KE by determining to use an interview. The interview is the most commonly used knowledge elicitation technique and takes many forms, from the completely *unstructured* interview to the formally-planned, *structured* interview. (For a full review of interview techniques see Sinclair in this volume.) The structured interview is a formal version in which the knowledge engineer has planned the whole session. The structured interview has the advantage that it provides structured transcripts that are easier to analyse than unstructured 'chat'. The relatively formal interview which we have specified here constrains the expert–elicitor dialogue to the general principles of the domain. Experts do not work through a particular scenario extracted from the domain by the elicitor; rather the experts generate their own scenarios as the interview progresses. The structure of a typical interview is as follows.

1. Ask the expert to give a brief (10 min) outline of the target task, including the following information:
   (a) an outline of the task, including a description of the possible solutions to the problem;
   (b) a description of the variables which affect the choice of solutions;
   (c) a list of major rules which connect the variables to the solutions.
2. Take each rule elicited in stage 1, ask when it is appropriate and when it is not. The aim is to reveal the scope (generality and specificity) of each existing rule, and hopefully generate some new rules.
3. Repeat stage 2 until it is clear that the expert will not produce any additional information.

It is important in using this technique to be clear and specific about how to perform stage 2. We have found that it is helpful to constrain the elicitor's interventions to a specific set of *probes*, each with a specific function. Here is a list of probes (P) and related functions (F) which will help in stage 2.

P1 Why would you do that?
F1 Converts an assertion into a rule.
P2 How would you do that?
F2 Generates *lower order* rules.

P3 When would you do that? Is <the rule> always the case?
F3 Reveals the generality of the rule and may generate other rules.
P4 What alternatives to <the prescribed action/decision> are there?
F4 Generates more rules.
P5 What if it were not the case that <currently true condition>.
F5 Generates rules for when current condition does not apply.
P6 Can you tell me more about <any subject already mentioned>.
F6 Used to generate further dialogue if expert dries up.

The idea here is that the elicitor engages in a type of slot/filler dialogue. The requirement that the elicitor listens out for relevant concepts and relations imposes a large cognitive load on the elicitor. The provision of fixed linguistic forms within which to ask questions about concepts, relations, attributes and values makes the elicitor's job very much easier. It also provides sharply focused transcripts which facilitate the process of extracting usable knowledge. Of course, there will be instances when none of the above probes are appropriate (such as the case when the elicitor wants the expert to clarify something). However, you should try to keep the interjections necessary in such situations to a minimum. The point of specifying such a fixed set of linguistic probes is to constrain the expert to giving you all, and only, the information you want.

The sample of dialogue below is taken from a real interview of this kind. It is the transcript of an interview by a knowledge engineer (KE) with an expert (EX) on VDU fault diagnosis*.

EX I actually checked the port of the computer.
KE Why did you check the port?
EX If it's been lightning recently then it's a good idea to check the port + because lightning tends to damage the ports.
KE Are there any alternatives to that problem?
EX Yes, that ought to be prefaced by saying do that if it was several keys with odd effects + not necessarily all of them, but more than 2.
KE Why does it have to be more than 2?
EX Well if it was only one or two keys doing funny things then the thing to do would be to check the keys themselves + check the contacts of the keys + check that they're closing properly + speed would affect all keys, parity would affect about half the keys.

This is quite a rich piece of dialogue. From this section of the interview alone we can extract the following rules:

IF there has been recent lightening
THEN check port for damage
IF there are two or fewer malfunctioning keys
THEN check the key contacts
IF about half the keyboard is malfunctioning
THEN check the parity

---

*In the transcripts we use the symbol + to represent a pause in the dialogue.

IF   the whole keyboard is malfunctioning
THEN   check the speed

Of course these rules may need refining in later elicitation sessions, but the text of the dialogue shows how the use of the specific probes has revealed a well-structured response from the expert[†].

In all the interview techniques (and in some of the other generic techniques as well) there exist a number of dangers that have become familiar to knowledge engineers. One problem is that experts will only produce what they can verbalize. If there are non-verbalizable aspects to the domain, the interview will not recover them. This can arise from two causes. It may be that the knowledge was never explicitly represented or articulated in terms of language (consider, for example, pattern recognition expertise). Then there is the situation where the knowledge was originally learnt explicitly in a propositional form but the experts may have *compiled* the knowledge to such an extent that they regard the complex decisions they make as based on hunches or intuitions; in fact, these decisions are based upon large amounts of remembered data and experience, and the continual application of strategies. In this situation they tend to give *black box* replies "I don't know how I do that . . . .", "It is obviously the right thing to do . . .".

Another problem arises from the observation that people (and experts in particular) often seek to justify their decisions in any way they can. It is a common experience of the knowledge engineer to get a perfectly valid decision from an expert, and then to be given a spurious justification. For these and other reasons we have to supplement interviews with additional methods of elicitation. Elicitation should always consist of a programme of techniques and methods. This brings us on to consider another technique much favoured by knowledge engineers.

Protocol analysis

Protocol Analysis (PA) (considered in detail by Bainbridge in this book) is a generic term for a number of different ways of performing some form of analysis of the expert(s) actually solving problems in the domain. In all cases the engineer takes a record of what the expert does—preferably by video or audio tape—or at least by written notes. Protocols are then made from these records and the knowledge engineer tries to extract meaningful rules from the protocols.

We can distinguish two general types of PA—*on-line* and *off-line*. In on-line PA the expert is being recorded solving a problem, and concurrently a commentary is made. The nature of this commentary specifies the two subtypes of the *on-line* method. The expert performing the task may be describing what they are doing as problem solving proceeds. This is called

---

[†]In fact, a possible *second-phase* elicitation technique would be to present these rules back to the expert and ask about their truthfulness, scope and so forth.

self-report. A variant on this is to have another expert provide a running commentary on what the expert performing the task is doing. This is called *shadowing*.

Off-line PA allows the expert(s) to comment retrospectively on the problem solving session—usually by being shown an audio-visual record of it. This may take the form of retrospective self-report by the expert who actually solved the problem, it could be a critical retrospective report by other experts, or there could be group discussion of the protocol by a number of experts including its originator. In the case in which only a behavioural protocol is obtained then obviously some form of retrospective verbalization of the problem solving episode is required.

Before PA sessions can be held, a number of pre-conditions should be satisfied. The first of these is that the knowledge engineer is sufficiently acquainted with the domain to understand the expert's tasks. Without this the elicitor may completely fail to record or take note of important parts of the expert's behaviour. A second requirement is the careful selection of problems for PA. The sampling of problems is crucial. PA sessions may take a relatively long time and only a few problems consequently can be addressed. Therefore, the selection of problems should be guided by how representative they are. Asking experts to sort problems into some form of order (Chi *et al.*, 1981, 1982) may give an insight into the classification of types of problems and help in the selection of suitable problems for PA (see also the next two sections on concept sorts and laddering).

A further condition for effective PA is that the expert(s) should not feel embarrassed about describing their expertise in detail. It is preferable for them to have experience in thinking aloud. Uninhibited thinking aloud has to be learned in the same way as does talking to an audience. One or two short training sessions may be useful, in which a simple task is used as an example. This puts the experts at ease and familiarizes them with the task of talking about their problem solving. In trying to decide when it is appropriate to use PA bear in mind that it is alleged that different KE techniques differentially elicit certain kinds of information. With PA it is claimed that the sorts of knowledge elicited include: the 'when' and 'how' of using specific knowledge, problem solving and reasoning strategies; evaluation procedures and evaluation criteria used by the expert; and procedural knowledge about how tasks and subtasks are decomposed.

When actually conducting a PA the following are a useful set of tips to help enhance its effectiveness:

1. Present problems and data in a realistic way; the way problems and data are presented should be as close as possible to a real situation.

2. Transcribe the protocols as soon as possible; the meaning of many expressions is soon lost, particularly if the protocols are not recorded.

3. In almost all cases an audio recording is sufficient, but videorecordings have the advantage of containing additional and disambiguating information.

4. Avoid long self-report sessions; thinking aloud is significantly more tiring for the expert, than is being interviewed, because of the need to perform a double task. This is one reason why shadowing is sometimes preferred.

5. In general, the presence of the knowledge engineer is required in a PA session. Although adopting a background role, the knowledge engineer's very presence suggests a listener to the interviewee, and lends meaning to the talking aloud process. Therefore, comments on audibility, or even silence by the knowledge engineer, are quite acceptable.

Protocol analyses share with the unstructured interview the problem that they may deliver unstructured transcripts which are hard to analyse. Moreover, they focus on particular problem cases and so the scope of the knowledge produced may be very restricted. It is difficult to derive general domain principles from a limited number of protocols. These are practical disadvantages of protocol analysis, but there are more subtle problems. Two actions, which look exactly the same to the knowledge engineer, may be the result of two quite different sets of considerations. This is a problem of impoverished interpretation by the knowledge engineer. They simply do not know enough to discriminate the actions. The obverse to this problem can arise in shadowing and the retrospective analyses of protocols by experts. Here the expert(s) may simply wrongly attribute a set of considerations to an action after the event. This is analogous to the problems of misattribution in interviewing.

A particular problem with self-report, apart from its being tiring, is the possibility that verbalization may interfere with performance. The classic demonstration of this is for a driver to attend to all the actions involved in driving a car. If one consciously monitors such variables as engine revs, current gear, speed, visibility, steering wheel position and so forth, the driving invariably gets worse. Such skill is shown to its best effect when performed automatically. This is also the case with certain types of expertise. By asking the expert to verbalize, one is in some sense destroying the point of doing protocol analysis—to access procedural, real-world knowledge.

Having pointed to these disadvantages, it is also worth remembering that context is sometimes important for memory—and hence for problem-solving. For most non-verbalizable knowledge, and even for some verbalizable knowledge, it may be essential to observe the expert performing the task, for it may be that this is the only situation in which the expert is actually able to 'communicate' knowledge.

Finally, when performing PA it is useful to have a set of conventions for the actual interpretation and analysis of the resultant data. Kuipers and Kassirer (1983) and Belkin *et al.* (1987) provide detailed guidelines for such analysis.

The two classes of generic technique discussed so far are *natural* and intuitively easy to understand. Experts are used to expressing their knowledge in these sorts of ways. The techniques that follow are what we might term *contrived*, and permit the expression of knowledge in ways that are likely to be unfamiliar to the expert.

Concept sorting

Concept sorting is a technique that is useful when we wish to uncover the different ways an expert sees relationships between a fixed set of concepts. In the version we will present an expert is presented with a number of cards on each of which a concept word is printed. The cards are shuffled and the expert is asked to sort the cards into either a fixed number of piles or else to sort them into any number of piles the expert finds appropriate. This process is repeated many times. One attempts to get multiple views of the structural organization of knowledge by asking the expert to do the same task over and over again, each time creating at least one pile that differs in some way from previous sorts. The expert should also provide a name or category label for each pile on each different sort.

Performing a card sort requires the elicitor to be not entirely naive about the domain. Cards have to be made with the appropriate labels before the session. However, no great familiarity is required as the expert provides all the substantial knowledge in the process of the sort. We now provide an example from our VDU domain to show the detailed mechanics of a sort.

The concepts printed on the cards were faults and corrective actions drawn from a structured interview with the expert. He had outlined 20 faults and outcomes:

(A) damaged VDU port	(B) faulty key contacts
(C) incorrect parity setting	(D) program is busy
(E) damaged tube	(F) terminal not switched on
(G) mains fuse blown	(H) fuse in VDU blown
(I) loose connection or break in line	(J) press control-c
(K) press control-q	(L) press control-z
(M) VDU power supply fault	(N) software fault
(O) video section fault on VDU	(P) incorrect terminal speed setting
(Q) terminal requires reset	(R) VDU transmission fault
(S) transmit line fault	(T) receive line fault

The expert was shown possible ways of sorting cards in a *toy* domain as part of the briefing session, and then asked to sort the real elements in the same way.

The dimensions/piles (P) which the expert used for the various sorts (S) were as follows:

S1 P1 hardware fault   P2 software fault

S2 P1 easy to clear   P2 difficult to clear

S3 P1 no component damage P2 component damage usually as symptom of another outcome

P3 component damage as an outcome

S4 P1 communication fault P2 not a communication fault

S5 P1 no output P2 no response
P3 garbled output P4 combination of no response and garbled output

The following table shows the pile number of each sort for each fault or corrective action (card label); almost all the fault concepts are distinguishable from one another—even with these few sorts.

	S1	S2	S3	S4	S5
A	1	2	3	1	4
B	1	1	3	2	3
C	2	1	1	1	3
D	2	1	1	2	1
E	1	2	3	2	1
F	2	1	1	2	1
G	1	1	2	2	1
H	1	1	2	2	1
I	1	2	3	1	4
J	2	1	1	2	2
K	2	1	1	2	2
L	2	1	1	2	2
M	1	2	3	2	1
N	1	2	1	1	2
O	1	2	3	2	1
P	2	1	1	1	3
Q	2	1	1	1	4
R	1	2	1	1	4
S	2	2	3	2	4
T	1	2	1	1	4

Using this information we can attempt to extract decision rules directly. An example of a rule extracted from the sortings is:

IF   it is a hardware fault (sort 1/pile 1)
AND   it is easy to clear (sort 2/pile 1)
AND   it is component damage (sort 3/pile 3)
AND   it is NOT a communication fault (sort 4/pile 2)
AND   there is garbled output (sort 5/pile 3)
THEN there are faulty key contacts (outcome B)

As can be seen from the example such sorts produce long and cumbersome rules. In fact many of the clauses may be redundant—once you have established that the fault is a hardware one, there is no need to check whether it is component damage. However, the utility of this technique does not reside solely in the production of decision rules. We can use it, as we have said, to explore the general interrelationships between concepts in the domain. We are trying to make explicit the implicit structure that experts impose on their expertise. Variants of the simple sort are different forms of *hierarchical* sort. One such version is to ask the expert to proceed by producing first two piles, on the second sort three, then four and so on. Finally we ask if any two piles have anything in common. If so you have isolated a higher order concept that can be used as a basis for future elicitation.

The advantages of concept sorting can be characterized as follows. It is fast to apply and easy to analyse. It forces into an explicit format the constructs which underlie an expert's understanding. In fact it is often instructive to the expert; a sort can lead the expert to see structure in his view of the domain which he himself has not consciously articulated before. Finally, in domains where the concepts are perceptual in nature (i.e. x-rays, layouts and pictures of various kinds) then the cards can be used as a means of presenting these images in an attempt to elicit names for the categories and relationships that might link them. There are, of course, features to be wary of with this sort of technique. Experts can often confound dimensions by not consistently applying the same semantic distinctions throughout an elicitation session. Alternatively, they may over-simplify the categorization of elements, missing out important caveats. One thing that we have found (Schweikert *et al.*, 1987) is that an expert's own opinion of the worth of a technique is no guide to its real value. In methods such as sorting we have a situation in which the expert is trying to demonstrate expertise in a non-natural or contrived manner. He or she might be quite used to chatting about their field of expertise, but sorting is different and experts are suspicious of it. Experts may in fact feel they are performing badly with such methods. However, on analysis one finds that the yield of knowledge is as good and sometimes better than for non-contrived techniques.

Laddered grids

Once again this is a fairly contrived technique, and it will be necessary to explain it fully to the expert before starting. The expert and the knowledge engineer construct a graphical representation of the domain in terms of the relations between domain elements. The result is a qualitative, two-dimensional graph where nodes are connected by labelled arcs. No extra elicitation method is used here, but expert and elicitor construct the graph together by negotiation.

In using the technique the elicitor enters the conceptual map at some point and then attempts to move around it with the expert. A formal specification

▸f how we use the technique is shown below together with an example of
ts use.

1. Start the expert off with a seed item.
2. Move around the domain map using the following prompts:
    1. To move DOWN the expert's domain knowledge:
        How can you tell it is <ITEM>?
        Can you give examples of <ITEM>?
    2. To move ACROSS the expert's domain knowledge:
        What alternative examples of <CLASS> are there to <ITEM>?
    3. To move UP the expert's domain knowledge:
        What have <SAME LEVEL ITEMS> got in common?
        What are <SAME LEVEL ITEMS> examples of?
        What is the key difference between <ITEM 1> and <ITEM
        2>?

The elicitor may move around the domain map of knowledge in any order
which seems convenient. As the elicitation session progresses, the elicitor
keeps track of the elicited knowledge by drawing up a network on a large
piece of paper. This representation allows the elicitor to make decisions (or
ask questions) about what constitutes higher or lower order elements in the
domain. In order to give the reader the flavour of the technique, there follows
an extract from a laddered grid elicitation session. Once again, the knowledge
domain is VDU fault diagnosis.

KE What examples of a 'no response' problem are there?
EX It's a software problem, unless a small set of keys is affected.
KE Can you give me an example of what you do if there's a small set of
    keys affected?
EX Yes. You check the key contacts.
KE Can you give me examples of actions for software problems?
EX Yes. You intervene or don't intervene. You don't intervene if there's
    a user error where they misinterpret a slow program as this fault.
KE Can you give me an example of when you intervene?
EX When there's an editor problem. Then you use control keys.
KE Can you give me an example of using control keys?
EX Press control-c, control-q and control-z.
KE What is the difference between control-c and control-q?
EX Control-q gets you out of control-s and control-c gets you out of the
    program. Control-q should be used before control-c.
KE What is the difference between control-c and control-z?
EX Control-c gets you out of the program and should be used before
    control-z. Control-z quits the system.

From this portion of a laddered grid interview the elicitor drew up a
hierarchical representation of the domain as shown in Figure 13.1.

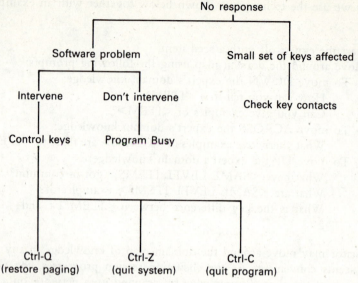

Figure 13.1.. Laddering in the VDU Domain.

This hierarchy gives rise to the following set of rules which could be included in the knowledge base of an expert system for VDU fault finding.

IF   there is no response
AND   it is a software problem
AND   intervention is needed
THEN   press control-q

IF   control-q doesn't work
THEN   press control-c

IF   control-c doesn't work
THEN   press control-z

We have found that this form of knowledge elicitation is very powerful for structured domains. As with other contrived techniques we have found that whilst an expert may think this technique is revealing little of interest, subsequent analysis provides good quality rules.

The limited information task

A technique which does not provide a spatial representation of the domain, but rather a set of hints or suggestions which may prove useful in expert system construction is a technique called the *limited information task* (Hoffman,

987) or *20 questions* (Grover, 1983). The expert is provided with little or no information about a particular problem to be solved, but must then ask the elicitor for specific information which will be required to solve the problem. The information which is requested, along with the order in which it is requested, provides the knowledge engineer with an insight into the expert's problem solving strategy. One difficulty with this method is that the knowledge engineer needs a good understanding of the domain in order to make sense of the experts' questions, and to provide meaningful responses. The elicitor should have forearmed themselves with a problem from the domain together with a *crib sheet* of appropriate responses to the questions.

In a version of the limited information task which we use, we tell the expert that the elicitor has a scenario in mind and the expert must determine what it is. The scenario might represent a problem, a solution or a problem context. The expert is told that they may ask the elicitor for more information, though what the elicitor gives back is terse and does not go much beyond what was asked for in the question. The expert may be asked to explain why each of the questions was asked.

An example of the kind of interaction produced by this technique is shown below. Here the problem domain is in the construction of lighting systems for the inspection of industrial products and processes.

EX   Is this in the manufacturing industry?

KE   Yes.

EX   So we've ruled out things like fruit, vegetables, cows?

KE   Yes.

EX   Is it the metal industry?

KE   The material is wood.

EX   So we could be dealing with a large object here like a chair or table.

KE   The object is large.

EX   It's likely to be a 3-D object, you've got to pick it up and turn it over.

KE   That's right.

EX   So what I need now are the dimensions of this object in terms of the cube that will enclose it.

KE   It would have similar dimensions to the table top.

EX   Do I inspect one surface or all the surfaces?

KE   All of them.

EX   Is the inspector looking for one or many faults?

KE   One particular fault.

EX   Can you describe it for me?

KE   It's pencil marks about half an inch long.

EX   What colour is the wood?

KE   Dark unfinished wood.

EX   We've got a contrast problem here. At this point I'd go and look at the job + to see if the graphite pencil marks reflect light + sometimes

it does, but it depends on the wood + if it does you can select th
light to increase the contrast between the fault and the background.
. . . . .

EX I'd be doing this in three phases: first a general lighting, then specifi
for surface lighting, and then some directional light [expert then give
technical specifications for these types of light].

This interview gives us an interesting insight into the natural line (
enquiry of an expert in this domain. Often expert systems gather the righ
data but the order in which they are gathered and used can be remote fror
how an expert works. This can decrease the acceptability of the system
other experts are to use it, and it also has consequences for the intelligibilit
of any explanations the system offers in terms of a retrace of its steps to
solution.
It will be seen that we can once again extract decision rules directly fror
the dialogue: e.g.

IF   fault colour is black
AND   object colour is dark
THEN contrast is a problem

The drawbacks to this technique are that the elicitor needs to hav
constructed plausible scenarios and to be able to cope with questions askec
The experts themselves are often uncomfortable with this technique; thi
may well have to do with the fact that, as with other contrived technique:
it is not a natural means of manifesting expertise. Whilst a few scenaric
may reveal some of the general rules in a domain the elicitation is very cas
specific. In order to get the range of knowledge for a sweep of situation
many scenarios would need to be constructed and used.

Automatic elicitation

As KE is acknowledged to be a time consuming and difficult process, th
idea of automated elicitation is particularly attractive. A number of progran
have been developed towards this goal, and we will briefly consider son
of them in this section.

There are two main types of automated elicitation: (1) those systems whic
are implementations of standard KE techniques; and (2) those systems whic
use machine learning techniques to induce rules from sets of worked exampl
and observed data. In addition to these categories, there are also systen
which use knowledge about the structure of a particular domain in order t
drive the elicitation. However, these are large-scale systems dedicated t
specific projects, and are not generally available. Readers interested in th
category of system are referred to Marcus *et al.* (1985) for a review.

Of those systems which implement standard techniques, the most successful are based on the *repertory grid*. This technique has its roots in the psychology of personality (Kelly, 1955) and is designed to reveal a conceptual map of a domain, in a similar fashion to the card sort as discussed above (see Shaw and Gaines, 1987a, for a full discussion; also, see Sinclair in this book). Briefly, subjects are presented with a range of domain elements and asked to choose three, such that two are similar, and different from the third. So, for example, people might be presented with a set of animals, and choose elephant and hippo as the two similar elements, and sparrow as the third. The subject is then asked for their reason for differentiating these elements, and this dimension is known as a construct. In our example 'size' would be a suitable construct. The remaining domain elements are then rated on this construct.

This process continues with different triads of elements until the expert can think of no further discriminating constructs. The result is a matrix of similarity ratings, relating elements and constructs. This is analysed using a statistical technique called *cluster analysis*. In KE, as in clinical psychology, the technique can reveal clusters of concepts and elements which the expert might not have articulated in an interview. However, the disadvantage of the technique is that it is very time-consuming to administer by hand, and the cluster analysis is complex to perform and interpret. This naturally suggests that an implemented version would be appropriate.

There are several repertory grid programs on the market. However, the best known programs are ETS (Boose, 1985), its successor, AQUINAS (Boose and Bradshaw, 1987), and KITTEN (Shaw and Gaines, 1987b). Although these are largely research tools rather than commercial products, they provide a good focus for discussion of KE using an automated repertory grid. With each of these systems the expert interacts directly with a computer. The programs are run in such a way that the repertory grid is built-up interactively, and the expert is shown the resultant knowledge. Experts have the opportunity to refine this knowledge during the elicitation process. The output of these systems is a set of machine-executable rules. In AQUINAS, these rules can take the form necessary for a wide range of expert system shells. These systems bypass the human elicitor altogether and have been used with considerable success in small-scale domains where solutions can be comfortably enumerated (see Boose, 1986 for a list of successful applications of ETS). However, the rep grid is not well-suited to more complex domains involving construction or planning.

We now turn to automated elicitation of the second kind. Machine induction is the process whereby a program is presented with many solved problems (or other appropriate large data sets) from the domain, and uses statistical regularities to infer underlying rules. An introductory overview to this technique can be found in Hart (1986). The best known algorithm for this is ID3 (Quinlan, 1979, 1983), and versions are now available in some small-scale shells. To conceptualize the process, imagine presenting sets of motor car symptoms alongside decisions about the fault diagnosis. It may be that a mechanic cannot readily articulate a rule relating a set of symptoms to a particular fault, but if they occur reliably together with the fault, the program induces a rule relating them. So, the next time this configuration of faults is encountered, the appropriate diagnosis is made.

Although this technique sounds attractive, it has several associated problems. Rules may be induced which do not have any basis in the domain. This can arise according to quite arbitrary decisions such as the order in which the learning examples are input. Furthermore, there is the problem of cause and effect—the fact that a particular set of faults co-occurs with a diagnosis on the learning set does not necessarily imply a causal relation. Research continues on more sophisticated machine learning techniques for KE (e.g. Michalski *et al.*, 1986; Mitchell, 1982). Clearly though, induction programs do not provide a complete solution to automatic elicitation.

Finally, in this section, we should mention that there are now several large-scale knowledge acquisition environments under construction. These typically provide a number of automated KE techniques, knowedge base editors, automated transcript analysis and various other support software for the knowledge engineer. These systems are currently at the research stage, and as yet are not generally available. However, the interested reader is referred to Anjewierden (1987), Motta *et al.* (1987) and Diederich *et al.* (1987) for accounts of KADS Power Tools, KEATS and KRITON, respectively. These systems indicate the shape and form of the next generation of knowledge engineering tools.

A taxonomy of KE techniques

We have attempted to sample some of the major approaches to elicitation and where appropriate give a detailed description of techniques that are likely to be of use. There are many variants on the methods we have described. Below we have provided a taxonomy of methods with which we are familiar together with a primary reference for each one.

    Non-contrived
      Interviews
        Structured
          Focused interviews (Hart, 1986)
          Forward scenario simulation (Grover, 1983)
          Teach back (Johnson and Johnson, 1987)
        Unstructured (Weis and Kulikowski, 1984)
      Protocol analysis
        Verbal
          On-line (Johnson *et al.*, 1987)
          Off-line (Elstein *et al.*, 1978)
          Shadowing (Clarke, 1987)

Behavioural (Ericsson and Simon, 1984)
Contrived
  Conceptual mapping
    Sorting and rating (Gammack, 1987a,b)
    Repertory grid (Shaw and Gaines, 1987a)
    Pathfinder (Schvaneveldt *et al.*, 1985)
  Goal decomposition
    Laddered grid (Hinkle, 1965)
    Limited-information task (Grover, 1983; Hoffman, 1987)
Automatic
  Rule induction (Shapiro, 1987)

Having discussed the principle methods of elicitation we should spend a little time reflecting on the nature of two other major components of the KE enterprise, namely the experts and the expertise they possess.

On experts

Experts of course come in all shapes and sizes. Ignoring the nature of your expert is another potential pitfall in KE. A coarse guide to a typology of experts might make the issues clearer. Let us take three categories we shall refer to as *academics, practitioners* and *samurai* (in practice experts may embody elements of all three types). Each of these types of expert differ along a number of dimensions. These include: the outcome of their expert deliberations, the problem solving environment they work in, the state of the knowledge they possess (both its internal structure and its external manifestation), their status and responsibilities, their source of information, and the nature of their training.

How are we to distinguish these different types of expert? The academic type regards their domain as having a logically organized structure. Generalizations over the laws and behaviour of the domain are important to them; theoretical understanding is prized. Part of the function of such experts may be to explicate, clarify and teach others, thus they talk a lot about their domains. They may feel an obligation to present a consistent story both for pedagogic and professional reasons. Their knowledge is likely to be well structured and accessible. These experts may suppose that the outcome of their deliberations should be the correct solution of a problem. They believe that the problem can be solved by the appropriate application of theory. They may, however, be remote from everyday problem-solving.

The practitioner class on the other hand are engaged in constant day to day problem-solving in the domain. For them specific problems and events are the reality. Their practice may often be implicit and what they desire as an outcome is a decision that works within the constraints and resource limitations in which they are working. It may be that the generalized theory of the academic is poorly articulated in the practitioner. For the practitioner heuristics may dominate and theory be thin on the ground.

The samurai is a pure performance expert—the only reality is the performance of action to secure an optimal performance. Practice is often the only training and responses are often automatic.

One can see this sort of division in any complex domain. Consider for example medical domains where we have professors of the subject, busy housemen working the wards, and medical ancillary staff performing many important but repetitive clinical activities.

The knowledge engineer must be alert to these differences because the various types of expert will perform very differently in KE situations. The academic will be concerned to demonstrate mastery of the theory. They will devote much effort to characterizing the scope and limitations of the domain theory. Practitioners, on the other hand, are driven by the cases they are solving from day to day. They have often *compiled* or *routinized* any declarative descriptions of the theory that supposedly underlies their problem-solving. The performance samurai will more often than not turn any KE interaction into a concrete performance of the task—simply exhibiting their skill. But there is more to say about the nature of experts and this is rooted in general principles of human information processing. Psychology has demonstrated the limitations, biases and prejudices that pervade all human decision making—expert or novice. To illustrate this, consider the following facts, all potentially crucial to the enterprise of KE.

It has been shown repeatedly that the context in which one encodes information is the best one for recall. It is possible then, that experts may not have access to the same information when in a KE interview, as they do when actually performing the task. So there are good psychological reasons to use techniques which involve observing the expert actually solving problems in the normal setting. In short, protocol analysis techniques may be necessary, but will not be sufficient for effective knowledge elicitation.

Consider now the issue of biases in human cognition. One well-known problem is that humans are poor at manipulating uncertain or probabilistic evidence. This may be important in KE for those domains which require a representation of uncertainty. Consider the rule:

IF   the engine will not turn over
AND   the lights do not come on
THEN   the battery is flat with probability X

This seems like a reasonable rule, but what is the value of X, should it be 0·9, 0·95, 0·79? The value which is finally decided upon will have important consequences for the working of the system, but it is very difficult to decide upon it in the first place. Medical diagnosis is a domain full of such probabilistic rules, but even expert physicians cannot accurately assess the probability values. In fact there are a number of documented biases in human cognition which lie at the heart of this problem (see for example Kahneman et al., 1982). People are known to undervalue prior probabilities, to use the ends and middle of the probability scale rather than the full range, and to *anchor* their responses around an initial guess. Cleaves (1987) lists a number of cognitive biases likely to be found in knowledge elicitation, and makes suggestions about how to avoid them. However, many knowledge engineers prefer to avoid the use of uncertainty wherever possible.

Cognitive bias is not limited to the manipulation of probability. A series of experiments has shown that systematic patterns of error occur across a number of apparently simple logical operations. For example, *Modus Tollens* states that if 'A implies B' is true, and 'not B' is true, then 'not A' must be true. However people, whether expert in a domain or not, make errors on this rule. This is in part due to an inability to reason with contrapositive statements. Also in part it depends on what A and B actually represent. In other words, they are affected by the content. This means that one cannot rely on the veracity of experts' (or indeed anyone's) reasoning.

All this evidence suggests that human reasoning, memory and knowledge representation is rather more subtle than might initially be thought.

On expertise

Clearly the expertise embodied by experts is not of a homogeneous type. In constructing expert systems it is likely that very different types of knowledge will be uncovered which will have very different roles in the system. There are a number of analyses available of the *epistemology* of expertise. Our analysis is based to a large extent on that of Breuker (1987).

First, we can distinguish what is called *domain level* knowledge. This term is being used in the narrow sense of knowledge that *describes* the concepts and elements in the domain and relations between them. This sort of knowledge is sometimes called *declarative*, it describes what is known *about* things in the domain. The propositions below can be seen as domain level knowledge in this sense:

    damaged VDU is a hardware fault
    transmit line fault is a software fault

There is also knowledge and expertise which has to do with what we might call the *inference level*. This is knowledge about how the components of expertise are to be organized and used in the overall system. This is quite a high level description of expert behaviour and may often be implicit in expert practice. The following is a description of knowledge about part of a systematic diagnosis system at the inference level.

To perform systematic diagnosis we will have knowledge about a complaint, and knowledge about observables from the patient or object. We select some aspect of the complaint and using a model of how the system should be performing normally we look to see if a particular parameter of the system is within normal bounds.

Another type of expert knowledge is the *task level*. This is sometimes called *procedural* knowledge. This is knowledge to do with how goals and sub-goals, tasks and subtasks should be performed. Thus in a classification task there may exist a number of tasks to perform in a particular order so as to utilize the domain level knowledge appropriately. This type of knowledge is present in the following extract.

First of all perform a general inspection of the object. Next examine the sample with a hand lens. Next use a prepared thin-section and examine that under a cross-polarizing microscope.

Finally, there is a level of expert knowledge referred to as *strategic* knowledge. This is information that monitors and controls the overall problem solving. This can have to do with the way resources are used; what to do if the proposed solution fails or is found to be inappropriate in some way; what to do when faced with incomplete or insufficient data. Such information is contained in the following extract from an interview.

If I had time I would always check the video driver board. If its a Zenith machine I'd always check that because they are notorious for going wrong . . .

Any field of expertise is likely to contain these various sorts of knowledge to greater or lesser extents. At any particular knowledge level the information may be explicit or implicit in an expert's behaviour. Thus in some domains the experts may have no real notion of the strategic knowledge they are following whilst in others this knowledge is very much in the forefront of their deliberations. Also, of course, the requirements on a system about how far it needs to implement these various levels will vary. It is almost universally acknowledged that significant reasoning about problem domains requires more than just modelling simple relationships between concepts in the domains—it requires causal models of how objects influence and affect one another, models of the processes in which objects participate. This is a hard problem, and often the limitations of first generation expert systems mean that sophisticated domain models cannot be supported.

This brings us to a final important feature of KE. Often the process of acquisition yields much more knowledge than can be implemented (Young, 1987). In this case the knowledge engineer ought to think about laying down the knowledge in a format that will allow it to be used when more powerful implementation systems arrive. Putting knowledge into *deep freeze* in this way requires using an expressive and unambiguous intermediate representation of the knowledge to be stored. A number of candidates exist for this; Young and Gammack (1987) provide a brief review of the alternatives.

Methodologies and programmes of KE

Are there any guidelines as to how techniques should be assembled to form a programme of acquisition? In other words when should we use the various techniques? The choice may depend on the characteristics of the domain, of the expert, and of the required system. Furthermore, it is clear that some techniques are going to be more costly in terms of time with the expert, or of the effort required for subsequent analysis of transcripts.

There are a number of articles and books available on 'how to do knowledge elicitation'. These often contain advice of the most general kind, and emphasize the pragmatic considerations of expert system development. General reviews can be found in Welbank (1983), Hoffman (1987), Kidd (1987) and Hart (1986). While these reviews are based on experience of the general kind, there have also been a number of attempts to make formal recommendations. Gammack and Young (1985) offer a mapping of knowledge elicitation techniques onto domain type. Their analysis requires that domain knowledge be separated into different categories, and provides suggestions about the particular techniques which are most likely to be effective in each category. Although the analysis alerts one to the fact that there are different types of knowledge with any domain, it does not provide engineers with guidelines about how to identify each type.

The most thorough attempt to integrate KE procedure is provided by KADS—Knowledge Acquisition and Domain Structuring (see Breuker, 1987 for an overview). KADS embodies seven principles for the elicitation of knowledge and construction of a system. These principles are stated below (following Breuker, 1987):

1. The knowledge and expertise should be analysed before the design and implementation starts.
2. The analysis should be model-driven as early as possible.
3. The content of the model should be expressed at the epistemological level.
4. The analysis should include the functionality of the prospective system.
5. The analysis should proceed in an incremental way.
6. New data should be elicited only when previously collected data have been analysed.
7. Collected data and interpretations should be documented.

Principle 1 is quite straightforward and requires no further explanation. Principle 2 requires that one should bring to bear a model of how the knowledge is structured early on in the process, and use it to interpret subsequent data. Principle 3 means that one should use an appropriate intermediate level representation device, and not try to force knowledge into a particular formalism before one knows how it may best be implemented. Principle 4 is a reminder that a complete analysis includes an understanding of how the system is to work, e.g. who will use it, and in what situation.

One cannot gain a full understanding of the problem simply by trying to map out an expert's knowledge without regard to how it will be used. Principle 5 emphasizes the fact that there is a wide variety of related topics within a domain. This means that construction of a model should be 'breadth-first', embodying all aspects at once, rather than attempting fully to represent one sub-part after another. Principles 6 and 7 are once again straightforward. Like many of the best recommendations, the utility of these statements is most apparent when they are not adhered to.

A basic insight from KADS is that knowledge acquisition should be viewed under the metaphor of model building, rather than the mining of information (Breuker, 1987). The principles above emphasize this point—in short, KE is not an independent part of constructing an expert system, and viewing it as such will lead to inefficient practice. Whether or not the adoption of such an approach makes for more efficient and effective KE is a moot point as these claims have not been formally evaluated. In so far as they impose a discipline on the KE process they are likely to be useful.

A rather less disciplined approach and yet one that is almost always associated with expert systems is *rapid prototyping* (Hayes-Roth et al., 1983). Indeed some see the approach as specifying a separate elicitation technique. The idea is that it is easier for experts to criticize a working system, than it is to specify the system in the first place. Initially, a prototype is built, without much regard to its weaknesses, and the expert makes suggestions about its performance. These suggestions are incorporated into the system by programmers, and at the next session there should be fewer errors. This cycle continues until the expert is satisfied with the behaviour of the system (see Christie and Gardiner in this volume).

Finally, there has been more formal evaluation of KE techniques (cf. for example, Cooke and McDonald, 1987; Schweickert et al., 1987; Burton e al., 1987a, 1988). This type of research is still in its infancy. Where there are concrete and robust findings from the work we have tried to subsume them when describing the pros and cons of the techniques themselves.

Conclusions

The problem of knowledge elicitation is a subtle and complex one. We have described some of the techniques that are used in this enterprise. But we have also sought to provide an indication of the difficulties inherent in doing this kind of work. At present knowledge elicitation is itself a form of expertise—experienced knowledge engineers come to recognize the subtleties of expert thinking, and can develop skills for capturing these. However, it is clear that a formal methodology is absent.

As expert systems technology becomes more readily available, more people will have to face the issue of knowledge elicitation. In order to provide support for this enterprise research is under way in areas as disparate as software engineering, artificial intelligence and psychology. Experience to date has identified interesting and difficult problems which form the research agenda for these groups. In the mean time, prospective knowledge engineers will have to rely on accumulated experience, and general readings such as this one.

References

Anjewierden, A. (1987). Knowledge acquisition tools. *AI Communications*, **0**, 29–39.

Belkin, N.J., Brooks, H.M. and Daniels, P.J. (1987). Knowledge elicitation using discourse analysis. *International Journal of Man–Machine Studies*, **27**, 127–144.

Boose, J.H. (1985). A knowledge acquisition program for expert systems based on personal construct psychology. *International Journal of Man–Machine Studies*, **23**, 495–525.

Boose, J.H. (1986). *Expertise Transfer for Expert System Design* (New York: Elsevier).

Boose, J.H. and Bradshaw, J.M. (1987). Expertise transfer and complex problems: using AQUINAS as a knowledge-acquisition workbench for knowledge-based systems. *International Journal of Man–Machine Studies*, **26**, 3–28.

Breuker, J. (Ed.) (1987). *Model-Drive Knowledge Acquisition Interpretation Models*. Deliverable task AI, Esprit Project 1098 (University of Amsterdam).

Burton, A.M., Shadbolt, N.R., Hedgecock, A.P. and Rugg, G. (1987). A formal evaluation of knowledge elicitation techniques for expert systems: domain 1. In *Research and Development in Expert Systems IV*, edited by D.S. Moralee (Cambridge: Cambridge University Press).

Burton, A.M., Shadbolt, N.R., Rugg, G. and Hedgecock, A.P. (1988). Knowledge elicitation techniques in classification domains. *ECAI-88: Proceedings of the 8th European Conference on Artificial Intelligence*, pp. 85–90.

Chi, M.T.H., Feltovitch, P.J. and Glaser, R. (1981). Categorisation and representation, physics problems by experts and novices. *Cognitive Science*, **5**, 121–152.

Chi, M.T.H., Glaser, R. and Rees, E. (1982). Expertise in problem solving. In *Advances in the Psychology of Human Intelligence: 1*, edited by R.J. Sternberg (Hillsdale, NJ: Erlbaum).

Cooke, N.M. and McDonald, J.E. (1987). The application of psychological scaling techniques to knowledge elicitation for knowledge-based systems. *International Journal of Man–Machine Studies*, **26**, 533–550.

Clarke, B. (1987). Knowledge acquisition for real time knowledge based systems. *Proceedings of the First European Workshop on Knowledge Acquisition for Knowledge Based Systems*, Reading University, UK.

Cleaves, D.A. (1987). Cognitive biases and corrective techniques: proposals for improving elicitation procedures for knowledge based systems. *International Journal of Man–Machine Studies*, **27**, 155–166.

Diederich, J., Ruhmann, I. and May, M. (1987). KRITON: a knowledge-acquisition tool for expert systems. *International Journal of Man–Machine Studies*, **26**, 29–40.

Elstein, A.S., Shulman, L.S. and Sprafka, S.A. (1978). *Medical Problem Solving: An Analysis of Clinical Reasoning* (Cambridge, MA: Harvard University Press). Cited in Kuipers and Kassirer (1983).

Ericsson, K.A. and Simon, H.A. (1984). *Protocol Analysis: Verbal Reports as Data* (Cambridge, MA: MIT Press).

Gammack, J.G. (1987a). Formalising implicit domain structure. *Proceedings of the SERC Workshop on Knowledge Acquisition for Engineering Applications.* SERC Report RAL-87-055.

Gammack, J.G. (1987b). Different techniques, and different aspects of declarative knowledge. In *Knowledge Acquisition for Expert Systems: A Practical Handbook*, edited by A.L. Kidd (New York: Plenum Press).

Gammack, J.G. and Young R.M. (1985). Psychological techniques for eliciting expert knowledge. In *Research and Development in Expert Systems*, edited by M. Bramer (Cambridge: Cambridge University Press).

Grover, M.D. (1983). A pragmatic knowledge acquisition methodology. *IJCAI-83: Proceedings 8th International Joint Conference on Artificial Intelligence.*

Hart, A. (1986). *Knowledge Acquisition for Expert Systems* (London: Kogan Page).

Hayes-Roth, F., Waterman, D.A. and Lenat, D.B. (1983). *Building Expert Systems* (Reading, MA: Addison-Wesley).

Hinkle, D.N. (1965). The change of personal constructs from the viewpoint of a theory of implications. Unpublished Ph.D. Thesis, University of Ohio.

Hoffman, R.R. (1987). The problem of extracting the knowledge of experts from the perspective of experimental psychology. *AI Magazine*, **8**, 53–66.

Jackson, P. (1985). *An Introduction to Expert Systems* (New York: Addison-Wesley).

Johnson, L. and Johnson, N. (1987). Knowledge elicitation involving teachback interviewing. In *Knowledge Elicitation for Expert Systems: A Practical Handbook*, edited by A.L. Kidd. (New York: Plenum Press).

Johnson, P.E., Zualkernan, I. and Garber, S. (1987). Specification of expertise. *International Journal of Man–Machine Studies*, **26**, 161–181.

Kahneman, D., Slovic, P. and Tversky, A. (Eds) (1982). *Judgement under Uncertainty: Heuristics and Biases* (New York: Cambridge University Press).

Kelly, G.A. (1955). *The Psychology of Personal Constructs* (New York: Norton).

Kidd, A.L. (Ed.) (1987). *Knowledge Acquisition for Expert Systems: A Practical Handbook* (New York: Plenum Press).

Kuipers B. and Kassirer, J.P. (1983). How to discover a knowledge representation for causal reasoning by studying an expert physician. *IJCAI-83: Proceedings of the 8th International Conference on Artificial Intelligence.*

Marcus, S., McDermott, J. and Wang, T. (1985). Knowledge acquisition for constructive systems. *IJCAI-85: Proceedings of the Ninth International Joint Conference on Artificial Intelligence.*

Michalski, R. (1983). A theory and methodology of inductive learning. In *Machine Learning: An Artificial Intelligence Approach*, edited by R. Michalski, J. Carbonell, and T. Mitchell. (Palo Alto: Tioga).

Michalski, R., Carbonell, J. and Mitchell, T. (Eds) (1986). *Machine Learning: An AI approach*, Volume 2 (Los Altos, CA: Morgan Kaufman).

Mitchell, T. (1982). Generalization as search. *Artificial Intelligence*, **18**, 203–226.

Motta, E., Eisenstadt, M., West, M., Pitman, K. and Evertsz, R. (1987). *KEATS: The Knowledge Engineer's Assistant*. HCRL Tech Report 20 (Milton Keynes: Open University).

Newell, A. and Simon, H.A. (1963). GPS, a program that simulates human thought. In *Computers and Thought*, edited by E. Feigenbaum and J. Feldman (New York: McGraw-Hill).

Quinlan, J.R. (1979). Rules by induction from large collections of examples. In *Expert Systems in the Micro-Electronic Age*, edited by D. Michie (Edinburgh: Edinburgh University Press).

Quinlan, J.R. (1983). Learning efficient classification procedures and their application to chess endgames. In *Machine Learning: An Artificial Intelligence Approach*, edited by R. Michalski, J. Carbonell and T. Mitchell (Palo Alto: Tioga).

Schvaneveldt, R.W., Durso, F.T., Goldsmith, T.E., Breen, T.J., Cooke, N.M., Tucker, R.G. and De Maio, J.C. (1985). Measuring the structure of expertise. *International Journal of Man–Machine Studies*, **23**, 699–728.

Schweickert, R., Burton, A.M., Taylor, N.K., Corlett, E.N., Shadbolt, N.R. and Hedgecock, A.P. (1987). Comparing knowledge elicitation techniques: a case study. *Artificial Intelligence Review*, **1**, 245–253.

Shadbolt, N.R. (1989). Knowledge representation in man and machine. In *Expert Systems: Principles and Case Studies*, 2nd edition, edited by R. Forsyth (London: Chapman and Hall).

Shapiro, A. (1987). *Structured Induction in Expert Systems* (New York: Addison-Wesley).

Shaw, M.L.G. and Gaines, B.R. (1987a). An interactive knowledge elicitation technique using personal construct technology. In *Knowledge Acquisition for Expert Systems: A Practical Handbook*, edited by A.L. Kidd (New York: Plenum Press).

Shaw, M.L.G. and Gaines, B.R. (1987b). KITTEN: Knowledge initiation and transfer tools for experts and novices. *International Journal of Man–Machine Studies*, **27**, 251–280.

Weiss, S. and Kulikowski, C. (1984). *A Practical Guide to Designing Expert Systems* (Towata, NJ: Rowman and Allanheld).

Welbank, M.A. (1983). *A Review of Knowledge Acquisition Techniques for Expert Systems* (Martlesham Heath: British Telecom Research).

Young, R.M. (1987). *The Role of Intermediate Representations in Knowledge Elicitation*. Keynote Address to 'Expert Systems 87', Brighton, UK.

Young, R.M. and Gammack, J. (1987). Role of psychological techniques and intermediate representations in knowledge elicitation. *Proceedings of the First European Workshop on Knowledge Acquisition for Knowledge Based Systems*, Reading University, UK.

Part IV

Assessment and design of the physical workplace

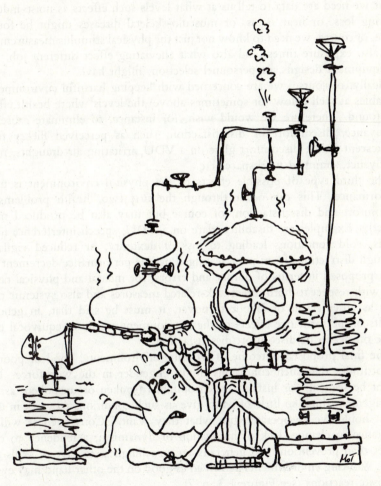

The physical environment comprises what must be the most widely explored of any group of ergonomics variables. These variables are components of most situations we investigate and affect people and their performance in a multitude of ways.

In spite of the long history of environmental assessments, the methods and techniques are still developing. This must be, in part, because of the impact of computers and information technology in general (see Drury's chapter on computerized data collection), but it is also because our understanding of the psychophysical and physiological effects of environmental stressors is increasing.

Moreover, the array of methods used reflects these various effects. Different aspects of the physical environment have been said to have effects—positive and negative—on all the facets of individual well-being described earlier in Figure 1.2 (p. 6). All environments at the extreme will affect workers' health; what we need are data to tell us at what levels such effects as noise-induced hearing loss, or heat stress, or musculo-skeletal diseases might be found. Here, of course, we need to know not just the physical stimulus measurements but also exposure times, and also what alleviating effect different job, task or equipment designs, or personnel selection, might have.

Ideally, of course, we are concerned with keeping harmful environmental variables at well below (or sometimes above) the levels where health effects are found. Therefore we would wish, for instance, to eliminate causes of annoyance, discomfort or dissatisfaction, such as perceived flicker from fluorescent lights, discomfort glare on a VDU, irritating air draughts, noise annoyance, cramped workspaces etc.

The third type of possible effects of the physical environment is upon performance. This can occur through the first two, health problems or discomfort and dissatisfaction, of course but may also be produced more directly. Examples are disability glare on a VDU, speech interference noise levels, cold conditions leading to loss of dexterity, or reduced vigilance through distractions from poor seating. Possible performance decrement has been proposed in terms of output and errors, for mental and physical tasks, and with respect to both direct task-related measures and also systemic ones such as absenteeism or labour turnover. It must be said that, in general, results on performance effects of the environment are more equivocal than those on health or discomfort/dissatisfaction.

The final proposed outcome of working environments, good or poor, is the attitudes that such conditions might engender in the workforce. This might be seen in, for instance, such opinions (spoken or unspoken) as: "If management think so little of us to give us such conditions to work in then why should we co-operate with what they want." Consequences will be resistance to change, lack of innovation or dynamism, a tendency to cure rather than prevention, and generally an unwillingness to give of skills and time. Working environments perceived as good on the other hand may evoke opposite reactions (see Figure 1.3, p. 7).

This section opens with three chapters reviewing all the effects described above in the context of the visual, climatic and auditory environments; consequently methods of assessment are described which will allow us to evaluate working conditions and to initiate the correct remedial action where necessary. Howarth (chapter 14), Parsons (chapter 15), and Haslegrave (chapter 16) explain for the unwary the multiplicity of physical characteristics and units of measurement of environmental stimuli, the different human responses of relevance, and the consequently diverse methods and techniques of measurement. They also describe evaluation criteria such that assessments can be made, and provide examples of practical working environment assessment.

Vibration, covered by Bonney in chapter 17, differs somewhat from the first three types of physical work environment; it will be a factor only in certain working conditions. It is relatively recently that hand–arm vibration has had attention from a major body of researchers, probably as a result of the widespread use of chain saws in forestry in the last twenty years. Measuring the extent of Vibration White Finger still does not embrace an agreed practice, the argument lying primarily between clinical judgement and the use of psychophysical threshold measures of spatial or vibratory perception. For some situations, e.g. chain saws or hand grinders, the assessment of vibration levels is quite structured and anti-vibration mountings can bring the levels down to those defined by ISO, keeping energy input to the hands such that damage is unlikely. Other work situations, e.g. riding motorcycles or hand fettling of castings, are less tractable and require changes to the process to overcome the problems.

Whole body vibration covers a range of experiences from seasickness to helicopter flying. Here the major area of motor vehicle riding has been emphasised since many of the techniques can be extended in either direction along the vibration spectrum. The use of mathematical models and computer simulations seems likely to extend in whole body vibration studies, at least on the design side. A greater understanding of the effects on people, however, will still require experimentation. Since only low stress levels are acceptable in research studies the importance of exploiting field evidence will be obvious.

The last three chapters in this section have close links, all having anthropometry as a major component. Pheasant (chapter 18) outlines the subject, the sources of data and the methods for applying them to real problems. Superficially the procedures are simple, but in fact there are many poor applications of anthropometry which arise to a great extent through insufficient recognition of the reasons for the choice of the various dimensions or percentiles. Pheasant gives clear guidance in this respect.

Computer graphics have made major strides in recent years and permit the simulation of workplaces at much less cost than physical mock-ups (Porter *et al.*, chapter 19). As is emphasised in the chapter, this does not obviate the need for physically testing the design, but it does bring a good design more quickly to realization and allows the rapid exploration of a greater number of different designs before investing in mock-ups or prototypes. As computer technology advances we can expect even more sophistication in these methods, including the calculation of environmental, physiological and psychological aspects in addition to the physical layouts.

The last chapter (20) by Corlett deals with a particular design or testing application, industrial seating, which has important anthropometric components. Many people believe they are competent to design seats, and the wide variety of seats available does, in part, support their view. The comments of seat users, particularly where the seat is a necessary part of a workplace, gives clear evidence that a lot of their confidence is misplaced.

The seat evaluation chapter brings into study the utility and acceptability of the seat, as well as its dimensions. In fact these two factors are relevant for any workplace, and variations on the methods in this last chapter could be used to test their effectiveness in these respects. One of the reasons for incorporating this chapter within the group is to point out these broader aspects, and to round off what might otherwise have been seen as primarily technical matters with a reminder that the ergonomist is interested more broadly than just in the relationships between physical dimensions.

Chapter 14

Assessment of the visual environment

Peter A. Howarth

Introduction

The scope of this chapter

Vision provides us with more information than all of our other senses combined. As a consequence, the environmental conditions necessary to optimize the eyes' performance are of paramount importance.

While the visual sense is in some ways exquisitely fine, in other ways it is dreadfully misleading. Although we can detect an offset in a line of as small as 5″ of arc (the width of a pencil viewed at a distance of 300 m), when indoors under artificial lights we can think that two pieces of cloth are the same colour—to discover on going outside that they are quite different. In the first of these examples, visual performance is dependent upon the *quantity* of light, whereas in the second it is the *spectral characteristics* of the light which limit performance.

There are many different aspects of the visual environment to consider when trying to produce the conditions necessary for good visual performance, not all of them immediately obvious. Consider an example: visual displays incorporating liquid crystal alphanumeric characters are to be installed in a self-service petroleum pump. What visual considerations are there? Some spring to mind immediately, such as character readability, and the positioning of the display so that it can be seen by drivers of different heights, sitting or standing. But the relevant elements of the visual environment are not simply those related directly to the visual task in isolation from its surroundings. Other questions need to be asked: Is the display going to be used both during the day and the night? If so, how is it going to be lit, and will supplementary lighting be needed? What is known of the spectral characteristics of the light that will fall on the display? Will it be lit by monochromatic sodium road lighting, and if so what account needs to be taken of this in the use of colour in the design? If supplementary lighting is going to be used, how should it be positioned to avoid producing veiling reflections in the display, and will discomfort be produced from glare? Is the display going to be covered with either glass or plastic, and if so what reflections will be produced in this cover by artificial lights, by car headlights, or by the sun as it crosses the sky?

This chapter discusses conditions of the visual environment that may be of concern to the ergonomist, and indicates why and how they may need to be assessed. Of necessity, the chapter is general in nature: the aim is to provide the reader with sufficient information to enable them to approach a wide range of problems with an adequate understanding of the key issues. With the background information provided here, texts dealing with specific issues—such as the Chartered Institute of Building Services (CIBS) Code of Interior Lighting (CIBS, 1984)—should be more readily understood. Finally, further information on the issues considered here, along with full references, can be found in standard texts such as Boyce (1988).

The nature of vision

It is often said that vision runs on light. While it is true that the eye's photoreceptors respond to light, they do not signal absolute light level—adaptation removes this information at the earliest stage of the visual process. Because of its adaptation systems the human eye has a tremendous total operating range, and the illuminance at midday on a Florida beach may be 10^9 times higher than on a dark and cloudy night on the Scottish moors. However, although the eye can operate under either condition, at any one time it is restricted to a small part of this range. It is this adaptation that makes us so poor at absolute visual judgements; on the other hand we are usually extremely good at relative judgements. Rather than responding to light level *per se*, the visual system responds to *variation* (temporal, spectral or spatial) in light. It is more appropriate to say that vision runs on contrast— a luminance or a chromatic change in either space or time.

Contrast describes a variation in light, and this chapter begins with a review of the nature of light and how it affects the visual system. This is followed by a description of the characteristics of natural and artificial light sources, and a discussion of which aspects of the visual environment deserve attention. Finally, those parameters which it is realistic and useful to assess in a practical situation are considered.

How light affects the visual system

Light is a small portion of the electromagnetic radiation spectrum. This spectrum includes radio waves, microwaves, ultraviolet radiation, infra-red radiation, and X-rays. What is special about light is that we can see it. While this seems an obvious statement, there is an important fact to be gleaned from it—namely, 'light' is defined by the human visual system, and not by the light source. If you cannot see it, it's not light. The sun, for example, radiates a wide range of the electromagnetic spectrum, some of which passes through the atmosphere and some of which is filtered out by it. However, the limits of what we call light are those of the human eye [approximately 380–760 nm (10^{-9} m)] not those of the sun. The spectral sensitivity of the eye, termed the Vλ function, is shown as a solid line in Figure 14.1. A photometer measuring *light* will, in effect, measure the total amount of radiation present over the spectrum as weighted by this function. This is done by passing the radiation through filters, so that the combination of the filter and the radiation sensor has the same response characteristics as the human eye.

The effectiveness of electromagnetic radiation in producing the sensation of vision depends upon how sensitive the eye is to the wavelengths present. The *spectral characteristics* of the light give it its characteristic appearance, and the weighted sum of the radiation present at each wavelength determines the *amount of light* present.

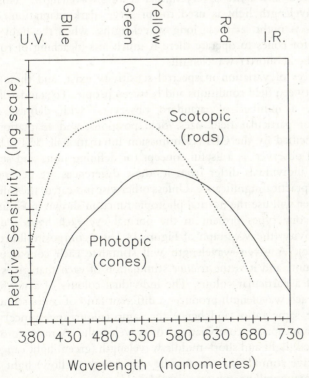

Figure 14.1. The spectral sensitivity of the eye for daytime light levels (solid line), known as the Vλ function, and the sensitivity for nightime light levels (dashed line), known as the V'λ function.

A potentially troublesome point to note here is that the spectral sensitivity of the eye is not totally independent of the light level. The eye has two kinds of photoreceptors, which have different operating ranges. One kind are called 'cones', of which there are three types (to give us colour vision), the other kind are called 'rods'. At normal light levels the cones are the active photoreceptors, and the rods are essentially saturated (their photopigment is highly bleached). However at lower light levels, e.g. around dusk, there is not enough light for cones to operate, and rods become the active photoreceptors. The spectral sensitivity of the rod system differs from that of the normal cone system, and the change in sensitivity of the eye as the light level changes from photopic (cone) to scotopic (rod) levels is termed the 'Purkinje shift'. Purkinje, a Czech physician, noticed in 1825 that while red and blue paint on signposts looked the same brightness during daylight, at night the blue looked much brighter than the red. There are two aspects of the Purkinje shift to note from Figure 14.1: (1) going from cone vision (solid curve) to rod vision (dashed curve) the peak of the function moves from 555 nm to 507 nm and (2) the overall sensitivity of the eye increases.

From the functions shown in Figure 14.1, we can see that the absolute sensitivity of rods and cones is very similar at long wavelengths. This shows why long-wavelength light is used to 'preserve' dark-adaptation, e.g. on ship's bridges at night: at these long wavelengths, when the light level is high enough for cones to operate there is much less bleaching of rods than there would be at shorter wavelengths.

Other sources of variation in spectral sensitivity exist, and these include differences between field conditions and between people. To standardize light measurement, a number of 'standard observers' with defined spectral sensitivities for particular field sizes, field positions, and adaptation levels have been specified by the CIE (Commission Internationale de l'Eclairage). The 'standard observer' is a useful concept for defining units and standards and although individuals differ from it, these differences are generally too small to be of practical significance. Unless otherwise indicated, the calibration of a light meter will use the normal photopic function shown in Figure 14.1.

The three cone types present in the normal eye each have a different spectral sensitivity (the $V\lambda$ graph of Figure 14.1 is a composite function for the whole eye). A given wavelength will stimulate each cone type by a different amount, and it is the *relative* stimulation of each that gives rise to the percept of a particular colour. The individual colours of a rainbow are seen because each wavelength produces a different ratio of cones stimulation. However, the same ratio (and hence the same colour appearance) can be produced by appropriate combinations of wavelengths. A mixture of long-wavelength (red) light and short–middle wavelength (green) light can provide the same relative stimulation as a middle wavelength (yellow) light, and in appearance the two will be indistinguishable*. However, the colour rendering

---

*This is a *metameric* match: two objects (in this case the light sources) appear to be identical in colour even though their spectral composition differs.

properties of these two light sources will be very different: under the mixture a rose will look red and a leaf will look green, whereas under the monochromatic light both would look yellow.

The *temporal characteristics* of the light source may be of interest when artificial lights are used, either because flicker is noticeable in the environment or because of interactions between the light source and the equipment being used. The visual system integrates light over a finite period of time, and a light flickering faster than the integration time will be perceived as steady rather than flickering. This integration time varies with light level, and so modern films, for example, appear to be continuous rather than flickering, even though their frequency is below the maximum that humans can detect under optimum conditions.

The *spatial characteristics* of tasks and lighting are the final aspect of the visual environment considered here. Luminance and chromatic variation can occur across a room, across a workplace, and at different heights from the ceiling. Directional lighting can facilitate tasks such as inspection. Also, lighting may be deliberately arranged within a room so that features stand out; this can be for safety (highlighting fire extinguishers and exits) as well as for aesthetic reasons.

Spatial variation is not always advantageous, however. At any given moment the eye has a limited operating range, and a large range of light levels within the environment is to be avoided because visual performance will be degraded at the extremes. This effect can be seen in large rooms when sunlight shines through a window, providing lots of light for the areas near the window but leaving the areas away from the window relatively gloomy. Here, without supplementary lighting the details in the shadows may be below threshold, and cannot be seen.

Light sources

It is appropriate here to consider light sources as being either natural or artificial.

Natural light

The sun is the main source of natural light. Direct sunlight, light which has been scattered by the atmosphere to become skylight, and light which has been bounced off the moon all originate from the sun. Three aspects of sunlight are considered here.

First, the wavelength spectrum of light from the sun is broad, and contains no large peaks. The spectrum of an overcast northern sky is illustrated in Figure 14.2(a). At different times of the day, and at different places in the sky, the relative amount of energy at each wavelength changes. The most noticeable example occurs late in the day, when the emission spectrum of the western sky is predominantly long-wavelength. For colour rendering purposes, daytime skylight has been accepted as 'normal' light, with a clear northern sky providing the spectral standard.

Second, the sun is giving off energy in a manner which can be considered to be continuous. There is essentially no fast temporal variation in sunlight, with slow natural variations being caused by clouds, by the earth's rotation, and by eclipses.

Third, the sun as a source is very small in angular terms, while the sky as a source is very large, unless walls and ceilings block it out. Although direct sunlight is a lot more intense than skylight, because of this size difference *the major component of natural lighting in buildings comes from the sky* and not from the sun, even on a cloudless day. This is particularly true in those parts of the World where the sun is reputed never to shine!

Incandescent light

When an object is heated, it gives off electromagnetic radiation. If you could take a metal ball and paint it perfectly matt black, then at normal room temperature it would neither reflect nor emit any light. These properties are what is meant by the term 'black body'. If you started heating the ball, it would eventually change colour and start to glow, i.e. it would emit radiation within the visible spectrum. An electric element of a cooker is like this—turn the cooker on and the black element starts glowing red (and also gives off lots of long-wavelength radiation as heat). In the same way, the metal ball would vary in colour as its temperature changed. At different temperatures it would have a characteristic (broad band) emission spectrum and, consequently, a characteristic colour. Because of this known variation in colour with temperature, light sources are often specified in terms of their 'colour temperature', which is the temperature at which a black body would have the same colour appearance as the source.

In the same way that a cooker element radiates light as well as heat, a thin piece of wire heated by passing electricity through it also gives off light. Again, the emission spectrum of the light will depend upon the temperature to which the filament has been heated. An incandescent light is simply a thin piece of tungsten wire, surrounded by inert gas in a glass container, heated to make it glow.

Figure 14.2(b) shows the spectral emission of a typical, commercial light bulb, which is yellow–white in appearance. The hotter the wire, the more it glows and the relatively greater amount of light it emits—particularly at short wavelengths (with a consequent rise in its colour temperature). However, the hotter the wire, the more evaporation there is from its surface and the shorter its life expectancy. So there is a trade off between lamp temperature (and hence light output) and lamp life. A recent improvement in this trade off has come with the use of halogen gases within the lamp. These combine with the evaporating tungsten during the lamp's operation, and the evaporated tungsten does not become deposited on (and blacken) the

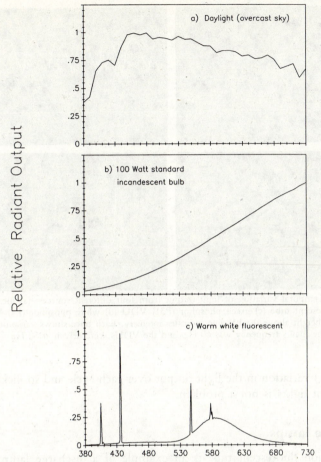

Figure 14.2. Emission spectra of (a) daylight, and two artificial sources: (b) an incandescent bulb, and (c) a fluorescent tube. Each graph has been normalised so that its peak equals 1.

inner wall of the bulb. Also, after evaporating into the halogen gas the tungsten is deposited back onto the filament. This increases the lamp's life expectancy, and it can then be run at a higher temperature with both a higher luminous efficacy (i.e. it gives off more light per unit of electricity) and also a spectrum more closely approaching that of skylight.

What about the temporal characteristics of incandescent lamps? An incandescent bulb run from the mains supply (50 Hz in the UK, Australia and Europe, 60 Hz in the USA) will show a temporal variation in its light output. However, Figure 14.3(a) shows that this variation is small—there is low luminance modulation. Although the filament will alter its temperature with the variation in the electricity, once the light is switched on the filament remains at a high temperature and never has a chance to cool down. Because the electrical variation produces only a small temperature change, there is

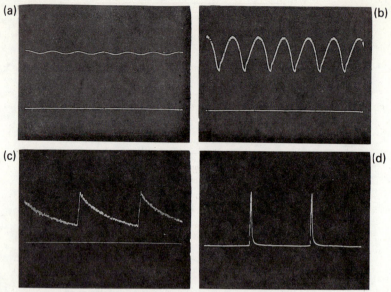

Figure 14.3. Temporal variation of light output from four different sources: (a) incandescent bulb (b) fluorescent tube (c) green phosphor (P31) VDU (d) white phosphor (P4) VDU. In each case the height above the baseline indicates intensity. Each trace shows variation over 1/20th second: the mains frequency was 60 Hz. and the VDUs were driven at 50 Hz.

little (<5%) variation in the light output over each cycle and so flicker from incandescent lights is not a problem.

Discharge lamps

The common fluorescent tube is an example of a discharge lamp. These work on a very different principle from that used in an incandescent lamp. Electrons emitted from a cathode at one end of the tube pass through a pressurized gas to an anode at the other end of the tube. Some are captured by atoms of the gas, thereby raising the atoms' energy level. However, like many high-energy objects these atoms are unstable. They subsequently release electromagnetic energy, and if this energy is within the visible spectrum, it's called light. Unlike the light from incandescent lamps, the emission from discharge lamps is usually at discrete wavelengths. The use of different gases, and different gas pressures, produces a variety of emission spectra.

The two gases most commonly used in commercial discharge lamps are mercury, which is found in the ubiquitous fluorescent tube, and sodium, found in road lights. Because discharge lamps generally emit at only a few discrete wavelengths, this emission by itself does not give satisfactory colour appearance. To improve the colour-rendering properties of fluorescent tubes, the inside of the tube is coated with a phosphor. This coating absorbs energy emitted by the lamp at discrete wavelengths (including the invisible ultraviolet) and re-emits light across a broad band of wavelengths within the visible spectrum. In Figure 14.2(c) the peaks are caused by the gas emission while the lower, continuous, background is produced by the phosphor. The choice of phosphor will determine the appearance of the emission spectrum, and lamp manufacturers are capable of producing green, blue, and even red fluorescent tubes. Normally tubes are produced with a warm appearance (yellowish-white) or with a cool appearance (blue-white), the use of which depends on the application and taste of the user. For example, to improve the product appearance, store meat counters are often lit by a 'warmer' redder, light than would be acceptable for normal interior applications.

Low pressure sodium lights have long been used for roadway lighting. These produce 'sodium yellow' light—almost monochromatic light of a wavelength just below 600 nm. These lamps are now being superceded by high pressure sodium lights which, although they have a broader spectral output, are still spectrally centred around the sodium lines, and appear yellow. These high pressure lamps have a higher luminous efficacy than normal incandescent bulbs, and can provide over 100 lm W^{-1} as opposed to 17 lm W^{-1}. For economic reasons they are being used increasingly in applications where accurate colour rendition is not necessary. However, because of the limited spectral output of these lamps they may not always be subjectively satisfactory as a sole light source. This situation may be remedied by providing both yellow-white high pressure sodium and blue-white cool fluorescent lamps, which together provide a broader overall spectrum. The combination can be subjectively very pleasant, mimicking as it does the combination of yellow sunlight and bluer skylight.

Discharge lamps may also differ from incandescent lamps in their temporal characteristics. As seen in Figure 14.3, the modulation of a discharge lamp is much higher (generally about 50%) than that of an incandescent light. With alternating current of 50 Hz the electrical signal driving the discharge lamp is, in effect, rectified when light is produced, and the light level rises and falls 100 times s^{-1}. This frequency is well above the maximum detectable visually, which for large fields is usually no higher than 60 Hz and, therefore, fluorescent light flicker should not be a problem. However, experience tells us otherwise: before the introduction of Visual Display Units (VDUs) into the workplace, fluorescent tubes were the principal topic of visual environmental complaints amongst office workers. One reason for these complaints is that sometimes other frequencies are also present in the emission. A mains frequency flicker can occur at the ends of the tubes; this can be remedied with appropriate shielding. Also, after thousands of hours of use, sub-harmonic flicker may be seen over the whole of the tube. This is remedied by tube replacement.

One solution to the problem of fluorescent tube flicker is to electronically stagger the phase of a bank of tubes. When some are fully on others are not, and this strategy reduces the overall modulation. However, this solution is being superceded and recent developments in ballast (the circuitry driving the tube) technology have enabled tubes to be driven at much higher frequencies (e.g. 20 kHz) than that of the mains, effectively eliminating all flicker. As well as reducing operating costs, (by increasing the luminous efficacy of the lights), these high-frequency lights have been reported to reduce complaints of headaches in offices lit by fluorescent lamps (Wilkins *et al.*, 1988).

The other main difference between incandescent lights and fluorescent tubes is their spatial extent. Although not all discharge lamps are large, some are the size of a normal light bulb, long tubes are commonly found in domestic, industrial and commercial applications. The amount of light given off by any point on the tube is far less than that given off by a point on a naked incandescent bulb providing the same illuminance. For this reason, we might expect fluorescent tubes to give rise to *fewer* complaints about glare (see p. 373). However, because it appears less bright a fluorescent tube will often be left exposed while an incandescent bulb will usually be incorporated into a glare-reducing fitting, and so the comparison is not a fair one.

Self-luminous tasks

Traditionally, the visual environment has been considered as consisting of two elements: the objects being viewed (such as paper copy, the road, instruments and dials) and their lighting (such as daylight, room lights or task-specific lights). However, the increasing use of self-luminous sources, such as VDUs, televisions, and LEDs, has led to a reconsideration of this viewpoint because the visual characteristics of the object are far less dependent upon the illuminating light than previously. As an illustration, VDUs are discussed briefly below.

VDUs essentially use the same technology as televisions and other cathode ray devices, where an electron beam scans rows of phosphor dots. The beam is turned on and off, and the phosphor glows or does not glow. Again, the spectral, temporal, and spatial characteristics of the display should be considered.

The main monochrome phosphors in present use are amber (Phosphor P134), white (Phosphor P4), and green (Phosphor P31 or P138). If luminance-matched, none has an intrinsic spectral advantage over the others*.

After the phosphor has been excited by the electron beam it emits radiation within the visible spectrum. The VDU phosphor, like that of fluorescent lights, does not decay instantaneously. Rather, it glows for a while, with ever decreasing intensity, after the beam has passed over it. The time course of this persistence varies between phosphors, so the temporal characteristics

---

*The claim that because green is closer to the peak of the human spectral sensitivity curve it is preferable from a user's viewpoint is incorrect if the screens are equiluminant. The advantage of having a phosphor which peaks where the eye is most sensitive is that less radiation is needed to produce a given luminance.

of all screens are not identical. For example, green <P31> phosphor has a slower decay than white <P4>. Whether or not this is preferable depends upon the application and upon personal choice. Figures 14.3(c) and (d) show the change in luminance over time of a single character on a green and on a white VDU, respectively. In each case, the discrete event of the phosphor excitation and subsequent decay is apparent. The greater persistence of the green phosphor [the long tail of Figure 14.3(c)] results in a lower average luminance variation, or modulation, with time when compared with the white phosphor. For an application where characters remain for a while in the same place on the screen this reduced modulation may be preferable because screen flicker will be less apparent. However, for an application where characters are being moved frequently the longer-persistence phosphor is often considered annoying, because of the 'ghost image' left briefly behind.

The *type* of ambient illumination, incandescent or discharge, has no spectral effect on the performance of screen-based tasks (but there may be spatial differences between them because the larger the area of a source the more likely it is to be reflected in a screen).

Apparent flicker (see later) is the major identifiable complaint arising from the temporal variation of VDUs. Often, you can see flicker from a VDU or a TV when looking to the side of the screen, but the flicker disappears when you look directly at it. This is because the eye is generally more sensitive to flicker in its periphery than in the centre of its visual field. Also, generally the larger the stimulus the more sensitive you are to flicker,* and so it is more apparent on a screen with dark characters and a bright background than on a screen with lit characters and a dark background. The flicker may be less obvious if the screen luminance is reduced, however this may be at the cost of decreasing the visibility of the screen characters, with a consequent decrease in visual performance. The long-term solution to this flicker problem is for the screen refresh rate to be raised.

As well as inherent flicker, older screens often show two problems which may be considered as spatial rather than temporal. These are 'jitter' and 'swim', and each is well-described by its name. The first is a fast oscillatory movement of a character around a central point, the second is a slow drift of the characters, as if the screen was being viewed through moving water. Both are caused by circuit instabilities and inadequate filtering.

Whilst it is possible to evaluate these problems objectively, a subjective assessment of the severity of the problem is generally a more realistic approach. There is no need here for intricate measurements, objective or subjective, as the only real issue to be faced is whether the screen is subjectively satisfactory for the user. This information may be determined from the operator by interview or questionnaire, and a satisfactory

---

*Sensitivity to flicker is context-dependant, and while frequencies as high as 90 Hz can be detected under optimum conditions, 60 Hz is generally taken as the maximum detectable under normal lighting conditions.

investigation of the issue could include both these user responses and an 'expert evaluation' of the severity of the problem. The solution to any substantial problem is to replace the screens.

Parameters of interest and their assessment

There are three main reasons why measurement of the visual environment is usually undertaken:

1. *Health and safety*. The lighting of an area should be adequate to ensure that people can live safely, and it should not in itself be a health hazard. Measurement of the visual environment can provide information as to whether or not these criteria are met.

2. *Visual performance*. There is a large body of knowledge telling us how visual performance is affected by variables in the visual environment, such as illuminance and contrast. Assessment of these variables can provide information about the expected performance in the location considered.

3. *Aesthetic reasons*. A pleasant environment is conducive to well-being, and will usually result in less stress and better task performance.

Laws, standards, and guidelines exist to provide information about the requirements for the first two reasons. The third is often a matter of personal preference, although many guidelines and recommendations for aesthetically pleasing lighting seem to have good general support.

The following section discusses the various parameters which might be relevant, and provides background information about each; the *taking* of measurements is discussed further in the final section.

Amount of light

In the recent past, measurement and specification of light has been complicated by the use of different units in different countries, and by different groups of people within the same country. The text below uses currently accepted SI units. Because alternative units will still occasionally be encountered, for example in instructions for older instruments, and in older texts, descriptions of some previous units, and their conversion factors, have been included here.

Because of the adaptability of the human visual system we are very poor at making *absolute* judgements (but very much better at making *relative* judgements) and the eye itself is not a particularly trustworthy meter. Hence, light levels are usually measured objectively (although their subjective appearance should not be ignored), and Figure 14.4 illustrates the different concepts involved. There are two distinct aspects to consider. The first is the question of how much light there is present to 'light things up'—how much light is falling on this book, or how much light goes into your eye. This is termed *illuminance*. The second is how much light is being given off

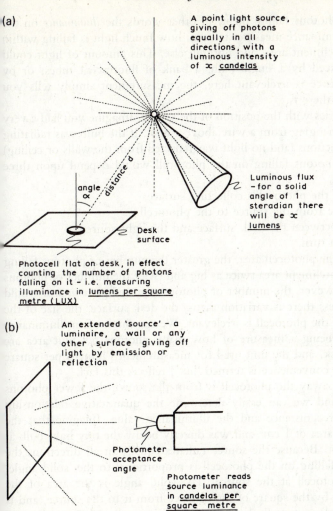

(a)

A point light source, giving off photons equally in all directions, with a luminous intensity of x <u>candelas</u>

Luminous flux - for a solid angle of 1 steradian there will be x <u>lumens</u>

angle α

distance d

Desk surface

Photocell flat on desk, in effect counting the number of photons falling on it - i.e. measuring illuminance in <u>lumens per square metre (LUX)</u>

(b)

An extended 'source' - a luminaire, a wall or any other surface giving off light by emission or reflection

Photometer acceptance angle

Photometer reads source luminance in <u>candelas per square metre</u>

Figure 14.4. The measurement of light; (a) illuminance, (b) luminance

by something, and this is termed *luminance* (or luminous intensity if the source is small). The difference between illuminance and luminance can be illustrated by closing this book (after reading the next sentence!). The amount of light falling on the book is constant—the illuminance doesn't change— but overall the page reflects more light than the book cover, and so the page has a higher average luminance.

Illuminance

If we took a photocell which faithfully recorded every photon which landed on it and placed it on a desk, then the photocell output would depend upon the number of photons falling on it, in other words the *illuminance* on the desk surface. Illuminance measures tell you how much light is falling within the photocell catchment area, and nothing else. This amount of light could have been produced by a single bulb, a bank of fluorescent tubes, or by daylight—the source is irrelevant here, the measurement simply tells you how much light there is.

Illuminance varies with the position of the source. Imagine you had a very small light bulb hanging from a wire above a desk. If the bulb was radiating equally in all directions (and no light was reflected from the walls or ceiling) the number of photons falling on the photocell would depend upon three factors:

1. The area of the photocell's collecting surface.
2. The distance from the source to the photocell.
3. The angle between the desk surface and the light source.

Consider these in turn.

1. The larger the photocell area, the greater the number of photons falling on it; make the catchment area twice as big and twice the number of photons are captured. However, the number of photons captured *per unit area* would be constant. Unless there is variation across the desk surface, the size of the collecting area of the photocell is irrelevant. We can think of an illuminance measurement as being a measure of how many photons per unit area are falling on the desk, and the unit used for measurement ['lumens per square metre' which for convenience is termed 'lux'] reflects this fact.

2. The further away the photocell is from the source the fewer photons will fall on it, and we can easily determine the quantitative relationship between the source distance and the illuminance value. Imagine that the photocell had an area of 1 cm^2 and was directly below the tiny light bulb at a distance of 1 m. Because the source radiates equally in all directions the amount of light falling on the photocell is proportional to the 'solid angle' made by the photocell at the source. The solid angle is the area of the photocell divided by the square of the distance from it to the source, and is measured in steradians (sr). So here the 1 cm^2 photocell at 100 cm distance would subtend $1/100^2 = 1 \times 10^{-4}$ sr.

The point to note here is that the solid angle varies inversely with the *square* of the distance, and from the previous deduction this means that the number of photons falling on the photocell varies inversely in proportion to the *square* of the distance from it to the source. Because the illumination on a surface is expressed in terms of the amount of light falling on it per unit area, the inverse square law follows: the illuminance on the surface varies with the square of the distance between the surface and the source.

To illustrate that *illuminance* depends on the distance from the source to the object being illuminated, imagine a torchlight with a diverging beam shone directly at a wall from a couple of inches away. If you walked away from the wall two things would happen: the circle of light on the wall would increase, and the amount of light falling at any point on the wall would decrease. Double the distance from the torch to the wall and you increase by four the area of wall illuminated, decreasing by four the illuminance (i.e. the amount of light per unit area of the wall) falling on the lit portion of the wall.

3. The above examples have assumed that the photocell was positioned perpendicularly to the source. What if the photocell were angled—which is what would happen if it was placed on an inclined document holder rather than flat on the desk—with the light source directly above it? Here the amount of light per unit area falling on the photocell would be less than if it were perpendicular to the source, and the illuminance varies with the angle that the photocell makes with the perpendicular. If we start with the photocell directly perpendicular to the source the illuminance will be at a maximum value. If we now rotate the photocell so that its face is no longer exactly perpendicular to the source, the illuminance on the face will be reduced by the *cosine* of the angle of rotation. For example, when the photocell is rotated 60° the illuminance will be reduced by a factor of two, as cos 60° = 0.5.

ILLUMINANCE UNITS

The amount of light falling on a given surface area, the luminous flux per unit area, is the illuminance at a point on the surface. The SI unit is the *lux*, which is one lumen per square metre (lm m^{-2}). That is to say, when luminous flux (see below) of 1 lm is spread over a surface area of 1 m^2, the illuminance is 1 lux. An alternative name, not in common use, for the lux is the 'metre candle', while the 'phot' is the illuminance when 1 lm is spread over 1 m^2.

The *foot-candle* is a unit still frequently encountered, especially on old illuminance meters; it is an illuminance of 1 lm ft^{-2}, and is equal to 10.76 lux.

The *troland* is a unit used in vision research, when an eye with a 1 mm^2 entrance pupil views a surface with a luminance of 1 cd m^{-2} (see below) then the retinal illuminance is 1 troland.

For illustration, Table 14.1 provides some representative indoor illuminance values recommended by the CIBS.

Luminous intensity (point sources)

In the previous section we considered the light source to be very small, and to give off light equally in all directions. If this source was so small that it effectively had no area, it would be called a 'point source'. In fact, this is a theoretical concept as all real sources have a finite area to them. However, the idea of radiation being emitted from a single point equally in all directions is a useful one in explaining the concepts of *luminous intensity* and *luminous flux*. The more intense a source is, the more light it gives off in all directions; luminous intensity is an attribute of the light source. Now consider a cone with a point source at its centre: the more intense the source the more photons there would be within this cone; in other words, the greater

Table 14.1. Examples of recommended illuminance values (from the CIBS Code, 1984)

Condition	Recommended value (lux)
Rarely visited locations, with limited perception of detail required, e.g. railway platforms	50
Continuously occupied areas, with limited perception of detail required e.g. waiting rooms	200
General offices Airport ticket counter	500
Drawing boards Bench and machine work (fine detail)	750
Cloth inspection Assembly work (fine detail, e.g. watchmaking)	1500
Inspection of extremely fine detail (e.g. small instruments)	2000

'luminous flux' there would be within the cone, see Figure 14.4(a). So luminous flux describes the light itself, and not the source.

An ideal point source gives off radiation equally in all directions. Although real sources can sometimes be assumed to have negligible area (consider the angular subtense of a motorcycle headlight viewed from a mile away) the property of directionality does not generally hold. Interior and exterior light fittings are often designed with reflectors or shades to produce a certain pattern of light, and the 'hot spots' of a headlight are a good example of this.

Given that the luminous flux varies with direction, it follows that the luminous intensity of a real (as opposed to a theoretical) source also varies with direction.

LUMINOUS INTENSITY UNITS

Luminous intensity is a measure which describes the amount of light emitted by a point source. The SI unit is the *candela* (cd). This basic unit is defined in terms of the emission of a black body at the freezing point of platinum 2040 K.

LUMINOUS FLUX UNITS

When light is emitted from a point source, the 'density of photons' within a cone centred at the source is termed the luminous flux. The number of photons within the cone will be a function of both the source intensity and the cone solid angle, and the SI unit of luminous flux is the *lumen* (lm). This is defined as the luminous flux emitted through a solid angle of 1 sr from a point source of intensity 1 cd. A point source with a luminance of 1 cd will emit overall a total of 12·57 (i.e. 4 π) lm.

Luminance (Extended sources)

In practice, one generally needs to consider extended sources rather than point sources. Usually a fluorescent tube, a light fitting, or some other source such as the sky subtends an appreciable angle at the eye, and cannot be considered to be infinitesimally small. Hence, luminance is specified in candelas per square metre of the source rather than in candelas. To take the measurement, a light meter with a small acceptance angle is used which will collect light within, e.g. 1°, 0.5°, or 1', depending on the instrument. It makes no difference how far away the source is, as long as it is larger than the meter's acceptance angle. Consider a luminance meter pointed at a wall, collecting and recording all of the light within its acceptance angle. If we move the meter closer to the wall more photons will be collected by the photocell from any given point. But because the acceptance angle of the meter is fixed, the portion of the wall from which photons are being collected by the photocell is also reduced. Because the increased number of photons collected from each point is balanced by the decreased area of the wall within the meter's acceptance angle the luminance reading remains constant. So luminance, unlike illuminance, is *independent* of the measurement distance.

LUMINANCE UNITS

Luminance describes how much light is emitted from a surface, and the SI unit is the *candela per square metre* (cd m^{-2}). (This unit in the past has been called a 'nit' but, to the chagrin of schoolchildren everywhere, this name is no longer in use.)

The other unit that may be encountered is the *stilb*, which is 1 cd cm^{-2}.

Reflectance

Illuminance and luminance are closely linked. The amount of light falling on a point on the wall is its illuminance, and the amount of reflected light coming back from the wall is its luminance. If all of the light that fell on the wall was reflected, then the values of the illuminance and the luminance would be the same, using appropriate units. If some of the light was absorbed, then the values would differ. The *reflectance* of the wall may be found by comparing the illuminance and the luminance values, as explained below.

An important point to consider before going further is the *directionality* of reflections. A perfect matt white surface will reflect light equally in all directions, irrespective of the direction of the ambient light, and is called a perfectly *diffuse* reflecting surface [Figure 14.5(a)]. On the other hand, a highly faithful reflector, like a mirror, will only reflect light at the same angle to the normal that the ambient light makes [Figure 14.5(b)]. Most surfaces lie somewhere between these two extremes, with a higher luminous flux occurring at the angle of reflection than elsewhere [Figure 14.5(c)]. (Try moving this book and see if you can detect specular reflection on the paper

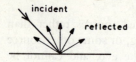

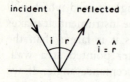

a) A perfect diffuse reflector (also known as a Lambert or as a cosine reflector) reflects equal luminous flux in all directions.

b) A perfect specular reflector reflects all light at an angle equal to the angle of incidence

c) Most surfaces reflect light in all directions, but with a greater proportion at an angle equal to the angle of incidence.

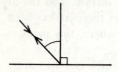

d) A retro-reflector reflects light back along exactly the same path.

Figure 14.5. The four types of reflection:
(a) diffuse, (b) specular, (c) mixed, (d) retro-reflection.

from any light sources.) Finally, if a surface reflects light back along its own path it is termed a retro-reflector.

The greater the proportion of unwanted specular reflection from a surface, the more likely a person is to experience annoyance, discomfort and degraded visual performance. Where specular reflection is likely to be a problem, the positioning of light fittings must take this into account: areas of relatively high luminance should be avoided at places where they might be reflected into the eye. Where new equipment is used in a previously-designed room, such as new VDUs in an old office, the aim should be to allow for adjustability in the equipment so that specular reflections from the screen can be reduced.

REFLECTANCE UNITS

Irrespective of its angle of incidence, all of the light falling on a *perfect diffuse* reflecting surface is emitted equally in all directions. No light is lost, and it is intuitively appealing to have units which are defined in such a way that the values for luminance and illuminance are identical here. This is the rationale behind the old (now deprecated) luminance units 'apostilb' and 'foot-lambert'. When the illuminance falling on a perfect diffuse reflector is 1 lux, the luminance of the surface will be 1 *apostilb* (asb). It follows that if the surface does not reflect all of the light, then the luminance in apostilbs will be equal to the reflectance of the surface, multiplied by the illuminance in lux. The same rationale applies for imperial units, and with an illuminance of one foot-candle (i.e. 1 lm ft^{-2}), the luminance of a perfect diffuse reflector will be one *foot-lambert*. It follows that 10·76 asb = 1 foot-lambert.

The accepted SI luminance unit is the candela per square metre, and for a perfectly diffuse reflecting surface there is a simple relationship between it and the apostilb:

$$1 \text{ cd m}^{-2} = \pi \text{ asb.}$$

So if the illuminance on a wall of reflectance 0·8 is 220 lx, then the luminance of the wall will be:

$$(220 \times 0·8) = 176 \text{ asb} \tag{1}$$

or

$$\frac{(220 \times 0·8)}{3·14} = 56 \text{ cd m}^{-2}. \tag{2}$$

The first calculation is more straightforward, so why is the use of the apostilb deprecated? The reason is that in practice most surfaces are *not* perfectly diffuse, but rather have some specular component. Because of this the wall luminance will vary, and the measured value will depend upon the observer's position.

Instead of using the reflectance of the wall to determine the wall luminance, a quantity called the 'luminance factor' is used. The luminance factor is the ratio of the luminance of a reflecting surface, viewed in a given direction, to that of a perfect white diffusing surface identically illuminated. If the reflecting surface is itself a perfect diffuser, then the value of the luminance factor is the same as the reflectance, is independent of the viewing position, and cannot be greater than one. On the other hand, if the surface does have specular reflections, then the luminance factor will vary with viewing position, and at the angle of reflection it could be greater than one. Hence:

$$\text{Luminance (cd m}^{-2}) = \frac{\text{illuminance (lux)} \times \text{luminance factor}}{\pi}.$$

The reflectance value of a surface is normally found for 'light', i.e. the whole of the visible spectrum. It is also possible (although not normally needed in practice) to determine the reflectance values for different wavelengths. If this were done, then it would be seen that different coloured walls had different reflectance spectra. For example, a wall which looks red might reflect a large proportion of long-wavelength light, but little medium- or short-wavelength light, while one which looks green might reflect a large proportion of medium-wavelength light, but little long- or short-wavelength light.

Flicker

Flicker, noticeable rapid fluctuations in light level, can be a serious problem in artificial environments. Unfortunately, objective measurement of flicker is not simple because it requires rapid-response equipment, normally available only to a lighting specialist.

Subjective assessment of flicker is, however, much more feasible and both the area of noticeable flicker and the degree of noticeable flicker can be adequately assessed by descriptive means. Also, because the periphery of the eye is more sensitive than the central area to flicker, subjective assessment may actually be a more relevant method. Hence, when dealing with flicker the precise circumstances under which it is seen, such as the luminance and the position of the source in the visual field, should be noted.

In considering the subjective assessment of flicker, we should take a lesson from noise assessment (see Haslegrave in this book). In the same way that the psychological effect of noise can be independent of the sound level (think about the dripping of a tap) flicker can have an annoyance or a distractive effect out of all proportion to its physical magnitude. A subjective assessment of the flicker should not only consider the physical aspects of the stimulus, such as the perceived flicker strength, but should also evaluate the psychological effect that the flicker is having on the person. The positive side to flicker is that because it is very attention-getting, its use is a very good visual method of conveying warning information.

Two further aspects need to be considered. First, some people are especially sensitive to flicker. Epileptics are an extreme example, and for them flicker (particularly at frequencies around 10 Hz) can provoke seizures. Second, flicker which is not visually detectable may still affect parts of the visual system. The human retina responds to flicker at high frequencies (over 100 Hz) even though the light appears steady, and no flicker is seen. It remains to be determined whether other parts of the human visual system are affected by high flicker frequencies, and whether performance or comfort are affected.

Colour

Again, a full spectral assessment of the environment requires specialized equipment. However, this assessment is very rarely needed.

Different light sources have different colour rendering properties, and the colour of an object is determined both by the spectral composition of the light source, and by the spectral reflectance properties of the object and its surround. In this way two objects, like a shirt and tie, can appear identical in colour under one set of lights (a metameric match) but can be noticeably different from each other under another set of lights. The eye is easily fooled under these circumstances. However, while changes in colour appearance like this do occur, the adaptability of the visual system is such that most objects will retain their correct appearance under a wide variety of light sources. This 'colour constancy' can be seen outdoors when a cloud passes over the sun—although the spectral composition of the light falling on the grass and the trees changes drastically, the colours of the objects don't seem to change.

When considering the chromatic aspects of the environment, both the colour rendering properties of the light sources and the pleasantness (or otherwise) of the lighting and the environmental colours must be included. The first of these is likely to have been considered by the lighting designer in an environment where it is important, e.g. where colour discrimination or colour matching are included in the tasks performed in that location. This occurs more often than casual thought would suggest, and text and instructions are often colour coded. There is an important point to stress here: colour coding is valueless if the ambient light does not reveal the colour differences. For example, because sodium light is monochromatic, in this light the three cone types are always stimulated in the same ratio. All objects appear to be the same colour, yellow, and depending on how much of the light they reflect, some are darker than others. Hence, the colour coding of road signs has to be revealed at night by means of supplemental lighting.

Good colour rendering is not always necessary, and other criteria may be important. In some situations (such as industrial exteriors and warehouse interiors) the lights may have a simple safety function. Good colour discrimination is not required, for example, if the lighting only has to reveal the presence or absence of objects. Here the cost of running the lights may be a more important criterion than their pleasantness.

In an environment where people have to spend a large proportion of their working time the colour rendering properties of the lights, or combination of the lights, plays an increasingly important role. Lamps with poor colour rendering are generally considered to provide a less pleasant environment than those with good rendering. Of particular importance here is the appearance of skin tones: the better the appearance of skin, the more subjectively preferable is the light. Although 'warm' lamps are generally considered preferable to those with a 'cool' appearance, lamps with a narrow spectral band (such as high pressure sodium) generally provide an unacceptable environment for long-term working. Also of interest here, and a factor where little research has been performed, is how people performing different occupations prefer different spectral combinations. Draughtspeople and designers, for example, may tend to use cool, blue–white light, while office work is more commonly performed under warmer, yellower light.

Daylight coming from a northern sky is broad-band (see Figure 14.2) and is the reference illuminant used for colour-vision testing. Different lamps are compared with this standard as far as their colour rendering properties are concerned, and the CIE have devised a 100 point scale, the 'colour rendering index' (CRI) for lamps. As a generality, the higher the value of the CRI the better the lamp performs [e.g. incandescent lamps may have a CRI of 99, an artificial daylight fluorescent lamp has a CRI of 93, while a white fluorescent lamp has a CRI of 56 (Boyce, 1988)]. However, these are *overall* values for the lamps, and a lamp with a high score does not necessarily perform well over all of the spectrum, although it should give good rendering of most colours.

Colours will often have specific culturally-based meaning associated with them (e.g. red = stop, green = go)*. Here assessment of the use of colour involves measuring the performance associated with the colour rather than the colour itself. Also, visual performance can sometimes be enhanced by using colour to provide a contrast between a task and its surround. Unfortunately, while the concept of colour contrast has some intuitive meaning, (most people would consider the colour contrast between red and blue to be greater than that between yellow and green), as yet there is no widely-accepted objective metric by which it can be evaluated. At present, this assessment has to be made on a subjective basis.

Glare

A number of quite distinct visual problems have been grouped together under the heading of 'glare'. These problems, such as discomfort and reduced visual performance, have in common that they are all associated with light levels that are *relatively* high when compared with the ambient light levels. Although different forms of glare may occur simultaneously, they are essentially independent because they do not have the same underlying physiological mechanism. It is not surprising, then, that it is possible to have discomfort without disability, and vice versa, even though both will be often found occurring together.

---

*Nature has developed colour coding, as seen in the colouring of wasps and bees: yellow and black hoops warn of a stinging insect (or an insect impersonating one). While 'yellow' is a human percept (the insect itself is not yellow/black, but rather we perceive it that way) we can infer from the colour coding that whatever might otherwise prey on these insects have some form of colour vision. Similarly, the use of colour in camouflage depends upon the colour vision of the observer: two objects may be metamerically matched to a person with normal colour vision, yet to someone with a colour vision defect (or a normal person looking through a coloured filter) they will look different. Because of this, people with colour vision defects have been used in the past to 'spot' camouflaged objects.

Discomfort glare

When a portion of the visual field has a much higher luminance than its surround, a feeling of discomfort may occur around the eyes and brow. While this effect has been studied empirically for many years, the physiological mechanisms involved are still unknown. The facial and the iris muscles have been suggested as the site of the discomfort, but possibly neither are involved. When a glare source is turned on, people tend to wrinkle their brow and partially close their eyelids; concurrently the iris sphincter muscle contracts and the pupil constricts. Whether these are simply *responses* to the increased light and the associated discomfort, or whether they *cause* the discomfort is not known.

Although the mechanism of discomfort glare is unknown, the conditions under which discomfort occurs have been well established for a number of years. The early work of Hopkinson, Petherbridge, and Guth (Hopkinson and Collins, 1970) revealed that:

Discomfort *increases* with:
an increase in the luminance of the glare source,
an increase in the angular size of the glare source at the eye.

Discomfort *decreases* with:
an increase in the luminance of the background,
an increase in the angular position of the glare source relative to the line of sight.

By definition discomfort is subjective, and discomfort glare is not easily quantified. A given physical configuration of lights will not only give rise to different reported amounts of discomfort from different people, but also to different reported amounts of discomfort from the *same* person on different occasions. Subjective assessment in this situation is not particularly reliable. On the other hand, the *physical* parameters of different lighting configurations are, in theory, easy to determine. If it is known how the above four factors (glare source luminance, size and position, and background luminance) interact, it should be possible to measure aspects of the environment and determine on an objective scale how good or how bad the environment is. This is the rationale behind the various glare indices established in different countries—they say little about how an individual will respond, but they do allow an objective evaluation to be made of the lighting configuration. Calculation of the Glare Index involves determining the luminance, size and position of the glare source(s). Full details for the calculations involved may be found in CIBS publications (1984 and further information list, p. 386).

Despite the apparent validity of these indices in carefully-controlled experimental conditions, the evaluation of the glare indices in practice is complicated. While the luminance and size of the glare source(s) are easily measured, there are real difficulties in determining the value of both the backgrouund luminance and the glare source position. First, the luminance of different walls and different parts of the ceiling will vary, and the problem arises of what value constitutes the 'true' background luminance. In practice, an 'average' value has to be estimated. Second, there is not one unique glare index for one room position, but rather one glare index for each position in the room *and* each position of the eyes.

As someone looks around a room, the glare index will vary. One of the parameters mentioned above is the angle between the line of sight and the glare source. So while a head and eye position may be specified for measurement purposes, e.g. the person may be assumed to be looking down at their desk, or perhaps to be looking straight ahead, each position is arbitrarily chosen and is not one which the person necessarily adopts or has problems with.

Not surprisingly, there is a poor correlation between subjective reports of discomfort and the glare index. The advantage of the method is that it does provide an objective description of the environment. The disadvantage of the method is that this figure does not in itself describe well the subjective discomfort of an individual subjected to that glare.

An alternative way of evaluating glare is to use an 'expert observer' approach, where an individual experienced in glare assessment can evaluate the degree of discomfort they feel in a certain situation. The simplest way of initially determining whether there is a problem is to shield the suspected glare source(s) and see if there is an improvement in comfort. If so, then you can safely assume that there was some discomfort present in the first place. This can then be evaluated more precisely. A number of descriptive scales have been produced, perhaps the most useful of which is that proposed by Hopkinson (Hopkinson and Collins, 1970). Four criterion points were used, each of which describes a transitory position of subjective feeling:

1. Just perceptible.
2. Just acceptable.
3. Just uncomfortable.
4. Just intolerable.

If necessary, the scale can be extended by including end points (imperceptible, and intolerable) and by allowing descriptions between the criterion points (e.g. not quite acceptable, but not yet uncomfortable). This will produce a nine point scale. An alternative method is to simply rate the amount of discomfort on a rating scale, with defined anchor points having specific, known, rating values, e.g. a rating scale where imperceptible = 0, just uncomfortable = 5, and intolerable = 10. (See Sinclair's chapter for general discussion of rating scales.)

There are a number of ways to reduce discomfort glare. Consider in turn the four factors mentioned previously. Reducing the glare source luminance will reduce the discomfort. This solution may not always be feasible, because the source could be providing illumination for a different part of the room. Increasing the angle between the viewers line of sight and the glare source, by moving either the source or the person, will decrease the discomfort. Another, less obvious, solution is to *increase* the background luminance. In seems counterintuitive that increasing the amount of light present reduces the discomfort, but this manoeuvre reduces the contrast between the source and the background. A practical way of doing this is by painting the walls and ceilings so that they reflect more light, thereby increasing the ambient light level. Finally, the solid angle of the glare source can be reduced by shielding the source. Changing the luminaire fitting to one with a narrower cut-off angle will decrease the visible extent of the source.

These solutions are not necessarily independent. For example, by reducing the size of the glare source the illuminance it provides for some other purpose may be decreased, necessitating an increase in the source luminance. In the same way, a bare light bulb or a luminaire with a small fitting or shade may be improved by increasing the size of the fitting. Although the glare source solid angle is increased, the source luminance is lowered, with a consequent reduction in discomfort. These are all solutions which are imposed on the person by outside manipulation of the light sources. Where feasible, it is preferable to allow individuals control over their immediate visual environment, with directional local lighting, and in this way enable them to position the light sources in a way that they are comfortable with.

As a final point, the long-term effects of small amounts of discomfort glare are unknown. It is reasonable to suppose that someone working all day under conditions where glare is just perceptible will have a build-up of discomfort over the day. The need for improvement here is greater than in the situation where someone is briefly exposed to 'just uncomfortable' glare once or twice per day. Similarly, someone's idea of how tolerable glare is will depend upon what they are doing, and how interested and involved they are in it. A person watching television may choose to have a low ambient light level but a bright screen—environmental conditions which in other circumstances would be considered intolerable. The point to remember here is that the glare assessment cannot be considered in isolation from the context in which it is made.

Disability glare

While the physiological mechanisms involved in discomfort are not understood, the ways in which an extraneous light source can affect visual performance are quite clear. All involve contrast reduction. The degradation of visual performance occurs in one of two ways: a glare source can act *directly* by reducing the contrast between an object and its background, or *indirectly* by affecting the eye.

'Direct' disability glare can occur because of a discrete reflection, such as the specular reflection of a light source from the surface of a VDU screen. Here the luminance of both the object (the characters being viewed) and the background (the surrounding screen) are raised by the addition of the extra light but the contrast is reduced. It can also occur because of a diffuse reflecting veil over the whole of the task, as is seen, for example, when a car windscreen mists up. The whole of the scene looks grey and washed out, and both luminance contrast and colour contrast are diminished. These two examples have in common that the contrast between the object and the background is decreased, with a consequent reduction in object visibility. Hence the disability will be reduced by raising the contrast.

'Indirect' disability glare affects the eye and not the visual task. It is seen, for example, when a car approaches at night with its headlights on full beam and your eyes get dazzled. The disability in this situation has two sources:

(a) there is scatter within the eye reducing the retinal image contrast, and
(b) the adaptation level of the eye is raised as the car approaches.

After the car has passed it takes a little while for the eye to re-adapt to the ambient light level.

Both sources of disability are present in other situations, such as when an object is seen against or near a bright window. They have been termed 'indirect' here to emphasize that they act by affecting the eye rather than affecting the scene. Indirect disability glare is usually found in circumstances where discomfort will also be present because in both cases the eye itself is affected by the glare. Like discomfort, the disability is often reduced by *raising* the light level. On a city street a car headlight, even when it is not dipped, causes less disability than the same light on an unlit road because the extra ambient light raises the eyes' adaptation level. In the city, the headlight is less bright in comparison with its surroundings, and we can take this reasoning further by considering that the same headlight would produce virtually no disability during daylight.

There is a problem encountered less often where performance is reduced, which can still be considered as glare, and this is where there is an area of *low* light as compared with the ambient light level to which the eye is adapted. The problem faced here by the visual system is exactly the same as that faced by a photographer encountering deep shadows. If the exposure is set manually for the ambient light the details in the shadows are lost, whereas if the camera is set to record these details everything else in the photograph will be overexposed. The eye's adaptation system acts like an automatic camera, but it isn't possible to vary it in the way you can with a manual override camera. The automatic setting the eye will adopt is appropriate for the ambient light, and not for the shadows, and so details in the shadows are lost to the visual system.

Contrast rendering, and directionality of light

The effectiveness of a lighting configuration is not dependent only upon whether any noticeable disability is present. Despite this, quantification of lighting effectiveness is an area in which laboratory-based research has been applied in the past largely to lighting design rather than to environmental assessment.

The relative positions of the light source, the visual task, and the observer determine how effectively the task contrast is rendered, and recently a measure of lighting effectiveness, the Contrast Rendering Factor (CRF) has been devised. The CRF has been used mainly in regard to paper-based tasks, which is where it is at its most useful. If the task lighting in an office is suspected to be deficient, then measurement of the CRF would be an appropriate way to investigate the problem. Ideally, the CRF is measured by comparing the contrast of the object under the ambient lighting with its contrast under reference lighting (completely diffuse, unpolarized illumination). It is important to realize that the CRF is specific to a particular target, a particular location, and a particular observer position and is not a measure which describes the lighting alone. So the CRF of writing on matt paper will be different from that of writing on glossy paper under otherwise identical conditions. The CIE Publication 19/2 (and Boyce, 1988) provides details of assessing the CRF if access to reference lighting is available.

If reference lighting is not available, do not despair, as two approximate methods are described in the CIBS (CIBS, 1984) code. For simplicity, these methods assume 'standard tasks', and, interestingly, because the task is no longer a variable, they *do* both provide a quantitative description of the lighting configuration. In the first method, the illuminance at different planes at the workplace is measured, and the value of the CRF is found from a calibration graph. In the second method, a CRF gauge is used at the workplace, and the CRF value is read directly.

Generally, the higher the CRF, the more acceptable the visual performance. This might lead one to suppose that the reference lighting conditions, uniform diffuse illuminance, could be considered as 'ideal'. However, this is not the case as CRF is not the only criterion by which the directionality of lighting is judged. Uniform, shadow-free lighting gives an extremely bland appearance to an interior, with the solidity of objects being less readily apparent because of the lack of relief. For example, facial features are far more pleasant when seen under directional lighting than under flat lighting.

The directionality and modelling of the lighting at any point in a room may be quantified by combining two measures. One is the scalar illuminance, the other is the illumination vector. These may be measured by suspending a cube (at any orientation) at the position of interest and measuring the illuminance value of each face. Incidentally, in the absence of a cube, this measurement may still be taken: improvise by imagining that you have a cube, and then hold the illuminance meter against each imaginary face. Averaging the values of the six cube faces gives the scalar illuminance. The illumination vector is then calculated by constructing a vector diagram of the differences within each of the three pairs of opposing faces, as illustrated in Figure 14.6.

Combining the illumination vector value with the scalar value gives the vector/scalar ratio. A value of 0·5 is considered essentially shadow-free, while

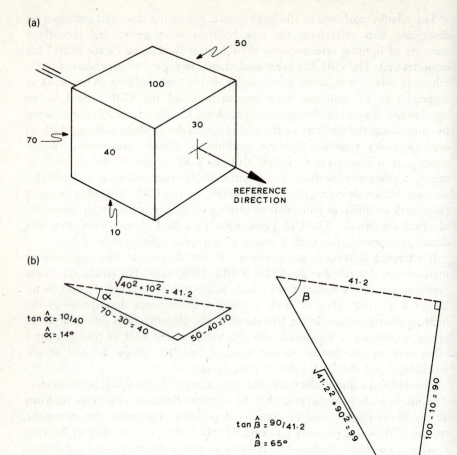

Figure 14.6. Calculation of the scalar and vector illuminances

(a) The illuminance values are measured on each face of a cube suspended in space, and the scalar illuminance is their arithmetic average (50 lux).

(b) A vector diagram can be constructed by first determining the difference between each pair of faces (90, 40, and 10 lux). Then the angle of the illumination vector in the horizontal meridian *relative to the reference direction shown in (a)* is determined as shown in the upper portion of (b); $\tan \alpha = 10/40$ so $\alpha = 14°$. The magnitude of this vector is the square root of $40^2 + 10^2$ (the two measurements taken in the horizontal plane) which equals 41.2. The angle in the vertical meridian, *relative to the direction just determined*, can now be found, as shown in the lower portion of (b). $\tan \beta = 90/41.2$ and so $\beta = 65°$. The *magnitude* of the illuminance vector is the square root of the sum of the squares of these two sides: the square root of $(41.2^2 + 90^2)$ which equals 99 lux.

ANSWER: In figure (a) the illumination vector has a magnitude of 99 lux, in a direction 14° to the left of the reference direction, and 65° below this direction. The vector: scalar ratio = 99/50 a value of 3·0 provides very strong modelling, with the details within the shadows often being invisible. Pleasant modelling of human facial features occurs between values of 1·2 and 1·8.

The directionality of lighting can also be used to reveal relief, and as a means of directing attention. For example, the inspection of textured items, such as cloth, may be enhanced by providing directional lighting which will cause shadows to be cast. The uniformity of these shadows can be easily assessed—defects showing up as an imprecision in their pattern. The elegant role that the lighting plays here is to facilitate performance by changing the nature of the visual task.

A further consideration in lighting uniformity is the illuminance distribution over a workplace. Variation in light level across a workplace will both differentially direct the attention to different areas, and will provide a more pleasing visual environment. For example, the whole surface of a desk may have an apparently-satisfactory illuminance level according to published guidelines, but if there is no spatial variation across the desk it appears uninteresting. This situation is greatly improved by providing slightly higher illumination on the area just in front of the person, where they put their immediate work. When the task is illuminated to a higher level than the surround, the appearance is much more interesting and pleasant.

Daylight

The need for windows in buildings, providing natural light, has come into question recently because of their cost in terms of heat-loss and energy conservation. The scientific evidence for a physiological need for windows is, at best, unproven, however the psychological evidence is clear. A small, windowless room can easily be considered cell-like and restricting, while the presence of a window provides visual access to the outside world. Larger rooms are considered less restrictive, and in a well-controlled environment the absence of windows becomes less important. However, in these windowless environments the information, such as the time of day, and the variety provided by the changing outside light is still absent.

As well as the illuminance variation over the day, the provision of daylight can also provide spatial and spectral variation within a room, again decreasing the monotony of the environment. The illumination variation may be quantified as a change in the 'daylight factor' across a room. The daylight factor is the ratio of the illuminance from the skylight measured on a horizontal surface within the room to the illuminance from the skylight (*not* direct sunlight) measured on a horizontal plane which has an unobstructed access to the hemisphere of the sky. At different positions within the room the daylight factor will vary, and if required this variation may be assessed over a room. For interiors where the daylight factor variation is not large, such as rooms with skylights, or rooms which are not too deep, when the average daylight factor is 5% or greater an interior will appear generally to be well day-lit. Also, if the illuminance from the sky is not known, the relative daylight factor at different positions within the room will give information about the spread of daylight within the environment.

The pleasantness of variation in spectral content and luminance level is not restricted to natural lighting. On the contrary, in some parts of the world the design of mood lighting provides a considerable source of revenue for interior designers and fixture designers alike.

Overview of measuring the visual environment

Standards and guidelines

Lighting guidelines and standards come in two forms, either statutory, where certain legal requirements have to be met, or recommendatory, where the recommendations provided are given as examples of 'good current practice'. In neither case should the values given be assumed to be 'ideal'. For example, in the case of a legal minimum requirement for a certain illuminance at a workplace, the *minimum* required is not necessarily the *optimum* value. Illuminance values which were considered 'good practice' 20 years ago are now, with the advent of more efficient lamps, often considered inadequate.

Most countries have their own standards and guidelines, and specific values will not be presented here. Rather, the reference section (p. 386) contains details of documents which contain the necessary information.

What measures are needed?

An assessment of the visual environment is going to be needed in one of two situations. The first is a design situation where the concepts discussed previously will have to be considered in a somewhat abstract manner. The example given in the introduction, that of designing a petroleum pump display, is a case in point. The second situation is a more immediately practical case, where an environment already exists and has to be evaluated in some way. Generally the assessment will be needed because problems are known to be present, and the appropriateness of the measures to be taken depends upon what these problems are.

The preceding sections have discussed the variety of factors of concern in the visual environment. Outside of the laboratory, many of these lend themselves to subjective assessment, and flicker is a good example. Objective physical quantification of the frequency and modulation of flicker is complicated, however it is usually unnecessary. The appropriate question may be 'Is there unpleasant flicker present?'—which has to be answered on subjective rather than on objective grounds.

In an assessment of an existing environment, the people who live and work in that environment are an important resource. As well as providing information about what environmental problems exist, their severity, and the appropriateness of the measures being taken to alleviate them, people familiar with the environment can also provide information which reveals problems which are not immediately obvious. Problems which may occur at different times of the day, or only under certain circumstances (such as when an infrequently used piece of equipment is operating) fall into this category. Once the problem has been established, then the relevant measures can be taken.

An example

Suppose you are approached by a client who feels that the average illuminance within an office is too low at the end of the day, and that some of the desks are inadequately lit. You are asked to make an assessment of the environment, and to suggest improvements. Off you go to the office armed with a luminance meter, an illuminance meter, and a tape measure. What do you then do?

Consider this problem in the following four stages:

Average illuminance within the room

The first issue to be considered in determining the average illuminance within the room is 'how many measurement points are needed, and where should the measurements be taken?' The answers to these questions depend upon the size and dimensions of the room, and the accuracy required. The following method will give accuracy to within 10%.

The first step is to determine the dimensions of the room, including the ceiling height, and to sketch a scale plan of a cross-section of the room. The positions and dimensions of the luminaires should be marked on this sketch. Next, the 'Room Index' (an overall size-measure of the room parameters) is found from the ceiling height and the length of one wall. For a square room, it is numerically half the ratio of the wall length to the ceiling height. If the room is almost square, then take the measurement for the largest wall in determining the wall:ceiling ratio. The minimum number of measurement points can then be found from Table 14.2.

If the room cannot be considered as square, an extra calculation is involved. First measure the *smaller* wall length, and the ceiling height. From these figures you can determine the minimum number of points needed within the square defined by the dimension of the smaller wall. This then tells you how far apart the measurements points need to be. By applying this value to the room *as a whole* the required number of measurement points may be calculated. Take the example of a room which is 20 m × 30 m with a 4 m ceiling. If the room were 20 m × 20 m then the wall:ceiling ratio would be 5, and from the above table a minimum of 16 measurement points would be needed. So in the 20 m × 30 m room, which is 1·5 times as large, 24

Table 14.2. Minimum number of measurement points needed for different sized rooms

Ratio wall length : ceiling height	Room index	Minimum number of points
<2	<1	4
2–4	1–2	9
4–6	2–3	16
>6	>3	25

measurement points would be needed to maintain the distance between measurement points.

Greater accuracy than 10% can be achieved by increasing the number of measurement points, and when time allows the more points taken the better. Doubling the number of points halves the error, and accuracy to within 5% can then be claimed.

Once you have determined the number of measurement points, the next step is to divide the plan of the room into a grid consisting of a number of squares each equal in area. If necessary, the number of measurement points should be increased to ensure that the grid is symmetrical. The idea is to take illuminance measurements at the centre of each of these squares. However, before doing this check from the plan that the measurements are not going to be biased by having a disproportionate number of measurements taken directly underneath luminaires. The illuminance measurements should be taken in the horizontal meridian, at around desk height, without shadows being cast on the meter and the values should be recorded directly onto the sketch plan of the office. In this way, you will also have a visual indication of the illuminance variation over the room to use in conjunction with the arithmetic average for the room as a whole.

The desk illumination

Again, it is a good idea to sketch a plan of what is being measured, in this case each desk. While doing this, note down what measurements need to be determined, and under what conditions they need to be taken. One position on the desk may be identified as being the location where the primary tasks are performed, with other areas being either where objects used less are placed, or simply storage space. If objects on the desk are inclined, in the way that a document holder may be, then the horizontal plane may not be the appropriate place to take the measurement and this should be noted on the sketch. Measurements should be taken with any occupants in place, so that any shadows usually cast by the person are present.

As well as the illuminance falling on the desk, the luminance of the various points of interest can be determined, as measured from the occupant's usual position. Knowing the illuminance on an object, and its luminance, the 'luminance factor' may be calculated (see earlier).

What other measurements are needed?

Depending on the circumstances, further measures could be appropriate. In the present example, the luminances and reflectances of the walls, floor and ceiling *should* be determined. Wall reflectance is usually important to the lighting of small rooms, but less so for large rooms where the area immediately adjacent to the wall is the only part of the room affected by light reflected from it. On the other hand, the larger the room the greater the importance of the ceiling reflectance. While a small room can have a satisfactory appearance with a darker ceiling, a large room needs to have a large proportion of the light incident on the ceiling reflected back into the room.

Optionally, a subjective 'expert observer' assessment of discomfort glare could be taken from each workplace; the relative daylight factor could be measured at each grid point; the contrast rendering factor could be established; measurements could be repeated to determine the variation over the day, and so on. While these measures might all be outside the original brief, where time and enthusiasm permits, they enable a much more complete description of the workplace to be made and provide a sounder basis on which to make recommendations about appropriate changes.

Assessment of the environment

Now that the measurements have been taken, the values obtained can be compared with values known to represent good practice, as detailed in the CIBS code. Here the recommendation for standard service illuminance (the mean illuminance over the lifetime of the lamps) for a general office is 500 lux, with a minimum of 200 lux for any working surface. The code also gives recommendations which are applicable here for illuminance ratios:

(a) The ratio of the minimum illuminance to the average illuminance over the task area should not be less than 0·8. While a greater degree of non-uniformity is acceptable between task areas and areas of the room where no tasks are performed, the illuminance on areas adjacent to the tasks should still not be less than 0·3 of the task illuminance.

(b) In an interior with general lighting, the ratio of the average illuminance on the ceiling to the average illuminance on the horizontal working plane should be within the range 0·3 to 0·9.

(c) In an interior with general lighting, the ratio of the average illuminance of any wall to the averge illuminance on the horizontal working plane should be within the range 0·5 to 0·8.

(d) In an interior with localized or local lighting, the ratio of the illuminance on the task area to the illuminance around the task should be not more than 3:1.

By comparing the measured values with the above recommendations, the extent of any problem present in the office may be evaluated, and remedial action may be suggested. Here the extra measurements suggested may be helpful in differentiating between the available solutions. For example, if the wall and ceiling reflectances are acceptable, then the solution may lie in modification of the luminaires, whereas if these reflectances are low the situation could be improved by painting the walls and ceiling.

This example is, of necessity, short and addresses a specific and relatively standard issue. Assessment of other visual environments will involve different circumstances and problems. For example, the criteria by which the design of visual aspects of a car interior is judged will be rather different from those applicable in an office. With this in mind, the next section contains a checklist to indicate the range of problems which might be present in *any* visual environment, and the questions that may need to be asked.

Checklist

What are the problems?
What measures are needed?
What guidelines are available?

Consider each of the following if appropriate:
Light levels
Luminance and illuminance:
  Where is the light needed?
  What variation is there?
    across the room?
    across the workplace?
  Is supplementary light needed anywhere?
    at any particular time of the day?
    at any particular time of the year?
    for particular purposes related to the immediate task?
    for purposes unrelated to the tasks, e.g. safety lighting.
Surfaces
What are the reflectances of the various surfaces?
  the walls,
  the ceiling,
  the floor.
What are the reflectance ratios between them?
Glare
  Discomfort glare:
    Is it a problem?
    Subjective or objective assessment needed?
    Can it be relieved by modifying the environment?
  Disability glare:
    Is it affecting performance?
    Can it be relieved by moving or shielding the lights?

Temporal aspects
  Is flicker apparent:
    in the visual task itself, e.g. a VDU?
    in the environment, e.g. from fluorescent tubes?
Chromatic considerations
  Is colour rendering of concern?
    Is the colour rendering of the lights acceptable?
    Is the colour appearance of the lights acceptable?
  Is colour discrimination a factor for concern?
    How good is the colour rendition of the present lights?
    Is supplementary light with better colour rendition needed for the task,
    to enhance colour discrimination?
    Is colour coding present, and if so are the lights adequate for the colours
    to be discriminated?
Spatial considerations
  Directionality:
    Is there a specific need for directional lighting?
    What variation is there over the room?
    What variation is there over the day? How is the daylight supplemented
    by artificial light, and how does this vary over the day?
    Are shadows a problem under working conditions?
  Highlighting:
    Are there any features that need special consideration to increase their
    conspicuity?
    Are there any special-purpose needs, such as safety lighting with a back-
    up power supply?
Users
  Are there particular user groups with specific requirements?
Non-visual considerations of the visual environment
  How can the environment be maintained under the desired conditions:
  what maintenance is needed?
  What non-visual effects occur, e.g. heat from luminaires?

References

Boyce, P.R. (1988). *Human Factors in Lighting* (London: Macmillan).
CIBS (1984). *Code for Interior Lighting*. London: Chartered Institution of
    Building Services.
Hopkinson, R.G. and Collins, J.B. (1970). *The Ergonomics of Lighting* (London:
    Macdonald).
Wilkins, A.J., Nimmo-Smith, I., Slater, I.A. and Bedocs, L. (1988).
    Fluorescent Lighting, headaches and eye-strain. Presented at the *National
    Lighting Conference*, Cambridge, March 1988.

Further information

Boyce, P.R. (1988). *See references* CIBS publications: These cover a wide variety of issues, e.g. glare, lighting for VDUs, lighting guides for various applications such as engineering, libraries, hospitals. Information can be obtained from: Chartered Institution of Building Services, 222 Balham High Road, London SW12 9BS.

CIE (Commission Internationale de l'Eclairage) publications: These also cover a wide range of issues, such as colour rendering, light measurement, and visual performance. Information on their availability can be obtained from lighting organisations of most countries (e.g. CIBS, IES).

Cronly-Dillon, J., Rosen, E.S. and Marshall, J. (1985). *Hazards of Light: Myths and Realities, Eye and Skin* (Oxford: Pergamon Press).

Egan, M.D. (1983). *Concepts in Architectural Lighting* (New York: McGraw Hill). A comprehensive book covering vision as well as lighting in an easily understood volume.

Galer, I. (Ed) (1987). *Applied Ergonomics Handbook: Chapter 9, The Environment—Vision and Lighting* (London: Butterworths). Although short, this is a useful reference chapter dealing with the principles of good lighting, and practical design considerations.

IES Lighting Handbook. (New York: Illuminating Engineering Society of North America). This handbook comes in two volumes, a Reference volume (1984) and an Applications volume (1987). The Society can be contacted at: Illuminating Engineering Society of North America, 345 East 7th Street, New York, NY 10017.

Interior Lighting Design Handbook (London: The Lighting Industry Federation). The Federation is a useful source of current information, and may be contacted at 207 Balham High Road, London SW17 7BQ.

Pritchard, D.C. (1985). *Lighting*, 3rd edition (London and New York: Longman). This inexpensive paperback is fairly up to date, and is highly recommended.

Overington, I. (1976). *Vision and Acquisition* (London: Pentech Press). A reference book about visual performance. Not easy reading, and no longer completely up to date, but a comprehensive collection of information.

Wyszecki, G. and Stiles, W.S. (1982). *Color Science: Concepts and Methods, Quantitative Data and Formulae,* 2nd Edition (New York: John Wiley). A standard reference work on colour: comprehensive, but not easy reading.

As well as the above sources, information can be obtained from the bodies which produce standards in individual countries (e.g. British Standards, American ANSI standards, German DIN standards). Also, International Standards have been established in many areas by the International Standards Organisation (ISO).

Chapter 15

Human response to thermal environments: principles and methods

Ken Parsons

Introduction

An integral part of applied ergonomics methodology is to consider how people will be affected by their environment. Traditionally, visual, acoustic (noise and vibration) and thermal environments are considered as well as spatial, psychosocial and chemical (e.g. air quality) environments. It is usual to consider effects in terms of separate components; however, it should always be remembered that people work in 'total' environments and interactions of environmental components may be important in some applications. Environmental effects are often considered in terms of workers' health, comfort and performance and this can contribute to human systems design and analysis in terms of overall system performance, worker safety, workstation design and so on.

Thermal environments can be divided conveniently into hot, neutral (or moderate) and cold conditions. Applied ergonomics methods for assessing thermal environments include objective methods, subjective methods and methods using mathematical (usually computer) models. Objective methods include measuring the physiological response of people to the environment. Responses in terms of sweat rate, internal body temperature, skin temperatures and heart rate are useful measures of body strain. Performance measures at simulated or actual tasks can also be useful. Subjective measures are particularly helpful when assessing psychological factors such as thermal comfort and satisfaction. They can also be useful in quantifying the effects of moderate cold or moderate heat stress. Mathematical models have become popular in recent years because, although often complex, they can be easily used in practical applications, employing digital computers. Some of the more sophisticated rational (or causal) models involve an analysis of the heat exchange between people and their environment and also include dynamic models of the human thermoregulatory system. Empirical models can provide useful mathematical equations which 'fit' data obtained from exposing human subjects to thermal conditions.

The aim of this chapter is to present the principles behind practical methods for assessing human response to hot, moderate and cold environments and to present a practical approach to assessing thermal environments with respect to human occupancy.

The principles

The following brief discussion provides the underlying principles behind assessing human response to thermal environments. For a fuller discussion and references the reader is referred to a standard text such as McIntyre (1980).

Relevant measures

It is now generally accepted that there are six important factors that affect how people respond to thermal environments. These are air temperature, air velocity, radiant temperature, humidity and the clothing worn by, and the activity of, the human occupants of the environment. In any practical assessment, instruments and methods for quantifying these factors may be used.

Thermoregulation

People are homeotherms, that is they react to thermal environmental stimuli in a manner which attempts to preserve their internal body ('core') temperature within an optimal range (around 37°C). If the body becomes too hot, vasodilation (blood vessel expansion) allows blood to flow to the skin surface (body 'shell') providing greater heat loss. If vasodilation is an insufficient measure for maintenance of the internal body temperature then sweating occurs resulting in increased heat loss by evaporation. If the body loses heat then vasoconstriction reduces blood flow to the skin surface and hence reduces heat loss to the environment. Shivering will increase metabolic heat production and can help maintain internal body temperature.

The physiological reaction of the body to thermal stress can have practical consequences. A rise or fall in internal body temperature can lead to confusion, collapse and even death in humans. Vasoconstriction can lead to a reduction in skin temperature and complaints of cold discomfort and a drop in manual performance. Sweating can cause 'stickiness' and warmth discomfort and 'mild heat' can provide a drop in arousal, and so on.

Thermal Indices

A useful tool for describing, designing and assessing thermal environments is the thermal index. The principle is that factors that influence human response to thermal environments are integrated to provide a single index value. The aim is that the single value varies as human response varies and can be used to predict the effects of the environment. A thermal comfort index for example would provide a single number which is related to the thermal comfort of the occupants of an environment. It may be that two different thermal environments (i.e. with different combinations of various factors such as air temperature, air velocity, humidity and activity of the occupants) have the same thermal comfort index value. Although they are different environments, for an ideal index identical index values would produce identical thermal comfort responses of the occupants. Hence environments can be designed and compared using the comfort index.

A useful idea is that of the standard environment. Here the thermal index is the temperature of a standard environment that would provide the 'equivalent effect' on a subject as would the actual environment. Methods of determining equivalent effect have been developed. One of the first indices using this approach was the effective temperature (ET) index (Houghten and Yaglou, 1923). The ET index was in effect the temperature of a standard environment (air temperature equal to radiant temperature, still air, 100% relative humidity for the activity and clothing of interest) which would provide the same sensation of warmth or cold felt by the human body as would the actual environment under consideration.

Heat balance

The principle of heat balance has been used widely in methods for assessing human responses to hot, neutral and cold environments. If a body is to remain at a constant temperature then the heat inputs to the body are balanced by the heat outputs. Heat transfer can take place by conduction (H), convection (C), radiation (R) and evaporation (E). In the case of the human body an additional heat input to the system is the metabolic heat production (M) due to the burning of oxygen by the body. Using the above, the following body heat equation can be proposed:-

$$M \pm C \pm R \pm H - E = S \tag{1}$$

If the net heat storage (S) is zero then the body can be said to be in heat balance and hence internal body temperature can be maintained. The analysis requires the values represented in equation (1) to be calculated from a knowledge of, for example, the physical environment, clothing and activity.

Rational indices

Rational thermal indices use heat transfer equations (and sometimes mathematical representations of the human thermoregulatory system) to 'predict' human response to thermal environments. In hot environments the heat balance equation (1) can be rearranged to provide the required evaporation rate (E_{req}) for heat balance $(S = 0)$ to be achieved, e.g.

$$E_{req} = (M - W) + C + R. \tag{2}$$

(H can often be ignored and W is the amount of metabolic energy that produces physical work.) Because sweating is the body's major method of control against heat stress, E_{req} provides a good heat stress index. A useful index related to this is to determine how wet the skin is; this is termed skin wettedness (w) where:

$$w = \frac{E}{E_{max}} = \frac{\text{actual evaporation rate}}{\text{maximum evaporation rate}}. \tag{3}$$
$$\text{possible in that environment}$$

In cold environments the clothing insulation required (I_{req}) for heat balance can be a useful cold stress index based upon heat transfer equations.

Heat balance is not a sufficient condition for thermal comfort. In warm environments sweating (or skin wettedness) must be within limits for thermal comfort and in cold environments skin temperature must be within limits for thermal comfort. Rational predictions of the body's physiological state can be used with empirical equations which relate skin temperature, sweat rate and skin wettedness to comfort.

Empirical indices

Empirical thermal indices are based upon data collected from human subjects who have been exposed to a range of environmental conditions. In hot environments curves can be 'fitted' to sweat rates measured on individuals exposed to a range of hot conditions. There has been little research of this kind for cold conditions, however a wind chill index was developed based upon the cooling of cylinders of water in outdoor conditions. Wind chill provides the 'trade off' between air temperature and air velocity. Comfort indices have also been developed entirely empirically from subjective assessments over a range of environmental conditions.

Direct indices

Direct indices are measurements taken on a simple instrument which responds to environmental components similar to those to which humans respond.

For example a wet, black globe with a thermometer placed at its centre will respond to air temperature, radiant temperature, air velocity and humidity. The temperature of the globe will therefore provide a simple thermal index which, with experience of use, can provide a method of assessment of hot environments. Other instruments of this type include the temperature of a heated ellipse and the integrated value of wet bulb temperature, air temperature and black globe temperature (WBGT).

Measuring instruments

Air temperature is traditionally measured using a mercury in glass thermometer, although more recently thermocouples and thermistors have been used. An advantage of electronic instrumentation is that values can be continuously recorded and fed into digital computers for later analysis. The dry bulb of a whirling hygrometer gives a value of air temperature. If there is a large radiant heat component in the environment then it will be necessary to shield the air temperature transducer (e.g. using a wide mouthed vacuum flask). Air humidity can be found from the wet and dry bulb of a whirling hygrometer. Other methods include capacitance devices and hair hygrometers.

Radiant temperature is usually quantified in the first analysis by measuring black globe (usually 150 mm diameter) temperature. Correcting the globe temperature for air temperature and air velocity allows a calculation of mean radiant temperature. If more detailed analysis is required then instruments for measuring plane radiant temperatures in different directions should be used. Correction factors may be necessary to allow for the shape of the human body; however use of a globe thermometer provides a satisfactory initial measurement method.

Air velocity should be measured down to about 0.1 m s^{-1} in indoor environments. Generally cup or vein anemometers (i.e. masses rotated by moving air) will not measure down to such low air speeds. Suitable instruments are hot wire anemometers, where the cooling power of moving air over a hot wire, corrected for air temperature, provides air velocity, or Kata thermometers, where air movement cools a thermometer and cooling time is related to air velocity. Kata thermometers are however cumbersome and time consuming to use in practical applications.

In more recent years integrating systems have been developed which detect all four environmental parameters and integrate measurements into thermal indices which predict, for example, thermal comfort. The instruments are usually simple to use and provide practical solutions for the non-expert. They are often relatively expensive however. Another development has been the use of transducers connected to digital storage devices, to allow recording of environmental conditions over long periods of time. The devices usually allow easy interfacing with digital computers where sophisticated analysis can be performed.

Subjective methods

Subjective methods range from simple thermal sensation votes to more complex techniques where semantic or cognitive models of human perception of thermal environments can be determined. In a simple practical assessment of thermal environments two types of scale are generally used. One type is concerned with thermal sensation and the other is concerned with acceptability (i.e. a value judgement). Specific questions regarding general satisfaction, draughts, dryness and open-ended questions asking for other comments provide a useful brief subjective form, especially for measuring the acceptability and comfort of an environment. There are biases and errors which can occur in taking subjective measures, but it should not be forgotten that the best judges of their thermal comfort are the human occupants themselves. (See the chapter on indirect observation by Sinclair.)

Human performance

Despite a number of studies having been carried out on the effects of thermal environments on human performance there is no specific information of practical value. Human performance can be considered in terms of physical (e.g. manual dexterity) and psychological (e.g. behavioural, cognitive) effects. If heat or cold stress is sufficiently severe that internal temperatures pass beyond limits at which major physiological effects occur (e.g. causing collapse or hallucinations) then clearly performance will be impaired. Within such limits, effects are influenced by factors such as motivation, level of proficiency at the task and individual differences.

The state of practical knowledge is such that it is not yet possible to predict reliably effects on manual or cognitive performance in hot environments or effects on cognitive performance in cold environments. Major effects do occur, however, on manual dexterity, cutaneous sensitivity and strength in cold environments. Hand skin temperature can be used to predict effects and it is generally agreed that to maintain manual dexterity the hands should be kept warm.

Clothing

Clothing can be worn for protection against environmental hazards and for aesthetic reasons as well as for thermal insulation. In thermal terms a microclimate is produced between the human body and the clothing surface. The applied ergonomist should ensure that the microclimate allows the body to achieve desirable physiological and psychophysical objectives. Physiological objectives may include the maintenance of heat balance for the body, and the presesrvation of skin temperatures and sweating at levels which allow for comfort. An interaction between thermal aspects and material type should also be considered (e.g. the effect of 'scratchy' materials being exacerbated by sweating).

The dry thermal insulation of clothing is greatly affected by how much air is trapped within clothing layers as well as within clothing. There are two common units which quantify dry clothing insulation; these are the CLO and the TOG. The CLO value is a clothing insulation value which is intended to be a 'relative unit', compared to a normal everyday costume necessary for thermal comfort in an indoor environment. For example, a 'typical' business suit (including underclothes, shirt, etc.) is often quoted as having a thermal insulation of 1·0 CLO. A nude person has zero CLO. In terms of thermal insulation 1 CLO is said to have a value of 0·155 m^2 °C W^{-1}. The TOG value is a unit of thermal resistance and is a property of the material. It can be measured on a heated flat plate, for example, in terms of heat transfer. It does not necessarily relate to the thermal insulation provided to a clothed person. For comparison, 1 TOG is equal to 0·1 m^2 °C W^{-1}. When clothing becomes wet, due to sweating or external conditions, then the clothing insulation is altered (usually greatly reduced). There are methods which estimate the thermal properties of wet clothing; however, these are crude and thermal insulation values for wet clothing are not well documented. Additional factors which can greatly affect the thermal insulation of clothing include pumping effects due to body movement, ventilation and wind penetration.

In application it is not usually necessary to have a detailed description of clothing insulation properties. The important point is whether the clothing achieves its objectives. The objectives for clothing will be determined in an overall ergonomics system analysis, involving a description of the objectives of the organization, task analysis, allocation of functions, job design and so on. This will include the objectives in terms of thermal insulation which will also involve the design of the clothing [to include pockets to keep hands warm, devices to keep workers cool in hot environments (e.g. ice jackets) etc.].

User tests and trials (described generally in chapter 10 by McClelland) will provide important information about whether clothing meets its objectives. Objective measures such as sweat loss, skin temperatures and internal body temperatures or subjective measures of thermal sensation, comfort or stickiness can be of great value. Performance measures at actual or simulated tasks can also be used to evaluate whether clothing has achieved its thermal and other objectives.

Safe surface temperatures

The applied ergonomist is also interested in the effect there will be on the body from physical contact between the human skin and surfaces in a workplace; for example, what sensation is caused by bare feet on a 'cold' floor or whether brief contact with a domestic product (e.g. a cooker, oven door, knobs or a kettle) will result in pain or burns. There are many factors involved in determining human response. These include the type and duration of contact, the material and condition of the surface and the condition of the human skin. However, a simple model based upon the heat transfer between two semi-infinite slabs of material in perfect thermal contact can provide a practical method of assessment.

The principal point is that there is a contact temperature which exists between the skin and the material surface which is dependent upon the physical properties of the skin and the material and which will influence the effect on the person. For example, if one touched a metal slab at 100°C then the contact temperature would be of the order of 98°C and the metal would be felt to be extremely hot. If, however, one touched a cork slab at 100°C then the contact temperature would be around 46°C and the cork would feel much less hot than the metal. Contact temperature can therefore be used to predict effects on the body, and can be calculated from the following equation:

$$T_{con} = (b_1 T_1 + b_2 T_2)/(b_1 + b_2) \qquad (4)$$

where T_1 and T_2 are the initial surface temperatures (°C); T_{con} is the contact temperature (°C); and b_1 and b_2 are thermal penetration coefficients calculated from the following equation:

$$b = (K\rho c)^{1/2} \, JS^{-1/2} \, m^{-2} \, °C^{-1}$$

where K is the thermal conductivity; ρ is density; and c is specific heat.
To calculate T_{con} the values of b for human skin and for different materials are required. These are provided in Table 15.1.

McIntyre (1980) argues that if one assumes a skin temperature of 34°C and a simple reaction time for an individual of 0·25 s then, from the work of Bull (1963), whose data suggests a partial burn at a skin temeprature of about 85°C for 0·25 s contact, one can estimate temperatures which would produce partial burns. Some of these temperatures are provided in Table 15.1 (from McIntyre, 1980).

A practical approach to providing maximum surface temperatures for heated domestic equipment is provided in British Standard 4086 (BSI, 1966).

Table 15.1. Thermal penetration coefficients (b)

Material	b $(JS^{-1/2} \, M^{-2} \, °C^{-1})$	Momentary contact surface temperature for burn threshold (°C)
Human skin	1000	
Foam	30	—
Cork	140	450
Wood	500	187
Brick	1000	136
Glass	1400	121
Metals	>10000	90

The Standard considers three categories of material type and three types of contact duration. A summary of the limits is provided in Table 15.2.

Because of the complexity of the problem and ethical considerations regarding experimentation involving pain and burns on human subjects, knowledge is incomplete in this area. In addition, specifications for safety of manufacturing products often involves other costs and benefits to be considered (see Simpson and Mason in this book). The limits provided in BS 4086 (BSI, 1966) have recently been reviewed and an additional document, BS PD 6504 (BSI, 1983), has been produced which provides background medical information and data in terms of discomfort, pain and burns. There is, however, little information concerning the inter- and intrasubject variation in response, effects of skin condition and effects on different populations (e.g. the aged, children) and other variables important for practical application.

Thermal models and expert systems

The increase in knowledge of the human response to thermal environments and methods of modelling the response, and the development of the digital computer, have allowed complex but useful methods to be easily used in practical application. Practical models which simulate how people respond to hot, moderate or cold environments can be used to assess and design thermal environments. They can also be integrated into larger computer based expert and knowledge based systems for use by the ergonomics practitioner.

Examples of models which simulate the human response to thermal environments are provided by Haslam and Parsons (1987). Models of the human thermoregulatory system controlling a passive body (e.g. made up of cylinders and a sphere with thermal properties similar to those of the human body) can be used to predict changes in temperature within and over different parts of a clothed body. These models can then simulate how persons could respond in terms of heat stress or cold stress in outdoor environments or in terms of thermal comfort indoors. Investigations of the nature of expertise used in assessing the human response to thermal environments, coupled with the requirements of ergonomics practitioners and simulations using such models as are described here, can provide the

Table 15.2. Maximum surface temperatures (°C) for heated domestic equipment (BSI, 1966)

	Handles (Kettles, pans, etc)	Knobs (not gripped)	Momentary Contact
Metals	55	60	105
Vitreous enamelled steel and similar surfaces	65	70	120
Plastics, rubber or wood	75	85	125

input to expert systems. An example of the structure of such a system as described by Smith and Parsons (1987) is provided in Figure 15.1. It is probable that such systems will become valuable tools in integrating the principles and knowledge involved in the assessment of human response to thermal environments for practical application. Techniques for the elicitation of knowledge in general for expert systems are discussed by Shadbolt and Burton elsewhere in this book.

The practice

Despite the lack of some information the principles mentioned above can provide a practical methodology which can be used to assess thermal environments with respect to effects on their human occupants. Almost by definition practical assessments will have factors specific to particular applications and one universal method is therefore difficult to provide. A general description of methods for assessing extreme environments is given later. More detailed guidance is also provided for the assessment of moderate environments.

Practical assessment of hot environments

The assessment of hot environments is particularly important as danger to health can occur rapidly. For environments where experience has been gained in monitoring workers, working practices can be developed and conditions monitored using simple thermal indices [e.g. wet bulb globe temperature (WBGT) index, wet globe temperature (WGT)].

The WBGT index is used in ISO 7243 (ISO, 1982) as a simple method for assessing hot environments. For conditions inside buildings and outside buildings without solar load:

$$WBGT = 0.7\,t_{nw} + 0.3\,t_g \tag{5}$$

and outside buildings with solar load:

$$WBGT = 0.7\,t_{nw} + 0.2\,t_g + 0.1\,t_a \tag{6}$$

where t_{nw} is the natural wet bulb temperature (i.e. not 'whirled'); t_g is 150 mm diameter black globe temperature; and t_a is the air temperature.

Acclimatization programmes, before workers begin work, are useful. It is particularly important that workers do not become unacceptably dehydrated (e.g. greater than 4% of body weight lost in sweat) or have an unacceptably elevated internal body temperature (e.g. greater than 38.0–38.5°C). A more detailed analysis can be provided by using rational assessments of the environment. Allowable exposure times based on such factors as predicted

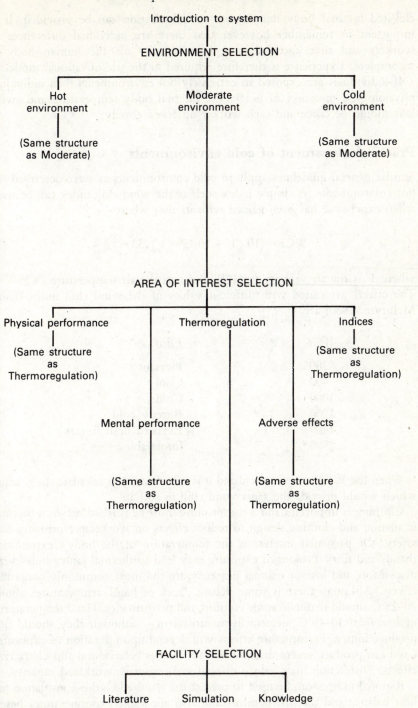

Figure 15.1. System tree diagram.

elevated internal body temperature or dehydration can be provided. It is important to remember however that there are individual differences in workers and that knowledge of heat transfer for the human body is incomplete. Experience is therefore required in the use of rational models.

If individuals are exposed to extremely hot environments then individual physiological measures of heart rate, internal body temperature and sweat loss should be taken and each worker observed closely.

Practical assessment of cold environments

Similar general guidelines apply to cold environments as were described for hot environments. A simple index such as the wind chill index can be used when experience has been gained with its use, where:

$$WCI = (10 \sqrt{V} + 10\cdot45 - V)(33 - t_a) \qquad (7)$$

where V is the air velocity (m s^{-1}); and t_a is the air temperature (°C). The effects associated with different values of the wind chill index (from McIntyre, 1980) are:

WCI	Effect
200	Pleasant
400	Cool
1000	Cold
1200	Bitterly cold
1400	Exposed flesh freezes
2500	Intolerable

When the WCI value is calculated it is often useful to calculate the t_a value which would provide the same wind chill in still air.

Clothing is important and a compromise must be reached between thermal insulation and clothing design to reduce effects on worker performance and safety. Of particular interest is the temperature of the body's extremities (hands and feet). Prolonged exposure may lead to thermal injury but severe discomfort and loss of manual dexterity are the most commonly occurring effects. Although there is some debate, 'back of hand' temperatures above 20–25°C should maintain some comfort and performance. Hand temperatures of less than 10–15°C are usually unsatisfactory, although they should not produce injury. Performance effects will depend upon duration of exposure. Cold can produce severe discomfort and this has behavioural and distractive effects. Distraction may reduce manual and cognitive workload capacity.

Rational indices can be used to predict the required clothing insulation for heat balance and thermal comfort and also allowable exposure times based upon a drop in body 'core' temperature. Physiological measures of heart rate, mean skin temperature and body core temperature should be used if assessing individuals in cold environments. Any drop in body core temperature is unsatisfactory, but 36°C is a lower working limit. A core temperature of below 35°C is defined as hypothermia. Medical screening of subjects should take place before exposure to either hot or cold environments.

Practical assessment of moderate environments

A common request to the environmental ergonomist is to assess an indoor climate such as found in an office. The practical method used in an actual case is outlined here, although some adjustment to the results has been made to illustrate points.

The workers in a large office were complaining that their thermal environment was unacceptable. The ergonomist was asked to assess the environment, quantify the problem and make recommendations for improvement if necessary. One day was allowed for the assessment and a total of four days for the whole project, including both analysis and final report, a not unusual time restriction.

Worker relations

Complaints about working environments can be stimulated by other work related problems and it is important for the ergonomist to gain an impression of the physical, social and organizational environment in general. In addition it is useful to have the co-operation and understanding of management and workers. The worker representative was therefore contacted and the ergonomist introduced. It was explained that the ergonomist was attempting to improve the thermal environment conditions. The physical and subjective measures which were to be taken were also demonstrated to the workers' representative who then passed on the information to the office occupants.

Where, when and what to measure

The question of where and when to measure is a question of statistical sampling. The thermal environmental conditions will vary throughout a space and also with time (during the day, night and seasonal variations). The more measuring points in the room and the more measuring times, in general, the more accurately the environment can be quantified. This then is a question of resources. Only one day was allowed for measurement so a plan of the office was obtained and individual workplaces identified. Measurements should be taken at the positions of the workers. Ankle, chest and head heights were chosesn as measuring points at each workplace. Ten workplaces were chosen as the sample, 'evenly spread' throughout the office. Measurements were taken over a 3-h period under what had been established as 'typical' conditions throughout the morning, a time when complaints had been received. The ventilation systems were identified and set to normal working. Outside weather conditions were noted.

A 150 mm diameter globe thermometer was placed at each workplace (only two were available so they had to be moved around) for at least 20 min before readings were taken. Using a hot wire anemometer, air velocity and air temperature were measured at ankle, chest and head height of the worker. A whirling hygrometer was used at chest height to measure wet and dry bulb temperatures (dry bulb was used as a cross check for air temperature with the air temperature sensor on the hot wire anemometer). The workers' clothing and activity were noted, and movements throughout the room were also noted.

Subjective assessment forms (see Appendix) were handed to each worker and collected centrally (i.e. the working position was noted but a degree of anonymity was maintained). The subjective forms allowed some information to be collected regarding time variations (i.e. outside the survey time) and general satisfaction.

Analysis

PHYSICAL MEASURES

Analysis of physical measurements takes place in two parts. The first part is to obtain for each measurement point air temperature, mean radiant temperature, relative humidity and air velocity from the instrument measures and also to determine metabolic heat production and clothing insulation values. The second part is to predict the degree of discomfort. The subjective measures are analysed separately and complement the physical measures.

The air temperature and air velocity were measured directly using the hot-wire anemometer. The mean radiant temperature (t_r) is obtained from globe temperature (t_g) corrected for air temperature (t_a) and air velocity (V). If the mean radiant temperature is within a few degrees of room temperature then McIntyre (1980) suggests that:

$$t_r = t_g + 2\cdot44 \sqrt{V} (t_g - t_a) \tag{8}$$

where temperatures are in °C and air velocity in m s^{-1}. Relative humidity is calculated from the dry bulb (air temperature) and aspirated (whirled) wet bulb of the whirling hygrometer. Table 15.3 provides typical values. Table 15.5 provides metabolic heat production values for typical activities and Table 15.4 provides clothing insulation values for typical clothing. Useful information is provided by presenting all physical data in a table, or on the office plan, in the final report.

PREDICTION OF WHOLE-BODY THERMAL DISCOMFORT

Despite some theoretical limitations, one of the most useful thermal comfort indexes is the predicted mean vote (PMV) of Fanger (1970) which is used in

Table 15.3. Relative humidity (%) from dry bulb and aspirated wet bulb temperature

Dry bulb temperature (°C)	Aspirated wet bulb temperature (°C)									
	12	14	16	18	20	22	24	26	28	30
12	100									
14	79	100								
16	62	81	100							
18	49	64	82	100						
20	37	51	66	83	100					
22	28	40	54	68	83	100				
24	20	31	43	56	69	84	100			
26	14	24	34	45	58	71	85	100		
28	9	18	27	37	48	59	72	85	100	
30	5	12	21	30	39	50	61	73	86	100

Table 15.4. Estimates of typical clothing insulation values (1 CLO = $0.155 \ m^2 \ °C \ W^{-1}$)

Type of clothing	Clothing insulation (CLO)
None	0
Light summer clothing (briefs, shorts, short sleeved shirt, light socks, light shoes)	0.3
Light work clothing (light underwear, cotton long sleeved workshirt, light long trousers, socks, shoes)	0.65
Light business suit (including underclothing etc)	1.0
Heavy business suit (including underclothing etc)	1.5

Table 15.5. Estimates of typical metabolic heat production values

Activity	Metabolic heat production ($W \ m^{-2}$)
Seated, at rest	58
Standing, relaxed	70
Standing, light arm work	100
VDU operation	70
Driving	70–100

ISO 7730 (ISO, 1984). Air temperature, mean radiant temperature, air velocity, humidity, clothing and activity values can be integrated to predict the mean thermal sensation vote of a large group of people on a seven point thermal sensation scale (as used on the subjective assessment form in the Appendix). The values range from PMV = 3 (hot) through PMV = 0 (neutral) to PMV = −3 (cold). PMV = 0 provides comfort conditions. From the PMV value a predicted percentage of dissatisfied (PPD) value can be calculated. This is related to the percentage of people likely to complain about the thermal conditions. Values of PMV and PPD are presented in Tables 15.6 and 15.7 for typical environmental conditions.

The thermal discomfort results for all ten workplaces will not be presented here; however the PMV and PPD values for each workplace were calculated and labelled on a copy of the plan of the office, for the final report. This showed the predicted whole-body thermal sensation (comfort) pattern over the office and those areas of likely complaint.

The following is an example of calculations for one workplace; the physical measurements were: $t_a = 18°C$; $t_r = 18°C$; $V = 0.15$ m s^{-1}; relative humidity = 50%; clothing insulation = 0.65 CLO; metabolic rate = 70 W m^{-2}.

Table 15.6. PMV values for air temperature, clothing and activity (assume: mean radiant temperature = air temperature, air velocity = 0.15 m s^{-1} and relative humidity = 50%)

Clothing (CLO)	Activity (W m^{-2})	Air temperature (°C)						
		16	18	20	22	24	26	28
0.65	58	—	−2.7	−2.0	−1.3	−0.6	0.0	0.8
1.0	58	−2.1	−1.6	−1.1	−0.5	0.0	0.6	1.2
1.5	58	−1.1	−0.7	−0.3	0.2	0.6	1.1	1.5
0.65	70	−2.2	−1.7	−1.2	−0.6	0.0	0.5	1.0
1.0	70	−1.3	−0.9	−0.5	0.0	0.4	0.9	1.3
1.5	70	−0.5	−0.2	0.2	0.5	0.9	1.2	1.6
0.65	100	−0.9	−0.5	−0.1	0.3	0.6	1.0	1.4
1.0	100	−0.3	0.0	0.3	0.6	1.0	1.3	1.6
1.5	100	0.3	0.5	0.7	1.0	1.3	1.5	1.8

Table 15.7. Interpretation of PMV values in terms of thermal sensation and Predicted Percentage Dissatisfaction (PPD)

Sensation	Cold	Cool	Slightly cool	Neutral	Slightly warm	Warm	Hot
PMV	−3	−2	−1	0	1	2	3
PPD (%)	—	75	25	5	25	75	—

Using Tables 15.6 and 15.7

$$PMV = -1.7 \text{ and } PPD = 62\%.$$

Therefore the prediction is that, on average, a person will be between slightly cool and cool at this position. Also it can be seen that for all other conditions remaining the same an increase in air temperature from 18 to 24°C will provide a PMV value of 0 required for comfort. This could be a recommendation, or a recommendation could be made in terms of increased CLO value, etc.

LOCAL THERMAL DISCOMFORT

As well as overall or whole-body thermal sensation thermal conditions can produce effects on local areas of the body. For example cold air moving around the workers ankles may cause a draught. The most common forms of local discomfort are caused by cooling due to air movement, heat losses due to asymmetric radiation (e.g. a radiant draught caused by workers sitting next to cold walls or windows) and thermal gradients. There is some debate about conditions which produce discomfort, but cool air movements (especially, if fluctuating) should be avoided above 0·15 m s^{-1}, and particularly for exposed skin areas and if the subject is already cool. Radiant asymmetry should not exceed 10°C (less in the case of heated ceilings) and vertical temperature gradients should not be greater than 3°C. General observation of the workplaces, air velocity measures at the three heights (ankle, chest and head), and mean radiant temperatures will provide an indication of possible local thermal discomfort.

Dryness is probably related to air velocity, humidity and air and radiant temperatures, and is usually due to the evaporation of fluids from the eyes, nose and mouth which can lead to various problems, for example, with contact lenses. Local discomfort and other factors such as dryness and overall satisfaction should also be examined using subjective methods.

SUBJECTIVE RESPONSES

Analysis of subjective responses involves determining the average of, and variation in, response. The responses of how workers felt at the time of measurement can be compared with predicted responses. In general subjects in the office example used, gave a wider range on the scale than the predicted measures. The subjective measures were also presented on a plan of the office in the final report. On average workers were between slightly cool and cool with some subjects cold and some neutral. Draughts were reported in some areas. Most workers wished to be warmer. Responses regarding general sensation at work were similar to responses made about the conditions when they were measured. Most people were generally dissatisfied with the thermal environment.

Concluding remarks and recommendations

The above measurement and analysis allowed recommendations to be made in a final report which were related to the original objectives. An average increase in air temperature was recommended with some specific recommendations about draughts for particular workstations. It was also noted that the high level of dissatisfaction indicated may be due to general work or workplace dissatisfaction and not simply related to thermal conditions.

References

B.S.I. (1966). *Recommendations for Maximum Surface Temperatures of Heated Domestic Equipment*. (London: British Standards Institution).

B.S.I. (1983). *Medical Information on Human Reaction to Skin Contact with Hot Surfaces*. (London: British Standards Institution).

Bull. J.P. (1963). Burns. *Postgraduate Medical Journal*, **39**, 717–723.

Fanger, P.O. (1970). *Thermal Comfort*. (Copenhagen: Danish Technical Press).

Haslam, R.A. and Parsons, K.C. (1987). A comparison of models for predicting human response to hot and cold environments. *Ergonomics*, **30**, 1599–1614.

Houghten, F.C. and Yaglou, C.P. (1923). Determining equal comfort lines. *Journal of American Society of Heating and Ventilation Engineering*, **29**, 165–176.

I.S.O. (1982). *Hot Environments—Estimation of the Heat Stress on Working Man, Based on the WBGT-index (Wet Bulb Globe Thermometer)*. ISO 7243 (Geneva: International Standards Organisation).

I.S.O. (1984). *Moderate Thermal Environments—Determination of the PMV and PPD Indices and Specification of the Conditions for Thermal Comfort*. ISO 7730 (Geneva: International Standards Organisation).

McIntyre, D.A. (1980). *Indoor Climate*. (London: Applied Science Publishers).

Smith, T.A. and Parsons, K.C. (1987). The design, development and evaluation of a climatic ergonomics knowledge based system. In: *Contemporary Ergonomics 1987*, edited by E.D. Megaw (London: Taylor & Francis), pp. 257–262.

Appendix

The following subjective form was used in a moderate office environment where workers had been complaining about general working conditions. Various details about workers' characteristics and location were collected separately. The form was handed to workers for completion at their workplace. Question 1 determines the workers' sensation vote on the ASHRAE/ISO scale. Note that this can be compared directly with the measured PMV. Question 2 provides an evaluation judgement. For example, question 1 determines subject's sensation (e.g. warm). Question 2 compares this sensation with how the subject would like to be. Questions 3 and 4 provide information about how workers generally find their thermal environment. This is useful where it is not practical to survey the environment for long durations. Questions 5 and 6 are catch-all questions about workers' satisfaction and any other comments. Answers to these questions will provide information about whether more detailed investigation is required. Answers will also indicate factors which are obvious to the workers but not obvious to the investigator.

Please answer the following questions concerned with YOUR THERMAL COMFORT.

1. Indicate on the scale below how you feel NOW.
   Hot
   Warm
   Slightly warm
   Neutral
   Slightly cool
   Cool
   Cold

2. Please indicate how you would like to be NOW
   Warmer   No change   Cooler

3. Please indicate how you GENERALLY feel at work:
   Hot
   Warm
   Slightly warm
   Neutral
   Slightly cool
   Cool
   Cold

4. Please indicate how you would GENERALLY like to be at work:
   Warmer   No change   Cooler

5. Are you generally satisfied with your thermal environment at work?
   Yes   No

6. Please give any additional information or comments which you think are relevant to the assessment of your thermal environment at work (e.g. draughts, dryness, suggested improvements etc.).

Chapter 16

Auditory environment and noise assessment

Christine M. Haslegrave

Introduction

Sound in our environment can be generated by transport of all forms, by equipment and machinery, and by people themselves. In the community this is mainly due to traffic noise, to neighbours, and to radios and other domestic equipment. In the working environment, office noise comes from people, typewriters, printers and telephones. In the manufacturing industry it comes mostly from motors, fans, pumps and compressors, moving machinery, contact between tool and workpiece, and resonant plates or housings. Traffic and machinery can also set up vibrations in the structure of the building. All these have to be considered when identifying the source of noise and investigating ways of reducing sound levels.

It is rare to experience silence in the modern world and our auditory environment is very complex. Sound can bring pleasure and information but unnecessary sound or too much sound is annoying, distracting and possibly harmful. Also, it is not always possible to separate these effects. It is quite possible for a sound to be wanted by one hearer and unwanted by many others, and conflicts may arise between differing interests in both living and working environments. For this reason, we tend to distinguish between sound and noise. Kryter (1985) has defined noise in terms of its effects on people as 'audible acoustic energy that adversely affects the physiological or psychological well-being of people.' As he says, this is consistent with the usual definition of noise as 'unwanted sound'.

Ergonomists therefore need to evaluate auditory environments which may include sounds which are unwanted although they are not so loud as to be harmful, or sounds which are enjoyed by one person while annoying to another. They need to measure noise levels to assess whether there is a problem in terms of safety, working performance, comfort or annoyance.

Further than this, they may need to identify the source and specify whether action or protection is needed.

In order to do this, they must be equipped to measure hearing ability, noise or sound levels of individual sources and of the environment, and to be able to assess the risk of damage to hearing. Some typical applications are:

1. Hearing assessment and screening.
2. Industrial health and safety.
3. Workplace (re)design.
4. Design and evaluation of the effectiveness of communication signals.
5. Design and performance of protective equipment.
6. Community noise assessment.
7. Noise suppression and shielding.
8. Machine design and testing.

Thus, the complexity of our auditory environment, and our highly subjective and personal responses to it, result in an equally wide range of measures which are used in its evaluation. This chapter presents the methodology for measuring sound and outlines some of the range of measures or criteria which have been proposed for assessing noise. The aim is to indicate some of the most useful techniques which are currently used by ergonomists, but without attempting to cover either specialist techniques or the concerns of acoustic engineers who deal with noise control, architectural acoustics or design of communication systems. Similarly there is no intention of covering the physics of sound or physiology of hearing, other than to introduce the necessary concepts and terms.

For a more detailed study of the ergonomic aspects introduced in this chapter the reader can consult texts dealing with hearing, perception of sound and the effects of noise on work and health, such as Jones and Chapman (1984), Kryter (1985), Loeb (1986) or Sanders and McCormick (1987). Guidance on acoustic treatments and techniques for reducing noise levels can be found in publications such as those by Brüel and Kjaer (1987) and the Health and Safety Executive (1983), as well as in noise control handbooks.

Units of measurement of sound

Sound is a variation of pressure in the air (or in any other elastic medium) which the human sense of hearing detects. It is transmitted in the form of pressure waves as a series of compressions and rarefractions travelling outwards from the source of the sound. The speed of the wave depends on the medium, but in air at 20°C sound travels at a velocity of 344 m s^{-1}. The amplitude of the sound pressure wave is the fluctuation above or below the ambient air pressure. Our sensation of loudness however does not come directly from the pressure, but from the intensity of the sound which is the rate at which energy is transmitted by the wave. This is defined in terms of the energy passing through a unit area in unit time or sound power per unit area. Pressure can be measured directly, but it is more difficult to measure sound power. Sound pressure and sound power are however closely related, as will be shown later. It is therefore usual to measure sound levels in terms of sound pressure and to use this to calculate sound power. The units used for noise measurements in fact normally take account of this.

Auditory stimuli at the ear result from the combined signals received from sound sources in the environment. The waveforms (or variation in intensity/ amplitude with time) of typical sound signals are shown in Figure 16.1.

A pure tone (which might be obtained from a tuning fork or a vibrating string) vibrates at a single frequency and can be represented as a sine wave, but most sounds are made up of complex tones containing many frequencies. Complex signals may be analysed into their component sine waves (with characteristic frequency, amplitude and phase) to understand the content of the signal. This is known as frequency or Fourier analysis. A frequency spectrum indicates the principal frequencies contained within the signal, and their relative intensities (energies). When a sound is made up of frequencies covering most of the sound spectrum, it is known as white or broad-band noise.

An impulsive signal (such as a door slamming or a hammer blow) is a single pressure pulse with a very fast rise time (around 35 ms or less) to the initial peak amplitude, followed by small pressure oscillations decaying over about 1 s. The audible range of sound levels is enormous, from a quiet whisper up to the level of a warning siren (which is around the pain threshold), representing a range in power of the sound signal of over one to one billion as shown in Table 16.1. Similarly, the ear is sensitive to a large range of frequencies—the audible range is approximately 20–20000 Hz, with greatest sensitivity between 2–5 kHz. Most speech is between 300 and 700 Hz, with all vowel sounds below 1000 Hz, but sibilant consonants may be higher than 5000 Hz. Low frequency sounds below 20 Hz are usually termed infrasound. The effects of infrasound on human listeners are not yet well understood, and this range is not usually considered when assessing the auditory environment.

Definition of units

Several units are used for sound measurement, and it is helpful to give some brief definitions to show the relationships between the various basic measures. A fuller description of the measures can be found in texts such as Kohler (1984).

Sound is described by intensity and by frequency. (The corresponding sensations in human hearing are termed loudness and pitch.) Intensity or

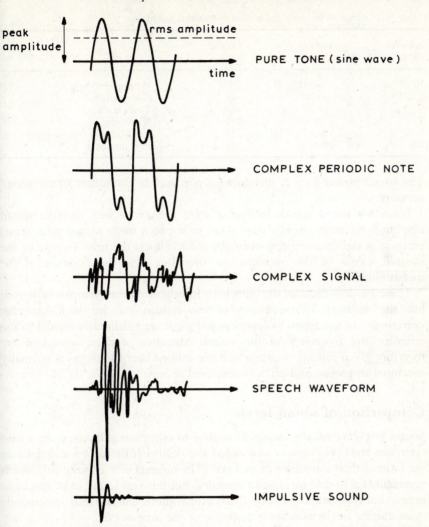

Figure 16.1. Waveforms of various sound signals.

power of the oscillations in the air is measured in watts per metre² (Wm⁻²), while pressure is measured in Newtons per metre² (N m⁻²); frequency is measured in Hertz (Hz). The total sound power of a sound source (such as a machine) can be measured and is expressed in watts (W).

In a sound wave emitted from a source, the power and pressure are related according to the following equation:

$$\text{Power/unit area} = \text{energy flow/unit area} = \frac{\text{pressure}^2}{\text{density of air} \times \text{speed of sound}}.$$

Table 16.1. Range of intensity and pressure of audible sound

Intensity (W m^{-2})	Pressure (N m^{-2})	Decibel level	
10^{-12}	0·00002	0	Hearing threshold
3×10^{-6}	0·04	65	Conversation
10^{-4}	0·2	80	Town traffic
10^{-2}	2	100	Workshop
3	36	125	Jet at take-off (60m away)
100	200	140	Pain threshold

The sound power level is therefore proportional to the square of the sound pressure.

From the sound signals in Figure 16.1, it may be seen that the sound amplitude fluctuates rapidly over time, so the root mean square value (rms) pressure is the measurement normally used. This is the time average of the squared values of the instantaneous pressures over the duration of the measurement.

These are definitions of the units used for physical measurements of sound, but the auditory characteristics of the human ear are such that the corresponding sensations of loudness and pitch are not linearly related to the intensity and frequency of the sound. Measures of these sensations are therefore given separate units (which are defined later): loudness is normally measured in phons, and pitch is measured in mels.

Comparison of sound levels

Sound levels are usually measured relative to other sound levels, or to a base reference level. For this, a unit called the decibel (dB) has been defined as the ratio of their intensities or powers. (The original unit was the bel, which represented a 10-fold increase in intensity, but this was found to be too large in practical applications.) The decibel is a logarithmic unit and thus corresponds quite closely to the non-linear response of the human ear.

The base reference sound intensity is chosen as the power of a standard vibration in the air which is just on the threshold of hearing, or one billionth of a watt per square metre (10^{-12} W m^{-2}). Thus, a sound level of N dB is related to its intensity of I W m^{-2} by

$$N \, dB = 10 \log_{10} \frac{I}{I_0}$$

where I_0 is the reference level intensity at threshold of hearing (10^{-12} W m^{-2}).

Given the relationship between sound intensity and pressure,

$$N \, dB = 10 \log_{10} \frac{I}{I_0} = 10 \log_{10} \frac{p^2}{p_0^2} = 20 \log_{10} \frac{p}{p_0}$$

where p is the sound pressure level and p_0 is the reference level amplitude at threshold of hearing (2×10^{-5} N m^{-2}).

The values of intensity, pressure and decibels for some typical sounds within the audible range are shown in Table 16.1. The relationships between the three scales are shown in Table 16.2. A sound which is 10 times louder than another has an intensity level of 10 dB relative to it. Since the decibel scale is logarithmic, a sound 100 times louder is said to be 20 dB louder. The smallest change in loudness that the human ear can discriminate is 1 dB, but in practice the minimum difference needed to recognize a sound above the background noise level is 3 dB, which represents a doubling in intensity or loudness. A doubling of the sound pressure level is equivalent to an increase of 6 dB in sound level.

Since the various units are related in this way (with loudness or sound power proportional to sound pressure^2) a useful guide to changes in sound levels is given by:

$$3 \text{ dB} = 1{\cdot}4 \times \text{ sound pressure level} = 2 \times \text{ sound power level,}$$
$$6 \text{ dB} = 2 \times \text{ sound pressure level} = 4 \times \text{ sound power level,}$$
$$20 \text{ dB} = 10 \times \text{ sound pressure level} = 100 \times \text{ sound power level.}$$

Effects of several sources of sound

What happens when two sounds are heard at the same time? How are their effects added? Sound levels are not additive when they are measured in logarithmic dB units. If two sounds are received at the ear simultaneously, the resultant is the sum of the energies ($I_1 + I_2$) in the two sounds and so the sound power levels are added:

$$\text{combined sound power level (W m}^{-2}) = I_1 + I_2 .$$

Table 16.2. Decibel scale

Ratio of intensities (powers) of two sounds		Ratio of pressures of two sounds	Relationship in decibels
1		1	0
10		3·16	10
100		10	20
	200		23
	400		26
	600		28
	800		29
1000		31·6	30
10000		100	40
100000		316	50
1000000		1000	60
1		1	0
1/10		1/3	−10
1/100		1/10	−20

Thus:

$$\text{combined sound level (dB)} = 10 \log \frac{I_1 + I_2}{I_0}.$$

The combined loudness of sounds from two or more sources can therefore be calculated from this formula, but a simple rule of thumb (accurate to 0·5 dB) is also given in Table 16.3.

If the two sounds are of equal intensity (difference 0 dB), their combined intensity is 3 dB higher than the intensity of either. If the difference is 5 dB, the combined sound is 1 dB higher than the greater. If the difference is 10 dB, the combined sound is approximately equal to the louder sound and there is no noticeable difference in intensity of the resultant sound. For example:

> two sounds of 75 dB combine to give a sound level of 78 dB,
> two sounds of 75 dB and 77 dB combine to give a sound level of 79 dB,
> two sounds of 75 dB and 85 dB combine to give a sound level of 85 dB.

As a practical consequence of this, the sound of a machine or piece of equipment can be measured even in the presence of background noise, provided that the background noise level is at least 10 dB below the machine noise level. If the background noise level is higher than this, a correction factor for the measured machine noise level can still be calculated as shown previously. It is also useful to note that the ratio of two sound levels is calculated by subtracting the levels in decibels: this gives the signal-to-noise ratio as the difference in decibel levels of the desired signal and the unwanted noise.

Measures of noise

It is obvious from this brief review that measurements of sound or noise levels will be influenced by the temporal, intensity and spatial characteristics of the signals. The temporal characteristics can include both the frequency

Table 16.3. Addition of two sound levels in decibels

Difference (dB) between sounds	Add to higher (dB)
0	3
1	2·5
2	2
4	1·5
6	1
8	0·5
10	0

spectrum and fluctuations in the overall sound level with time. A variety of measures have therefore been developed for specific purposes.

Subjective measures

Since the response characteristics of the ear are non-linear, both frequency and intensity affect our perception of loudness of a sound. We do not perceive a sound arriving at the ear to be equally loud at 20 Hz and at 10 kHz—a sound at 20 Hz will appear very much quieter. Curves of equal subjective loudness can be plotted, giving the sound levels which provide constant perceived loudness at various frequencies. These indicate the combinations which appear equally loud to a human listener. This shows, for example, that a 50 Hz tone must have a loudness of about 85 dB to give the same subjective loudness as a 1000 Hz tone at 50 dB.

Subjective loudness is measured in a unit called a phon. This unit is equivalent to the decibel at 1000 Hz and is also logarithmic. While the phon measures the subjective equality of sounds, a unit called the sone was defined to measure the relative loudness of sounds. One sone is defined as the loudness of a 1000 Hz tone of 40 dB (40 phons). A sound of 2 sones is one judged to be twice as loud. The phon and sone scales are related by the following formula:

$$phons = 40 + 10 \log_2 sones.$$

Every increase of 10 phons then doubles the loudness in sones, so that (for example) 50 phons is equivalent to 2 sones.

There is also a subjective measure of pitch, which corresponds to frequency for a pure sinusoidal tone and to the fundamental frequency for complex waveforms. Perceived pitch is given a unit called a mel—defined by a pure tone of frequency 1000 Hz at a sound pressure level of 60 dB, which is said to have a pitch of 1000 mels. Any two tones separated by a given number of mels appear equally far apart in pitch, regardless of their frequency.

Weighted measures

In view of the non-linearity of hearing response, most instruments are designed to measure sound levels on a decibel scale which is weighted to match the characteristic of the ear. Several scales have been developed for different purposes and their response characteristics have been standardized internationally. The A, B, and C-weighted scales were developed to match the responses for sounds of low, moderate and high intensity. The most commonly used is the dBA scale, which gives the best correlation with subjective tests of perceived loudness, and also with ratings of noise annoyance. The A-scale was in fact designed to match the 40 phon equal loudness contour.

This means that the actual measurement of the sound pressure level is converted to a weighted dBA sound level, in accordance with the response characteristic shown in Figure 16.2 which takes account of the middle range of frequencies to which the human ear is most sensitive. The other scales shown in Figure 16.2 are less commonly used. The B-scale was designed to match the equal loudness contour at 70 dB, and the C-scale for a flat rating across the frequency range. There are also D-weighted scales which were designed as equal noisiness scales for assessing aircraft noise. The unweighted sound levels are normally only used in frequency analysis.

Empirical measures

In most everyday situations noise contains sounds from various sources, of differing frequencies and intensities, and also extending over different periods of time. In order to describe the noise environment, various statistical distribution measures are used. These are based on noise levels measured on the dBA scale. The most important are defined as follows:

Equivalent level of sustained noise (L_{eq}). This is the average level of sound energy over a given period of time, which integrates all the fluctuating noises to represent them as an average steady level. It therefore takes account of short but high peaks.

Median noise level (L_{50}). This is the noise level which is exceeded for 50% of the time period.

Background noise level (L_{90}). This is the 10th percentile level—the level that is exceeded for 90% of the time.

Peak noise level (L_1 or L_{10}). This is the 99th or 90th percentile level—the level that is exceeded for 1% or 10% of the time.

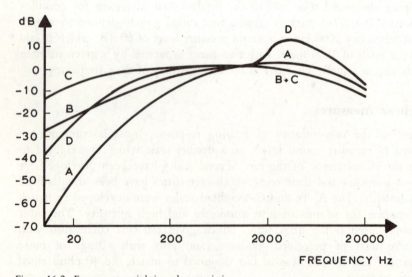

Figure 16.2. Frequency weighting characteristics.

All these indices can be used to give a measure of the total environment in, say, an office or a factory. The last three give a feel for the range of the noise—the 'average' level, the low background level and the highest levels heard over a period of time. They are measured by taking a continuous recording of the dBA level over a known time period and then performing a statistical analysis of the record.

Day–night equivalent level (L_{dn}). This is a 24-h L_{eq} used for community noise exposure, where the value of L_{eq} is measured over 24 h but the readings between 22.00 and 06.00 hours are increased by 10 dB.

Sound exposure level (SEL or L_{AE}). This is useful for comparing unrelated noise events in terms of their total acoustic energies: the energy is integrated over the duration of the event and expressed as the equivalent level over 1 s. It is often used to describe the noise energy of a single event such as a passing car.

Choice of instrumentation

In order to measure the auditory environment instruments are needed to measure sound levels within the range of 0–150 dB and over the frequency range up to 20 kHz. A variety of types of instruments are used and the more generally used are described briefly below. Further information on measurement, calibration and analysis techniques can be found by consulting manufacturers' handbooks or by reference to texts such as Kohler (1984), Hassall and Zaveri (1979) or Peterson and Gross (1978).

Sound level meters

Sound level meters are the most commonly used instruments for measuring the acoustic environment and most general purpose meters contain a variety of signal measures which are suitable for different applications. The basic measuring system is shown diagrammatically in Figure 16.3. At its simplest, the sound level meter consists of a microphone which converts pressure variations into electrical signals. These are then amplified by an electronic network and displayed in some form, either on a digital or analogue instrument, or as a continuous reading which is recorded on tape, computer or hard copy. The sound level displayed is usually the instantaneous rms value of the signal in decibels.

Sound pressure level (SPL) meters normally have an appropriate frequency weighting network (with a response matching the A, B, C or D-scale) in addition to providing the unweighted (linear) decibel level. When used with the A-weighted network, the response of the instrument is similar to that of the human ear. Portable filter sets giving octave band or 1/3 octave band frequency analysis are commonly used in conjunction with an SPL meter to give the frequency content of the sound. These record the sound level in

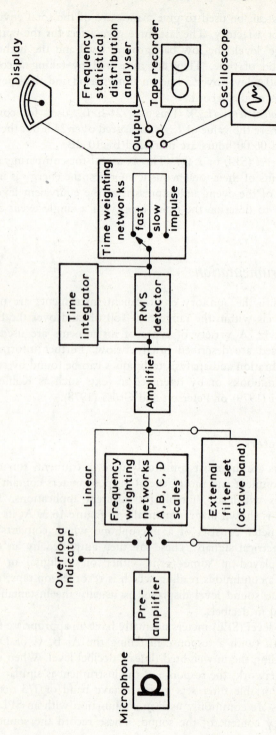

Figure 16.3. Diagram of a typical sound level meter.

each filter band. Figure 16.4 shows a sound level meter in use coupled to a filter set. In laboratory applications, wave analysers may be used for more detailed narrow band frequency analysis.

Integrating sound level meters are capable of measuring over a longer time period and provide the average sound level and L_{eq} value. These integrate and average the sound over a period determined by the response characteristic of the instrument: for 200 ms on the FAST setting or 500 ms on the SLOW setting. The SLOW setting allows the user to read the overall sound level displayed even when the signal is fluctuating rapidly. The measured level will depend on the time weighting setting, which should of course be appropriate to the type of source and purpose of the measurement. Additional analysis of the temporal characteristics over a longer period of time may require an external statistical distribution analyser. The noise signals may also be recorded on tape for further analysis at a later date, but in this case a reference signal level must be recorded for calibration during analysis.

Impulsive sounds cannot be measured accurately on most normal SPL meters because the response time of the instrument is not sufficiently fast. On some meters a 'hold' facility permits a display of either peak level or maximum rms level, which is useful for measuring impulsive noise. Some meters may have impulse time weighting with a time constant of 35 ms, instead of peak hold. Detailed analysis of the waveform of impulsive sound would normally be made using an oscilloscope.

Sound level meters should also have an input amplifier overload indicator, for use when measuring on the weighted scales, because the difference between the weighted and unweighted levels can be several decibels. The reading displayed on the weighted scale (say dBA) may be below the limiting amplitude while the unweighted decibel level exceeds the maximum limit, and it is therefore possible to overload and clip the peak amplitude of the input signal.

Sound level meters require calibration before use, either by means of built-in calibrating networks or with an external acoustic calibrator held over the microphone. The calibrator is a reference sound source of accurately specified sound level. It is useful also as a reference level when recording noise for later analysis.

Dosemeters

Dosemeters are small integrating sound level meters which are used to measure the total personal exposure to noise during a period such as a working day, and are usually carried in a pocket or attached close to the wearer's ear. These give the total A-weighted sound energy received during the measurement period, and may be used to calculate the L_{eq} value. The dose is often displayed as the proportion of the maximum permitted 8-h dose (usually 90 dBA). The instrument may also indicate whether a standard peak noise level has been exceeded.

Probe microphones

Probe microphones are used for measuring sound in the ear or other small space for instance for evaluating the effect of hearing protectors. These are coupled to the sound level meter by means of a small probe tube, and it is necessary to apply a correction for loss of sound pressure in the tube. An alternative method of measuring the sound in earphones is to use an acoustic coupler (artificial ear), which is a standard cavity having an acoustic impedance similar to that of the ear.

Sound generation

Noise sources are often required in experimental work, for instance in the measurement of hearing ability or in testing hearing protectors. Recordings of speech and other sounds may be used, but other noise generators are available to produce single tones, continuous or intermittent signals. These can be played either through earphones or in free-field using loudspeakers. Pure tones can be generated using a system incorporating an oscillator, amplifier and attenuator. Complex tones can be generated by waveform generators. The frequency of the signal is controlled by an oscillator and the intensity by an attenuator, while the frequency range may be limited by the use of filters. Several types of noise are commonly used:

Figure 16.4. Measurement of machine noise in the field.

1. *Wide-band noise* containing a wide range of frequencies (within the bandwidth limits specified).

2. *Narrow-band noise* containing only a small range of frequencies.

3. *White noise* containing all the audible frequencies and with an essentially flat spectrum.

4. *Pink noise* also containing the range of audible frequencies, but weighted towards the lower frequencies so that the sound pressure spectral density is inversely proportional to frequency.

Impulse or impact noise can be generated by, for instance, using high voltage spark plugs or solenoid activated hammers. It is more difficult to produce impulsive sounds with specified characteristics. If recordings are used, the speaker is likely to 'ring', which extends the duration of the signal, while the signal itself may be distorted by limitations in the dynamic range of the speaker.

Analysis techniques

Detailed analysis of noise signals and environments can be complex and theoretical. A short review of the techniques can be found in Kohler (1984), and noise control handbooks should be consulted for the more technical details.

Amplitude analysis

A continuous reading from a SPL meter can be analysed to give an amplitude–time plot, showing the variation in intensity over a long period. This may be used to compute statistical distribution parameters, such as median or background noise levels. The waveform of a signal over a short duration can be displayed by means of an oscilloscope.

Frequency analysis

Frequency analysis is used when detailed information is needed about a complex sound signal. It can help to isolate possible sources of noise in machinery or in a complex auditory environment. It is also used to evaluate the relative contributions of different frequency components when assessing the risk of damage to hearing. The most usual analysis is the plot of a frequency or power spectrum, which displays the sound energy across the range of frequencies. The frequency range can be split into frequency bands (usually one octave or one third octave wide) which are analysed separately.

An octave is an interval which represents a doubling in frequency: the upper frequency bound is twice the lower frequency bound and the centre frequency of the band is taken as the geometric mean of the two bounds. The octave bandwidths used for industrial sound analysis are normally similar to those forming the musical scale, but the two sets of octave bands do not coincide. The centre frequencies of the two sets are:

Musical octaves   32   64   128   256   512   1024 2048 4096 Hz.

Industrial sound octaves

  37·5   75   150   300   600   1200 2400 4800   9600   19200 Hz.

When more accurate frequency analysis is required, one third octave bands can be used by logarithmically dividing each octave band into three. Narrow band analysis (using a spectrum analyser) gives more detail. The bandwidth is usually defined as a fixed percentage of the frequency to be analysed. For example, a 6% frequency band at centre frequency 500 Hz would be a bandwidth of 30 Hz covering the range between 485 Hz and 515 Hz. Filter sets are used in conjunction with SPL meters for octave band analysis in field use, but narrow band analysis would normally be performed using a recording of the noise.

Further analysis may be used to investigate the temporal characteristics of the noise. For instance, a spectrogram or sonogram is a three-dimensional representation of the frequency, intensity and duration of a signal. This is a plot of frequency against time for a signal or sample of relatively short duration (e.g. speech), in which the density (darkness) of the lines represents the intensity of the sound at each frequency.

Noise measurement procedures

Noise of individual machines or products may be measured in the laboratory or in an anechoic chamber, if precise measurements are required. For most purposes, however, this is not necessary and it is sufficient to use a quiet period in the workplace itself. Environmental measures obviously have to be carried out in the field, under normal conditions. The procedures adopted for measuring noise levels under these different conditions will be described separately.

Since the intensity of sound waves emitted from a noise source is attenuated with distance, noise level decreases with the square of the distance from the source. A measurement of sound level is therefore meaningless unless the location of the measurement is specified. This is normally chosen as the position or positions at which people are likely to be present.

Field surveys

A useful checklist of the procedures which need to be adopted in a field survey is given in Beranek (1971). There are obviously problems in measuring individual noise sources in a workplace or in a town environment, due to the presence of other noise sources. It may be possible to choose a quiet period at night or at the weekend, when very little other machinery is working, but it is unusual in modern environments to have no background sources of noise. Even at night there is likely to be traffic noise, and in the built environment there is frequently central heating, air conditioning or other plant operating. However, reference to Table 16.3 shows that this will have a small or negligible effect on the noise measurement, provided the noise level is at least 10 dB above the background level.

If the difference is less than 3 dB (in any frequency band), or if the source noise level is below that of the background, it cannot be measured reliably. If the difference is between 3 dB and 10 dB, the effect of the noise source may usually be calculated to an acceptable degree of accuracy by comparison with the background level measured with the source switched off. This is more difficult when background noises are intermittent or fluctuating, and care has to be taken to monitor these during the measurement period.

Laboratory measurements

Measurements in the laboratory are liable to be affected by the enclosed environment (by walls or other objects in the acoustic field). Sound may be reflected off hard surfaces (such as steel or concrete), or alternatively may be absorbed by fabric surfaces or acoustic tiles. When sound is reflected, sound level distribution in the space will depend on the phase relationships of the incident and reflected waves. This may increase noise levels and interfere with speech and other signals. In rooms, sound may be reflected several times from the walls and cause reverberation. (Reverberation time is the time taken for the resultant sound to decay.) A discussion of the effects of sound fields can be found in Beranek (1971) or Peterson and Gross (1978).

Briefly, in enclosed spaces, there are three distinct regions around a noise source:

1. *Near field* where there may be interference between the emitted sound wave and reflected waves, so that the sound level varies with slight changes in meter location. The near field extends over a distance approximately equal to the wavelength of the lowest frequency emitted or twice the greatest dimension of the source machine (whichever is the greater).

2. *Far field*, which approximates to free-field conditions. The noise intensity decreases according to the inverse square law. This region can be identified by noting whether the sound level measurements obey this law.

3. *Reverberant field* close to reflecting surfaces, where the reflected noise is diffuse and levels are high. In this case, measurements of the source will not be accurate as the noise levels depend on the room geometry and absorption properties of the surfaces.

For accuracy, measurements should be made in the far field region if possible.

Two extreme conditions are used for laboratory measurements: the anechoic chamber and the reverberation chamber. In an anechoic chamber (such as the one shown in Figure 16.5), the surfaces are covered in highly sound absorbent material, so that there are no reflections or echoes, simulating the free-field conditions experienced outdoors. A reverberant room is one in which the walls completely reflect sound energy and where no walls or surfaces are parallel. This gives a diffuse sound field in which sound energy is uniformly distributed. The sound pressure at any point is an average value due to the many reflections. This is most suitable for measuring total power output of a noise source.

Both the anechoic chamber and the reverberation chamber can be used to determine the sound power of a source, but an anechoic chamber can also be used to determine the reflectivity of surfaces. Most measurements are in fact made in a semi-reverberant room—typical of most normal working conditions—where there is a mixture of direct and reflected sound, and it is necessary to ensure that the measurements are not made in the near field.

Location of microphone

The microphone and observer can both interfere with the acoustic field being measured by blocking or reflecting sound waves, causing significant errors

Figure 16.5. Anechoic chamber measurement of the noise level of a ship's whistle (Reproduced with permission from the Motor Industry Research Association, Nuneaton, UK).

which may be as large as 6 dB. If possible, the sound level meter should be left mounted on a stand. When it is handheld, the observer should hold it at arm's length.

In general, whether outdoors or in an anechoic chamber, a directional (free-field or frontal incidence) microphone is used and should be pointed towards the noise source. A random-incidence microphone is designed to record sound from all directions and is normally used in reverberant or diffuse sound fields. However, some standards differ from this in their prescribed test conditions. If a random-incidence microphone is used in a free-field, its readings are most accurate when it is orientated at an angle of 70–80° to the source.

The microphone is normally located in a position representative of a hearer's ear, but without the person present. For a standing operator, ear height is usually assumed to be 1·5 m. In seated workplaces, the measurement might be taken at the height of the individual operator's ear. The measurements must be made at all positions at which people may be exposed to the noise, and also taking account of normal operating conditions and working practices. For instance, measurements of the noise levels of machinery should include locations used during activities such as maintenance.

In the working environment, it is sometimes necessary for the machine operator to be present while the noise measurements are made. The microphone must then be positioned as close to the operator's head as possible, while avoiding reflections from the head or body or absorption in clothing, at least 50 mm from the side of the head (and preferably further away). Where miniature instruments are attached to the hearer's collar or helmet, or inside noise protectors, it is necessary to take account of such factors and to make corrections to the measured values.

Noise mapping

Noise mapping is useful for an initial survey of a complex environment such as a machine workshop. The noise levels in a working area or around a machine are measured systematically and plotted in the form of a noise map, as shown in Figure 16.6. Maps of this form can be used to identify zones of noise danger (particularly where people may be moving around), and to indicate where preventive measures are required. They can also be used to isolate noise sources. Such noise maps are the auditory equivalent of the room illuminance assessment method described for the visual environment by Howarth in this book.

The noise levels are measured at a sufficient number of locations and plotted on a sketch of the layout of the machinery and workplace. The measurements should be taken at regular intervals, either as a grid covering the area of interest or around the circumference of a machine. If the sound field is complex, it will be necessary to take measurements at closely spaced intervals.

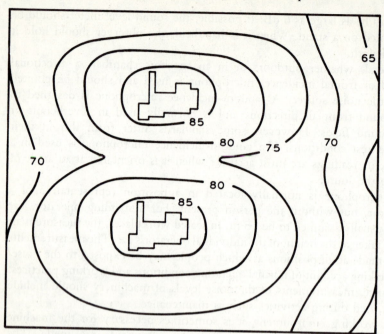

Figure 16.6. Noise map in machine shop.

Accuracy of measurements

Regular calibration of instruments is essential to ensure the accuracy of measurements. A discussion of some of the methods used is given in Kohler (1984).

Outdoors several factors may affect the measurements. Wind is the most common problem, since this causes air turbulence and low frequency noise at the microphone. It is usual to shield the microphone with open-cell foam when taking measurements outside, although this is not fully effective at higher wind speeds. Wind, temperature and humidity can also affect the actual sound levels, as they change attenuation over distance and can create shadow zones. For a discussion of these atmospheric effects, and the attenuation produced by surroundings and barriers or walls, see Beranek (1971).

Identifying noise sources

Various methods of isolating noise sources are used. Where possible the components of the machine or environment should be switched on separately and investigated in turn. Sound intensity analysers are now available which have highly directional probe microphones capable of determining the direction of sound propagation and identifying noise sources, but simpler methods can also be used.

In isolating noise sources of machinery it may be necessary to consider whether reflection or reverberation are contributing to the noise level. In some cases housing and panels may act as a sounding box, either through vibration or by reflecting noise from other sources. It may help in investigating these effects to use acoustic materials to fill the air spaces and damp the reverberations. Weights may be added to increase damping and test the effects of different components.

Potential sources can be isolated by surrounding them with a suitable absorbent material, such as lead sheet, which can easily be shaped around components. Alternatively, they can be shielded by barriers or temporary walls of attenuating materials such as boards sandwiched with acoustic foam.

Standard test methods

The general principles of noise measurement have been outlined in the preceding part of this chapter, but standardized test procedures also have been developed to give accurate and comparable measurements for specific applications. Many are covered by international standards (ISO), but there are also national and industrial standards which should be consulted.

ISO 2204 (ISO, 1979) is a guide to noise problems, covering the general procedures used in the measurement of noise and evaluation of its effects on human beings. This document lists the ISO standards which are applicable to specific problems. Current ISO standards dealing with acoustic measurements are also listed in Brüel and Kjaer (1987), which contains guidance on arrangements of instruments which are suitable for testing to these standards.

Procedures for measurement of vehicle or traffic noise are given in ISO/R 362 (ISO, 1964), and also in ISO 1996 (ISO, 1982) which deals more generally with community response to noise. There is a British Standard BS 4142 (BSI, 1967) dealing with the measurement of environmental noise. Other applications of measurements of communication and annoyance effects are covered later.

The procedures adopted for measurement of noise emitted by machinery or other equipment depend on the purpose of the measurements, and particularly on whether this is to determine the sound power output of the machine or the sound pressure levels to which operators are exposed in the workplace. ISO 3740 (ISO, 1980) is a guide to the measurement of sound pressure levels of machines and equipment, which helps in deciding which methods should be used for different test environments [the more detailed methods being given in the subsequent series of standards ISO 3741 to ISO 3746, which are identical to those of BS 4196 (BSI, 1981)].

ISO 3744 (ISO, 1981a) covers testing in free-field conditions over a reflecting surface, which corresponds to many industrial workplaces. By this method, measurements are made over a hypothetical surface (defined as either a hemisphere or a rectangular parallelepiped) which envelops the noise source under investigation. These are used to calculate the sound power level of the source, and may also be used to compare machines or to rate equipment on its sound power output, or for prediction of the sound pressure level at any given location around the machine/source.

If the workplace is not free-field (for instance where it is close to a wall or surrounded by other machinery), the survey method of ISO 3746 (ISO, 1975) should be used to determine the weighted sound pressure level (SPL) at prescribed microphone positions close to the sound source. This may also be used to calculate the sound power level of the source, and is useful for rating the sound output of a source producing steady noise when it cannot be moved to an acoustic chamber for more accurate measurements.

Both ISO methods require large numbers of measurements, and some of these may be difficult for access on larger machines, especially in crowded machine shops. British Standard 4813 gives a simpler method for measuring the sound power level from machine tools (BSI, 1972). However, it cannot be used to estimate the total power output of a machine, and is less accurate for identifying highly directional noise sources. Following BS 4813, noise levels are measured at a height of 1·5 m and at intervals around the machine at a distance of 1 m from its surface. The microphone positions should not be more than 1·5 m apart, to obtain sufficiently detailed analysis, with a minimum of five measurements and one of these should be at the point of highest sound level. These measurements can be used to plot a noise map of the environment, but can also be used to calculate the mean noise level at the 1 m distance. In addition a measurement should be taken at the operator's position, and any other positions occupied by personnel for a significant period of the working day.

Measurement of hearing

A person's hearing ability is measured in terms of the minimum threshold of perception, using the normal techniques for threshold measurements such as the 'method of limits'. The measurements are usually made with an instrument called an audiometer, and the measurements presented in the form of an audiogram. 'Normal' thresholds have been established (see for instance ISO, 1972) and loss or impairment of hearing is usually defined as a minmum change of 10 or 15 dB in the threshold. Hearing level is taken as the amount by which the average threshold is raised in an individual, so that a positive hearing level represents hearing ability worse than the defined 'normal' level at any frequency.

Audiometer

An audiometer consists of a calibrated oscillator and amplifier which presents pure tones over the range of audible frequencies. At each frequency, the amplifier can be scanned through the range of intensities to measure the hearing threshold. The tones are normally presented through headphones, and the signal can be directed to one of the pair of earphones, so that the two ears can be tested separately. Two forms of audiometer are used: either manual or automatic (Bekesey), as described later.

In an audiogram, the reference level of 0 dB represents the normal threshold of hearing (i.e. for people who have no hearing disability or age deterioration) at each frequency over the auditory range.

Measuring hearing thresholds

The audiometer is operated by the experimenter who alters signal frequency and sound level. The hearing threshold is measured at discrete frequencies across the hearing range, usually as the average of the ascending and descending thresholds. The absolute threshold for a sound is the minimum level which is detected, when presented in the absence of other sounds. The audiometer and operator should both be screened from the direct view of the subject, so that no cues are given to the presentation of the test signal. In order to avoid any influence from other sounds on the measurements of the hearing threshold, no words should be spoken to the subject and the subject should be asked to indicate the presence or absence of a noise by a silent signal such as the movement of a finger.

Bekesey audiometers give a semi-automatic procedure similarly based on the 'method of limits'. They automatically vary the sound level, decreasing while the subject holds down a switch (which controls the direction of the motor driving the attenuator), and increasing when it is released. The subject is asked to hold the switch pressed down as long as the sound is audible, and therefore maintains it at threshold level. A continuous trace of the hearing level is recorded at each frequency, and the mean value indicates the auditory threshold. Frequency is scanned across the range in a programmed sequence so that the tone is maintained at each frequency for a short period to establish the threshold.

Whichever technique is used, measurements should be carried out in a quiet room, and a period allowed for the subject to adapt to the low noise level and recover from any temporary threshold shift due to previous exposure to noise. The US regulations on occupational noise exposure (OSHA, 1983) specify that the first baseline audiograms taken in a hearing conservation programme should be taken after at least 14 h without exposure to workplace noise, although subsequent annual audiograms are permitted at any time during the working day.

Sounds may be presented through speakers instead of earphones, but there may be some differences in the measured threshold (perhaps up to 6 dB) due to the differences in the acoustic field. According to Loeb (1986), the threshold in free-field conditions is likely to be lower than when presented through earphones, and the binaural threshold is likely to be lower than the monaural threshold.

Measuring masking thresholds

Absolute hearing threshold is measured in a quiet environment, but it is often important to know the hearing threshold for a signal or for speech in a noisy environment, which is termed the masked threshold. This defines the threshold of detection (not intelligibility) against the background level of noise. The effect is measured by determining the absolute threshold of the sound when presented alone, then measuring it in the presence of the masking sound. The amount of masking is then defined as the difference in decibel level by which the threshold of audibility is raised above the absolute threshold.

The probability that a signal will be detected increases with the level of the signal above the background. So the masked threshold is sometimes defined as the level at which there is a given probability (say 75%) of correct detection of the signal. This is not an absolute measure since it depends on factors such as the level of expectancy of the hearer and the rise time, duration and temporal shape of the signal. For a detailed discussion of the factors involved, the reader may consult texts such as Sorkin (1987).

Assessing the risk of hearing damage

Although a single violent sound can cause damage to the ear-drum, this is rare and hearing loss is commonly caused by long-term exposure to noise. The exposure does not need to be continuous, since the effects of intermittent exposure are cumulative. The main indicator used to assess hearing damage is a change in the hearing threshold of an individual. If this is measured before and after exposure to loud noise the threshold shift may be determined by the difference in the two thresholds. More usually hearing loss is assessed by the drop below the population norm. In industrial hearing conservation programmes, the operators at risk are assessed at the start of their employment and changes in hearing are monitored by annual tests (OSHA, 1983).

Temporary threshold shift (TTS) is a short-term and reversible change experienced after exposure to loud noise, which may persist for minutes or hours depending on the exposure. Since recovery starts as soon as the noise ceases, it is necessary to specify the time at which the TTS is determined. This is normally measured 2 min after exposure.

Noise-induced permanent threshold shift (NIPTS) is the long-term effect of exposure to noise and is not reversible. The mechanisms of TTS and NIPTS are not necessarily identical, but Kryter (1985) has suggested that the TTS of a group of workers after 8-h exposure can be used as a criterion for the risk of long-term damage.

Risk of hearing damage has to be assessed in terms of duration of exposure as well as noise level. ISO Standard 1999 (ISO, 1972) gives a risk table in relation to age, duration of exposure and intensity of noise, which shows that intensities above 90 dBA risk hearing damage.

Since industrial workers are exposed to noise which may vary considerably over the working period, it is important to calculate the noise levels over the whole time of exposure (usually over an 8 h working day). The noise dose is calculated from the measured L_{eq} value (equivalent level of sustained noise). In measuring this, it should be remembered that operating conditions, especially communication signals, may contribute to the total noise dose. This occurs for occupations such as aircraft pilots, where the signals are presented through headphones. In these cases the noise levels can be recorded at the ear, if necessary using a miniature or probe microphone, and the noise dose obtained by analysis of the recording (Glen, 1976).

According to Davies and Jones (1982), most standards apply the 'equal-energy principle', which suggests that exposure to higher intensities can be permitted for short periods, providing that the total energy within an 8 h period does not exceed the normally permitted dose [as in ISO 1999 (1972)]. This allows for quiet periods during the working day, but does not take account of recovery which may occur in these periods. For example, the equal energy principle assumes that a continuous 4 h exposure followed by 4 h quiet has the same effects as four 1 h exposures to the same level of noise separated by quiet periods of 1 h. Thus on this assumption, halving the duration of exposure permits the sound level to be increased by 3 dB. The American Standard (OSHA, 1983) uses a less conservative 5 dB halving rule.

For a discussion of assessment of occupational noise-induced hearing loss and standards for noise exposure limits, the reader should consult Kryter (1985) or Davies and Jones (1982). Kryter includes comments on the methodology of audiometry and an extensive review of the evidence for the relative influence of presbycusis, sociocusis and nosocusis on hearing thresholds.

The complex effects of impulsive noises are reviewed in both Loeb (1986) and Kryter (1985), which quote damage risk contours giving an indication of acceptable maximum peak SPL values for impulsive noises of different durations and repetition frequencies.

Measurement of effects of noise

People vary enormously in their responses to noise, being influenced by factors such as intrusion into privacy or whether the sound is intermittent or unexpected, and probabilistic or statistical measures are needed to assess both annoyance and performance effects of noise. Some of the reasons for individual differences are discussed by Jones and Davies (1984). People can adapt well to a continuous background of sound, and experiments therefore have to be designed carefully to include an adequate degree of realism in the experimental conditions. A corollary to this is that measurements of the level of annoyance may be due to factors other than the noise levels which are the subject of the experiment: noise annoyance may just be a symptom of poor morale or stress.

Annoyance

There is an enormous number of measures of noisiness, annoyance and intrusiveness, which have been developed for assessment of environmental community and transport noise.

The annoyance level of noise is frequently measured by means of survey and questionnaires. The wording of questionnaires and of instructions given to experimental subjects has a considerable influence on the rating. Loudness or noisiness cannot be equated with level of annoyance, and investigators have found large differences between noise sources judged on these criteria (Kryter, 1985). A discussion of the design of rating scales may be found in Loeb (1986). Studies measure many different aspects of the response to noise such as judged noisiness or loudness of the neighbourhood, dissatisfaction with present noise level, frequency with which annoyance is felt, degree of annoyance, and interference with activities (Jones and Davies, 1984). It is important to distinguish between these, since simple reduction in noise levels may not reduce the level of annoyance.

Noise annoyance is obviously a multi-factorial problem. At least 13 primary and derived measures have been used to assess community noise (Sander and McCormick, 1987). The following indicates a few of the methods of evaluation which have been used.

Kryter (1985) produced equal annoyance curves (pN dB) analogous to the equal loudness curves (phons), and also scales of annoyance or noisiness in units called noys. However annoyance also has social, economic and psychological dimensions, and Kryter indicated that the threshold of perceived annoyance varies with previous exposure, location indoors or outdoors, time of day, and the impulsive or startle character of the noise. His scales have been used to develop measures of aircraft and traffic noise. In fact, many indices have been produced to quantify disturbance by fluctuating noise (e.g. noise rating (NR), noise and number index (NNI), traffic noise index (TNE), noise pollution index (NPL)), and these are discussed in Loeb (1986) or Kryter (1985). They are often recorded on noise level analysers, which perform statistical analyses of noise levels over long duration measurement periods.

Schultz (1978) produced a dosage response curve as a predictive tool. This relates noise exposure (in terms of a day–night average sound level L_{dn} in dBA) to the level of community annoyance, based on large scale surveys from various countries (mostly related to transport noise). Others suggest that short duration noises can be particularly annoying and have proposed measures using peak levels of noise events, as for aircraft overflights (Fidell 1984). Other special measures have been developed to evaluate aircraft noise exposure (ISO, 1970), and a discussion of some of these may be found in Ollerhead (1973).

Preferred noise criteria (PNC) curves were introduced as design criteria for background noise in offices, rooms or halls (Beranek, 1971). They were based on the speech interference level (SIL), which is a measure of speech intelligibility and is described in the next section. The noise criteria curves are a set of arbitrary sound spectra for steady noises, serving as a reference for rating noise environments. They are also useful in deciding where in the frequency spectrum the greatest benefits could be obtained by noise reduction treatments. PNC curves are based on the need for acceptable speech communication, and specify that the loudness in phons should not exceed SIL by more than 22 units. The spectrum of the measured background noise is plotted over the noise criteria curves, and the noise is rated by the number of the curve which equals or just exceeds the noise spectrum at any point. This rating is then compared to a table of recommended ratings to evaluate the suitability of the noise level for the particular room environment.

It is not only loud noises that can be annoying. The effects of intrusiveness of low-level (or infrequent) noise exposure have been investigated by Fidell and others (Fidell *et al.*, 1979; Fidell and Teffeteller, 1981). They found that the L_{eq} measure is insensitive to these types of noises and have developed methods of predicting the level of intrusion or annoyance from measures of signal detectability.

Communication effects and speech intelligibility

Noise can mask speech, but the degree of masking depends on various factors since there is a large measure of redundancy in speech, especially when the hearer is familiar with the subject matter. Thus, the assessment of speech intelligibility is not a simple question of measuring the signal-to-noise ratio. Speech intelligibility can be measured most directly by listening tests, presenting different types of standardized material, such as nonsense syllables, phonetically balanced word lists which contain all speech sounds (or phonemes), and sentences. The A-weighted decibel scale is normally used to measure sound level in these applications.

Several methods have been developed to predict speech intelligibility from physical measures of the speech signal, noise environment and task parameters (e.g. speaker–hearer distance). Three of the methods are given here, and discussions of their relative merits can be found in Kryter (1985) and other textbooks.

1. *The Articulation Index* (ANSI, 1969) predicts the likelihood of difficulties with speech communication, and is calculated from the differences between the SPLs of the speech and masking noise in various frequency bands. These are weighted according to their importance in the intelligibility of speech. Originally, 20 bands were used between 250 and 7000 Hz, but the articulation index is now calculated from octave or one third octave analysis. The signal

and noise levels are measured in each band and the differences weighted and summed to give the articulation index.

The relationship between the articulation index and intelligibility has been determined experimentally for different types of spoken material, and is expressed in a series of graphs in the ANSI Standard which indicate the percentage of syllables, words or sentences which will be correctly understood. Corrections are given in the standard to take account of factors such as reverberation or noise interruption. An indication of the difficulty of speech communication is given from the following values of articulation index:

< 0·4	difficulties likely
0·4–0·7	some difficulties may occur
> 0·7	good speech communication possible

2. *Speech Interference Level* (SIL) is a simpler method (also specified in ANSI (1969)) which does not require direct measurement of the speech level. The SIL is calculated from the mean of noise SPL for octave bands centred at 500, 1000, 2000 and 4000 Hz. Again, this value is compared with tables relating it to speech intelligibility for different voice efforts (normal, raised, shouting) and for varying distances between speaker and listener. Preferred speech interference level is a similar measure which is related to the maximum distance over which speech communication is possible (ISO, 1974).

3. *Direct Measurements of SPL* (in dBA) have also been used as an index of speech interference, predicting SIL. Loeb (1986) suggests a relationship of

$$SIL = SPL - (9 \text{ or } 10 \text{ dB}).$$

Webster (1979) produced a chart relating dBA, voice effort and speaker–hearer distance to quality of communication, so that either SPL or SIL could be used in assessing this.

Design of warning and information signals for safety and audibility

The effects of masking on warning and alarm signals are discussed in Webster (1984), who gives a method for predicting the level at which a pure tone signal will be audible. However, additional criteria are needed to ensure that the signal is also attention-getting and recognizable. The perception of masked signals varies with frequency, signal-to-noise ratio and absolute level of background noise.

Four main factors need to be taken into account in designing auditory warnings and signals: audibility, noise dose, startle and discriminability (Coleman et al., 1984). Reports from Patterson and Milroy (1979) and Coleman et al. (1984) describe procedures which have been used to design the appropriate sound level for auditory warnings.

The first stage in this process is the measurement or prediction of the masked threshold (and modelling techniques for this are mentioned later). As described by Coleman *et al.*, 'The method employed . . . involved the tape recording of the particular background noise and a detailed narrow-band spectral analysis followed by computerised calculation of masked thresholds, incorporation of a 97·5th percentile absolute threshold criterion, and a graphical output showing the complete design window with its constituents.'

This 'design window' defines the upper and lower boundaries of frequency and intensity within which the signal components must lie in order both to be audible and to attract attention without causing startle or risk of hearing damage. Figure 16.7 shows how several factors are considered in specifying the 'design window' for a warning signal. This is constructed from knowledge of the hearing abilities of the workforce, the nature of the masked threshold due to the background noise and the levels above threshold which are necessary for the signal to be audible without causing startle. It is also possible in this to take account of the attenuation effects when using hearing protectors.

As seen in Figure 16.7, the thresholds may also be set for different population criteria: the maximum level to avoid startle is determined for

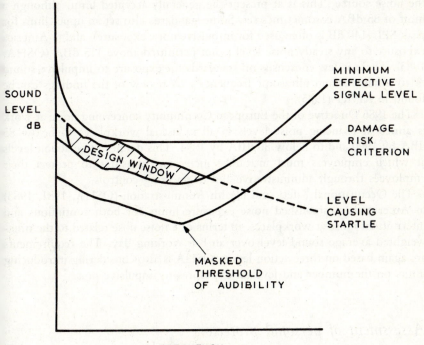

Figure 16.7. 'Design window' method for specifying signals to be used in noisy environments (after Coleman *et al.*, 1984).

people with normal hearing, while the minimum effective signal level is set for the people in the population who have the lowest hearing ability.

Standards and guidelines for noise exposure

International, national and industry wide standards have been drawn up for assessing noise levels in different environments. British legislation is reviewed by Acton *et al.* (1984), covering community and occupational noise exposure, as well as road traffic and aircraft noise, and by the Health and Safety Commission (1987) consultative document dealing with the European Regulation (Commission of the European Communities, 1986) which will be introduced in 1990.

The effects of noise on performance in different types of task are more difficult to determine, and are task specific. These are often assessed by experimental evidence of the effects of noisy environments on performance or comfort, and guidelines on acceptable and hazardous noise levels can be found in most ergonomics textbooks.

The evidence on the risk of hearing damage [for instance from ISO (1972)] shows that exposure to noise of over 90 dB for 8 h a day will produce significant deafness, while the damage potential increases with frequency of the noise source. This is at present the generally accepted limit, although a limit of 85 dBA is sometimes set. Some standards also set an upper limit for peak SPL (140 dB is often used for impulsive noise exposure), and in America exposure to any steady noise level is not permitted above 115 dBA (OSHA, 1983). There is less consensus on standards for exposure to impulsive sound or to infrasonic and ultrasonic frequencies. A review of the findings can be found in Kryter (1985).

The 1986 Directive of the European Community concerning noise at work is aimed at reducing noise levels in all industrial workplaces to below 85 dBA by 1990 and to below 80 dBA by 1994. This defines three action levels at which employers must introduce increasing measures to protect their employees through administrative and engineering controls.

The Occupational Safety and Health Administration (OSHA, 1981, 1983) in America has established noise exposure limits for both continuous and intermittent noise at workplaces, in terms of a noise dose related to the time-weighted average sound level over an 8-h working day. The requirements are again based on three action levels. OSHA is also considering introducing limits on the number and level of exposures to impulsive noise.

Assessment of personal protectors

Hearing protectors have to be assessed for their effect on the audibility of communication signals, as well as for the attenuation they provide.

Evaluating attenuation

Noise attenuation curves are normally supplied with hearing protection equipment from the manufacturers. The attenuation may be tested by methods such as ANSI S3.19-1974 (ANSI, 1974). The American Environmental Protection Agency (1979) requires manufacturers to test hearing protectors to this standard and to label them with a noise reduction rating (NRR). The NRR is calculated from the attenuation provided in 1/3 octave band analysis. This can be used to estimate the noise exposure (dBA) of the wearer: either as the workplace sound level in dBC less NRR, or as the sound level in dBA plus 7 dB less NRR (OSHA, 1983).

Sutton and Robinson (1981) review the various procedures which can be used to estimate the level of protection given to the wearer (usually defined as the reduction in dBA at the ear, although this of course varies between wearers) from a knowledge of the frequency response of the protector and the spectrum of background noise. The most accurate method of calculating the protection provided is given in ISO 1999 (ISO, 1972), where the attenuation (minus a variance correction) is subtracted from the workplace sound levels measured with octave band analysis. It should be noted, however, that these methods indicate the optimum protection provided, not allowing for the poor fitting of the protector or leakage due to hair or other factors, and various studies have shown that the protection is overestimated under workplace conditions (Berger, 1983; Lempert, 1984).

A subjective method of measuring attenuation (by threshold shift) can also be used, as outlined in ISO 4869 (ISO, 1981b). A minimum of 10 subjects should be tested (using normal audiometric techniques), although more should be used if possible. The subjects must have a hearing threshold level in either ear which is not worse than 15 dB at frequencies below 2 kHz and not worse than 25 dB at frequencies above 2 kHz. The test signals specified are pink noise filtered through 1/3 octave bands from 63–8000 Hz. This method can be used to compare or rank different models of protectors, and to evaluate design features which may affect performance.

The Health and Safety Commission (1987) consultative document then gives a method of calculating the 'assumed protection' from the attenuation values obtained from testing a group of subjects. The 'assumed protection' is taken as the mean attenuation at each frequency minus the standard deviation over the tests (probably of the order of 5 dB), in order to allow for the variation in protection between wearers.

Dummy heads and miniature microphones can also be used to measure attenuation, although the results may be expected to differ from subjective measurements because of the effects of bone conduction leakage. A semi-objective technique, in which a miniature microphone is attached to a subject's ear, can be used to investigate the effects of hair, glasses or helmets with different models of hearing protector.

Evaluating audibility of communications

Speech communication needs to be considered when assessing the effectiveness of hearing protectors. This can be done in terms of the articulation index, speech interference level or by direct experimental testing of intelligibility. Although the articulation index generally assumes normal hearing, it can be calculated against reduced hearing thresholds, and a measure of AIIHA (articulation index incorporating hearing ability) is suggested (Coleman *et al.*, 1984). Coleman *et al.* also propose a technique for selecting protectors, which takes into account the interaction between the attenuation characteristics and hearing ability, the range of noise and the speech conditions.

Simulation and modelling of the acoustic environment

A few mathematical models of the auditory environment have been developed to assist ergonomic analyses. Three such simulations are described here related to the design of auditory warning signals, the reduction of noise levels in factories and the acceptability of traffic noise in the urban environment. The reader should also refer to the thermal models and expert systems described by Parsons in this book, and also to the chapter on modelling and simulation by Meister.

Models of the auditory filter have been used to demonstrate the effects of masking of signals in a noisy environment, and to predict the audibility and discriminability of signals used in the workplace (Patterson *et al.* 1982; Coleman *et al.* 1984). The models assume that the ear acts as a set of bandpass filters, and predict the masked thresholds of signals from the measured spectrum of the noise environment by modelling the characteristic of the auditory filter centred on the signal frequency to be heard, then integrating the noise power within the filter passband.

Shield (1980) describes a computer model called Noiseshield, which was developed to predict noise levels during the design of a factory. This uses information on the layout of the area (dimensions of the room and construction materials of walls, floor and roof), and location and sound power output of all machines and other noise sources. The package can also be used to evaluate noise problems in existing factories, by investigating the effects of treatments such as erecting barriers or enclosures around machines, or the introduction of sound absorbing materials on floor or walls.

Two approaches to modelling road noise have been reported by Clayden *et al.* (1975), who described the early development of a model to predict traffic noise levels in the urban environment given a plan of the buildings and the locations of vehicles, and by Nelson (1973) who used a model to predict the temporal distribution of noise from freely flowing traffic (using speed, traffic mix, and distance from the road).

Concluding remarks

As can be seen from the variety of techniques introduced in this chapter, the most important consideration in any noise assessment is the choice of appropriate criteria to use in the evaluation. It is not usually sufficient simply to measure the sound pressure level or acoustic power at a particular location in order to make an adequate assessment of the effects of noise on hearers. Both the perception and the preference of the hearer are influenced by the auditory characteristics of the human ear, while there is a further influence in the social or occupational context in which the sound is heard.

Ergonomists often need to consider complex effects concerning safety, annoyance, distraction, information or pleasure for different groups of hearers in a noisy environment. Different measures and criteria may be needed to evaluate these aspects. It is hoped that the brief discussion in this chapter will give some guidance on the most appropriate methods to choose in noise assessments for various community, transport and industrial environments.

References

Acton, W.I., Grime, R.P. and Ratcliffe, K. (1984). Legal aspects of noise. In *Noise and Society*, edited by D.M. Jones and A.J. Chapman, (Chichester: John Wiley).

ANSI (1969). *American National Standard for Calculation of the Articulation Index*. ANSI S3.5. (New York: American National Standards Institute).

ANSI (1974). *American National Standard Method for the Measurement of Real-ear Protection of Hearing Protectors and Physical Attenuation of Ear-muffs*. ANSI S3.19-1974. (New York: American National Standards Institute).

Beranek, L.L. (1971). *Noise and Vibration Control*. (New York: McGraw-Hill).

Berger, E. (1983). Using the NRR to estimate the real world performance of hearing protectors. *Sound and Vibration*, **18**, 5, 26–39.

Brüel and Kjaer (1986). *Noise Control, Principles and Practice*. (Naerum, Denmark: Brüel and Kjaer).

Brüel and Kjaer (1987). *Acoustic Measurements According to ISO Standards and Recommendations*. (Naerum, Denmark: Brüel and Kjaer).

BSI (1967). *Method of Rating Industrial Noise Affecting Mixed Residential and Industrial Areas*. BS 4142. (London: British Standards Institution).

BSI (1972). *Method of Measuring Noise from Machine Tools Excluding Testing in Anechoic Chambers*. BS 4813. (London: British Standards Institution).

BSI (1981). *Sound Power Levels of Noise Sources*. BS 4196. (London: British Standards Institution).

Clayden, A.D., Culley, R.W.D. and Marsh, P.S. (1975). Modelling traffic noise mathematically. *Applied Acoustics*, **8**, 1–12.

Coleman, G.J., Graves, R.J., Collier, S.G., Golding, D., Nicholl, A.G. McK., Simpson, G.C., Sweetland, K.F. and Talbot, C.F. (1984). *Communications in Noisy Environments*. Technical Memorandum TM/84/ 1 (EUR P.74). (Edinburgh: Institute of Occupational Medicine).

Commission of the European Communities (1986). *Council Directive of 1.
    May 1986 on the Protection of Workers from the Risks Related to Exposur
    to Noise at Work*. Council Directive 86/188/EEC. Official Journal of th
    European Communities, NO L 137/29, 24 May 1986.

Davies, D.R. and Jones, D.M. (1982). Hearing and noise. In *The Body a
    Work*, edited by W.T. Singleton. (Cambridge: Cambridge Universit
    Press).

Environmental Protection Agency (EPA) (1979). Noise labeling requirement
    for hearing protectors. *Federal Register,* **42**(190), 56139–56147.

Fidell, S. (1984). Community response to noise. In *Noise and Society*, edite
    by D.M. Jones and A.J. Chapman. (Chichester: John Wiley).

Fidell, S. and Teffeteller, S.R. (1981). Scaling the annoyance of intrusiv
    sounds. *Journal of Sound and Vibration*, **78**, 291–298.

Fidell, S., Teffeteller, S.R., Horonjeff, R.D. and Green, D.M. (1979)
    Predicting annoyance from detectability of low-level sounds. *Journal c
    the Acoustical Society of America*, **66**, 1427–1434.

Glen, M.C. (1976). The contribution of communications signals to nois
    exposure. *Applied Ergonomics*, **7**, 197–200.

Hassall, J.R. and Zaveri, K. (1979). *Acoustic Noise Measurements*. (Naerum
    Denmark: Brüel and Kjaer).

Health and Safety Commission (1987). *Prevention of Damage to Hearing fror
    Noise at Work: Draft Proposals for Regulations and Guidance*. Consultativ
    Document (London: HMSO).

Health and Safety Executive (1983). *100 Practical Applications of Noise Reductio
    Methods* (London: HMSO).

ISO (1964). *Measurement of Noise Emitted by Vehicles*. ISO/R 362 (Geneva
    International Standards Organization).

ISO (1970). *Procedure for Describing Aircraft Noise Around an Airport*. ISO R50
    (Geneva: International Standards Organization).

ISO (1972) *Assessment of Occupational Noise Exposure for Hearing Conservatio
    Purposes*. ISO 1999 (Geneva: International Standards Organization).

ISO (1974) *Assessment of Noise with Respect to its Effect on the Intelligibility c
    Speech*. ISO Technical Report TR 3352-1974 (Geneva: Internationa
    Standards Organization).

ISO (1975). *Acoustics—Determination of Sound Power Levels of Noise Sources—
    Survey Method*. ISO 3746 (Geneva: International Standards Organization)

ISO (1979). *Guide to International Standards on the Measurement of Airborn
    Acoustical Noise and Evaluation of its Effects on Human Beings*. ISO 220
    (Geneva: International Standards Organization).

ISO (1980). *Acoustics—Determination of Sound Power Levels of Noise Sources—
    Guidelines for the Use of Basic Standards and for the Preparation of Nois
    Test Codes*. ISO 3740 (Geneva: International Standards Organization)

ISO (1981a). *Acoustics—Determination of Sound Power Levels of Noise Sources—
    Engineering Methods for Free Field Conditions over a Reflecting Plane*. ISC
    3744 (Geneva: International Standards Organization).

ISO (1981b). *Acoustics—Measurement of Sound Attenuation of Hearing Protectors—
    Subjective Method*. ISO 4869 (Geneva: International Standards Organiz
    ation).

ISO (1982). *Acoustics—Description and Measurement of Environmental Noise*. ISC
    1996 (Geneva: International Standards Organization).

ones, D.M. and Chapman, A.J. (1984). *Noise and Society*. (Chichester: John Wiley).

ones, D.M. and Davies, D.R. (1984). Individual and group differences in the response to noise. In *Noise and Society*, edited by D.M. Jones and A.J. Chapman (Chichester: John Wiley).

Kohler, H.K. (1984). The description and measurement of sound. In *Noise and Society*, edited by D.M. Jones and A.J. Chapman (Chichester: John Wiley), pp. 35–76.

Kryter, K.D. (1985). *The Effects of Noise on Man*, 2nd edition. (London: Academic Press).

Lempert, B.L. (1984). Compendium of hearing protection devices. *Sound and Vibration*, **18**, 26–39.

Loeb, M. (1986). *Noise and Human Efficiency*. (Chichester: John Wiley).

Nelson, P.M. (1973). *A Computer Model for Determining the Temporal Distribution of Noise from Road Traffic*. TRRL Laboratory Report 611 (Crowthorne: Transport and Road Research Laboratory).

Ollerhead, J.B. (1973). Noise: how can the nuisance be controlled? *Applied Ergonomics*, **4**, 130–138.

OSHA (Occupational Safety and Health Administration) (1981). Occupational noise exposure; hearing conservation amendment. *Federal Register*, **46**, 4078–4179.

OSHA (Occupational Safety and Health Administration) (1983). Occupational noise exposure; hearing conservation amendment; final rule. *Federal Register*, **48**, 9738–9783.

Patterson, R.D. and Milroy, R. (1979). *Existing and Recommended Levels for Auditory Warnings on Civil Aircraft*. Civil Aviation Authority Contract Report (Contract No 7D/S/0142). (Cambridge: Medical Research Council, Applied Psychology Unit).

Patterson, R.D., Nimmo-Smith, I., Weber, D.L. and Milroy, R. (1982). The deterioration of hearing with age: frequency selectivity, the critical ratio, the audiogram, and speech threshold. *Journal of the Acoustical Society of America*, **72**, 1788–1803.

Peterson, A.P.G. and Gross, E.E. (1978). *Handbook of Noise Measurement*, 8th edition. (Concord, MA: General Radio Company).

Sanders, M.S. and McCormick, E.J. (1987). *Human Factors in Engineering and Design*, 6th edition (New York: McGraw-Hill).

Schultz, J. (1978). Synthesis of social surveys on noise annoyance. *Journal of the Acoustical Society of America*, **64**, 377–405.

Shield, B.M. (1980). A computer model for the prediction of factory noise. *Applied Acoustics*, **13**, 471–486.

Sorkin, R.D. (1987). Design of auditory and tactile displays. In *Handbook of Human Factors*, edited by G. Salvendy, (New York: John Wiley), pp. 549–576.

Sutton, G.J. and Robinson, D.W. (1981). An appraisal of methods for estimating effectiveness of hearing protectors. *Journal of Sound and Vibration*, **77**, 79–91.

Webster, J.C. (1979). Effects of noise on speech. In *Handbook of Noise Control*, 2nd edition, edited by C.M. Harris (New York: McGraw-Hill).

Webster, J.C. (1984). Noise and communication. In *Noise and Society*, edited by D.M. Jones and A.J. Chapman, (Chichester: John Wiley).

Chapter 17

Human responses to vibration: Principles and methods

Rosemary A. Bonney

Introduction

Vibration can affect people's health, comfort and performance. The methods used to assess the effects of vibration include objective measurement, subjective measurement and mathematical modelling. This chapter defines the principles of hand/arm and whole body vibration as well as outlining the methods used to evaluate the effects they may have on the individual. It is not intended as a full and deep review of the subject; for those requiring more information the reader is referred to detailed texts such as Wasserman (1987) and Brammer and Taylor (1982).

General principles

Vibrations are mechanical oscillations produced by regular or irregular movements of a body about its resting position. The direction in which this motion occurs must be defined in terms of its three orthogonal components:

'x'—front to back,
'y'—side to side,
'z'—up and down.

Both hand/arm and whole body vibration measurements are obtained with respect to internationally agreed biodynamic coordinate systems, as defined by ISO 2631 (ISO, 1985). (See Figure 17.1.)

Once the direction of the motion has been defined, its other characteristics, in terms of repetition rate or frequency (number of cycles per unit time) and amplitude (distance displaced from resting position) must also be specified (see Figure 17.2). The frequency (in hertz) is the number of times a body

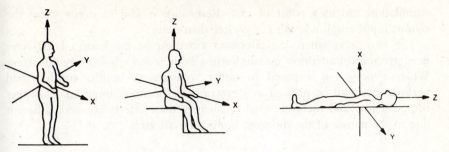

Figure 17.1. International vibration coordinate system for a standing, sitting and lying person (ISO 2631, 1985).

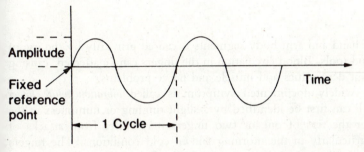

Figure 17.2. Frequency (number of cycles/unit time) and amplitude for sinusoidal vibration.

vibrates in a specific time period, normally 1 s. Amplitude is the maximum amount a point of interest is displaced from its central position. Since achievement of a given amplitude at a chosen frequency requires a certain acceleration (which, of course, implies a certain maximum velocity), a wave form can be defined in terms of frequency and acceleration alone. Acceleration is more often used in place of amplitude to define a wave form, not least because the acceleration can be measured more readily than amplitude or velocity in most practical cases.

Many structures which have elastic properties will tend to vibrate freely at a particular frequency, called the natural frequency. If vibration is applied to a structure at or near this frequency then it will resonate, or vibrate at an amplitude greater than that applied to it. At other frequencies the opposite of resonance occurs so that the body absorbs or reduces the input intensity. This is known as damping or attenuation.

The human body is extremely complex, composed of organs, bones, joints and muscles. Each of these parts can be affected in the ways described above. At some frequencies therefore, they might vibrate at greater amplitudes than the vibration applied to them and at others they may absorb and attenuate the inputs. The resonant characteristics of a particular body system can be measured by comparing the vibration intensity of the system at the point of stimulation and at a point of exit. Resonance is said to occur when the output/input amplitude ratio is greater than unity.

The two areas where the effects of vibration on the human body have been given most attention are hand/arm vibration and whole body vibration. When a person is exposed to either, changes in health, comfort and performance will be noticed as a reaction to the stress imposed. However the principles and methods used to identify these effects are very different due to the nature of the different body parts affected.

Hand arm vibration

Principles

Vibration of hand and arm body segments is caused primarily by the use of vibrating hand tools. Blood circulation in the fingers can e affected, causing bone and joint deformities and muscle and nerve problems.

The most widely documented symptom is called *vibration white finger* (VWF). VWF can first be identified by a slight tingling or numbness in the fingers. Later the tips of one or two fingers may suffer from attacks of blanching particularly in the morning and in cold conditions. The fingers affected and distribution of damage are initially dependent on the tool type and the method by which it is held. VWF attacks usually last less than 1 h and are ended by a red flush and pain, when the blood rushes back to the finger tips. During an attack, sensitivity to touch, pain and temperature are reduced.

The time taken for VWF to appear depends on the level of exposure for the individual. For very high levels of vibration this may be only a few months, however for most occupations it is usually 5–10 years. The period elapsing before symptoms appear is called the *latent* interval. VWF as well as being extremely painful, can reduce finger sensitivity and manual dexterity. Thus work requiring fine finger movement may no longer be possible. It is also socially handicapping, as the symptoms can be brought on by cold, which makes many outside activities unpleasant and difficult. Employees may be forced to change their jobs as a result of the disease and at the moment there is no universal cure. Some remission from early stages of the disease has been noted on people who have been withdrawn from exposure (Riddle and Taylor, 1982).

Of the population, 10% exhibit VWF symptoms, even though they are not actually exposed to a vibrating environment (Griffin, 1982). This makes control and diagnosis of the disease more difficult. Recommendations as to the acceptable limits of vibration exposure below which the disease is unlikely to occur, are provided in ISO standard DIS 5349.2 (ISO, 1986) (see Figure 17.3).

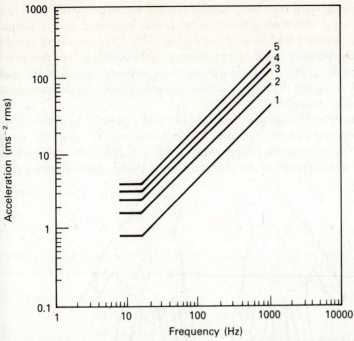

Figure 17.3. Acceptable limits of vibration exposure of the hand (ISO/DIS 5349, 1986). Curves 1 to 5 refer to multiplying factors, associated with exposures per 8 hour shift of 4–8, 2–4, 1–2, ½–1, and up to ½ hour respectively.

Equipment and processes which are regularly associated with the causes of VWF are:

(a) hand fettling of castings and forgings,
(b) grinders or other hand-held rotary tools,
(c) pneumatic hammers, drills and other oscillating tools,
(d) chain-saws, and
(e) motor cycles.

These equipments vibrate at frequencies between 5–1000 Hz, the range of frequencies within which VWF is most likely to occur. The actual risk resulting from use of these tools depends on the length of exposure, the thermal environment in which they are used, and how well the tools have been maintained, as well as on the physical state of the user.

Practical assessment techniques for vibrating hand tools

Where vibrating hand tools are being used the ergonomist may be called in to identify whether there is a risk of VWF, quantify its severity and make recommendations for design improvements if necessary.

The first requirement is to use a reliable method for assessing the incidence of VWF amongst the work-force. Griffin (1982) suggested that a simple categorization system could be used (see Figure 17.4). This assessment should be done by an experienced medical practitioner. The possible actions described in Figure 17.4 are protective measures for those exposed. Ergonomists and engineers will recognize that further actions towards eliminating the hazardous situation are also necessary.

An assessment by Taylor and Pelmear (1975) of a number of clinical tests for VWF proposed that the Renfrew Ridge Aesthesiometer gave the most reliable separation between people with VWF and those not exposed to vibration. A development of the device, using a rising ridge around a disc

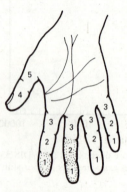

Digit	Th.	1	2	3	4
Right Hand					
Possible Score	4+5	1+2+3	1+2+3	1+2+3	1+2+3
*Actual Score	0	1	3	0	0
Total Score	4/33				

Digit	Th.	1	2	3	4
Left Hand					
Possible Score	4+5	1+2+3	1+2+3	1+2+3	1+2+3
*Actual Score	0	1	3	6	6
Total Score	16/33				

(Note T (tingling) or N (numbness) may be inserted as appropriate where score = 0)

*Example   TOTAL SCORE : 4ᵣ, 16ₗ

Maximum Score on
Either Hand   Possible Action
1   Vibration-exposed person to be regularly warned
           of problem and informed of possible
           consequences
3   Vibration-exposed person to be advised not to
           continue with vibration work
5   Vibration-exposed person to be removed from
           vibration work
9   Compensation of vibration-exposed worker

Possible Use of Proposed Categorisation of Vibration-induced White Finger
(The action appropriate to each score should be defined by the appropriate authorities: the actions listed are only intended to illustrate the possible use of the proposed method of categorisation)

Figure 17.4. Categorization system proposed for vibration induced white finger, with actions of worker warning and protection (Griffin 1982).

was developed by Corlett *et al.* (1981) (see Figure 17.5). This provided closer control over experimental errors. Studies by Marsh (1986) suggest that increased discrimination can be achieved by using as a measure the degree of rotation of the disc between the just noticeable appearance of the ridge and the 'just noticeable disappearance' of the ridge.

The other most generally well-considered method suitable for field use is a vibration aesthesiometer. This has its detractors, in that concern is expressed about the level of adaptation experienced by the fingers which may bias the measurements.

Apart from these objective measures, discussion with the workers concerned will provide information on symptoms and when they occur, which will give additional evidence for identification. Methods of working, characteristics of tools and equipment and the circumstances under which they are used, as well as the various uses to which they are put, will give the investigator knowledge to help in recognizing the likely paths for the vibration transmission. Such directions of enquiry will increase the opportunity to understand the particular situation and help to reduce the effects of preconceptions and over-simplifications, such as can all too frequently occur if the investigator is unfamiliar with the practical working activities.

Recording the vibration from tools is not just a matter of screwing accelerometers onto the handle. Many tools create shocks during use, presenting overloads which can cause a DC shift in the record, and which can be misinterpreted as a low-frequency vibration. It is customary to interpose a resiliant pad between accelerometer and tool, which acts as a mechanical filter to vibrations over 2000–3000 Hz. Care must also be taken to see that the mounting does not have its own natural frequency which superimposes onto the tool frequencies.

Figure 17.5. Adapted Renfrew Ridge Aesthesiometer used for detecting VWF.

Piezoelectric accelerometers are most commonly used for measuring tool vibration, together with the appropriate recording equipment. Where tape recording of the results is being used, the response of the recorder must be sufficient for the purpose, and this normally means using a high quality one. If piezoresistive accelerometers are used, it is necessary to employ damped accelerometers; the undamped version is too easily damaged by shock. Some workers mount the accelerometer on a plate which is gripped by the operator, seeking to record the vibrations transmitted to the hand. Rasmussen (1982) reported studies of various forms of such mountings which gave transmissibility values close to unity, i.e. the transfer function was flat over the range of frequencies of interest for hand/arm vibration.

The intensity of a worker's grip on the tool is an important aspect; grip is almost always necessary for control and the accelerations should be recorded at least in the direction in which the grip by the fingers is being exerted. It is the transmission of vibration to the fingers which is of major importance, and both frequency and acceleration are required. The measure of acceleration is related to the energy present in the system.

Apart from measuring the vibration characteristics, the extent of exposure to vibration must also be recorded. Activity sampling (discussed as 'occurrence sampling' in Drury's chapter 2 on direct observation) to give the proportion of the working period over which the person is exposed may be sufficient, but it may be desirable also to record the lengths of work and rest periods. Especially where working sessions are long and breaks short, it is advisable to maintain the record of actual work and rest intervals, since a general statement of percentage exposure per day could be misleading.

When investigating hand/arm vibration the objective is usually to determine if the levels to which the person is exposed will induce VWF during a working lifetime. Current practice is to specify how to assess the energy input to the hand rather than just limiting values of frequency and acceleration. The method is described in an appendix to ISO DIS 5349.2 (ISO, 1986).

Where exposures cannot be reduced to the recommended levels and engineering changes cannot reduce the severity of the vibration, recourse is usually made to limiting total operating times. Brammer (1982a,b) proposed a method for estimating the latent interval, and standard deviation of that interval, for a given level of the frequency weighted r.m.s. component acceleration. This latter value is the one defined in the ISO appendix (ISO, 1986) and represents the energy input to the hand. The method does provide some measure which is of use in the design of new products as well as being useful for those concerned with the improvement of current practices.

For test purposes various ways of mounting vibration recording equipment, particularly on chain-saws, have been proposed, which may be specified for standards purposes. These methods endeavour to record vibration without any confounding by the variability introduced by the human operator, and they are of particular value when assessing the effects of various designs of vibration isolators on, for example, chain-saw handles (Reynolds and Wilson, 1982).

Results obtained through using the techniques described provide the data for assessing the level of exposure against the appropriate standard, which in this case is ISO 5349.2 (ISO, 1986).

Recommendations to reduce the effects of hand/arm vibration

Some general (non-medical) measures to reduce the effects of hand/arm vibration are recommended in ISO 5349.2 (ISO, 1986), which is summarized as follows:

Technical recommendations

Measure vibration in all three orthogonal directions on the handles (or gripped parts of tool or component) and compare with ISO recommended levels. If they exceed these levels:

1. Reduce the vibration by anti-vibration handle/mountings, or where appropriate, eliminate any out of balance components in the equipment.
2. Where grinding on pedestal grinders is concerned, improve the wheel dressing to reduce wheel irregularities and introduce softer wheels.
3. Improve the maintenance of equipment.
4. Ensure that the tools *are* appropriate for the task.
5. Where rotary hand-held grinders are in use, provide flexible suspension to reduce the forces needed to manipulate the tool.

Advice to the employer

1. Ensure that adequate training and instruction are provided in the proper use of the tool.
2. The worker should let the tool do the work, grasping it as lightly as is possible consistent with safe work practice and tool control.
3. Vibration hazards are reduced when continuous vibration exposure over long periods is avoided. Therefore work schedules with adequate rest breaks are essential.
4. The worker should use the tools only when absolutely necessary.
5. Investigate for the incidence of VWF at least annually.

Advice to employee

1. Wear adequate clothing to keep body temperature at an acceptable level, and wear gloves whenever practically possible during use of the tools.
2. Before starting a job warm the hands, and thereafter keep them as warm as possible.

3. Reduce smoking when using vibrating hand tools; smoking can act as a vasoconstrictor, i.e. it reduces blood supply to the fingers.

4. Should signs of tingling, numbness or blueness in the fingers occur, see a physician and possibly seek a change of job.

Whole body vibration

Principles

Unlike hand/arm vibration there is no particular injury which is due specifically to *whole body vibration* (WBV). However, it is recognized that exposure can be detrimental to health, comfort and performance and that methods should therefore be sought to minimize WBV effects. Vibration can be sensed in different parts of the body across a very wide range of frequencies, from 0.1–10000 Hz (see Figure 17.6). However, it is generally agreed that human sensitivity to WBV is greatest around 4–8 Hz in the 'z' (up and down) direction and 1–2 Hz in the 'x' (front to back) and 'y' (side to side) directions. Furthermore, there is strong epidemiological evidence to suggest that there is an increased level of low-back pain and gastrointestinal

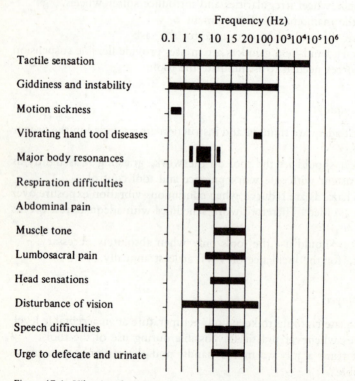

Figure 17.6. Vibration frequencies at which different physiological effects occur.

problems amongst people who are exposed to WBV at these frequencies for long periods of time (Rosegger and Rosegger, 1960; Kelsey, 1975; Heliovaara, 1987).

Human reaction to WBV has been studied extensively, mainly in relation to transportation industries; these studies have been aimed primarily at improved comfort and performance for crew and passengers. Other studies have investigated the actual amount of vibration reaching the operator through the seat surface for land vehicles, aircraft and ships (e.g. Coermann, 1960; Lovesey, 1975).

It is difficult to establish simple and acceptable principles for measuring exposure to vibration since the acceptable level of exposure is very much dependent on the environment within which the operator is exposed. Furthermore, most criteria which seek to limit vibration exposure are based on studies using sinusoidal vibration. ISO 2631 (ISO, 1985) defines the limits of exposure to vibration in both the 'z' and 'y' directions (1–80 Hz range) in terms of three criteria:

(a) preservation of health, *exposure level* (EL),
(b) working efficiency, *frequency decreased proficiency boundary* (FDPB), and
(c) comfort, *reduced comfort boundary* (RCB).

The levels for each criterion are defined in terms of the maximum time for which an operator should be exposed to vibration. The levels range from 1 min to 24 h. For any exposure time:

$$EL = 2FDPB \text{ and}$$
$$RCB = FDPB/3 \cdot 14 \text{ (see Figure 17.7)}.$$

Oborne (1983) has criticized the 1978 version of ISO 2631 (which is very similar to the 1985 amended version) and claimed that the experimental evidence for it is lacking, both in quantity and quality. The standard however is in wide use, giving guidelines for exposure limits which should not be exceeded.

Practical assessment techniques for whole body vibration

Most situations of exposure to WBV occur whilst people are in a seated position. As with hand/arm vibration it is necessary to record the amplitudes and accelerations of the vibrations to which the person is exposed, together with the periods of exposure. It is also important to note the posture of the person, since it has been found that the way the person sits in, say, a driving seat can affect the amount of vibration reaching the upper part of the body (Wilder *et al.*, 1982). Posture changes and time spent in different postures should be noted during observation.

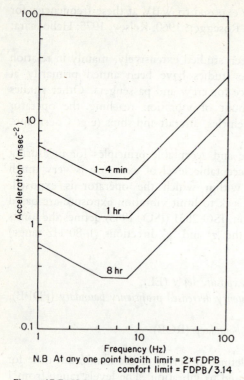

N.B At any one point heaith limit = 2 × FDPB
        comfort limit = FDPB/3.14

Figure 17.7. ISO acceptable limits of whole body vibration exposure for working efficiency.

Major groups who are exposed to WBV, and on whom much research attention has been focused, are vehicle drivers and pilots. In the following discussion the vehicle driver will be used as the example.

As the seat itself may be flexible, it is necessary to measure its transmissibility. This is the ratio of the acceleration appearing at the seat surface to that imposed on the floor to which the seat is attached. By running a series of tests using accelerometers attached to the floor and to a plate on which the subject sits, and gathering data from a sequence of frequencies, the natural frequency of the loaded seat will be found. This is the imposed frequency for which the transmissibility ratio is greatest. Where the ratio is less than unity, the seat is providing damping.

As the body is a complex organism, the various segments of which have their own natural frequency and damping characteristics, it is not sufficient to assume that the frequency at the seat surface is experienced all over the body. If the head under vibration is being investigated, a lightweight bite bar can be used, carrying a triaxial accelerometer.

Basic instrumentation for obtaining vibration measurements is:

an accelerometer,
a preamplifier,

a multichannel FM tape recorder, and
a Fourier analyser.

Vibration will be measured using accelerometers, which produce a voltage directly proportional to the acceleration value of the motion. Piezoelectric accelerometers are the most commonly used because they are small, light and have a wide operating frequency range from about 0·2 Hz to several kilohertz. Most accelerometers are designed to measure vibration in one direction only; thus one accelerometer is required for each of the directions of motion being measured. For the seated subject of our example, to measure the vibration at the seat surface, a triaxial accelerometer can be used embedded in the centre of a hard rubber disc and positioned where the operator's buttocks would be located. Before recording, each accelerometer must first be calibrated.

Once the accelerometers have been attached they must then be connected to a preamplifier so that the vibration signal can be recorded with a magnetic tape recorder. The data can then be analysed using commercially available systems (for example, as supplied by Brüel and Kjaer) designed to display the predominant frequencies and amplitudes of the particular wave form. This method of separating a vibration signal into a number of individual components is called *Fourier* or *spectral analysis*. By comparing the results to ISO 2631 it should be possible to identify whether the level of vibration to which the subject is being exposed is too high. It should be emphasized that to obtain the frequencies and amplitudes of vibrations at different points on a human body is by no means as easy as these simple descriptions might make it appear, and the experimental controls are complex. In addition the equipment required for vibration analysis is costly and sophisticated. Further advice should be sought before embarking upon such experiments and analyses. A discussion of the measurements related particularly to heavy transport vehicles will be found in SAE J1013 (SAE, 1980).

In addition to measuring the actual amount of vibration to which the operators are exposed, it is also important to question them about any discomfort they may feel as a result. Use of subjective measures can provide useful information about the environment which would not be available if only objective measures were used. Whitham and Griffin's (1978) semantic scaling technique may be used as an indicator to describe those areas of the body which are most severely affected by the vibration environment. Rating scales, reviewed by Oborne (1978), can be used to find acceptable tolerance levels within the constraints of the system to which the operator, in this case the driver, is exposed. The response to the subjective questions, however, depends on the interviewee's expectations; for example, an army tank driver would obviously have different expectations of what is comfortable compared to a driver of a family saloon car.

It is known that poor postures and heavy lifting, as well as vibration, are related to back injury problems. Thus in any vibration assessment it is important to consider not only the level of vibration to which the drivers are exposed but also the constraints placed on them within the confines of their equipment, as well as the tasks carried out before and following driving. Only in this way will important parameters relating to the problem be properly identified.

Recommendations to reduce the effects of whole body vibration

1. Where feasible, for example, for rail travel, reduce vibration at source by minimizing the undulations of the surface over which the vehicle must travel.

2. Reduce the transmission of vibration to the driver by improving vehicle suspension, and altering the position of the seat within the vehicle.

3. Decrease the amount of vibration to which the driver is exposed by reducing the speed of travel, minimizing the exposure period, and increasing the recovery time between exposures.

4. Modify the seat and control positions to reduce the incidence of forward or sideways leaning of the trunk; maintain back rest support. Eliminate awkward postures due to difficulty of seeing displays or reaching controls.

Methods for evaluating the effects of vibration on performance have not been mentioned. Body parts tend to vibrate in sympathy with vibrating machinery; vibration can therefore reduce motor control, producing hand unsteadiness, or can cause oscillations of the eyeballs resulting in difficulty in focusing. There is little evidence to suggest that vibration can affect central information processes.

Conclusions

The overall objectives of this chapter have been: first, to summarize the principles and methods for evaluating the effects of vibration in situations where exposure to the human operator is an unnecessary and unexpected problem; and secondly, to provide some recommendations as to how the effects might be reduced. This has been done for both hand/arm and for whole body vibration.

Methods employed to assess vibration environments are dependent upon whether we are interested in effects in terms of whole body or hand/arm vibration. For both situations a major problem with exposure to vibration is that often neither work-force nor management realizes the potential dangers and they do not take the recommended precautions at the right time. The overall aim of vibration assessment should be to determine the extent of vibration exposure, and to compare this with recommended limits, but in any case to redesign equipment and jobs such that exposure is minimized in terms of degree and time.

References

Brammer, A.J. (1982a). Relations between vibration exposure and the development of vibration syndrome. In *Vibration Effects on the Hand and Arm in Industry*, edited by A.J. Brammer and W. Taylor (New York: John Wiley), pp. 283–290.

Brammer, A.J. (1982b). Threshold limit for hand-arm vibration exposure throughout the work day. In *Vibration Effects on the Hand and Arm in Industry*, edited by A.J. Brammer and W. Taylor (New York: John Wiley), pp. 291–301.

Brammer, A.J. and Taylor, W. (eds) (1982). *Vibration Effects on the Hand and Arm in Industry* (New York: John Wiley).

Coermann, R.R. (1960). The passive mechanical properties of the human thorax-abdomen system and of the whole body system. *Journal of Aerospace Medicine*, **31**, 915–924.

Coermann, R.R. (1962). The mechanical impedance of the human body in sitting and standing position of low frequencies. *Human Factors*, **4**, 227–253.

Corlett, E.N., Akinmayoa, N.K. and Sivayoganathan, K. (1981). A new aesthesiometer for investigating Vibration White Finger (VWF). *Ergonomics*, **24**, 603–630.

Griffin, M.J. (1982). *The Effects of Vibration on Health*. Memorandum 632. (University of Southampton: Institute for Sound and Vibration Research).

Heliovaara, M. (1987). Occupation and risk of herniated lumbar intervertebral disc or sciatica leading to hospitalization. *Journal of Chronic Diseases*, **40**, 259–264.

ISO (1985). *Guide for the Evaluation of Human Exposure to Whole Body Vibration*. ISO 2631. (Geneva: International Standards Organization).

ISO (1986). *Guidelines for the Measurement and Assessment of Human Exposure to Hand Transmitted Vibration*. ISO/DIS 5349.2 (Geneva: International Standards Organization).

Kelsey, J.L. (1975). An epidemiological study of acute herniated lumbar intervertebral discs. *Rheumatic Rehabilitation*, **14**, 144–159.

Lovesey, T. (1975). The helicopter: some ergonomic factors. *Applied Ergonomics*, **6**, 139–149.

Marsh, D.R. (1986). Use of a wheel aesthesiometer for testing sensibility in the hand. *Journal of Hand Surgery*, **11-B/2**, 182–186.

Oborne, D.J. (1978). The stability of equal sensation contours for whole body vibration. *Ergonomics*, **21**, 651–658.

Oborne, D.J. (1983). Whole body vibration and International Standard ISO 2631: a critique. *Human Factors*, **25**, 55–69.

Rasmussen, G. (1982). Measurement of vibration coupled to the hand–arm system. In *Vibration Effects on the Hand and Arm in Industry*, edited by A.S. Brammer and W. Taylor, (New York: John Wiley), pp. 89–96.

Reynolds, D.D. and Wilson, F.L. (1982). Mechanical test stand for measuring the vibration of chain saw handles during cutting operation. In *Vibration Effects on the Hand and Arm in Industry*, edited by A.S. Brammer and W. Taylor, (New York: John Wiley), pp. 211–224.

Riddle, H.F.V. and Taylor, W. (1982). Vibration-induced white finger among chain sawyers nine years after the introduction of anti-vibration measures. In *Vibration Effects on the Hand and Arm in Industry,* edited by A.S. Brammer and W. Taylor, (New York: John Wiley), pp. 169–172.

Rosegger, R. and Rosegger, S. (1960). Health effects of tractor driving. *Journal of Agricultural Engineering Research,* **5**, 241–275.

SAE (1980). *Measurement of Whole Body Vibration of the Seated Operator of Off Highway Work Machines.* SAE J1013. SAE Handbook Volume 4 (Warrendale, PA: Society of Automotive Engineers, Inc.).

Taylor, W. and Pelmear, P.L. (1975). *Vibration White Finger in Industry.* (London: Academic Press).

Wasserman, D.E. (1987). *Advances in Human Factors/Ergonomics,* Volume 8, *Human Aspects of Occupational Vibration.* (Amsterdam: Elsevier).

Whitham, E.M. and Griffin, M.J. (1978). The effects of vibration frequency and direction on the location of areas of discomfort caused by whole body vibration. *Applied Ergonomics,* **9**, 231–239.

Wilder, D.G., Woodworth, B.B., Frymoyer, J.W. and Pope, M.H. (1982). Vibration and the human spine. *Spine,* **7**, 243–254.

Chapter 18

Anthropometry and the design of workspaces

Stephen T. Pheasant

Introduction

Anthropometry is the branch of the human sciences which deals with body measurements—particularly those of size, shape and body composition. Biomechanics is the application of mechanical principles to the study of the structure and function of the human body. As applied to ergonomics, the two are closely linked, since the science of biomechanics commonly provides the *criteria* for the application of anthropometric data to the problems of design.

This chapter is concerned with anthropometry and the methods for using anthropometric data. For the latter it will be necessary to consider criteria, but a fuller discussion of biomechanics will be found in a further chapter by Tracy. For the wider applications of these subjects to design, the reader is referred to Pheasant (1986) or Clark and Corlett (1984).

Anthropometric data

The variability of most bodily dimensions may be described, to a tolerable degree of accuracy, by a mathematical function known as the *normal distribution*. This name does not imply that it describes the distribution of 'normal people'—whatever that might mean anyway—but rather as meaning 'the distribution which you will find most useful in practical affairs'. To deal with this slight semantic difficulty, it is sometimes referred to as the Gaussian distribution, after Johann Gauss (1777–1855), the German mathematician and physicist who first explored its mathematical properties in detail. Gauss's concern in this respect was with random errors in physical measurement. The fact that bodily characteristics such as stature (i.e. standing height) are also normally distributed is an empirical observation due to the English anthropologist and geneticist Sir Francis Galton (1822–1911). Geneticists have in fact contrived certain plausible mathematical models to account for the normal distribution of anthropometric characteristics, but the details of these need not concern us here.

The distribution of stature in adult British men is shown in Figure 18.1. Frequency is plotted vertically and stature is plotted horizontally. Beneath the horizontal axis is a second scale showing percentiles of stature. In any particular characteristic, *n*% of the population concerned are smaller than the *n*th percentile, i.e. the *n*th percentile is the value which is exceeded in $(100-n)$% of cases. Note that the percentiles are close together in the centre of the distribution and widely spread in the tails.

The curve is symmetrical about the mid-point. The 50th percentile value is also the most common value and the frequency declines systematically as you enter the tails of the distribution.

Figure 18.2 shows exactly the same data, plotted in a slightly different way. This is the cumulative form of the normal distribution, sometimes known (because of its shape) as the *normal ogive*. Plotted horizontally is stature; plotted vertically are percentiles of stature. This is a particularly useful way of presenting the data for our present purposes because it allows us to read off directly the percentage of people who will be *accommodated* (i.e. satisfied) with respect to a particular anthropometric criterion. To take a slightly trivial example, Figure 18.2 tells us the percentage of men who would be able to pass under a doorway of a given height, without running the risk of banging their heads.

The normal distribution is completely described by two parameters, the mean and the standard deviation. The mean is the same thing as the familiar arithmetical average and for normally distributed variables it is equal to the

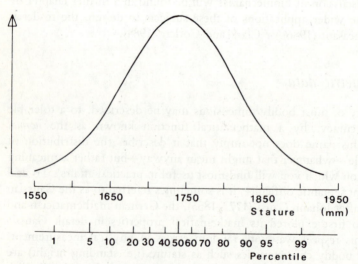

Figure 18.1. The normal distribution of the stature of British men. After Pheasant (1986).

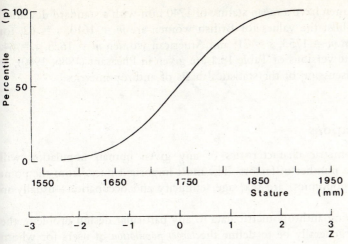

Figure 18.2. The normal distribution of the stature of British men, plotted in cumulative form. After Pheasant (1986).

50th percentile. The standard deviation is a measure of dispersion; it describes the extent to which an individual might be expected to differ from the mean. So we might say, for example, that the mean stature of a carefully selected sample of high-jumpers was greater than the mean stature of the population at large, but that their standard deviation was less.

If the mean (m) and the standard deviation (s) of a normally distributed variable are known; then any percentile (X_p) which we might happen to require may be calculated from the following equation:

$$X_p = m + s\,z$$

where z (the standard normal deviate) is a factor for the percentile concerned. Values of z for some commonly used percentiles (p) are given in Table 18.1 which is the cumulative version of the normal distribution plotted in tabular

Table 18.1. Selected values of z and p

p	z	p	z
1	−2·33	99	2·33
2·5	−1·96	97·5	1·96
5	−1·64	95	1·64
10	−1·28	90	1·28
15	−1·04	85	1·04
20	−0·84	80	0·84
25	−0·67	75	0·67
30	−0·52	70	0·52
40	−0·25	60	0·25
50	0·00	50	0·00

form. British men have a mean stature of 1740 mm with a standard deviation of 70 mm, whilst the values for British women are $m = 1610$, $s = 62$; for American men $m = 1755$, $s = 71$; and American women $m = 1625$, $s = 64$. More complete versions of Table 18.1 are given in Pheasant (1986, 1990), as is a further discussion of the statistical basis of anthropometrics.

User populations

The anthropometric characteristics of any given human population will depend upon a number of factors. The most important ones, from the point of view of ergonomics, are sex, age, ethnicity and occupation—usually in that order.

When applying anthropometric data to any particular design problem, the first step will generally be to define the *target population* of users for whom the product (workstation, environment) is intended, and to locate a source of anthropometric data for the population concerned (or for one which resembles it as closely as possible in relevant respects). The consequences of using inappropriate data will be that fewer people will be satisfied than intended—sometimes quite drastically so—or even worse, the performance of the eventual users will be impaired.

In most adult populations a difference of about 7% between the average heights of men and women (and a somewhat larger difference in their standard deviations) will be found. On average, men will also be larger in most other respects, although the magnitude of the difference will vary from dimension to dimension. The most important exception to this rule is hip breadth. In addition to the more obvious differences in shape between men and women, it is worth noting that men have proportionally greater limb lengths. That is, if we were to compare a man and a woman *of equal stature*, we should expect the man to have longer arms and legs, bigger hands and feet, and so on. There will also be proportional differences in dimensions which have a substantial soft tissue component. This is partly because men have (on average) greater muscle bulk whereas women have greater bodily fat; and it is partly due to sex differences in fat distribution.

People also change in shape as they get older. In our society at least, they tend to put on weight; and after the age of about 55, muscle bulk begins to decrease and the spine begins to shorten due to changes in the properties of the intervertebral discs.

Anthropometric differences due to the ageing process itself are confounded with differences due to long-term historical processes known as secular trends. In Britain, Europe and North America, people have been getting steadily taller. For the last century or so, there has been an upward trend in adult stature at the rate of about 10 mm/decade (an inch/generation). This trend was first noticed by Galton in the latter part of the last century. In Britain and in North America though there is some evidence that the trend ıas now come to a halt, but in Japan it is still under way at a very rapid ate.

The ethnic groups of the world differ in both size and shape. Figure 18.3 hows sitting height (i.e. the distance from the seat to the crown of the head) ›lotted against stature. The oblique lines on the chart show sitting height livided by stature. Each of the major ethnic divisions of the world include ›oth tall and short ethnic groups, but they tend to have characteristic body ›roportions. Black Africans have relatively long legs for their stature; far :astern peoples have relatively short legs (particularly so the Japanese). Europeans and Indo-Mediterraneans (who together make up the so-called Caucasoid division of mankind) are somewhere between the two extremes.

A compilation of basic anthropometric data for the British population, ₊ith some advice on its use, will be found in Pheasant (1990), whilst a wider ₊ange of information, including adult dimensions for certain other nationalities, ₌s given in Pheasant (1986). The largest collection of anthropometric data ›ublished in any one place at present is held by NASA (1978) although the nternational Ergonomics Association is seeking to draw together an nternational data set appropriate for ergonomics applications.

Criteria

Before pursuing some of the methods for using anthropometric data we nust look at how we decide whether what we do will be adequate for the ›urpose. We need criteria both to guide us in our applications and to provide ₊ framework against which we can test our decisions.

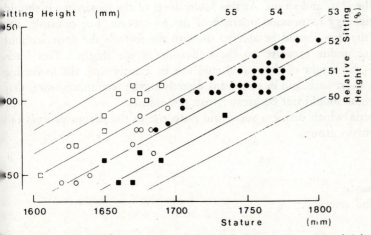

Figure 18.3. Ethnic differences in the relationship between average sitting height and average stature in samples of adult men. (● = European, ○ = Indo-Mediterranean, □ = Far Eastern, ◼ = African) After Pheasant (1986).

In the context of ergonomics, a criterion is a standard of judgement which defines the extent to which a particular product (workstation, environment) is appropriately matched to its human users. For a criterion to have any practical (or scientific) value, it must be possible to specify the operations which an investigator would have to perform, in order to determine whether the criterion had indeed been satisfied in any particular case. A criterion which may be defined in this way is known as an *operational criterion*. There is not much point in saying that a product should be 'easily usable' unless we can define usability in terms of how it could be measured. Some product standards work in this way; BS 6652, *Packages Resistant to Opening by Children*, defines a set of procedures for conducting an experiment to determine how easily people are able to open a particular container. An experiment of this kind is called a *user trial* (see the chapter by McClelland in this book for a full review of user trials). An ergonomic criterion of this kind indicates whether an existing product is to be regarded as satisfactory but it does not reveal much about how the product should be designed. This has both advantages and disadvantages.

Design criteria are hierarchical. At the highest level there are very general concepts: comfort, efficiency, safety, usability and so on. In themselves, these are not easy to define in operational terms—at least not in ways which are directly useful to the designer. To get round this problem, we need to break down these high-level criteria into subordinate criteria at successively lower levels in the hierarchy. Consider the criteria which might define an 'ergonomically-designed chair'. One of these would obviously be comfort. Subordinate to this would be more specific design principles, such as the provision of adequate postural support and the avoidance of pressure hot-spots on supporting surfaces. These in turn could be broken down into component parts dealing, for example, with the angle of the backrest, the height of the seat, and so on. At this lower level of the hierarchy, it should be relatively easy to provide operational design criteria. For example, the criterion, 'the user should be able to sit with the feet on the floor without experiencing undue pressure on the underside of the thighs'. This leads directly to 'the height of the seat should not be greater than the lower leg length of a short user' and in turn leads directly to the recommendation that the seat height should not be greater than 400 mm.

The criteria which define a successful outcome to the design process fall into three main groups:

comfort,
performance,
health and safety.

(We will see that these groupings recur in assessment of the physical environment, in the chapters by Howarth, Haslegrave and Parsons.)

Case studies show that those ergonomics measures which increase comfort and well-being are also likely to improve productivity (and vice versa). For example Ong (1984) studied a group of data entry operators at an airline computer centre in Singapore, before and after ergonomic improvements to the design of their workstations. He found both an increase in productivity and a dramatic reduction in the symptoms of visual and muscular fatigue and discomfort. Performance as measured by keystrokes per hour increased by 25%—but at the same time, the error rate dropped from an overall 1·5% (1 character in 66 incorrect) to 0·1% (1 character in 1000 incorrect). Dainoff and Dainoff (1986) describe experiments on data-entry operators in which comparisons were made between a workstation which was designed according to commonly accepted ergonomics guidelines and one which deliberately broke most of the rules. (The workstations differed with respect to seat design, keyboard height, screen location, task lighting, glare, etc.) Performance was 25% better at the ergonomically-designed workstation. When the differences in lighting were eliminated there was still a performance difference of 17·5%. The subjects also expressed a preference for the ergonomically-designed workstation and experienced less back and shoulder pain. Aside from any humanitarian considerations involved, performance differences of this magnitude amply justify the costs of the ergonomics intervention.

In many problems of workstation design, the immediate objective will be to achieve appropriate muscular efforts for the performance of a given task. We may reasonably assume that this will have both short- and long-term benefits. In the short-term it will reduce fatigue—and by so doing, improve both performance and subjective comfort. In the long-term, it will reduce the incidence of conditions such as back pain, neck pain and repetitive strain injuries. The sickness absence which results from these common musculoskeletal disorders is economically costly, both for the organization concerned and for society as a whole.

Physiological and biomechanical principles

To achieve the desirable outcomes described earlier we need some rational procedures and principles. To help us to decide the broad arrangements of equipments, displays and controls, and layouts in general, there are four principles, stated by McCormick (1970), which might be seen as a formalization of common sense; because they are so evidently ignored in many situations they should not be underrated.

1. *Importance principle.* Those components which are most essential to safe and efficient operation should be in the most accessible positions.
2. *Frequency of use principle.* Those components which are used most frequently should be in the most accessible positions.
3. *Function principle.* Components with closely related functions should be located close to each other.

4. *Sequence of use principle.* Components which are often used in sequence should be located close to each other and their layout should relate logically to the sequence of operation.

Note that the term 'accessible' relates not only to physical accessibility (such as ease of reach) but also to visibility and to other more abstrac characteristics.

Further sets of principles are needed to deal with the direct relationship between the person and the workplace. These will arise from consideration o the physiological, biomechanic and sensory (information transfer) relationship required if the people concerned are to maintain the comfort, performance and health aspects which were stated earlier to be the basis of our desig requirements. Reference to ergonomics texts, appropriate to the aspects being designed, will give information on the requirements to be met if good performance is to be possible.

Within industry it has been common to rely on concepts such as the principles of easy movement, still listed in many methods texts, for decision on workspace arrangements. These principles are limited in how they accord with the needs of human physiology and psychology, and Corlett (1978 proposed a set which was more in line with current knowledge (Table 18.2) These provide more specific guidance for the choice of anthropometri dimensions to meet our basic principles.

Table 18.2. New principles for workplace layout

---

1. The worker should be able to maintain an upright and forward facing posture during work

2. Where vision is a requirement of the task, the necessary work points must be adequatel visible with the head and trunk upright or with just the head inclined slightly forward.

3. All work activities should permit the worker to adopt several different, but equally health and safe, postures without reducing capability to do the work.

4. Work should be arranged so that it may be done, at the worker's choice, in either a seate or standing position. When seated, the worker should be able to use the backrest of th chair at will, without necessitating a change of movements.

5. The weight of the body, when standing, should be carried equally on both feet, and foo pedals designed accordingly.

6. Work activities should be performed with the joints at about the mid-point of their rang of movement. This applies particularly to the head, trunk and upper limbs.

7. Where muscular force has to be exerted it should be by the largest appropriate muscl groups available and in a direction co-linear with the limbs concerned.

8. Work should not be performed consistently at or above the level of the heart; even th occasional performance where force is exerted above heart level should be avoided. Wher light hand work must be performed above heart level, rests for the upper arms are requirement.

9. Where a force has to be exerted repeatedly, it should be possible to exert it with either o the arms, or either of the legs, without adjustment to the equipment.

10. Rest pauses should allow for all loads experienced at work, including environmental an information loads, and the length of the work period between successive rest periods.

Application of anthropometric data and criteria

Design limits

The slope of the normal ogive is steepest at the mean and it decreases steadily in the tails of the distribution. This has an extremely important consequence for ergonomics in that it is increasingly difficult to accommodate extreme individuals. Let us consider a specific example: the desirable range of adjustment for the height of a working chair. For the purposes of argument we shall ignore things like the table with which the chair will be used and the task which will be performed (both of which are, in practice, very important) and we shall consider the chair taken in isolation. A seat which is too high causes undue pressure on the undersides of the thighs; one which is too low makes standing up and sitting down needlessly difficult and encourages a slumped position of the spine. Most ergonomics books would recommend therefore that the height of a chair should be a little below the popliteal height of its user. (Popliteal height is the vertical distance from the floor to the crease at the back of the knee.) It so happens that the popliteal height of British adults (ignoring sex differences) has a mean of 455 mm and a standard deviation of 30 mm. If we assume the optimal height of a seat to be 40 mm less than popliteal height, then the distribution of optimal heights will have a mean of 455 − 40 = 415 mm, with a standard deviation of 30 mm. By calculating percentiles of this distribution, we can calculate the percentage of the target population who would be matched by any given range of height adjustment. From Table 18.1 we find that the 25th and 75th percentiles are 0·67 standard deviations below and above the mean, respectively. Hence to satisfy the 50% of users who are between these limits, we would need to make our chair adjustable by 0·67 × 30 = 20 mm, on either side of the mean value, i.e. from 395 to 435 mm. Repeating this calculation for other percentiles, the following results are obtained:

Per cent satisfied	50	60	70	80	90	95	98
Millimetres adjustment	40	50	62	77	98	118	140

We are in a situation of diminishing returns in which each additional unit of adjustment yields less benefit in terms of the percentage of satisfied users.

In practical terms we have to set limits on our attempts to satisfy the largest possible number of users. Where should we set these *design limits*? By convention we usually choose to accommodate (i.e. satisfy) the range of users who are between the 5th and 95th percentile of whatever characteristic we are dealing with. In some problems (like the one we have just considered) this will accommodate 90% of people; in others (as we shall see shortly) it will accommodate 95%. This is an arbitrary choice based on expediency. Beyond these limits the situation of diminishing returns begins to tell against us rather dramatically. In some cases however (such as when dealing with health and safety issues) we will need to set wider limits. The rule of thumb for making such decisions is to consider the worst possible consequences of a mismatch.

Fitting trials

A fitting trial is an experimental investigation of the relationships between the dimensions of a product (workstation, environment) and the dimensions of its users. In general, the experimental subjects will be asked to try out an adjustable mock-up of the product concerned. Critical dimensions of the product will be adjusted through a range of values, and the subjects will be asked to express their preferences with respect to comfort, ease of use, and so on. A fitting trial is therefore a special kind of psychophysical experiment. It is important that the subjects in the experiment should be a representative sample of users of the product—both with respect to their body dimensions and with respect to their general fitness and anything else which might be relevant.

Suppose we wished to conduct an experiment to determine the narrowest gap between two obstructions, of a given height, that subjects could pass through without experiencing undue inconvenience. We could do the experiment simply by shifting furniture around, so as to create gaps of different widths. We could start with a gap which was so narrow that no one could get through; and then systematically widen it (perhaps in 100 mm increments) until we had a gap which everyone could get through. If we plotted a graph of gap width (on the horizontal axis) against the percentage of people who could get through (on the vertical axis), we should expect to get a relationship which looked something like Figure 18.2, i.e. the cumulative form of the normal distribution. We could make it look exactly like Figure 18.2 by calculating the mean and standard deviation of the minimum gap widths through which each individual subject could pass, and then use these parameters to calculate percentiles, using the z values in Table 18.1. (In essence we are smoothing our data by fitting a normal distribution.) The experiment could be made more sophisticated by observing whether the subjects passed through crab-wise or head on, or by asking them whether it was easy or difficult, and so on. In fact, many different criteria could be used, and if the data plotted for each of these criteria in turn, a series of normal distributions, spaced out along the horizontal axis, would be expected.

To conduct a fitting trial to determine the optimal height for a control (like a door handle or a light switch) or for a working surface would be slightly more complicated. A gap between obstacles can be too narrow for convenience, but it cannot reasonably be too wide; whereas a door handle or a work bench can be too low, too high or just right. We could start with a position which was too low for everyone; then as we raised the level of the object, we would expect the percentage of subjects who judged it to be too low to steadily decrease. As it did so, the percentage of people saying it was just right would climb to a peak value, before beginning to fall again—as an increasing number of people began to say that it was too high. The characteristic form which the results of such experiments are generally found to take, is shown in Figure 18.4.

The latter experiment has a considerably greater element of subjective judgement than the former (which is much more of a go/no-go situation, in which you can either get through the gap or else you cannot). This has a number of important consequences. Characteristically, it is found that the range of optima reported in ascending trials (i.e. from low to high) is placed higher than the range reported in descending trials (i.e. from high to low). This is true of psychophysical experiments in general.

If we conduct our fitting trial with each subject running through the range of possible positions of the door handle, from low to high and back to low again, and plot the results, a graph similar to that of Figure 18.5 would result. Subjects have experienced all the positions from 'too low' to 'too high', and a section in between which they say is satisfactory. The test is conducted, for each person, with the height changing up and down, to incorporate the effects of ascending and descending trials as described above. We use subjects who are at the extremes of the body dimensions which are relevant to the dimension of the object which we are testing, in the example

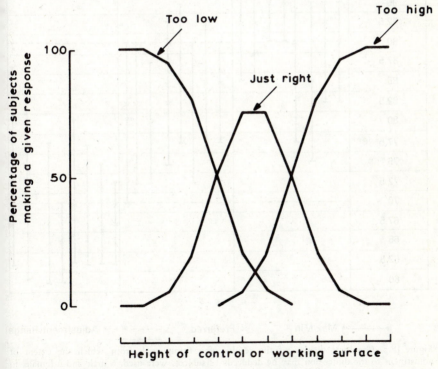

Figure 18.4. Characteristic form taken by the results of a fitting trial.

of Figure 18.5, the 5th and 95th percentiles. We may include some others if we wish, and 50th percentiles were also used here.

If a line can be drawn through all the 'satisfactory' dimensions, then this indicates that there is one level of this dimension which suits everyone. If this cannot be done, then a line through the bottom of the highest 'satisfaction line', and one through the upper end of the lowest, demonstrate the minimum range of adjustment needed to suit the population concerned. The method,

SUBJECT

DIMENSION (cm)	5th				50th				95th			
	M		F		M		F		M		F	
	1	2	3	4	5	6	7	8	9	10	11	12
110												
107.5												
105												
102.5												
100												
97.5												
95												
92.5												
90												
87.5												
85												
82.5												
80												
77.5												
75												
72.5												
70												
67.5												
65												
62.5												
60												

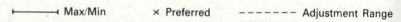

├────────┤ Max/Min   × Preferred   ------- Adjustment Range

Figure 18.5. Graphical display of the results of a fitting trial, from which the extent of adjustment to the dimension may be deduced. 12 subjects were used, 6 male and 6 female, in the 3 percentile groups shown.

due to Jones (1963), finally needs all the separately selected dimensions putting together and checking with the sample of subjects, since dimensions may interact with each other, which would only be revealed by their final grouping and checking.

It is also worth noting that the form of results shown in Figure 18.4 also turns up in quite different areas of ergonomics, for example, in studies of thermal comfort (Fanger, 1970; Grandjean, 1988). In fact it is likely to be relevant to studies of subjective preference in general. The lines on the graph could represent all sorts of things: too soft and too hard; too cool and too warm; too light and too dark; too young and too old, etc.

The anthropometric method of limits

The process by which we establish final design recommendations with respect to anthropometric and biomechanical criteria, is known as the *method of limits*. The term is borrowed from psychophysics. The anthropometric method of limits is essentially a model or analogue of the fitting trial, in which the anthropometric data stand as substitutes for the experimental subjects. You could say that, in applying this method, we are attempting to predict, using pencil and paper methods, what the result of a fitting trial would be if we were to perform one. (Note that the fitting trial is a special case of the psychophysical method of limits.)

In applying this method to any particular design problem, we are attempting to establish those boundary conditions, which make an object 'too big', 'too small', and so on, with respect to certain design criteria. In the simplest cases we may do this by inspection. This requires us to identify the *limiting user*— that is a hypothetical individual, who by virtue of his or her extreme bodily characteristics, is particularly difficult to accommodate with respect to the criteria concerned. If the limiting user is accommodated, it necessarily follows that the majority of the population, who are less demanding in their requirements, will be accommodated as well.

Anthropometric criteria fall into three principal categories: clearance, reach, and posture.

Clearance

Clearance problems include those which relate to head room, knee room, elbow room and so on, as well as those concerned with access through passageways, around and between equipment and into equipment for maintenance purposes. These are amongst the most important issues in workspace design since mismatches in these respects may be particularly hazardous. Unresolved clearance problems may also have knock-on effects in terms of unsatisfactory working postures, which the user is forced to adopt. For example, if the distance between the surface of the seat and the underside of a table does not provide adequate clearance for the thighs and knees of the users, then they may have to adapt to the situation by pushing the chair backwards (and leaning forward excessively to perform the task) or by perching on the front of the seat (and hence losing the support of the backrest). This is quite a common problem at service counters and cash tills.

Clearance should be adequate for the largest user. For practical purposes, it will usually be expedient to set the design limits at the 95th percentile, hence by definition accommodating 95% of the target population.

As an example, suppose we wish to find the minimum clearance required between the arms of a chair. We look up a suitable table of anthropometric data. Searching through this table we find the dimension 'hip breadth', which for our target population (British adults) has 95th percentile values of 405 mm for men and 435 mm for women. We adopt the larger figure. This would be an exact fit for the hips of the limiting user, but it would not allow for her clothing (since anthropometric data are usually quoted for unclad people) nor would it allow her any leeway. Making a commonsense correction for these we arrive at a round figure of 500 mm. (We could, if we wished, formalize this last stage, by specifying exactly what clothing ensemble she is likely to be wearing, and how much leeway we wish to give, but in reality (and for this problem) this degree of precision is rarely realistic.)

The process we have gone through could be summarized as follows:

Criterion: seat breadth ≥ hip breadth.
Limiting user: 95th percentile woman = 435 mm hip breadth.
Corrections: clothing and leeway = 65 mm.
Design recommendation: seat breadth ≥ 500 mm.

This application of the method of limits is in many respects equivalent to the first fitting trial we discussed in the previous section; and if we plotted a graph of the seat breadth (horizontally) against the percentage of users accommodated (vertically) we should again get a normal ogive, as shown in Figure 18.2.

Reach

Reach problems include those which are concerned with the location of controls in the workspace as well as things like the height of a seat (where it is necessary for the feet to reach the floor) and the height of visual obstructions. Seat depth also falls into this category since it is necessary for the user to 'reach' the backrest without undue pressure on the backs of the knees.

The procedure for dealing with reach problems is the same as the one used for clearance problems, except in this case the limiting user will be a small member of the target population, usually a person who is 5th percentile in the relevant characteristic.

Note that both the clearance and the reach criteria impose limits in one direction only. The clearance criterion indicates when an object is too small, but not when it is too large—and conversely for reach. Both clearance and reach therefore are *one-tailed constraints*. There may of course be other constraints acting in the opposite direction. In the case of clearance and access problems, these might include economy of space, or the reduction of distances travelled. In laying out working areas, we often find that clearance problems (e.g. elbow room) interact with reach problems (e.g. the accessibility of controls). The interaction of two opposing sets of constraints, creates a situation which is equivalent to the one which is described in Figure 18.4.

Note also that there is an important class of design problems (concerned with the safeguarding of machinery) in which the conventional criteria of clearance and reach are reversed. In these cases we actively wish to *prevent* access to hazardous areas, or to place the hazards *out of reach*. So the limiting user might, for example, be a person with long slender limbs who could reach the furthest distance through the smallest aperture.

Posture

Posture problems are inherently more complicated. For example, a working surface which is too high is just as undesirable as one which is too low. This limits the design in two directions to give a *two-tailed constraint* of the kind shown in Figure 18.4. In this situation there are two options open to us. We either have to provide an adjustable workstation, so that each user may set it to his or her own optimum dimensions (as discussed previously under 'Design limits'); or else we have to settle on a single overall compromise value which will maximize the number of users who are accommodated and minimize the inconvenience suffered by the remainder. Supposing our problem concerns the height of a work bench to be used by a standing person for performing a certain manipulative task. (It is assumed that the bench will be used by men and women.) A suitable height is between 50 and 100 mm below elbow height. It follows from the shape of the normal distribution that the single overall compromise value which will accommodate the greatest number of people is 75 mm below the average (or 50th percentile) elbow height. Looking up an appropriate table of data, we find the 50th percentile elbow height is 1090 mm for men and 1005 mm for women. This gives an overall average (i.e. for both sexes) of 1048 mm, to which we should add 25 mm for shoes, giving 1073 mm. Subtracting the 75 mm and rounding up, gives us a final recommendation of exactly 1000 mm. We may express this formally as follows:

Criterion: (elbow height − 100 mm) < working surface height < (elbow height − 50 mm).
Optimal compromise: 50th percentile elbow height − 75 mm = 973 mm.
Corrections: 25 mm for shoes.

Design recommendation: working surface height = 998 mm, rounded = 1000 mm.

At this point it will be pertinent to consider how many users will be mismatched with respect to the criteria and how seriously inconvenienced they will be. This will enable us to decide whether a single compromise height will indeed be satisfactory, or whether an adjustable workstation will be necessary.

For a further discussion of the method of limits and further examples of its application see Pheasant (1986, 1990). For a compilation of design standards and guidelines concerned with anthropometric and ergonomic issues, see Pheasant (1987).

Conclusions

In this chapter some of the basic uses of anthropometric data have been described, together with an introduction to sources. When choosing anthropometric data it is necessary to read the small print. Who have been measured? What ages? Both sexes? How many in each sample? How old are the data? Listings of anthropometric data and their diagrams look very impressive, but this is no substitute for reliability and statistical validity.

When preparing a design for a particular situation, e.g. in a factory or office, and you are not sure whether or not tabulated data are suitable, it is valuable to measure some major dimensions for a sample of potential users (e.g. their stature). By comparing the mean and standard deviations of relevant dimensions for the situation you are concerned with, to the tabulated dimensions in published data, simple statistical tests will confirm whether or not they are from the same populations. If so, then you can use the tabulated values with confidence.

It must be emphasized that ergonomics is not synonymous with anthropometrics; it is an unfortunate belief in some design offices that this is the case, and that an ergonomic situation is achieved by choosing dimensions in conjunction with anthropometric tables. Anthropometrics are a necessary, but by no means sufficient, contribution to ergonomic design and evaluation. If the reader keeps in mind that we are dealing with a whole person, with needs for comfort, performance, interest and all those other things which make people so fascinating, then we are unlikely to reduce them to a simple matter of linear dimensions, no matter how vital the statistics.

References

Clark, T.S. and Corlett, E.N. (1984). *The Ergonomics of Workspaces and Machines: A Design Manual*. (London: Taylor and Francis).

Corlett, E.N. (1978). The human body at work: new principles for designing workspaces and methods. *Management Services*, May, 20–52.

Dainoff, M.J. and Dainoff, M.H. (1986). *People and Productivity—A Manager's Guide to Ergonomics in the Modern Office* (Toronto: Holt, Reinhardt and Winston).

Fanger, P.O. (1973). *Thermal Comfort* (New York: McGraw-Hill).

Grandjean, E. (1988). *Fitting the Task to the Man—An Ergonomic Approach*, 4th edition (London: Taylor and Francis).

Jones, J.C. (1963). Fitting trials. Architects Journal Information Library, February, p. 321.

McCormick, E.S. (1970). *Human Factors Engineering*. (New York: McGraw-Hill).

NASA (1978). *Anthropometric Source Book*. US National Aeronautics and Space Administration.

Ong, C.N. (1984). VDT workplace design and physical fatigue: a case study in Singapore. In *Ergonomics and Health in Modern Offices,* edited by E. Grandjean. (London: Taylor and Francis), pp. 484–94.

Pheasant, S.T. (1986). *Bodyspace—Anthropometry, Ergonomics and Design*. (London: Taylor and Francis).

Pheasant, S.T. (1987). *Ergonomics—Standards and Guidelines for Designers*. (London: British Standards Institution).

Pheasant, S.T. (1990). *Anthropometrics—An Introduction*, 2nd edition. (London: British Standards Institution).

Chapter 19

Computer workspace modelling

J. Mark Porter, Keith Case and Maurice C. Bonney

Man*-modelling CAD systems

Computer aided design (CAD) methods are becoming very popular with engineers as they provide considerably more flexibility than conventional techniques. Although they are now commonplace in manufacturing industries the great majority of CAD systems completely ignore the most important component of the human–machine system being designed—humans themselves.

The importance of an ergonomics input to a design is now recognized by many industries as being essential. The increasing complexity of modern systems and the social, economic and legislative pressures for good design have led to the demand for the ergonomics input to be made available as early as possible in the design programme, starting preferably at the concept stage. Traditionally, ergonomists have had to wait until the mock-up stage before being able to perform a detailed evaluation of a prototype design. This delay has several consequences, which will be discussed later in this chapter, all of which are detrimental to the design process.

Clearly, the optimum solution is to provide a means of supplying the ergonomics input in a complementary fashion to the engineering input; the logical conclusion being to develop CAD systems with facilities to model both equipment and people. Recognizing the potential of this solution, in some cases as early as the late 1960s, several research teams have developed man-modelling CAD systems. These have met with varying degrees of success but, essentially, they are design tools which enable evaluations of postural comfort and the assessment of clearances, reach and vision to be conducted on the earliest designs, and even from sketches. In order to achieve these predictions, the systems need: three-dimensional modelling of equipment and workplaces which can be displayed on a computer graphics screen; three-

---

* *Editors' note*: 'Man' is used in this chapter as a generic term in preference to human or people in the context of modelling systems, since this is the terminology employed in the area.

dimensional man models (representations of the human form which can be varied in size, shape and posture for a variety of populations); evaluative techniques, based around the man model, to assess reach, vision, fit and posture; and a highly interactive user interface which allows the user to tailor design evaluations to their own requirements.

Existing systems

Existing man-modelling CAD systems show considerable differences in the extent to which the above facilities have been developed, and the breadth of their potential applications. Brief descriptions of the most established man-modelling systems are given below.

BOEMAN
Developed by the Boeing Corporation, Washington in 1969 for use in checking cockpit layout (see Figure 19.1), the system was complex to use and it was not designed for interactive use as graphics terminals were not commonly in use at that time.

BUFORD
Developed by Rockwell International, California (see Figure 19.2), it offers a simple model of an astronaut, with or without a space suit. Body segments can be selected separately and assembled to construct any desired model, although these segments must be moved individually to simulate working postures. The model does not predict reach but a reach envelope of two-

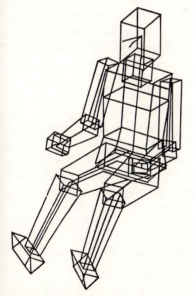

Figure 19.1. BOEMAN (reproduced from Dooley, 1982).

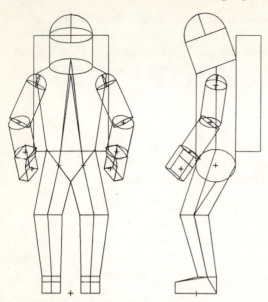

Figure 19.2. BUFORD (reproduced from Dooley, 1982).

handed functional reach can be defined and displayed around the arms. It is not generally available.

CAR (Crew Assessment of Reach)

Developed by Boeing Aerospace Corporat... 1 for use by the Naval Air Development Centre in the USA, the system is designed to estimate the percentage of users (i.e. aircrew) who will be able to be accommodated physically in a particular workstation. The analysis is purely mathematical and the system has no graphical display. It is only available for in-house assessment of reach in aircraft crew stations.

COMBIMAN (Computerized Biomechanical Man Model)

Developed by the University of Dayton for the US Air Force in 1973 to assist in the design and evaluation of aircraft crew stations (see Figure 19.3), the equipment modelling assumes that the work space is made up of panels and controls and the model is only available in the seated position. The use of this system is mainly limited to the prediction of vision and hand reach, and is anyway restricted to in-house users.

CYBERMAN (Cybernetic Man Model)

Developed by the Chrysler Corporation in 1974 for use in design studies of car interiors (see Figure 19.4). There are no constraints on the choice of joint angles so the man model's usefulness for in-depth ergonomics evaluations is rather limited. This system is also not generally available.

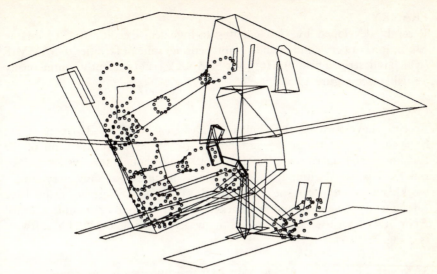

Figure 19.3. COMBIMAN (reproduced from Dooley, 1982).

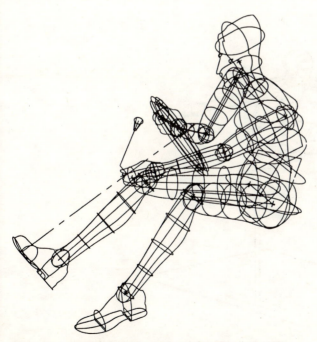

Figure 19.4. CYBERMAN (reproduced from Dooley, 1982).

FRANKY

Recently developed by Gesellschaft fur Ingenieur-Tecnick (GIT) mbH in Essen, it has a very similar (and comprehensive) suite of facilities to SAMMIE (which is described below). However FRANKY is not presently commercially available (see Figure 19.5).

OSCAR

Recently developed by the Hungarian Design Council in Budapest, it has been adapted for use in Western Europe by the SOMACAD team of the Fachhochschule Darmstadt (see Figure 19.6). Although this system can be run on a personal computer, its usefulness is limited as only very simple workplace models can be constructed. This West German team have subsequently developed and are marketing a '3D ergonomic template' called ANYBODY for use as a module in the widely available CADKEY software (see Figure 19.7).

SAMMIE (System for Aiding Man–Machine Interaction Evaluation)

Developed originally at Nottingham University in the late 1960s, and more recently at Loughborough University of Technology (see Figure 19.8), the general purpose nature of the system makes it suitable for a wide range of applications (described later in this chapter). In addition, the system permits the modelling of any special or logical relationships between components of the models, allowing the models to be functional; for example, the operational

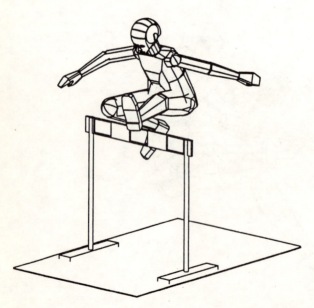

Figure 19.5. FRANKY (reproduced from Elias and Lux, 1986).

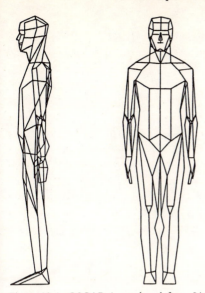

Figure 19.6. OSCAR (reproduced from Lippmann, 1986).

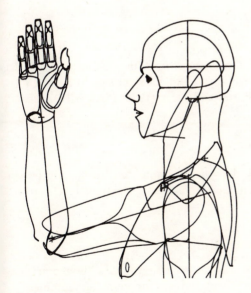

Figure 19.7. ANYBODY.

movements of pedals, doors, seats or levers can be easily specified and executed. SAMMIE is currently the only system being marketed world-wide which provides sophisticated ergonomics facilities and a powerful workplace modelling system.

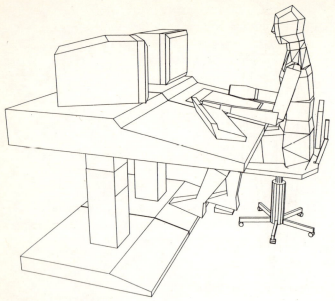

Figure 19.8. SAMMIE.

Further information

Dooley (1982) and Rothwell (1985) both present surveys of man-modelling systems, including most of the above systems. Other sources include: McDaniel (1976) (COMBIMAN); Elias and Lux (1986) (FRANKY); Lippmann (1986) (OSCAR); and Porter and Freer (1987) (SAMMIE).

A man-modelling CAD system

The SAMMIE system will now be described in more detail to demonstrate how a man-modelling CAD system can be used as an extremely effective ergonomics tool.

Equipment and workplace modelling

The workplace modelling system is used to generate full-size 3D geometric representations of a working environment and specific items of equipment. A boundary representation form of solid modelling is used to enable the system to be highly interactive whilst maintaining a sufficiently accurate 3D model. This method requires that solid shapes are constructed from a description of the location of their vertices, a knowledge of which vertices are joined together to form edges and which edges form plane polygon faces. Models of considerable complexity can be quickly built from the range of parametrically defined primitive shapes available such as cuboids, prisms and cylinders (see Figure 19.9). These primitives require only a brief specification

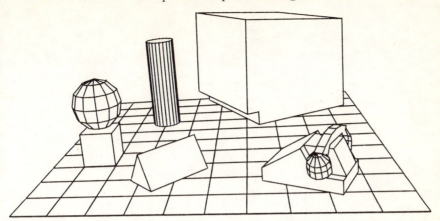

Figure 19.9. Examples of simple model types available in the SAMMIE system. The telephone is an example of how models are formed from these basic types.

[i.e. cuboid name, width (mm), depth (mm), height (mm)], whereas non-regular solids need the complete description of vertices, edges and faces. Solids of revolution (e.g. a sphere) can be created by defining the axis of revolution and the desired profile. Although truly curved surfaces are not available, this has never been a cause for concern from an ergonomics point of view, as sufficient accuracy can be obtained from a suitably configured faceted model. A reflection facility is also available so that mirror images of solids can be constructed automatically; for example, only one side of a car needs to be defined manually.

As mentioned earlier, the SAMMIE modeller is particularly strong in its ability to specify logical or functional relationships between items in the model. This is achieved using a hierarchical data structure, an example of which is shown in Figure 19.10. This hierarchy allows the designer to move the whole car as one unit or to open individual doors or the boot (see Figure 19.11), to rotate the steering wheel or to adjust the tilt of the driver's seat cushion. To achieve this selectivity, users need to travel across and up or down the data structure until they reach the level which will control the particular item(s) to be adjusted.

The data describing the 3D models are normally prepared away from the computer terminal using engineering drawings or sketches, although it is possible to create models interactively at the graphics screen during a design session. Another important feature of the system is its ability to interactively modify the geometry of an item in ways relevant to the design situation. For example, if a table was modelled as a top and four legs, then increasing the width of the table would automatically reposition the legs to maintain a valid model.

Man modelling

The man model is a 3D representation of the human body with articulation at all the major body joints. Limits to joint movement can be specified and

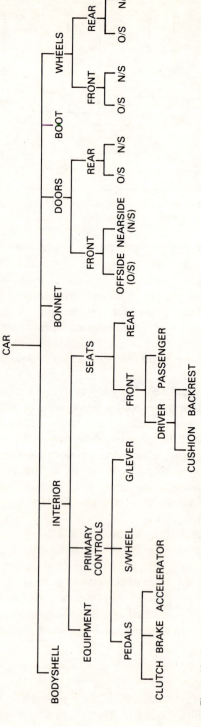

Figure 19.10. An example of the hierarchical data structure used in the SAMMIE system. This structure allows the model to be functional as shown in Figure 19.11.

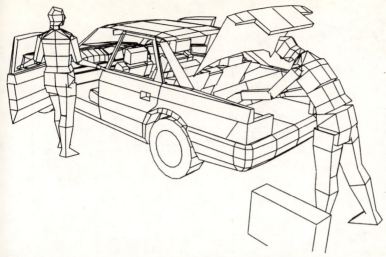

Figure 19.11. A complex car model. SAMMIE's hierarchical data structure enables functional as well as geometric relationships to be modelled, thus all moving parts of the model can be made to function. For example the car's doors, bonnet and boot can be made to open and close. Inside the car it is possible to adjust the seat and steering wheel within the design specification.

the dimensions and body shape of the man model can be varied to reflect the ranges of size and shape in the relevant national and/or occupational populations (see Figure 19.12).

The man model is displayed as a set of 17 pin joints and 21 straight rigid links structured hierarchically to represent the major points of articulation and the body segment dimensions (see Figure 19.13). The hierarchical structure is similar to that shown for the car, so that when the man model's right upper arm is raised, then the right forearm and hand follow accordingly. By dropping down the hierarchy users can control just the forearm and hand together or just the hand, at their discretion.

The size, shape and range of postures permitted are a function of the anthropometric and biomechanical databases chosen by the user. The data required consist of the linear dimensions between adjacent joints (e.g. from elbow to wrist), the body segment parameters of weight and centre of gravity, and the absolute and 'normal' limits for each joint in each of the three degrees of freedom (i.e. flexion–extension, abduction–adduction and medial–lateral rotation).

The limb length data can be stored as either a set of mean dimensions together with standard deviations, or as a set of dimensions explicitly defining the anthropometry of an individual. The displayed man model can be interactively amended by changing the overall body percentile, individual link percentile, explicit link dimension and the use of correlation equations to relate internal link dimensions to external anthropometric dimensions.

Figure 19.12. Shown, from left to right, are male models of 95th, 50th and 5th percentile stature from a chosen population. The system also enables changes to be made to individual limbs allowing representation of specific users or groups of users. The shape of the models' flesh envelope can be varied in accordance with somatotypes providing a useful evaluative technique for situations involving work in confined spaces.

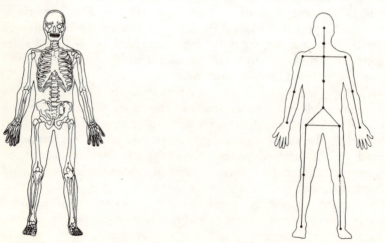

Figure 19.13. The link structure of the man model is a simplification of the human skeletal frame, with pin-joints suitably constrained to simulate human movement capabilities. The rigid links between the joint centres are defined by use of anthropometric data and are usually displayed with 3D flesh shapes.

The flesh shape is controlled by a classification system known as somatotyping (Sheldon, 1940) which enables the extent of endomorphy, (plumpness), mesomorphy (muscularity) and ectomorphy (leanness) to be specified; the somatotype number and the height and weight enables 17 body dimensions to be obtained from Sheldon's experimental data.

The joint constraints prevent the man model being positioned in an unattainable posture. For example, it is impossible to abduct the elbow. The system indicates whether a selected joint angle is within the 'normal' range of movement, within the maximum range, or infeasible. The limb dimensions and somatotype can be interactively altered to construct 3D man models to the user's unique specification if desired. Additionally, the joint constraints can be limited to represent disability, the effects of bulky clothing or unusual working conditions, for example where high gravity forces may severely limit arm movement.

The variable anthropometry of the man model is clearly advantageous for the evaluation of body clearances (fit) and reach. In addition, the 'man's view' facility allows the user to display the man model's field of view on the graphics screen. These facilities allow the user to predict the likely work postures that a given design will enforce. For example, a tall and fat model of a driver might be shown to adopt a slouched posture to gain sufficient headroom with arms at full stretch to the steering wheel under which the thighs are trapped. The view to the main driving displays may be obscured by the steering wheel, causing the driver to slouch to an even greater extent. This posture can be visualized by the designer and specified in terms of joint angles which can be compared with recommended angles in the literature (e.g. Rebiffe, 1966 for the driving task). The ergonomist would be able to comment upon such a posture saying that tall drivers of that particular car would suffer considerable discomfort in the neck, shoulders, lower back and thighs. Furthermore, the design can then be interactively modified by lowering the seat or raising the roof-line, and re-positioning or providing adjustment to the steering wheel.

Ergonomics facilities

The system has several facilities to help the user assess the ergonomics of a particular design.

A clasher routine

This facility automatically detects whether two solids are intersecting and, if this is the case, it flashes the appropriate items to attract the user's attention. This feature can be used to check clearances with the man model set to an appropriate size and shape, say 99th percentile limb lengths and an extreme endomorph. Alternatively, visual inspection from a variety of angles will achieve the same result.

Reach algorithms

Reach can be assessed simply by positioning the arms or legs so that the hands or feet either contact, or fail to contact, a specified control or point in space (see Figure 19.12). This method could become tedious for a large number of controls so an algorithm has been developed which predicts a feasible posture for the arms or legs given a specified model item or co-ordinates to be reached. Generally, there will be a large number of feasible postures for any successful reach attempt. The algorithm selects the limb posture to be displayed by attempting to minimize the extension of the joints away from their neutral positions and by preferring the greater extension of distal links to those that are more proximal. This feature does not ensure that the displayed limb posture is the likely posture adopted by a human, but it does confirm whether or not the reach attempt will be successful. If a reach attempt fails, the system displays this fact together with the distance by which it failed.

There are two other automated methods to define reach: reach areas and reach volumes. Both methods are especially suited to concept design as they are generated without specifying control locations or co-ordinates. The first method enables envelopes of reach areas to be overlaid on any surface of the design as an aid to assessing suitable positions for control locations. The second method is an extension of this whereby reach is assessed over a number of imaginary surfaces parallel to either the frontal, saggital or transverse planes of the man model. An example of a reach volume in the transverse plane is shown in Figure 19.14; such information is particularly useful for locating controls above head height. A major study was conducted using this facility to determine both hand and foot reach zones for drivers of agricultural tractors (Reid *et al.*, 1985).

Vision tests

The view 'seen' by the man model (man's view) is under the full control of the user (see Figure 19.15). For example, one can select left, right or a mean eye position, 60 or 120° cone of vision and specify the angle of vision using the eyes and/or head as appropriate. Constraints limit the maximum angles of vision from the eyes. As with reach, the testing of vision can be achieved manually by directing the head and eyes or else the user can specify the model item or co-ordinates to be viewed; the resulting view, together with the visual angle and viewing distance, will be displayed automatically.

Further developments include 2D visibility plots whereby vision can be determined at any given surface (e.g. checking vision of the fascia of the vehicle and, in particular, through the steering wheel) and 3D visibility charts which describe all-round visibility (e.g. checking external visibility from a vehicle through all the windows). Simple calculations allow one to calculate the maximum vertical visibility at any given point on the ground so, for

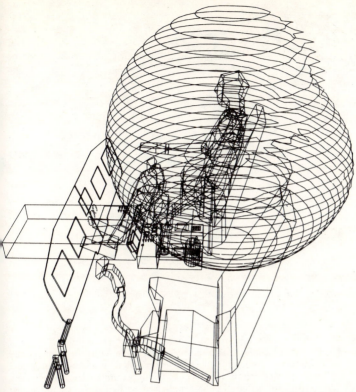

Figure 19.14. This plot illustrates volumetric reach facility (available for both hands and feet). In this example the right hand reach for a 50th percentile male helicopter pilot is being assessed.

example, the user can check whether a tall driver would be able to see signposts and traffic lights without leaning forward. These charts are described in detail in Porter *et al.* (1980).

Mirrors and reflections

The mirror modelling facility can be used to design mirrors for vehicles (see Figure 19.15) or to determine whether reflections will be a problem in windscreens or computer screens. The mirror parameters of focal length, convexity/concavity, size and orientation are all variable and can be interactively adjusted to provide the required field of view displayed on the mirror surface, as seen by the man model.

Saving postures

Having selected an appropriate size and shape of man model and adjusted his/her posture to suit the task demands and physical constraints of the

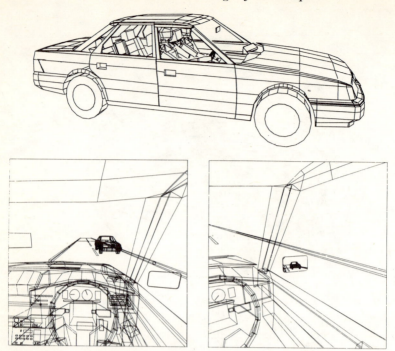

Figure 19.15. SAMMIE's viewing facilities enable the evaluation of the visual field for the full range of operator sizes. Left shows a 95th percentile male driver's view of the car controls and displays and the road through the windscreen. Right shows a 95th percentile male driver's view in his off-side exterior mirror. Reflections in the windscreen at night due to unshielded illuminated displays can also be identified at an early stage in the design.

workplace, it is important that this posture can be stored and recalled at a later date. This facility exists and it enables the user to run through a sequence of typical work postures in rapid succession, for example driving forwards, depressing the clutch and engaging first gear, depressing the clutch, engaging reverse gear and looking rearwards (see Figure 19.16).

User dialogue

The system is highly interactive and allows designers to proceed through the design process in a manner determined by their own requirements rather than in a predetermined manner. The user communicates with the system via a menu based dialogue using either keyboard, light pen, or mouse. Each menu, of which there are nearly 40, contains commands grouped according to their functions. A brief description of the main menus is given below.

View menu

The status of the graphics display is governed by four main parameters. The first is the centre of interest, basically what the user is looking at, either

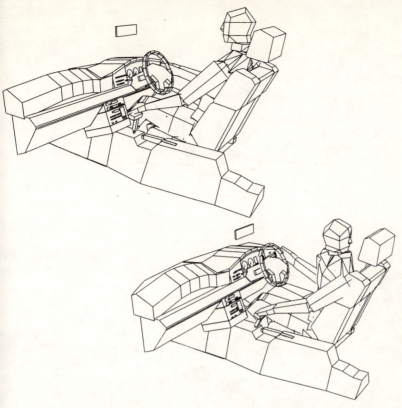

Figure 19.16. SAMMIE can be used to evaluate fit, postural comfort (by reference to joint angle data) and reach to controls in the car interior. As well as the appraisal of static reach and comfort it is also possible to consider use sequences.

directly or through the man's view. The second is the viewing point, which can be set at the man model's eyes or any other point in 3D space around or inside the models that have been constructed. The third parameter is the choice between displaying view in plane parallel projection (e.g. engineering drawing style) or in perspective and the fourth is the size of the displayed model, which is set by the scale factor in plane parallel projection and by the acceptance angle (i.e. the viewing angle) in perspective. The 'view menu' contains a variety of ways of interactively changing these parameters and it also provides a directory of 'saved views' which the user has set up for future use.

Workplace menu

These commands allow the interactive positioning of models or component parts of models in the workplace. Items can be shifted or rotated about either their own (local) axis system or the global axis system. An example of this important distinction is illustrated in Figure 19.17.

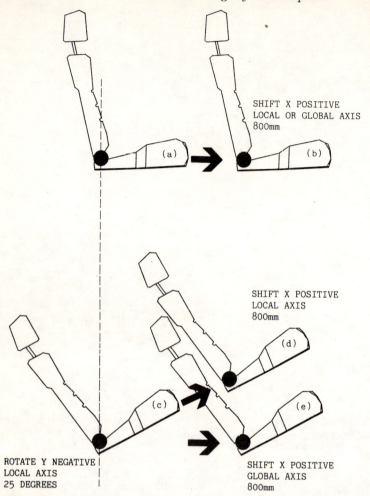

SHIFT X POSITIVE
LOCAL OR GLOBAL AXIS
800mm

(a)   (b)

SHIFT X POSITIVE
LOCAL AXIS
800mm

(d)

(c)   (e)

ROTATE Y NEGATIVE
LOCAL AXIS
25 DEGREES

SHIFT X POSITIVE
GLOBAL AXIS
800mm

Figure 19.17. An example of the use of the local and global axis systems available in the SAMMIE system. In some orientations these axis systems are identical, as shown in (a) and (b) where the seat is shifted 800mm along the global or local X axis. In (c) the car seat has been rotated about its local Y axis to produce seat tilt. (If it had been rotated about the global Y axis then it would have pivoted around the centre of the available workspace.) Examples (d) and (e) show how a subsequent 800 mm shift along the local and global axis systems, respectively, can produce different results. If the intended movement is to simulate fore and aft adjustment of the seat, then only (e) is appropriate.

A commonly used alternative to specifying the shift distance in millimetres is to 'drag' the chosen item(s) to a desired location on the screen using the light pen, keyboard cursor keys or mouse. This method can be faster because the location can be changed in two axes simultaneously and the accuracy can be maintained by increasing the scale of the model.

Display menu

Complex models take longer to be drawn on the graphics screen and sometimes these models appear confusing. The 'display menu' allows the user to select which items need to be displayed as required.

Man menu

This menu contains a variety of sub-menus including the 'anthropometry menu' for changing the anthropometry of the man model, the 'joint movement menu' for postural changes, the 'man's view menu' for displaying the view seen by the man model and the 'reach menu' for producing reach areas and reach volumes.

Geometry editor menu

When evaluating a design it is useful to be able to expand or shrink the dimensions of some items. This menu enables the X, Y or Z dimensions (width, depth and height) to be modified in isolation or concert. This feature is very useful at the concept stage because a large number of small cubes can be built and interactively edited to form appropriate sized building blocks for the construction of the early models.

Hidden lines menu

Models are usually displayed on the graphics screen in wire frame form (see figure 19.14) so that all the edges of the model are visible, even though some in reality would be totally or partially obscured by solid objects. This type of display is easily interpreted by an experienced user although, for extra clarity or presentations, the 'hidden lines' can be automatically removed (e.g. Figures 19.8, 19.11 and 19.16).

Plot menu

The end result of a design and/or evaluation will usually be in the form of a variety of views taken from the graphics screen and drawn on a pen plotter. The 'plot menu' provides a standard format for these views with the option of including a title and several lines of comments.

Case studies using SAMMIE

Two projects carried out using SAMMIE are described here to give the reader an insight into the way in which such systems are used; the typical length of a project using SAMMIE is around ten days.

Computer workstation design

The aim of this project was to design an integrated workstation to be used in the computer aided design of printed circuit boards. The original workstation was purely a grouping together of the hardware needed to perform the required functions, which resulted in a three-sided configuration comprising an alphanumeric VDT on the left, an A0 digitizer board in the centre and a graphics VDT on the right. Not unexpectedly, this arrangement was far from satisfactory with a high incidence of physical discomfort reported by the users. The manufacturers then designed two prototype integrated workstations where the graphics VDT and a much reduced digitizer, which was sunk into the worksurface, were placed directly in front of the user. However, both these designs were found to cause problems for the user for several reasons, including lack of thigh clearance, forward leaning over the worksurface, difficult reach to the keyboard and an excessive viewing distance to the graphics VDT. The manufacturers were both surprised and disappointed when these problems came to light within the first few days of testing, as they had invested considerable time and expense to produce the prototypes. However, most of their attention had been directed at the engineering problems and the interface design had suffered as a consequence.

Following initial discussions with the manufacturers, it was decided to develop three alternative designs using SAMMIE, covering a range of manufacturing costs. These designs are illustrated in Figure 19.18 and are now briefly described:

(a) This was the cheapest design with all the components free standing on the fixed height worksurface. Whilst this option may appear satisfactory as a paper specification, the visualization of the workstation clearly shows its shortcomings, such as the lack of space for paperwork, the likely wrist and arm discomfort arising from the raised digitizer board, and the generally clumsy layout.

(b) This was the most expensive design as it offered both worksurface height and tilt adjustment. The digitizer was sunk into the worksurface and the workstation could be set up for either left- or right-handed use as it was divided into two modules; this feature also made it considerably more portable.

(c) This was the medium cost design which had all of the features of (b) above except the adjustable tilt angle. The VDTs were adjustable. An evaluation using a man model is shown in Figure 19.8.

These designs were presented to the manufacturers in the form of slides as reproduced here. The SAMMIE plots were visually enhanced by an industrial designer who was closely involved in the project. The manufacturers were able to visualize accurately the concept workstations knowing that the SAMMIE system had been used to evaluate the designs in terms of fit, reach, vision and posture. The chosen design was (c) because of several factors, namely its aesthetic appeal, ease of manufacture, cost and sound ergonomics.

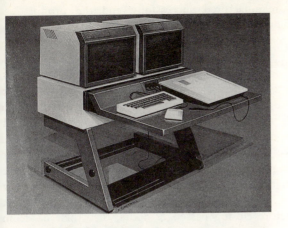

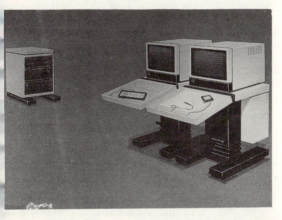

Figure 19.18. Three alternative designs of computer workstations; (a), (b) and (c) were the low cost, expensive and medium cost alternatives respectively (alternative (c) is the same design as shown in Figure 19.7). The SAMMIE plots were enhanced by an industrial designer.

This workstation was manufactured successfully and the product wa nominated for a design award the following year.

Train driver's workstation

This project was conducted on behalf of London Underground and it aros because of their policy to change some trains to OPO (one person-drive operation). The guard's main function had been to open and close th passenger doors at stations, having checked that it was safe to do so. I order for drivers to take on this extra responsibility, it was necessary fc them to leave the driving workstation (desk) and walk to the appropriat side of the cab to open the door and check that it was safe to open th passenger doors. The drivers would then wait for all the passengers t disembark or embark before closing passenger doors, then their door, an return to the desk, before pulling out of the station. This additional workloa delayed the train from leaving each stop by about 8 s which was considere unacceptable by London Underground's management.

The proposed solution to this problem was to modify the driver's des by adding a set of passenger door controls to the front edge (see Figu 19.19). In order for this to be a satisfactory solution it was necessary to chec that the reduced clearance did not make the workstation too cramped particularly when getting in and out of the seat. This was assessed using 95th percentile male man model with an extreme endomorph somatotyp and was found not to be a problem as long as the seat cushion was reduce in length by 2 cm. This was achieved by moving the pivot point (the cushio was able to pivot vertically for when the driver wished to stand whil driving) forwards by 2 cm and cutting off 2 cm of the seat frame at the re of the cushion. This recommendation was checked for sitting comfort usin a full-size mock-up with human subjects; in fact the majority of subjec preferred it to the original design.

One other aspect of the workstation needed investigation before recom mending the fitting of new door controls to the desk and this was to ensu that the driver could clearly see the passengers leaving and entering the trai without moving from the desk. Obviously, it would be impossible for th driver to see the passengers by direct vision, so underground stations ar now fitted with mirrors or video monitors situated on the platform just i front of the train when it is stationary. These displays enable the driver t see the complete length of the train, but only if the displays are complete visible through the driver's windscreen. This was assessed by defining a 3 volume within which all the displays would appear for all the undergroun stations. These boxes are shown in Figure 19.20 for a 95th percentile mal driver. The views show the driver's view of the desk, incorporating the ne door controls, the windscreen, the track and the nearside and offside displa boxes. The windscreen is shown with a mesh superimposed over it to allo recommendations to be made regarding the swept area of the wiper blad

Figure 19.19. A model of a London Underground train; the principal evaluations carried out here concerned with the driver's workstation which was evaluated for ease of access, reach of controls and vision.

The top plot shows the poor view that the driver has when sitting upright; only half of the nearside box is visible and very little of the offside box. The left hand plot shows the view with the driver adopting a 40° forwards lean, whereupon the nearside box becomes completely visible as the driver's eyes are closer to the windscreen enabling a wider angle of view. A further lean of 20° to the left permits the clear view of the offside box. Clearly, any new cab design should enable the driver to see the displays without any leaning at all. However, these postures were considered to be acceptable in the existing cab as they will be fairly infrequent and maintained for short periods of time, and then only when the train is stationary.

The SAMMIE work showed clearly that the proposed modification to the driver's desk was acceptable in terms of both fit and vision, subject to the

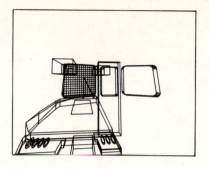

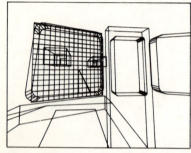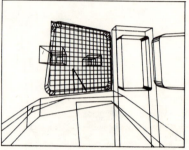

Figure 19.20. Selected views for a 95th percentile male driver in an underground train ca
Vision through the windscreen was evaluated to determine whether the mirrors or vid
monitors mounted on the platform were obscured by the bodywork of the cab. The top p
shows the view, when sitting upright, of the control panel, windscreens, track and two cubo
depicting the envelopes of possible display locations for all stations. Clearly, the displ
envelopes are only partly visible .However, if the man model is instructed to lean forward
40° (bottom left) then the left hand envelope becomes completely visible; the right hand envelo
becomes visible (bottom right) when the man model leans to the left in addition (20°). The
infrequent postures are considered satisfactory when the train is stationary.

minor alteration to the seat cushion length. This project highlights the val
of CAD in assessing compromises in design.

Details of the above two projects can be found in Porter (1981) and Port
and Porter (1987). Other projects have been described in Bonney *et a*
(1979a,b), Case and Porter (1980), Levis *et al.* (1980), and Porter and Ca
(1980).

The advantages of using CAD

There are several important advantages to using 3D man-modelling CA
systems in design and these are now briefly discussed.

Reduced timescale

This clearly can be a major factor and it may often decide whether or n
the project receives any ergonomics input at all. Time can be saved in sever

eas, for example, the construction of a computer-based mock-up might ke between 1 (simple) to 5 (complex and large) days compared to as many onths using wood, glass fibre or other materials. Subject selection can be time consuming process when conducting user trials, whereas the athropometric database of the computer system can be used to select the quired man models in seconds. For example, when designing driving ackages it is important to consider people with long legs and short arms ecause they will have a personal conflict between positioning the seat arwards for good leg posture, whilst having the steering wheel at full retch, or having the seat further forwards for good arm posture at the crifice of leg posture. The best solution is to provide steering wheel ljustment but this requirement may not be apparent if user trials are rushed sing only a small handful of subjects who may have similar percentile reach ith their hands and feet. Another saving is made at the evaluation stage as ly a few man models are examined compared to 20–30 subjects, with the nsuing lengthy data analysis.

arly input of ergonomics expertise

ecause of the rapid modelling facilities it is possible to start the ergonomics put right at the beginning of the project. This is particularly necessary as ngineers are using CAD systems themselves and the design might be rtually finished from their point of view by the time the first full size ock-ups are ready for traditional user trials.

erative design

arly commencement coupled with reduced timescale make it very easy to tablish an iterative design programme and to promote the exploration of wide range of design solutions. Compromises are an essential feature of esign and the above features are important ingredients in developing e optimum trade off between, for example, cost and the ergonomics ecification.

D analysis

part from user trials, other traditional techniques involve using anthropo- etric data or 2D manikins. Both of these methods are unsatisfactory for mplex tasks, for example driving a tractor and ploughing a field (see Figure .21). The driver will have both feet operating foot controls, one hand will e on the steering wheel and the other will be on a hydraulic control lever adjust the height of the plough. The driver will be looking both in front d, twisting the spine, over the right shoulder to the furrows behind. This osture cannot be assessed without using 3D analysis.

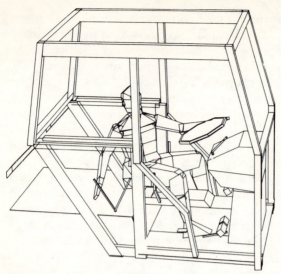

Figure 19.21. Being three-dimensional, the man models can assume complex postures. Fo example, the tractor driver shown above must be able to reach the hydraulic control and watc. the plough as well as operating the normal driving controls.

Improved communication

Computer graphics provide an excellent means of presenting ergonomic input to design committees. The visual impact of the ergonomics specification is far stronger and easier to grasp than numerous recommendations in report. Additional realism can easily be supplied using the services of a industrial designer or stylist (see Fig. 19.22) and this collaboration improve communication within the design team.

Cost effective ergonomics

The use of CAD is cost effective because of the advantages described above If the ergonomics input lags behind the engineering, then the end result i often last minute modifications which take time and money to implemen or a product that does not meet the full ergonomics specification. Both o these are undesirable; the first because it increases the development anc production costs, whilst the second is likely to reduce the success of the product or service.

There are few disadvantages, and these are more to do with restricting the potential advantages. One problem is that CAD is a powerful tool and, lik any tool, it can be dangerous in the wrong hands. The selection of relevan and accurate databases and decisions concerning workstation design anc posture require the skills of an experienced ergonomist or a designer/enginee with suitable training. The systems are designed to supplement an ergonomist' skills, not replace them. It would be short-sighted to think that such system

Figure 19.22. A concept model of a helicopter cockpit interior. The combined strengths of the ergonomist and stylist are clearly shown in the above photograph. The ergonomics contribution to the design can be communicated powerfully using 3D graphics.

can replace totally user trials; they should only be used to explore alternative designs, to eliminate the poor ones and select and, if possible, improve upon, the promising ones. The results of the CAD evaluation should lead straight to an in-depth user trial with working prototypes, especially if the tasks are complex and performed under adverse conditions.

Future developments

The future of man-modelling CAD systems looks very promising as manufacturing organizations are always looking for ways to reduce development times and costs, whilst producing good quality design for the increasingly 'design aware' public. With regard to the development of SAMMIE, the following useful enhancements to the system are being considered.

Control of the man-model's posture

The current methods for setting the posture are limited by the fact that it is often difficult to predict the actual posture that people would adopt in some circumstances. For example, could you specify exactly how you would get out of a car without taking mental notes as you do it? Even if you do this, it would be quite tedious to set up such complex postures for the man model. One interesting solution to this problem, currently being investigated, is the use of a catsuit worn by the user with strain gauges at the major body joints. This device enables the user's posture to be recorded in the form of voltages which could be linked directly to the control of the man model's posture. Another use of the catsuit would be to collect postural data from a sample of people performing a variety of tasks and use the findings as a database for SAMMIE.

Anthropometric database

Very few anthropometric surveys take sufficient measurements to define an accurate 3D model of people. In addition both external dimensions and the location of joint centres, including ranges of movement, are required. It has been suggested (Bonney *et al.*, 1980) that surveys should take into account these requirements and take more comprehensive measurements to maximize the potential applications of their data. The major problem with this request is the time and cost required. However, developments in recording methods may allow the automated collection of thousands of measurements that define points all over the body in seconds.

Other areas of future interest include the implementation of a static strength modeller and the development of a SAMMIE 'expert' system.

References

Bonney, M.C., Blunsdon, C.A., Case, K. and Porter, J.M. (1979a). Man–Machine Interaction in Work Systems. *International Journal of Production Research*, **17**, 619–629.

Bonney, M.C., Case, K., Porter, J.M. and Levis, J.A. (1979b). Design of mirror systems for commercial vehicles. *Applied Ergonomics*, **11**, 199–206.

Bonney, M.C., Case, K. and Porter, J.M. (1980). User Needs in Computerised Man Models. In *Anthropometry and Biomechanics: Theory and Application*, edited by R. Easterby, K.H.E. Kroemer and D.B. Chaffin (New York: Plenum Press), pp. 97–101.

Case, K. and Porter, J.M. (1980). SAMMIE: a computer aided ergonomics design system. *Engineering*, **220**, 21–25.

Dooley, M. (1982). Anthropometric modelling programmes—a survey. *IEEE Computer Graphics and Applications*, **2**, 17–25.

Elias, H.J. and Lux C. (1986). Gestatung ergonomisch optimierter Arbeitsplatze und Produkte mit Franky und CAD. [The design of ergonomically optimized workstations and products using Franky and CAD.] *REFA Nachrichten*, **3**, 5–12.

Levis, J.A., Smith, J.P., Porter, J.M. and Case, K. (1980). The impact of computer aided design on pre-concept package design and evaluation.

In *Human Factors in Transport Research,* Volume 1, edited by D.A. Oborne and J.A. Levis (London: Academic Press), pp. 356–364.

Lippmann, R. (1986). Arbeitsplatzgestaltung mit Hilfe von CAD. [Workstation Design with help from CAD.] *REFA Nachrichten,* **3,** 13–16.

McDaniel J.W. (1976). Computerised biomechanical man-model. *Proceedings of the 6th Congress of the International Ergonomics Association and the 20th Annual Meeting of the Human Factors Society,* pp. 384–389.

Porter, J.M. (1981). *Ergonomic Aspects of CAD Workstations.* Report No. CAS.027. (Loughborough: SAMMIE CAD Ltd.).

Porter, J.M. and Case, K. (1980). SAMMIE can cut out the prototypes in ergonomics design. *Control and Instrumentation,* **12,** 28–29.

Porter, J.M., Case, K. and Bonney, M.C. (1980). Computer generated three-dimensional visibility chart. In *Human Factors in Transport Research,* Volume 1, edited by D.J. Oborne and J.A. Levis (London: Academic Press), pp. 365–373.

Porter, J.M. and Freer, M.T. (1987). *The SAMMIE System, Information Booklet,* 5th edition (Loughborough: SAMMIE CAD Ltd).

Porter, J.M. and Porter, C.S. (1987). *An Ergonomics Study of the C69 Stock Cab.* Unpublished report for London Underground Ltd.

Rebiffe, R. (1966). An ergonomic study of the arrangement of the driving position. *Ergonomics and Safety in Motor Car Design, London Symposium,* 27 September, pp. 26–33.

Reid, C.J., Gibson, S.A., Bonney, M.C. and Bottoms, D. (1985). Computer Simulation of Reach Zones for the Agricultural Driver. In *Proceedings of the 9th International Congress of the International Ergonomics Association,* edited by I.D. Brown, R. Goldsmith, K. Coombes and M.A. Sinclair (London: Taylor and Francis), pp. 646–648.

Rothwell, P.L. (1985). Use of man-modelling CAD systems by the ergonomist. In *People & Computers: Designing the Interface,* edited by P. Johnson and S. Cook (Cambridge: Cambridge University Press), pp. 199–208.

Sheldon, W.H. (1940). *The Varieties of Human Physique* (New York: Harper and Bros.).

Chapter 20

The evaluation of industrial seating

E. Nigel Corlett

Introduction

From an ergonomics perspective it is necessary to recognize that the work seat is as much a tool to achieve the work objectives as any other piece of equipment in the workplace. Its design and functioning will be influenced by the tasks to be done, the other equipment to be used, the environment and, of course, the individual human differences. It will be evident that there is unlikely to be one seat suitable for all jobs and the concept of an 'ergonomic chair' independent of the tasks is not possible. Since what we try to do affects our postures, e.g. what we look at or reach for, then it is evident that the contributions of the seat to comfort and support should be developed in relation to these activities.

Seat requirements

From such considerations as these, Table 20.1 provides some of the important requirements for a work seat. They are based on the seat as a full body support, rather than as a temporary perch. There is utility in a perch, where most of the body weight is still on the feet, but it is not part of the present discussion.

Maintaining one third or less of the body weight on the feet was shown by Eklund *et al.* (1982) to be necessary if people were not to complain of leg discomfort. More than this amount on the legs requires continuous muscular activity for body support, a major source of discomfort. The requirement to allow changes of posture is a good ergonomic one and also a practical necessity in many jobs. The use of a high seat at, for example, a supermarket checkout may increase the sitter's reach by moving the legs, whilst still being fully supported by the seat.

In many cases a work seat has to resist other forces than just the body weight of the sitter. Where movements of the arms are needed, or forces

Table 20.1. Functional factors in sitting

The task
    Seeing
    Reaching

The sitter
    Support weight
    Resist accelerations
    Under-thigh clearance
    Trunk–thigh angle
    Leg loading
    Spinal loading
    Neck/arms loading
    Abdominal discomforts
    Stability
    Postural changes
    Long-term use
    Acceptability
    Comfort

The seat
    Seat height
    Seat shape
    Backrest shape
    Stability
    Lumbar support
    Adjustment range
    Ingress/egress

exerted by arms or legs, these forces must be transmitted through the body and the seat to the ground. It is evident that a backrest is a channel for such forces on many occasions, otherwise the musculature of the trunk must be tensed to provide a semi-rigid path for the force transmission. The generated muscle tensions will increase the load on the spinal column, particularly in the lumbar spine, which is the major channel for load transmission from the upper to the lower part of the body. A backrest can also reduce loads on the lumbar spine by transmitting part of the gravity forces due to the head, arms and upper trunk (Corlett and Eklund, 1984).

The evaluations of seating

A chair is for a sitter, not for itself, and there is no doubt that this is a case where form must follow function. Thus the priorities in any evaluation must include the responses of sitters, both in their behaviour and in their subjective judgements. It also follows, as noted in the literature, that to evaluate a chair on its dimensions alone, and how they relate to anthropometric data, is inadequate; it should not be necessary to repeat this, but unfortunately there is still a widely held view that anthropometric data are sufficient for seating selection, and even workspace design.

It might be as well to begin this section with almost the last words from the Shackel *et al*. (1969) paper: 'Seating comfort is still a very complex problem and the only valid approach is the experimental method'. This does not rule out the need for dimensional criteria, but certainly it calls for a better understanding of them than we currently possess. The Shackel *et al* paper demonstrates no significant correlations between the BSI recommended dimensions and the reported comfort. In a study by Langdon (1965) of a group of key punch operators, average stature 64 inches, 85% reported their seat as being comfortable, even though the average seat height was 19.2 inches. This highlights a point made by Branton (1969) and others, that chair comfort has many dimensions, including the task and possibly the appearance. These are discussed by Shackel *et al*. (1969) and further by Lueder (1983) who comments, after an extensive review 'little insight is available into the meaning of comfort'.

Table 20.2 indicates what methods have been used to assess many of the sitting aspects which are important for chair users. It is evident that people can be asked their views on all the sitting aspects listed. The choice of methods will, to some extent, depend on the aims of the investigations and hence whether or not laboratory equipment and conditions are appropriate. In what follows some examples of practical evaluations are presented demonstrating the selections of test methods which have been found useful. These are based around the seat model of Table 20.3, which links functions effects and the required measures.

A more comprehensive overview of seating evaluations would extend the 'methods' section of Table 20.3 to include research techniques, including X rays, optical posture recording systems and a wider range of physiological areas of study. Much seating research, complementary to the areas discussed here, is concerned with understanding the reasons why discomforts arise spinal discs suffer damage or particular postures are adopted. This fundamental work, on which the practical evaluations of working seats are based, require its own review but is not further discussed here.

Dimensional evaluations

Although body sizes are not sufficient data for the design of seating, it is obvious that they are essential. Pheasant (1984) has provided up-to-date data for the UK population, together with useful notes on their use. The clearances which are desirable in fixing the length and height of the seat have been described by Akerblom (1954), Floyd and Roberts (1958) and Murrell (1965). Briefly, for a chair in use at a table, the seat length must clear the calf of a 5th percentile female user, and the height allow light contact under the thigh when the sitter's feet are flat on the floor. This latter requirement makes some adjustment necessary if a 95th percentile male and a 5th percentile female are to use the same chair.

Associated with the question of dimensions is the requirement for adjustment. Jones (1963) has put forward a simple and effective fitting trial

Table 20.2. Some of the methods used in assessing the functional qualities of industrial seating

Functional factors \ Methods	Dimensional measurements	Fitting trials	Force or pressure measurements	Bio-mechanics calculations	Observations or timing of behaviours	Subjective judgements overall	Subjective judgements body parts	Check-lists	Cross-modality	Reach/force/stability	Stature changes
Seeing	✓					✓					✓
Reaching	✓	✓		✓		✓				✓	✓
Seat	✓		✓	✓		✓					✓
Backrest	✓			✓		✓					✓
Adjustment	✓					✓					
Ingress/egress		✓			✓	✓				✓	
Stability		✓		✓		✓	✓	✓		✓	✓
Support weight			✓			✓	✓	✓	✓		✓
Under-thigh clearance	✓						✓				
Trunk–thigh angle	✓			✓		✓	✓	✓			
Leg load			✓	✓		✓	✓				✓
Spinal load				✓		✓	✓				
Neck/arm load						✓	✓				
Posture changes					✓	✓					
Long-term use					✓	✓		✓			
Acceptability						✓	✓		✓		
Comfort				✓		✓	✓			✓	✓
Lumbar support	✓		✓			✓	✓				

Table 20.3. Seating model for assessment of industrial seats (adapted from Eklund, 1986)

Functional factors		Responses and effects	
		Initial	Subsequent
The task	(detailed	Postures	Discomfort
The sitter	items as	Loads	Pain
The seat	in Table	Pressures	Disease
	20.1)	Influences on blood flow	Reduction in performance
		Discomfort	
		Preferences	
Measures			
Workplace dimensions		Biomechanical load	Rating
Work weights		EMG	Ranking
Work forces		Stature change	Clinical examination
Work reaches		Rating	Epidemiological studies
Work time patterns		Ranking	Performance
Anthropometry		Dilations of body parts	
Strength		Linear measurements	
		Posture	

procedure, using subjects from the extremes of the population. (See Pheasant's chapter on anthropometry and the design of workspaces.) In running such trials we might note the comments of Branton (1969) that people older than 30–35 years are likely to be more sensitive to discomfort than younger people. Again, the evidence is in favour of experimentation since, particularly for work chairs the variability in the tasks will require evidence for the facility to see, to reach, to exert forces effectively and to maintain a stable position on the seat.

By using the fitting trials procedure, direct evidence is obtained for the need for adjustments on the chair, and the necessary extent of these adjustments. LeCarpentier (1969) noted the variability in dimensions recorded by the same subjects on different days when studying comfort. Although these findings are important for recognizing the flexibility inherent in dimensions for comfort, the same variability might not exist for some constraining work activities, and there is clearly need for further investigation into the precision to be expected from fitting trials conducted to assess a seat for various functions.

Ingress and egress

These factors are especially important in two cases in particular, seating for the elderly and sit-stand seats. The former is not part of this paper, but studies by Shipley et al. (1969) showed that accessibility of chairs for the elderly was best assessed by the times taken to get into and out of them, coupled with observations of the subjects' behaviours during these manoeuvres. However, for people without disability, time might not be such a good means of separating seats.

For sit-stand seats the advantages of a saddle are reduced when their use is observed. For women in skirts they are awkward. The times of use for what is often a hard seat surface which supports only a part of the buttocks, as well as observations of the manoeuvres to get on and off, are probably the best methods. Added to this must be stability requirements, which are discussed later. As will be a constant refrain in this chapter, the experiences of the users must also be part of the evaluations.

Observational methods

It is an obvious point that the investigator should watch what the user does, the other side of the coin to gathering their experiences of using the chair. Kember (1976) utilized an adaptation of Benesh notation, a choreographer's tool, to record in detail the postures and movements of a chair user. It is capable of recording on a time base and, providing the three months' intensive training suggested by Kember is not prohibitive, will record sitting activities in great detail.

For most studies however this investment in training will prove excessive, for there is still the need, after recording, for a detailed analysis.

Most workers have used simpler methods, of which that by Branton and Grayson (1967) is typical. It may be modified to include actions, and so on, relevant to the problem under investigation with little difficulty. For their easy chair comfort study they selected positions of the head, trunk, arms and legs, with separate numbers in each body part for the important postures (see Table 20.4). If, for example, the head was also supported by the hands, or the legs were stretched forward, the relevant number on the recording sheet was ringed. The resulting records are easy to enter into, and analyse by, computer and the coding is easy to learn.

Clearly the whole range of observational techniques can be drawn on and modified to suit the investigators. Analysis, however, usually takes longer than recording, and even activity sampling methods (Grandjean, 1980; Branton, 1969), which are very useful to record the postural behaviour over several hours and are simple to apply in many cases, can condemn the investigator to many hours of tedious analysis. If body markers can be used, electronic analysis methods are becoming more practical and in the laboratory are well used (Corlett *et al.* 1986). In the field, however, body markers are not often practical and video recording not always welcomed.

The length of time, and time of day are both important in observational studies. The length of time must be relevant to the task. Branton and Grayson (1967) used 4–5 h, since that was the length of a long train journey, and hence the period of use of the seat. Shipley *et al.* (1969) studied people in homes for the elderly over the whole day. On the other hand, where work

Table 20.4. Coding of sitting postures (Branton and Grayson, 1967)

Head
Free of support	1.
Against headrest	2.
Against side wing	3.
Supported by hands	ring appropriate no.

Trunk
Free from backrest	1.
Against backrest	2.
Lounging/slumped back	3.

Arms (one or both)
Free from armrest	1.
Upon armrest	2.

Legs
Free, both feet on floor	1.
Crossed at knee	2.
Crossed at ankle	3.
Stretched forward	ring appropriate no.

imposes a variety of activities during the day, as for many office workers, randomly chosen periods of 1 or 2 h, or half a day may be appropriate. Sampling methods should cover the whole day, however. Both Branton and Grayson and Shipley *et al.* report significant differences between morning and afternoon sitting behaviours. There can also be differences between men and women in the postures adopted (Branton and Grayson, 1967) and in the length of the sitting periods (Shipley *et al.* 1969), both factors which may apply in situations other than those studied by these workers.

Subjective methods

This heading embraces a wide range of methods, requiring an understanding of psychophysics. Comfort is one of the variables which is most often examined by ranking or rating scales. Allen and Bennett (1958), for an air pilot's seat, used a forced choice method of ranking all body areas in decreasing levels of comfort. Corlett and Bishop (1976) preferred to focus on discomfort, asking subjects either to rank, or to rate, body areas perceived as suffering discomfort. There was no requirement to cover all body areas. (See the chapter on static muscle loading in this book.)

The recognition of discomfort appears to change with circumstances. Branton (1969) proposed a model in which people traded off stability against relaxation, indicating that no posture will remain comfortable for long periods. Shipley (1980) suggests that discomfort varies with arousal and attention; periods when attention 'turns inward towards the condition of the self' cause what has been ignored to receive attention, producing fluctuations in subjective discomfort.

Several workers echo the view of Habsburg and Mittendorf (1980) that people seem to attempt to produce a general view of comfort unless asked to consider 'is it for me/not for me?'. This question focuses the response more sharply and the instruction to subjects to consider how it is for them at the time the questions are asked, rather than to leave this aspect of the survey implicit, would appear to be good practice.

A *general comfort rating* scale, developed by Shackel *et al.* (1969), is given in Figure 20.1. It is fine enough to evaluate even small differences in discomfort, and was administered by Drury and Coury (1982) every half hour of their trials. Shackel *et al.* (1969) provide details of the range of comfort of the 10 chairs they tested, and Drury and Coury (1982) also show their chairs on the same metric. It is thus possible to compare results from this scale with other workers, and get some feel for the relative effectiveness of the chair under test.

The question of focus is important, Wachsler and Learner (1960) found that the only factors which correlated highly with feelings of comfort in their aircrew study was comfort of the buttocks and back. If these were accommodated, subjects were willing to overlook other discomforts when rating a seat. In later studies the use of a 'body map' to obtain *body part discomfort* (BPD) has been widely used, often based on Corlett and Bishop (1976).

The use of BPD scales, either with or without scales looking at somatic conditions, gives 'before and after' comparisons based on personal data which are very compelling evidence of changes and their effects, providing the controls or intervening variables are maintained. Shackel *et al.* (1969) are not the only researchers to draw attention, for example, to the effects of appearance on judgements of seat quality. Extended periods of use were

General comfort rating
  Please rate the chair on your feelings *now*

— I feel completely relaxed
— I feel perfectly comfortable
— I feel quite comfortable
— I feel barely comfortable
— I feel uncomfortable
— I feel restless and fidgety
— I feel cramped
— I feel stiff
— I feel numb (or pins and needles)
— I feel sore and tender
— I feel unbearable pain

Figure 20.1. General comfort rating scale of Shackel *et al.* (1969).

employed by Bendix *et al.* (1988) prior to investigating their subjects' responses to two chairs under comparison. Drury and Coury (1982) report a comparison between several hours' use and a rapid evaluation procedure, the two studies shared closely similar results. These workers used small groups (12 subjects). Jones (1969) has suggested that economies can be obtained by training testers, quoting that teaching testers to discriminate different levels of discomfort can give highly repeatable results. These however are within-testers; the between-testers comparison gave considerable differences.

Apart from focusing attention on body parts, subjects can be asked to focus attention on chair details, such as the *chair feature check list* (CFCL) (Shackel *et al.* 1969; Drury and Coury, 1982) or on aspects of the task, which could be rated for difficulty. This last presupposes a good task analysis, as called for by Lueder (1983), which is not often done.

The CFCL shown in Figure 20.2, is again by Shackel *et al.* (1969), but modified by Drury and Coury (1982). It provides an opportunity to get the mean and distribution of the effects of the various aspects of a chair as experienced by the sample of sitters. Typically, those dimensions which the

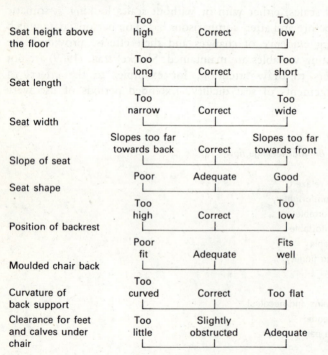

Chair feature checklist

Figure 20.2. Chair feature checklist of Shackel *et al.* (1969), modified by Drury and Coury (1982).

subjects can adjust prior to the test will give a smaller range, and a mean closer to the optimum, than the scores on other features.

When evaluating seating, the length of time during which the subjects are exposed to the conditions is important. Jones (1969) illustrated examples where certain boundary levels of discomfort, just distinguishable by untrained subjects, did not appear until after 3 h or more. Branton (1966) identified a pattern of sitting behaviour in less time than this, whilst LeCarpentier (1969) reported that after 20 min he found no changes in preferred easy-chair sittings even though the subjects sat for 2 h.

Shackel *et al.* (1969) used brief (5 min) periods of sitting, under carefully controlled sitting, and a ranking procedure, to separate a range of chairs so that they were suitable for a group of people. Drury and Coury (1982) used a short period for adjustment of the chair by each subject to an initial position and, in common with several other studies, then evaluated its effectiveness against longer periods of sitting.

Cross-modality matching (CMM)

Discomfort has been related to pressure distribution measured on the seat and backrest (Wachsler and Learner, 1960; Habsburg and Mittendorf, 1980). Lueder (1983) comments that 'the lack of an accepted measure of comfort [. . . has] frequently caused comfort to be relegated to a low priority in comfort (sic) decision making'. Whilst accepting the first point, chair researchers appear to have put comfort high on their list of criteria, as well as giving it extensive discussion in their papers.

In the discussion on comfort by Branton (1969), he reports an attempt to use a hand dynamometer in a cross-modality match with feelings of bodily tension. Although unsuccessful in his trials to relate the responses to seat features, he considered the method had some potential, and that it would facilitate the subject's ability to evaluate responses to the seat.

One attempt to quantify perceptions of pressure and their links with discomfort is reported by Gregg (personal communication). Using a sphygmometer cuff as the other modality, he asked subjects to use magnitude estimation to estimate perceived pressure, and demonstrated a highly significant reliability between repeated trials for estimating perceived pressure. With this information Gregg used the pressure cuff to determine the pressure values which characterize a discomfort scale with five points lying between 'no noticeable discomfort' and 'extremely uncomfortable'. He found repeatability to be good.

Extending these studies he then compared the results of BPD scales, and pressure recordings from a cuff on the subject's arm, in a number of trials, in which he compared seat heights, pressure on the ischii and discomfort experienced in various body parts, whilst sitting for 2 h. He demonstrated close agreement between the two methods. By combining a CFCL with the body part assessments by CMM using the cuff, a recognition of some of the causes of discomfort were obtained. In particular it was found possible to identify pressure distributions across a seat surface by requiring the pressure judgements to be made via the comparison with a cuff.

Posture changes and stability

An aspect of behaviour which researchers have sought to use as an indicator of seat comfort is the amount of posture change which occurs during sitting. The argument is that if the subject changes position frequently (fidgets), then the seat is not comfortable. Difficulties have been found with using the concept; for example, it is generally agreed that some changes in posture are desirable, and of course some are necessary due to the task's demands. In looking at automobile seating, Rieck (1969) found that although a questionnaire separated one seat from the other four, the measurement of small movements in the seats, using a force platform, did not. Branton (1966) found distinct changes of posture in relation to seat shapes, although he was looking more at gross changes than 'fidgets'.

The requirement for stability has usually been discussed in terms of the chair tipping over during use, or of its resistance to movement during work activities. Branton (1969) raised the point of the stability of the sitter, in the seat, in the context of easy-chair comfort. He pointed out that 'if the seat does not allow postures which are both stable and relaxed, the need for stability seems to dominate that for relaxation', requiring muscular activity to secure stability. It may appear self evident that people will not relax on a seat if, by so doing, they would fall out of it, but some proposals for forward-sloping seats have certainly increased the efforts needed to maintain stability at the expense of relaxation. This is not to say that, in some cases, the trade off has not been satisfactory; it is just necessary to note that muscular effort is a concomitant of the design.

Changes in stature

Having devised a seating model which, *inter alia*, proposed that one requirement of a good industrial seat was that it should reduce the load on the spine, Eklund (1986) examined several methods for evaluating a seat in these terms. Biomechanic analysis in conjunction with a force platform (Eklund *et al.*, 1983) or an instrumented chair, (Eklund *et al.*, 1987) could assess the loads on the spine, but provided no direct evidence of the validity of the values, or their actual effects on the sitter.

Eklund and Corlett (1986) reported a study to compare BPD, biomechanic analysis and stature change measures for industrial seats. The study was conducted in a laboratory, and involved a force production task, a sideways viewing task (as in some fork-lift truck driving) and a sit-stand seat. They noted that the three methods were all effective but contributed different areas of information. The theoretical evaluation of the load, using biomechanics, was quick and inexpensive. The stature change method needed at least 30 min exposure by each subject, may need several subjects but was better for major loading situations. It had the advantage of giving a numerical measure of the actual effects of the load on the spine. Discomfort assessment was inexpensive, sensitive and suitable for field work but it, too, required long exposure by the subjects. Trials with the precision stadiometer had demonstrated its practicability in an industrial setting.

Comprehensive evaluation procedure

1. Starting with the need to evaluate a range of chairs for their utility in various situations, and no comprehensive assessment procedure available, the procedure of Shackel *et al.* (1969) involved:

   (a)   ranking of the chairs for preference, whilst not being allowed to see or touch them;
   (b)   long-term sitting whilst working, with regular completion of a general comfort rating and BPD ranking; and
   (c)   completion of a CFCL at the end of the session.

This procedure showed where chairs were inadequate, either in dimensions or in relation to tasks, and which chairs were preferred.

2. Where one chair has to be tested, for example when a prototype has been built or a company is exploring the purchase of a particular chair, Drury and Coury (1982) proposed a variation on this set of tests. They utilized the data from Shackel *et al.* (1969) as a basis for part of their evaluation, also bringing in a modified general comfort questionnaire and the CFCL from their methods. They used LeCarpentier's (1969) concept of initial adjustments and the BPD procedure of Corlett and Bishop (1976).

   The procedures adopted, in summary, were:

   (a)   comparison with anthropometric data, standards and principles;
   (b)   opportunity to adjust the chair until the feeling of comfort is maximized for the sitter, typically taking about 5 min;
   (c)   a sitting and working period of 2.5 h, with general comfort and BPD scales given every half hour; and
   (d)   CFCL given at the end of the sitting session.

The results of the *general comfort* and CFCL measures were compared with Shackel *et al.* (1969), to identify how the chair lay in the general field. They confirmed by further studies the reliability of the battery of methods, and demonstrated its sensitivity by noting the discrimination which occurred, and the opportunities available for the interpretations of differences between the different measures.

Drury and Coury (1982) report that the method enabled a single chair to be evaluated both for its use in given jobs and to provide information to the maker on its weak points. It would certainly be desirable to have a broader base for comparison than the 10 chairs from the Shackel *et al.* (1969) study, but even without this section of the procedure, the method is probably the most economic procedure available which covers the major features in sitting which are important to the user.

3. Recognizing the complexity of any seating assessment, Yu *et al.* (1988) used a fractional factorial experiment to evaluate seven variables in seat design for a sewing task. These were seat height, seat angle, whether or not the seat rocked, whether or not it swivelled, backrest distance, backrest height and backrest angle. The statistical procedure is one that minimizes experimental costs and demonstrated the practicality of this form of experimental design. Their dependent variables were general discomfort, BPD, stature change and electromyography (EMG).

Two females, spanning 90% of female stature, were used as subjects. In a summary table of results the use of BPD and stature change provided most information, and their results supported each other. The other two methods did not contribute additional information so far as the selection criteria for the chair were concerned.

4. To develop a seat for a university auditorium, Wotzka *et al.* (1969) first did an activity sampling study of student activities in four large auditoria, accompanied by a questionnaire to the students about the seat and desk characteristics. From this they deduced a range for the variables of importance and built five seats to test their effectiveness. A number of subjects attended one test session of 20–30 min. First, the seat was adjusted for each subject 'until the most comfortable posture had been achieved', when each subject was given a questionnaire to assess six body parts for 'uncomfortable, medium or comfortable'. From this trial, a set of seats, including some modified as a result of the previous experiment, were ranked by paired comparisons, followed by the use of the same body part questionnaire. The final test was in an auditorium, using the same seats as in the paired comparison study. Here, again the same questionnaire was used.

This study links the questionnaire with observed behaviours and body part judgements in a long study, which required a large number of subjects. This has the advantage of increasing the confidence one may have in the seat's acceptability, but increases the time and expense of the study.

The study is typical of those which associate questionnaires with other measures. Hünting and Grandjean (1976) have used activity sampling and discomfort bi-polar questionnaires for selected body parts in the comparison of seats. They have deduced preferences, and desirable changes, from the results, although subject numbers have been large.

Conclusions

A list of the functional factors relevant to seat use is given in Table 20.3. When evaluating a seat, those factors relevant to the purposes of the seat must be studied. To define which are relevant may need a task analysis, as Lueder (1983) has pointed out. An activity sampling study of postures linked with the task analysis, will give some insight into the ways that people are coping with the job, helped by, or in spite of, the chairs provided.

From such a study, combined perhaps with measurements of the seat and workplace and backed by occupational health evidence and/or a questionnaire or BPD study, the present state of a situation can be documented. This would provide data for a focused attack on the seating and its associated factors, as well as a baseline for comparing the results of any changes. These data would come particularly from the combination of measurements, observations and BPD measures. But then, having made the design decisions and the related tests which appear to be relevant, what must be done about the testing of the final results?

A two-stage procedure seems appropriate. With only one or a few chairs, but with data derived from previous studies, the methodology used by Drury and Coury (1982) is appropriate. Briefly, this is a dimensional comparison with standards or other validated data, a short self-adjustment period for each subject, then a long (2 h plus) exposure when the seat is used for its designed purposes, during which BPD data are gathered, combined at the same sessions with a general comfort scale if this is seen as desirable. At the end of the test period, a CFCL is administered.

This test procedure locates the chair within a spectrum of other chairs used for similar purposes and identifies weaknesses in the new design, as subjectively recognized by users. However, if certain additional requirements are needed, e.g. a given reach envelope, reduced back load compared with existing chairs, or similar, then additional measures must be added to the above procedure. In the light of Yu *et al.* (1988), measures of stature change, the BPD for discomfort and the CFCL may well give all the necessary data plus clear evidence of spinal loading.

The wide variation in chair users, as well as the wide range of uses to which a chair will be put, require a further stage after the installation of a number of the modified chairs. It is at this final stage that the activity sampling and general questionnaire are useful. The procedure is not necessarily expensive but it does allow the modes of use to be assessed against what was expected in the design and in the first experimental stage. It also assesses the acceptability of the design, giving a comparison with the initial study and indications of any gains made.

The reader will recognize that there is still much to do in chair research and application. Our knowledge of comfort factors is still small and our ability to provide an effective working seat has much room for improvement. This is in no way to discredit the studies which have demonstrated good designs for particular cases. But a walk around an office, factory or supermarket, or a discussion with an orthopaedic surgeon, would reveal how far there is to go. If one thing is in the ergonomist's favour, it is the increasing recognition of the health and safely aspects of seating. A major weakness in the computerized office is the operator's back, and it is this 'discovery' that has again put seating firmly on the ergonomist's agenda.

Note

This chapter is largely based upon the Ergonomics Society's Lecture 1989, reproduced in *Ergonomics*, **32**, 257–269.

References

Allen, P.S. and Bennett, E.M. (1958). *Forced Choice Ranking as a Method of Evaluating Psychological Feelings.* Technical Report No. 58 (USAF. WADC).

Akerblom, B. (1954). Chairs and sitting. In *Symposium on Human Factors in Equipment Design*, edited by W.F. Floyd and A.T. Welford (London: H.K. Lewis).

Bendix, A., Jensen, C.V. and Bendix, T. (1988). Posture, acceptability and energy consumption on a tiltable and knee support chair. *Clinical Biomechanics*, **3**, 66–73.

Branton, P. (1966). *The Comfort of Easy Chairs.* Interim report. Furniture Industry Research Association, UK.

Branton, P. (1969). Behaviour, body mechanics and discomfort. *Ergonomics*, **12**, 316–327.

Branton, P. and Grayson, G. (1967). An evaluation of train seats by observation of sitting behaviour. *Ergonomics*, **10**, 35–51.

Corlett, E.N. and Bishop, R.P. (1976). A technique for assessing postural discomfort. *Ergonomics*, **19**, 175–182.

Corlett, E.N. and Eklund, J.A.E. (1984). How does a backrest work? *Applied Ergonomics*, **15**, 111–114.

Corlett, E.N., Wilson, J. and Manenica, I. (eds) (1986). *The Ergonomics of Working Postures, Section 5: Seats and sitting.* (London: Taylor and Francis).

Drury, C.G. and Coury, B.G. (1982). A methodology for chair evaluations. *Applied Ergonomics*, **13**, 195–202.

Eklund, J.A.E. (1986). Industrial seating and spinal loading. Ph.D thesis University of Nottingham. Distributed by Dept of Industrial Ergonomics. University of Technology, Linköping, Sweden.

Eklund, J.A.E. and Corlett, E.N. (1986). Experimental and biomechanical analysis of seating. In *The Ergonomics of Working Postures*, edited by E.N. Corlett, J. Wilson and I. Manenica. (London: Taylor and Francis).

Eklund, J.A.E., Corlett, E.N. and Johnson, F. (1983). A method for measuring the load imposed on the back of a sitting person. *Ergonomics*, **26**, 1063–1076.

Eklund, J.A.E., Houghton, C.S. and Corlett, E.N. (1982). *Industrial Seating, A Report of Some Pilot Studies.* Internal report. University of Nottingham, published in Eklund (1986).

Eklund, J.A.E., Örtengren, R. and Corlett, E.N. (1987). A biomechanical model for evaluation of spinal load in seated work tasks. In *Biomechanics XB*, edited by B. Jonsson. (Champaign, IL: Human Kinetics Publishers).

Floyd, W.F. and Roberts, D.F. (1958). Anatomical and physiological principles in chair and table design. *Ergonomics*, **2**, 1–16.

Grandjean, E. (1980). *Fitting the Task to the Man*. (London: Taylor and Francis).

Habsburg, S. and Mittendorf, L. (1980). Calibrating comfort: systematic studies of human responses to seating. In *Human Factors in Transport Research*, edited by D.J. Oborne and T.A. Levis. (New York: Academic Press).

Hünting, W. and Grandjean, E. (1976). Sitzverhalten und subjektives Wohlbefinden auf Schwenkbaren und fixierten Formisitzen. *Z. Arbeitswissenschaft*, **30**, 161–164. (Quoted in report by E. Grandjean and W. Hünting from Federal Institute of Technology, Zurich of 20.10. 1976.)

Jones, J.C. (1963). Fitting trials. *Architects Journal*, **137**, 321–325.

Jones, J.C. (1969). Methods and results of seating research. *Ergonomics*, **12**, 171–181.

Kember, P. (1976). The Benesh movement notation used to study sitting behaviour. *Applied Ergonomics*, **7**, 133–136.

Langdon, F.J. (1965). The design of card punches and the seating of operators. *Ergonomics*, **8**, 61–65.

LeCarpentier, E.F. (1969). Easy chair dimensions for comfort—a subjective approach. *Ergonomics*, **12**, 328–337.

Lueder, R.K. (1983). Seat comfort: a review of the construct in the office environment. *Human Factors*, **25**, 701–711.

Murrell, K.F.H. (1965). *Ergonomics*. (London: Chapman and Hall).

Pheasant, S.T. (1984). *Anthropometry, An Introduction for Schools and Colleges*. BSI Education No. PP7310. (London: British Standards Institution).

Rieck, A. (1969). Über die Messung des Sitzkomforts von Autositzen. *Ergonomics*, **12**, 206–211.

Shackel, B., Chidsey, K.D. and Shipley, P. (1969). The assessment of chair comfort. *Ergonomics*, **12**, 269–306.

Shipley, P. (1980). Chair comfort for the elderly and infirm. *Nursing* supplement **20**, *Sleep and comfort*.

Shipley, P., Haywood, J., Furness, W. and Rose, J. (1969). *Testing Easy Chairs for the Elderly*. Report to the Research Institute for Consumer Affairs, London.

Wachsler, R.A. and Learner, D.B. (1960). An analysis of some factors influencing seat comfort. *Ergonomics*, **3**, 315–320.

Wotzka, G., Grandjean, E., Burandt, U., Kretschmar, H. and Leonhard, T. (1969). Investigations for the development of an auditorium seat. *Ergonomics*, **12**, 182–197.

Yu, C.-Y., Keyserling, W.M. and Chaffin, D.B. (1988). Development of a work seat for industrial sewing operations: results of a laboratory study. *Ergonomics*, **31**, 1765–1786.

Part V

Analysis of work activities

Having examined and explained the methods and techniques basic to all areas of ergonomics investigation, looked in more depth at techniques developed or adapted to study some particular design and evaluation applications, and then devoted a number of chapters to the broad area of physical environment assessment, we turn now to analysis techniques for work activities. Each of the chapters in this section takes either a specific, if broad, area of work activity, such as visual performance, or a specific consequence of certain types of work, as in stress assessment. The individual contributions draw upon any number of the methods discussed earlier in the book in more basic and general form.

Work involves both physical and mental activities and this section considers both of them. The physiological costs of work are a reality to many, and are not necessarily alleviated by modern technology. Hence it is proper that we should be informed about the major methods for evaluating physical effort, including simple methods of field investigation to recognise when this effort is high or excessive.

Although dynamic work (Kilbom, chapter 21) is obvious when it is performed, static and postural load (Corlett, chapter 22) is often overlooked. It is easy to assume that if someone is sitting down there is no load of any significance, and forget that postural muscles are still active. If the job restricts the opportunity for change between muscles, then the situation is exacerbated. People will often discount the discomforts of postural loading, expecting to 'feel tired after work'. However these manifestations give guidance to the ergonomist on where to look for the sources of problems, and methods to assess the effects on the person are important in very many situations. It will have been noted, in the chapter on seating at the end of the last section, how subjective methods were utilised, whilst many previous chapters all use people's responses as primary data for analysis.

When risks, and boundaries for levels of exposure, have to be assessed, we need numbers. Chapter 23 by Tracy on biomechanics is cautious about the acceptance of figures but strongly in favour of biomechanic analyses for 'before and after' comparisons. If figures are to be used as absolute measures and matched against criteria, the availability of a range, or measure of the likely spread of results, together with the proposals for using 'worst-case' values, contributes to the assurance that the decisions taken will be conservative.

If the assessment of physical workload is difficult and contentious, then the situation for mental workload (Meshkati *et al.*, chapter 24) is no different. The methods used do parallel some of those for physical workload—especially performance (or primary task measures), subjective assessment and physiological measurement. Even the fourth group of techniques—secondary task measures—can be used for physical workload too, for instance checking the effects of extended keyboard use by testing subsequent performance on one of the many tests of manipulative ability.

Of course, one way in which mental workload assessment differs from physical workload assessment is in the less overt and visible nature of behaviour involved. Direct observation of task performance or worker response can often at least confirm or illuminate investigations of physical work. This is less easy for mental work, and techniques such as protocol analysis (see Bainbridge, chapter 7) may be needed. A similar problem, i.e. that the causative factors are not easily or directly observable, is also found in the assessment of stress. In chapter 25, Cox discusses the various models or definitions of stress—as with many topics in this book the model or definition chosen can affect our selection of methods as well as of preventive strategies. He then addresses the identification of stress potential, and recognition and measurement of stress in individuals.

The final two chapters in this section are complementary reviews of the assessment and consequences of performing visual work, appropriate since over 90 per cent of work task-related input is visual. Bullimore *et al.* (chapter 26) are concerned with the performance of visual tasks, whereas Megaw (chapter 27) concentrates upon one possible outcome or group of outcomes of visual performance, namely visual fatigue. As well as the types of method that may be adapted for use in almost any situation, visual processes and their consequences may be assessed by use of some particular techniques also, including specific performance tests (e.g. for colour vision) and observed behavioural measures such as eye movements. Indeed, visual task performance is a very rich topic as far as methods and technique development is concerned.

In conclusion, if there were one theme which connects all the work activity analysis in this section, it is that measurement and assessment of effects on people are made doubly difficult because of the problems in defining unambiguously the phenomena concerned. For instance, workload, physical or mental, is made harder to understand and evaluate if we are unsure of its causes; disagreement about what visual fatigue is does not make the task of assessing it any easier; and so on.

Chapter 21

Measurement and assessment of dynamic work

Åsa Kilbom

Introduction

The human body is continuously required to perform physical work. Three main types of demands must be met:

(a) Moving the body or its parts, e.g. in walking and running,

(b) transporting or moving other objects, e.g. in carrying, lifting, hitting, cranking, and

(c) maintaining the body posture, e.g. in forward stoop of the body, twisted trunk, raised arms.

When exposed to these demands the human body responds with a complex series of events, leading to the performance of muscular exercise. Thus the muscle contraction is the end point of events taking place in the sensory organs, the brain, nervous system, lungs, heart and blood vessels and musculoskeletal systems (see Figure 21.1).

The term *physical stress* is often used to describe the demands, while *strain* is used to describe the response in the human body. The assessment of these physical stresses and strains is an important component of ergonomics. It is used to identify excessive physical stresses and to design external demands so that they fit the capacity of the workers. This chapter will deal with the quantitative evaluation of (1) physical stress, and (2) physiological strain on the cardiovascular and pulmonary systems.

Sometimes physical stress can be accurately predicted. For example, the energy and the power needed to transport the body up a ladder can be calculated, knowing the weight of the body, the vertical distance and the time available. In a similar way the force or torque necessary to maintain body posture or to hold an object can be calculated. In mixed tasks, encompassing body movements, exertion of forces and maintaining body posture in a complex time sequence, it is often impossible to predict the demands, although attempts have been made to model them. In these cases the demands must be measured, using the methods described below.

Obviously the response—*the strain*—will be influenced by the capacity of the individual and not only by the demands. For optimum performance all systems of the body must function efficiently. However, any of the organs participating in the events leading to muscle contractions can have a low functional capacity or small dimensions, thereby limiting the capacity for muscular work. Those systems that most commonly limit the rate of physical work are the cardiovascular system and the muscles.

Different types of exercise

During *dynamic* exercise muscles are shortening and lengthening rhythmically, e.g. in running and walking. The shortening phase is also called *concentric* exercise, while a contraction with simultaneous lengthening is called *excentric* exercise. During *static* exercise the muscles maintain a contraction with unchanged force and length for a period of time varying from a few seconds to several hours. In practice purely static contractions hardly ever occur; the most common being a low intensity contraction with small variations both in muscle length and force.

Other terms used to describe muscle exercise are *isometric* (i.e. unchanged muscle length) and *isotonic* (i.e. unchanged muscle force). In *intermittent* work muscular exercise (either dynamic or static) is performed for a duration of a few seconds up to several minutes, interrupted by rest, then resumed, again followed by a pause and so on.

Muscle metabolism during exercise

The discharge of a nerve impulse on the motor end plate of the muscle is the signal for a fast conversion of chemically bound energy, in the form of ATP (adenosine triphosphate), to mechanical energy:

$$ATP \rightleftarrows ADP + \text{phosphate} + \text{energy}$$

where ADP is adenosine diphosphate.

Available stores of ATP are very limited, and therefore they have to be built up continuously from energy obtained by the oxidation of glucose and fatty acids. Protein is also metabolized but at a much lower rate, its main function being to provide material for tissue repair and growth.

The metabolism can take place either aerobically (with oxygen) or anaerobically (without oxygen). Oxygen is transported to the muscles by the circulation. If enough oxygen is available the aerobic pathway is chosen, because it is more efficient as it leads to a more complete metabolism of energy-rich nutrients and gives less fatigue. The fatigue perceived in

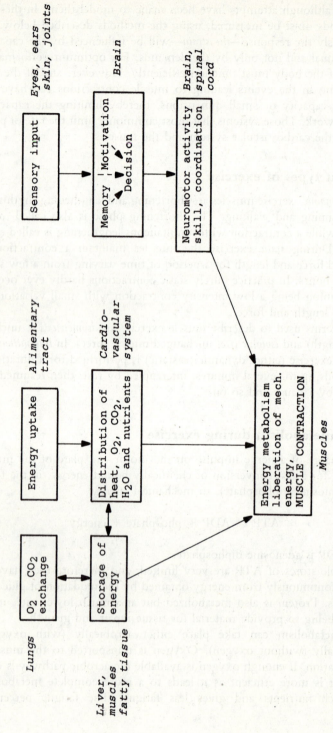

Figure 21.1. Pathways leading to muscular contractions.

conjunction with anaerobic muscle metabolism is probably caused by the lowering of pH that takes place when lactate is produced.

aerobic pathway $\left\{ \begin{array}{l} \text{free fatty acids} \\ \text{glucose} \end{array} \right.$ $+ O_2 \rightarrow CO_2 + H_2O +$ energy

anaerobic pathway $\qquad\qquad$ glucose $\rightarrow$ lactate + energy

Note that the breakdown of ATP is also anaerobic, but no lactate is produced.

Role of the cardiovascular system during exercise

The main role of the cardiovascular system during exercise is to transport:

(a) heat from exercising muscles to the body surface,
(b) nutrients (fatty acids and glucose) from their stores in liver and fatty tissue to exercising muscles,
(c) oxygen from lungs to muscles,
(d) CO_2, H_2O and lactate from muscles to lungs and liver for excretion and metabolism.

In order to meet these demands the circulation can increase the transporting capacity by a factor of 100 within a few minutes after the onset of exercise. This is achieved by:

(a) distribution of more blood to exercising muscles,
(b) a more efficient uptake of O_2 and excretion of CO_2,
(c) increasing the total blood flow, by increasing the cardiac stroke volume and heart rate.

Physical working capacity

The ability of any individual to perform physical work varies within very wide limits. These variations are mainly due to genetic factors which influence both body dimensions and functional capacity. The ability to perform aerobic work is best evaluated through measurements of the *maximum aerobic power*, measured as the highest uptake of oxygen per minute (max $\dot{V}O_2$) that can be achieved during dynamic exercise with large muscle groups. Physical inactivity, e.g. a few days in bed, leads to a fast reduction of the maximal $\dot{V}O_2$, whereas training gives nearly as fast an increase. Thus the aerobic power achieved in an individual is the effect of an adaptation process. Variations in average male and female values by age are presented in Figure 21.2 (adapted from Åstrand, 1960).

Fatigue and recovery

Aerobic metabolism leading to lactate accumulation and muscular fatigue takes place:

(a) at the onset of dynamic exercise,

Maximal
oxygen uptake
ℓ/min

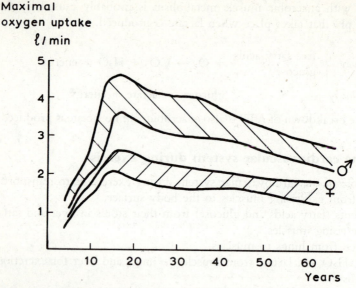

Figure 21.2. Maximal aerobic power in physically active men and women by age. The solid
lines indicate mean values, the upper shaded area represents +2 S.D. for men and the lower
shaded area represents −2 S.D. for women. Modified from Åstrand (1960).

(b) during heavy dynamic exercise, i.e. when the energetic demands
exceed 50% of the individual maximal aerobic power,

(c) during static exercise exceeding about 10% of the maximal muscle
strength.

Some of the lactate produced during exercise can be metabolized in the
muscles and/or removed. However, during very heavy dynamic work or
during static exercise blood circulation cannot keep up with the demands on
oxygen supply and lactate removal, and this leads to lactate accumulation,
lowered pH, perception of fatigue and reduced endurance. The endurance
during static contractions decreases as the intensity of the contraction increases.
For a relatively unlimited endurance, without subjectively perceived fatigue,
contraction intensities of 5–10% of maximal muscle strength must not be
exceeded.

The cycle of fatigue and recovery is demonstrated in Figure 21.3. The rate
of recovery after fatiguing exercise depends on:

(a) the duration of exercise,
(b) the intensity of exercise, and
(c) the physical fitness of the individual.

The rate of recovery depends to a large extent on the ability of the blood
circulation to supply oxygen to tissues (repayment of the 'oxygen debt').
Therefore recovery has usually been studied by measuring heart rate after
exercise.

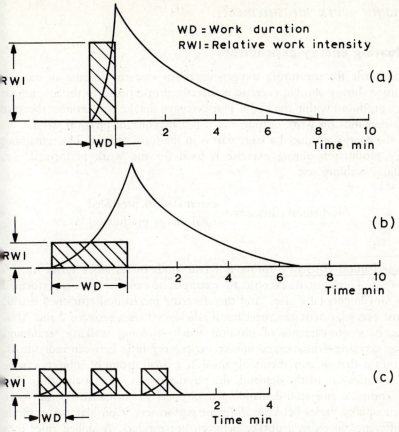

Figure 21.3. Schematic representation of build-up of fatigue, and recovery in three types of exercise. (a) Very high intensity exercise during 1 min, followed by recovery which in this case took around 9 min. (b) Exercise at ⅓ of previous intensity, performed during 3 min, followed by recovery which took around 7 min. (c) Intermittent exercise consisting of three exercise periods at ⅓ of initial intensity, each performed during 1 min with 1-min pauses interspersed. Note the brief recovery periods.

Recent research indicates that recovery of muscle function may not be complete even though all circulatory variables have returned to pre-exercise levels. After static contractions heart rate may normalize, but on repeated contractions endurance may still be reduced. After prolonged static or excentric exercise to exhaustion full recovery may take several hours and maybe even days. The most likely cause of this reduced performance capacity is damage to the structure of the muscles (including ruptures and oedema formation) and maybe also depletion of energy-rich nutrients in the muscles.

Dynamic work measurements

Evaluating energy expenditure

The rationale for measuring oxygen uptake is that the amount of oxygen consumed during aerobic exercise is directly proportional to the amount of energy produced within the body. Thus, oxygen uptake is an indirect measure of the demands on work output, i.e. it is a measure of physical stress.

Of the energy produced a large part is in the form of heat. The remaining energy production during exercise is used for the work performed, i.e. pedalling, walking, etc.

$$\text{Mechanical efficiency} = \frac{\text{external work produced}}{\text{total energy production}}.$$

The mechanical efficiency during different kinds of activities varies from 0 to 50–60%. During static exercise, for example, no external work is performed as no movements take place, and therefore the mechanical efficiency is 0%. In most everyday activities mechanical efficiency varies between 0 and 20%. In the basic constituents of physical work—cycling, walking, cranking, hitting, carrying—the oxygen uptake varies very little between individuals. This means that measurements obtained in a small group of subjects can be used as a measure of the demand, the physical stress, in that situation. The more complex and skill-demanding a physical performance is, the more oxygen uptake varies between different performers. Consider for example the difference in oxygen uptake between horse riders. A skilled rider uses less oxygen per minute to master the horse than an unskilled one, who uses additional muscle contractions for balancing on horseback and in preparation for unexpected actions from the horse.

For most purposes, including manual labour, oxygen uptake is a good indicator of physical stress. Important exceptions are tasks which induce a heavy heat stress, tasks with a large static component, and activities which demand a large proportion of anaerobic metabolism. For such tasks oxygen uptake gives valuable information about the aerobic component, but the measurements must be supplemented with others (see later under heart rate and body temperature measurements). For example, short-lasting, very demanding athletic tasks (e.g. short distance running) should not be evaluated only on the basis of oxygen uptake. However, if the oxygen uptake is measured over the entire *exercise plus recovery* period the total demands or energy consumption can be calculated. Because occupational activities usually go on for several hours, it is very uncommon that they contain an appreciable degree of anaerobic exercise. Exceptions may be all-out life saving operations (fire-fighting, diving, emergency maintenance) but in such cases the tasks are usually short lasting and time is given afterwards for recovery. Such situations can be potentially hazardous especially when combined with high heat load (see later). Performers of such tasks should therefore be selected for good physical performance capacity, and the task time may have to be controlled in order to avoid heat stroke or accidents caused by fatigue.

Thus measurement of oxygen uptake during work aims at assessing the physical stress during a work operation. It is used in order to:

(a) identify the most demanding tasks in an occupation. Such tasks often exclude the weaker members of the work-force from taking a certain job. The most strenuous tasks can be redesigned to make them less demanding, job rotation or pauses can be introduced, or the task can be reduced in pace (see Figure 21.4),

(b) compare the demands of alternative ways of performing a task, for example with improved tools or work routines (see Figure 21.5),

(c) evaluate the component of dynamic exercise in a complex work situation. Many jobs have components of static and dynamic exercise as well

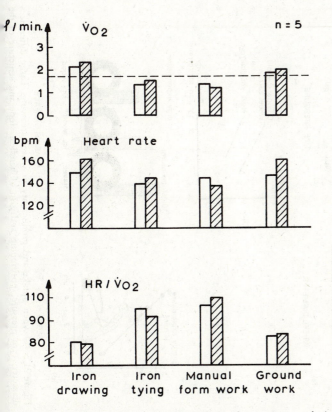

Figure 21.4. Oxygen uptake and heart rate during four common tasks in the building industry. All tasks were performed at a normal (unfilled columns) and an impeded (filled columns) pace. The level for 'very heavy work' = $\dot{V}O_2 > 1.75$ l/min, has been marked. Observe the high $HR/\dot{V}O_2$ quotient in some tasks.

Source: Kilbolm *et al.* 1989.

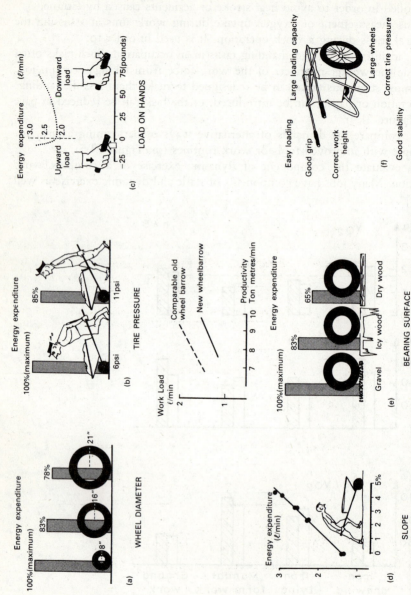

Figure 21.5. Oxygen uptake and productivity in wheelbarrowing. Large differences were obtained with changes in wheel diameter, tyre pressure, positioning and grip of handles, slope and structure of the terrain. The optimal design improved

as heat stress. Strain measures like heart rate (see later) and rectal temperature indicate the total physiological response to a number of different stressors. Oxygen uptake however increases appreciably only as an effect of dynamic exercise (see Figure 21.4).

The task demands can also be expressed in relation to the maximal aerobic power of the individual. Consider Table 21.1, which gives oxygen uptake during a number of occupational or everyday tasks. By tradition tasks demanding oxygen uptake above 1.75 l/min are considered very heavy, those demanding 1.5–1.75 l/min are considered as heavy, those demanding 1.0 − 1.5 l/min are considered moderately heavy and those demanding less than 1.0 l/min are classified as light. However, the physiological response, the strain, will obviously vary markedly between different individuals when performing the same task. For an individual with a maximal oxygen uptake of 1.75 l/min, tasks with an oxygen uptake of 1 l/min demand 60% of his/her maximal capacity, while the same task requires only 30% of the maximal capacity of an individual with a maximum of 3.0 l/min. Thus the energetic strain, expressed in relation to the individual's maximal aerobic power, may vary considerably between individuals doing the same task.

Many investigations have demonstrated that the demands during an 8-h work day should not exceed about 35–40% of the individual maximal capacity. For shorter time periods a somewhat higher proportion can be used. For very short-lasting tasks—a few minutes—close to maximal demands can be met provided a sufficiently long rest period is given after the exercise to allow for lactate metabolism (see also earlier under 'Fatigue and recovery').

Measuring and analyzing oxygen uptake

Oxygen uptake is calculated after analysis of expired air for its content of oxygen and (usually) carbon dioxide.

Table 21.1. Oxygen uptake during various physical tasks and activities

Activity	Oxygen uptake (l/min)
Running, skiing, swimming (male elite)	> 5·0
Running, skiing, swimming (female elite)	> 4·0
Cycle ergometer, external load 200 W	2·8–2·9
Fire-fighting, manual work in forestry and mining	2·0–3·0
Cycle ergometer, external load 150 W	2·1–2·2
Heavy industrial work, heavy gardening and agriculture	1·5–2·0
Cycle ergometer, external load 100 W	1·5–1·6
Heavy cleaning and manufacturing, fast walking or slow running	0·8–1·5
Cycle ergometer, external load 50 W	0·9–1·0
Walking 4–5 km/h, nursing, catering, light manufacturing, changing between sitting, standing and walking	0·6–1·0
Passive standing	0·4–0·5
Sitting assembly work, driving, office work	0·3–0·6
Passive sitting	0·2–0·4
Supine	0·2–0·3

$$\dot{V}O_2 = \dot{V}_E \, (CO_2i - CO_2e)$$

where $\dot{V}O_2$ is oxygen uptake in litres per minute, $\dot{V}_E$ is expired pulmonary ventilation in litres per minutes, CO_2i is concentration of oxygen in inspired air and CO_2e is concentration of oxygen in exspired air. Oxygen uptake is usually expressed in litres per minute, STPD (standard temperature, pressure and saturation, i.e. 0°C, 760 mm Hg and dry).

Expired air volume is either measured, or calculated, from the volume of inspired air. The volume of inspired and expired air is not the same, as expired air volume is slightly expanded through heating and the addition of carbon dioxide and humidity. Therefore CO_2i must be corrected for differences in temperature, humidity and carbon dioxide content. Thus a complete calculation of oxygen uptake includes the analysis of volume, carbon dioxide and oxygen concentrations of expired air, and temperature of ambient inspired air. Carbon dioxide and oxygen content of inspired air are constant (0·03 and 20·94%, respectively), and the temperature of expired air is assumed to be 37°C.

In order to convert oxygen uptake to energy consumption, an energy coefficient of 20·2 kJ at rest and 20·6 kJ at exercise should be used. Thus, the combustion of one litre of oxygen with a mixture of carbohydrates, fat and protein yields on the average 20·2 kJ at rest. (Previously energy consumption was often expressed in calories, where 1 kcal = 1000 cal = 4·186 kJ.)

Measurements of oxygen uptake require relatively expensive and complex instrumentation. For the laboratory, fully automated systems which measure ventilatory air flow and temperature, current barometric pressure and humidity, ventilation rate and concentrations of oxygen and carbon dioxide in expired air are available. Large errors in the calculation of oxygen uptake can be introduced if the volume of expired air is not correctly measured, or if the air sample for oxygen analysis is mixed with ambient air. Therefore the system for collection and analysis must fit tightly to the subject's mouth via a mouth piece, while a nose clip is used to eliminate leakage from the nose. A whole face mask can also be used but it is less reliable due to risk of leakage (especially for bearded persons) and a larger 'dead space'.

For field studies simplified automated instruments are used. They usually analyse only inspired air flow and expired oxygen concentration, making assumptions for other factors. Thus a small error is introduced, but within the most common range of oxygen consumption values this error is usually less than 5%.

Instead of this complex procedure a simple measurement of pulmonary ventilation can be used to predict oxygen uptake. For light to moderately heavy exercise intensities pulmonary ventilation is closely related to oxygen uptake. During heavy exercise (oxygen uptake over 2·0 l/min and pulmonary ventilation above 40–50 l/min) pulmonary ventilation is an unreliable measure of energy expenditure.

Measuring physical working capacity

The maximal aerobic power (max $\dot{V}O_2$) can be accurately measured during dynamic exercise with large muscle groups, e.g. cycling or running. Other physical activities which can produce maximal or near maximal oxygen uptakes are skiing, rowing and swimming, but such activities are for practical reasons not often used for testing purposes.

Choice of cycle ergometer vs. treadmill

The advantages of cycle ergometers are:

(a) the subject is more stationary than on the treadmill, and therefore all measurements are performed more easily and accurately,

(b) the mechanical efficiency during cycling varies little between individuals. This means that a given workload on the cycle ergometer can be expected to yield very similar values for oxygen uptake in a group of subjects. On a treadmill, however, running at a certain speed will give a large range of oxygen uptakes for a group of subjects, because of their varying body weights,

(c) elderly subjects usually find maximal exertion on the cycle ergometer less frightening than on the treadmill.

The advantage of treadmill exercise is that maximal oxygen uptake is up to 8% higher than on the cycle ergometer.

Test protocol for cycle ergometer tests

The subject is usually tested on two to three submaximal and one maximal workload (Figure 21.6). The submaximal workloads are chosen to correspond to 35–40, 50–60 and 70–80% of the individual's expected maximal aerobic power. In practice this often coincides with a power output of 50, 75 and 100 W for female subjects and 50, 100 and 150 W for males. For small or old individuals and for well-trained persons, the test loads must be adjusted. The testing on submaximal workloads serves several purposes:

1. The investigator can observe the subject's reaction to exercise, register pulmonary ventilation, heart rate (see below) and subjective responses and thereby more easily estimate which maximal workload should be used.

2. Contraindications to further work stress can be identified, i.e. symptoms and physiological signs indicating cardiovascular, pulmonary or musculoskeletal disease.

3. Submaximal workloads serve as a warm-up, i.e. they increase muscle temperature. Thereby exchange of oxygen and carbon dioxide is facilitated, and the risk of strains and sprains is decreased through improved muscle co-ordination.

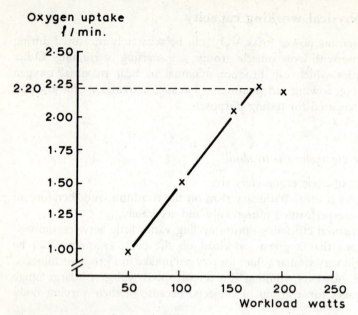

Figure 21.6. Relation between workload and oxygen uptake during submaximal and maximal exercise on the cycle ergometer. The maximal oxygen uptake was in this case reached at a workload of 175 W and it was 2·20 l/min. On another day the subject was also tested on a workload of 200 W, but could not further increase the oxygen uptake.

Exercise on each submaximal workload is maintained for 5–6 min. Oxygen uptake should not be measured until during the last 1–2 min of this exercise period, as it takes 1–3 min after the start of exercise for oxygen uptake to stabilize at a steady state.

The maximal workload is chosen so as to exhaust the subjects completely within 3–7 min. Usually oxygen uptake is measured continuously after 2 min and the highest value obtained thereafter is used as max $\dot{V}O_2$.

Test protocol for treadmill

The workload is gradually increased through increments of slope and speed of the treadmill. Usually subjects are tested for 2–3 min on each workload without pauses between, and oxygen uptake is measured during the last minute of each workload. As the subject approaches exhaustion, continuous measurements are made. A number of different test protocols for treadmill exercise have been described. If oxygen uptake is not measured during the test, it can be estimated from the number of metabolic units, METS (1 MET = oxygen uptake at rest, i.e. 3.5 ml × kg × min) required for a certain increment in slope and speed of the treadmill.

An alternative to cycle ergometer and treadmill exercise tests is the so-called *step test*, which may be suitable in field studies, where sometimes no other testing equipment is available. In the Harvard step test, the physical working capacity is estimated by letting the subject step up and down a stool at a given pace. In modified step tests the height of the stool may be varied to suit the subject's estimated capacity. Gradational tests, using submaximal and maximal intensities, have also been described. Step tests, however, are difficult to standardize, measurements of, for instance, heart rate are difficult to perform, and the tests often lead to muscle soreness.

For further reading on exercise testing the reader is referred to Åstrand and Rodahl (1986, pp. 354–390).

Contraindications to exercise testing

Before the test the subject must be asked about cardiovascular or pulmonary disease, ongoing infections and medications, and exercise habits. Ongoing infections and chest pain are absolute contraindications, whereas previous cardiac insufficiency, angina pectoris, myocardial infarctions and hypertension are relative contraindications. In such cases exercise testing must only be done under electrocardiograph (ECG) surveillance, by trained medical staff and with resuscitation equipment available. Symptoms like chest pain, excessive breathlessness and certain arrhythmias that occur during exercise are indications to immediately stop the test. In subjects above 40 years of age it is advisable to register ECG before, during and after the exercise test. With the above precautions the risks of exercise testing are exceedingly small. A further reduction of risk is obtained by doing a submaximal work test (see 'Heart rate during exercise' later) using the heart rates at submaximal work loads to predict the maximal oxygen uptake.

Heart rate during exercise

Heart rate increases during both static and dynamic exercise, during heat exposure, and as an effect of psychological stress. Thus a heart rate increase is an *unspecific* cardiovascular strain response, and the interpretation of heart rate recordings must always be made against a background knowledge of the circumstances of the recording. Other factors that can influence heart rate are tobacco smoking, certain types of medication, ongoing infections, and so on.

During moderate dynamic exercise at a constant workload (e.g. on a cycle ergometer) heart rate increases during the first 1–3 min and then reaches a steady state. Steady-state heart rates are linearly related to workload or oxygen uptake (Figure 21.7). A very unfit individual will have his/her heart rate–oxygen uptake relationship shifted to the left, and a well-trained individual will have it shifted to the right. During static exercise steady-state levels are usually not reached.

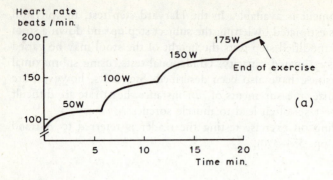

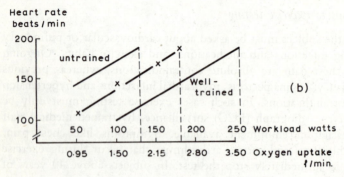

Figure 21.7. Heart rate during a standardized cycle ergometer test. (a) Heart rate increase and steady-state levels during exercise on increasing workloads. Note that no steady state was reached at the maximal workload. (b) Crosses indicate steady-state heart rate levels from Figure 21.7(a). The maximal heart rate, 180 beats/min, was reached at workload 175 W, at an oxygen uptake of 2.5 l/min. Heart rate – oxygen uptake relationships for an untrained and a well-trained individual with their respective maximal oxygen uptakes, are also indicated.

Submaximal exercise testing

The relationship between heart rate and oxygen uptake is the basis for the method to predict maximal oxygen uptake from submaximal heart rates. By measuring heart rate at, say, 3 submaximal workloads, the maximal oxygen uptake can be predicted, either by extrapolation to the maximal heart rate or by using nomograms. These methods require knowledge of the maximal heart rate, which gradually decreases with age, from around 220/min in young children to around 160/min at age 60. However, there is also a relatively large variation in maximal heart rate between different individuals, even at the same age. Thereby an error of about 10% is introduced in the prediction of maximal oxygen uptake, compared to direct measurements.

The same safety precautions as during maximal testing must be used, i.e. a brief medical history regarding cardiovascular and pulmonary disease and infections must be taken, and the subjects must be instructed to report any symptoms apart from normal breathlessness and general fatigue. Usually the test is interrupted when heart rates 30 beats/min below the expected maximal heart rate are reached, i.e. at heart rate 170/min for a 20-year-old and 130/min for a 60-year-old. In order to avoid some other factors which influence heart rate, testing should not be performed within 1 h after smoking or after a large meal, and the air temperature in the laboratory should be around 18°C.

Heart rate as a strain measure during occupational work

The continuous measurement of heart rate during work is a common method to evaluate cardiovascular strain. The measurements are relatively simple to perform, and the results are usually reliable. The most commonly used system is telemetry, i.e. the ECG impulse is registered via chest electrodes and transmitted to a receiver. The receiver identifies the R-waves of the ECG signal and stores them in a microprocessor, where the number of beats is counted for given time periods (usually 1 min). The receiver can be positioned either on the wrist of the subject or in an external station, where the signals from several transmitters, using different frequencies, can be stored. The circuitry of the receiver can be constructed to identify the R-wave with good accuracy, so that no artefacts are recorded. Another method is to record the ECG signal continuously using miniaturized tape recorders. This also permits analysis of arrhythmias and is often used for clinical purposes.

If no measurement system is available, heart rate can nevertheless be recorded manually with reasonably good accuracy. Usually the work tasks to be studied are interrupted regularly at 1–5 min intervals and the time for 10 beats is recorded with a stopwatch.

As already emphasized heart rate is an unspecific measure of cardiovascular strain. Therefore measurements must be supplemented with activity recordings, i.e. the type of physical activity, psychologically stressful situations, simultaneous heat exposure, etc., must be noted. Often the investigation is supplemented with measurements of oxygen uptake, skin and deep body temperatures, subjective ratings of exertion and environmental measurements.

Analysis of heart rate recordings during work

The occupational recordings should be supplemented with a submaximal or maximal exercise test. The purpose of this testing is to expose the subject to a standardized activity for reference, and also to measure the maximal capacity. At least some of the exercise workloads should be designed so that they resemble the type of exercise performed occupationally. For example, if the occupational tasks studied are mainly performed walking, the exercise test should preferably include testing at one or two walking speeds on the treadmill. Or if the occupational tasks are mainly performed by the upper body, an exercise test including arm cranking may be most appropriate.

The analysis includes (see Figure 21.8):

1. Calculation of average heart rate. This average value is related to the oxygen uptake (or workload) at the same heart rate during the standardized

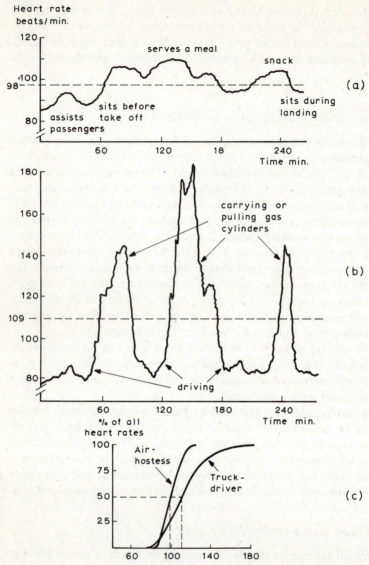

Figure 21.8. Examples of heart rate curves recorded during occupational work. (a) Air-hostess. (b) Truck driver delivering heavy gas cylinders to customers. (c) Cumulative heart rate curves. Note the very steep slope and narrow range of heart rates in the air-hostess and the much wider range of heart rates in the truck driver.

exercise test. This will permit an estimation of the proportion of the maximal oxygen uptake used.

2. Identification of the most demanding tasks.

3. Analysis of the mean heart rate for each of the different activities performed.

4. A comparison between the distribution of heart rates for the studied job (or work task) and that of another job.

Recovery heart rate: the Brouha method

If no automatic system for heart rate measurement is available, and if the work cannot be interrupted for manual recordings, the cardiovascular strain can nevertheless be estimated using recovery heart rate measurements done according to Brouha (1960). This method has even been claimed to be more reliable than measurements during exercise.

Immediately after the termination of a work task, the subject is seated and the heart rate is measured for three minutes after cessation of work. Recovery heart rate measurements can be performed repeatedly during a work day in order to evaluate whether the workload is too high, and whether recovery is incomplete between the different tasks. Brouha (1960, pp. 107–8) has outlined the practical procedure for assessing work and recovery. Pulse rate, P, during the first, second and third minutes after exercise is obtained by counting the beats during the last 30 s of each of the minutes, and doubling the value. $P_{AV\ 1,2,3}$ is the average of these three values, and is highly correlated with total cardiac cost.

1. If $P_1 - P_3 \geq 10$, of if P_1, P_2 and P_3 are all below 90, then recovery is normal.

2. If the *average* of P_1 over a number of recordings is ≤ 110, and $P_1 - P_3 \geq 10$, the workload is not excessive.

3. If $P_1 - P_3 < 10$, and if $P_3 > 90$, then recovery is inadequate for the task requirements.

The rate of recovery of heart rate is influenced by the absolute level of heart rate at the interruption of work, and by the fitness of the individual. In addition, heat exposure influences the rate of recovery. If recovery is unsatisfactory, the work must be redesigned in such a way as to reduce the physical stress. The intensity of exercise is usually difficult to influence, especially in industrial tasks where the pace is often set by machines. The fitness of the individual too is difficult to influence, and even in well designed, intensive training experiments, improvements of 10–20% are the best that can be achieved. The best way of securing sufficient recovery is usually to limit the duration of each task, or, if this is not possible, to increase the duration of the pauses between tasks (see also Figure 21.3).

Rating of perceived exertion

There is a well-known, curvilinear relationship between the intensity of a range of physical stimuli and our perception of their intensity. A positively accelerating relationship has been found between physical workload and perceived exertion. Thus, in a given individual, there is a highly reproducible relation between, e.g. the workload on a cycle ergometer and the perceived exertion. In the so-called Borg, or RPE (rating of perceived exertion) scale (Borg, 1985), the scale steps have been adjusted so that the ratings, from 6 to 20, are linearly related to the heart rate divided by ten. The scale (Figure 21.9) is presented to the subject before the start of the exercise test and the 'endpoints'—6 and 20—are thoroughly defined. The scale is then shown to the subject at the end of each exercise intensity and he/she is asked for a rating. The verbal explanations are used as support information. Recently a non-linear scale, which also has ratio properties, was developed.

These scales can be used to supplement physiological measurements during exercise testing. They often provide valuable additional information about subjective responses, especially in cases where the heart rate response is unreliable (e.g. patients with atrial fibrillation, or medication which influences heart rate). Similar scales have also been developed to quantify intensity of pain.

The RPE scales have been used to a limited extent in industry. The idea of substituting physiological measurements by subjective ratings is attractive, as ratings do not require any instrumentation. However, recent findings in industry and in industrial tasks suggest that ratings are influenced not only by the overall perception of exertion, but also by previous experience and motivation of the subjects. Thus highly motivated subjects tend to underestimate their exertion.

BORG'S RPE-SCALE

6	NO EXERTION AT ALL
7	
8	EXTREMELY LIGHT
9	VERY LIGHT
10	
11	LIGHT
12	
13	SOMEWHAT HARD
14	
15	HARD (HEAVY)
16	
17	VERY HARD
18	
19	EXTREMELY HARD
20	MAXIMAL EXERTION

Figure 21.9. Borg scale for rating of perceived exertion.

Body temperature during heat exposure and exercise

Body temperature increases during exercise, since a large proportion of the energy produced is converted to heat. To some extent the exercise performance benefits, since the metabolic processes work faster in higher temperatures. Thus deep body temperature is adjusted in relation to the relative workload, and although this adjustment is slow and takes at least 30 min, it is highly accurate. For example, during exercise corresponding to 50% of maximal aerobic power, deep body temperature is 38.0°C.

However, human tissues have a limited range of temperature tolerance, so the major part of the heat produced must be dissipated. In low environmental temperatures radiation and convection are the main ways of heat dissipation, whereas sweating is the only possibility in high environmental temperatures. With an ambient air temperature above 37°C, or in intense radiant heat, external heat is transferred to the body and must also be dissipated through sweating. Since sweating demands increased blood circulation to the skin, heat exposure puts large demands on the cardiovascular system. Hence exercise in hot environments induces increments in heart rate far above those obtained at the same exercise intensity in a cold environment (Figure 21.10).

Prolonged heavy sweating also leads to loss of fluid, which can severely reduce the blood volume and thereby further increase the cardiovascular strain. A loss of 1% of body weight through sweating leads to a deterioration in the physical working capacity and reduced orthostatic tolerance. The corresponding heart rate increase is around 10 beats/min. The World Health Organisation recommends that sweat production should not exceed 4 l in an 8-h work period. Thus physical work in hot environments imposes large stresses on the human physiology. If uptake and production of heat cannot be balanced by heat dissipation, deep body temperature increases with a concomitant risk of heat exhaustion and heat stroke. Deep body temperatures above 38.0°C indicate that this balance may be upset.

Through environmental measurements of wet bulb temperature, radiant heat and air velocity, an index, WBGT, can be calculated. This index can be used to predict the risk of heat exhaustion during exercise at different intensities. A more accurate prediction can only be obtained through physiological measurements of heart rate, deep body and skin temperatures, and fluid balance. A more extensive discussion on the measurement of the body's responses to heat will be found in the chapter by Parsons in this book.

Conclusions

Physiological measures of physical activity have a long history of contributions to ergonomics. They provide readily measured indications of the dynamic efforts people are exerting, and well recognized and reliable limits can be

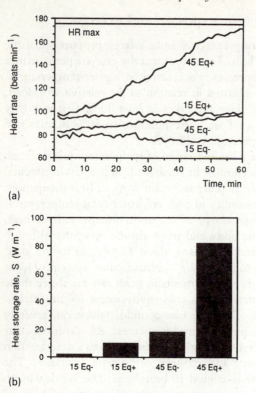

(a)

(b)

Figure 21.10. Heart rates (a) and heat storage (b) during standard work at increasing workloads at 15°C and 45°C, with (Eq+) and without (Eq−) heavy equipment. From Sköldström (1987).

used to ensure that people are not overloaded. As pointed out at the end of the chapter, there are confounding factors which must be watched for. Emotional stresses, including anxiety, can add to the physiological costs of human effort, whilst adverse thermal environments will detract in a major way from the available work capacity of a person.

Subjective measures, epitomized by the Borg scale, give a cross comparison to the physical measures, and allow the investigator to explore how the job feels to the job holder. The combination of the two types of measurement can be revealing as well as protecting the investigator from dropping into the error of relying solely on numbers, as if the people in the work situation were merely biological machines. As with all ergonomics the measures are to understand the effects of the situation on people—who are multidimensional. In situations where a single extreme stressor is acting, a one-dimensional measure could be appropriate. In most situations though, we need more than one measure in order to understand how best to modify, to design, or to assess a situation.

References

Åstrand, I. (1960). Aerobic work capacity of men and women with special reference to age. *Acta Physiologica Scandinavica* (Suppl. 169) **49**.

Åstrand, P.O. and Rodahl, K. (1986). *Textbook of Work Physiology* (New York: McGraw-Hill).

Borg, G. (1985). *An introduction to Borg's RPE-Scale* (Ithaca, NY: Movement Publications).

Brouha, L. (1960). *Physiology in Industry* (Oxford: Pergamon).

Hansson, J.E. (1970). *Ergonomics in the Building Industry*. Research Report No. 8. Byggforskningen, State Council of the Building Industry.

Kilbom, Å., Jörgensen, K. and Fallentin, N. (In press). Recording workload in occupational work – a comparison between observation methods, physiological measurements and subjective estimates. *Arbete och Halsa*.

Sköldström, B. (1987). Physiological responses of fire fighters to workload and thermal stress. *Ergonomics*, **30**, 1589–97.

Chapter 22

Static muscle loading and the evaluation of posture

E. Nigel Corlett

Introduction

The maintenance of postures and the support of loads are particular examples of the performance of static work. Although these are both quite common, what can be overlooked, presenting particular difficulties for the analyst, are those cases where postures *and* other physical activities intermingle. If one or the other situation predominates it may be sufficient to assess the whole situation on the basis of its worst aspect, but if serious levels of effort are required whilst posture is maintained, this procedure is unsatisfactory.

In general we can say that, whilst the limitations on dynamic physical activities felt by a person would be high heart rate and shortage of breath, the limits to static work will be the experience of muscular pain. As described in chapter 21 by Kilbom this arises particularly from the anaerobic metabolic activity of the muscles, whose blood supply is restricted due to the increased intramuscular pressure. A consequence of this pressure is that the heart rate does not represent the static effort involved, although post-effort heart rate can be an important indicator of the existence of static load (see Brouha, 1960).

Methods for direct measurement of the effort involved in posture, as well as its effects, are less common than for dynamic work. The range of methods embraces estimation techniques (biomechanics and estimates from maximum voluntary contraction methods); direct measures of muscular activity (analysis of the electromyographic–EMG–signal); measures of the resultant effects (e.g. spinal shrinkage); subjective measures (e.g. discomfort recordings) and a wide range of interpretive methods. These last range from epidemiological studies to estimates of likely effects from posture recordings (using such as OWAS, NIOSH Nordic questionnaire, posture targetting) and posture measurements (e.g. goniometers, SELSPOT or CODA). The measurement of posture and its effects is more extensively discussed in Corlett et al. (1986).

Maximum voluntary contraction

Perhaps the earliest scientific approach to estimating the appropriateness of static loads was to evaluate, experimentally, the holding times for various loads, expressing the result as a proportion of the maximum load which a person could hold. The force required to achieve maximum load is referred to as maximum voluntary contraction (MVC).

Whilst MVC may be measured quite simply (in many cases using a spring balance) there are some essential controls. An impulse force is not required for the measurement; the instruction to a subject is usually of the form, 'build up your maximum force gradually, over a period of 2–3 s, and hold it for 3 s'. The value used is the mean force over the last, relatively constant, period.

The relationship between force and holding time, demonstrated by Monod and Scherrer (1965) and by Rohmert (e.g. 1973a, b) is a logarithmic one (Figure 22.1). Today it is accepted that a long-term *constant* static effort greater than 2–3% of MVC is unacceptable, although at one time 15% was believed to be possible. Knowledge of the force holding-time relationship, which appears to hold for all skeletal muscle, does allow us to estimate the effects of some postures, and provide guidance as to their appropriateness.

The maximum holding time for a posture is not, by itself, a very useful measure, since we usually wish to know the frequency with which the posture may be held, and the consequent likelihood of damage. Hence recovery from static work loads is of interest. In experimental work, evidence

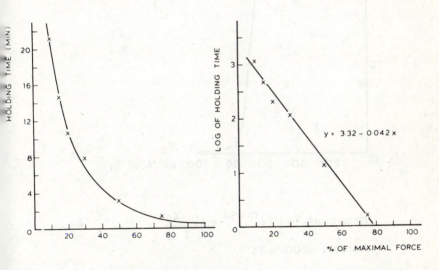

FORCE – TIME RELATIONSHIP FOR PULL CONDITION

Figure 22.1. Subjects held various percentages of their maximum force, exerted by pulling at shoulder level, for as long as they could. The log-normal relationship gives a means for calculation of intermediate values

of recovery has been taken as being when the same posture can be held again for the same maximum time. In gathering such data we must seek to achieve the same level of motivation for each test and treat subjects as their own controls. Thus we calculate the forces as a percentage of the subject's own MVC, and provide rest periods as a proportion of each subject's own holding time. A typical recovery curve is shown in Figure 22.2. It arises from a number of different forward bending postures held for as long as possible (T_1), after which the subject had a rest interval equal to 12 times T_1 and then repeated the posture again (T_2). An important consequence of the relationship shown in this curve is that if the maximum holding time is considerable, e.g. the posture is such that discomfort builds up over a long period, then recovery also takes a very long time.

If we are seeking MVC for a particular situation it is important that the posture adopted, including foot positions and any constraints due to the workplace, are repeated during the tests, so that as nearly as possible the

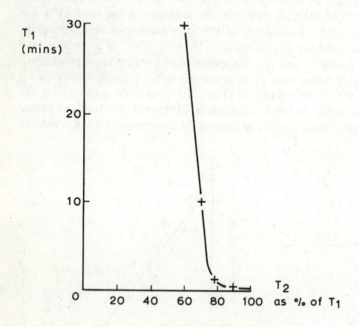

Graph of $T_2 = T_1{}^{0.854} \, e^{\frac{-0.152}{I}}$

For $I = 1200 \% \, T_1$

Forward   bending   posture

Figure 22.2. The graph shows the recovery (T_2) after a rest pause equal to twelve times the first holding time (T_1). Data for the formula came from 42 subjects experiencing five different postures and five rest intervals

same muscle groups are used. Few *practical* force exertions are undertaken by just a single group of muscles so, if a number of different muscles is recruited to do the task, unless the posture is identical, the group of muscles could be operating differently, or else other groups of muscles may be called into play.

Subjective methods

Two subjective methods can contribute to the evaluation of static work. Borg's scale has already been described, together with its rationale, in Kilbom's chapter on dynamic work. Although its use in static work is not valid in terms of the relationship of the numbers to the person's heart rate, the judgements of severity do give important information. An example of this arose from a demonstration of the difficulties involved in butchery. A rig was available which permitted subjects to exert single-handed forces on a set of knife handles, pulling them in directions across and down the body in the vertical plane and exerting as much force as possible. Although differences in the forces were found when working at different heights, they were relatively modest. However, when subjects used Borg's scale to judge the difficulty of pulling in each of the directions, large differences were identified in the ease of use in the different directions, which separated them more clearly than the analyses of the imposed forces would do.

The second subjective method uses muscular pain as a measure. Because 'pain' is sometimes seen as a specific and localized experience, the term 'discomfort' is used. Experiments (Corlett and Bishop, 1976) demonstrated that, if a force was exerted for as long as possible, until the pain was unbearable, and estimates of the discomfort levels made on a scale (5 or 7 points) at intervals during the holding time, the growth of discomfort was linearly related to the holding time regardless of the level of force being exerted (Figure 22.3). So discomfort itself can be used as a linear scale. There are deviations from linearity at the top end of the scale, where subjects will sometimes near the end of the scale before reaching their own discomfort limits. A magnitude estimation technique can be adopted if the extreme end of the scale is important (see a text on psychophysics).

To specify the site(s) of discomfort, a body map is used, divided into segments (Figure 22.4) depending on the sites of discomfort experienced by those engaged in the tasks being investigated. This information is found from preliminary enquiries or pilot trials. The procedure for mapping the development of discomfort can proceed in two ways.

1. At intervals during the whole working day, people are asked to point to the site(s) of current discomfort on the body map. Then they are asked to rate the intensity of discomfort at each identified site on a 5 or 7 point scale, preferably by marking a paper scale which is 'anchored' at the 0 and 5 (7) points by 'no discomfort' and 'extreme discomfort', respectively. These

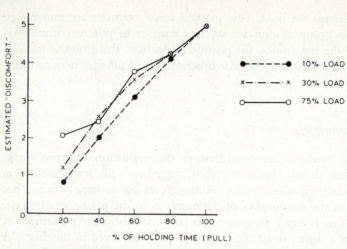

Figure 22.3. Mean values for overall discomfort ratings when different amounts of MVC are exerted for as long as possible

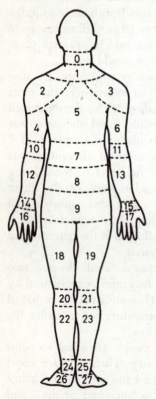

Figure 22.4. The Body Map for evaluating Body Part Discomfort, either by rating or ranking

scores are plotted against time of day for each body site, dividing the scale during analysis at the 1/2 points as well to give an effective 10 (or 14) point scale. Since it is likely that differences in body size or person–equipment relationships will cause changes in the distribution of discomfort around the body, the effect of adding together the scores from several subjects should be considered carefully; it is usually unwise.

The reason for urging the collection of data throughout the working day is that recovery from static load is slow (see above), and a lunch break is often not sufficient to achieve full recovery. A study of engravers (Figure 22.5), where four workers were studied and the average result taken, illustrates the point forcibly. It is evident from the high discomfort levels reached that the posture is extreme, and the curves representing the most heavily loaded body parts appear as one curve spread across the whole day, rather than the morning curve repeated in the afternoon.

2. The second way in which the body-mapping procedure can be used is recommended when rating of individual sites will take up too much of the subject's time. The person can be asked to point out the site(s) which are most uncomfortable, then those next most uncomfortable and so on until no more sites are reported. We have asked the person to identify a sequence of just-noticeable differences in discomfort, so it is unlikely that more than five or six will be recognized, as any text on psychophysics will confirm. Again these results are plotted against time of day, but a numerical value is obtained by counted back the number of levels of discomfort reported and numbering them, using the no-discomfort sites as zero, the last reported sites

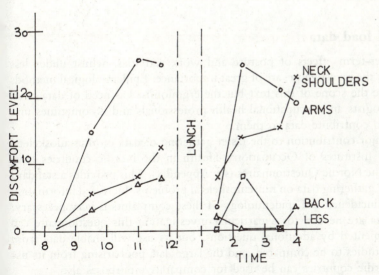

Figure 22.5. Discomfort scores, rating on a 7-point scale by four workers. The relative continuity of the neck and arms graphs over the lunch period will be noted

as 1, and so on. Although less detailed than the previous method, it is much quicker, involves less explanation to the subjects and reveals the most heavily loaded body parts equally as well.

Of course, all the other aspects of experimental control apply for this method as well, as described in (1) earlier.

Posture recording

As a first thought, it may be proposed that photographs or videos, perhaps in pairs of views orthogonal to each other, would be enough to record postures. It is true that for some situations this is adequate, if for example just a visual record is needed or some particular angles are to be measured. In other cases, however, it will be realized that, although the posture may have been recorded using a video or still photography, the data for analysis still have to be retrieved from the recorded images. Where accuracy is needed, problems of parallax in a plane transverse to the optical axis of the lens can be overcome, but inaccuracies along the optical axis can be considerable.

However, rarely is a record of posture of use on its own; it is necessary also to have data on task activities, the loads moved, conditions of the people concerned, and the workplace. It is fitting, therefore, before continuing with a number of methods for recording postures, to comment on the wider aspects of postural loading. It is also appropriate to note that postural data are frequently used in biomechanical analysis, and the chapter by Tracy on biomechanics should be consulted to assess what data are required, so that the method most appropriate for the need is selected.

Postural load data

The longer-term effects of posture and work activities, whilst under less control by the investigator, are of great importance. Epidemiological methods are outside the scope of this text but the ergonomist has need of data from epidemiologists and occupational health professionals and is sometimes in a position to contribute data to them.

As a major contribution to the faster gathering of data on musculoskeletal problems, Institutes of Occupational Health in the Nordic countries have designed the Nordic Questionnaires (see Appendix). This provides a standard format for gathering data on musculoskeletal problems. Increased information about the incidence and epidemiology of these complaints is very necessary. Where data are needed for a particular investigation this questionnaire can be supplemented by additional questions, but its use will enable data from different studies to be compared, and the large data pool arising from its use in the Nordic countries can be used for comparative purposes also.

Another tool developed at the Swedish National Board of Occupational Safety and Health is the single sheet analysis for identifying musculoskeletal stress factors (see Appendix). This is self-explanatory and uses the site of discomfort or injury to focus attention on a number of possible workplace faults which could be their causes. The list of possible causes is equally applicable to the body mapping procedure described earlier, enabling a direct link to be made to the sources of the problems. After changes have been introduced, it is clear that these same methods can be used to demonstrate any improvements which have been achieved.

Goniometers

Individual angles can be measured using a simple goniometer, of the type shown in Figure 22.6, which will give angles to the vertical, or the angle between adjacent body segments. As will be realized, accuracy is not high but is perfectly adequate for many purposes. The pendulum goniometer gives quick measures and is very simple to use. Where spinal measures are being made, care is needed since flexion of the spine introduces a pronounced curvature. The goniometer can be used at about L3 to L5, to get an indication of the angle at the lumbar–sacrum junction, and on the lower part of the thoracic spine for an approximate estimation of the spinal angle.

A combination of instruments in conjunction with a flexible rod, to record spinal shape, can be used for a better estimate (Burton, 1986). The changes in spinal flexibility have been well documented by use of such a system, and it could be used to demonstrate the difficulties in performance faced by some workers with limited mobility, due either to their condition or to workplace restrictions.

Recordings for segment or whole-body postures are possible. The ease of recording from small electronic goniometers can allow several to be mounted at selected points on a subject, sufficient to reconstruct from the recordings the postures adopted. Since the recording could cover a whole waking period, gathering the data as analogue signals on magnetic tape, computerized analysis is readily possible (see Drury's chapter on computer data collection and analysis).

Observational methods

Detailed records of postures have been used to record ballet for some two centuries. Two popular methods in wide use are the Benesh Notation, and Labanotation (Hutchinson, 1970). An example of the use of the former is given by Kember (1976). Both methods incorporate the record of timing, which is a major advantage, but each requires about three months of training in order to become proficient in even a simplified form of the methods.

A less comprehensive procedure, which does not incorporate timing data, is posture targetting (Corlett *et al.*, 1979). This makes use of a diagram (Figure 22.7), which has 'targets' located alongside each of the major limb segments, and on head and trunk. As shown in the positive target record of

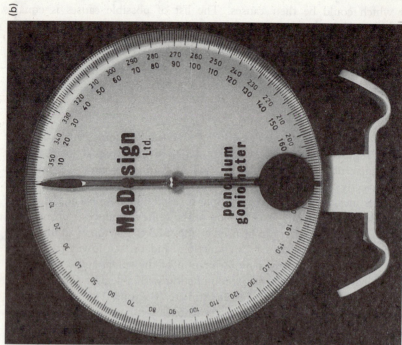

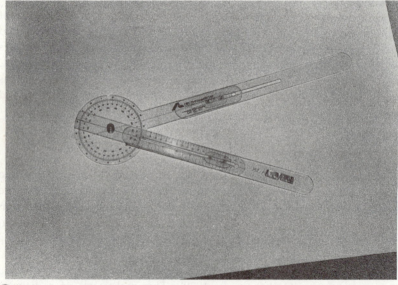

Figure 22.6. Goniometers. The radiating arms can be set either side of a joint, or used to assess displacement from a datum direction. The circular goniometers operate by gravity on the pendulum

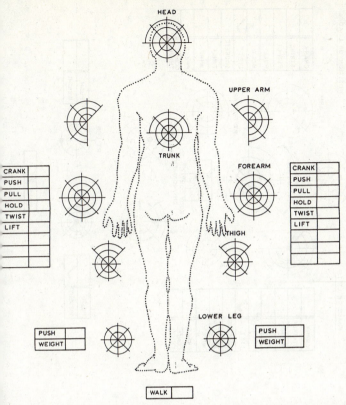

Figure 22.7. Posture targets, with the datum position shown by the dotted figure

Figure 22.8, the same position of the targets, but without the human shape, can be used, saving space on the recording form.

Marking the form requires the user to take the posture of Figure 22.7 as the normal or zero position. For a person in this position a mark would be made on the centre of each target. Movements forward by limbs, trunk or head (in the sagittal plane) would require a mark along the vertical axis of the target ('up' for forward) and positioned on that axis according to the estimated angle of the displacement. Each concentric circle marks off a 45° step. Displacements to the side of the body would be marked on the horizontal axis, again in a position according to the estimated angle. Directions (in a horizontal plane) are marked along the appropriate radius or between appropriate radii. An example of the use of the system is shown in Figure 22.8.

There are cases where the trunk may be twisted rather than bent, or where a bend and a twist occur together. In most of these the recording of the positions of arms and legs will result in the trunk record adopting the appropriate position. Where it is felt desirable to record the angle of twist of the trunk with respect to the hips, the arc between the head and trunk

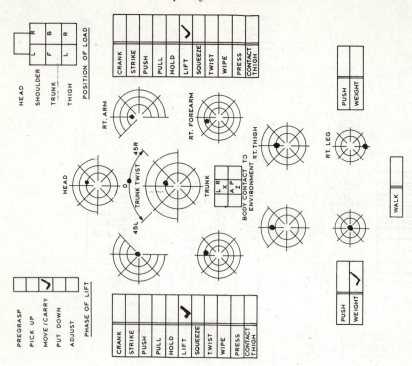

Figure 22.8. Posture target records taken from a photograph

targets—marked "trunk/hips"—may be marked at the appropriate point, using the same angular scale as for the targets.

With only a modest amount of practice this procedure can be learnt, its accuracy can be tested with goniometers during the learning period. The posture can be reconstructed from the diagram, and with simple mathematics the angles of all segments can be related to each other, to permit the whole posture to be redrawn, or input to a computer. With measures of body weight, body sizes and exerted forces, a biomechanic program can then be used to calculate required torques and loads.

For a rapid assessment of the adequacy of working postures, Ovako Oy, in conjunction with the Finnish Institute of Occupational Health* developed the OWAS method. Postures are observed, and recorded as shown in Figure 22.9(a). An accompanying assessment sheet, similar to Figure 22.9(b), enables each posture to be assessed for acceptability or else appropriate remedial action.

It will be evident that OWAS seeks to identify postures which put the body in positions where force exertions can be dangerous. Balanced and symmetrical postures are the ones which are, in general, acceptable. Pushing, pulling or moving loads when the person is twisted, or the body is in other ways asymmetrically loaded, are recommended for change.

Procedures such as OWAS have some limitations. They take limited account of external loads or time, for example, and were devised for heavy industrial work (in a steelworks). Their great strength is the facility they provide for the rapid identification of most of the major inadequate postures. Furthermore, as they are easy to learn and use, they can be used by a wide spectrum of the work-force, and alert people to those aspects of activity which may be hazardous.

Optical methods

There are several commercial methods which use three or more video cameras, together with software, to track markers on a subject moving in the cameras' field of view. Advances in computer software are making it increasingly possible for laboratories to develop their own systems using, e.g. 'frame-grabbing' techniques and taking data from video recordings.

Video methods often use markers on the body to provide measurement points. These can be stripes placed, for example, along the long axes of the limbs, reflective spots on appropriate body parts (e.g. over the joints of limbs) or small lights similarly mounted. An optical system which uses reflective markers is the Cartesian Optoelectronic Dynamic Anthropometer (CODA)† device. This equipment is usable on-line to a computer. It scans a number (up to 12) of retroreflective pyramidal markers with light from

---

*Institute of Occupational Health, Topeliuksenkatu 41 a A, SF-00250 Helsinki, Finland.
†Charnwood Dynamics Ltd, Loughborough, Leicestershire, UK.

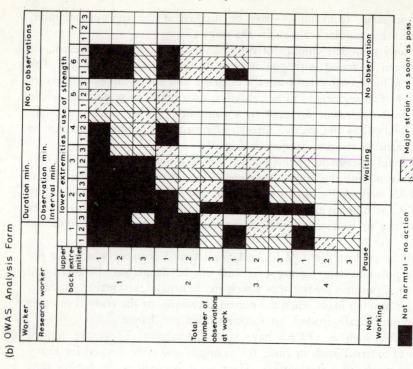

(a)

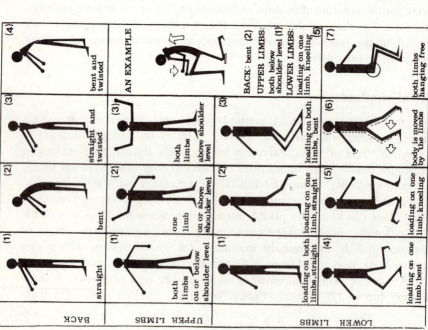

(b) OWAS Analysis Form

Figure 22.9. (a) The OWAS posture set (b) A typical posture evaluation chart. Other versions are available which incorporate the support of

three segmented mirrors at a rate of 300 times/s. One mirror rotates in the vertical plane centrally between two horizontally rotating mirrors which are about 1 m apart. The system, which is illustrated in Figure 22.10, will record the position of each marker (which is identified by its colour) to ±0·1 mm in the plane parallel to the mirrors and ±0·3 mm perpendicular to this plane, within a workspace of 4 m³.

(a)

(b)

Figure 22.10. The CODA optical posture recording instrument (a), with examples of the retroreflective pyramidal markers (b)

Figure 22.11 shows the use of an optical pointer to locate and record points outside the vision of the instrument. Two different markers are spaced a known distance apart and at a known distance from the point of the rod. The computer is programmed to recognize the markers when both are visible, identify their positions and calculate the position in space of the point of the rod from this information. In use, one marker is covered and the point of the rod located on the required anatomical point. The concealed marker is then revealed to CODA, the computer giving a 'bleep' when the recording is complete. To cancel a false reading, the computer may be programmed so that when a marker of a different colour is displayed the previous reading is cancelled. This feature overcomes a major restriction of all vision systems, that they cannot see the other side of the subject (many use multiple cameras to compensate for this). The optical pointer is usable only for stationary postures.

However, by the use of this technique a posture can be introduced into the computer. Force data from strain gauged instruments can also be input, and calculations done virtually on-line. Figure 22.12 shows a stick man displayed on a screen, developed from the CODA readings, with body load data presented alongside. Applications of this method are discussed in Tracy's chapter on biomechanics.

Electromyography

Electromyography (EMG), the recording of myoelectric signals which occur when a muscle is in use, can be used to assess the level of activity occurring

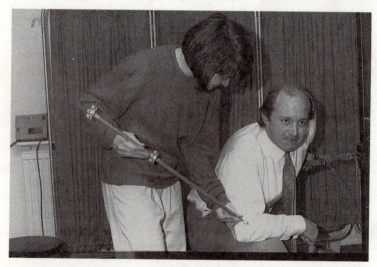

Figure 22.11. The optical pointer in use, providing direct input for reference points, even when the point itself is invisible to CODA

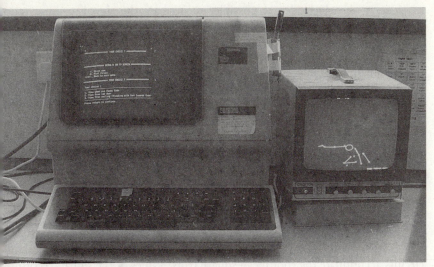

Figure 22.12. On-line display of the posture via a "stick man", and immediate calculation of forces and torques at the desired joints

over a period of time. It can also be used to show the presence of muscle fatigue, a state when a skeletal muscle is unable to maintain a required force of contraction (Hagberb, 1981). A high correlation has been shown between EMG activity and muscular force, for both static and dynamic activities (Hagberg, 1981). This relationship was once thought to be linear, but is now proposed as exponential (Lind and Petrofsky, 1979; Hagberg, 1981).

When a muscle begins to fatigue, there is an increase in the amplitude in the low frequency range and a reduction in the amplitude in the high frequency range of EMG activity (Petrofsky *et al.*, 1982). There is also a shift in the frequency spectrum towards the lower end of the spectrum as fatigue occurs.

Although needle electrodes, entering specific muscles, are used for medical research, occupational EMG records are usually taken from surface electrodes. These are stuck over the central part of the muscle and leads taken, via pre-amplifiers, to amplification and recording equipment. Telemetering can be done, but the existence of miniaturized circuitry and recorders has rendered this less necessary.

Skin preparation is necessary for reliable recordings. This involves the use of fine sandpaper to remove layers of dead skin prior to fixing the electrodes. Many modern electrodes, usually small silver discs, have a central hole into which electrode jelly, a saline grease, may be inserted with a syringe after the electrode is in place. The placement is generally over the central part of the muscle, where most of the active fibres will lie, and the electrodes will be from 3–5 cm apart. An 'earthing' electrode is sometimes included, placed away from the muscle being recorded and where other muscular activity is unlikely to be picked up. Signals from the electrodes should be pre-amplified as close to the electrodes as possible, to increase the signal-to-noise ratio, before passing them via a low pass filter to a recorder or analyzer.

As opposed to clinical EMG, where the quality of the signal is of importance, in occupational EMG the quantity of the signal is usually the important factor. The signal is analysed with respect to its frequency or amplitude, and the amplitude is usually analysed with the signal given as a percentage of that from a standard MVC taken prior to the investigation. This conversion permits comparison across different tasks and different people.

There are three major methods for EMG analysis: the integrated EMG (IEMG), Fourier analysis and *amplitude probability distribution function* (APDF) analysis.

IEMG

The integrated EMG gives a measure of the power in the signal. Integrating circuits accumulate the root mean square (rms) values of the signal, recording when a certain selected total value has been reached and starting the addition again. The visual record shows a series of triangular waves, with equal peaks but spaced more closely where the EMG signal was greater. Counting the peaks per unit time or calculating the rms value, again per unit time, provides values representative of the muscular activity.

Fourier analysis

The speed of response of muscle fibres varies, and they are conventionally divided into fast and slow twitch fibres. As a muscle is used, slow twitch activity becomes more evident, and is taken to be a sign of increased muscular fatigue. The frequencies bound up in a raw EMG signal are identified by Fourier analysis, a procedure which breaks down the signal into its component sine waves. Usually the analysis is done by taking successive short samples of the EMG, analysing them and presenting the results as a frequency spectrum, showing the frequency and amplitude of the component waves.

Amplitude Probability Distribution Function

Work by Jonsson (1976) and Hagberg (1979) demonstrated the utility of analysing EMG in terms of the amplitudes present in the signal. Each excursion of the signal represents the innervation of muscle fibre(s) to exert the force. Large excursions are related to the exertion of external force or rapid movement; small and frequent excursions can be interpreted as indicating the maintenance of static work, e.g. for holding a position.

The analysis is relatively simple in concept. Again short, successive samples of the signal are taken and the amplitudes of all the peaks counted and grouped. They are plotted as an amplitude spectrum or a cumulative amplitude distribution function. If the latter plot is adopted, the 10th, 50th and 90th deciles can be identified. The 10th decile is proposed as the level which demonstrates the static work load.

Although close correlations have been reported between EMG analyses and force, Hagberg (1981) has noted some sources of variance in experiments. The relationship will change with temperature, such as might arise from high levels of work activity, with fatigue, with whether the contraction is concentric or eccentric, and with changes in velocity of contraction. For much occupational work these factors may not be serious influences on results, but should be considered in relation to the quoted literature for any extensive studies. Where defined test contractions are easy to apply and the muscle action substantially isometric, the APDF can be a useful measure of the muscle performance.

Comment

Each method has its uses, depending on the problems under study. As will be recognized, all methods may be used from the same EMG recordings. The amount of data collected in occupational EMG is usually very large as longer recording periods are required. This creates difficulties in storage and analysis. Also artefacts are more common, such as movement artefacts, changes in recording due to temperature changes or muscle isolation.

Occupational EMG is a good technique for assessing which muscles are used in a task but is of more limited use in accurately assessing the fatigue process. It is, however, of value to establish the changes in the amplitude and frequency domains, with respect to time, to gain some understanding of how a muscle is operating. It is a tool which should be used in conjunction with other assessment techniques for a good understanding of a work situation.

Spinal loading

Investigations of posture are frequently concerned with loads on the spine. Where these loads are of short duration a biomechanic analysis may be the most suitable method. Where the exposure is over a longer period, or is frequently repeated, an assessment of changes in the work situation can be made by measuring changes in total stature (Eklund and Corlett, 1984).

The forces imposed on the lumbar spine during the day are the gravity loadings from thorax, head and arms, together with the components of forces exerted by the arms which have to be transmitted to the pelvis. These loads cause a reduction in stature over the day of 15 mm or more due to a slow decrease in spinal disc height; this will vary with age. Such shrinkage is recovered when lying down. To compare the effects of different workplaces, postures or work regimes on spinal loading, the use of a precision stadiometer is required (Figure 22.13). By close control of the experiment and measurement

Figure 22.13. Precision stadiometer

protocol, changes in stature of about 0·5 mm can be identified. As with many biological response measures, it is preferable to use subjects as their own controls, since averaging across subjects—due to differences in responding—increases the variance considerably.

There are several points to note when using the technique, which has been used in the workplace as well as in the laboratory. As with all precision measurements, tight experimental control is required.

If repeated tests are to be made, the time of day and prior activities of the subjects should be consistent. As recovery of height loss is quite rapid, the periods between exposure to load and subsequent measurement should be kept short and, again, consistent. Rest pauses between any sequence of measurements should also be controlled so that no extra increase or major decrease of load arises, say from a major change in posture. Thus if, at one point in an experimental sequence a subject lay down, the resultant change in disc condition would make the effects of a subsequent test condition very different from earlier trials (Abu Amin *et al.*, 1988).

Some instruction to subjects is needed on how to position themselves on the stadiometer, to help in maintenance of a consistent posture. The experimenter must also check and control weight distribution between heels and soles, location of the spine on the micro-switched pads set into the backboard of the instrument, head position (which is assessed by the nose marker), and ensure that the subject has folded arms. About five recordings over 2 or 3 s are taken. The subject then steps off, and back on the stadiometer immediately, is repositioned and more readings are taken. The repetition gives measures for the estimation of error as well as evidence for the consistency of the measuring posture.

For any study it is desirable to differentiate stature changes due to load from those which would normally arise under gravity loadings in the postures associated with the situation under investigation. Thus it will often be advisable to run a number of preliminary measures to establish the rate of shrinkage prior to loading, following by the trials. Extrapolation of the initial measures to the time of the final measures under load will allow assessment of a loading effect which will be independent of the expected rate of shrinkage at that time of the day. Recent work by Brinkmann (personal communication) has demonstrated the importance of this point, as well as the difficulty of establishing it in some cases.

A study by Foreman (1989) noted that compressibility of the heel is a major influence on total stature. He showed changes ranging between 2 and

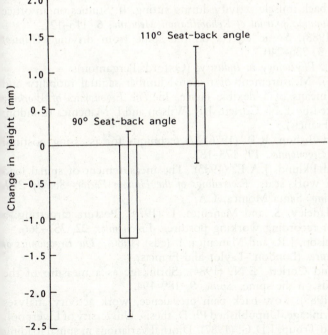

Figure 22.14. Changes in height during simulated driving, for two different seat-back angles

6 mm in a group of 20 subjects over a period of about 1·5 min. Hence it i. desirable that, where experimental conditions reduce the load on the heel some control is exercised to take this change into account. This might b done by recording initial calibration curves for heel compression anc correcting results in relation to the time from standing on the stadiometer.

Some results from the use of the method are shown in Figure 22.14 Comparison was made between seat-back angles for drivers of a car. Subject 'drove' a video driving game for 1-h periods. The changes in stature for sea geometries support the work of Andersson and Örtengren (1974), whc showed that spinal load was least for a seat-back angle of about 110°, whils the reduction in load for the lorry driving arrangement can be supported by biomechanical analyses.

The technique has been used to study nursing activities (Foreman anc Troup, 1987), design of working seats (Corlett and Eklund, 1983), the effect of circuit weight training and running (Leatt *et al.*, 1986) and the effects o equipment arrangement and vibration in vehicle driving (Bonney, 1988).

References

Abu Amin, A., Corlett, E.N. and Bonney, R.A. (1988). Does wearing a seat belt alter the load on the back whilst driving. In *Contemporary Ergonomics*, edited by E.D. Megaw (London: Taylor and Francis).

Andersson, B.J.G. and Örtengren, R. (1974). Lumbar disc pressure and myoelectric back muscle activity during sitting. II. Studies on an office chair. *Scandinavian Journal of Rehabilitation Medicine*, **6**, 115–121.

Bonney, R.A. (1988). Some effects on the spine from driving. *Clinical Biomechanics* **3**, 236–240.

Brouha, L. (1960). *Physiology in Industry*. (Oxford: Pergamon).

Burton, K. (1986). Measurement of regional lumbar sagittal mobility and posture by means of a flexible curve. In *The Ergonomics of Working Postures*, edited by E.N. Corlett, J.R. Wilson and I. Manenica (London: Taylor and Francis).

Corlett, E.N. and Bishop, R.P. (1976). A technique for assessing postural discomfort. *Ergonomics*, **19**, 175–182.

Corlett, E.N. and Eklund, J.A.E. (1983). The measurement of spinal load arising from work seats. *Proceedings of the Human Factors Society 27th Annual Meeting*, Santa Monica, CA.

Corlett, E.N., Madeley, S. and Manenica, I. (1979). Posture targetting: a technique for recording working postures. *Ergonomics*, **22**, 357–366.

Corlett, E.N., Wilson, J.R. and Manenica, I. (eds) (1986). *The Ergonomics of Working Postures* (London: Taylor and Francis).

Eklund, J.A.E. and Corlett, E.N. (1984). Shrinkage as a measure of the effect of loads on the spine. *Spine*, **9**, 189–194.

Foreman, T.K. (1989). Low back pain prevalence, work activity analyses and spinal shrinkage. Unpublished Ph.D. thesis, University of Liverpool.

Foreman, T.K. and Troup, J.D.G. (1987). Diurnal variations in spinal loading and the effects on stature. *Clinical Biomechanics*, **2**, 48–54.

Hagberg, M. (1979). The amplitude distribution of surface EMG in static and intermittent static muscular exercise. *European Journal of Applied Physiology*, **40**, 265–272.

Hagberg, M. (1981). An Evaluation of Local Muscular Load and Fatigue by Electromyography. *Arbete och Hälsa*, **24**, Solna, Sweden.

Hutchinson, A. (1970). *Labanotation* (London: Oxford University Press).

Jonsson, B. (1976). Evaluation of the myoelectric signal in long-term vocational electromyography. In *Biomechanics V*, edited by A.P.V. Komi (Baltimore: University Park Press), pp. 509–514.

Kember, P.A. (1976). The Benesh movement notations used to study sitting behaviour. *Applied Ergonomics*, **7**, 133–136.

Kemmlert, K., Kilbom, Å., Nilsson, B., Andersson, R. and Bjurvald, M. (1987). Prevention of injuries related to physical stress through intervention by labour inspectors. In *Musculoskeletal Disorders at Work*, edited by P. Buckle, (London: Taylor and Francis), pp. 146–152.

Kuorinka, I., Jonsson, B., Kilbom, Å., Vinterberg, H., Biering-Sørenson, F., Andersson, G. and Jørgensen, K. (1987). Standardised Nordic questionnaires for the analysis of musculo-skeletal symptoms. *Applied Ergonomics*, **18**, 233–237.

Leatt, P., Reilly, T. and Troup, J.D.G. (1986). Spinal load during circuit weight training and running. *British Journal of Sports Medicine*, **20**, 119–124.

Lind, A. and Petrofsky, J.S. (1979). Amplitude of the surface EMG in fatiguing isometric contractions. *Muscle and Nerve*, **2**, 257–264.

Monod, H. and Scherrer, J. (1965). The work capacity of a synergic muscle group. *Ergonomics*, **8**, 329–338.

Petrofsky, J.S., Glaser, R.M. and Phillips, C.A. (1982). Evaluation of the amplitude and frequency components of the surface EMG as an index of muscle fatigue. *Ergonomics*, **25**, 213–223.

Rohmert, W. (1973a). Problems in determining rest allowances. 1. Use of modern methods to evaluate stress and strain in static work. *Applied Ergonomics*, **4**, 91–95.

Rohmert, W. (1973b). Problem in determining rest allowances. 2. Determining rest allowances in different tasks. *Applied Ergonomics*, **4**, 158–162.

Appendix

The Nordic Questionnaires

(i) Survey of musculoskeletal complaints

These questionaires were designed and tested in Norway, Sweden, Denmark and Finland and have been translated into English. More extensive information can be obtained from: Berit Ydreborg, Department of Occupational Medicine, Medical Centre Hospital, S-701 85 Orebro, Sweden.

Continued on p. 569

LEAD-IN PAGE

CODE NUMBER: | | | | | | —| | |

for the questionnaires

FHV 013   Trouble with the locomotive organs
FHV 014:1   Neck trouble
FHV 014:2   Shoulder trouble
FHV 015   Low back trouble

> NB: Shaded areas are to be filled in by the occupational health staff.

Reason 1	Date of inquiry year month day 2—7	User 8—12	County 13—14	Occupation 15—19	Sector specific code 20—25	Company or authority 26—29 30—32	Signature

TO BE FILLED IN BY THE EMPLOYEE

What year were you born? | 1 | 9 | | | 33—34

Sex   ☐ 1 Male   ☐ 2 Female   35

How many years and months have you been doing your present type of work? |___| years + |___| months   36—39

On average, how many hours a week do you work? |___| hours   40—41

How much do you weigh? |___| kg   42—44

How tall are you? |___| cm   45—47

Are you right-handed or left-handed?   ☐ 1 right-handed ☐ 2 left-handed   48

Name:

Company/authority:

TROUBLE WITH THE LOCOMOTIVE ORGANS*

FHV 013 D

Year of birth: ☐ 1 Male ☐ 2 Female
1 9 | |
Code number: | | | | — | |

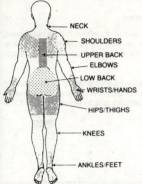

NECK
SHOULDERS
UPPER BACK
ELBOWS
LOW BACK
WRISTS/HANDS
HIPS/THIGHS
KNEES
ANKLES/FEET

How to answer the questionnaire:

In this picture you can see the approximate position of the parts of the body referred to in the questionnaire. Limits are not sharply defined, and certain parts overlap. You should decide for yourself in which part you have or have had your trouble (if any).

Please answer by putting a cross in the appropriate box – one cross for each question. You may be in doubt as to how to answer, but please do your best anyway. Note that the questionnaire is to be answered, even if you have never had trouble in any part of your body.

To be answered by everyone	To be answered only by those who have had trouble	
Have you at any time during the **last 12 months** had **trouble** (ache, pain, discomfort) in:	Have you at any time during the **last 12 months** been **prevented from doing your normal work** (at home or away from home) because of the trouble?	Have you had **trouble** at any time during the **last 7 days**?
14 Neck 1 ☐ No 2 ☐ Yes	**15** 1 ☐ No 2 ☐ Yes	**16** 1 ☐ No 2 ☐ Yes
17 Shoulders 1 ☐ No 2 ☐ Yes, in the right shoulder 3 ☐ Yes, in the left shoulder 4 ☐ Yes, in both shoulders	**18** 1 ☐ No 2 ☐ Yes	**19** 1 ☐ No 2 ☐ Yes
20 Elbows 1 ☐ No 2 ☐ Yes, in the right elbow 3 ☐ Yes, in the left elbow 4 ☐ Yes, in both elbows	**21** 1 ☐ No 2 ☐ Yes	**22** 1 ☐ No 2 ☐ Yes
23 Wrists/hands 1 ☐ No 2 ☐ Yes, in the right wrist/hand 3 ☐ Yes, in the left wrist/hand 4 ☐ Yes, in both wrists/hands	**24** 1 ☐ No 2 ☐ Yes	**25** 1 ☐ No 2 ☐ Yes
26 Upper back 1 ☐ No 2 ☐ Yes	**27** 1 ☐ No 2 ☐ Yes	**28** 1 ☐ No 2 ☐ Yes
29 Low back (small of the back) 1 ☐ No 2 ☐ Yes	**30** 1 ☐ No 2 ☐ Yes	**31** 1 ☐ No 2 ☐ Yes
32 One or both hips/thighs 1 ☐ No 2 ☐ Yes	**33** 1 ☐ No 2 ☐ Yes	**34** 1 ☐ No 2 ☐ Yes
35 One or both knees 1 ☐ No 2 ☐ Yes	**36** 1 ☐ No 2 ☐ Yes	**37** 1 ☐ No 2 ☐ Yes
38 One or both ankles/feet 1 ☐ No 2 ☐ Yes	**39** 1 ☐ No 2 ☐ Yes	**40** 1 ☐ No 2 ☐ Yes

* Nordiska Ministerrådets projekt nr 170.21–1.15
Yrkesrelaterade muskeloskeletala sjukdomar och deras prevention.

NECK TROUBLE*

FHV 014 D:1

Code number: ⊔⊔⊔⊔⊔–⊔⊔⊔

How to answer the questionnaire:

By neck trouble is meant ache, pain or discomfort in the shaded area. Please concentrate on this area, ignoring any trouble you may have in adjacent parts of the body. There is a separate questionnaire for shoulder trouble.

Please answer by putting a cross in the appropriate box – one cross for each question. You may be in doubt as to how to answer, but please do your best anyway. Note that question 1 is to be answered, even if you have never had neck trouble.

---

1. Have you **ever** had neck trouble (ache, pain or discomfort)?

   1 ☐ No   2 ☐ Yes

If you answered **No** to Question 1, do not answer the questions 2—8.

---

2. Have you **ever** hurt your neck in an **accident**?

   1 ☐ No   2 ☐ Yes

---

3. Have you **ever** had to **change jobs or duties** because of neck trouble?

   1 ☐ No   2 ☐ Yes

---

4. What is the **total** length of time that you have had neck trouble during the **last 12 months**?

   1 ☐ 0 days
   2 ☐ 1—7 days
   3 ☐ 8—30 days
   4 ☐ More than 30 days, but not every day
   5 ☐ Every day

If you answered **0 days** to Question 4, do not answer the questions 5—8.

---

5. Has neck trouble caused you to **reduce** your activity during the **last 12 months**?

   a. Work activity (at home or away from home)?
      1 ☐ No   2 ☐ Yes
   b. Leisure activity?
      1 ☐ No   2 ☐ Yes

---

6. What is the **total** length of time that neck trouble has **prevented** you from doing your normal work (at home or away from home) during the **last 12 months**?

   1 ☐ 0 days
   2 ☐ 1—7 days
   3 ☐ 8—30 days
   4 ☐ More than 30 days

---

7. Have you **been seen** by a doctor, physiotherapist, chiropractor or other such person **because of neck trouble** during the **last 12 months**?

   1 ☐ No   2 ☐ Yes

---

8. Have you had neck trouble at any time during the **last 7 days**?

   1 ☐ No   2 ☐ Yes

---

* Nordiska Ministerrådets projekt nr 170.21—1.15
Yrkesrelaterade muskeloskeletala sjukdomar och deras prevention.

SHOULDER TROUBLE[*]   FHV 014 D:2   Code number:

How to answer the questionnaire:

By shoulder trouble is meant ache, pain or discomfort in the shaded area. Please concentrate on this area, ignoring any trouble you may have in adjacent parts of the body. There is a separate questionnaire for neck trouble.

Please answer by putting a cross in the appropriate box – one cross for each question. You may be in doubt as to how to answer, but please do your best anyway. Note that question 9 is to be answered, even if you have never had shoulder trouble.

9. Have you **ever** had shoulder trouble (ache, pain or discomfort)?

   1 ☐ No   2 ☐ Yes

If you answered **No** to Question 9, do not answer the questions 10—17.

10. Have you **ever** hurt your shoulder in an **accident**?

   1 ☐ No   2 ☐ Yes, in my right shoulder
   3 ☐ Yes, in my left shoulder
   4 ☐ Yes, in both shoulders

11. Have you **ever** had to **change jobs or duties** because of shoulder trouble?

   1 ☐ No   2 ☐ Yes

12. Have you had shoulder trouble during the **last 12 months**?

   1 ☐ No   2 ☐ Yes, in my right shoulder
   3 ☐ Yes, in my left shoulder
   4 ☐ Yes, in both shoulders

If you answered **No** to Question 12, do not answer the questions 13—17.

13. What is the **total** length of time that you have had shoulder trouble during the **last 12 months**?
   1 ☐ 1–7 days
   2 ☐ 8–30 days
   3 ☐ More than 30 days, but not every day
   4 ☐ Every day

14. Has shoulder trouble caused you to **reduce** your activity during the **last 12 months**?

   a. Work activity (at home or away from home)?
      1 ☐ No   2 ☐ Yes

   b. Leisure activity?
      1 ☐ No   2 ☐ Yes

15. What is the **total** length of time that shoulder trouble has **prevented** you from doing your normal work (at home or away from home) during the **last 12 months**?

   1 ☐ 0 days
   2 ☐ 1–7 days
   3 ☐ 8–30 days
   4 ☐ More than 30 days

16. Have you **been seen** by a doctor, physiotherapist, chiropractor or other such person **because of shoulder trouble** during the **last 12 months**?

   1 ☐ No   2 ☐ Yes

17. Have you had shoulder trouble at any time during the **last 7 days**?

   1 ☐ No   2 ☐ Yes, in my right shoulder
   3 ☐ Yes, in my left shoulder
   4 ☐ Yes, in both shoulders

[*] Nordiska Ministerrådets projekt nr 170.21–1.15
Yrkesrelaterade muskeloskeletala sjukdomar och deras prevention.

LOW BACK TROUBLE*

FHV 015 D

Code number: ☐☐☐☐☐—☐☐

How to answer the questionnaire:

In this picture you can see the approximate position of the part of the body referred to in the questionnaire. By low back trouble is meant ache, pain or discomfort in the shaded area whether or not it extends from there to one or both legs (sciatica).

Please answer by putting a cross in the appropriate box – one cross for each question. You may be in doubt as to how to answer, but please do your best anyway. Note that question 1 is to be answered, even if you have never had low back trouble.

1. Have you **ever** had low back trouble (ache, pain or discomfort)?

   1 ☐ No   2 ☐ Yes

If you answered **No** to Question 1, do not answer the questions 2—8.

2. Have you **ever** been hospitalized because of low back trouble?

   1 ☐ No   2 ☐ Yes

3. Have you **ever** had to **change jobs or duties** because of low back trouble?

   1 ☐ No   2 ☐ Yes

4. What is the **total** length of time that you have had low back trouble during the **last 12 months**?

   1 ☐ 0 days
   2 ☐ 1—7 days
   3 ☐ 8—30 days
   4 ☐ More than 30 days, but not every day
   5 ☐ Every day

If you answered **0 days** to question 4, do not answer the questions 5—8.

5. Has low back trouble caused you to **reduce** your activity during the **last 12 months**?

   a. Work activity (at home or away from home)?
      1 ☐ No   2 ☐ Yes

   b. Leisure activity?
      1 ☐ No   2 ☐ Yes

6. What is the **total** length of time that low back trouble has **prevented** you from doing your normal work (at home or away from home) during the **last 12 months**?

   1 ☐ 0 days
   2 ☐ 1—7 days
   3 ☐ 8—30 days
   4 ☐ More than 30 days

7. Have you **been seen** by a doctor, physiotherapist, chiropractor or other such person **because of low back trouble** during the **last 12 months**?

   1 ☐ No   2 ☐ Yes

8. Have you had low back trouble at any time during the **last 7 days**?

   1 ☐ No   2 ☐ Yes

* Nordiska Ministerrådets projekt nr 170.21—1.15
  Yrkesrelaterade muskeloskeletala sjukdomar och deras prevention.

Continued from p. 563

There are five sheets, the first of which asks for basic information about the subject. The second sheet gives an overview of the distribution of the musculoskeletal troubles and their extent. Then the three subsequent sheets seek more detailed information about the neck, the shoulder and the low back.

Although this set of questionnaires can provide important evidence of musculoskeletal problems, it requires complementing by ergonomic work analyses and estimates of exposure times. From the questionnaires themselves only the job title is available to identify the work activities, which is usually insufficient information for assessing exposure and workload.

(ii) Identifications of musculoskeletal stress factors

This technique was developed at the National Board for Occupational Safety and Health, Solna, Sweden, to provide a simple instrument for widespread use for workplace assessments amongst the Labour Inspectorate.

The form reproduced on p. 570 is used by entering in it the column or columns associated with the site of the injury. Numbered areas in each column should be marked if the question at the right hand end of its line is relevant to the workplace being examined. A list of other important factors is given, which should be reported on with the analysis provided from the questions.

The procedure gives an analysis of likely stress factors which may have given rise to the injury. To implement appropriate changes requires more knowledge than the form provides, what it does give is a case for making changes to reduce the likelihood of further injuries, and points where an attack should be mounted.

Method for the identification of musculo-skeletal stress factors which may have injurious effects.

Kemmlert, K. Kilbom, A. (1986) National Board of Occupational Safety and Health, Research Department, Work Physiology Unit, 171 84 Solna, Sweden

Method of application.

* Find the injured body region
* Follow while fields to the right
* Do the work tasks contain any of the factors discribed?
* If so, tick where appropriate

	neck/shoulders, upper part of back	elbows, forearms hands	feet	knees and hips	low back
1. Is the walking surface uneven, sloping, slippery or nonrasilient?			1.	1.	1.
2. Is the space too limited for work movements or work materials?	2.	2.	2.	2.	2.
3. Are tools and equipment unsuitably designed for the worker or the task?	3.	3.	3.	3.	3.
4. Is the working height incorrectly adjusted?	4.				4.
5. Is the working chair poorly designed or incorrectly adjusted?	5.				5.
6. (If the work is performed whilst standing): Is there no possibility to sit and rest?			6.	6.	6.
7. Is fatiguing foot-pedal work performed?			7.	7.	
8. Is fatiguing leg work performed eg: a) repeated stepping up on stool, step etc.? b) repeated jumps, prolonged squatting or kneeling? c) one leg being used more often in supporting the body?			8. a b c	8. a b c	8. a b c
9. Is repeated or sustained work performed when the back is: a) flexed forward, more than 20°? b) severely flexed forward, more than 60°? c) bent sideways or twisted, more than 15°? d) severely twisted, more than 45°?	9. a b c d				9. a b c d
10. Is repeated or sustained work performed when the neck is: a) flexed forward, more than 15°? b) bent sideways or twisted, more than 15°? c) severely twisted, more than 45°? d) extended backwards?	10. a b c d				
11. Are loads lifted manually? Notice factors of importance as: a) periods of repetitive lifting b) weight of load c) awkward grasping of load d) awkward location of load at onset or end of lifting e) handling beyond forearm length f) handling below knee height g) handling above shoulder height	11. a — c b — f c — g d				11. a — c b — f c — g d
12. Is repeated, sustained or uncomfortable carrying, pushing or pulling of loads performed?	12.	12.			12
13. Is sustained work performed when one arm reaches forward or to the side without support?	13.	13.			
14. Is there repetition of: a) similar work movements? b) similar work movements beyond comfortable reaching distance?	14. a b	14. a b			
15. Is repeated or sustained manual work performed? Notice factors of importance as: a) weight of working materials or tools b) awkward grasping of working materials or tools	15. a b	15. a b			
16. Are there high demands on visual capacity?	16.				
17. Is repeated work, with forearm and hand, performed with: a) twisting movements? b) ... c) uncomfortable hand positions?	17. a — c	17. a — c			

Also take these factors into consideration:
a) the possibility to take breaks and pauses
b) the possibility to choose order and type of work tasks or pace of work
c) if the job is performed under time demanded or psychological stress
d) if the work can have unusual or unexpected situations
e) presence of cold, heat, draught, noise or troublesome visual conditions
f) presence of jerks, shakes or vibrations

Chapter 23

Biomechanical methods in posture analysis

Moira F. Tracy

Introduction

Biomechanics is the study of forces on the human body. It is used in the ergonomics field most often to assess manual handling tasks. This chapter outlines the types of tasks for which biomechanical analysis is or is not an appropriate method, details the tools required to carry out an analysis, and describes various methods and models used in the field, with their advantages and disadvantages. When interpreting results of a biomechanical analysis we come to the question of how reliable are the simplifications and approximations made in the analysis. And finally, what are our criteria for 'safe' levels of force and how reliable are they?

Relevant tasks

Biomechanics is a useful tool in most manual handling situations, whether people are lifting, pushing, pulling, or even when no load is handled but the body's own weight is creating postural stress. The human body is complex and biomechanics cannot at this stage give very fine detail. For instance it is, to our knowledge, not possible to show through calculations which of two backrest shapes would be superior: the posture or force from the backrest would need to change quite noticeably for calculated results of body loadings to show differences.

Biomechanics is best used as a comparative method because, as will be shown throughout this chapter, results rest on simplifications and approximations. The method is particularly successful at demonstrating possible improvements obtained from redesigning a task. It can also be used to identify the most stressful parts of a job.

Interpretation of results

Having calculated forces at joints such as elbows or shoulders, and forces on the spine or within trunk muscles, results for various tasks or designs can be compared. It is also possible to assess the feasibility and safety of tasks even at the design stage, by comparing these forces with recommended limits. However, as will be discussed in this chapter, these limits are not absolute guarantees of safety, so results of a biomechanical analysis should not be used in isolation, but combined with other assessment methods. Direct observation, discomfort charts or questionnaires may identify sources of discomfort which force calculations cannot. For repetitive tasks a physiological assessment is often necessary as well. Injury statistics will help to identify problems for which biomechanics can give relevant answers and help towards an assessment of the cost-effectiveness of redesign.

Drury *et al.*'s (1983) evaluation of a palletizing aid provides a good example of the use of biomechanics with other methods. Very simple calculations based on video recordings showed to what extent the load on the lumbar spine was reduced when a palletizing aid was used. The heart rate was also shown to go down, and the authors used these factors and injury statistics to evaluate the cost-effectiveness of introducing the aids. The interesting point is that if this study had based itself only on accepted 'safe' limits of spinal load, it would have found with this particular biomechanical model that they were quite acceptable without an aid. The use of several methods to evaluate the task ensured that problems were not overlooked and biomechanics demonstrated the benefits of introducing an aid.

How detailed should the analysis be?

The simplest biomechanical calculations are those relating to a static posture and force in the sagittal plane, i.e. where there is neither twisting nor lateral bending. The problem is purely two-dimensional (2D), and there are no extra forces caused by accelerations and inertia. It is easy to record the posture in that one plane and all the calculations can be done with a small calculator.

If the task is characterized by much lateral bending or twisting, problems or possible improvements would be overlooked with a 2D analysis. Recording posture in three dimensions (3D) requires special methods outlined later. Calculations are quite lengthy and a computer program is usually necessary. A 3D analysis is therefore quite time consuming unless the posture-recording method and computer program have already been set up.

If the task is not static but dynamic, extra forces resulting from the accelerations have to be added to the calculations. The analysis now requires continuous monitoring of posture, along with the value and direction of the acceleration of each limb. This may restrict dynamic analyses to the laboratory and to our knowledge this has not yet been done in 3D. In 2D it is possible sometimes to avoid recording posture and acceleration continuously by using simulations. This has been done for lifting, the accuracy of the results depending on how similar the lifting technique is to the one used as a data base by the computer program. Finally it is quite common to analyse dynamic tasks statically, freezing at the beginning, middle, and end of the task, for instance. If the task was done slowly and smoothly, this is a good approximation. Otherwise this may underestimate forces during the acceleration phase up to two or three times (Garg *et al.*, 1982).

Equipment required

A biomechanical analysis requires measurements of posture, hand-force (and any other forces, such as the force from a backrest) and, in the case of dynamic tasks, accelerations. Hand-forces can be measured with a spring balance, but in some cases a tool equipped with strain-gauges may have to be constructed. This can be interfaced to a computer for on-line recordings. One quick-and-ready method is to press onto bathroom scales. Sometimes it is practical to place subjects on bathroom scales or a force plate and take one reading when they are applying a force and one reading when they are not. The difference is due to the force applied.

The choice of a posture recording method will depend on whether the analysis is 2D or 3D, in the field or in the laboratory. It may not be necessary to record posture when using some 2D computer programs: the posture is input via a stick-man on the screen. However it is best to base this on a photograph in order to input a realistic posture. Photographs or videos can be digitized for 2D or 3D posture recordings. This is time consuming in 3D but often used as a simple field method in 2D. Another simple method for static 2D tasks is to take measurements from the subject with a tape measure and plumb bob (Schultz *et al.*, 1983). This is more of a laboratory method as it would interrupt real work in many cases.

For 3D analyses or dynamic tasks, scanners like Selspot, Vicon, Coda or Elite are practical when on-line to a computer, though costly and usually not very portable (Tracy *et al.*, 1987). A method for field work in the future could be to interface a posture-recording suit based on light-emitting diodes (Samuelson *et al.*, 1987) or strain-gauges. Also for field work, a pen and paper method such as Corlett *et al.*'s (1979) posture targetting, or the Labanotation used in dance recording, may also be of interest as posture inputs to a computer evaluation (see the chapter on static work evaluation by Corlett).

Human variability

Biomechanics is sometimes used with the aim of determining safe limits for most of the population. At this stage, various aspects of human variability must be borne in mind. The most obvious one is the large range in strength found in the population. Men are usually stronger than women, but there is considerable overlap, some women exceeding some men. (Strength testing though does not necessarily identify people who will be the most capable of doing a task, if the test requires the use of different muscle groups from those used in the task (Keyserling *et al.*, 1980).)

Human variability in body weight and stature also comes into biomechanics. The posture people choose to adopt for a given task is another factor, but given a particular posture, the loads on the body are greatest for larger body weights and limb lengths. It is therefore recommended to use 95th percentile body weight and stature in order to err on the side of safety.

Finally, when results of biomechanical calculations are evaluated against 'safe' limits, it must be borne in mind that people differ in their susceptibility to back pain or injury. Hutton and Adams (1982) have shown that as a rule women's spines are more susceptible to fracture than men's, and age also is a weakening factor. Troup *et al.* (1987) found that people who had experienced back pain chose to lift lower loads. The mechanism of back pain is still uncertain in many cases: tests on cadaveric spines do not necessarily produce the same effects as those observed *in vivo* (Brinckmann, 1986). The best predictor of susceptibility to back pain seems, at the moment, to be a history of previous low-back trouble (MacDonald, 1984).

Principles of biomechanical calculations

The aim in this section is to demonstrate how loads at any body joint are calculated. These results can be used in a comparative way, for instance to quantify the improvements obtained from redesigning a task. On the other hand they can be compared to population data on maximum strength capabilities, in order to assess how strenuous the task is.* The present section will also demonstrate how forces within the low back can be evaluated with a simple 2D model.

The calculation of forces rests on the principle that all forces must balance each other if the body is to be in equilibrium. If there is a resultant force in any direction, the body will move in that direction. The calculation of loads at body joints which follows is in fact the calculation of moments, or turning forces around a point. Moments must also balance, so that the sum of moments around any point is zero, if there is to be no rotation.

Moments and lever arms

The moment, or torque, of a force about a point, is a measure of the turning force round the point. For instance holding a weight in the hand creates a

---

* This last aspect is discussed in the next main section. Readers who do not need to know the details of calculations can turn directly to that section after reading the paragraph on moments and lever arms.

moment around the elbow, tending to make it extend. Muscles spanning the elbow provide the opposite moment by contracting, so that the elbow is able to support the weight. The greater the weight, the larger the elbow moment.

In general terms, the moment of a force about a point is the product of the force and the perpendicular distance between the point and the line of action of the force. This is shown in Figure 23.1. A weight of 100 N* held in the hand creates a moment of $100 \times 0.20 = 20$ Nm (Newton metres) for the position shown. If the weight was held with the arm hanging down, the lever arm of the force would be zero, and so the moment about the elbow would be zero. There would, of course, still be a force at the elbow, resisting the downward pull of the 100 N weight.

So far we have not taken into account another force exerting a moment round the elbow: the weight of the hand and forearm. The location of the centre of gravity of the hand and forearm, and their weight, can be estimated from tables, presented later in the chapter (see section 'Inputs to biomechanical calculations').

A simple example with a 2D low-back model

To evaluate forces in the lumbar region, the moment around a point of the low back is calculated in the same way as has been demonstrated for the elbow. Then a model of the muscular layout is used to calculate how much force the back muscles need to exert to counteract this moment. This enables

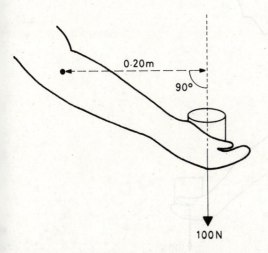

Figure 23.1. The moment at the elbow due to the 100 N weight held at the hand is the product of the force (100 N) and the perpendicular distance (0.20m) through which it acts.

---

* Newtons (N) are units of force or weight. A 1 kg mass weighs approximately 10 N. More on this and other units appears at the end of this section.

the compression force on the spine itself to be evaluated, which is a much used criterion for safety.

What follows is a simple example in 2D (see Figure 23.2): the posture and hand-force are in the sagittal plane. Calculations in 3D, for lateral bending and twisting, will follow after that.

A subject is depicted holding a 100 N weight and moments are calculated around the point indicated by a star, situated on a point of the lumbar spine. The weight of the part of the body above this point is 400 N in this example, acting with a lever arm of 0·20 m. It creates a moment of 80 Nm. Add to this the effect of the weight at the hands, acting with a 0·60 m lever arm, and the total moment created around the starred point is 140 Nm. Thus the weight at the hands and the subject's own body weight tend to flex the trunk: trunk muscles and ligaments must counteract this so that the posture is held.

Many studies have been carried out to record the trunk extension moment capabilities of subjects. We will return to this later in the chapter but note that a range of 100–700 Nm has been found in the literature, for male subjects without back pain. The subjects whose maximum strength was 100 Nm

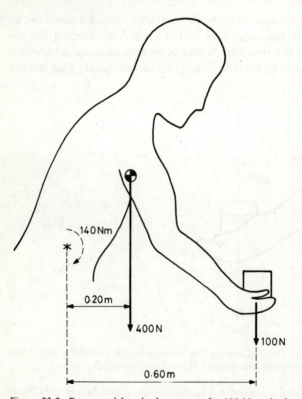

Figure 23.2. Forces and low-back moment for 100 N at the hands.

would therefore not be capable of exerting the 140 Nm of our example. For the strongest subjects, those capable of 700 Nm, it would be an easy task.

Forces within the trunk can be evaluated at this stage, using a model of the low back. Such models vary in detail and complexity, as will be seen later, but the principle can be shown here. In a simple 2D model, the 140 Nm trunk flexion moment is resisted by back muscles alone. The greater the leverage those muscles have from the spine, the smaller the force needed from them. Let us suppose the line of action of the back muscles is 5·8 cm posterior to the spine (this is the average from a recent study: more information is given in the section 'Inputs to biomechanical calculations'). The force these muscles need to exert is 140/0·058 = 2414 N. This is much larger than the body weight or the hand-force, because the muscles are balancing the moments through a very small lever arm.

As the back muscles pull to counteract moments, they compress the lumbar spine. The weight of the body above the lumbar spine and the weight at the hands also compress it, so finally the total compression force is the sum of all these components: 100 + 400 + 2414 = 2914 N. This again may seem large, but compression tests have shown that the spine can, in general, withstand this type of force. Further discussion on this is given in the section 'Forces on the low back'.

Figure 23.3 shows all the forces involved in this simple problem: the sum of all these forces is zero, and so is the sum of all moments around any point. The point indicated by a star on the lumbar spine was chosen only to simplify calculations, as the compression force does not create a moment round this point. In this example the centre of mass of the upper part of the body is shown as being 0·20 m from the lumbar spine. In fact this would not be known in an ordinary problem—what is estimated is the location of the centre of mass of each body segment, and so the moments created by each segment can be summed.

We will now go on to calculations in 3D static or dynamic tasks. However, in many cases very valuable information can be obtained from the simple calculations described so far (for instance Drury *et al.*'s (1983) evaluation of palletizing aids discussed earlier).

Moments in 3D space

The following, on moments in 3D space, contains details which need not be read by users who will not actually be performing calculations.

Previous examples were restricted to postures and forces in one plane, the sagittal plane. For asymmetric postures, or forces not contained in this plane, the calculation of the moments can be made by the following method. Figure 23.4 shows a force **F** (bold type indicates a vector) acting at the hand, with **r** the vector running from the elbow to the hand. The simplest way to determine the moment, **M**, around the elbow is to record the x, y, z components of vectors **r** and **F**, along a set of perpendicular axes. Most

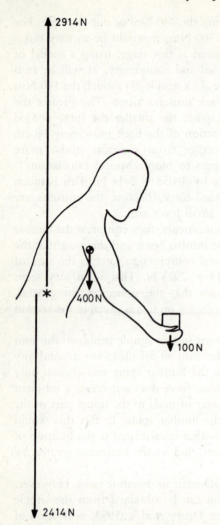

Figure 23.3. Equilibrium of forces when holding a 100 N weight.

measuring equipment will allow this. Thus the components of **r** are r_x, r_y, r_z along the x, y, z axes, and those of **F** are F_x, F_y, F_z. The moment **M** around the elbow is also a vector and it is simple to calculate its components M_x, M_y, M_z. M_x is the turning force in the (y, z) plane, M_y is the turning force in the (z, x) plane, and M_z is the turning force in the (x, y) plane.

The resultant of M_x, M_y, M_z, is the size of the vector **M**. From Pythagoras' theorem:

$$M^2 = M_x{}^2 + M_y{}^2 + M_z{}^2.$$

M_x, M_y and M_z are obtained through the following equations:

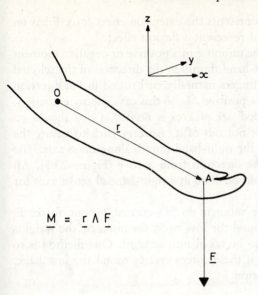

Figure 23.4. The moment of a force about O is the vector product (noted $\wedge$) of the vector r running from O to A, with the force vector F.

$$M_x = r_y F_z - r_z F_y$$
$$M_y = r_z F_x - r_x F_z$$
$$M_z = r_x F_y - r_y F_x.$$

A short-hand notation for these is:

$$\mathbf{M} = \mathbf{r} \wedge \mathbf{F}$$

$\mathbf{M}$ is described as the vector product (noted $\wedge$) of $\mathbf{r}$ and $\mathbf{F}$ (the order is important).

For calculations of moments in a 2D situation we have seen that the moment was the product of the force and its perpendicular distance (lever arm). This gives the same result as the equations above, so the method chosen depends on which is easiest to record: the lever arm, or components along the x, y and z axes.

Each component of the moments, M_x, M_y, M_z, represents the turning force round the x, y, or z axis. For instance in Figure 23.4, the y axis is directed into the paper, and M_y is the moment round that axis, and represents the flexion/extension moment about the elbow.

This is the only moment in the case of Figure 23.4: the reader can verify from the above equations that M_x and M_z are zero. (The only component of $\mathbf{F}$ is along the z axis and $\mathbf{r}$ and $\mathbf{F}$ are in one plane: $r_y = 0$, $F_x = 0$, $F_y = 0$.) This means that there are no twisting or abduction/adduction requirements on the elbow.

The sign (positive or negative) of a moment indicates the direction of the turning force. In the example of Figure 23.4, F_z is negative, so from the equations, M_y is positive. This represents the extension effect force **F** has on the elbow. A negative M_y would represent a flexion effect.

One method of working out the meaning of a positive or negative moment is as follows: stick out your right-hand thumb in the direction of the selected axis, y in this case. Your other fingers naturally curl round in the direction of the rotation corresponding to a positive M_y, in this case, elbow extension.

For this system, a 'right-handed' set of axes is needed, i.e. y should go into the paper, as in Figure 23.4, not out of it. An easy trick to ensure the axes are right-handed is to point the right-hand thumb along the z axis. The fingers curl round to indicate the direction from x to y (Figure 23.5). All force and posture recording should be done in a right-handed set of axes for consistency.

So far we have looked at the moment in 3D created by one force, **F**. When calculating the moment round the low back, for instance, the weights of several body segments have to be taken into account. One method is to add up the moments that each of these forces creates round the low back. This is summarized by the equation:

$$\mathbf{M} = \mathbf{r} \wedge \mathbf{F} + \mathbf{r}_1 \wedge m_1\mathbf{g} + \mathbf{r}_2 \wedge m_2\mathbf{g} + \ldots$$

where **F** is an external force acting on the body, such as a weight at the hands; **r** is the vector running from the joint to the point of application of **F**; m_1, m_2 are the masses of body segments (kg); $\mathbf{r}_1$, $\mathbf{r}_2$ are the vectors running from the joint to the centres of mass of body segments; and **g** is the acceleration due to gravity ($9 \cdot 81$ m/s², downwards).

A second method to calculate **M**, which is useful when moments at several joints of the body are already known, is to calculate the moment round the wrist, then use this result and add the effect of the forearm weight to work up to the elbow, and so on to the shoulder, till the low back is reached. This second method, although it may seem less immediate, is more economical if the moments at the wrist, elbow and so on were required anyway. The following equation is used (symbols are shown on Figure 23.6).

$$\mathbf{M} = \mathbf{r}_{cm} \wedge m\mathbf{g} + \mathbf{M}_{adj} + \mathbf{r}_{adj} \wedge \mathbf{R}_{adj}$$

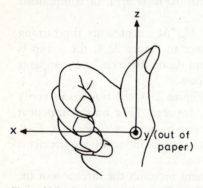

Figure 23.5. Conventions for a system of axes.

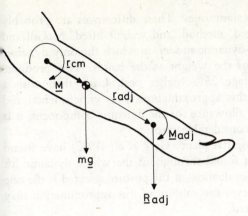

Figure 23.6. The moment M_{adj} at the wrist can be used to calculate the moment M at the elbow.

where **M** is the moment at the selected joint; $\mathbf{M}_{adj}$ is the moment at the adjacent joint; $\mathbf{r}_{adj}$ is the vector running from the selected joint to the adjacent one; m is the mass of the segment beween these two joints; $\mathbf{r}_{cm}$ is the vector running from the selected joint to the centre of mass of the segment; and $\mathbf{R}_{adj}$ is the resultant force calculated at the adjacent joint.

For example:

$$\mathbf{R}_{adj} \text{ at the wrist is } \mathbf{F} + m_{hand}\mathbf{g}$$

$$\mathbf{R}_{adj} \text{ at the elbow is } \mathbf{R}_{wrist} + m_{forearm}\mathbf{g}$$

and so on, if **F** is a force on the hand.

Moments in dynamic tasks

Previous sections have focused on calculations for static tasks, but if body segments are going through accelerations this will add the effect of inertial forces to the moments at the joints. The main difficulty in performing an analysis of a dynamic task is recording instantaneous accelerations throughout the task. The problem has therefore been tackled mainly for lifting in the sagittal plane, where acceleration data is obtained from video recordings, say. Ayoub and El Bassoussi (1978) have included a prediction of accelerations in a dynamic computer model: it is approximated as a function of the angle of each limb at the end of the lift and the duration of the lift.

The most common simplification is to ignore accelerations, and treat the problem as a static one. This may lead to errors if the task is performed quickly. McGill and Norman (1985) have tested this on lifting tasks: they carried out both static and dynamic evaluations of the load on the L4–L5 intervertebral joint. Results with the dynamic analyses were on average 19% higher than with the static approximation, and could go up to 52% higher. A similar investigation by Garg *et al.* (1982) resulted in dynamic evaluations two to three times higher than static ones. These differences are probably due to differences in lifting speed, method, and weight lifted. McGill and Norman (1985) propose a quasi-dynamic model, in which the only dynamic component is the acceleration of the weight at the hands. This produces conservative estimates: on average 25% higher peak loads than with a dynamic evaluation. Although this approximation saves experimental and computer time and makes some allowance for the dynamic component, it is still an approximation liable to considerable error.

Still related to lifting in the sagittal plane, Garg *et al.* (1982) have found that the peak low-back moment found throughout the whole dynamic lift can be approximated by a static evaluation, if the posture selected is the one at the initiation of the lift. However the validity of this approximation may depend on the type of lift analysed.

In order to interpret results obtained from a dynamic analysis, more research is needed on voluntary dynamic moments, and on responses of tissue and intervertebral disc material to high loads applied for a short instant. At the moment, the greatest value of a dynamic analysis is for comparative purposes. The rest of this section will describe the calculations required for a dynamic analysis.

The moment at a particular joint varies throughout the motion due to changes both in lever arms and in accelerations. At any instant, the moment depends on the value of the linear acceleration of the centre of mass of each segment, on the direction of this acceleration, and also on the angular acceleration of each segment. The resistance to rotation that an object has depends on its mass and shape and is described by its moment of inertia. This parameter differs for different axes of rotation, but some moments of inertia for movement in the sagittal plane can be found in the literature (Winter, 1979).

The general equation for the moment **M** at a joint at a particular instant is:

$$\mathbf{M} = \mathbf{r}_{cm} \wedge \mathbf{mg} + \mathbf{M}_{adj} + \mathbf{r}_{adj} \wedge \mathbf{R}_{adj} + \mathbf{r}_{cm} \wedge \mathbf{ma} + \mathbf{I} \wedge \ddot{\boldsymbol{\theta}}$$

The first three terms have already been described (see Figure 23.6) in the previous section on static moments. Symbols used in the additional terms are: **a** is the linear acceleration of the centre of mass of the segment (ms^{-2}); $\ddot{\boldsymbol{\theta}}$ is the angular acceleration of the segment about its centre of mass (degrees s^{-2}); and **I** is the moment of inertia of the segment about its centre of mass (kg m^2).

For 3D tasks, Ito *et al.* (1980) provide detailed equations for a dynamic analysis of a man-model made of several links, each with a given position, orientation, velocity and acceleration.

A note on units for moments

In the previous example, the lever arm was expressed in metres (m), the force in Newtons (N) and so the moment was obtained in Newton-metres

(Nm). These are SI units and therefore recommended; however the following units also can be found in the literature.

The kg-force or kilopond is the force exerted by a mass of 1 kg due to gravity; it is equal to 9·81 N. Thus if a 10 kg object is held at the hand, it exerts a force of 98·1 N. Its weight is 98·1 N, while its mass is 10 kg.

One can also come across the pound (lb). This is equal to 0·4536 kg, and multiplying by 9·81 one obtains 4·450 N as the force exerted by 1 lb. Sometimes moments are expressed in inch-pounds, or foot-pounds. With 1 inch = 0·025 m and 1 foot = 0·30 m, one obtains the equivalence: 1 inch-pound = 0·1112 Nm and 1 foot-pound = 1·3349 Nm. One can also encounter kg-cm or kg-inch but, luckily, SI units are increasingly being used.

Prediction of strength and task feasibility using moments

Many experimental studies have been carried out to measure the maximum voluntary static strength of men and women. This can be overall body strength or else the strength of individual joints. The first category includes data on lifting strength in various postures. This type of information is very useful when it is directly applicable to a task; however maximum strength can be very different if the posture adopted is slightly different from the one tested, with joints not at their best angles and body weight contributing a different moment due to the different posture. In spite of these reservations, maximum voluntary static strength tests are frequently used for comparison with task loads.

Another approach to task assessment is to evaluate the moments created at each joint by a task and compare them with joint strength data from the population. If the moment evaluated for a given joint is higher than the estimated maximum capabilities within the population, one can predict that the task will not be feasible for most people.

A practical way of expressing the feasibility of a task is to state what percentage of the male or female working population is likely to be capable of it. Chaffin (1988a) gives examples of this method. One of these applies to pulling carts for moving stock, and it is shown how the percentage of women capable of this drops as the force required increases. In another example, it is shown that most of the working population is capable of performing a particular lifting task (however evaluation of loads on the spine shows the task actually is hazardous).

The method of comparing evaluated moments with population maxima is sometimes used to predict the maximum strength possible in a particular posture: one raises the hand-force in the calculations until the maximum capability of one joint is reached. That joint is the limiting factor, the weak link or bottleneck. If this method of raising the hand load is used to predict maximum strength, results may be inaccurate because as the hand-force is increased, a subject's real posture is likely to change to ensure body balance is maintained. Subjects may also change their posture to use other muscles and avoid the restrictions arising from weaker muscle groups.

Reliability of maximum joint strength data

Another problem is the reliability of the data on a particular joint's maximum strength. A very wide range of results can be found in the literature, and this may be due to different testing methods as well as to human variability. For instance elbow strength results are sometimes given as the force subjects are able to exert using their hand. This method is not satisfactory as the wrist may be the weak link in this task. To avoid this problem there have been experiments in which the subjects exert a force on a device attached proximal to the wrist. These results are incomplete if the distance between this point and the elbow is not quoted. The most useful data are those where elbow strength is expressed directly as a moment and the method for its determination is noted. Low-back strength is particularly difficult to define, as there is no obvious point from which to define moments when testing for strength.

Dependence of joint strength with joint angles

The next limitation concerns joint angles. The force a muscle can exert depends on its length, so moment capabilities depend on joint angles. For example elbow strength depends not only on the elbow angle but also on the shoulder angle, as muscles span across both joints. Yet many results on elbow moments do not report the shoulder angle.

For a compilation from the literature of moment-angle curves of major joints, the reader is referred to Svensson (1987). Figure 23.7 has been adapted

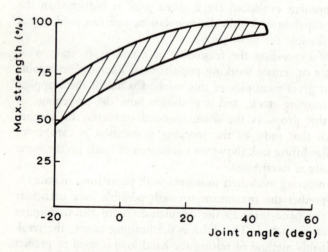

Figure 23.7. Back extensor strength as a function of trunk angle. Results from 4 studies, normalised by denoting the top value of each curve as 100%. (From Svensson, 1987).

rom one of these results and shows the dependence of trunk extensor
noment and trunk angle. The hatched area includes curves from four different
studies.

A word of caution is in order when referring to published results concerning
oint angles: the field of biomechanics does not appear to have any angle
conventions and the position of the zero angle varies across studies. One
useful standard may be that set by the British Orthopaedic Association (1966)
n their booklet describing terminologies used in joint motion. Their method
s the 'Zero Starting Position': to accept the 'anatomical position' of a limb
as zero degrees. For instance (see Figure 23.8) the elbow angle is zero for
he extended straight arm, and its range of movement is from about 150°
flexion to 10° hyperextension.

Interpretation of results

Readers can refer to prediction equations in Chaffin and Andersson (1984,
p. 224) for maximum moments as a function of joint angles. However, as a
wide range of values can be found from other sources, a rough compilation
of these ranges is presented in Table 23.1. It is only intended to give the
reader an order of magnitude with which to compare evaluated moments,
and further work is needed in this area. The ranges given in Table 23.1
include all those found in some of the literature, and they all refer to the so-
called 'fit and healthy' volunteer. The angle notation in Table 23.1 follows
he British Orthopaedic Association convention described earlier.

Results refer to moments along a single axis, for instance pure flexion or
pure abduction, so caution must be exercised in using them for a task in
which moments in several directions are combined.

An additional word of caution is given by Grieve (1987) who showed that
at some joints, antagonistic muscles (i.e. muscles that have opposite effects)

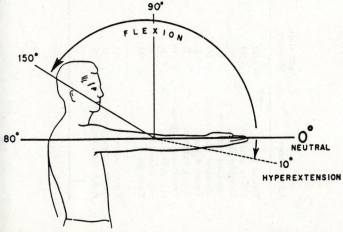

Figure 23.8. The elbow—flexion and hyperextension (from British Orthopaedic Association,
1966).

Table 23.1. Maximum voluntary joint strengths (Nm) from some of the literature. The ranges presented include the ranges from these studies

Joint strength	Joint angle (degrees)	Range of moments (Nm) of subjects from several studies		Variation with joint angle
		Men	Women	
Elbow flexor	90	50–120	15–85	Peak at about 90°
Elbow extensor	90	25–100	15–60	Peak between 50° and 100°
Shoulder flexor	90	60–100	25–65	Weaker at flexed angles
Shoulder extensor	90	40–150	10–60	Decreases rapidly at angles less than 30°
Shoulder adductor	60	104	47	As angle decreases, strength increases then levels at 30° to -30°
Trunk flexor	0	145–515	85–320	Patterns differ among authors
Trunk extensor	0	143	78	Increases with trunk flexion
Trunk lateral flexor	0	150–290	80–170	Decreases with joint flexion
Hip extensor	0	110–505	60–130	Increases with joint flexion
Hip abductor	0	65–230	40–170	Increases as angle decreases
Knee flexor	90	50–130	35–115	In general, decreases with knee flexion but some disagreement with this, depending on hip angle
Knee extensor	90	100–260	70–150	Minima at full flexion and extension
Ankle plantarflexor	0	75–230	35–130	Increases with dorsiflexion
Ankle dorsiflexor	0	35–70	25–45	Decreases from maximum plantar flexion to maximum dorsiflexion

ontract simultaneously. For example at some elbow angles, voluntary elbow exion recruits not only the biceps (flexor), but also to some extent the ·iceps (extensor). This co-contraction, found also at the knee and shoulder, ; believed to have a joint stabilizing role. Therefore, it is possible that rediction models overestimate a subject's strength, because co-contraction τ a joint reduces the net moment provided by it.

At this stage there are very few data on maximum moments exerted ynamically, probably because of the problem in recording moments that over a wide enough maximum dynamic range of angles, velocities and cccelerations.

In conclusion, comparing the moments required by a task with data on naximum capabilities provides some useful guidance in terms of orders of nagnitude, but the method should be used with caution and complemented vith other methods, especially for repetitive tasks which may cause fatigue.

ʼorces on the low back

ι study by Chaffin (1988a) was mentioned earlier, in which a particular .fting task was analysed. From the moments at the joints it was estimated nat most of the working population would be capable of performing the ιsk; however evaluation of loads on the spine indicated that the task could ut the back at risk. This section discusses what criteria are available for ıssessing such risk. A simple example of a 2D model to calculate the ompression force on the spine was given earlier in the chapter (Figure 23.3). ʼhis section will describe other models, including 3D ones for asymmetrical ostures.

A relatively direct method to evaluate forces on the spine is to insert a ressure-measuring needle into the intervertebral disc (Nachemson and Morris, 1964). Results have been used to evaluate various seated or lifting ιsks (e.g. Nachemson and Elfström, 1970; Andersson and Örtengren, 1974). ι recent studies, intradiscal pressure measurements have been used mostly ι conjunction with electromyography (EMG) and intra-abdominal pressure neasurements to verify the validity of low-back models (Schultz *et al.*, 1982). t is presumably more pleasant for subjects to have their spine compression ιlculated than their disc pressure measured. Nachemson and Morris (1964) nd Aspden (1989) discuss how the compression force between vertebrae can ·e evaluated from intradiscal pressures.

ʒuidelines from low-back forces

ʼhe most commonly used guideline for task assessment is the value of the ompression force between vertebrae. Experiments on cadaveric spines have hown that fractures appear above certain levels of compression. The level ; lowest for older people; female spines are, as a general rule, weaker than male spines (Hutton and Adams, 1982). The National Institute for Occupational Safety and Health (NIOSH, 1981) have agreed on the following guidelines: tasks causing a compression on the lumbo-sacral joint greater than 6400 N are above the 'maximum permissible limit': they are unacceptable and engineering controls are required. On the other hand compressions under 3400 N can be tolerated by most young, healthy workers (over 75% of women and over 99% of men). It must be noted these guidelines relate to lifting in the sagittal plane, and the spine may be much more vulnerable under axial rotation or hyperflexion (Adams and Hutton, 1981). The majority of compression tests have sought the ultimate compression strength, but work by Brinckmann *et al.* (1987) is promising, for data relating to repetitive tasks and the strength of intervertebral joints under cyclic loading. For instance they have shown that for a cyclic load of about half the ultimate compression strength, the probability of a fatigue fracture after 100 cycles is nearly 50%.

The value of the compression force may not be the most relevant parameter related to back injury, and guidelines such as the two NIOSH limits should be used with their limitations in mind, especially at extremes of trunk motion and in dynamic tasks. Additional indications of the severity of a task are given by the forces evaluated for the muscles of the low back. As a general rule, the maximum strength of a muscle is proportional to its largest cross sectional area, and is approximately 50–100N/cm² (Schultz and Andersson, 1981). (Some cross-sectional areas can be found later in Table 23.5.)

Applications

The value of lumbar spine compression is the most frequently used criterion in the evaluation of tasks that may put the back at risk. It has been used extensively in the analysis of lifting tasks, for instance to determine a good lifting method (Bejjani *et al.*, 1984) or to determine maximum acceptable weights (Hutton and Adams, 1982; Jäger and Luttmann, 1986).

Gagnon *et al.* (1986) used low-back estimates to compare three methods to lift a patient out of a wheelchair. Their paper points out the low-back model's limitations and how this puts uncertainty on absolute values for forces; however modelling did allow comparison between the three methods. Energy expended to lift patients was also taken into account in the choice of a lifting method.

More examples of the use of low-back modelling in the analysis of industrial tasks can be found in Jäger and Luttmann (1986).

Models of the low back

In the simple example of Figure 23.3, one set of back muscles, situated posterior to the spine, was used to resist a trunk flexion moment. This created a compression force on the spine. This is only one model of the low ack: many others, varying in complexity, have been developed. Some, used n the orthopaedics field, take into account a large number of muscles and gaments attached to several points of each vertebra. For the ergonomics eld, the main guideline is obtained from the value of the compression force etween two lumbar vertebrae, and a less detailed model is usually considered dequate.

The rest of this section describes how to evaluate muscle and compression rces with 2D and 3D models. It has been written for readers who wish to arry out calculations themselves.

A simple 2D model

We will now complete the simple 2D model used at the beginning of the hapter (Figure 23.3). It allowed back muscles (representing the erector spinae group) to resist a trunk flexion moment. If trunk extension is to be resisted, hese muscles can relax and abdominal muscles (rectus abdominis) take over. Figure 23.9 summarizes the results of this model. The lumbar spine

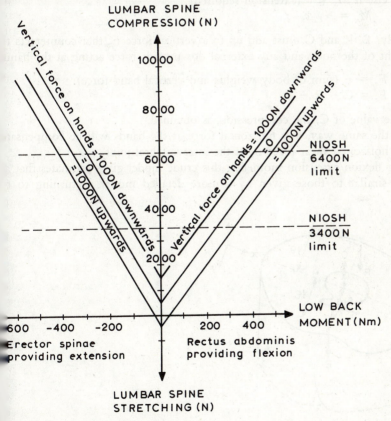

Figure 23.9. Lumbar spine compression as a function of low-back moment, with a simple 2D model.

compression is shown as a function of the low-back moment. The vertic
force on the hands also compresses the spine, as shown from the parall
lines for 0 N, 1000 N upwards and 1000 N downwards.

The equations describing this 2D model, and from which Figure 23.9 w
obtained are now described.

A flexion moment is provided by the rectus abdominis (R), and a
extension moment is provided by the erector spinae (E) (see Figure 23.10
E acts at a distance y_E from the centre of the spine, and the lever arm for
is y_R.

If M_x is the flexion or extension moment which these muscles mu
provide,

if $M_x > 0$ (flexion required)
$M_x = (-y_R) \times (- R)$
(both the y and z axes are in an opposite direction to y_R and
hence the minus signs)
so $M_x = y_R R$
else if $M_x < 0$ (extension required)
$M_x = -y_E E$

Finally, E, R and C must add up to a vertical force F_z that counteracts t
weight of the body and any external downwards force acting at the hand

$F_z = -$ (sum of body weights and vertical hand force)
$F_z = C - R - E$

so the value of C, the compression, is obtained.

In the same way, any horizontal force at the hands will be compensat
by a horizontal shear force S_y at the intervertebral joint.

For flexion/extension moments, this crude model gives estimates that a
very similar to those given by a more detailed model. Returning to t

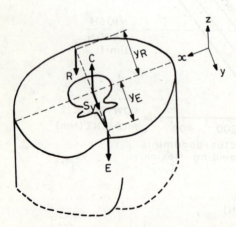

Figure 23.10. A simple 2D model, represented on a section of the low-back (R = rect
abodominis, E = erector spinae, C = compression, y_R, y_E = lever arms).

graphical summary in Figure 23.9 for flexion/extension, the slopes on the graph are $1/\gamma_R$ for positive moments and $1/\gamma_E$ for negative moments, with $\gamma_R = 8 \cdot 0$ cm and $\gamma_E = 5 \cdot 8$ cm. (These numbers are average values: see the later paragraph on 'low-back geometry'.)

Simple 3D models

The previous model did not have any muscles accounting for lateral flexion or axial rotation of the trunk. A model described by Chaffin and Andersson (1984, p. 206) will be briefly summarized here.

It consists of six muscles (Figure 23.11): the rectus abdominis (R) and erector spinae (E), provide flexion and extension, respectively; the vertical components of the left and right obliques (VL and VR) provide lateral flexion to the left and to the right. Finally, the horizontal components of the left and right obliques (HL and HR), provide respectively positive (anti-clockwise) and negative axial moments round the z axis. Other low-back forces in the model are the compression force C on the intervertebral joint (if C < 0, the force is on the contrary an extension force), and the lateral and antero-posterior shear forces, S_x and S_y on the intervertebral joint. One more force not mentioned so far is the force P due to intra-abdominal pressure: it is believed that the rise in pressure in the abdominal cavity, that occurs during heavy manual handling, supports the trunk and effectively produces an extensor moment. This moment is equivalent to a force P acting on the centre of the diaphragm. This topic will be discussed in more detail later.

The equations for this model are presented in Table 23.2. Antero-posterior lever arms are denoted by y, lateral lever arms by x, so y_E is the distance of the erector spinae force E behind the spine, for instance. x_O is the lateral lever arm of the obliques. M_x, M_y, M_z are the low-back moments to be provided by the muscles. F_x, F_y, F_z are the forces provided by the low back to counteract body weight and hand-forces.

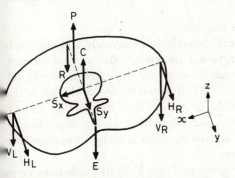

Figure 23.11. A schematic diagram of a simple 3D model, adapted from Chaffin and Andersson 1984. (R, E, C = rectus abdominis, erector spinae and compression; P = force due to intra-abdominal pressure; VL, HL are the vertical and horizontal components of the obliques on the left side of the body; VR, HR on the right side).

Table 23.2. Equations for the simple 3D model (adapted from Chaffin and Andersson, 1984)

IF	M_x	$\geq$	0	THEN E	=	0	(flexion required)	
	M_x	$\leq$	0	THEN R	=	0	(extension required)	
IF	M_y	$\geq$	0	THEN VR	=	0	(flexion to left required)	
	M_y	$\leq$	0	THEN VL	=	0	(flexion to right required)	
IF	M_z	$\geq$	0	THEN HR	=	0	(anti-clockwise rotation required)	
	M_z	$\leq$	0	THEN HL	=	0	(clockwise rotation required)	

$$F_z = S_x$$
$$F_y = S_y + HL + HR$$
$$F_z = C + P - R - E - VL - VR$$
$$M_x = -\gamma_R P + \gamma_R R - \gamma_E E$$
$$M_y = x_o \ (VL - VR)$$
$$M_z = x_n \ (HL - HR)$$

One limitation of this model is that the vertical and horizontal components of the obliques are made to act independently, whereas oblique muscles in reality always pull simultaneously in the vertical and horizontal directions. As this model allows an oblique to provide purely a horizontal force, the compression on the spine may be underestimated for tasks involving axial rotation. We will discuss later a 10-muscle model by Schultz and Andersson (1981), which models the obliques in a more realistic way, with internal obliques acting posteriorly downwards and external obliques acting anteriorly downwards.

This 10-muscle model, and others involving more muscles require a computer to carry out a particular mathematical procedure (linear optimization). This is a handicap to some users wanting a simple program quickly written on a microcomputer, or even just worked out on a calculator. Accordingly a 'micro-model' is proposed (Tracy, 1988), to model the obliques more realistically and produce results that are closer to those of models requiring linear optimization.

The rule for oblique action is as follows:

Internal and external obliques pull respectively posteriorly and anteriorly downwards, as in Figure 23.12. Suppose a clockwise axial rotation moment must be provided ($M_z < 0$): this can be done by the external on the left (XL) and by the internal on the right (IR). If lateral flexion to the left is also required ($M_y > 0$), XL and IL will be in action. With this model, XL acts strongly to provide both M_z and M_y, and either IR or IL act, depending on which moment is the largest. So if axial rotation is more important than lateral flexion, XL and IR are active.

The equations for this model are given in Table 23.3. Its predictions come close to those of the 10-muscle model which is to follow, but a computer is not needed. Its main limitations are that the erector spinae and the rectus abdominis (E and R) are placed in the mid-sagittal plane, whereas in reality they are groups of muscles situated to the left and the right. They contribute

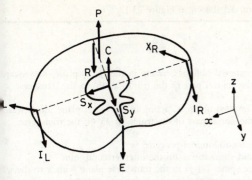

Figure 23.12. Schematic diagram of the micro-model – a simple 3D model with more realistic representation of the obliques. (Symbols described in Table 23.3.)

lateral flexion moments, whereas the simple models seen here only allow obliques to do this.

3D models requiring optimization

If the erector spinae and the rectus abdominis are placed as separate forces on either side of the sagittal plane, and if any other trunk muscles are also represented, special techniques are required to decide how several muscles than can all do the same job are going to share it out. Lateral flexion to the right, for instance, can now be provided by the right obliques, the right rector spinae, or the right rectus abdominis. Mathematically, there are more variables (forces) than equations. The indeterminacy must be solved by making assumptions, and giving rules as in Tables 23.2 or 23.3 is not possible or practical when many muscles are involved.

A technique is then to use a computer library for linear programming: this optimizes one variable while making sure a number of equations are satisfied. Schultz and Andersson (1981) and Schultz *et al.* (1983) have established models with 10 to 22 muscles, based on the assumption that the spinal compression C is to be at a minimum. A linear programming routine will ensure that all muscle forces provide the required moments and do not exceed maximum capability of 100 N/cm^2 and that at the same time the forces have been distributed so that C is as low as possible. Although the problem may seem complicated, linear programming routines are in principle straightforward to use. For asymmetrical tasks, models such as the ones developed by Schultz *et al.* (1983) are far superior to simple calculator-based models, as shown by Schultz *et al.* in validation experiments. The reader is referred to Schultz and Andersson (1981) and Schultz *et al.* (1983) for details of these models.

The condition to minimize the compression force will have the effect that muscle with a larger lever arm will provide a force in preference to one with a smaller lever arm. Muscles close to the spine only act if other muscles have reached their maximum capability.

Table 23.3. Equations for the micro-model shown in Figure 23.12.

Conventions

R	rectus abdominis
E	erector spinae
XL, XR	left and right external obliques, acting in the (y, z) plane, downward and towards the ventral part of the trunk, at 45° to the transverse plane
IL, IR	left and right internal obliques, acting in the (y, z) plane, downward and towards the dorsal part of the trunk, at 45° to the transverse plane
P	force due to intra-abdominal pressure
C, S_x, S_y	compression and shear forces on the intervertebral joint
Axes (x, y, z)	centered on the spine, (x, y) in the transverse plane with x to the left of the body, y directed posteriorly, z directed upwards
F_x, F_y, F_z	resultant reaction forces at the level of the section
M_x, M_y, M_z	resultant reaction moments at the level of the section
ABS (x)	absolute (positive) value of x
SUM =	$\dfrac{M_y + M_x}{2\,(x_o \cos 45)}$
DIFF =	$\dfrac{M_y - M_z}{2\,(x_o \cos 45)}$

Equations

$F_z = C + P - E - R - (IL + IR) \cos 45 - (XL + XR) \cos 45$
$F_y = (IL + IR) \sin 45 - (XL + XR) \sin 45 + S_y$
$F_x = S_x$
$M_x = -y_E\, E + y_R\, R - y_P\, P$
$M_y = x_O\, (IL - IR) \cos 45 + x_O\, (XL - XR) \cos 45$
$M_z = x_O\, (IL - IR) \sin 45 + x_O\, (XR - XL) \sin 45$
If $M_x \geq 0$ then $E = 0$
If $M_x < 0$ then $R = 0$

This assumption produces good predictions but it must be pointed out that validation experiments have only been carried out on small loads. Bea et al. (1988) have proposed a more sophisticated assumption: first use linear programming to minimize not C but muscle intensities (force per cross sectional area): the largest of all the muscle intensities, I^* must be as small as possible. This ensures that no muscle is giving its maximum while other muscles which could also contribute are inactive. The next step is to solve the problem all over again, minimizing C, but with the condition that no muscle intensity exceeds I^*. This recent method seems to provide more realistic modelling than when to minimize C is the only criterion, and validation with experimental data is expected.

Minimizing C also implies that antagonistic muscles do not co-contract for instance that if trunk extension is required the rectus abdominis are inactive. This is a simplification of reality, as co-contraction can be important in some tasks. Nachemson et al. (1986) for instance observed co-contraction when a Valsalva manœuvre (voluntary raising of intra-abdominal pressure was performed. This implies a higher value of spine compression than predicted by the previous models.

Table 23.3. Continued

If $M_y \geq 0$ and $M_z \geq 0$		
and ABS $(M_y) \geq$ ABS (M_z) then	IR	= 0
	XR	= 0
	IL	= SUM
	XL	= DIFF
and ABS $(M_y) <$ ABS (M_z) then	IR	= 0
	XL	= 0
	IL	= SUM
	XR	= −DIFF
If $M_y \geq 0$ and $M_z < 0$		
and ABS $(M_y) \geq$ ABS (M_z) then	IR	= 0
	XR	= 0
	IL	= SUM
	XL	= DIFF
and ABS $(M_y) <$ ABS (M_z) then	IL	= 0
	XR	= 0
	IR	= −SUM
	XL	= DIFF
If $M_y < 0$ and $M_z < 0$		
and ABS $(M_y) \geq$ ABS (M_z) then	IL	= 0
	XL	= 0
	IR	= −SUM
	XR	= −DIFF
and ABS $(M_y) <$ ABS (M_z) then	IL	= 0
	XR	= 0
	IR	= −SUM
	XL	= DIFF
If $M_y < 0$ and $M_z \geq 0$		
and ABS $(M_y) \geq$ ABS (M_z) then	IL	= 0
	XL	= 0
	IR	= −SUM
	XR	= −DIFF
and ABS $(M_y) <$ ABS (M_z) then	IR	= 0
	XL	= 0
	IL	= SUM
	XR	= −DIFF

Inputs to biomechanical calculations

Posture input

Biomechanics is sometimes used as a predictive tool, on a posture that has not been observed but has been estimated as a likely posture for a task. However, the posture may not be realistic and it may be worth ensuring that the body is in balance by checking that the resultant of all external force lies in the area between the two feet. If the posture is asymmetric, each leg can take a different proportion of the resultant force at the feet, and this needs to be measured with a force plate, which brings us back to the laboratory. Therefore biomechanics must be used with caution when used as a predictive tool.

The reader is referred back to the beginning of the chapter for an overview of methods to record posture.

Body segment weights

Table 23.4 summarizes masses and the locations of the centre of gravity of body segments, compiled by Pheasant (1986). Other anthropometric data used in modelling, such as link lengths, can be found in the same reference. Segment mass data originate from very small, poorly representative samples, so may be a source of error in biomechanical calculations. This and other sources of error are discussed in the last section of this chapter.

Low-back geometry

There have been recent improvements in the data available for low-back models. Both *computed axial tomography* (CAT) and *magnetic resonance imaging* (MRI) have been used to measure muscle lever arms and cross-sectional

Table 23.4. Segment masses and locations of centre of gravity, from Pheasant (1986)

Segment	Mass (percent body mass)	Location of centre of gravity
1. Head and neck	8.4	57% of distance from C7 to vertex
1a. Head	6.2	20 mm above tragion
2. Head and neck and trunk	58.4	40% of distance from hip to vertex
2a. Trunk	50.0	46% of distance from hip to C7
2b. Trunk above lumbo-sacral joint	36.6	63% of distance from hip to C7
2c. Trunk below lumbo-sacral joint	13.4	Approximately at the hip joint
3. Upper arm	2.8	48% of distance from shoulder to elbow joints
4. Forearm	1.7	41% of distance from elbow to wrist joints
5. Hand	0.6	40% of hand length from wrist joint (at centre of an object gripped)
6. Thigh	10.0	41% of distance from hip to knee joints
7. Lower leg	4.3	44% of distance from knee to ankle joints
8. Foot	1.4	47% foot length forward from the heel (half height of ankle joint above the ground)—mid-way between ankle and ball of foot at the head of metatarsal III

| Total Body mass* (kg) | | | | | | | |
| | Men | | | | Women | | |
Percentiles	5th	50th	95th	S.D.	5th	50th	95th	S.D.
British (19–65 years)	55.3	74.5	93.7	11.7	44.1	62.5	80.9	11.2

* Masses (in kg) to be multiplied by 9.81 to obtain weights (or forces in N) for the calculation of moments.

areas, whereas previously values came from a limited sample of cadaveric data. Some of the results needed for the models described in this chapter can be found in Table 23.5. These are lever arms and areas at the L3–L4 level for 26 males (Tracy *et al.*, 1989). Data for more muscles and at other lumbar sections can be found in this same study.

Lever arms for both sexes have been measured by Nemeth and Ohlsen (1986) at the lumbo-sacral joint, and by Kumar (1988) at L3 and L4 (as well as T7 and T12). Chaffin (1988b) has announced that results on larger samples will soon be published.

At the moment it is common to use muscle lever arms observed at a particular cross-section, but ideally the values used should take into account the line of action of the muscles and their points of attachment. More information will no doubt become available with the recent developments in CAT and MRI scanning.

Role of intra-abdominal pressure

The most widespread theory of the role of intra-abdominal pressure (IAP) in low-back force production is that the pressure supports the trunk, and its action on the diaphragm and pelvic floor is equivalent to a force for trunk extension. According to this model, the force produced by IAP is calculated by multiplying the pressure by the area of the diaphragm. The extensor moment created by this force is the product of the force with the lever arm of the centroid of the area on which IAP acts. Using this model, IAP reduces lumbar compression by 4–30% according to Schultz *et al.* (1982), 2–8% according to Leskinen and Troup (1984). This range is large because the percentage reduction depends on the value of IAP and on the value of compression. The reduction in spinal compression due to IAP may be underestimated, because calculations of moments and lumbar loads, ignoring IAP, sometimes still result in excessive compression values although no structural failure is observed (Jones, 1983; Chaffin and Andersson, 1984). On the other hand, both Krag *et al.* (1985) and Nachemson *et al.* (1986) have argued from experimental evidence that IAP does not reduce lumbar compression; EMG readings showed that trunk extensor muscle action was not reduced when the abdominal cavity was voluntarily pressurized.

Table 23.5. Lever arms and cross-sectional areas* of some muscles at L3–L4 level, from an MRI study of 26 males (standard deviations in parentheses)

	Antero-posterior lever arm (mm)		Lateral lever arm (mm)		Area (cm²)	
Erector spinae	57.6	(4.6)	38.2	(3.2)	26.0	(3.3)
Rectus abdominis	79.5	(17.6)	33.8	(9.9)	6.6	(2.4)
Obliques	17.2	(12.0)	122.1	(10.8)	35.1	(5.1)

* Areas are for the muscles on one side of the body.

A number of theories for the role of IAP have been put forward, and the reader is referred to other texts (e.g. Aspden, 1987) for a review of these various theories. Until the controversies on the role of IAP are resolved, the most common approach is to represent it as an extensor force, as in Figure 23.11. Shown in Table 23.6 are some values found in the literature for various tasks, but many authors choose to ignore IAP and set the force to zero. It is possible to measure IAP with a swallowed radio-pill (Davis and Stubbs, 1977), but some experience is required to use this technique. Chaffin and Andersson (1984, p. 192) have published a prediction equation for IAP, using the hip moment and angle, for lifting in the sagittal plane.

It must be added that the measure of IAP is sometimes used directly to evaluate manual handling tasks. Davis and Stubbs (1977) conducted some studies in which they found IAP increases that were proportional to weights lifted. They found higher incidences of back pain among males undertaking heavy physical work, during which peak abdominal pressures frequently exceeded 13 kPa (100 mmHg). Accordingly they developed a set of recommendations for safe limits of force from values of forces which produce 12 kPa (90 mmHg) in a 5th percentile man.

However, there is some disagreement on the validity of the relationship between IAP and spinal load. For instance Andersson et al. (1977) and Örtengren et al. (1981) found a linear relationship between intradiscal and intra-abdominal pressures, whereas Schultz et al. (1982) and Nachemson et al. (1986) did not. Schultz et al. (1982) found a poor correlation between estimated spine compression and IAP values, Chaffin and Andersson (1984) report a relationship between IAP and hip moment which is not linear but nearly quadratic, while Mairiaux et al. (1984) found it to be linear.

Table 23.6. Intra-abdominal pressure for various tasks (1 kPa = 7.6 mmHg)

Task	IAP (kPA)	Force (N) developed by IAP over 299 cm^2
Schultz et al. (1982)		
Relaxed standing	1.0	30
Upright, arms in, holding 8 kg in both hands	1.5	45
Flexed 30°, arms out	4.2	125
Flexed 30°, arms out, holding 8 kg in both hands	4.4	130
Davis and Stubbs (1977)		
Breathing	1	30
90 mmHg 'safe limit'	12	560
Grieve and Pheasant (1982)		
Competitive weight lifting	40	1196
Nachemson et al. (1986)		
Valsalva manoeuvre	4	120

Estimate for diaphragm area: 299 cm (Leskinen and Troup, 1984)
Estimate for IAP lever arm: 48 mm (Schultz et al., 1982)

Angle of discs

So far we have referred to forces on 'the lumbar spine' without specifying a particular vertebra or disc. The low-back models discussed are too crude to differentiate between different vertebrae, and the main difference between calculations at L3 or at L5–S1 is the weight of the trunk above it. There is one other difference, though, and that is the angle of the intervertebral discs. The force referred to as the compression force earlier on is actually partly compression and partly shear if the intervertebral joint is not perpendicular to the line of action of the erector spinae or rectus abdominis. Unfortunately there is very little information on disc angles for various postures, and as there are large variations in the degree of lordosis in the population, predictions on disc angles are associated with a large uncertainty. Chaffin and Andersson (1984, p. 192) obtain the L5–S1 angle from hip and thigh angles, for postures in the sagittal plane. Another approach is to infer the shape of the spine from the shape of the surface of the back (Stokes and Moreland, 1987; Tracy *et al.*, 1989) but this work is still somewhat inconclusive. Until more information is available, one solution is to make the approximation that L3 remains perpendicular to the line of action of the erector spinae and rectus abdominis, whatever the posture (Schultz *et al.*, 1983, for instance), or to use Chaffin and Andersson's (1984) relationship for the L5–S1 angle. Either way the uncertainty will mean that some of the compression force evaluated may in fact be a shear force, and vice versa.

Uses and limitations of biomechanics

This chapter has surveyed the application of biomechanics in the ergonomic field. Calculating moments at joints provides an estimate of the severity of the task. Moments calculated in 3D will also highlight possible twisting efforts which could be eliminated. Biomechanical calculations allow applied forces or posture to be varied, so that problems and solutions can be identified.

Repetitive work and fatigue

Unfortunately, biomechanics cannot on its own answer questions of the type: 'What force can be applied safely and without fatigue *x* times a minute for *y* hours, given *n* rest pauses of *m* minutes are provided?'

Biomechanics can only give some indications of the effect of fatigue if the moment required of a joint is close to its maximum capability, for the onset of fatigue is near. Rohmert (1973) provides information for rest allowances in static work as a function of the percentage of maximum strength a task requires. However Rohmert *et al.* (1986) question the universality of these relationships; it appears that postures where passive structures (skeleton and ligament) are playing the key role can be held longer than those requiring mainly active muscular force.

For intermittent static work a rough guideline is to keep the force exerted under 15–30% of the maximum capability (Bjorksten and Jonsson, 1977; Pheasant and Harris, 1982) to avoid fatigue. Even less can be said at the moment about dynamic work, presumably because of the large number of variables in the problem. In general, whether the work is static, intermittent or dynamic, biomechanics cannot on its own give reliable answers, except in extreme cases where the task can be shown to be so strenuous it can only be performed occasionally.

Safe limits

Human variability is the main problem when determining acceptable limits, especially where back pain is concerned. Limits on lumbar spine compression, such as those used by the NIOSH (1981) guidelines, are based on results of ultimate compression strength tests performed on cadaveric spines. However, as discussed in this chapter, these do not usually produce the effects observed in real back injuries; also it is not possible in many back-pain patients to define the precise source of back pain (Wells, 1985).

There may be some confusion on how often a task can be repeated if it creates a spine compression of the order of magnitude of the ultimate compression strength of cadaveric specimens. These tests imply that one single exertion would damage the spine, yet the NIOSH guidelines allow this type of task to be done quite frequently and these tasks are indeed regularly performed in industry. Studies on fatigue fractures from repetitive loading should eventually bring some light on the matter. A further discussion of these problems can be read in Jones (1983).

Sources of inaccuracy

Results of low-back forces depend on the model used, so it is useful to bear in mind the assumptions and simplifications a model employs, and if several models are available to compare their predictions. Some inputs to low-back calculations are subject to uncertainty. These include the following parameters:

1. *Intra-abdominal pressure.* If IAP is not measured, an estimate needs to be made, and the contribution of IAP to forces in the low back depends on the model. Pressures range from about 1 kPa (relaxed standing) to about 40 kPa (competitive weight lifting) (Table 23.6). If the spine compression calculated without IAP is around 3400 N (first NIOSH limit), including IAP could reduce the estimate by up to 60%, depending on the value of IAP. At the second NIOSH limit (6400 N), the reduction is up to 30%.

2. *Angles of discs.* There are individual variations in spinal shape between subjects, and there is further uncertainty on disc angle changes with trunk

motion. The largest compression estimates will be obtained with discs that are perpendicular to the line of action of the muscles.

3. *Geometry of the low back.* There are quite large inter-subject variations in some muscle lever arms—for instance in Table 23.5, the lever arm of the erector spinae has a standard deviation of nearly 10% of the mean value. Accordingly the uncertainty about the force provided by the erector spinae is also represented by a standard deviation of about 10% of the force calculated with the mean lever arm.

Other sources of uncertainty in biomechanical calculations have already been mentioned. How accurately were the posture and force recorded? Was an asymmetric task evaluated with a 2D model? Were accelerations ignored in a static model? There are also uncertainties on limb weights and their centres of mass. These have the greatest effect when the trunk and arms are held out at large lever arms. In this case the spine compression on a person with a 95th percentile body weight is nearly 30% greater than that for a male with a 50th percentile weight.

All these sources of uncertainty are not usually important when results are used to compare tasks, but are useful to bear in mind when results are used as absolute numbers, and perhaps evaluated against guidelines. There is a mathematical method of evaluating the effect of all the uncertainties on the final result (Barford, 1985), but another way is to experiment with different values of the input parameters.

Conclusion

Biomechanics is a useful tool to evaluate manual handling tasks, highlight problems, and test out possible improvements. Models used can be more or less sophisticated; some require computers while a lot can be achieved just with a calculator. Results should be interpreted with a basic knowledge of the simplifications and uncertainties that have been involved in the calculations. Other methods usefully drawn in to complement biomechanics include physiological measurementse, EMG, injury statistics, discomfort charts and questionnaires.

Acknowledgment

The author is indebted to Diva Ferreira for her contribution to the development of this chapter.

References

Adams, M.A. and Hutton, W.C. (1981). The effect of posture on the strength of the lumbar spine. *Engineering in Medicine,* **10**, 199–202.
Andersson, B.J.G. and Örtengren, R. (1974). Lumbar disc pressure and

myoelectric back muscle activity during sitting. II. Studies on an office chair. *Scandinavian Journal of Rehabilitation Medicine*, **3**, 115–121.

Andersson, G., Örtengren, R. and Nachemson, A. (1977). Intradiscal pressure, intra-abdominal pressure and myoelectric back muscle activity related to posture and loading. *Clinical Orthopaedics and Related Research*, **129**, 156–164.

Aspden, R.M. (1987). Intra-abdominal pressure and its role in spinal mechanics. *Clinical Biomechanics*, **2**, 168–174.

Aspden, R.M. (1989). The spine as an arch. A new mathematical model. *Spine*, **14**, 266–274.

Ayoub, M.M. and El Bassoussi, M.M. (1978). Dynamic biomechanical model for sagittal plane lifting activities. In *Safety in Manual Materials Handling*, edited by C.G. Drury (OH: US Department of Health, Education and Welfare), pp. 88–95.

Barford, N.C. (1985). *Experimental Measurements: Precision, Error and Truth*, 2nd edition (New York: John Wiley).

Bean, J.C., Chaffin, D.B. and Schultz, A.B. (1988). Biomechanical model calculation of muscle contraction forces: a double linear programming method. *Journal of Biomechanics*, **21**, 59–66.

Bejjani, F.J., Gross, C.M. and Pugh, J.W. (1984). Model for static lifting: relationship of loads on the spine and the knee. *Journal of Biomechanics*, **17**, 281–286.

Björksten, M. and Jonsson, B. (1977). Endurance limit of force in long-term intermittent static contractions. *Scandinavian Journal of Work and Environmental Health*, **3**, 23–27.

Brinckmann, P. (1986). Injury of the annulus fibrosus and disc protrusions. An *in vitro* investigation on human lumbar discs. *Spine*, **11**, 149–153.

Brinckmann, P., Johannleweling, N., Hilweg, D. and Biggemann, M. (1987). Fatigue fracture of human lumbar vertebrae. *Clinical Biomechanics*, 2, 94–96.

British Orthopaedic Association (1966). Joint motion. Method of measuring and recording. Published by the American Academy of Orthopedic Surgeons, reprinted by the British Orthopaedic Association, 1966.

Chaffin, D.B. (1988a). A biomechanical strength model for use in industry. *Applied Industrial Hygiene*, **3**, 79–86.

Chaffin, D.B. (1988b). Biomechanical modelling of the low back during load lifting. *Ergonomics*, **31**, 685–697.

Chaffin, D.B. and Andersson, G. (1984). *Occupational Biomechanics* (New York: Wiley-Interscience).

Corlett, E.N., Madeley, S.J. and Manenica, I. (1979). Posture targetting: a technique for recording working postures. *Ergonomics*, **22**, 357–366.

Davis, P.R. and Stubbs, D.A. (1977). Safe levels of manual forces for young males. *Applied Ergonomics*, **8**, 141–150; **8**, 219–228; **9**, 33–37.

Drury, C.G., Roberts, D.P., Hansgen, R. and Bayman, J.R. (1983). Evaluation of a palletising aid. *Applied Ergonomics*, **14** (4), 242–246.

Gagnon, M., Sicard, C. and Sirois, J.P. (1986). Evaluation of forces on the lumbo-sacral joint and assessment of work and energy transfers in nursing aides lifting patients. *Ergonomics*, **29**, 407–421.

Garg, A., Chaffin, D.B. and Freivalds, A. (1982). Biomechanical stresses from manual load lifting: a static vs dynamic evaluation. *IIE Transactions*,

14, 272–281.

Grieve, D.W. (1987). Demands on the back during minimal exertion. *Clinical Biomechanics*, **2**, 34–42.

Grieve, D.W. and Pheasant, S.T. (1982). Biomechanics. In *The Body at Work—Biological Ergonomics*, edited by W.T. Singleton. (Cambridge: Cambridge University Press), pp. 71–161.

Hutton, W.C. and Adams, M.A. (1982). Can the lumbar spine be crushed in heavy lifting? *Spine*, **7**, 586–590.

Ito, K., Minamizaki, Y. and Ito, M. (1980). *Computer-aided dynamic analysis of multi-link system for biomechanical applications*. Research Reports of Automatic Control Laboratory, Faculty of Engineering, Nagoya University, Volume 27.

Jäger, M. and Luttmann, A. (1986). Biomechanical model calculations of spinal stress for different working postures in various workload situations. In *The Ergonomics of Working Postures*, edited by E.N. Corlett, J.R. Wilson and I. Manenica (London: Taylor and Francis), pp. 144–154.

Jones, D.F. (1983). Back injury research: have we overlooked something? *Journal of Safety Research*, **14**, 53–64.

Keyserling, W., Herrin, G., Chaffin, D.B., Armstrong, T. and Foss, T. (1980). Establishing an industrial strength testing program. *American Industrial Hygiene Association Journal*, **41**, 730–736.

Krag, M.H., Gilbertson, L. and Pope, M.H. (1985). Intra-abdominal and intra-thoracic pressure effects upon load bearing of the spine. *31st Annual Meeting Orthopedic Research Society*, Las Vegas, Nevada.

Kumar, S. (1988). Moment arms of spinal musculature determined from CT scans. *Clinical Biomechanics*, **3**, 137–144.

Leskinen, T.P.J. and Troup, J.D.G. (1984). The effect of intra-abdominal pressure on lumbosacral compression when lifting. *Computer-aided Biomedical Imaging and Graphics Physiological Measurement and Control: Proceedings*, Aberdeen, PMCS: 4.

MacDonald, E.B. (1984). Back pain, the risk factors, and its prediction in work people. Occupational aspects of low back disorders. *Society of Occupational Medicine, Symposium Proceedings*, pp. 1–17.

McGill, S.M. and Norman, R.W. (1985). Dynamically and statically determined low-back moments during lifting. *Journal of Biomechanics*, **18**, 877–885.

Mairiaux, P., Davis, P.R., Stubbs, D.A. and Baty, D. (1984). Relation between intra-abdominal pressure and lumbar moments when lifting weights in the erect posture. *Ergonomics*, **27**, 883–894.

Nachemson, A. and Elfström, G. (1970). Intravital dynamic pressure measurements in lumbar discs. *Scandinavian Journal of Rehabilitation Medicine* (Suppl. 1), 1–40.

Nachemson, A. and Morris, J. (1964). *In vivo* measurements of intradiscal pressure. *Journal of Bone and Joint Surgery*, **46A**, 1077–1092.

Nachemson, A.L., Andersson, G.B.J. and Schultz, A.B. (1986). Valsalva maneuver biomechanics: effects on lumbar trunk loads of elevated intra-abdominal pressures. *Spine*, **11**, 476–479.

Nemeth, G. and Ohlsen, H. (1986). Moment arm lengths of trunk muscles to the lumbosacral joint obtained *in vivo* with computed tomography. *Spine*, **11**, 158–160.

NIOSH (National Institute for Occupational Safety and Health) (1981). A work practices guide for manual lifting. Cincinnati: DHHS (NIOSH) publication no 81–122.

Örtengren, R., Andersson, G.B.J. and Nachemson, A.L. (1981). Studies of relationships between lumbar disc pressure, myoelectric back muscle activity, and intra-abdominal (intragastric) pressure. *Spine*, **6**, 98–103.

Pheasant, S. (1986). *Bodyspace. Anthropometry, Ergonomics and Design* (London: Taylor and Francis).

Pheasant, S. and Harris, C.M. (1982). Human strength in the operation of tractor pedals. *Ergonomics*, **25**, 53–63.

Rohmert, W. (1973). Problems in determining rest allowances. *Applied Ergonomics*, **4**, 91–5; **4**, 158–162.

Rohmert, W., Wangenheim, M., Mainzer, J., Zipp, P. and Lesser, W. (1986). A study stressing the need for a static postural force model for work analysis. *Ergonomics*, **29**, 1235–1249.

Samuelson, B., Wangenheim, M. and Wos, H. (1987). A device for three-dimensional registration of human movement. *Ergonomics*, **30**, 1655–1670.

Schultz, A.B. and Andersson, G.B.J. (1981). Analysis of loads on the lumbar spine. *Spine*, **6**, 76–82.

Schultz, A., Andersson, G., Örtengren, R., Haderspeck, K. and Nachemson, A. (1982). Loads on the lumbar spine. *Journal of Bone and Joint Surgery*, **64A**, 713–720.

Schultz, A., Haderspeck, K., Warwick, D. and Portillo, D. (1983). The use of lumbar trunk muscles in isometric performance of mechanically complex standing tasks. *Journal of Orthopaedic Research*, **1**, 77–91.

Stokes, I.A.F. and Moreland, M.S. (1987). Measurement of the shape of the surface of the back in patients with scoliosis. *Journal of Bone and Joint Surgery*, **69A**, 203–211.

Svensson, O.K. (1987). On quantification of muscular load during standing work. A biomechanical study. Dissertation from the Kinesiology Research Group, Department of Anatomy, Karolinska Institute, Stockholm, Sweden.

Tracy, M.F. (1988). Strength and posture guidelines: a biomechanical approach. Ph.D. Thesis, University of Nottingham.

Tracy, M., Haslegrave, C.M. and Corlett, E.N. (1987). Automating the measurement and biomechanical analysis of posture. In *New Methods in Applied Ergonomics* edited by J.R. Wilson, E.N. Corlett and I. Manenica (London: Taylor and Francis), pp. 267–272.

Tracy, M.F., Gibson, M.J., Szypryt, E.P., Rutherford, A. and Corlett, E.N. (1989). The geometry of the lumbar spine determined by magnetic resonance imaging. *Spine*, **14**, 186–193.

Troup, J.D.G., Foreman, T.K., Baxter, C.E. and Brown, D. (1987). The perception of back pain and the role of psychophysical tests of lifting capacity. *Spine*, **12**, 645–657.

Wells, N. (1985). *Back Pain* (Office of Health Economics).

Winter, D.A. (1979). *Biomechanics of human movement.* (New York: Wiley Interscience).

Chapter 24

Techniques in mental workload assessment

Najmedin Meshkati, Peter A. Hancock and Mansour Rahimi

Introduction

The question of mental workload (MWL) assessment is relatively new and important; new in comparison to companion techniques for the assessment of physical load, whose origins are the contemporary of the Industrial Revolution, and important in that an increasing proportion of work taxes the information processing capabilities of operators, rather than their physical capacity. It is the load placed upon such cognitive capabilities that mental workload assessment is designed to measure. The techniques used to measure this load are the primary focus of this chapter. Four contemporary groups of methods, each comprising several techniques, are evaluated below from the view point of their practicality and utility for the working ergonomist:

1. *Primary task measures*. These are probably the most obvious method of mental workload assessment. For example, if we want to know how driving is affected by differing task demands, e.g. traffic conditions, fatigue or lane width, we should be able to utilize the driving performance itself as a criterion (Hicks and Wierwille, 1979).

2. *Secondary task measures*. A secondary task is a task which the operator is asked to do in addition to his/her primary task. If he/she is able to perform well on the secondary task, this is taken to indicate that the primary task is relatively easy; if he/she is unable to perform the secondary task and at the same time maintain the primary task performance, this is taken to indicate that the primary task is more demanding (Knowles, 1963). The difference between the performances obtained under the two conditions, with and without inclusion of the primary task, is then taken as a measure, or index, of the workload imposed by the primary task.

3. *Subjective rating measures*. These include direct or indirect queries of the individual for their opinion of the workload involved in a task. The easiest

way to estimate the mental workload of a person who performs a certain task is to ask him/her what he/she feels about the mental load level of the task.

4. *Physiological (or psychophysiological measures).* Individuals who are subjected to some degree of mental workload commonly exhibit changes in a variety of physiological functions. As a result, several researchers have advocated the measurement of these changes to provide an estimate of the level of workload experienced.

Primary task measurement

There are several methodological approaches to the measurement of performance, or system output measures (Chiles and Alluisi, 1979). From the practical standpoint the *analytical approach* appears most appropriate. Welford's (1978) concept of the analytical approach looks in detail at the actual performance of the task to be assessed, examining not only overall achievement, but also the way in which it is attained. The advantage of this method is that the various decisions and other processes that make up performance are considered in the context in which they normally occur, so that the full complexities of any interaction between different elements in the task can be observed. For instance, it has been shown that the presence in a cycle of operations of one element which has to be carried out more deliberately than the rest slows the performance of all the elements in the cycle, so that a prediction of the time taken made on the basis of the time required to carry out each element in isolation would be too low.

The analytical approach requires that several scores be taken of any one performance. For example, the component parts of a complex cycle of operations should be measured separately, errors may be recorded as well as time taken, and different types of errors need to be distinguished. Welford (1978) has argued that the greatest value of this approach is probably that it enables the more subtle effects of workload to be examined by showing the strategies in use, such as maintaining a balance between speed and accuracy or between errors of omission and commission, and methods of operation which in various ways seek to increase efficiency and to reduce excessive load. This approach has two difficulties. First, the detailed scores required may be difficult to obtain for tasks such as process monitoring in which most of the decisions made do not result in any overt action, and second that even where there is sufficient observable action, recording may have to be elaborate and analysis of results laborious.

Synthetic methods comprise another major approach to performance measurement as a MWL assessment technique. According to Chiles and Alluisi (1979) these methods start with a task analysis of the system, in which the proposed operating profile is broken down into segments or phases that are relatively homogeneous with respect to the way the system is expected to operate. For each phase, the specific performance demands placed on the operator are then identified through task-analytic procedures. Peformance times and operator reliabilities are assigned to the individual tasks and subtasks on the basis of available or derived data. The information on performance time is then accumulated for a given phase and the total is compared with the predicted duration of the phase. This comparison of required time with available time can be employed as an index of workload.

Welford's (1978) approach to this method is relatively similar. He considered loads imposed by task demands (e.g. data gathering, choices, actions) which are separate, in terms of time taken or other measures, either in the laboratory or in artificially simplified work conditions. The total load is then assessed by adding the components together. This is the approach on which standard time analysis of manual work is based. For the assessment of mental workload based on this approach, the work of Kitchin and Graham (1961) is considered of seminal importance. They questioned the applicability of conventional work measurement techniques to mental workload and were able to show that those techniques could give a satisfactory quantitative expression to three types of mental activity on a time basis. The first type of mental activity involves the direction and co-ordination of muscular activity during all defined physical movements of the body, including highly manipulative work. The second type of mental activity is that of perception (taken to mean the actual receipt of information by any of the senses), while the final activity is that of the senses searching for a random (but likely) signal demanding instant action. In summary, work-measurement, although predominantly a technique for measuring work which can be observed as being carried out physically, satisfactorily recognizes for practical purposes those mental activities which are an integral part of some physical activity and which are defined by the physical activity which they accompany in time.

The third major approach to mental workload assessment via performance analysis is the *multiple measurement of primary task performance*, which is a composite technique. These techniques might be considered useful for workload assessment when individual measures of primary task performance do not exhibit adequate sensitivity to operator workload because of operational adaptivity due to perceptual style and strategy. According to Williges and Wierwille (1979), using multiple measures in a combined analysis has the beneficial effect of reducing the likelihood that important strategy changes will go unnoticed. There are numerous studies that report both positive and negative results for the application of multiple measures which can be found in this paper.

Although use of the multiple measures approach is potentially advantageous, it may have a detrimental effect similar to noise amplification. By measuring a large number of variables, it becomes more likely that some will not change reliability as a function of workload. Another area of concern is the differential sensitivities of the individual members of the multiple measures to the different aspects of the task. For instance, there might be two or more different measures which appear almost equal in ability to discriminate change in operator workload, but which may in fact have large differences in sensitivity. Consequently, this might cause disarray in scaling of the different workloads.

The lack of sensitivity of performance measures to changes in mental workload levels is one of the major problems of these methods. This issue has been addressed by many authors such as Gaume and White (1975) and Gartner and Murphy (1975). Gaume and White argue that the level of mental workload may increase while performance is unchanged so that performance may not be a valid measure of workload. Gartner and Murphy propose that an operator may show equal performance for two different configurations, but in reality effort on one system may greatly exceed the effort on the other. Generalization to different task situations poses an additional problem in the application of performance measures, since for each experimental situation a unique measure must be developed (Hicks and Wierwille, 1979). Williges and Wierwille (1979) also refer to this point and argue that the measures of performance of the primary task are task-specific. Each time a new situation is examined, new measures must be developed and tested. In several other techniques of workload assessment, the same measures can be used regardless of the application.

Williges and Wierwille (1979) cite numerous studies which support the same concept; namely, that no substantial change occurs in the primary task as a function of workload. In general, the studies cited appear to have been performed at workload levels where the operator had sufficient reserve capacity to adapt to the increased load. Rouse (1979) also refers to long-term performance measures as the indicator of relative workload, although this would seem to provide at best an ordinal scale of workload measurement. Further, unless one is willing to assume that humans always operate to capacity and that all humans have the same capacity, the performance-based workload scales would only reflect the states of particular individuals for which the data were collected. In other words, interindividual comparisons may not be valid for this measure.

Secondary task measures

The concept of using a secondary task as a measure of mental workload is grounded on the assumption of the limited channel capacity of the human information processing system (Welford, 1959; Kalsbeek, 1968, 1973). This approach assumes that an upper limit exists on the ability of a human operator to gather and process information. The secondary task is a task which the operator is asked to do in addition to the primary task. There are two types of secondary tasks; 'loading' and 'subsidiary' (or 'non-loading'). If the subject is instructed to aim for error-free performance on the secondary task at the expense of the primary task, the secondary task is called a loading task

(Rolfe, 1976). In this case the operator must always attend to the secondary task which may cause performance degradation on the primary task (Sheridan and Johannsen, 1976). If the subject is instructed to avoid making errors on the primary task, the secondary task is called a non-loading or subsidiary task. In this case the operator attends to the secondary task when time is available.

According to Knowles (1963), one of the best ways of measuring operator load is to have the operator perform an auxiliary or secondary task at the same time as performing the primary task under evaluation. If the operator is able to perform well on the secondary task, this is taken to indicate that the primary task is relatively easy. If he/she is unable to perform the secondary task, and at the same time maintain primary task performance, this is taken to indicate that the primary task is more demanding. The difference between the peformance obtained under the two conditions is taken as a measure or index of the workload imposed by the primary task. There are two related, but different, reasons for using the secondary or loading task. The first one, according to Knowles (1963), is to compensate for any deficiency in the loading of the primary task and to stimulate aspects of the total job that may be missing. Therefore, the secondary task is used simply to bring pressure on the primary task with the idea that as the operator becomes more heavily stressed, performance on difficult tasks will deteriorate more than performance on easy tasks. In the first application of the secondary task, the emphasis is upon stressing the primary task. Differences in operator workload are indicated by differences in the primary task performance measures taken under the stress induced by the auxiliary task. In the second application of the secondary task as a subsidiary task, the auxiliary task is used not so much with the intention of stressing the primary task as with intention of finding out how much additional work the operator can undertake while still performing the task to meet satisfactorily the system criteria.

Some examples of secondary tasks are arithmetic addition, repetitive tapping, choice reaction time, critical tracking tasks and cross-coupled dual tasks. Varying combinations of these tasks have been employed by many investigators in their measurement of operator workload. Ogden *et al.* (1978) reviewed 144 experimental studies which used secondary task techniques to measure, describe or characterize operator workload. In addition to the secondary task techniques that have been reported by Ogden, his colleagues and Williges and Wierwille (1979), there are two other related (secondary task) techniques. The first of these is occlusion and the second is handwriting analysis.

Occlusion is actually a time-sharing technique. However, in occlusion, the time-sharing is forced rather than voluntary. The operator is given samples over time of the visual information required to perform the primary task; the time-sharing is thus accomplished by suppressing the information input (Hicks and Wierwille, 1979). The usual method is to block the operator's visual input from the display. Senders *et al.* (1967) and Farber and Gallagher

(1972) used this technique in measuring the attentional demand of driving an automobile and reported some positive results. However, according to Hicks and Wierwille (1979) the occlusion method is not particularly sensitive and is more intrusive compared to other techniques in a simulated automobile driving situation. *Handwriting analysis* is another potential measure of mental workload because of its deterioration due to distraction of the individual by other tasks. Kalsbeek and Sykes (1967) utilized handwriting as a secondary task while the primary task was to respond, via pedal depression, to a random series of binary choice signals which were presented to either the auditory or visual senses. They were able to show a step-by-step disintegration of writing performance provoked by the increasing number of binary choices.

The secondary task as a mental measurement technique has many shortcomings. Perhaps the most difficult aspect of secondary task methodology for assessing workload is intrusion. When the secondary task is introduced, performance on the primary task is known to be modified and usually degraded (Williges and Wierwille, 1979). This problem has been addressed by other authors such as Welford (1978), who regarded the extra load imposed by the secondary task as a factor that might produce a change of strategy in dealing with the primary task and consequently distort any assessment of the load imposed by the primary task alone. Brown (1978) argued that since the dual task method is essentially a resource-limiting device (human-processing resources being limited), interference should occur within the processing mechanisms, rather than at sensory input or motor output. He claimed there is empirical support for the idea that interference is maximal at the level of response selection. Brown (1978) indicated that the dual task interference is greater when the tasks share the same response modality than where responses occupy different modalities.

The nature of the primary task and its informational load and structural characteristics can cause problems of efficiency and a reduction in the utility of the secondary task. Workload may be largely a function of the structural characteristics of a task rather than of the informational load imposed by its component parts (Brown, 1978). Therefore, the more interesting tasks (i.e. those which most closely assimilate real-life situations) may be relatively inaccessible to study by the dual task methodology. This fact and other expected and unexpected interactions between certain tasks, drove Ogden *et al.* (1978) to point out that the choice of the secondary task is problematic (see also McCormick and Sanders, 1982).

The question of individual differences in secondary task performance has been addressed by some investigators through association with personality constructs. The introduction of an additional task may increase arousal which has been shown to affect differing personality types in contrasting ways (Gibson and Curran, 1974; Huddleston, 1974). Motivation is another related factor which may play an important role in secondary task performance (Kalsbeek and Sykes, 1967). Knowles (1963) attempted to provide a set of criteria against which to judge the desirability of a secondary task. These criteria included unobtrusiveness with respect to the primary task, ease of learning, self-pacing (in order for the secondary task to be neglected in maintenance of primary task performance) and compatibility with the primary task. Ogden *et al.* (1978) added sensitivity and representativeness to the above set in order to address a wide range of human abilities and functions. Kalsbeek (1971) suggested the dual task method for use in two ways. First, in the traditional way, measuring the so-called spare mental capacity, and second, in experiments where the main task, to which preference has to be given, is a simple or repetitive one. For instance, a binary choice task can be regarded as a stress condition in the performance of a secondary task.

The various problems outlined above point to Brown's (1978) conclusion that:

> The dual task method should be used only for the study of individual difference in processing resources available to handle workload . . . If it is so used, it should probably be in the form of an additional, secondary task, presenting discrete stimuli of constant load, on a forced paced schedule, and competing for processing resources only.

Subjective rating measures

As mentioned in the introduction, the easiest way to estimate the mental workload of a person who performs a certain task is to ask what he/she subjectively feels about the load of the task. Sometimes a list of key words or definitions describing different levels of load can be given. The subject then has to rate the load with reference to these levels (Sheridan and Stassen, 1979). Sheridan (1980) has also argued that mental workload should be defined in terms of subjective experience, and he continued: 'subjective scaling is the most direct measure of such subjective experience.'(For a general discussion of subjective assessment methods see Sinclair in this book.)

Subjective estimates of load have often been obtained through either the use of *rating scales* or *interviews/questionnaires*. A widely used rating scale in systems evaluation is the Cooper and Harper (1969) scale, originally developed to measure the handling characteristics of aircraft by using the subjective reports of test pilots. The Cooper–Harper scale was designed primarily to assess flight characteristics and the descriptors of this scale pertain to 'flyability' of an aircraft. Therefore, it can be applied to manual control tasks (Moray, 1982). However, according to Williges and Wierwille (1979), if this scale was used for workload assessment the assumption must be made that handling difficulty and workload are directly related. The assumption is that if a pilot states that an aircraft is difficult (or impossible) to fly, this is equivalent to saying that the task of flying imposes a very heavy or unsupportable load (Moray, 1982). The modified Cooper–Harper scale (Rahimi and Wierwille, 1982; Wierwille and Casali, 1983; Wierwille *et al.*, 1985b), which is a modified version of the original Cooper–Harper, is considered as further development for the subjective measurement of mental workload. This scale is applicable to a wider variety of task workloads, especially for systems which load perceptual, mediational and communication activities (Wierwille *et al.*, 1985a).

Wewerinke (1974) has confirmed the validity of the Cooper–Harper scale as an indicator of MWL. He was able to report an extremely high correlation coefficient (0·8) between subjective difficulty rating and objective workload level in his study. Gartner and Murphy (1975) echoed the above idea and referred to the positive qualities of the Cooper–Harper scale as operational relevance, convenience and unobtrusiveness. Some of the other examples of application of rating scales can be found in the work of Williges and Wierwille (1979).

Another widely used subjective rating technique is SWAT (*Subjective Workload Assessment Technique*), which was designed specifically to measure operator workload in a variety of systems for a number of tasks. It uses the conjoint measurement technique (Nygren, 1982) to combine ratings on three different dimensions of workload: time load, mental effort load, and stress load. It should be noted that SWAT, like the Cooper–Harper scale, can be applied to workload in a number of different settings, although cockpit evaluation has been most common.

Application of subjective-rating technique in strictly cognitive tasks has been addressed by Borg *et al.* (1971) who were able to achieve a high correlation between subjective and objective measures of difficulty. The strictly cognitive tasks, unlike the manual control tasks and time-stressed tasks (e.g. some signals must be processed before the processing of predecessors is completed), are single-trial tasks. The subject is not under pressure associated with a continuous or arbitrary stream of signals that may arrive before he/she has finished dealing with an earlier signal. However, Phillip *et al.* (1971) argued that in time-stress tasks it is not possible to differentiate unambiguously between the criteria of stress time and difficulty of the control task, based upon subjective rating methods. Therefore, the subjective feeling of difficulty in work processing seems to be essentially dependent on the time-stress involved in performing the task. Gaume and White (1975), in their study of mental workload evaluated the subjective estimates of stress levels obtained during the experiment, and they considered them as potentially valid and reliable indicators of mental workload.

The second approach to subjective rating is through the application of interviews/questionnaires. The procedures used in this approach are not as structured as rating scales. They range from completely open-ended debriefing sessions to self-reporting logs of stressful activities, from carefully chosen questionnaire items (Williges and Wierwille, 1979). Usually, interviews and questionnaires have been used primarily as supplementary measures to other techniques. For example, Sherman (1973) demonstrated a high correlation between subjective measures of workload and various physiological measures.

The advantages of the approach stem from its unobtrusiveness and extreme ease of application.

Since the subjective rating of the difficulty of a task is primarily a function of the raters' perception, the concept of perceived difficulty has to be given importance and analysed directly. Audley *et al.* (1979) proposed that the perceived difficulty of scaling the task demands, should be the primary consideration and attempts should be made to dissociate this from other facets of subjective aspects of workload. The perceived difficulty of a task might alter the human operator's attitude to it. This in turn could affect the time operators would be prepared to spend and the level of confidence in their decisions (Moray, 1982). The perceived difficulty for the individual is influenced by at least three groups of factors. The first group deals with the content of long-term memory including both general experience and memories of similar tasks. The second group is of background factors such as personality traits, habits and general attitudes including likes and dislikes, aspiration and expectation levels. The third group of factors represents momentary conditions, e.g. one's emotional state, general fatigue, motivation and the importance ascribed to the task, as well as actual anticipated success or failure (Borg *et al.*, 1971).

The subjective rating of task difficulty could also be affected by the situation and job as a whole rather than by only the task induced or individual rater's factors. Borg (1978) proposed that it is necessary to point out the set of factors which seem to cause the experience of the difficulty in one job which may be different from those in another job. Finally, it should be noted that general individual differences and differences in the adaptivity of the operator to the system, the task and the resultant impression of the task as viewed by the operator, causes higher than normal rating (Williges and Wierwille, 1979).

Physiological measures

Individuals engaged in cognitive activities provide indirect indices of their level of effort through changes in the status of a number of physiological systems. Ursin and Ursin (1979) recognized that these physiological methods do not measure the imposed load but rather they give information concerning how the individuals themselves respond to the load and, in particular, whether they are able to cope with it. In what follows we have attempted to differentiate the main current physiological approaches on the basis of their validity as a measure of workload and their applicational utility. From this analysis, the most practical method to emerge is heart rate or one of its derivatives (e.g. Kalsbeek, 1968) and the most valid measures are changes in central nervous system (CNS) activity (e.g. Wickens, 1979). We suggest that a tympanic temperature measure, taken in the ear canal, can provide a potentially useful compromise in the trade-off between the concern for validity and practicality (see Hancock and Brainard, 1981; Hancock and Dirkin, 1982; Hancock, 1983, 1984). The interested reader can find specific details of differing physiological measures in the review articles of Ursin and Ursin (1979), Williges and Wierwille (1979), and Hancock *et al.* (1985).

Use of physiological measures as indicators of mental workload is influenced by a combination of several factors, such as the cost of both hardware and software to operate the equipment, the training level of the personnel who administer the physiological tests, environmental conditions of the workplace and the willingness of the employees to be connected to a physiological recording mechanism. Due to problems caused by various combinations of these factors at the present time, physiological measures are among the least practical methods of mental workload assessment for use in complex machine–person systems. We should note, however, that technological innovations such as telemetric monitoring, have reduced some of the problems associated with these measures and use of these techniques in the working environment may be realized in the very near future. Two frequently used physiological methods of mental workload measurement are *event-related potentials* and *heart rate variability*.

Event-related potentials

Event-related potentials (ERPs) are fluctuations in the activity of the nervous system recorded in response to environmental stimulation in association with psychological processes, or in preparation for motor activity (Martin and Venables, 1980). Repeated presentations of a physical stimulus elicit waves which are subsequently signal-averaged to reduce or eliminate random variation and to yield a stimulus locked wave or ERP. Various ERP components have been taken to reflect information processing activity, and change in mental workload. Workload inferences are based upon the amplitude and latency elements of the elicited wave. For example, it has been observed that the amplitude of the wave at a latency of 200 ms following stimulus onset (P2) and the overall maximum power in the evoked response provided a metric of subjective difficulty of task performance (Spyker *et al.*, 1971).

The major advantage of ERPs is their representation as direct reflections of the information processing activity of the operator. In addition, ERPs are multivariate measures characterized by differing latency peaks which provide a considerable amount of information per observation. Also, as ERPs are elicited by discrete events in the environment there is more specificity between stimulus and response compared to other more global methods. Wierwille (1979) acknowledged that the dependence of the ERP upon the operator's perceived utilities, attitude and understanding, in addition to the imposed load, represents both an advantage and a limitation with respect to its power as a workload assessor.

The major disadvantages of this approach include: first, the single recorded trial response contains a high noise-to-signal ratio (Wickens, 1979). Consequently, either multiple trial recording and subsequent signal averaging, or filtering and application of analysis such as template matching, is required to extract meaningful information from single observations (Squires and Donchin, 1976). Second, the response may be contaminated by motor artifacts. Also, because of individual differences, 'calibrations' must be undertaken for each different operator. In practical terms, the technique requires considerable supporting instrumentation (e.g. computer facilities) and trained personnel for operation and interpretation. These limitations, however, represent technical barriers which are subject to constant change. Therefore, ERPs potentially represent the most promising physiological measure of mental workload for future exploitation.

Heart rate variability

If ERPs represent the most valid physiological measure, then measures pertaining to heart rate and its derivatives are currently the most practical physiological method of assessing imposed mental workload. Among such measures perhaps none is more thoroughly investigated than that of heart rate variability (HRV).

In research involving this measure, Kalsbeek (1971) noted a gradual suppression of the heart rate irregularity due to increases in the difficulty of a task. In consequence, it was posited that such a measure could be used to reflect mental workload. Several empirical investigations attest to the strength of this assertion (Kalsbeek, 1968, 1973). Since such observations were first made, there have been many experimental studies in which the connection between HRV and mental workload has been observed. Detailed reviews of these efforts are available (Wierwille, 1979; Meshkati, 1983).

Measures of HRV have been assessed through the use of three major calculational approaches (see Meshkati, 1988a): (1) scoring of the heart rate data or some derivative (e.g. standard deviation of the R–R interval); (2) through the use of spectral analysis of the heart rate signal; and (3) through some combination of the first two methods. There are two advantages to such a measure—a relative and an absolute advantage. First, the absolute advantage refers to the sensitivity of the measure to change in mental workload as demonstrated in the previously cited investigations. The second advantage is its practical utility and relative simplicity in both administration and subsequent interpretation, when compared to alternate physiological techniques. However, it should be acknowledged that since HRV is a particularly sensitive physiological function, it is vulnerable to potential contamination from the influence of both stress and the ambient environment (Kalsbeek, 1971). For the interested reader, further information on the details of each of these measures can be found in reports in Meshkati *et al.* (1984), and in Moray (1979).

Some other physiological means that may be of interest to ergonomics practitioners are pupil diameter, body chemical analysis and auditory canal temperature (ACT). For instance, Wierwille and Connor (1983) examined five different physiological measures (mean pulse rate, pulse rate variability, respiration rate, pupil diameter, and voice pattern) elicited by digit shadowing and mental arithmetic tasks. According to their results, only the mean pulse rate demonstrated some limited 'sensitivity' to some of the differences in the psychomotor load conditions. In a related study, Casali and Wierwille (1983) monitored respiration rate, heart rate mean, heart rate standard deviation, pupil diameter, and eye blinks and concluded that the sole physiological measure to display sensitivity to changes in communications load is the pupil diameter measure.

There are studies which utilized relatively unconventional and novel physiological approaches to assess human mental workload. Hyyppa *et al.* (1983) investigated psychoneuroendocrine responses to mental workload. They were able to find a significant decline of the cortisol and prolactin levels of subjects undergoing psychologically demanding achievement-oriented tasks. Loewenthal (1983) proposed alveolar gas concentration level could be a 'cleaner' physiological measure than the others (e.g. respiratory arrhythmia). In his extensive study, Loewenthal cited several studies that tried to demonstrate a relationship between alveolar gas pressures and mental workload.

Hancock (1983) and Hancock *et al.* (1985) considered tympanic temperature or, more correctly, deep auditory canal temperature (ACT) as an alternative measure which circumvents certain problems associated with other physiological measures. It has been observed that subjects beginning work on a simple mental task, after a period of quiescence, exhibit small but constant increases in ACT (Hancock, 1983). Also, subjects encountering a number of different computational problems embedded in a series of simple mathematical additions show an increase in ACT (Hancock and Brainard, 1981).

Relevance and coping with individual differences

A common and important aspect of all mental workload assessment methods is their relative sensitivity to individual differences. Moray (1984) asserted that:

> Individual differences in workload research is far more important than has hitherto been acknowledged. Without taking this into account we are seriously delaying the development of a useful measure.

Many investigators who failed to obtain significant results in applying mental workload measurement techniques have suggested that either the sample population must be homogenized or, alternatively, personality traits, individual differences, and other related factors should be incorporated into the experimental design. The following is a summary of such expert recommendations (for further discussion see Meshkati and Loewenthal, 1988a).

Kitchin and Graham (1961) refer to the character of the human operator as 'a very important area of concentration,' without which the assessment of operator's physical and mental abilities are of little value to industry. Mulder and Mulder–Hajonides Van Der Meulen (1973) acknowledged the large differences among subjects and therefore recommended single subject analysis, and Leplat (1978) stated that the characteristics of personality could intervene in a far from negligible manner in regard to workload. Hamilton *et al.* (1979) tried to analyse the activation responses as a function of the task characteristics. Furthermore, they acknowledged that the subject's active information processing involves personality traits. Hopkin (1979) considered personality variables as potentially relevant to mental workload.

According to Firth (1973), in the real-life working environment individual differences in operators' characteristics very much influence information processing of the individuals. These differences arise from a combination of past experience, skill, emotional state, motivation and the estimation of risk and cost in a task. The influence of these individual differences is important, since many of these factors have been shown to directly influence cardiac responses. There are some other indications of the relationship between personality traits and physiological reaction parameters, e.g. Rotter's (1966) internal-external locus of control and heart rate control. Ray and Lamb (1974) and Gatchel (1975) found that internal locus of control subjects were better able to increase their heart rate as compared with their external counterparts.

Duffy (1962) reported that individual differences in responsiveness have been observed in many forms, in the frequency and amplitude of rhythms in the EEG, in the occurrence of 'spontaneous' changes in skin resistance, peripheral blood flow, heart rate, muscle tension and other functions. The author referred to the work of Armstrong (1938) who detected correlation between cardiovascular reactivity and emotional stability in 700 Army Corps candidates. Offerhaus (1980), based upon his study of hospital staff (normal subjects) and psychiatric patients, concluded that by employing the concept of heart rate variability, it is possible to differentiate between two pairs of groups of subjects: first, the high anxiety group from the anxiety one (i.e. psychotic patients from non-patients), and second, the stress reactor group from the non-stress reactor group (i.e., acute patients and neurotic staff from chronic patients and stable staff).

The issue of individual differences and psychological variables and their substantial effects on automatic responses has been addressed by Cleary (1974), Van Egeren *et al.* (1972) and Sutton and Tueting (1975). The concept was experimentally evaluated and confirmed by Bryson and Driver (1969). They found that 'cognitively complex' subjects manifest higher GSRs in attending to stimuli. Lykken (1968), in the same regard, referred to two additional areas of consideration of individual differences, as the tonic psychophysiological level and phasic response to specific stimuli. The effect of personality traits and individual differences on the performance of a mental task bears a great amount of significance. Hopkin (1979) stated that on many occasions, individual differences have precluded general judgements on whether the task induced workload is excessive as distinct from high. He considered this typical inability to generalize the findings as mostly due to the fact that any given individual characteristic becomes a pertinent factor in workload only insofar as the task being peformed brings that characteristic into play. Schroder *et al.* (1967) also reiterated this fact by arguing that if the task requires the processing of large amounts of descriptive information, and if this information must be integrated into a flexible, comprehensive system, then it can be expected that the 'integratively complex' person would perform better than integratively simple persons. They also postulated and later demonstrated that superior performance may be expected of a simple person, in an open situation, if the environment is complex and the criterion is simple.

Thackray *et al.* (1973) studied the role of personality in performance decrement and attention. Their results indicated that individuals scoring high on a distractibility scale (i.e. extrovert) found it difficult to maintain a uniform mode of performance. This group of subjects exhibited increasing lapses in attention, while introverted ones failed to show any evidence of a decline in attention.

According to an experimental study of mental workload by Meshkati and Loewenthal (1988b), operators' individual information processing behaviour affects their sinus arrythmia and subjective rating of task difficulty.

Wickens (1979) also regarded the relatively large differences among subjects in time sharing abilities as the cause of substantial variance in dual-task performance. His proposal to tackle this problem was to 'calibrate' particular workload measurement techniques for different operators. Furthermore, with reference to Pew (1970), who recommended that at the same time these individual differences might actually be exploited to enhance system performance by employing them to provide guidelines for merging operators to specific systems, or by modifying systems to the limitations and strengths of individual operators, Kahneman (1973) rates a system possessing these qualities as 'perfect'.

Guidelines for use

In measuring human mental workload it is extremely important to define the measure and the nature of the loading task as accurately as possible. Suppose, for example, that an investigator uses the secondary task of time estimation as a means of assessing mental workload. Time estimation is a secondary task technique for measuring the spare mental capacity of the human operator while performing a primary task. Estimating a 10-s time interval is a popular approach for this measure. First, the investigator has to define the actual measure. Next, the procedure and equipment by which this measure is to be taken should be precisely defined. Otherwise, another investigator may obtain different results because of a difference in the type of measure and/or procedure.

In time estimation, to continue the example, the procedure includes many aspects, such as instructed interval, method of prompting, method of responding, amount of training, instructions regarding counting, and instructions regarding relative importance of the estimation task compared with the primary task. Measure definition includes type of estimates to be included (e.g. only those actively initiated and actively completed; or perhaps those actively initiated whether completed or not, with all incompleted estimates scored as the maximum interval between prompts). Measure definition also includes the types of computation, such as mean, median, standard deviation, or root mean square value of the estimated intervals. For example, our experience with these experiments has shown that standard deviation is by far the most sensitive type of computation for time estimation. Also, there are no guarantees that the most sensitive computation remains viable for all task-loading situations. The following is a short description of guidelines and pitfalls in application of some mental workload measures.

Primary task measures

One difficulty in application of these measures is determining which component of the task is paramount to the overall task performance. In particular, tasks which are complex and multidimensional may need multiple primary task measures. It may even be necessary to divide the task components into different classes of performance (e.g. Berliner *et al.*, 1964) and assign primary measures to each category. That is, to break the global operator tasks into individual task elements and to determine what subset of those elements drives the workload metric.

A potential problem with application of primary task measures is intrusiveness, particularly in field evaluations. Another problem is in the interpretation of the results. Primary task data may reflect a wide variety of influences such as motivation and learning effects. The effects of training and the ability of the operators to muster more effort may reduce the sensitivity of this measure to the real changes in mental workload.

Secondary task measures

In general, secondary task measures are relatively more expensive and difficult to administer. One reason is that both primary and secondary task performances need to be monitored and measured at the same time. One study even found a significant effect of secondary task measurement on the primary task performance of pilots while flying a simulator (Wierwille *et al.*,

1985b). This may indicate a certain degree of danger in measuring mental workload for real-world tasks that require critical operator attention. Also, the secondary task measures should not be combined with subjective measures, because operators may inadvertently include loading due to the secondary task in their ratings.

Subjective rating measures

There has been significant attention given to the use of subjective measures, mostly in the form of opinion scales. Subjective measures are used to assess the mental difficulty of the assigned task. One of the most useful opinion scales, the modified Cooper–Harper scale, shows a consistent level of sensitivity in tasks which have perceptual and cognitive components. Yet it requires after-the-fact evaluation of the task difficulty, using a guide scale. It is applicable to a wide variety of critical tasks (Wierwille and Casali, 1983). Most opinion scales are easy and inexpensive to administer. However, care must be given to the interpretation of these scales since they contain words that may be interpreted differently under different task situations.

Physiological measures

In using these measures, care must be exercised in their selection and application. In general, the sensitivity of these measures is task dependent (Wierwille *et al.*, 1985b). For example, heart rate is one of the simplest physiological measures used in a number of early studies, yet the results indicate lack of generalizability, sensitivity and reliability across some mentally demanding tasks. Another difficulty with this measure is the nonmonotonicity of some physiological measures. That is, the measure may indicate increase in mental activity when task difficulty increases from low to medium, but it may show decrease after the workload continues to increase (Noel, 1974).

It should be mentioned, however, that physiological measures might be more useful in assessing the effects of time on task and task 'strain'. Also, in regards to intrusion, physiological measures have been fairly successful indicators without influencing the primary task performance of the operators.

Evaluation criteria

The following criteria are useful in evaluating which mental workload assessment technique is appropriate for which particular setting. The listing below provides general guidelines which are elaborated in the sections on specific techniques (Meshkati, 1988b).

1. *Validity.* The chosen mental workload measure should satisfy three validity constraints, those of content, predictability and construct.

2. *Reliability*. The measure should provide stable and repeatable results across repeated administrations.

3. *Sensitivity*. Sensitivity refers to the capability of a technique to discriminate significant variations in the workload levels imposed by a task or group of tasks (Eggemeier, 1985).

4. *Diagnosticity*. Diagnosticity refers to the capability of a technique to discriminate the amount of workload imposed on different resources or capabilities of the human operator (Eggemeier and O'Donnell, 1982; Wickens 1984).

5. *Intrusion*. The mental workload measurement technique should not interfere with and/or cause degradations in the task being performed.

6. *Focus*. The measurement technique should be focused only on the changes in mental workload levels and should not reflect changes in status due to artifacts from variation in environmental conditions.

7. *Ease of field utilization*. The chosen technique should be robust and easy to administer within the constraints of the work environment under consideration. Factors which contribute to making a technique cumbersome include the instrumentation, analyst and operator training and data recording and analysis.

8. *Operator acceptance*. The success of a mental workload measurement technique is largely dependent upon operator acceptance and co-operation. This implies the necessity to understand the psychological profile of the typical end-user population prior to the application of any measurement technique.

By referring to the above criteria, some investigators have advocated a multiple battery approach, using representative elements from each of the four major groups of workload assessment methods. In such an approach, however, the advantage of non-interference is sacrificed for the greater specificity of workload information obtained. In the operation of dynamic person–machine systems, the limits of human mental capabilities need to be considered. The limit of such abilities and the operators' approaching of their own individual load tolerances are of particular importance with respect to safety and efficiency of overall action.

In this brief review, four major groups of mental workload assessment methods, their applications, advantages, and disadvantages and feasibility, have been presented. Subjective rating measures, due to their apparent simplicity, ease of administration and interpretation, seem to be the most widely used by practitioners for the measurement of non-physical workload. However, as mentioned previously, the problem of perceived task difficulty and the resultant artifacts inherent in operators' responses are as yet unresolved issues (cf. Meshkati and Driver, 1984). Until these questions are answered, subjective ratings of task difficulty should be normalized for individual differences and analysed with extreme caution. Also, the validity and reliability of this method could be enhanced by the use of multiple scaling techniques which are validated through cross-checking procedures.

References

Armstrong, H.G. (1938). The blood pressure and pulse rate as an index o emotional stability. *American Journal of Medical Science*, **195**, 211–220.

Audley, R.J., Rouse, W., Senders, J. and Sheridan, T. (1979). Final repor of mathematical modeling group. In *Mental Workload: Its Theory an Measurement*, edited by N. Moray (New York: Plenum Press).

Berliner, C., Angell, D., and Shearer, D.J. (1964). Behaviors, measures, an instruments for performance evaluation in simulated environments Paper presented at the *Symposium and Workshop on the Quantification (Human Performance*, Albuquerque, NM.

Borg, G. (1978). Subjective aspects of physical and mental load. *Ergonomics* **21**, 215–220.

Borg, G., Bratfisch, O. and Dorinc, S. (1971). On the problem of perceive difficulty. *Scandanavian Journal of Psychology*, **12**, 249–260.

Brown, I.D. (1978). Dual task methods of assessing workload. *Ergonomics* **21**, 221–224.

Bryson, J.B. and Driver, M.J. (1969). Conceptual complexity and interna arousal. *Psychonomic Science*, **17**, 71–72.

Casali, J.G. and Wierwille, W.W. (1983). A comparison of rating scale secondary task, physiological and primary task workload estimatio techniques in a simulated flight task emphasizing communications load *Human Factors*, **25**, 623–641.

Chiles, W.D. and Alluisi, E.A. (1979). On the specification of operator c occupational workload with performance-measurement methods. *Huma Factors*, **21**, 515–528.

Cleary, P.J. (1974). Description of individual differences in autonomi reactions. *Psychological Bulletin*, **81**, 934–944.

Cooper, G.E. and Harper, R.P. (1969). *The Use of Pilot Rating in the Evaluatio of Aircraft Handling Qualities*. Report No. ASD-TR-76-19. (Moffe Field, California: National Aeronautics and Space Administration).

Duffy, E. (1962). *Activation and Behavior* (New York: John Wiley).

Eggemeier, F.T. (1985). Workload measurement in system design an evaluation. In *Proceedings of the 29th Annual Meeting, the Human Facto Society*, (Santa Monica, CA: Human Factors Society).

Eggemeier, F.T. and O'Donnell, R.O. (1982). A conceptual framework fo development of a workload assessment methodology. In *Text of th Remarks made at the 1982 American Psychological Association Annue Meeting*. (Washington, D.C.: American Psychological Association).

Farber, E. and Gallagher, V. (1972). Attentional demands as a measure c the influence of visibility conditions on driving task difficulty. *Highwa Research Record*, **414**, 1–5.

Firth P.A. (1973). Psychological factors influencing the relationship betwee cardiac arrhythmia and mental load. *Ergonomics*, **16**, 5–16.

Gartner, W.B. and Murphy, M.R. (1975). *Pilot Workload and Fatigue: , Critical Survey of Concepts and Assessment Techniques* Report Nc ASD-TR-76-19 (Washington DC: National Aeronautics and Spac Administration).

Gatchel, R.J. (1975). Change over training sessions of relationships between locus of control and voluntary heart rate control. *Perceptual and Motor Skills*, **40**, 424–426.

Gaume, J.G. and White, R.T. (1975). *Mental Workload Assessment. III. Laboratory Evaluation of One Subjective and Two Physiological Measures of Mental Workload*. Report MDC-J7024/01. (Long Beach, CA: McDonnell-Douglas Corp.).

Gibson, H.B. and Curran, J.B. (1974). The effect of distraction on a psychomotor task studied with reference to personality. *Irish Journal of Psychology*, **2**, 148–158.

Hamilton, P., Mulder, G., Strasser, H. and Ursin, H. (1979). Final report of the physiological psychology group. In *Mental Workload: Its Theory and Measurement*, edited by N. Moray (New York: Plenum Press), pp. 367–385.

Hancock, P.A. (1983). The effect of an induced selective increase in head temperature upon performance of a simple mental task. *Human Factors*, **25**, 441–448.

Hancock, P.A. (1984). An endogenous metric for the control of perception of brief temporal intervals. *Annals of the New York Academy of Sciences*, **423**, 594–596.

Hancock, P.A. and Brainard, D.M. (1981). *Tympanic Temperature: A Non-invasive Physiological Measure of Workload*. Technical Report (MA: Endeco).

Hancock, P.A. and Dirkin, G.R. (1982). Central and peripheral visual choice reaction time under conditions of induced cortical hyperthermia. *Perceptual and Motor Skills*, **54**, 395–402.

Hancock, P.A., Meshkati, N. and Robertson, M.M. (1985). Physiological reflections of mental workload. *Aviation, Space, and Environmental Medicine*, **56**, 1110–1114.

Hicks, T.G. and Wierwille, W.W. (1979). Comparison of five mental workload assessment procedures in a moving-base driving simulator. *Human Factors*, **21**, 129–143.

Hopkin, V.D. (1979). General discussion based upon interactive group sessions. In *Mental Workload: Its Theory and Measurement*, edited by N. Moray (New York: Plenum Press), pp. 484–487.

Huddleston, H.F. (1974). Personality and apparent operator capacity. *Perceptual and Motor Skills*, **38**, 1189–1190.

Hyyppa, M.T., Aungola, S., Lahtela, K., Lahti, R. and Marniemi, J. (1983). Psychoneuroendocrine responses to mental load in an achievement-oriented task. *Ergonomics*, **26**, 1155–1162.

Kalsbeek, J.W.H. (1968). Measurement of mental workload and of acceptable load: possible applications in industry. *International Journal of Production Research*, **7**, 33–45.

Kalsbeek, J.W.H. (1971). Standards of acceptable load in ATC tasks. *Ergonomics*, **14**, 641–650.

Kalsbeek, J.W.H. (1973). Do you believe in sinus arrhythmia? *Ergonomics*, **16**, 99–104.

Kalsbeek, J.W.H. and Sykes, R.N. (1967). Objective measurement of mental load. *Acta Psychologica*, **27**, 253–261.

Kahneman, D. (1973). *Attention and Effort* (Englewood Cliffs, NJ: Prentice-Hall).

Kitchin, J.B. and Graham, A. (1961). Mental loading of process operations: an attempt to devise a method of analysis and assessment. *Ergonomics*, **4**, 1–15.

Knowles, W.B. (1963). Operator loading tasks. *Human Factors*, **5**, 155–161.

Leplat, J. (1978). Factors determining workload. *Ergonomics*, **21**, 143–149.

Loewenthal, A. (1983). *Alveolar Gas Concentration and Mental Workload*. Technical Report 83−2 (Department of Industrial and Systems Engineering, University of Southern California).

Lykken, D.T. (1968). Neuropsychology and psychophysiology in personality research. In *Handbook of Personality Theory and Research*, edited by E.F. Borgatta and W.W. Lambert (Chicago: Rand McNally), pp. 413–509.

Martin, I. and Venables, P.H. (1980). *Techniques in Psychophysiology* (New York: John Wiley).

McCormick, E.J. and Sanders, M.S. (1982). *Human Factors in Engineering and Design* (New York: McGraw-Hill).

Meshkati, N. (1983). A conceptual model for the assessment of mental workload based upon individual decision styles. (Dissertation). University of Southern California.

Meshkati, N. (1988a). Heart rate variability and mental workload assessment. In *Human Mental Workload*, edited by P.A. Hancock and N. Meshkati (Amsterdam: North Holland).

Meshkati, N. (1988b). Toward development of comprehensive theories of mental workload. In *Human Mental Workload*, edited by P.A. Hancock and N. Meshkati (Amsterdam: North Holland).

Meshkati, N. and Driver, M.H. (1984). Individual information processing behavior in perceived job difficulties: A decision style and job design approach to coping with human mental workload. In *Human Factors in Organizational Design and Management*, edited by H.W. Hendrick and O. Brown, Jr. (Amsterdam: North Holland).

Meshkati, N., Hancock, P.A. and Robertson, M.M. (1984). The measurement of human mental workload in dynamic organizational systems: an effective guide for job design. In *Human Factors in Organizational Design and Management*, edited by H.W. Hendrick and O. Brown, Jr. (Amsterdam: North Holland).

Meshkati, N. and Loewenthal, A. (1988a). An eclectic and critical review of four primary mental workload assessment methods: a guide for developing a comprehensive conceptual model. In *Human Mental Workload*, edited by P.A. Hancock and N. Meshkati (Amsterdam: North Holland).

Meshkati, N. and Loewenthal, A. (1988b). The effects of individual differences in information processing behavior on experiencing mental workload and perceived task difficulty: an experiemental approach. In *Human Mental Workload*, edited by P.A. Hancock and N. Meshkati (Amsterdam: North Holland).

Moray, N. (1979). *Mental Workload: Its Theory and Measurement* (New York: Plenum Press).

Moray, N. (1982). Subjective mental workload. *Human Factors*, **24**, 24–45.

Moray, N. (1984). Mental workload. *Proceedings of the 1984 International Conference on Occupational Ergonomics*, Toronto, Canada, pp. 41–46.

Mulder, G. and Mulder-Hajonides Van Der Meulen, W.R.E.H. (1973). Mental load and the measurement of heart rate variability. *Ergonomics*, **16**, 69–83.

Noel, C.E. (1974). Pupil diameter versus task layout. Master's Thesis. Monterey, California, Naval Postgraduate School.

Nygren, T.E. (1982). *Conjoint Measurement and Conjoint Scaling: a User's Guide (AFAMRL-TR-82-22)*. DTIC No. ADA 122579 (Wright-Patterson Air Force Base, OH: Air Force Aerospace Medical Research Laboratory).

Offerhaus, R.E. (1980). Heart rate variability in psychiatry. In *The Study of Heart Rate Variability*, edited by R.I. Kitney and O. Rompelman (Oxford: Clarendon Press), pp. 225–238.

Ogden, G.D., Levine, J.M. and Eisner, E.J. (1978). *Measurement of Workload by Secondary Tasks: A Review and Annotated Bibliography*. Prepared under contract no. NAS2-9637 to National Aeronautical and Space Administration. (Ames Research Center, Washington DC: Advance Research Resources Organization).

Pew, R.W. (1970). Comments on promotion of man: challenges in sociotechnical systems: design for the individual operator. *Proceedings of the Global Systems Dynamics International Symposium*, Charlottesville, NJ, pp. 59–65.

Phillip, V., Reiche, D. and Kirchner, J. (1971). The use of subjective ratings. *Ergonomics*, **14**, 611–616.

Ray, W.J. and Lamb, S.B. (1974). Locus of control and the voluntary control of heart rate. *Psychosomatic Medicine*, **36**, 180–182.

Rahimi, M. and Wierwille, W.W. (1982). Evaluation of the sensitivity and intrusion of workload estimation techniques in piloting tasks emphasizing mediational activity. *Proceedings of the IEEE International Conference on Cybernetics and Society*, pp. 593–597.

Rolfe, J.M. (1976). The measurement of human response in man vehicle control situations. In *Monitoring Behavior and Supervisory Control*, edited by T.B. Sheridan and G. Johannsen (New York: Plenum Press).

Rotter, J.B. (1966). Generalized expectancies for internal versus external control of reinforcement. *Psychological Monographs*, **80** 1–22.

Rouse, W.B. (1979). Approaches to mental workload. In *Mental Workload: Its Theory and Measurement*, edited by N. Moray (New York: Plenum Press).

Schroder, H., Driver, M. and Streufert, S. (1967). *Human Information Processing* (New York: Holt Rinehart and Winston).

Senders, J.W., Kristofferson, A.B., Levision, W.H., Dietrich, C.W. and Ward, J.L. (1967). The attentional demand of automobile driving. *Highway Research Record*, **195**, 15–33.

Sheridan, T.B. (1980). Mental workload, what is it? why bother with it? *Human Factors Society Bulletin*, **23**, 1–2.

Sheridan, T.B. and Johannsen, G. (1976). *Monitoring Behavior and Supervisory Control* (New York: Plenum Press).

Sheridan, T.B. and Stassen, H.G. (1979). Definitions, models and resources of human workload. In *Mental Workload: Its Theory and Measurement*, edited by N. Moray (New York: Plenum Press).

Sherman, M.R. (1973). The relationship of eye behavior, cardiac behavior and electromyographic responses to subjective responses of mental fatigue and performance on a doppler identification task. Masters Thesis, Naval Postgraduate School, Monterey, California.

Spyker, D.A., Stackhouse, S.P., Khalafalla, A.S. and McLace, R.C. (1971). *Development Techniques for Measuring Pilot Workload*. Technical Report NASA CR-1888.

Squires, K.C. and Donchin, E. (1976). Beyond averaging: the use of discriminant functions to recognize event related potentials elicited by single auditory stimuli. *EEG Clinical Neurophysiology*, **41**, 449–459.

Sutton, S. and Tueting, P. (1975). The sensitivity of the evoked potential to psychological variables. In *Research in Psychophysiology*, edited by P.H. Venables and M.J. Christie (New York: John Wiley).

Thackray, R.I., Jones, K.N. and Touchstone, R.M. (1973). *Personality and Physiological Correlates of Performance Decrement on a Monotonous Task Requiring Sustained Attention*. Report No. AM-73-14 (Washington, DC: FAA Office of Aviation Medicine).

Ursin, H. and Ursin, R. (1979). Physiological indicators of mental workload. In *Mental Workload: Its Theory and Measurement*, edited by N. Moray (New York: Plenum Press).

Van Egeren, L.F., Headrick, M.W. and Hein, P.L. (1972). Individual differences on autonomic responses: illustration of a possible solution. *Psychophysiology*, **9**, 626–633.

Welford, A.T. (1959). Evidence of a single-channel decision mechanism limiting performance in a serial reaction task. *Quarterly Journal of Experimental Psychology*, **11**, 59–66.

Welford, A.T. (1978). Mental workload as a function of demand, capacity, strategy, and skill. *Ergonomics*, **21**, 157–167.

Wewerinke, P.H. (1974). Human operator workload for various control conditions. *Proceedings of the 10th Annual NASA Conference on Manual Control*, Wright-Patterson AFB, Ohio, pp. 167–192.

Wickens, C.D. (1979). Measures of workload, stress and secondary tasks. In *Mental Workload: Its Theory and Measurement*, edited by N. Moray (New York: Plenum Press).

Wickens, C.D. (1984). *Engineering Psychology and Human Performance* (Columbus, OH: Charles E. Merril Publishing Co.).

Wierwille, W.W. (1979). Physiological measures of air crew mental workload. *Human Factors*, **21**, 575–593.

Wierwille, W.W. and Casali, J.G. (1983). A validated rating scale for global mental workload measurement applications. In *Proceedings of the 27th Annual Meeting of the Human Factors Society* (Santa Monica, CA: Human Factors Society), pp. 129–133.

Wierwille, W.W. and Connor, S.A. (1983). Evaluation of 20 workload measures using a psychomotor task in a moving-base aircraft simulator. *Human Factors*, **25**, 1–16.

Wierwille, W.W., Casali, J.G., Connor, S.A. and Rahimi, M. (1985a). Evaluation of the sensitivity and intrusion of mental workload estimation techniques. *Advances in Man Machine Systems Research*, **2**, 51–57.

Wierwille, W.W., Rahimi, M. and Casali, J.G. (1985b). Evaluation of sixteen measures of mental workload using a simulated flight task emphasizing mediational activity. *Human Factors*, **27**, 499–502.

Williges, R.C. and Wierwille, W.W. (1979). Behavioral measures of air crew mental workload. *Human Factors*, **21**, 549–574.

Chapter 25

The recognition and measurement of stress: conceptual and methodological issues

Tom Cox

Introduction

Medical and psychological sciences have long been interested in a wide range of phenomena and issues attracting the common label 'stress'. Over the last 10 years this interest has increased dramatically, partly fuelled by the popular media's promotion of stress and health as related issues of societal importance. Much of the concern expressed has focused on the workplace, and stress has now become a major aspect of occupational health and safety.

If the experience of stress is accepted as an occupational hazard, then a series of questions need to be raised about (a) the identification of potentially stressful situations, (b) the recognition of the experience of stress in others, and (c) its subsequent measurement. Together these questions represent the first steps in a problem solving approach to the management and control of stress at work.

Problem solving paradigm

The problem solving paradigm is derived, in part, from general systems theory (see, for example, Checkland, 1972), and has been used in relation to stress in several different ways (Cox, 1987). For example, it has been used to provide an action plan not only for research into stress (Hingley and Cooper, 1986, in relation to nurses), but also for the management of stress within organizations (Cox, 1988; Cox et al., 1989, in relation to schools). It has also been discussed, at a more theoretical level, in relation to decision making and personal coping (Cox, 1987), and briefly in relation to counselling.

Essentially, the problem solving paradigm is based on a cyclical process of problem analysis, selection of actions, implementation and evaluation, feeding back into a re-analysis of the problem. The entry point is often the declaration or recognition of a problem promoting its further analysis. This analysis is then an iterative process comprising a cycle of deconstruction (taking apart) and reconstruction (putting together) until an adequate working description (model) is arrived at. In most circumstances it is necessary, as a prerequisite for the creation of an action plan, to agree the problem description with those involved in that scenario. The process of analysis and the subsequent negotiation and agreement of the problem description require a language and a framework (Cox, 1987). Contemporary stress theory offers the concepts necessary for both. Thus the definition of occupational stress is fundamental to the process of problem analysis and the subsequent process of selecting actions. It also provides a further framework for considering the conceptual and methodological issues which arise with respect to the subsidiary questions of recognition and measurement.

The definition of stress

Sometime ago, the author reviewed the different approaches which then existed to the definition and study of stress. It was concluded (Cox, 1978; Cox and Mackay, 1981) that there were essentially three approaches, each with its associated baggage of concepts, methods, theories and prejudices: (1) engineering and ergonomics, (2) medicine and physiology, and (3) psychology. Somewhat similar accounts and conclusions were offered by other authors at this time (for example, Lazarus, 1966; Appley and Trumbull, 1967; McGrath, 1970).

Engineering approach

The engineering approach treated stress as a stimulus characteristic of the person's environment, usually conceived in terms of the load or level of demand placed on the individual, or some aversive (threatening) or noxious element of that environment. Stress, so defined, produced a strain reaction in the individual which, although often reversible, could on occasions prove to be both irreversible and damaging. The concept of a stress threshold grew out of this approach. Sadly, the simple equation of demand and stress has given rise to the notion that 'a certain amount of stress is good for the person' which in turn has justified many dubious management practices (see later).

Physiological approach

The physiological approach considered stress in terms of a generalized and non-specific physiological response to noxious or aversive environmental stimuli. This response was very much seen to be invested in two neuroendocrine systems, the anterior pituitary–adrenal cortical system (PAC) and the sympathetic–adrenal medullary system (SAM). The origin of such an approach has been largely identified with the work of the Canadian physiologist, Hans Selye (1950).

Criticisms

Particular theories and models characteristic of these two approaches have attracted specific criticisms, for example that by Mason (1968) of Selye's work. However, in addition to these, there are several general problems associated with these approaches, not least that they tend to treat the existence of stress within a simple and somewhat dated stimulus–response paradigm. They fail to recognize the importance of individual differences in the experience of and response to stress, or consider the role and the importance of cognitive and perceptual processes in determining those experiences and responses. The person is treated as a passive vehicle linking stimulus and response in a simple linear system. Such ideas are inconsistent with current thinking in all three of the contributing disciplines, and cannot easily account for the wide range of individual and situational differences which are known to exist in the experience of and response to stress.

These approaches have also led to various unfortunate confusions, such as those of (a) demand with stress (engineering model) and (b) arousal with stress (physiological model). Furthermore, the ideas promoted by Selye and his more recent advocates have imposed false expectations on the nature of the relationship between different markers of stress, and very much dictated a 'medicalization' of the subject area, with an attendant narrowing of focus within the practical field of stress management. This approach encourages strategies which concentrate on the individual and their responses to stress independent of the organizational context within which the problem occurs. Partly as a result of this, we have witnessed the development of 'band aid' solutions (relaxation, for example) to stress problems in the workplace. Such views tend to encourage the attribution of responsibility for 'breakdown' to the individual.

Attempts to rework these approaches have not resulted in a more coherent or clearer account of stress research or stress management. The development of notions of positive stress (eustress) and negative stress (distress) has not resolved any of the existing problems, but have added to the semantic confusion. They have further encouraged the misguided view that 'some stress is good for you'.

A good example of this approach is provided by the guide notes issued by HMSO on behalf of several different government departments (Berry, 1987). These publications neatly encapsulate all the difficulties referred to previously.

Psychological approach

The third approach to the definition of stress is a psychological one, which offers what is an essentially cognitive model of stress. In doing so it attempts to overcome some of the criticisms that the other two models have attracted, and there is now some consensus developing around this psychological approach to stress.

As an aside, it is interesting to reflect on its development, which owes much to the work of Lazarus (1966) and McGrath (1970), because it appears to mirror the cognitive revolution within psychology (see Dember, 1974).

Stress as a psychological phenomenon

Stress has been defined as a cognitive state (Cox, 1985a) which reflects the person's perception of an adaptation to the demands of their (work) environment. This approach emphasizes the person's cognitive appraisal (Lazarus, 1966) of their situation, and treats the whole process of perceiving and reacting to stressful situations within a problem solving context (Cox, 1987). Stress is not a dimension of the physical or psychosocial environment; it cannot be defined simply in terms of workload or the occurrence of events determined by consensus to be stressful. Equally, it cannot be defined in terms of responses that are sometime correlates of stress, such as physiological mobilization or performance dysfunction. Framed in this way, the study of stress is about normal people coping with, and failing to cope with, the problems that face them.

Stress resides in the person's perception of the balance or 'goodness of fit' between the demands on them and their ability to cope with those demands. The absolute level of demand is therefore not the important factor in determining the experience of stress at work. What is important is the *discrepancy* that exists between the person's perception of those demands and that of their ability to cope with them.

A central feature of such 'transactional' approaches to stress is the process of cognitive appraisal (Lazarus, 1966), which Holroyd and Lazarus (1982) have defined in terms of being "the evaluative process that imbues a situational encounter with meaning". Cognitive appraisal appears to take account of the person's perceptions of:

(1) the demands on them, matched against
(2) their personal characteristics and coping resources—their knowledge, attitudes, behavioural and cognitive skills, and behavioural style,
(3) the constraints under which they have to cope, and
(4) the support they receive from others in coping.

The situation which is typically perceived and experienced as stressful is one in which the person's resources are not well matched to the level of demand placed on them, and where there are constraints on how they can cope and little social support for coping.

Demands

Demands are requests for action or adjustment, whether cognitive or emotional, behavioural or physiological. They require some degree of decision making and the exercise of skill. They may be imposed by the external environment, say as a function of work or the work–home interface, or may be internal reflecting the person's needs, material, social or psychological. There may be several important dimensions describing external (job) demands, for example, 'pleasantness/unpleasantness' and 'ease/difficulty' (see Cox, 1985b). Demands usually have a time base, the effects of which may be amplified by an acute sense of time urgency (type A behaviour: Zyzanski and Jenkins, 1970).

The absolute level of demand would not appear to be the important factor in determining the experience of stress. More important is any discrepancy which exists between the level of demand and the person's ability (personal coping resources) to meet that demand. The size of this discrepancy appears to be an important determining factor in the stress process. However, the relationship between the discrepancy and the intensity of the stress experience may be curvilinear rather than linear (Cox, 1978). Within reasonable limits stress can arise through either overload (demand>abilities) or through underload (demand<abilities), or through some combination of the two.

Ability and personal coping resources

There are many different ways in which the person's coping resources might be conceptualized; however, it could be useful to think of them in terms of energy, knowledge, attitudes, behavioural style (or personality) and skills. The idea of skill has to be extended beyond traditional conceptualizations in terms of psychomotor and technical skills, to include social and cognitive skills.

The attitudes and behavioural style, as well as personal knowledge and skills, can be developed both through formal education and training and more informally through untutored experience. Furthermore, several elements of this 'package' of resources are subject to change, influenced by factors such as time of day, energy and fatigue, and state of health.

The person's ability to cope with such an imbalance may be constrained or supported in different ways and to different extents.

Constraints

Constraints operate as restrictions or limitations on free action or thought, reflecting a loss or lack of discretion and control over actions. These may be imposed externally. For example, constraints may be imposed by the requirements of specific jobs or by the rules of the organization. They may also be role related or reflect the beliefs and values of the individual.

Support

Support can be made available in different ways, most essentially through 'social interaction'; through advice and information, through practical assistance or by providing understanding and by declaring empathy. It is possible that women need, and are more sensitive to, social support than men (Cox *et al.* 1983, 1984).

Cognitive appraisal and coping

Two somewhat different processes exist side by side: first, the person's appraisal of their ability to cope with the demands made of them, which underpins the stress state and initiates coping, and second the way in which the person attempts to cope with that stress. An obvious question arises: How are the cognitive elements of these two processes related?

Folkman *et al.* (1980) have offered a distinction between primary and secondary appraisal processes (Lazarus, 1966) in terms of their foci: 'Am I okay or in trouble?' compared to 'What can I do about it?'. Such a distinction fits the present discussion. Primary appraisal, an ongoing process, can give rise to the recognition that a state of stress exists. It establishes stress in the existence of a problem situation, albeit at a superficial level of analysis. It involves a continual *monitoring* of the four different aspects of the person's transaction with their environment and a continual *evaluation* of the balance between them (Cox, 1987). Secondary appraisal is not an ongoing activity, but is contingent on the recognition that a problem exists. It offers a more detailed analysis of the problem and the generation of possible coping strategies and actions. Accepting this form of relationship between the recognition that stress exists and subsequent problem solving requires that in discussing the way in which stress arises the term 'primary appraisal' is used. The term 'secondary appraisal' relates to the cognitive aspects of coping.

In addition to any consideration of its cognitive elements, the state of stress is often defined by the person's experience of negative emotion, unpleasantness and general discomfort.

Identification of stressful situations

According to contemporary stress theory, as described above, stressful situations have, at least, three characteristics:

(1) The person is faced with demands and pressures which are not matched to their personal resources: they have difficulty in coping with those demands.

(2) The person is constrained in the way they carry out their work and cope with its demands: they have little control over their work.

(3) The person is relatively isolated and receives little support from colleagues, supervisors, friends or family.

Together these characteristics describe the archetypal stressful situation, and provide the basis for its identification.

Responses to stress

Together, an awareness that an unmanageable problem exists and an associated negative emotional experience, normally initiate a cycle of changes in the person's perceptions and cognitions, and in their behavioural and physiological function. Some of these changes are attempts at mastering the problem, or attenuating the experience of stress, and have been termed 'coping' by Lazarus (1966). Coping usually represents either an adjustment *to* the situation or an adjustment *of* the situation. Elsewhere the author has described the process of coping with stress within a 'problem solving' framework (Cox, 1987, and earlier).

In addition to these psychological responses to stress, there may be significant changes in physiological function, some of which might facilitate coping, at least in the short-term, but in the longer term may threaten physical health.

Psychophysiology of stress

The stress state is usually accompanied by a characteristic mood change (see SACL below), and possibly by a more intense and focused emotional experience: the person feels tense or anxious, worn out or fatigued. Such moods and emotions are unpleasant, and on most occasions serve to define the stress state for the individual.

The physiological correlates of stress have been studied largely in terms of the activities of two neuroendocrine systems: the sympathetic—adrenal medullary system, and the pituitary—adrenal cortical system (Cox and Cox, 1983). For many people, and on many occasions, changes in the function of these systems with the experience of stress do not have implications for pathology. However, for some, that experience can be of aetiological or prognostic significance.

The recognition of stress

It should be obvious from a consideration of the nature of stress, that there is no simple checklist of 'symptoms' that can be sensibly applied to the recognition of the experience of stress in employees at work. The whole notion of such an approach represents a gross oversimplification of the underlying theory. It attempts to apply defined standards, such as the presence or absence of diagnostic symptoms, to the relevant decision making processes, simply conceived in terms of crossing a threshold on an additive unidimensional scale.

What then is the process by which stress can be recognized in individuals? It is essentially one of building up the picture of the person under stress, seeking information from several different domains and deliberately cross-referencing what is known in an attempt to validate the final conclusion. From our understanding of stress theory there would appear to be three different domains of important information:

(1) Situational information relating to antecedent conditions (see Identification of stressful situations, earlier).

(2) Self report data on perceptions and emotions.

(3) Observed changes in the person's behaviour and self presentation, that are uncharacteristic and inappropriate (see Responses to stress, earlier).

Information from all three domains needs to be taken together and interpreted and cross-referenced within two rather different frameworks: the theoretical and the organizational. Decision making, with regard to the recognition of stress, depends on the interplay of this information, and on the picture that emerges from that interplay. The recognition that a person is experiencing stress is essentially an emergent property of this information system, and is not simply vested in any one of its single components. The rules that need to be followed in this decision making process are, in themselves, relatively simple, and require that consistent information be available in at least two of the three domains. For example, the person is seen to be in a high risk situation, as defined by contemporary stress theory, and to be reporting feeling 'stressed', or the person is seen to be in a high risk situation and although not reporting, and even denying feelings of stress, is showing changes in their behaviour consistent with that experience. This later case is important beyond providing another example. It represents the situation which can occur in organizations with 'macho' cultures, where it is not acceptable to report such feelings, and where such reports may be 'punished' by their implications for future career development.

The measurement of stress

In accepting the definition of stress as a psychological state, it becomes obvious that following recognition the measurement of stress should logically tap into this state or the accompanying emotional experience. Measures thus need to generate state-dependent subjective data. Subjectivity should not, however, imply a lack of proper development of the measurement instruments, and attention needs to be paid to the issues of reliability, validity and fairness. Together, these three criteria are of great importance for *all* methods of data capture, not only those which are subjective. Research in this general area must thus be dependent on a good knowledge of psychometrics (see for example, Oppenheim, 1966).

Some discussion of the nature of such measures has already been offered (Cox, 1985a, b). The relevant measures should focus on (a) the person's overall perception of situations as 'stressful', or on (b) the various elements of the appraisal process, that is, degree of perceived demand, personal resources, nature and extent of support of and constraints on coping. Alternatively, instruments have been developed which appear to tap into the emotional (or mood) response to stressful situations, for example, the stress arousal checklist (SACL: Mackay *et al.*, 1978; Cox and Mackay, 1985).

The stress arousal checklist (SACL)

The measurement of mood may offer one direct method of tapping the individual's experience of stress, and there has been a recent resurgence of interest in this issue as witnessed by a series of articles in the *British Journal of Psychology* (King *et al.*, 1983; Cruickshank, 1984; Cox and Mackay, 1985) and elsewhere (Burrows *et al.*, 1977; Russell, 1979, 1980; Ray and Fitzgibbon, 1981; Watts *et al.*, 1983). Most of these studies have employed the 'stress arousal checklist' (SACL) developed by Cox and Mackay and originally published in the *British Journal of Social and Clinical Psychology* (Mackay *et al.*, 1978).

This adjective checklist (SACL) was developed at Nottingham, using factor analytical techniques (Mackay *et al.*, 1978; Cox and Mackay, 1985), for the measurement of self reported mood. It presents the respondent with 30 relatively common mood describing adjectives, and asks to what extent they describe their current feelings.

The model of mood which underpins the checklist is two-dimensional. One dimension appears to relate to feelings of unpleasantness/pleasantness or hedonic tone (stress) and the other to wakefulness/drowsiness or vigour (arousal). Such a model is well represented in the relevant psychological and psychophysiological literature (see, for example, Mackay, 1980; Russell, 1979, 1980). It was suggested by Mackay *et al.* (1978) that the stress dimension may reflect the perceived favourability of the external environment, and thus have a strong cognitive component in its determination. This view is consistent with the author's psychological model of stress. Arousal, it was suggested, might relate to ongoing autonomic and somatic activity, and be essentially psychophysiological in nature. It has now become obvious that stress may partly reflect how appropriate the level of arousal is for a given situation, and the effort of compensating for inappropriate levels (Cox *et al.* 1982).

The split half reliability coefficients for the two scales which tap into these dimensions have proved acceptable: arousal 0·82 and stress 0·80 (Watts *et al.*, 1983). Both were conceived of and developed as state measures, and are thus seen as transient in nature. Together the two dimensions can be used to describe a four quadrant model of mood within which characteristic emotions and related states may be identified: high arousal and high stress (anxiety), high arousal and low stress (pleasant excitement), low arousal and high stress (boredom), and finally, low arousal and low stress (relaxed drowsiness).

There are now many reported studies using the SACL, and reporting data from its two scales (Burrows *et al.*, 1977; Ray and Fitzgibbon, 1981; Cox *et al.*, 1982, 1983; King *et al.*, 1983; Watts *et al.*, 1983). There have also been a number of studies which have used modified versions of the checklist (Cruickshank, 1982, 1984), although locally inspired changes in the instrument cannot always be defended (Cox and Mackay, 1985).

A third scale has been suggested based on the use of a '?' category on the response scale associated with the different mood adjectives. This category signifies, in part, uncertainty about whether the adjective given currently describes the respondent's mood. A score based on the frequency of '?' responses might reflect an inability to report feelings, and this may be symptomatic of a disordered psychophysiological state. Such a scale has an acceptable split half reliability coefficient of 0·89 (Cox and Mackay, 1985).

A recent compilation of the available British and Australian data has allowed the publication of mean levels for different groups, broken down by country of origin, age, sex and occupation (Cox *et al.*, 1988). Parts of these 'normative' data are presented in Table 25.1.

While the SACL offers some assessment of mood state, another psychometric instrument developed at Nottingham (UK) offers a related assessment of health or general well-being (Cox and Brockley, 1980; Cox *et al.*, 1983, 1984).

Table 25.1. Some normative data for the SACL (derived from Cox *et al.*, 1988*)

Sample	Dichotomized Scores						Q		
	Stress			Arousal					
	x	SD	*n*	*x*	SD	*n*	*x*	SD	*n*
Mixed population	6·0	4·6	1027	6·4	3·2	1040	4·2	4·2	1079
Males: mixed sample	6·0	4·7	296	6·6	3·2	297	4·9	4·6	266
Females: mixed sample	6·0	4·6	731	6·3	3·3	743	3·9	4·1	584
Students	6·3	4·9	515	5·7	3·6	518	4·7	4·1	535
Ages 16 to 30	6·2	4·6	466	6·0	3·2	469	5·0	4·3	379
Ages 31 to 45	5·9	4·9	344	7·2	3·3	353	3·5	4·2	334
Age more than 45	5·1	4·2	122	6·4	3·3	123	3·7	4·1	1132

*More complete normative data are being published as part of a manual for the SACL (Cox *et al.*, 1988). Further information can be obtained from the author.

The nature and measurement of health

The definition of health has been no less a subject for debate than that of stress, although the broad view espoused by the World Health Organization (1946) is often cited as a starting point for further discussion and development. In its constitution, the WHO offered a positive definition of health in terms of psychological and social as well as physical well-being, and emphasized that it is both dynamic and changeable (like the weather). Somewhat later, Rogers (1960) proposed that the health state was a function both of the individual's heredity, and the accumulated and current effects of the person's environment as 'they act upon the psyche and body'. This offered an alternative approach to the traditional medical emphasis on constitutional and genetic factors. Rogers (1960) also suggested that health might be usefully viewed as a continuum, the opposing poles of which are 'complete well-being' and 'death', with a significant watershed existing at the point where the person is recognized as being obviously ill or injured. Accepting this simple model immediately highlights the area between complete well-being and obvious illness; an area which Rogers (1960) referred to as suboptimum health, and which the advertising world of the early 1960s discussed in terms of being 'one degree under'.

For most people, the day to day variation in their state of health occurs within this 'grey' area of suboptimum health, and therefore it may be changes in suboptimum health which reflect similar variation in the impact of the environment on the person, and any subsequent experience of stress.

Suboptimum health and well being

There are several different questionnaire instruments which have been used to tap into subjects' suboptimum health, and which by the nature of their scales and internal structure offer some description of that area of health (Gurin, *et al.*, 1960; Crown and Crisp, 1966; Goldberg, 1972; Derogatis *et al.*, 1974). Studies at Nottingham have also attempted to map suboptimum health using self reported symptoms of general malaise. Initially a compilation of general non-specific symptoms of ill health was produced from existing health questionnaires (see above) and from diagnostic texts. These symptoms included reportable aspects of cognitive, emotional, behavioural and physiological function, none of which were clinically significant in themselves. From this compilation, a prototype checklist was designed with each symptom being associated with a five point frequency scale ('never' through to 'always') which referred to a 6-month response window. In a series of classical factor analytical studies, on British subjects, now variously reported (Cox *et al.*, 1983, 1984; Cox, 1988), two clusters of symptoms or factors were identified (see Table 25.2). These factors were derived orthogonal.

The first factor (GWF1) was defined by symptoms relating to tiredness, emotional lability, and cognitive confusion; it was colloquially termed 'worn

Table 25.2. Items defining the GWBQ Scales (international version)

GWF1

Have your feelings been hurt easily?
Have you got tired easily?
Have you become annoyed and irritated easily?
Has your thinking got mixed up when you have had to do things quickly?
Have you done things on impulse?
Have things tended to get on your nerves and wear you out?
Has it been hard for you to make up your mind?
Have you got bored easily?
Have you been forgetful?
Have you had to clear your throat?
Has your face got flushed?
Have you had difficulty in falling or staying asleep?

GWF2

Have you worn yourself out worrying about your health?
Have you been tense and jittery?
Have you been troubled by stammering?
Have you had pains in the heart or chest?
Have unfamiliar people or places made you afraid?
Have you been scared when alone?
Have you been bothered by thumping of the heart?
Have people considered you to be a nervous person?
When you have been upset or excited has your skin broken out in a rash?
Have you shaken or trembled?
Have you experienced loss of sexual interest or pleasure?
Have you had numbness or tingling in your arms or legs?

out'. The more cognitive items would appear to imply difficulties in decision making (in the specific context of feeling 'worn out'): (a) Has your thinking got mixed up when you have had to do things quickly? (b) Has it been hard for you to make up your mind? and (c) Have you been forgetful? These may have implications for personal problem solving and coping (see Cox, 1987). The second factor (GWF2) was defined by symptoms relating to worry and fear, tension and physical signs of anxiety; it was colloquially termed 'uptight and tense'. This model of suboptimum health appeared to have some face validity in that it was acceptable to a conference audience of British general practitioners and medical and psychological researchers (see Cox *et al.*, 1983).

A questionnaire was derived from this factor model, and has now been used in a number of studies conducted by the Stress Research group at Nottingham, and elsewhere by other researchers. Scores on both scales have been shown to be determined by the nature of the person, and by the nature of their work and work environment. For example, a study of 300 school teachers revealed that 'neuroticism' scores on the Eysenck Personality Inventory were significantly related to scores on the general well-being questionnaire, concurrently administered. Between 37 and 41% of the variance in well-being was accounted for by 'neuroticism' (emotional instability).

However, there was no significant relationship between 'extraversion' and well-being (Cox et al., 1983). Significant sex differences have been reported for workers engaged in semi-skilled and unskilled work (Cox et al., 1984). Working women were shown to report poorer well-being than working men, controlling for the age of the worker. Within this sample, well-being scores were shown to be related to the nature of the work on which the person was employed: e.g. repetitive vs. non-repetitive (Cox, 1985b).

New data have been collected by Cox and Gotts through a series of linked studies in Britain and Australia. These data have recently been re-analysed and the model and its associated scales have been slightly amended to increase their robustness in relation to this international sample (and also to diverse homogenous samples). A number of symptoms have now been deleted from the two original scales, although no new symptoms have been added. The two new 'international' scales are now defined by 12 symptoms (See Table 25.2), but retain their essential nature: 'worn out' and 'uptight and tense'. The deleted symptoms were among the weaker ones in terms of scale definition (and item loadings). New norms have been computed for the 'international' scales and are to be published elsewhere (Cox and Gotts, unpublished data); Table 25.3 presents some of those new data.

Table 25.3. Some normative data for the GWBQ (derived from unpublished data of Cox and Gotts*) for mixed populations

SAMPLE	INTERNATIONAL VERSION (1987)					
	'Wornout' (12 Items)			'Uptight' (12 Items)		
	x	S.D.	n	x	S.D.	n
All	16·7	8·3	2300	10·7	7·4	2312
Males	15·9	7·8	1031	8·2	6·5	1042
Females	17·4	8·6	1262	12·8	7·4	1262
British sample by age (years)						
16–20	16.5	8.7	141	11·5	7·9	141
21–25	16.9	9.2	147	11·3	7·6	147
26–30	15.6	8·4	236	10·2	7·5	236
31–35	17·2	8·6	239	9·0	6·5	239
36–40	16·1	8·1	201	9·2	7·5	201
41–45	15·5	8·6	199	10·4	7·7	199
46–50	16·0	8·3	175	9·7	7·7	175
51–55	14·5	8·0	174	9·1	7·4	174
56–60	13·7	8·0	127	7·7	6·6	127
> 60	13·5	6·4	26	4·8	5·8	26

*More complete normative data are being published as part of a manual for the GWBQ (Cox and Gotts, unpublished). Further information can be obtained from the author.

It is thus suggested that suboptimum health, the 'grey area' between complete well-being and obvious illness, is made up of two states, one related to being 'worn out', and the other related to being 'uptight and tense'. The former has an interesting cognitive component, possibly related to decision making and coping, while the latter is partly defined by physical symptoms of anxiety and tension. It has been shown that people vary in the extent to which they report these feelings, both between individuals and across time, and it has been suggested that this variation may not only (a) reflect the experience of stress, but also (b) affect other responses to stress, such as self reported mood (see Mackay *et al.*, 1978; Cox and Mackay, 1985).

Study paradigms

Together the two instruments, the SACL and the GWBQ, have been used in many different studies of people's reactions to stressful situations. The following section considers some of the issues which are important to the design and analysis of such studies when conducted in simulated or 'real' settings. In doing so it considers the nature of the quasi-experiment, using recent studies on routine computer based work to provide examples of the different possible designs. The reader should also refer to the chapter by Drury in this book on designing ergonomics studies and experiments.

Experiments

The word 'experiment' denotes a test, and the test is usually one of a causal proposition: for example, 'Does allowing VDU operators to take rest breaks reduce the experience of stress?'. The notion of a 'trial' or deliberate manipulation is also part of the concept of the experiment. Deliberate trials have long been used to test causal propositions or hypotheses, although this often follows from observations made about more natural occurrences. From these observations, hypotheses are formed about causality, and these are then tested out and modified in a cyclical manner.

Causality relationships, within this experimental framework, involve treatments and outcomes (independent and dependent variables, respectively), and to infer treatment effects some form of comparison is needed. In order to ascribe the level of self reported stress, measured at a VDU based task, to the presence of a rest break, logically one has to be able to compare levels of stress in the presence and absence of rest breaks, and demonstrate a reliable difference. Even better would be to show concomitant variation between stress or performance levels and the number of breaks taken or their duration, thus establishing some analogy to a 'dose response curve' or 'titration'.

Random assignment

One of the great breakthroughs in the development of the experimental paradigm was the realization that random assignment of subjects to treatment groups provided a means of comparing the effects of different treatments in a manner that ruled out most alternative explanations. Given a sufficient number of subjects relative to the variability between subjects, the random assignment procedure makes the average subject in any one treatment group equivalent to the average subject in any other treatment group, before the treatments are applied. Random assignment is the basis of causal inference.

Thus, all experiments involve at least a treatment, an outcome measure, subjects, and some comparison from which change or difference can be inferred. The random assignment of subjects to conditions allows for causal inferences to be drawn about the effects of treatment.

Quasi-experiments

Unfortunately, it is usually more difficult to (randomly) assign subjects to treatment groups in 'real' settings than it is in simulation or laboratory situations. This implies that random assignment will be less common in field studies than in laboratory studies.

Although the term was not coined until later, both Stouffer (1950) and Campbell (1957) placed a special emphasis on *quasi-experiments* in their discussions of the methodological issues involved in studies conducted in social settings. Essentially, quasi-experiments are those that have treatments, outcome measures and subjects, but which cannot use random assignment to create the comparisons from which any changes caused by treatments can be inferred. Instead the comparisons depend on *non-equivalent* groups that may differ from each other in many ways other than the presence of the treatment whose effects are being tested. The fundamental problem is how can the effects of interest be disentangled from those due to 'non-equivalence'? There are at least three strategies which can be used to overcome this problem of non equivalence; all require that the specific threats to valid causal inferences which exist be identified and made explicit. These can be dealt with at three different levels: (a) through trade offs in control, (b) through the use of statistical analyses which take those threats into account, such as analysis of covariance, and (c) through increased care in describing and interpreting results and in generalizing from them.

Greater control is often available in simulations and laboratory studies than in 'real' situations, and relatively more control can be exercised in some 'real' situations (such as schools or prisons) than in others (such as the home or factory). What is at issue is really the degree of control and not whether it exists or not. Few study environments offer no control, and none offer complete control. The reduced possibilities for the random assignment of subjects to treatments in 'real' situations has led to the refined specification of the other controls needed in such studies.

It should be obvious already that the term control can be used in several different ways in relation to research design, and we need to be precise regarding its usage. In addition to control over treatments (random assignment), it also refers to environmental control, and control over selection of subjects. Thus, amongst other things, it refers to the ability to control or determine the situation or environment in which an experiment or study takes place, so as to rule out the influence of extraneous factors and the generation of spurious causal inferences. The third sense in which the term control is used refers to attempts to take out known differences between subjects which might also threaten valid causal inference. Such differences, although irrelevant to the focus of the experiment, may affect the outcome measures. Thus the threat of 'non-equivalence' may be reduced through the design of the experiment and selection of subjects. If the threats to equivalence can be identified and measured in some way, it is also possible to take them into account during the analysis of the data (*statistical control* using, say, analysis of covariance).

Essentially, all three senses of the term control involve ruling out threats to valid causal inference. Different types and degrees of control will be possible in different settings, and often it is necessary to attempt to compensate for lack of control in one domain by activity in another. For example, it may not be possible to ensure that all VDU operators entering a study have the same level of general well-being (lack of control over subjects). However, it may be possible to reliably and validly measure the differences in well-being that do exist between operators, say using the GWBQ (Cox *et al.*, 1983), and take them into account during the analysis (statistical control) of the study data, and in the interpretation of the results of that analysis.

There has been a belief, in some quarters, that statistical control can replace control over subjects in the design of experiments. However, the difficulties inherent in fully modelling initial interindividual and intergroup differences, reliably and validly measuring each element of that model, and then removing the variance attributable to those elements are now known to be substantial (see for example, Rivlin, 1971; Duncan, 1975; Heise, 1975).

Types of study

Controlled settings often make it easier to implement factorial designs where all levels of one treatment are crossed with all levels of another. For example, one might wish to compare VDU work with and without breaks and whether it is supervised or not. If the treatments are factorially combined then four groups would be created. Factorial designs are important because they allow *interactions* between factors to be tested. While experimental control is not absolutely necessary for factorial studies, it certainly facilitates inferences about the form of causal relationships.

Several distinctions are traditionally made among types of quasi-experiments: (a) non-equivalent group designs, (b) interrupted time series designs, and (c) correlational designs.

Non-equivalent groups

Non-equivalent group designs are typically those in which responses of existing groups, designated treatment and control, are measured and compared before and after a treatment. This would be the case where VDU operators in two departments were compared and measures of self reported stress and task performance are taken at the beginning and end of each shift.

Interrupted time series designs

Interrupted time series designs are those in which the effects of a treatment are inferred from comparing measures of performance taken at many time intervals before a treatment with those taken at many intervals afterwards. For example, operators' self report of stress at the VDU might be sampled many times before rest pauses were introduced and then many times afterwards. Interrupted time series designs are improved if this longitudinal comparison is combined with the cross-sectional comparability of non-equivalent group designs.

Correlational designs

The term correlational design occurs in the older methodological literature. It has been used to refer to efforts after causal inference based on measures of both effects and exposures taken at the same point in time, often in natural situations without any experimental intervention.

Summary of study paradigms

Problems with quasi-experimental designs may be approached in a variety of ways. However, efforts must always be made to provide the most adequate design possible: if not a true experimental design, then the most adequate quasi-experimental design. The shortcomings of the latter approach must be recognized and dealt with in one of three ways: the study environment and selection of subjects must be tightly controlled, as far as possible; the sources of non-equivalence must be identified and measured, as far as is possible, and those data used in the statistical analysis; and, finally, the limitations of the design and analysis must be taken into account when the data are interpreted and reported. This often affects the detail or preciseness with which data are described, and the way in which they can be generalized.

Given that these guidelines are followed then there is value in conducting quasi-experimental research, particularly where there is no alternative form of data collection and analysis possible.

Concluding comments

The definition of stress at work is central to its recognition, measurement and management. A psychological approach has been adopted in this chapter, which is given form in the transactional approach of Cox and Mackay. The implications of this approach for recognition and measurement were then briefly explored. The methodology of the quasi-experiments which have characterized so many field studies on stress were then discussed.

It is hoped that the reader will at least have obtained a sense of coherence from what is written, and come to understand that despite much of what happens in research, there is a theoretical base available which is capable of driving further developments and studies in both the pure and applied aspects of our concern.

Acknowledgement

The author acknowledges the support of the University of Nottingham; however, the views expressed here are those of the author.

References

Appley, M.H. and Trumbull, R. (1967). *Psychological Stress* (New York: Appleton-Century-Crofts).

Berry, D. (1987). *Stress Management* (London: HMSO).

Burrows, G.C., Cox, T. and Simpson, G.C. (1977). The measurement of stress in a sales training situation. *Journal of Occupational Psychology*, **50**, 45–51.

Campbell, D.T. (1957). Factors relevant to the validity of experiments in social settings. *Psychological Bulletin*, **54**, 297–312.

Checkland, P.B. (1972). Towards a system based methodology for real world problem solving. *Journal of Systems Engineering*, **3**, 2–28.

Cox, T. (1978). *Stress* (London: Macmillan).

Cox, T. (1985a). The nature and measurement of stress. *Ergonomics*, **28**, 1155–1163.

Cox, T. (1985b). Repetitive work: occupational stress and health. In: *Job Stress and Blue Collar Work*, edited by C.L. Cooper and M. Smith, (Chichester: John Wiley).

Cox, T. (1987). Stress, coping and problem solving. *Work and Stress*, **1**, 5–14.

Cox, T. (1988). Stress in organizations: meeting the challenge of work. *BUPA Symposium, The Management of Health*, Edinburgh.

Cox, T. and Brockley, T (1984). The experience and effects of stress in teachers. *British Journal of Educational Research*, **10**, 83–87.

Cox, T. and Cox, S. (1983). The role of the adrenals in the psychophysiology of stress. In *Current Issues in Clinical Psychology*, edited by E. Karas (London: Plenum Press).

Cox, T. and Mackay, C.J. (1981). A transactional approach to occupational stress. In *Stress, Work Design and Productivity*, edited by N. Corlett and J. Richardson (Chichester: John Wiley).

Cox, T. and Mackay, C.J. (1985). The measurement of self reported stress and arousal. *British Journal of Psychology*, **76**, 183–186.

Cox, T., Boot, N. and Cox, S. (1989). Stress in schools: a problem solving approach. In *Stress and Teaching*, edited by M. Cole and S. Walker (Milton Keynes: Open University Press).

Cox, T., Gotts, G. and MacKay, C.J. (1988). The Stress Arousal Checklist: A manual. Maxwell & Cox Associates, Sutton Coldfield.

Cox, T., Thirlaway, M. and Cox, S. (1982). Repetitive work, wellbeing and arousal. In *Biological and Psychological Bases of Psychosomatic Disease*, edited by H. Ursin and R. Murison (Oxford: Pergamon Press).

Cox, T., Thirlaway, M. and Cox, S. (1984). Occupational well being: sex differences at work. *Ergonomics*, **27**, 499–510.

Cox, T., Thirlaway, M., Gotts, G. and Cox, S. (1983). The nature and assessment of general well being. *Journal of Psychosomatic Research*, **27**, 353–359. (Paper based on presentation to the 26th Annual Conference of the Society for Psychosomatic Research, Royal College of Physicians, London.)

Crown, S. and Crisp, A.H. (1966). A short clinical diagnostic self rating scale for psychoneurotic patients. The Middlesex Hospital Questionnaire (MHQ). *British Journal of Psychiatry*, **112**, 917–923.

Cruickshank, P.J. (1982). Patient stress and the computer in the waiting room. *Social Science and Medicine*, **16**, 1371–1376.

Cruickshank, P.J. (1984). A stress and arousal mood scale for low vocabulary subjects. *British Journal of Psychology*, **75**, 89–94.

Dember, W.N. (1974). Motivation and the cognitive revolution. *American Psychologist*, **29**, 161–168.

Derogatis, L.R., Lipman, R.S., Rickels, K., Uhlenhuth, E.H. and Convi, L. (1974). The Hopkins Symptom Checklist (HSCL). In *Modern Problems in Pharmacopsychiatry*, Volume 7, edited by P. Pichot (Basel: Karger).

Duncan, O.D. (1975). *Introduction to Structural Equation Models* (New York: Academic Press).

Folkman, S., Schaefer, C. and Lazarus, R.S. (1980). Cognitive processes as mediators of stress and coping. In *Human Stress and Cognition*, edited by V. Hamilton and D.M. Warburton (Chichester: John Wiley).

Goldberg, D.P. (1972). *The Detection of Psychiatric Illness by Questionnaire*, Maudsely Monograph No. 21 (London: Oxford University Press).

Gurin, G., Veroff, J. and Feld, S. (1960). *Americans' View of Their Mental Health* (New York: Edinburgh).

Heise, D.R. (1975). *Causal Analysis* (New York: John Wiley).

Hingley, P. and Cooper, C.L. (1986). *Stress and the Nurse Manager* (Chichester: John Wiley).

Holroyd, K.A. and Lazarus, R.S. (1982). Stress, coping and somatic adaptation. In *Handbook of Stress*, edited by L. Goldberger and S. Breznitz (New York: Free Press).

King, M.G. Burrows, G.D. and Stanley, G.V. (1983). Measurement of stress and arousal: validation of the stress arousal checklist. *British Journal of Psychology*, **74**, 473–479.

Lazarus, R.S. (1966). *Psychological Stress and the Coping Process* (New York: McGraw-Hill).

McGrath, J.E. (1970). *Social and Psychological Factors in Stress* (New York: Holt).

Mackay, C.J. (1980). The measurement of mood and psychophysiological activity using self report techniques. In *Techniques in Pscychophysiology*, edited by I. Martin and P. Venables (Chichester: John Wiley).

Mackay, C.J., Cox, T., Burrows, G.C. and Lazzerini, A.J. (1978). An inventory for the measurement of self reported stress and arousal. *British Journal of Social and Clinical Psychology*, **17**, 283–284.

Mason, J.W. (1968). A review of psychoendocrine research on the pituitary–adrenal cortical system. *Psychosomatic Medicine*, **30**, 576–607.

Oppenheim, A.N. (1966). *Questionnaire Design and Attitude Measurement* (London: Heinemann).

Ray, C. and Fitzgibbon, G. (1981). Stress, arousal and coping with surgery. *Psychological Medicine*, **11**, 741–746.

Rivlin, A.M. (1971). *Systematic Thinking for Social Action* (Washington DC: Brookings Institution).

Rogers, E.H. (1960). *The Ecology of Health* (New York: Macmillans).

Russell, J.A. (1979). Affective space is bipolar. *Journal of Personality and Social Psychology*, **37**, 345–346.

Russell, J.A. (1980). A circumplex model of affect. *Journal of Personality and Social Psychology*, **39**, 1161–1178.

Selye, H. (1950). *Stress* (Montreal: Acta Incorporated).

Stouffer, S.A. (1950). Some observations on study design. *American Journal of Sociology*, **55**, 355–361.

Watts, C., Cox, T. and Robson, J. (1983). Morningness–eveningness and dirunal variations in self reported mood. *Journal of Psychology*, **113**, 251–256.

World Health Organization (1946). Constitution of the World Health Organization (3). (Geneva: WHO).

Zyzanski, S.J. and Jenkins, C.D. (1970). Basic dimensions within coronary prone behaviour pattern. *Journal of Chronic Diseases*, **22**, 781–795.

Chapter 26

Assessment of visual performance

Mark A. Bullimore, E. Jane Fulton
and Peter A. Howarth

Introduction

This chapter is concerned with the assessment of visual performance. Let us begin by considering what we mean by visual performance and why we might want to assess it. By visual performance we mean how well people can deal with visual inputs. This can involve a range of levels of complexity from simply detecting a light to integrating complex qualitative and quantitative information from a display or visual scene. Visual performance is a function of:

(a) the abilities of the observer—the inherent capacities and limitations of the human visual system, individual idiosyncracies and special characteristics like levels of arousal and fatigue;

(b) the characteristics of the observed objects—'displays'—how bright, how much contrast, how big and for how long viewed; and

(c) the characteristics of the visual environment in which viewing takes place.

We deal principally with the first two aspects: characteristics of the observer and of the viewed objects. For each we will consider the important aspects which affect visual performance and describe ways of assessing it. Environmental factors affecting visual performance are dealt with in the chapter by Howarth. We recommend strongly that these two chapters be read together. It is rarely possible to deal with a practical issue of visual performance without some concern also for environmental characteristics. Also, we should consider the consequences of performing a visual task and amongst these may be visual fatigue, discussed in the chapter by Megaw.

Except for a few special circumstances, such as camouflage, the ultimate purpose of assessment will most often be to optimize a visual task. In an ergonomics framework we are concerned with issues such as:

Design: How should this display be designed so that, say, bleary-eyed night-shift nurses will be able to read, from the other side of the bed, how much of the drug has been infused into a patient's arm?

Trouble shooting: Why do quality controllers/inspectors continually miss flaws in the seal of this milk powder packaging? How can we improve their performance?

Evaluation: It may be important to know about the visual function of specific people in consideration of the needs of a particular population or to check an experimental sample, e.g. what proportion of a group show colour vision deficiencies and of what type?

These scenarios might lead an ergonomist to ask questions about visual performance, such as:

How accurately can people absorb this kind of visual information. How much information do they miss?

Would performance improve if we changed the way information was displayed?

What kinds of error are common?

Does this performance deteriorate over time or improve with practice or experience?

In this chapter we will cover basic information about the human visual system and the important characteristics of objects/displays which affect visual performance. This will provide the reader with information about what might be important in particular contexts and help in deciding what to measure and how. We aim to clarify what is feasible for the general ergonomist to attempt in the way of measurement; some kinds of visual performance assessment can be carried out fairly readily using easily obtainable equipment but other kinds are complex and are more appropriately performed by vision experts. The chapter will also refer to other sources which give more detail about particular issues raised here.

The human visual system

In this section we will give an overview of the human visual system and discuss how its physiology can influence visual performance. The basic structure of the human eye is shown in Figure 26.1. Light enters the eye through the transparent cornea (where the majority of the refraction occurs) and, passing through the pupil (the circular aperture defined by the iris), it is refracted further by the lens to form an image on the retina. (For a detailed treatise of retinal image formation see Charman, 1983.) The pupil alters in size in order to moderate the amount of light reaching the retina. Constriction of the ciliary muscle modifies the shape of the lens, and hence its power, such that objects at various distances from the eye can be brought into focus on the retina, a process termed *accommodation*. Young adults and children

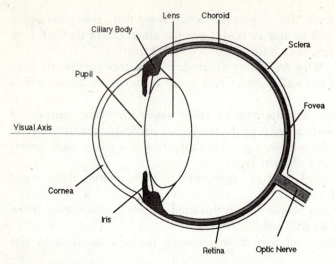

Figure 26.1. Horizontal cross-section through the human eye.

possess large amounts of accommodation and hence have no difficulty in focusing on objects as close as 10 cm, although sustained viewing at such close distances may cause fatigue. An observer's ability to accommodate will, however, decrease with age. This is termed *presbyopia* and above the age of 40–45 years most people require spectacles for near vision.

Sensitivity to light

Light falling on the retina stimulates light sensitive cells called *photoreceptors*. These convert the light energy into electrical signals which are transmitted to the visual cortex, in the rear portion of the brain. The complex processing which occurs in the retina, visual pathways and the cortex is discussed elsewhere (e.g. DeValois and DeValois, 1988). The photoreceptors are divided into two kinds, *rods* and *cones*, which have different characteristics and properties. The distribution of rods and cones across the retina is shown in Figure 26.2. The cones, which are responsible for vision at higher light levels, discrimination of fine detail and the perception of colour, are most abundant in the central or foveal region. The rods, which are responsible for vision at low levels of illumination, are found in greater numbers in the peripheral retina. The implications of these relative distributions will be considered later. The visual system is unable to detect light at levels below 10^{-6} cd m^{-2} and in the *scotopic* range, between 10^{-6} and 10^{-3} cd m^{-2}, only rods are functioning. At light levels above 3 cd m^{-2}, the *photopic* range, cones play the major role in vision. The area between 10^{-3} and 3 cd m^{-2} is called the *mesopic* range wherein both rods and cones are operating.

In the dynamic visual environment the eye has to adapt to changing light levels and does so in three different ways. First, the pupil can change size;

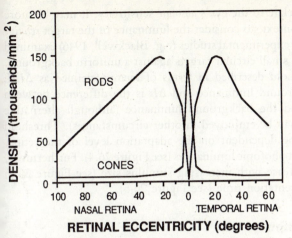

Figure 26.2. Density of rods and cones across the human retina in the horizontal meridian. The gaps in the functions are due to the optic nerve.

however the maximum area change is only 10–20 times—far too small to account for the eye's immense change in sensitivity. Secondly, small rapid changes in neural sensitivity take place in the retina. These occur in milliseconds and compensate for small changes in light levels, e.g. walking in and out of shade. The third mechanism involves slow changes in the photopigments in the rods and cones and is seen, for example, in the slow adaptation after entering a cinema or a photographic darkroom. The process of dark adaptation may be observed by measuring the eye's increasing ability to detect a dim light over a period of time in darkness. The dark adaptation curve is a bi-phasic function (see Figure 26.3) where the first portion represents changes in the sensitivity of cones and takes around 10 min, and the second portion shows changes in visual sensitivity mediated by rods.

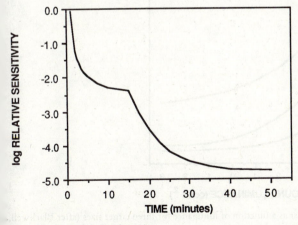

Figure 26.3. The dark adaptation curve.

The processes above relate to the eye's *absolute* sensitivity. It may be more relevant, in a practical context, to consider the luminance of the target *relative* to the background. Early experimental studies (e.g. Blackwell, 1946) examined the eye's ability to detect small circular targets against a uniform background, with the detection threshold described in terms of *contrast*, defined as $\Delta L/L$ —where L is the background luminance and ΔL is the difference between the target luminance and the background luminance (although alternative definitions of contrast may be employed in other circumstances). Threshold contrast was found to be dependent on the adaptation level of the retina, thresholds being lowest at photopic luminances (see Figure 26.4). Furthermore, detection thresholds decrease with increasing stimulus size (see Figure 26.4) and with increasing presentation time (see Figure 26.5).

Spatial aspects of vision

We are often concerned not just with detecting an object but also with discriminating detail. This attribute of the visual system is normally referred to as *visual resolution* or *visual acuity* (VA). Visual acuity is usually defined as the minimum angular separation between two lines which are perceived as two lines rather than one. A variety of targets can be employed in its measurement, such as letters, Landolt Cs or gratings (see Figure 26.6). Visual acuity values may be expressed in terms of minutes of arc (min arc) or as a Snellen fraction, e.g. 6/6 (or 20/20 in the USA). The fraction is more commonly used by clinicians where the numerator refers to the test distance in metres (or feet) and the denominator signifies the distance at which the limbs of the letter would subtend 1 min arc. Under optimal conditions the range of normal visual acuity is 6/4 to 6/6 (0·67 to 1·00 min arc). Not

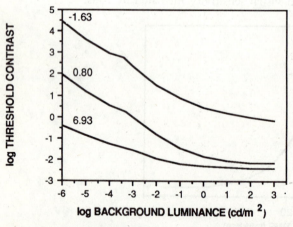

Figure 26.4. Threshold contrast as a function of luminance for three target sizes (after Blackwell, 1946). Target sizes are displayed in log mrad^2.

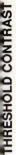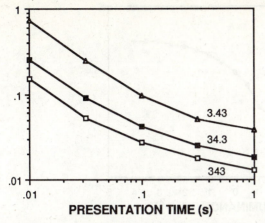

Figure 26.5. Threshold contrast as a function of presentation time for a 4 min arc target and for three background luminances (after Blackwell, 1959).

Snellen   Landolt C   Grating   Vernier
Letter   Acuity

Figure 26.6. Various targets which may be used in the measurement of visual acuity.

urprisingly a reduction in luminance or contrast will result in a decrease in visual acuity (see Figure 26.7).

A more complete picture of the visual system's spatial capabilities may be determined by testing people's ability to detect luminance sine wave gratings. The grating is defined in terms of its contrast and its spatial frequency—the number of cycles (light or dark bars) per degree. Threshold contrast is determined as a function of the spatial frequency to yield the *contrast sensitivity function* (CSF)—a graph of (the reciprocal of) threshold contrast as a function of the spatial frequency. Although such a function may appear of little practical interest, any object or scene can be represented as a series of sine waves of different contrast and spatial frequency and hence its visibility can be predicted from known contrast sensitivity values. (For further reading see Cornsweet, 1970.)

A further important aspect of the eye's spatial sense is its extraordinary ability to detect the misalignment of two lines. Berry (1948), Westheimer (1979a) and others have shown that observers can identify misalignment to an accuracy of 1 *second* of arc. This threshold is referred to as *vernier acuity* and is relevant to the reading of micrometers, slide rules and other tasks where the precise judgement of alignment is required.

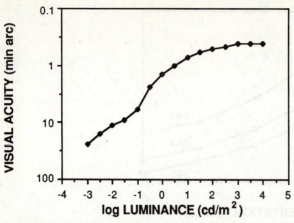

Figure 26.7. Visual acuity (min arc) as a function of luminance.

Temporal aspects of vision

The human visual system is fairly good at detecting a target that is changing with respect to time. Observers can detect that an object is moving for velocities as low as 7 min arc sec^{-1} with no frame of reference (Boyce, 1965) or 1 min arc sec^{-1} with a reference frame (Salaman, 1929). Further work has shown that target velocities of up to 5 deg sec^{-1} have little influence on visual acuity or vernier acuity (Westheimer and McKee, 1975).

The visual system is also very sensitive to detection of flicker and two thresholds are important. The *critical fusion frequency* (CFF) is the maximum temporal frequency (in Hz) at which flicker can be detected. Under photopic conditions the CFF is around 60 Hz and hence the typical 100 Hz flicker of fluorescent lights is undetectable (see the visual environment chapter by Howarth). Like visual acuity, CFF declines with luminance. The second threshold which may be of interest is the minimum modulation (or change) in luminance required for the detection of flicker. This is termed *temporal contrast sensitivity* and has been shown to be a function of both temporal frequency and luminance (de Lange, 1958).

Colour vision

An important feature of our visual system is the ability to discriminate colour and colour can be a powerful tool in the design of visual displays. Colour may be defined in terms of hue and saturation; hue describes the perceived colour, e.g. red, violet, while saturation describes paleness or how 'deep' the colour is.

The nature of human colour vision is discussed more fully in specialized articles and texts (e.g. Hurvich, 1981; Adams and Haegerstrom-Portnoy, 1987) but it suffices to say that there are three types of cone in the retina with peak sensitivity to short (S-cones), medium (M-cones) and long wavelengths (L–cones). These receptors are sometimes called blue, green and red cones respectively, but these terms are misleading. Our ability to discriminate between different colours arises from the fact that a given wavelength of light will stimulate each cone type to a different extent, in the same way that colour televisions produce a range of colours by varying the luminance ratio of the blue, green and red pixels. Optically, we are able to discriminate between colours as close as 2 nm in wavelength (see Figure 26.8) although our ability to discriminate between desaturated colours is poorer. Colour vision is, however, defective in some individuals. This will be discussed later.

The optical power of the eye is dependent upon the wavelength of light and objects of different wavelengths are focused at different points within the eye. This phenomenon is called 'chromatic aberration', and while its practical consequences are generally not severe, focusing difficulties can occur when wavelengths from extremes of the spectrum are viewed together. To generalize, this means that if a red and a blue object are adjacent, one may seem to be blurred compared with the other.

The visual field and visual search

The majority of the aspects of visual performance discussed previously have concerned optimal or foveal viewing. Each eye has, however, a wide field of vision extending 100 degrees temporally, 50 degrees nasally, 60 degrees superiorly and 90 degrees inferiorly from the visual axis. Our visual capabilities vary across the visual field; for example, visual acuity and colour discrimination are best at the fovea (see Figure 26.9).

Our ability to detect a static object is in part a function of its position within the visual field. The probability of detection, within a single fixation pause, may be plotted against eccentricity to yield the characteristic *visual*

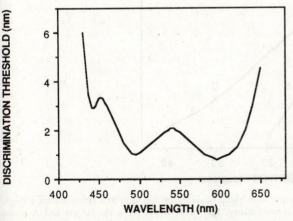

Figure 26.8. The variation in wavelength discrimination with test wavelength.

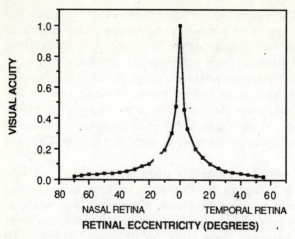

Figure 26.9. The variation in visual acuity with retinal eccentricity (after Wertheim, 1891).

detection lobe (see Figure 26.10). The visual lobe varies with exposure time and target size, hence the detectibility of a peripheral target can be improved by increasing its size (see Figure 26.10). The visual lobe is an important concept in visual search and inspection tasks since most detection takes place away from the visual axis.

Unlike static visual performance, detection of a dynamic target can actually be better in the periphery. CFF and temporal contrast sensitivity do not decline rapidly with eccentricity. On the contrary, CFF is actually higher in the peripheral visual field than the central field. This can be demonstrated easily with a conventional TV or VDU. If you look directly at the screen you can probably not perceive the flicker whereas if you shift your gaze to the left or right of the screen it will appear to be shimmering.

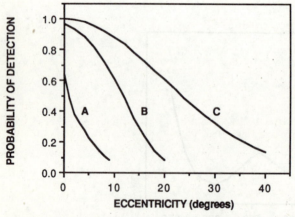

Figure 26.10. The visual detection lobe, the probability of detection of a target within a single fixation pause as a function of eccentricity, for three targets (C is the largest and A is the smallest).

Stereopsis and eye movements

One of the most important attributes of the human visual system is that we have two eyes which can move together. Possessing two eyes enables us to perceive the world in three dimensions—an ability termed stereopsis—with which we can make extremely accurate judgements of the relative distance of objects from ourselves. Stereopsis is usually described in seconds of arc, and thresholds of 10 sec arc or less are possible: that is, an object 1 m away can be perceived to be closer than another which is 1·00075 metres away.

The muscles which move the eyes are controlled by visual feedback so that with both eyes open they remain pointed towards the object of interest. Covering one eye will break the feedback loop—there is no visual information as to where that eye is directed—and the eye may take up a different position. This change in eye position is termed *heterophoria* and while it is quite normal for someone to have a small degree of heterophoria, for some individuals this can lead to discomfort and symptoms such as headaches.

The eyes can move rapidly—a saccade—to enable the object of interest to be imaged on the most sensitive part of the retina, the fovea. This is particularly important, for example, in the context of visual search. Alternatively the eyes can track a moving object—a pursuit movement—in order to keep the image on the fovea. Furthermore, the eyes can move rapidly in order to compensate for voluntary and involuntary movements of the head. For all of these 'version' movements the eyes move left or right together. The eyes can also make 'vergence' movements, which alter the angle between the visual axes, in order to look at objects at different distances. The eyes converge in order to view a near object and diverge to view a more distant object. An important characteristic of the vergence system is that it fatigues relatively easily and hence sustained convergence or frequent changes in vergence may produce discomfort. Most people will be able to converge closer than 10 cm: hold a pencil in front of your nose and bring it towards you. When it appears double, you've passed your *near point of convergence*.

Inter-subject variations in visual performance

We have now considered the major characteristics of the normal human visual system. We must also consider factors which will decrease the visual capabilities of the observer. Disease and poor health, for example, may influence visual performance, as will the intake of tobacco, prescribed (and non-prescribed) drugs and alcohol (see Adams *et al.*, 1978; Gilmartin, 1987).

In many individuals the optical components of the eye do not form a clear image on the retina due to a *refractive error*. There are three types of refractive error, the most commonly considered being *myopia* or near-sightedness. Myopia affects 20–25% of the working population and, because the cornea and lens are too powerful or the eye is too long, the image of a distant object is brought to focus in front of the retina. Because of this, distant objects appear blurred. Myopes can, however, see near objects clearly and hence uncorrected myopia may not decrease visual performance in the near environment. Myopia is corrected with concave spectacle or contact lenses. In 10–15% of people the eye is too short, or the optical components are not powerful enough, and the image will be focused behind the retina, a condition termed *hyperopia* (hypermetropia) or far-sightedness. Unlike myopia, the visual effects of this refractive error are often not obvious. This is because many hyperopes can exert their accommodation in order to bring distant objects and, if the hyperopia is not too severe, near objects into focus. Convex spectacle or contact lenses will be required for clear and comfortable vision in the older hyperope and the younger hyperope performing sustained visual tasks. The third class of refractive error, which affects the majority of the population is *astigmatism*. Like myopia, this produces a decrease in visual performance which cannot be compensated for by accommodation, but unlike either myopia or hyperopia it is equally detrimental for distance and near vision. In the astigmatic eye, lines of different orientations are focused at different positions relative to the retina. For example, an astigmat may see the horizontal poles of a scaffold clearly while the vertical poles appear blurred. Virtually everyone has *some* astigmatism, and it is only when the amount is large that visual performance is affected. As for myopia and hyperopia, astigmatism may be corrected with spectacles or contact lenses.

The visual performance of an observer may change considerably with age. Not only does the over 45-year-old have to come to terms with their decreased ability to accommodate (presbyopia) but also with a reduction in pupil size and changes in the crystalline lens which will result in less light reaching the retina. Hence people over 50 may require higher levels of illumination in order to perform as well as their younger colleagues and take longer to adapt to changes in illumination. Furthermore, changes in the crystalline lens may increase their susceptibility to disability glare (see the chapter by Howarth). Finally, the correction of presbyopia with bifocals or trifocals may cause focusing problems if people are looking through the wrong part of the spectacle lens.

Earlier we discussed the characteristics of normal colour vision. It should be acknowledged, however, that some 8% of males and 0·5% of females have defective colour vision. The relative frequencies and characteristics of the various types of defect are given in Table 26.1. All congenital colour deficiencies are due to anomalies in the retina, the most common type being anomalous trichromacy where one of the three cone types in the retina is abnormal. A more severe defect is dichromacy where one of the cone types is absent. The most dramatic defect occurs in the rod monochromat who has no colour discrimination, a scotopic spectral sensitivity function and reduced visual acuity, but these people make up a minute proportion of the population.

Table 26.1. Prevalence and properties of colour vision defectives in the male population. Although the prevalence of each type is much less in the female population, the *relative* proportions are similar

Type of defect		Spectral colour discrimination	Prevalence (%)
Anomomalous	*Trichromacy*		
	Protanomalous	Reduced for green, yellow, orange and red	1·0
	Deuteranomalous		5·0
	Tritanomalous	Reduced for blue-green cyan, and blue	0·001 (?)
Dichromacy			
	Protanope	Absent for green, yellow, orange and red	1·0
	Deuteranope		1·0
	Tritanope	Absent for blue-green cyan, and blue	0·001 (?)
Rod Monochromacy		Little or no discrimination	0·003 (?)

Characteristics of tasks and viewed objects which affect visual performance

In the visual working environment we are generally concerned with more than the detection of simple spots of light, distinguishing single characters, or discriminating two colours. We are concerned with the acquisition of visual information from various sources. These are usually complex rather than simple stimuli and are most often well above threshold levels for visual detection.

This section describes briefly characteristics of visual tasks and viewed objects which affect how well they can convey information to people through the visual system. Here 'viewed objects' means all those things from which people receive visual information. These may be 'displays' in the traditional sense, they may be the focus of an inspection task, or they may be any other kind of visually apprehended material such as printed documents or vehicles on the road.

Types of visual task

What constitutes good visual performance depends on the requirements of the task. In thinking about assessment of visual performance it is important to appreciate the nature of the tasks being carried out—this may affect the type of assessment which is appropriate. Three examples will illustrate task differences:

1. *Detection.* Some visual tasks require simply that an observer detects the presence or absence of something or finds out where something is. Examples are detecting that a warning light has come on, checking a manufactured unit for breaks in a seal, finding the cursor on a computer screen. Here good

visual performance requires only that the observer see the object against its background—no other discrimination is needed.

2. *Recognition*. Most often a visual task will require that an observer detect *and* recognize what something is—this demands a higher level of discernment because there has to be discrimination between stimuli. This is the case, for example, in obtaining information from graphic displays and text or carrying out complex inspection tasks*. Here good performance involves being correct in the judgement of what it is that has been detected. In counting out change, for example, it is important to distinguish between different coins.

3. *Interpretation*. Most tasks also require observers to interpret what they have seen in terms of what it means for their subsequent actions. Examples include establishing which of a row of warning lights has come on, the significance of a blemish on a photograph, what a dial is indicating and what a text message means.

Here we are not considering this cognitive level of extracting meaning from visual information, but rather the sensory capacity to obtain information such that cognitive factors can begin to play. We should not, however, lose sight of the fact that both sensory and cognitive factors are important in the design and evaluation of visual material; no matter how lucid and interesting the prose, it will be without value if it is written in tiny grey characters on a grey page; conversely, no matter how legible the message it is useless if it makes no sense. Beware!—cognitive problems can be mistaken for problems of visual performance.

Types of visual display

Visual information comes to us in the form of light either reflected by objects or emitted by them. Nowadays more and more visual information, specifically from displays in the working environment, comes to us via sophisticated emissive technologies in the form of cathode-ray tubes (CRTs), light emitting diodes (LEDs), backlit liquid crystal displays (LCDs), plasma displays, etc. Before new technologies were widely available, visual information was displayed principally by the traditional reflective technologies of inks and paper, printed labels for electro-mechanical dials and some simple emissive displays like warning lights. There is a rather large body of knowledge associated with these traditional media in terms of guidelines for good design for performance (e.g. McCormick and Sanders, 1983; Helander, 1987; Boff and Lincoln, 1988).

Visual inspection. Visual inspection, usually associated with monitoring product quality, represents a specific kind of visual task notable for its sustained and invariable nature. The same general principles affecting visual performance and assessment of other tasks also apply to visual inspection. Visual inspection is, however, an area of industrial ergonomics which, because of the direct impact of its performance on profitability, has received special attention over the last 20 years. While we make reference to inspection in a general way, for more detailed discussion we refer the reader to writings dedicated to the subject (Smith and Lucaccini, 1977; Drury, 1973; Drury and Addison, 1973; Megaw, 1979).

The important special characteristics of the new emissive displays with respect to visual performance are those concerned with the stability of images and their resolution. These include raster screen refresh rates, pixels density, phosphor persistence, dot sizes, brightness and spacing on matrix displays. There are reference books (e.g. Knave, 1983; National Research Council, 1983) which give guidelines for visual information presented on these kinds of displays, but these need to be interpreted circumspectly because the new technologies are themselves developing very rapidly in terms of their ability to support good quality images.

It is more difficult to produce robust standards and guidelines about this class of displays. This makes *assessment of visual performance* more important since it is often not possible to refer with confidence to guidelines about the physical characteristics of images on the displays; the chances are that there are none researched sufficiently for the particular quality of display with which you are concerned. This fact is reflected in the current emphasis of the International Standards Organization's (ISO) efforts to produce ergonomics standards for visual displays which attempt to define standard test procedures for visual performance rather than physical characteristics (e.g. the 1988 ISO draft standard for VDUs). Nevertheless, although existing guidelines may be inappropriate for state-of-the-art displays there will always be work environments where people are exposed to examples of outdated technology.

Basic principles of appropriate design for visual performance

It is the intention in this section to explain underlying principles which govern the suitability of visual information for the human visual system, rather than to present exhaustive guidelines for design or evaluation. The section is divided into two. In the first part, simple physical aspects of the display are considered, while the second part considers aspects which involve some cognitive component.

Physical aspects

As a general rule, visual performance is better the brighter the ambient lighting, the greater the contrast between object and background, the larger the object and the longer the viewing time. The influence of these physical parameters on visual performance can be seen in a general model presented in Figure 26.11. If the parameter value is too low (e.g. the size is too small) then the task will be below threshold. As the parameter increases threshold is reached and subsequent increases will improve performance until the optimal level is reached. In some cases though, if the parameter continues to increase performance will eventually decline. We shall now consider the relevant parameters.

ILLUMINATION
Within the normal range of illumination levels which we encounter naturally, or produce artificially, visual performance is improved by increasing

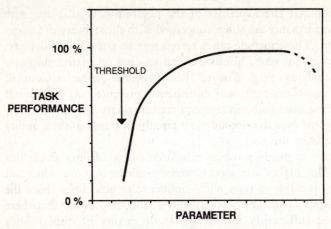

Figure 26.11. The influence of parameter on visual performance. The 'parameter' may be size, contrast, luminance or time. The broken portion of the line signifies a potential decrease in performance.

illumination. This is because the eye is relatively more sensitive to change at higher illumination levels. Elsewhere in this book Howarth discusses the importance of maintaining a relatively constant illumination level within the visual field so that visual performance is not affected by adaptation to either of the extreme levels. There are some rare occasions when illumination levels can become too high, for example, in a visual environment combining bright sunshine and wide expanses of snow, visual performance deteriorates.

CONTRAST

While it is generally true that the greater the luminance contrast the better the task performance, it is important to qualify this statement. For a moment consider, as an example, driving at night. Performance in the task of detecting oncoming vehicles is enhanced by their displaying bright headlights, but the contrast between these and the rest of the scene commonly causes discomfort and disability and hence a reduction in performance of the visual driving task as a whole. This example demonstrates the need to think about the whole task context rather than simply parts of it. It is also important to think of contrast both in terms of bright objects against dark backgrounds and of dark objects against light backgrounds. For some tasks there are advantages in illuminating the background to improve the observer's ability to see a stimulus—for example, in checking for flawed items using backlighting or shadows. Conversely, for tasks involving written character recognition it seems that *in general* dark text and symbols on a light background ('positive contrast') is preferable to light text on a dark background ('negative contrast') (Gould *et al.*, 1987a).

SIZE

Generally the larger the object (the greater the angular subtense at the eye) the more easily it will be seen and discriminated from other objects. For a resolution task this generalization holds for sizes greater than 1 min arc (the normal resolution threshold) up to the point where optimum performance is reached—the value for which depends also on other factors discussed here. As demonstrated by the model in Figure 26.11, above a certain point increasing the size will not improve performance and might in fact degrade it. Imagine standing directly in front of a large advertising hoarding and trying to read it! Applied to alphanumeric characters, size recommendations are that character heights should be large enough to subtend an angle of around 20 min arc at the observer's eye, i.e. about 4 mm height at a viewing distance of 600 mm.

EXPOSURE TIME
Visual performance is better the longer an observer gets to look at or look for something. This has implications both for the design of tasks—it is important to ensure that presentations are for an adequate length of time— and for the measurement of performance. Indeed, 'time to detect' can be used to assess the visibility of a stimulus.

Aspects which include a cognitive component

All the above physical parameters apply to the basic sensory processes of the human visual system. The parameters considered now all have, in addition, some cognitive aspects to them, and require a more sophisticated appreciation of their interactions. Consider Figure 26.12 where the physical aspects of the stimulus do not easily reveal the way the visual system responds to the pattern elements in this illustration.

DYNAMIC ASPECTS
Most often when we consider criteria for the design of visual material we are concerned with stable and static images—for most tasks this is an optimum condition. Images might, however, be unstable due to vibration of the observer or the viewed object or characteristics of the display itself. Furthermore, dynamic displays are becoming more widespread. Care must

Figure 26.12. The text is more conspicuous because of the pattern of lines, an effect which is not predictable from simple parameters of luminance, size or contrast.

be taken to compensate for the effects of these movements by ensuring that speed of movement is controlled and that illumination, contrast, image size and viewing time are increased above that which permits adequate performance with static and stable images. The conspicuity of an object, i.e. its capacity to attract our visual attention, is in part a function of its dynamic characteristics—movement or intermittency. A moving or flashing stimulus is more conspicuous than a static one.

CHANGE AND COMPARISON OF VISUAL STIMULI

Relative judgements are easier to make than absolute judgements. For example, you can easily detect that a vehicle brake light has come on if you notice the increase in intensity; however, if you miss the brightness change then because they are the same colour it is difficult to know whether you are looking at brake lights or rear lights. In this example there is no external brightness reference to compare the lights with, and the task is extremely difficult. A reference item for comparison will improve performance in many circumstances, such as inspection tasks and monitoring tasks, and the use of reference lines and markers are of significant value in assisting visual search and judgements.

PATTERN RECOGNITION AND CODING

We are adept at pattern recognition and tend to group visual information in terms of similarities of its physical appearance; colour, brightness, shape, size and orientation. These factors can be used to enhance visual performance by helping the viewer organize visual information. Designing a bank of dials so that the pointers all line up in the same direction (particularly either horizontally or vertically) when status is normal makes it easier to detect when one of them is registering an abnormal condition. We are better able to see and distinguish objects if they have unbroken lines and boundaries and whole regular shapes. These factors are exploited in the design of camouflage, where a major principle is to break up boundaries and outlines with colour or shading. As another example, in the design of alphanumeric characters the implication, and empirically supported wisdom, is that it is important to use clear, non-slanting and simple fonts without serifs.

REDUNDANCY

Visual performance can sometimes be enhanced by providing observers with redundant information. The detection and discrimination of warning lights for different functions, for example, can be improved by making them a different colour *and* a different shape. Similarly, by being colour- and size-coded, British paper money scores over its US counterpart for visual discriminability. It is sometimes appropriate to use other senses as a redundant cue to aid visual performance, for example auditory cues will improve the detection of visual warnings.

Use of colour

Colour has an important role as a coding device in separating and grouping elements in a design. When colour is used for coding, or for grouping information, it is important not to use too many colours, although authorities differ in the maximum number advisable (e.g. Grether and Baker (1972) recommend using no more than 10 colours but preferably 3!). Of course, many more colours and shades can be used to render form and depth in visual displays. The saturation level of colour can be important and desaturated (pastel like) colours should be avoided. In using colour for coding the colour discrimination abilities of the user population must be considered. An example of failure to do so is the use of self-administered glucose tests for diabetics, where the test involves colour matching even though diabetes is known to cause colour vision defects.

Colour is also important in providing contrast and it can be used to make objects conspicuous. The success of a colour used as a highlight depends upon the visual context in which it is used; all else being equal, orange has better contrast with green, for example, than it does with red. The concept of generic 'high visibility' colours can be misleading. Those colours which we generally refer to as 'high visibility', such as bright and fluorescent yellows, oranges and yellow-greens are effective in many environments because on average they contrast well with their backgrounds. In other specific background circumstances these colours can also be 'low visibility'. For example, red flags by the roadside will be highly conspicuous as few natural scenes are bright red whereas soccer linesmen no longer use red flags because they merge in with red garments in the crowd and with red seats in some stadiums.

Display/viewed objects location, size and area

The best location for display of visual material is roughly perpendicular to the observer's line of sight, unobstructed and preferably requiring a minimum of eye movement. The most frequently accessed information is, therefore, best placed centrally. Standardizing location of specific types of information is useful in reducing search time. The appropriate size to make a display depends on characteristics of the task, mainly the amount of information which must be displayed and its relative importance. If the display area is too small and too dense this will increase search time and decrease legibility. Clutter and complexity in layout reduce performance, and issues such as these are receiving increasing attention in the user-interface design literature (e.g. Shneiderman, 1987).

Conclusion on task and object characteristics

This section has reviewed basic general principles of the design of visual displays and tasks which affect people's visual performance. It is worth bearing in mind, in the following section about methods, that a widely used and economical method of assessing visual performance is to compare the particular circumstances of interest with standards and guidelines of good practice. Some sources for these guidelines have been referenced here. Applying these principles, with discretion, can often save a great deal of time and energy by preventing assessments of performance which effectively repeat other people's work.

Methods for assessment of visual performance

Assessment of visual performance may be important in a number of different circumstances. Just what it is appropriate to measure, and how, will depend on these circumstances. Assessments are generally necessary either to troubleshoot unsatisfactory conditions or as a tool in design and research.

When something is wrong there is often objective and quantifiable evidence of poor performance. It may be evident from people making errors (e.g. confusing alphanumeric characters or failing to detect flaws) or performing more slowly than anticipated (e.g. taking longer to do specific tasks or spending more time idle). Unsatisfactory conditions can also become evident as the result of subjective complaints from people about fatigue[†], discomfort (e.g. glare, 'eyestrain'), or general matters (e.g. ill-health). (See Megaw, chapter 27 in this book for some clarification.) In all cases the assessor's job is to try and identify the causes of poor performance and usually to devise and evaluate ways of improving it. This might involve assessment of:

1. *Individual's visual functional abilities.* For example, are specific individuals displaying decrements in particular aspects of vision related to their work? Is their visual system deficient in any way?

2. *Characteristics of the viewed objects and visual task.* For example, are the contrast and luminance values too close to threshold levels for optimum performance? Does spatially reorienting the task improve the situation?

3. *Visual environmental factors.* For example, is the spectral output of overhead lighting adversely affecting colour discrimination?

4. *Non-visual factors.* For example, is the general health of employees good? How satisfactory are social and organizational factors within the workplace? Are other environmental factors, heat and humidity or vibration for example, having a detrimental influence on visual performance.

[†]*Visual Fatigue.* One of the problems in considering visual fatigue is that the term itself is used in a variety of contexts. Some authors use visual fatigue to describe subjective complaints of discomfort while others apply the term to changes in visual function. This led the National Research Council's Committee on Vision (1983) to conclude that:

'The terms *visual fatigue* and *eyestrain* are frequently used in ill-defined and differing ways. These terms do not correspond to known physiological or clinical conditions. We suggest instead that researchers and others use terms that specifically describe the phenomena discussed, such as *ocular discomfort, changes in visual performance* and *changes in oculomotor functions.*'

In design and research it is sometimes important to assess what level of visual performance can be expected from a specific group of people or from a specific design of task or display. In these circumstances the most appropriate assessment will involve:

(a) *measuring abilities of people* doing tasks typical of those they might be required to do,

(b) *reference to guidelines* for comparison of the characteristics of a task or display,

(c) *evaluation tests* of the display/task with a sample of people typical of the likely user population.

In sampling from populations, *assessment of visual function* is sometimes necessary to describe or screen a group of people and decide whether they represent the abilities of a specific population. This is important in selecting people to take part in empirical assessments of visual stimuli. (See also McClelland's chapter for a discussion of user trial sample selection.) Whenever assessment is necessary the common elements are the person, the task and the environment. The assessment of the visual environments and of non-visual factors is dealt with in other chapters of this book. The other two elements, namely assessment of personal visual function and assessment of tasks and displays, are discussed separately here, although some overlap will be evident.

Assessment of task/display

Performance based measures

Performance-based measures involve the observation and measurement of people's performance on visual tasks, either in their natural environment (e.g. factory, driving cab or office) or in laboratory settings where selected attributes of the task can be simulated and examined under more controlled circumstances. These measures can and have been employed extensively in the evaluation of lighting conditions, display quality and the effects of prolonged visual performance. However, there are a number of problems in designing and interpreting these measures, and a variety of approaches have been taken to account for these problems.

First, performance usually involves both speed and errors and people often make complex trade-offs between them. For tests that allow subjects to establish their own criteria for time and errors, a slight change in error rate could be reflected in a relatively large change in speed. This can make the use of performance-based measures extremely difficult unless an underlying model of the trade-off is available. The problem can sometimes be overcome in the design of the tests themselves. For example, people can be allowed to take as long as they want, and the performance measure will then be accuracy alone; this approach is seen in the reading of the optometrist's letter chart.

As an alternative, the performance measure could be the time taken to achieve a certain level of accuracy: an example of where the criterion is 100% accuracy is how long it takes someone to locate a particular town on a map. Similarly, other accuracy levels could be fixed by rejecting trials where the subject's error rate is greater or less than a predetermined level, and then using speed alone as the performance measure.

A second problem with performance-based measures is that, in short-term studies, visual performance may differ in unpredictable ways from when the task is performed on a prolonged or permanent basis. This is particularly of concern in long-term inspection tasks when vigilance and tiredness may be involved. While this is a general problem in ergonomics, particular difficulties can come about in visual tasks because of the long-term demands on accommodation and convergence.

Finally, there is a complex relationship between the visual demands imposed by the task, the amount of effort and attention allocated and the resultant performance levels. Again, it may be possible for the design of tests to control subjects' arousal and attention to some extent, for example by rewarding good and penalizing poor performance or by employing secondary tasks.

Analytical and empirical approaches

Two types of performance methods can be identified, the 'analytical' approach and the 'empirical' approach (Hopkinson and Collins, 1970; Boyce, 1988). In the analytical approach the performance of simple contrived tasks is observed so that a quantitative model may be developed to relate visual performance to visual conditions. For example, Blackwell (1946) describes a method which involved detecting a spot of light against a darker or lighter background. In this way the relationship between contrast, luminance and visual performance can be modelled, and this model can subsequently be applied to more complex tasks such as the legibility of characters on a given background. This approach has a great deal of merit when comparing between tasks or displays which differ only with respect to one or two variables.

In the empirical approach, the speed and accuracy with which a task is performed is measured under real or simulated conditions. Weston (1945), like Blackwell (1946), investigated the relationship between task contrast and visual performance. He used a large number of Landolt Cs ('C's oriented in various directions with the subject's task being to identify the location of the gap; see Figure 26.6) to test people's speed and accuracy under a variety of contrast and light levels. Various other types of tasks have been employed such as reading text (Carmichael, 1948; Kruk and Muter, 1984; Nordqvist et al., 1986; Gould et al., 1987b), simulated inspection (Brozek et al., 1950; Murch, 1983), and search (Bodmann, 1962; Neisser, 1964). Modifications of these tasks have been used to examine the effects of contrast, luminance and size (Khek and Krivohlavy, 1966; Boyce, 1974; Stone *et al.*, 1980).

In many real-life situations the empirical approach has great practical value because it allows for comparison between a number of options where multiple variables are involved and where there are not resources to develop a complex model to help predict performance. For example, suppose that a choice must be made between three different liquid crystal displays for use on a chemical analysis machine. If these displays vary in a single dimension, say the sizes of character that they can support, then a good decision can be made confidently on the basis of an analytical approach. Knowing the range of distances from which chemists will need to read results it is possible to select the most appropriate display sizes. However, if the choice had to be made between three different sized displays, one of which was liquid crystal, one was a vacuum fluorescent display and the other an LED display, there are many more variables differentiating them: e.g. display colour, luminance, contrast, character form, size and effective viewing angles. There is no ready model to help make the decision about which would be best. Who knows what the appropriate weightings are for each variable? This choice can be made empirically in a user test by comparing the legibility of characters on each of the displays in ambient lighting conditions and from angles and distances to the display which cover the range expected in the machine's use. The advantage of a performance based experiment like this is that it provides useful and sound predictive information for the specific application. The disadvantage of the approach is that it adds little to the body of theoretical knowledge about visual performance, since it has compared the performance of discrete complex objects but revealed nothing quantitative about the interactions between the many variables which were involved.

The empirical approach is also particularly useful when visual performance is being affected by higher level factors beyond the simple physical attributes of the task (such as size and contrast) or the physiological attributes of the visual system (e.g. Figure 26.12). These factors range from the legibility of characters to the organization of visual information in certain ways to capitalize on our visual pattern recognition abilities. For example, an analytical approach could help in improving inspection performance in the detection of stitching irregularities in the seams of jeans, by increasing ambient illumination levels and changing the lights' spectral characteristics to enhance colour contrast between stitching and cloth and/or allowing inspectors longer to look at each pair of jeans. On the other hand, there might be vast scope for improvement in the visual performance of railway timetable-enquiry clerks, even if they are using full-colour high-resolution visual displays. For example, it might be helpful to organize the listings graphically and to introduce different grouping and colour coding conventions on to the screens. Here an analytical approach would not be a suitable way to assess different ways of organizing the visual layout. An empirical approach, simulating the clerks' search tasks and measuring visual performance with each of several layout options, would be a more appropriate way to assess potential improvements.

Task evaluation

How can we assess the effects of changing the physical attributes of a task? Suppose we know from the analytical approach that an increase in illumination level might be expected to improve the performance of someone reading documents. How can we assess first, whether there is any need for improvement, and second whether the strategy we have adopted has been successful?

In assessing visual tasks and performance, the approach taken by the Commission Internationale de l'Eclairage (1972, 1981) was to use the parameter of contrast to define a measure they termed 'visibility level'. By determining what contrast reduction is necessary to reduce the task to threshold you effectively determine how far above threshold the task was in the first place. The higher above threshold, the more 'visible' the task. The approach has been successfully applied to a variety of lighting situations and to paper-based tasks (Boyce, 1988). To use this approach you need to have some means of reducing contrast without affecting overall luminance, and this is the function of a 'visibility meter'. In assessing a visual task, a vision or a lighting specialist would probably either use this approach or would measure the physical attributes of the task and then apply the values obtained to an existing visual performance model.

But what if you haven't got a visibility meter or a sophisticated photometer? As discussed earlier in this chapter there is a general relationship between task performance, and each of the physical attributes of size, contrast, illumination and viewing time. The relationship between performance and any of these parameters is shown graphically in Figure 26.11, and knowledge of this function provides us with a simple, yet elegant, means of assessing whether a visual task is adequate. If we knew for a given parameter how far above threshold the performance is optimal, then we could devise a strategy to reduce the parameter by that amount. If the visual task was originally above the optimum level, then this reduction would still leave the task above threshold. On the other hand, if the task was sub-optimal, then this reduction would leave the task below threshold!

Size is an appropriate candidate for this approach, and as a good rule of thumb, if the task is reduced in size by a factor of three and can still be performed, then the initial conditions were acceptable for adequate visual performance. The integrative aspect of this simple approach can be seen by considering that by reducing *any* parameter such as contrast, task luminance or illumination, the whole curve relating performance to size (size being the 'parameter' in Figure 26.11) will be altered. With this alteration the threshold size will increase. The strategy of reducing the task size could then take the task below threshold, and it could not be seen. This size reduction can be achieved easily by increasing the distance from the eye to the task by a factor of three (Bailey, 1987).

A major advantage of this simple strategy is that the match between the task and the individual performing it can be assessed by using the person themself as the observer. Alternatively, the task alone can be assessed by using an observer with good eyesight. A word of caution is in order though! It is important that size is the only parameter that is being varied and that the measurement procedure itself should not affect the task. Since, in this example, the task is moved further away, the focusing demands on the observer are less. However, supposing the person who normally performs the task wears spectacles designed to focus at the task distance and not at further distances. The task itself could be quite acceptable but when the viewing distance is increased detail could become unclear due to focusing rather than image size reasons. This could lead to an incorrect conclusion that at the normal working distance the task was inadequate.

Subjective reports

Another class of methods involves the use of subjective measures based on questionnaires, interviews or informal discussion. Typically, this approach has been used both to assess visual performance and to investigate complaints of visual discomfort. The advantage of these methods for environments and tasks outside the laboratory (where testing procedures can be strictly controlled) is that the effect of complex variables can often be rapidly assessed. These are relatively easy and economical methods but are not without their drawbacks. Subjective reports should not always be taken at face value—people are often mistaken in their assessment of their own visual system and its performance. (Sinclair's chapter provides a review of subjective assessment and advantages and disadvantages.)

Subjective reports are also likely to be biased by popular beliefs and topical misconceptions. A complaint about glare on screens, for example, could be prompted by a belief that VDUs would be better if provided with a special anti-glare screen rather than because there is a real performance problem. The investigator needs to develop methods to avoid being misled by the subjects' analysis; a brief investigation with placebo treatments or use of subjective reports from people other than those who were party to the original analysis would be useful techniques to adopt as controls.

Subjective reports are of different kinds; they can be more or less structured and are often most reliable when they are most structured and specific. For example the choice between specific options—which of these fonts is more legible—is more likely to yield useful results than asking an open-ended question. On the other hand, open-ended questions can often reveal unforeseen problems which might affect performance, for example, with equipment cleaning and maintenance practices or seasonal variations in light levels.

A good example of the use (and misuse) of subjective reports is the literature concerning reports of ocular discomfort and visual display units. The increasing use of VDUs in the early 1970s brought with it a plethora of studies reporting a high incidence of complaints of visual discomfort. However, reviews of these early studies are invariably critical: Helander *et al.* (1984), for example, stated that 21 of the 28 studies they surveyed had serious design faults. These flaws included a lack of control groups and biased samples (there is anecdotal evidence that in at least one early study subjects were encouraged to over-report difficulties by one of the participants because this would bring problems to the attention of the management). Subsequently, Howarth and Istance (1986) suggested that in many of these studies of visual discomfort the use of one-off questionnaires was inappropriate. If groups are well-matched and appropriate measures used (such as *change* in discomfort over the day, rather than simply discomfort at the end of the day), then no difference is found between VDU users and non-users (Howarth and Istance, 1985). This finding does not negate results of studies which show significant differences between VDU users and non-users (e.g. Knave *et al.*, 1985), but rather indicates that these reports are demonstrating problems other than the use of VDUs *per se.*

Assessment of personal visual function

In certain circumstances we may wish to assess a person's visual capabilities. This could be because we suspect that an individual's poor visual performance has a physiological basis. Alternatively, we may wish to evaluate a task and prior to doing so check that the group of observers to be used are 'normal'. In the same way, assignment of subjects to experimental groups may be on the basis of their visual capabilities. Many people will be able to tell you something about visual problems they have; however, they may be totally unaware of visual disabilities like colour vision defects. Finally, a change in visual function could itself be the metric of interest. This section is divided into two parts. In the first, basic tests of visual function are described which a competent ergonomist should be able to perform. In the second, visual functions needing elaborate (and expensive) equipment not generally available to the non-specialist are reviewed.

Basic tests of visual function

Test charts are available that allow visual acuity measurements to be made easily. Distance visual acuity charts contain rows of letters which decrease in size down the chart. These letter sizes are labelled by the distance at which they would subtend 5 min arc (and the limbs, 1 min arc) at the eye. Hence, an '18m' letter on the chart would be three times as large as a '6m' letter. Most charts are designed for use at 6 m and this distance should be adhered to wherever possible in order to avoid confusion. To use the standard Snellen chart, the observer is instructed to read as far down it as possible and the lowest line in which most of the letters are read may be taken as the threshold (the visual acuity values are marked clearly on most charts). Visual acuity may also be measured for near vision with appropriate charts, which usually employ lower case Times Roman print. An observer with normal visual acuity should have no difficulty in reading 5 point (N5) print at 40 cm. This raises an important point, which is that 6/6 distance visual acuity does not itself guarantee good intermediate or near vision, particularly in observers over 40 years of age. Hence visual acuity should always be assessed at a distance relevant to the task or display that the observer is or will be using. Also, as well as measuring both eyes together visual acuity should be measured for each eye separately since an imbalance may be contributing to any reported symptoms. Careful attention should also be paid to the luminance of the test chart (see Figure 26.7): there are a variety of international standards for chart luminance and as a guideline we recommend a value of between 80 and 300 cd m^{-2}.

The standard Snellen chart described earlier has been used for many years with little change. Over the last 15 or so years a number of new charts have been developed. These range from charts consisting of luminance sine waves at various orientations to letters embedded in random-dot noise. The type of test-chart we recommend was introduced by Bailey and Lovie (1976) and consists of rows of five black letters on a white background. The size of the letters on each row is related logarithmically to the rows above and below, and with this chart we record the *logarithm of the minimum angle of resolution* (logMAR). The work of Westheimer (1979b) and Hallden (1972) suggests that a logMAR scale is a perceptually equal-interval scale. Being logarithmic the scale does not have a 'true' zero, although conventionally 6/6 is recorded as a logMAR of zero, and so the scale can be taken to be at an interval level of measurement, but not at a ratio level. The chart has five letters on each line, and each letter correctly read increases the person's score by 0·02 log units. The person reads as much of the chart as they can, and their vision is then scored according to the number of lines and letters they correctly identified. Despite their scientific advantages, these logMAR charts are not yet widely available and the standard Snellen chart is more likely to be encountered. This latter chart is quite adequate for most purposes, however keep in mind that we *cannot* assume that measurements using this chart are at a measurement level higher than ordinal (or possibly ordered metric). The practice of averaging vision scores is, therefore, incorrect.

The ergonomist should be aware of the contribution of accommodation and vergence problems to visual discomfort. A subject's near point of accommodation (NPA) can be measured with a near vision chart (or even a newspaper!). The print is moved slowly towards the observer until they report it beginning to blur—this point is the near point of accommodation. As mentioned earlier the person's accommodative ability declines with age, and so while a NPA of <10 cm might be normal for a teenager, a 35-year-old might not be able to focus much closer than 20 cm from their eyes. It is desirable that an individual should have a near point of accommodation significantly closer than their required viewing distance for a display. Reading glasses and bifocals alter a subject's near point, and it will be more appropriate to take this measurement with the subject wearing their spectacles. The near point of convergence can be measured in a similar fashion using no more than a pencil. This is held vertically and moved towards the observer until it first appears 'double'—this represents the near point of convergence. Most people, irrespective of their age, should have near points of convergence no further than 8–10 cm and any value much beyond this range may give rise to symptoms. Do not confuse the near points of accommodation and convergence: in the one you are looking for *blurring* of the target, while in the other you are looking for the target to appear *double*.

Colour vision may also be assessed relatively simply by the ergonomist. The simplest and most common type of test uses pseudo-isochromatic plates. These are book tests of numbers, letters or symbols in which the background camouflages the task for the colour defective. The Ishihara Plates are an example of this type of test. These tests are fairly efficient at detecting colour defectives and will often differentiate between protan- and deutan-type defects. It is important, if the correct standard illuminant (Illuminant C) for which the tests were designed is not available, that daylight is used to illuminate such tests. If daylight is not available either, then cool fluorescent tubes can be used. Other light sources, e.g. incandescent lights, will unacceptably alter the apparent colour of the plates, (see the chapter by Howarth), possibly producing incorrect results.

Stereopsis is the final visual function that one can realistically assess without a large amount of equipment. Inexpensive tests are readily available, such as the Titmus Fly Test and the TNO Test. In these tests, the two eyes are dissociated with either crossed polarizing filters or red and green filters, and a composite picture or pattern (e.g. of random dots) is placed in front of the person. Because of the filters employed the two eyes will see different images, in the same way that the two eyes see slightly different views of a 3D object. If the person has stereopsis, a 3D pattern will be seen coming out from, or going into, the page.

Instead of using the above techniques, the ergonomist may use a 'vision screener' in order to evaluate the vision of an observer. These instruments are based on the principle of the Wheatstone stereoscope, and eyes are tested either singly or together. In most instruments the following aspects of vision are assessed:

1. Distance and near visual acuity.
2. Colour vision.
3. Heterophoria.
4. Stereopsis.

Several vision screeners are commercially available including the Bausch and Lomb Ortho-Rater and the Mavis Vision Screener. Although instrument norms are available, the quantitative results obtained from these machines should be treated with caution since their false alarm rate is generally high. As a screening instrument they are, however, generally excellent and will usually detect people who should be referred for expert evaluation. Individuals should normally be referred to an optometrist for a visual examination, who, on request, will provide a written report. Although there may be a charge for this service, it may be the most economical way to solve problems.

Specialized tests

A variety of visual functions have been assessed in the evaluation of visual workload. We shall examine briefly some of the techniques described in the literature although the practising ergonomist may not have the resources to perform most of them. It is important that when measuring these functions we understand the relevance of any recorded changes. Indeed, there is clearly a need to distinguish 'fatigue', as described in the literature, from an adaptation process.

Malmstrom *et al.* (1981) employed an objective optometer to measure the accommodative response to a far and near sinusoidally moving target. They found that the accommodative response diminished significantly over a 6·5-min period and propose that this is due to fatigue of the accommodation system. There is an abundance of literature demonstrating that a period of sustained near vision can induce changes in the accommodation and vergence systems (e.g. Ostberg, 1980; Pigion and Miller, 1985; Fisher *et al.*, 1987; Gilmartin and Bullimore, 1987; Owens and Wolf-Kelly, 1987). Ostberg (1980) demonstrated that 2 h of close work induced a proximal shift in both the resting focus and the far-point of accommodation, although Murch (1983) could not replicate these findings. There is, however, no evidence that such changes represent fatigue rather than simply the adaptability of the human visual system. Fisher *et al.* (1987) found that although symptomatic and asymptomatic individuals showed accommodative adaptation of similar magnitudes, there were significant differences in the baseline measures and the temporal characteristics of the adaptation.

Haider *et al.* (1980) demonstrated that distance visual acuity decreased from 0·93 to 1·22 min arc following 3 h of near work whereas Dainoff *et al.* (1981) found no change in distance visual acuity for 23 subjects who undertook near work. Jaschinski-Kruza (1984) demonstrated that contrast sensitivity for high spatial frequency gratings presented at 5 m was significantly reduced after 3 h of near work and that the results of these studies were due to optical effects. It should be noted that these changes are for distance visual acuity and may not imply any change in visual function for near work nor explain any associated discomfort. Conversely, Lunn and Banks (1986) showed that after reading text presented on a VDU, contrast sensitivity was reduced for a limited range of spatial frequencies. In this instance the reduction was neural rather than optical in origin, hence it can be seen that a change in contrast sensitivity does not itself tell us anything about causal factors.

It has been suggested that visual fatigue can result in a change in eye movement behaviour. Megaw (1986) and Megaw and Sen (1984) were able to demonstrate effects of continuous VDU viewing on some eye movement parameters, although the effects also showed significant intersubject variations. Wilkins (1986) has shown that the presence of 50 Hz flicker causes the eye to overshoot its target and therefore increases the frequency of corrective saccades. Wilkins is unable, however, to explain any relationship between these changes and reports of visual discomfort. Leermakers and Boschman (1984) have shown that fixation times and the length of primary saccades are determined by the contrast of the text and that these effects are correlated with subjective reports of comfort. It is unlikely, therefore, that changes in saccadic behaviour reflect anything other than the difficulty in extracting information from the display.

A variety of other methods have been used in an attempt to evaluate visual performance. These include measuring changes in critical fusion frequency, pupil size and blink rate. The plethora of discrepant tests suggested for assessing visual workload indicates that no single visual function adequately reflects visual work. (Megaw provides some comparative assessment of different tests in the context of visual fatigue in chapter 27 of this book).

Summary—an example

Along with assessment methods, this chapter has discussed aspects of the human visual system and of viewed objects and tasks which are relevant to visual performance. By way of conclusion we present an example which illustrates the variety of factors which may need to be considered.

A client informs you that someone in their office keeps complaining of difficulty seeing information on their VDU and is definitely getting through work more slowly than expected. The office has 10 other people, performing similar work, who seem to be symptom-free. How should you approach this problem?

The person and task are apparently not well-matched; the question is whether you can assess how to improve the situation. You visit the office on a Monday morning and interview the person about matters such as the nature of the work, how it varies through the day, whether any other visual problems are experienced at home or when driving. From this discussion you establish that the difficulty actually occurs only at the end of the day. Given that no-one else in the room is having problems, even though they are all using similar equipment, the most likely explanation is that the complication is related to this person's visual system. The onset of the difficulty, towards the end of the day, suggests that there may be muscular problems either of co-ordinating the two eyes (vergence problems) or focusing (accommodation problems). To investigate these physiological aspects you could measure them yourself using a vision screener or, preferably, refer the person to their optometrist. This person is in their mid-thirties and if they were far-sighted (hyperopic) they could well be approaching presbyopia. You try out this possibility by having them move a sheet of printed text towards them until the characters begin to blur. You discover that their near point of accommodation appears to be around 15 cm which is fairly normal. At this point you decide to refer them for an eye examination; as well as their vision and muscular co-ordination, the health of their eyes will also be checked.

In the meantime, you decide to examine whether the display is adequate by using two strategies. First you suggest that the person having problems changes places temporarily with someone else in the office and that you will check back in a few days to see whether either of them are experiencing any difficulty; this will act as a check on both the VDU and the individual. Second, you assess the VDU and its image quality yourself. You note that the characters are stable, they are dark on a light background (positive contrast), glare does not appear to be a problem; the VDU is facing away from windows and even moving the screen around on its adjustable base does not lead to the reflection of luminaires in the office. Also, with your good eyesight you can read the screen at three times the normal viewing distance, not easily, but you can do so.

You re-visit the office a few days later. The optometrist's assessment was that the person's eyes were healthy, vision good, and muscular co-ordination excellent. Furthermore, neither the person themselves, nor the person now using their workplace is having difficulty seeing their VDU!

The person's vision seems normal, the physical task seems acceptable and has not altered since your first visit, and the person's job is the same as before. This seems to leave you with two possibilities: either there is a 'Hawthorne effect' operating here, i.e. the mere fact that attention is given to a perceived problem is itself a temporary alleviation of it, or there is an environmental problem which you may not yet have uncovered. You are concerned that the first of these seems more likely and so test the idea by returning the operators to their original workstations. A day later the person reports difficulties seeing the screen just as before, and so you decide to visit the office that afternoon to find out what changes have been made in the environment, and this reveals the real problem. Late in the afternoon the temperature in the office becomes too high for comfort and the practice is to open the door to the corridor which is situated behind the problem workstation. The corridor is painted white and is brightly lit with fluorescent striplights—the open door causes a veiling glare to be cast over this one VDU, reducing its visibility. The small increase in luminance caused by the veiling glare is hardly noticeable on the light background display, and looking at the screen as a whole, the operator had not been aware of glare as a problem. However, the veiling effect on the dark characters is sufficient to significantly reduce the contrast of its dark characters and hence affect the operator's performance. The other operator had been intolerant of noise from the corridor while seated by the door and had kept it closed.

Having discovered that the screen contrast has been reduced by veiling reflections successful remedies you can recommend include fitting an anti-reflection cover over the screen, moving the screen on the desk, or fixing the office heating system!

This example illustrates the range of aspects which need to be considered in the assessment of visual performance. Visual performance encompasses a variety of person-, task- and environment-related issues, each of which the ergonomist must be aware of, and any of which can be detrimental to a person's health, well-being and performance.

References

Adams, A.J. and Haegerstrom-Portnoy, G. (1987). Color deficiency. In *Diagnosis and Management in Vision Care*, edited by J.F. Amos (Boston: Butterworths), pp. 671–713.

Adams, A.J., Brown, B., Flom, M.C., Jampolsky, A. and Jones, R. (1978). Influence of socially used drugs on vision and vision performance. *AGARD Conference Proceedings*, No. 218, C5. 1–11.

Bailey, I.L. (1987). Mobility and visual performance under dim illumination. In *Night Vision: Current Research and Future Directions*, National Research Council Committee on Vision (Washington DC: National Academy Press) pp. 220–230.

Bailey, I.L. and Lovie, J.E. (1976). New design principles for visual acuity letter charts. *American Journal of Optometry and Physiological Optics*, **53**, 740–745.

Berry, R.N. (1948). Quantitative relations between vernier, real depth, and stereoscopic depth acuity. *Journal of Experimental Physiology*, **38**, 708–15.

Blackwell, H.R. (1946). Contrast thresholds of the human eye. *Journal of the Optical Society of America*, **36**, 624–643.

Blackwell, H.R. (1959). Specification of interior illumination levels. *Illumination Engineering*, **54**, 317–353.

Bodmann, H.W. (1962). Illumination levels and visual performance. *International Lighting Review*, **13**, 41–47.

Boff, K.R. and Lincoln, J.E. (1988). *Engineering Data Compendium: Human Perception and Peformance*, Volumes I, II and III, (New York: John Wiley).

Boyce, P.R. (1965). The visual perception of movement in the absence of a frame of reference. *Optica Acta*, **12**, 47–52.

Boyce, P.R. (1974). Illumination, difficulty, complexity and visual perform-ance. *Lighting Research and Technology* **6**, 222–226.

Boyce, P.R. (1988). Human Factors in Lighting (New York: Macmillan).

Brozek, J., Simonson, E. and Keys, A. (1950). Changes in performance and in ocular functions resulting from strenuous visual inspection. *American Journal of Psychology*, **63**, 51–66.

Campbell, F.W. and Durden, K. (1983). The visual display terminal issue: a consideration of its physiological, psychological and clinical background. *Ophthalmic and Physiological Optics*, **3**, 175–192.

Carmichael, L. (1948). Reading and visual fatigue. *Proceedings of the American Philosophical Society*, **92**, 41–42.

Charman, W.N. (1983). The retinal image in the human eye. In *Progresses in Retinal Research*, Volume 2, edited by N. Osborne and G. Chader (Oxford: Pergamon), pp. 1–50.

Commission Internationale de L'Eclairage (CIE) (1972). *A Unified Framework of Methods for Evaluating Visual Performance Aspects of Lighting*. Publication CIE 19 (TC 3.1) (Paris: International Commission on Illumination).

Commission Internationale de L'Eclairage (CIE) (1981). *An Analytic Model for Describing the Influence of Lighting Parameters Upon Visual Performance*. Publication CIE 19/2.1 (TC 3.1) (Paris: International Commission on Illumination).

Cornsweet, T.N. (1970). *Visual Perception* (London: Academic Press).

Dain, S.J., McCarthy, A.K. and Chan-Ling, T. (1988). Symptoms in VDU operators. *American Journal of Optometry and Physiological Optics*, **65**, 162–167.

Dainoff, M.J., Happ, A. and Crane, P. (1981). Visual fatigue and occupational stress in VDU operators. *Human Factors*, **23**, 421–428.

de Lange, H. (1958). Research into the dynamic nature of the human fovea-cortex systems with intermittent and modulated light: I. Attenuation characteristics with white and coloured light. *Journal of the Optical Society of America*, **48**, 777–784.

DeValois, R.L. and DeValois, K.K. (1988). *Spatial Vision* (New York: Oxford University Press).

Drury, C.G. (1973). The effect of speed working on industrial inspection accuracy. *Applied Ergonomics*, **4**, 2–7.

Drury, C.G. and Addison, J.L. (1973). An industrial study of the effects of feedback and fault density in inspection performance. *Ergonomics*, **16**, 159–169.

Fisher, S.K., Ciuffreda, K.J., Levine, S. and Wolf-Kelly, K.S. (1987). Tonic adaptation in symptomatic and asymptomatic subjects. *American Journal of Optometry and Physiological Optics*, **64**, 333–343.

Gilmartin, B. (1987). The Marton Lecture: ocular manifestations of systemic medication. *Ophthalmic and Physiological Optics*, **7**, 449–459.

Gilmartin, B. and Bullimore, M.A. (1987). Sustained near-vision augments inhibitory sympathetic innervation of the ciliary muscle. *Clinical Vision Sciences*, **1**, 197–208.

Grether, W.F. and Baker, C.A. (1972). Visual presentation of information. In *Ergonomic Aspects of Visual Display Terminals*, edited by H.P. Van Cott and R.G. Kincade (Washington DC: American Institutes for Research), pp. 41–121.

Gould, J.D., Alfaro, L., Finn, R., Haupt, B. and Minuto, A. (1987a). Reading from CRT displays can be as fast as reading from paper. *Human Factors*, **29**, 497–517.

Gould, J.D., Alfaro, L., Barnes, V., Finn, R., Grischkowsky, N. and Minuto, A. (1987b). Reading is slower from CRT displays than from paper: attempts to isolate a single-variable explanation. *Human Factors*, **29**, 269–299.

Haider, M., Kundi, M. and Weisenbock, M. (1980). Worker strain related to VDUs with differently coloured characters. In *Ergonomic Aspects of Visual Display Terminals*, edited by E. Grandjean and E. Vigliani (London: Taylor and Francis), pp. 53–64.

Hallden, U. (1972). Notes on the statistical treatment of the visual resolution. *Acta Ophthalmologica*, **50**, 47–57.

Helander, M.G. (1987). Design of visual displays. In *The Handbook of Human Factors*, edited by G. Salvendy (New York: John Wiley), pp. 507–549.

Helander, M.G., Billingsley, P.A. and Schurick, J.M. (1984). An evaluation of human factors research on visual display terminals in the workplace. In *Human Factors Review: 1984*, edited by F.A. Muckler (Santa Monica: The Human Factors Society), pp. 55–129.

Hopkinson, R.G. and Collins, J.B. (1970). *The Ergonomics of Lighting*. (London: MacDonald).

Howarth, P.A. and Istance, H.O. (1985). The association between visual discomfort and the use of visual display units. *Behaviour and Information Technology*, **4**, 131–149.

Howarth, P.A. and Istance, H.O. (1986). The validity of subjective reports of visual discomfort. *Human Factors*, **28**, 347–351.

Hurvich, L.M. (1981). *Color Vision* (Sunderland, MA: Sinauer Associates).

Jaschinski-Kruza, W. (1984). Transient myopia after visual work. *Ergonomics*, **27**, 1181–1189.

Khek, J. and Krivohlavy, K. (1966). Variation of incidence of error with visual task difficulty. *Light and Lighting*, **59**, 143–145.

Knave, B.G. (1983). The visual display unit. In *Ergonomic Principles in Office Automation* (Stockholm: Ericsson Information Systems), pp. 11–41.

Knave, B.G., Wiborn, R.I., Voss, M., Hedstrom, L.D. and Berqvist, U.O. (1985). Work with video display terminals among office employees: 1. Subjective symptoms and discomfort. *Scandinavian Journal of Environmental Health*, **11**, 457–466.

Kruk, R.S. and Muter, P. (1984). Reading of continuous text on video screens. *Human Factors*, **26**, 339–345.

Leermakers, M.A.M. and Boschman, M.C. (1984). Eye movements, performance and visual comfort using VDTs. *IPO Annual Progress Report*, **19**, 70–75.

Lunn, R. and Banks, W.P. (1986). Visual fatigue and spatial frequency adaptation to video display of text. *Human Factors*, **28**, 457–464.

Malmstrom, F.V., Randle, R.J., Murphy, M.R., Reed, L.E. and Weber, R.J. (1981). Visual fatigue: the need for an integrated model. *Bulletin of the Psychonomic Society*, **17**, 183–186.

McCormick, E.J. and Sanders, M.S. (1983). *Human Factors in Engineering and Design* (New York: McGraw-Hill).

Megaw, E.D. (1979). Factors affecting inspection accuracy. *Applied Ergonomics*, **10**, 27–32.

Megaw, E.D. (1986). VDUs and visual fatigue. In *Contemporary Ergonomics 1986*, edited by D.J. Oborne (London: Taylor and Francis), pp. 254–258.

Megaw, E.D. and Sen, T. (1984). Changes in saccadic eye movement parameters following prolonged VDU viewing. In *Ergonomics and Health in Modern Offices*, edited by E. Grandjean (London: Taylor and Francis), pp. 352–357.

Murch, G.M. (1983). Visual fatigue and operator performance with DVST and raster displays. *Proceedings of the Society for Information Display*, **14**, 53–61.

Muter, P., Latremouille, S.A., Treurniet, W.C. and Beam, P. (1982). Extended reading of continuous text on television screens. *Human Factors*, **24**, 501–508.

National Research Council Committee on Vision (1983). *Video Displays, Work and Vision* (Washington: National Academy Press).

Neisser, U. (1964). Visual search. *Scientific American*, **210**, 94–100.

Nordqvist, T., Ohlsson, K. and Nilsson, L. (1986). Fatigue and reading text on videotext. *Human Factors*, **28**, 353–363.

Oborne, D.J. (1982). *Ergonomics at Work* (New York: John Wiley).

Ostberg, O. (1980). Accommodation and visual fatigue in display work. In *Ergonomic Aspects of Visual Display Terminals*, edited by E. Grandjean and E. Vigliani (London: Taylor and Francis), pp. 41–52.

Owens, D.A. and Wolf-Kelly, K. (1987). Near work, visual fatigue and variations in oculomotor tonus. *Investigative Ophthalmology and Visual Science*, **28**, 743–749.

Pigion, R.G. and Miller, R.J. (1985). Fatigue of accommodation: changes in accommodation after visual work. *American Journal of Optometry and Physiological Optics*, **62**, 853–863.

Salaman, M. (1929). *Some Experiments on Peripheral Vision*. MRC Special Report 136 (London: HMSO).

Shneiderman, B. (1987). *Designing the User Interface: Strategies for Effective Human-Computer Interaction*, (Reading, MA: Addison-Wesley).

Smith, R.L. and Lucaccini, L.F. (1977). Vigilance research: its application to industrial problems. In *Human Aspects of Man–Machine Systems*, edited by S.C. Brown and J.N.T. Martin (New York: Open University Press).

Stone, P.T., Clarke, A.M. and Slater, A.I. (1980). The effect of task contrast on visual performance and visual fatigue at a constant luminance. *Lighting Research and Technology*, **12**, 144–159.

Wertheim, T. (1891). Peripheral visual acuity, translated by I.L. Dunsky, 1980. *American Journal of Optometry and Physiological Optics*, **57**, 915–924.

Westheimer, G. (1979a). The spatial sense of the eye. *Investigative Ophthalmology and Visual Science*, **20**, 893–912.

Westheimer, G. (1979b). Scaling of visual acuity measurements. *Archives of Ophthalmology*, **97**, 327–330.

Westheimer, G. and McKee, S.P. (1975). Visual acuity in the presence of retinal image motion. *Journal of the Optical Society of America*, **65**, 847–850.

Weston, H.C. (1945). *The Relation between Illumination and Visual Performance*. Report No. 87 (London: Industrial Health Research Board).

Wilkins, A. (1986). Why are some things unpleasant to look at? In *Contemporary Ergonomics 1986*, edited by D.J. Oborne (London: Taylor and Francis), pp. 259–263.

Chapter 27

The definition and measurement of visual fatigue

Ted Megaw

Definition of visual fatigue

The early literature is full of attempts to arrive at an acceptable definition of visual fatigue, beginning with the frequently referenced phrase of Plautus quoted by Ramazzini (1700): 'Sitting hurts your loin, staring your eyes'. This implies that the subjective feeling of pain is one of the essential features of visual fatigue. Weber (1950) claimed that ocular fatigue is a form of 'occupational neurosis' like writer's cramp. The term neurosis was taken more seriously by Adler (1962) who believed that many ocular complaints are really psychosomatic in origin, a view that would not be very popular with today's work-force.

The results of early attempts to define visual fatigue can be seen generally to have failed as reflected in the following two quotes. Simmerman (1950) concluded that: 'The subject of fatigue has been investigated from many angles by many research workers. However, the results of all investigations have failed to give us a clear and concise definition of the meaning of fatigue' and Collins (1959): 'In speaking about visual fatigue, we are speaking about something which has no precise and unique definition. Visual fatigue means different things to those concerned with different aspects of vision'. There is, in fact, a strong temptation to take the view expressed by Muscio who in 1921, after considering whether any kind of general fatigue test were possible, concluded that the term fatigue should be totally banished from precise scientific discussion and that attempts to develop a fatigue test be abandoned. Rather, he suggested that one looked for direct correlations between task conditions and various measures of human functioning so that so-called tests of fatigue could be used to determine the effects of the task factors rather than to determine the presence or absence of fatigue itself.

Many early studies have shown that several measures one might have suspected of being sensitive to various visual task factors are not so. In an early study of Carmichael (1948), subjects read one of two books continuously for 6 h while their eye movements were recorded. One of the books was Adam Smith's *Wealth of Nations* which was considered by many of the subjects to be dull while the other, Blackmore's *Lorna Doone*, was often considered interesting! The text was presented either as hard copy or projected on a microfilm device. The lighting levels in both cases were around 170 lux. Results showed no effects of continuous reading on any of the eye movement measures and on comprehension scores. There was no change in visual acuity measures taken before and after the reading task. However, some subjects did show an increase in reported feelings of fatigue as the experiment progressed including 'wishing to stop'. All these results were independent of the reading material and the method of presentation.

While the low accuracy and reliability of early methods of recording and analysis should not be forgotten, this typical lack of significant results on objective measures of fatigue does explain why so much importance is attributed to subjective measures or, to give them their proper title, symptoms of asthaenopia. The symptoms can be grouped into *ocular symptoms* such as discomfort of the eyes and feelings that the eyes are tired, itchy, dry and burning, *visual symptoms* such as difficulty in focusing and blurred vision and, finally, *systemic symptoms* reflected in reports of headaches, postural fatigue and general tiredness. It is these frequently reported symptoms of asthaenopia associated with the introduction of visual display units (VDUs) over the last 10 years that have rekindled the interest in the subject (Matula, 1981; Stellman *et al.*, 1987). Moreover, as a result of improved recording and analysis methods and a closer understanding of the visual mechanisms involved, we are now in a better position to look at the possible relationships between task factors and human functioning along the lines suggested by Muscio (1921).

Not wishing to avoid totally giving a definition of visual fatigue, it is suggested that the following points should be included if and when a more formal definition is proposed:

(a) Visual fatigue does not occur instantaneously but involves people being subjected to the same visual task conditions over a period of time during the course of which the fatigue tends to build up. If a person suffers from some visual defect such as uncorrected refractory errors, the time course may be comparatively rapid.

(b) Visual fatigue should be distinguished from mental workload which is associated with the information and cognitive demands of the task.

(c) Visual fatigue can be overcome, often very rapidly, either by rest or by changing the task conditions. The presence of fatigue does not have any long-term harmful effects except possibly if an individual is prone to some related disease.

(d) Visual fatigue should be distinguished from an adaptive response of the visual system. This is to say, like any other biological system, the functioning of the visual system adapts to the particular conditions it meets.

Sometimes this adaptation process takes a comparatively long time as reflected, for example, by the typical dark adaptation curve. On other occasions, it is very fast. The neural adaptation in the retina to small changes in luminances takes only a matter of milliseconds (Rushton and Westheimer, 1962).

(e) Symptoms of asthaenopia are the main reasons for assuming the existence of visual fatigue.

(f) Individual factors, as well as task and environmental ones, can contribute to fatigue. If a person has uncorrected refractive errors, particularly say if they are presbyopic as well as already being hypertropic (see the chapter by Bullimore *et al.* for more explanation of terms), they are bound to experience feelings of discomfort when performing a reasonably demanding visual task. The same applies to those suffering from uncorrected abnormal muscle balance.

(g) Symptoms of asthaenopia can result from conditions which are not visual in origin. To take a very obvious example, a dry environment containing noxious particles can lead to complaints of visual discomfort (Messite and Baker, 1984).

Origins of visual fatigue

The origins of visual fatigue are probably three-fold. Most likely is fatigue of one or more of the oculomotor control systems. The others relate first to fatigue of the neural processes involved in visual processing ranging from effects at the retina, the optic nerve, the lateral geniculate body through to the striate cortex and second, to relatively non-specific effects such as arousal levels and the amount of so-called effort exerted by subjects. There are obvious problems in disassociating the latter from effects related to mental workload. Moreover, it is obvious that these three sources of fatigue are not independent. A lowering of arousal may affect any of the oculomotor control systems, as exemplified by the effects of alcohol and various drugs on the execution of rapid eye movements (Gentles and Llewellyn Thomas, 1971; Wilkinson *et al.*, 1974). Similarly, a lowering of the resolution powers of the retina will have effects on the accommodation demands as blur is one of the main stimuli for accommodation (Phillips and Stark, 1977). A further problem is that the presence of other factors can obscure the origins of fatigue. For example, the pupil response is sensitive to short-term memory load and arousal level (Kahneman and Beatty, 1966) as well as to visual factors such as levels of illumination and depth of focus requirements. It is not difficult to appreciate that in these circumstances the use of well controlled experimental procedures is essential.

The main oculomotor control systems for each eye can be listed as follows:

Accommodation control achieved by the ciliary muscle which is intermediate between striated and non-striated muscles.

Vergence control and muscle balance achieved by six extraocular eye muscles: the four recti and two oblique muscles.

Version control in relation to both saccadic and pursuit eye movements achieved by the same muscles responsible for vergence.

Pupil control achieved by two muscles which are non-striated, the sphincter pupillae causing constriction of the pupil and the dilator pupillae causing dilation.

Blink control achieved mainly by a sphincter muscle, the orbicularis palpebrarum, although the upper eyelid can be raised by the action of the levator palpebrae superioris to expose the eye ball.

The extent of the interaction between these control systems means that it is often very difficult to ascertain whether any changes in the functioning of one of them reflects fatigue or whether it is an adaptive change whereby the emphasis on the control of the visual system is moved from one oculomotor system to another. The most often quoted example of such an interactive adaptive response is reflected by the close interdependence of the accommodation, vergence and pupillary mechanisms. In particular, pupil constriction, which increases the depth of focus, can reduce the demands on the accommodation system, especially when viewing blurred material as on a VDU.

Measurement of visual fatigue

Methods for measuring visual fatigue can be divided into four classes. First, there are those methods which attempt to measure, more or less directly, changes in functioning of the various oculomotor systems that have been described earlier. Then there are methods based on measures of visual acuity which may reflect changes in visual processing ability. Third, there are methods which look more generally at visual task performance. Naturally, any changes in performance may reflect effects on both visual acuity and oculomotor functioning. Finally, there are methods based upon reported symptoms of asthaenopia.

Measures of oculomotor functioning

Eye movements

Many different types of eye movements are executed by the extraocular muscles. They range in amplitude from less than 1 minute of arc (min arc) to 40 degrees (deg) or more and in acceleration from a very few deg s^{-2} to 40 000 deg s^{-2}. They are often accompanied by head movements. The essential eye movement types are listed in Table 27.1.

There is a variety of methods to record eye movements and these have been reviewed very thoroughly by Young and Sheena (1975). Apart from direct observation, the following methods are commonly used:

Table 27.1. Types of eye movements

Saccadic eye movements	Rapid voluntary conjugate movements resulting in changes of fixation observed during visual search and reading executed in order to bring the retinal image of the object being viewed onto the fovea where there is high spatial resolution
Corrective saccades	Conjugate saccadic movements to compensate for target undershoot or overshoot by the main saccadic eye movements
Tremor, slow drift and microsaccades	A variety of very small amplitude movements, less than one degree, associated with visual fixation of a target
Pursuit or slow tracking movements	Slow smooth conjugate movements executed to stabilize the retinal image as a moving target is tracked, normally not under voluntary control
Compensatory eye movements	Smooth movements, closely related to pursuit movements and the slow phase of vestibular nystagmus which compensate for active or passive movements of the head or trunk
Vergence eye movements	Movements of the two eyes in opposite directions to facilitate fusion of the retinal images of the two eyes when objects are viewed at different distances
Optokinetic and vestibular nystagmi	A mixture of smooth image stabilizing movements to either a continuously moving visual field or stimulation of the semicircular canals and rapid 'return' saccadic eye movements
Dynamic overshoot	Small amplitude saccadic-like eye movements which sometimes occur in the opposite direction to the saccadic eye movements immediately preceding them, usually monocular
Glissades	Slow movements which sometimes occur in either the same or opposite directions as the saccadic eye movements immediately preceding them, often monocular

Electro-oculography. By placing surface electrodes around the eye, changes in the corneoretinal potential can be recorded as the eye rotates in respect of the head.

Corneal-reflection. Because of the geometry of the cornea, the angle of reflection of a light source directed onto the surface of the cornea changes as the eye rotates in relation to the head. The reflected light source is often recorded using a video system. The most well known commercial system based on this method is the NAC Eye Mark Recorder, the most recent version of which (V) also registers pupil size [see Figure 27.1(a)].

Limbus tracking. A light source, often an infra-red one, is directed at the boundary between the iris and the sclera (the limbus). As the eye rotates, there is a change in the amount of reflected light due to the differences in the reflection characteristics of the iris and the sclera which can be detected using photosensitive diodes or some other method of recording the amount of light [see Figure 27.1(b)].

Contact lens methods. There are several different contact lens methods. With the original method, a mirror was attached to the cornea so that a beam of light could be directed onto the mirror and reflected from it to the recording device, the angle of the reflected beam changing as the eye rotates. A more recent development involves attaching a small induction coil onto the limbus. Eye movements are detected by changes in magnetic induction as the coil moves in relation to a uniform magnetic field generated by a series of coils arranged around the head. Such a system is manufactured by Skalar and is illustrated in Figures 27.1(c) and (d).

Pupil-centre corneal-reflection. If one looks at a remote source of light reflected in the pupil, its position in relation to the centre of the pupil will vary as a function of head or eye movements. By video recording the pupil, it is possible to identify where somebody is looking by measuring the relative position in the pupil of the reflected source. This is the basis of the remote oculometers manufactured by Applied Science Laboratories, ISCAN and by Demel [see Figure 27.1(e)]. These systems also permit pupil size to be monitored.

Double Purkinje image method. This method is employed by the Stanford Research Institute and takes advantage of the relative changes in angle of the first (front of the cornea) and fourth (rear of the lens) reflections from the eye as it rotates.

The choice of eye movement recording methods is influenced by many factors, the most important of which will often be cost. This can vary from a few hundred pounds to over £40 000. The questions to consider are:

1. Are you mainly interested in where somebody is looking (point of regard) or in the characteristics of the eye movements themselves or both?

2. Is it necessary to permit head movements or is a fixed-head system acceptable?

3. Do you wish to record movements other than horizontal ones?

4. What are the prevailing lighting conditions?

5. Do you wish to record eye movements from people wearing glasses or contact lenses?

6. Should the system be unobtrusive?

7. What is the smallest amplitude of eye movement you wish to register?

8. What accuracy of recording is required?

9. Do you require measurements of pupil size?

Since the review of Young and Sheena (1975), developments in computerized recording and analysis have had an enormous effect on the quality of the data that can be obtained. Not only has accurate data calibration become easier to achieve, but by using high sampling rates and digital filtering, the undesirable effects of system noise are greatly reduced (McConkie, 1981; Harris *et al.*, 1984).

That eye movement parameters can be affected by fatigue to the saccadic eye movement system was suggested by the results of Bahill and Stark (1975). They showed that the longer subjects performed a step-tracking task, the lower was the peak velocity of their saccades, the more likely it was for saccades to overlap and the greater was the probability of glissades being

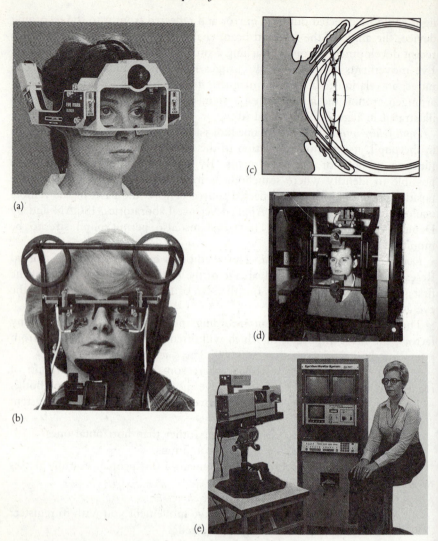

Figure 27.1. Illustration of some of the common methods used to register eye movements and point of eye fixation. (a) A corneal-reflection system manufactured by NAC which permits free head movements and is suitable for registering point of regard. (b) A limbus tracking system manufactured by Applied Science Laboratories which requires the head to be fixed. (c) and (d) A magnetic induction system manufactured by Skalar which requires the head to be fixed (c illustrates the attached scleral induction coil). (e) A pupil-centre corneal-reflection system manufactured by Applied Science Laboratories which permits head movements.

executed (see Table 27.1 for definitions). Using a standard infra-red limbus tracking technique, Megaw and Sen (1984), Sen and Megaw (1984) and Megaw (1986) were unable to demonstrate similar effects as a result of subjects performing tasks requiring continuous VDU viewing. On the other hand, they did observe changes in the probabilities of execution of glissades and dynamic overshoot in some subjects but not always in the direction predicted by the results of Bahill and Stark (1975). It should be said that the laboratory task demands imposed by Bahill and Stark are unlikely to be met in the course of day to day visual behaviour.

Wilkins (1986) has shown that the occurrence of 50 Hz display flicker can cause the eye to overshoot its target and, therefore, increases the frequency of corrective saccades. However, Wilkins is unable to offer an explanation for any relationship between this change in oculomotor behaviour and reported symptoms of visual discomfort. Leermakers and Boschman (1984) have shown that fixation times and the length of primary saccades are determined by the contrast of reading material and that these effects are correlated with reports of visual comfort. However, it is extremely unlikely that the changes in saccadic behaviour reflect anything other than the increased difficulty in extracting information from the display as contrast is reduced and does not reflect fatigue to the saccadic system.

If subjects continuously track a sinusoidally moving target, the amplitude of the accompanying pursuit movements decreases (Malmstrom *et al.*, 1981). However, when it came to comparing pursuit eye movement performance before and after reading aloud from either a normal or high resolution display terminal, Miyao *et al.* (1988) were unable to demonstrate any effect on root mean square tracking error as a result of performing the intervening reading task. Nor were they able to demonstrate any differences for the two displays. On the other hand, there was an increase in reported symptoms of visual fatigue following the reading task.

Studies on vergence eye movements and blink rate are described later.

Accommodation

Because of the limitation in recording methods, most of the early studies relating to accommodation concentrated on measures of accommodation time or of the nearest point of accommodation rather than measurements of the accommodation power of the lens while viewing a particular stimulus. Results from the early studies are inconclusive. Both Collins and Pruen (1962) and Krivohlavy *et al.* (1969) have reported an increase in accommodation time following performance on interpolated tasks, while Brozek *et al.* (1950) found no increase in the near point of accommodation following strenuous inspection work. Similar techniques have been used more recently by Stone *et al.* (1980), Mourant *et al.* (1981) and Gunnarsson and Soderberg (1983). Stone *et al.* (1980) failed to find any consistent differences in accommodation time before and after performing 1 h of various visual tasks with different levels of contrast. On the other hand, Mourant *et al.* (1981) reported that accommodation time increased significantly after carrying out visual search on a video display but no such increase occurred when the task was in hard copy form. These results were supported by the results from Gunnarsson and Soderberg (1983) who reported a decrease in the near point of accommodation following strenuous VDU work, accompanied by increases in subjective symptoms of visual discomfort.

There are some doubts about the validity of all of these results, mainly because of the relative inaccuracy of the recording methods. On more specific grounds, the results of Mourant *et al.* (1981) have been criticized on the grounds that the near and far targets they used to measure speed of accommodation were sufficiently close together not to require any accommodative response because of the available depth of perception. It should also be remembered that measures of speed of accommodation and near point of accommodation typically involve a perceptual element in that subjects have to perform some kind of target detection task such as detecting the gap in a Landolt ring; it may be, therefore, that the reported findings reflect perceptual rather than oculomotor fatigue.

As a result of recent developments, it has been possible to monitor more or less directly the accommodative power of the eyes. Three main techniques have been used: laser optometry (Hennessey and Leibowitz, 1972), infra-red optometry (Cornsweet and Crane, 1970) and polarized vernier optometry (Simonelli, 1980). With some optometers, subjects have to make subjective estimates in order that the accommodative state can be assessed while with others, the state is assessed directly. It has been asserted that, for example, the necessity to make judgements about the direction of speckle movement when using traditional laser optometers may influence the recorded states of accommodation (Post *et al.*, 1984).

Using a tracking task where subjects had to alter their accommodation between targets moving sinusoidally between 0·0 dioptres (D) (infinity) and 0·4 D (25 cm), Malmstrom *et al.* (1981) were able to show a decrease in the amplitude of accommodation over time. Iwasaki and Kurimoto (1987) used the low-frequency component of accommodative oscillations as a possible indicator of fatigue. They demonstrated that these oscillations significantly increased after performing a VDU based task, but not if the task was presented in traditional hard copy form. The results were supported by subjective reports of visual discomfort.

In a later study, Iwasaki and Kurimoto (1988) have reported an increase in the accommodation time when changing focus from a near to a far target from subjects who performed 1 h of VDU-based work. They claimed that this increase was accompanied by some changes in the latencies and amplitudes of visually evoked potentials (VEPs).

Most of the recent interest in the use of accommodation measures has been in interpreting accommodative power for a particular task in relation to a person's so-called dark-focus point. This point, often referred to as night myopia, is the accommodative state that the eye takes when there is an absence of any distance cues. Typically this varies between subjects from 0·6 D (167 cm) to 2·3 D (44 cm), accommodative powers considerably greater than zero (infinity). Since the influential results of Leibowitz and Owens (1975), there have been numerous studies which have shown that when viewing a task at a particular distance the accommodative state of the eye tends to be some way intermediate between the task viewing distance and the dark-focus point for the subject (Ostberg, 1980; Murch, 1983). Moreover, Ostberg (1980) and Ostberg and Smith (1987) report that as a result of performing visually demanding tasks, there is a shift in the dark-focus point and in the accommodation to targets presented at distances greater than the dark-focus point, in the direction of greater myopia. Because this increase in myopia is accompanied by increased reports of visual discomfort, the authors suggest there is reasonable evidence to suggest that accommodation state can provide a useful measure of objective visual fatigue. However, neither Murch (1983) nor Hedman (1988) were able to confirm the results of increased myopia following intense VDU work.

The interpretation of many of these results is debatable, particularly as little control has been taken over possible accompanying variations in pupil size which might reduce the demands on the accommodation system. Some researchers have suggested that one reason for the reports of visual fatigue from those people who work at VDUs is because the viewing distance is usually less than their dark-focus point. In support of this conclusion, Jaschinski-Kruza (1987) reported that subjects reported fewer feelings of visual fatigue with a viewing distance of 100 cm compared with one of 50 cm, particularly in the case of subjects whose dark-focus distance was relatively long. There is, however, a danger in adopting this approach. People can perform many different kinds of visual tasks, such as reading from hard copy or driving at distances, which do not coincide with their dark-focus distance, but without experiencing fatigue. Clearly, there are other important task features, such as image quality and the amount of variation in viewing distance that determine the ideal viewing distance.

Convergence and phorias

Convergence can be measured by the same methods used to measure accommodation. Traditional methods usually involve measuring the near point of convergence. For example, Luckiesh and Moss (1935a) developed a method whereby a pair of prisms were automatically rotated in front of the subject's right eye in order to gradually bring a stimulus closer to a subject until the motion was stopped by the subject when (s)he reported seeing the stimulus as double. After reading for 1 h under a low level of illumination (11 lux), results showed that the amplitude of convergence as measured by prism strength was reduced from around 16 to 13 dioptres. Gunnarsson and Soderberg (1983) measured the near point of convergence of VDU operators by moving a pointer along a ruler towards the base of the nose until they reported seeing a double image. An increase in the near point of convergence was observed (i.e. a reduced ability to converge) as the working day progressed. There are, however, doubts over the reliability of these results suggesting a fatigue to the convergence system. Stone *et al.* (1980), based on similar methods of measurement to Gunnarsson and Soderberg, reported that subjects showed a greater ability to converge after performing visually demanding tasks, irrespective of the contrast of the visual material. In addition, using laser optometry to measure the point of task convergence, Murch (1983) was not able to find any effects on the point of task convergence following 2·5 h of intense display work.

Rather than using measures of convergence, some authors have looked for changes in muscle balance following visually demanding work. Muscle balance can be measured approximately by industrial visual screening equipment such as the Bausch and Lomb Vision Tester or the Keystone Ophthalmic Telebinocular. To measure the balance, separate stimuli are displayed to the two eyes and because there is no meaningful way that the two images can be merged, it is possible to measure the tendency of the eyes to deviate inwards (esophoria) or outwards (exophoria) when binocular fusion is impossible. Some studies have found no changes in phoria following strenuous visual work (e.g. Collins, 1959) while others (Stone *et al.*, 1980) have shown a trend to esophoria. What seems likely is that whether or not changes occur in near point convergence or phoria will depend upon the actual task viewing distance. Recently, Marek *et al.* (1988) have reported an increase in the instability of phorias as VDU work progressed.

Pupil size

Pupil size has been monitored by Zwahlen *et al.* (1984) and by Geacintov and Peavler (1974). The former study used the pupil-centre corneal-reflection technique and failed to find any effects on pupil diameter as a result of performing a lengthy VDU task. This result confirms Campbell and Whiteside's (1950) conclusion that it is probably difficult, if not impossible, to fatigue the sphincter muscle of the iris. On the other hand, the study of Geacintov and Peavler (1974) based on infra-red photographic records of the pupil did show a gradual constriction of the pupil with continuous work when using material either printed on paper or displayed on a microfilm reader. It was difficult to conclude if the extent of the constriction varied between the two presentation methods. There is the possibility that the increased constriction reflected either a lowering of arousal or an adaptive response to increase the depth of field and thus reduce the demands on accommodation control.

Blinking

Blinking can be measured comparatively easily by either direct filming techniques or using electromyographic (EMG) recording techniques. Bitterman and Soloway (1946) compared both methods and failed to find any effects of glare on blink rate. Results of Brozek *et al.* (1950) failed to show any consistent reduction in blink rate as a result of performing a strenuous visual task. On the other hand, Poulton (1958) on re-analysing the results of Carmichael and Dearborn (1947), found evidence of increased blink rate as a function of the time spent on reading. These results were confirmed by Krivohlavy *et al.* (1969) who were concerned with close machine work. It seems likely that blink rate is affected by mental workload (see the chapter by Meshkati *et al.*) and this is supported by the results of Stern and Skelly (1984), which showed that fewer blinks were executed by pilots when controlling an aircraft compared with when they were not controlling the aircraft. In this context, Stark (National Research Council, 1983) has suggested that blinking be considered as a 'cybernetic windshield wiper' for the retina. Because a reduction in blink rate leads to drier eyes, it is not unexpected to find that more demanding tasks should yield complaints of irritation to the eyes. In the same way, a dry atmosphere is likely to cause similar complaints or an increase in blink rate.

Rather than use blink rate as an index of fatigue, Berg *et al.* (1988) have used the electrical activity originating in the muscles surrounding the eye socket: paraorbital electromyography. These muscles, in addition to controlling blinking, control the muscles which determine the facial expressions, including squinting, often associated with eyestrain, and particularly those said to result from specular glare. Berg *et al.* (1988) found significant changes in the spectral power distribution of the recorded muscle activity as a result of prolonged VDU work and they also showed that the distribution was affected by whether subjects performed the proof-reading task under what could be described as either good or poor VDU viewing conditions.

Visual acuity measures

Using the so-called 'li' test developed by Ferree and Rand (1927), Tinker (1939) reported decreases in clear seeing after reading text under fairly low levels of illumination (around 10 lux). With improved recording techniques, Haider *et al.* (1980) have proposed transient myopia to far targets as a possible indicator of fatigue having observed significant decreases in acuity following 3 h of continuous VDU work. Apart from the fact that other investigators have failed to replicate these results (Dainoff *et al.*, 1981), it is difficult to decide whether to attribute reduced acuity to fatigue of one of the oculomotor control systems such as the vergence or accommodation systems, to a reduced sensitivity of the visual processing processes, or to a general reduction in arousal.

By using measures of contrast sensitivity, rather than traditional measures of acuity, Jaschinski-Kruza (1984) was able to suggest that the source of the observed decrease in sensitivity was probably due to optical factors because it could be compensated for by the addition of an appropriate lens. He was also able to show that contrast sensitivity recovered to the pre-work value after only 1·5 min of rest. Lunn and Banks (1986) were able to demonstrate that after reading text presented on a VDU, spatial sensitivities in the range of 2–6 Hz were reduced. This range is largely responsible for controlling the accommodative response and therefore, they argue, adaptation at these frequencies could account for the changes in accommodation that have been described earlier. However, the same arguments would apply to reading from hard copy. In addition, Woo *et al.* (1987) were unable to demonstrate any consistent changes in contrast sensitivity at the end of a day of VDT work.

As well as spatial acuity, there have been studies which have used temporal measures. Aoki *et al.* (1984) failed to find any effect on critical fusion frequency (CFF) of continuous VDU work but Osaka (1985) reported a reduction in CFF for both foveal and peripheral tests particularly if the VDU task was performed with either blue or red characters. As with measures of spatial acuity, it is difficult to establish what is responsible for this decrease in CFF. That general arousal levels are affected by continuous performance of a visual task was demonstrated as early as 1935 by Luckiesh and Moss (1935b). They reported that heart rate level progressively decreased as the subject read text under an illuminance of 11 lux. Since then, CFF has frequently been interpreted as a measure of general arousal (e.g. Weber *et al.*, 1980).

Visual performance measures

Numerous studies have reported the results on performance measures in relation to the main fatiguing task. Apart from reading tasks (Carmichael, 1948; Nordqvist *et al.*, 1986) and simulated inspection tasks (Brozek *et al.*, 1950; Murch, 1983), a number of specially designed laboratory tasks have been used. These include Weston's original Landolt rings test (Weston, 1953), Bodmann's search tasks (Bodmann, 1962) and Neisser's search tasks (Neisser, 1964). These tasks were originally used to compare the effects of various task factors including target illuminance, contrast and size. Modifications to these tasks have been used to examine the same factors (Khek and Krivohlavy, 1966; Boyce, 1974; Stone *et al.*, 1980). Occasionally, these specially designed tasks are given in addition to the main fatiguing tasks. Thus Nordqvist *et al.* (1986) presented subjects with a Neisser-type search task after completing every 15 min of the main reading task. The results in relation to scanning speeds and errors on the Neisser-type search task were inconclusive and showed little evidence of a decline in performance as a result of the intervening demanding tasks presented either as hard copy or displayed on Videotex.

There are three major limitations to performance measures. First, any decline in performance over time may reflect factors unrelated to visual fatigue, factors such as boredom, low arousal and possibly mental fatigue. Second, it is possible for people to prevent performance from falling off either by supplying increased 'effort' or as a result of learning. Finally, it is often difficult to obtain valid measures, particularly as most tasks involve a speed/error trade off. For example, Wilkinson and Robinshaw (1987) reported that more errors were committed and reading speed increased over time as subjects performed a 50-min proof-reading task. Thus there was the possibility that subjects changed their speed/accuracy trade off as time proceeded rather than that there was a decline in performance. Using a somewhat *ad hoc* index of change in performance, the authors were able to conclude that the increase in errors could not be totally attributed to an increase in reading speed. Comparing hard copy with VDU presentation of the stimulus material, they also demonstrated that the decline in performance over time, indicative of fatigue they claim, was much greater for the VDU presentation. (Bullimore *et al.*, in the preceding chapter, provide a wider review and critique of visual performance assessment generally.)

Symptoms of asthaenopia

As mentioned before, it is the high frequency of reported symptoms of asthaenopia accompanying the introduction of VDU-based work into the workplace that has led to the revival of interest in the topic of visual fatigue. It is, therefore, worth considering the possible reasons for these complaints. The introduction of VDUs obviously alters the physical characteristics of the displayed information. Amongst the obvious differences from printed material are the size of the characters, their shape, their contrast and colour and their luminance profile, the presence of flicker, the viewing angles and distances, and the general lighting conditions under which the material is viewed. In addition, the introduction of VDUs alters the job characteristics. Not only are working methods and workload levels changed but the amounts of pacing and autonomy are altered. The effects of all these factors are reflected in measures of job attitudes.

When it comes to considering the characteristics of the display, Campbell and Durden (1983) have implied that there is often little *a priori* evidence why several of the characteristics of VDU-based work should lead to visual fatigue, such is the adaptability of the visual system. In fact, if one looks at those studies which have used well designed control groups, there is not as much evidence as commonly thought to suggest that working at a VDU itself does lead to more complaints of visual fatigue. For example, Muter *et al.* (1982), who made a direct comparison between continuous reading of material presented as hard copy and on a VDU, found no significant differences in subjective measures of comfort although subjects did read more slowly in the case of the VDU group. In the study of Nordqvist *et al.* (1986), there were again no significant differences in reported visual fatigue between those subjects who read from hard copy and those that read from a VDU and, unlike the previous study, there were no differences in the number of pages read under the two conditions.

Turning to less controlled field studies where, nevertheless, care was taken to balance control and experimental tasks in respect of job characteristics, results often fail to show any differences in the frequency and severity of reported symptoms of visual fatigue (Starr *et al.*, 1982; Howarth and Istance, 1985). However, against this, Knave *et al.* (1985) have reported that groups of subjects exposed to over 5 h of VDU work per day recorded more eye discomfort symptoms than the control groups. Unfortunately, the authors do not tell us the extent to which the groups were matched. More interesting, and less speculative, are their results concerning gender effects. Generally, the women tended to report a higher frequency and severity of symptoms of eye discomfort than the men, particularly under the VDU condition, a result that has been reported in several other studies (e.g. Levy and Ramberg, 1987). There is no obvious reason why there should be such gender effects solely arising from the way visual information is presented and these results strongly suggest that the tendency to report symptoms of visual discomfort must, therefore, often be related to factors which are not visual in origin. Such a suggestion is strongly supported by the analysis of subjective data reported by Howarth and Istance (1986). They used both pre-survey reports of symptom prevalence based on a single retrospective questionnaire given before the real survey and reports of symptoms obtained during the course of the survey given before and after each working day for one week. They found that if subjects were matched for their pre-survey reports, there were still significant differences in the frequency of reports obtained during the course of the survey and vice versa. The authors propose that this indicates that groups may differ in their reporting behaviour when comparisons are based on comparing group responses to single questionnaires. Megaw (1986) suggested that an increase in reported symptoms of visual fatigue may reflect a general dissatisfaction with VDU work as well as a fear over the employment prospects arising from the introduction of new technology. Summing up the results based on subjective reports, and as noted by Bullimore *et al.* in the chapter on visual performance also, Helander *et al.* (1984) went so far as to state that of the 28 studies they surveyed, 21 of which were field studies, a majority had serious design faults to the extent that the results were often useless. Returning for a moment to gender effects, there are, for instance, doubts with some studies that the men and women were carrying out identical tasks.

Conclusions

Just because some doubts have been cast over the validity of reported symptoms of fatigue and because of the absence of any coherent evidence of more objective indications, the existence of visual fatigue should not be denied. Laboratory studies have demonstrated that the vergence, version and accommodation mechanisms can be fatigued, but that usually this requires subjects to perform tasks that can be considered 'unnatural', that is to say, tasks that are unlikely to be performed during the course of normal visual behaviour. Moreover, recovery from fatigue induced by such laboratory methods is extremely rapid, less than minutes and often in the order of a few seconds.

At the same time, there is some evidence to suggest that certain display characteristics such as stimulus contrast and screen flicker can modify oculomotor performance and that these modifications are accompanied by subjective reports of visual discomfort. However, this does not necessarily mean that these changes in oculomotor performance reflect fatigue of the oculomotor control systems. One cannot overemphasize the close interplay that takes place between the accommodation, convergence and pupil control systems, and that, therefore, any changes in oculomotor performance may only reflect an adaptive response to the prevailing visual conditions where the burden of control is redistributed amongst the different control systems.

To establish the presence of fatigue more unequivocally, it will be necessary to examine functioning of the control systems in relation to factors which would on *a priori* grounds be expected to lead to fatigue. For example, there is the possibility that both the accommodation and vergence systems might become fatigued as a result of a 'hunting' action when viewing displays that are untextured which, therefore, provide few distance cues or that have images reflected in them. Similarly, the idiosyncratic effects from VDU viewing on certain eye movement parameters might reflect a disturbance to the vergence and version control mechanisms. Until our knowledge of the various oculomotor systems and their complex interactions are extended, it will remain difficult to decide whether or not the reported symptoms of visual discomfort do reflect fatigue of the oculomotor mechanisms or whether they are more a consequence of non-visual factors related to job design and job satisfaction.

References

Adler, F.H. (1962). *Textbook of Ophthalmology*, (Philadelphia: W.B. Saunders).

Aoki, K., Yamonoi, N., Aoki, M. and Horie, Y. (1984). A study of the change of visual function in CRT display tasks. In *Human-Computer Interaction*, edited by G. Salvendy (Amsterdam: Elsevier), pp. 465–468.

Bahill, A.T. and Stark, L. (1975). Overlapping saccades and glissades are produced by fatigue in the saccadic eye movement system. *Experimental Neurology*, **48**, 95–106.

Berg, W.K., Krantz, J.H., McGovern, J.B., Donohue, R.L. and Silverstein, L.D. (1988). The development of an objective electromyographic measure of VDT-induced visual stress. In *Proceedings of the 1988 Society for Information Display International Symposium Digest of Technical Papers*, Volume XIX, edited by J. Morreale. (Playa del Rey, CA: Society for Information Display) pp. 348–351.

Bitterman, M.E. and Soloway, E. (1946). Frequency of blinking as a measure of visual efficiency: some methodological considerations. *American Journal of Psychology*, **59**, 676–681.

Bodmann, H.W. (1962). Illumination levels and visual performance. *International Lighting Review*, **13**, 41–47.

Boyce, P.R. (1974). Illuminance, difficulty, complexity and visual performance. *Lighting Research and Technology*, **6**, 222–226.

Brozek, J., Simonson, E. and Keys, A. (1950). Changes in performance and in ocular functions resulting from strenuous visual inspection. *American Journal of Psychology*, **63**, 51–66.

Campbell, F.W. and Durden, K. (1983). The visual display terminal issue: a consideration of its physical, psychological and clinical background. *Ophthalmic Physiological Optics*, **3**, 175–192.

Campbell, F.W. and Whiteside, T.C.D. (1950). Induced pupillary oscillations. *British Journal of Ophthalmology*, **34**, 180–189.

Carmichael, L. (1948). Reading and visual fatigue. *Proceedings of the American Philosophical Society*, **92**, 41–42.

Carmichael, L. and Dearborn, W.F. (1947). *Reading and Visual Fatigue* (Boston: Houghton Mifflin).

Collins, J.B. (1959). Visual fatigue and its measurement. *Annals of Occupational Hygiene*, **1**, 228–236.

Collins, J.B. and Pruen, B. (1962). Perception time and visual fatigue. *Ergonomics*, **5**, 533–538.

Cornsweet, T.N. and Crane, H.D. (1970). Servo-controlled infra-red optometer. *Journal of the Optical Society of America*, **60**, 548–554.

Dainoff, M.J., Happ, A. and Crane, P. (1981). Visual fatigue and occupational stress in VDU operators. *Human Factors*, **23**, 421–438.

Ferree, C.E. and Rand, G. (1927). An investigation of the reliability of the 'li' test. *Transactions of the Illuminating Engineering Society*, **22**, 52–75.

Geacintov, T. and Peavler, W.S. (1974). Pupillography in industrial fatigue assessment. *Journal of Applied Psychology*, **59**, 213–216.

Gentles, W. and Llewellyn Thomas, E. (1971). Effect of benzodiazepines upon saccadic eye movements in man. *Clinical Pharmacology and Therapeutics*, **12**, 563–574.

Gunnarsson, E. and Soderberg, I. (1983). Eye strain resulting from VDT work at the Swedish telecommunications administration. *Applied Ergonomics*, **14**, 61–69.

Haider, M., Kundi, M. and Weisenbock, M. (1980). Worker strain related to VDUs with differently coloured characters. In *Ergonomic Aspects of Visual Display Terminals*, edited by E. Grandjean and E. Vigliani (London: Taylor and Francis), pp. 53–64.

Harris, C.M., Abramov, I. and Hainline, L. (1984). Instrument considerations in measuring fast eye movements. *Behavior Research Methods and Instrumentation*, **16**, 341–350.

Hedman, L.R. (1988). VDT users and eyestrain. *Displays*, **9**, 131–133.

Helander, M.G., Billingsley, P.A. and Schurick, J.M. (1984). An evaluation of human factors research on visual display terminals in the workplace. In *Human Factors Review: 1984*, edited by F.A. Muckler, (Santa Monica:

Human Factors Society), pp. 55–129.

Hennessey, R.T. and Leibowitz, H.W. (1972). Laser optometer incorporating the Badal principle. *Behavior Research Methods and Instrumentation*, **4**, 237–239.

Howarth, P.A. and Istance, H.O. (1985). The association between visual discomfort and the use of visual display units. *Behaviour and Information Technology*, **4**, 131–149.

Howarth, P.A. and Istance, H.O. (1986). The validity of subjective reports of visual discomfort. *Human Factors*, **28**, 347–351.

Iwasaki, T. and Kurimoto, S. (1987). Objective evaluation of eye strain using measurements of accommodative oscillation. *Ergonomics*, **30**, 581–587.

Iwasaki, T. and Kurimoto, S. (1988). Eye-strain and changes in accommodation of the eye and in visual evoked potential following quantified visual load. *Ergonomics*, **31**, 1743–1751.

Jaschinski-Kruza, W. (1984). Transient myopia after visual work. *Ergonomics*, **27**, 1181–1189.

Jaschinski-Kruza, W. (1987). Is the resting state of our eyes a favourable viewing distance for VDU-work? In *Work with Display Units 86*, edited by B. Knave and P.G. Wideback (Amsterdam: North-Holland), pp. 526–538.

Kahneman, D. and Beatty, J. (1966). Pupil diameter and memory load. *Science*, **154**, 1583–1585.

Khek, J. and Krivohlavy, J. (1966). Variation of incidence of error with visual task difficulty. *Light and Lighting*, **59**, 143–145.

Knave, B.G., Wiborn, R.I., Voss, M., Hedstrom, L.D. and Berqvist, U.O. (1985). Work with video display terminals among office employees: 1. Subjective symptoms and discomfort. *Scandinavian Journal of Work, Environment and Health*, **11**, 457–466.

Krivohlavy, J., Kodat, V. and Cizek, P. (1969). Visual efficiency and fatigue during the afternoon shift. *Ergonomics*, **12**, 735–740.

Leermakers, M.A.M. and Boschman, M.C. (1984). Eye movements, performance and visual comfort using VDTs. *IPO Annual Progress Report*, **19**, 70–75.

Leibowitz, H.W. and Owens, D.A. (1975). Night myopia and the intermediate dark focus of accommodation. *Journal of the Optical Society of America*, **65**, 1121–1128.

Levy, F. and Ramberg, I.G. (1987). Eye fatigue among VDU users and non-VDU users. In *Work with Display Units 86*, edited by B. Knave and P.G. Wideback, (Amsterdam: North-Holland), pp. 42–52.

Luckiesh, M. and Moss, F.K. (1935a). Fatigue of convergence induced by reading as a function of illumination intensity. *American Journal of Ophthalmology*, **18**, 319–323.

Luckiesh, M. and Moss, F.K. (1935b). The effect of visual effort upon the heart-rate. *Journal of General Psychology*, **13**, 131–138.

Lunn, R. and Banks, W.P. (1986). Visual fatigue and spatial frequency adaptation to video display of text. *Human Factors*, **28**, 457–464.

Malmstrom, F.V., Randle, R.J., Murphy, M.R., Reed, L.E. and Weber, R.J. (1981). Visual fatigue: the need for an integrated model. *Bulletin* of the *Psychonomic Society*, **17**, 183–186.

Marek, T., Noworol, C., Pieczonka-Osikowska, W., Przetacznik, J. and Karwowski, W. (1988). Changes in temporal instability of lateral and vertical phorias of the VDT operators. In *Trends in Ergonomics/Human Factors V*, edited by F. Aghazadeh (Amsterdam: North-Holland), pp. 283–289.

Matula, R.A. (1981). Effects of visual display units on the eyes: a bibliography (1972–1980). *Human Factors*, **23**, 581–586.

McConkie, G.W. (1981). Evaluating and reporting data quality in eye movement research. *Behavior Research Methods and Instrumentation*, **13**, 97–106.

Megaw, E.D. (1986). VDUs and visual fatigue. In *Contemporary Ergonomics 1986*, edited by D.J. Oborne (London: Taylor and Francis), pp. 254–258.

Megaw, E.D. and Sen, T. (1984). Changes in saccadic eye movement parameters following prolonged VDU viewing. In *Ergonomics and Health in Modern Offices*, edited by E. Grandjean (London: Taylor and Francis), pp. 352–357.

Messite, J. and Baker, D.B. (1984). Occupational health problems in offices: a mixed bag. In *Human Aspects in Office Automation*, edited by B.G.F. Cohen (Amsterdam: Elsevier), pp. 7–14.

Miyao, M., Allen, J.S., Hacisalihzade, S.S., Cronin, S.A. and Stark, L.W. (1988). The effect of CRT quality on visual fatigue. In *Trends in Ergonomics/Human Factors V*, edited by F. Aghazadeh, (Amsterdam: North-Holland), pp. 297–304.

Mourant, R.R., Lakshmanan, R. and Chantadisal, R. (1981). Visual fatigue and cathode ray tube display terminals. *Human Factors*, **23**, 529–540.

Murch, G.M. (1983). Visual fatigue and operator performance with DVST and raster displays. *Proceedings of the Society for Information Display*, **14**, 53–61.

Muscio, B. (1921). Is a fatigue test possible? *British Journal of Psychology*, **12**, 31–46.

Muter, P., Latremouille, S.A., Treurniet, W.C. and Beam, P. (1982). Extended reading of continuous text on television screens. *Human Factors*, **24**, 501–508.

National Research Council (1983). Committee on Vision. *Video Displays, Work and Vision*, (Washington: National Academy Press).

Neisser, U. (1964). Visual search. *Scientific American*, June, 94–100.

Nordqvist, T., Ohlsson, K. and Nilsson, L. (1986). Fatigue and reading text on videotex. *Human Factors*, **28**, 353–363.

Osaka, N. (1985). The effect of VDU colour on visual fatigue in the fovea and periphery of the visual field. *Displays*, **6**, 138–140.

Ostberg, O. (1980). Accommodation and visual fatigue in display work. In *Ergonomic Aspects of Visual Display Terminals*, edited by E. Grandjean and E. Vigliani (London: Taylor and Francis), pp. 41–52.

Ostberg, O. and Smith, M.J. (1987). Effects of visual accommodation and subjective discomfort from VDT work intensified through split screen technique. In *Work with Display Units 86*, edited by B. Knave and P.G. Wideback, (Amsterdam: North-Holland), pp. 512–521.

Phillips, S. and Stark, L. (1977). Blur: a sufficient accommodative stimulus.

Documenta Ophthalmologica, **3**, 65–89.

Post, R.B., Johnson, C.A. and Tsuetaki, T.K. (1984). Comparison of laser and infrared techniques for measurement of the resting focus of accommodation: mean differences and long-term variability. *Ophthalmic and Physiological Optics*, **4**, 327–332.

Poulton, E.C. (1958). On reading and visual fatigue. *American Journal of Psychology*, **71**, 609–611.

Ramazzini, B. (1700). *De Morbis Artificum—Diseases of Workers*, (English edition, New York: Hafner Publishing, 1964).

Rushton, W.A.H. and Westheimer, G. (1962). The effect upon the rod threshold of bleaching neighbouring rods. *Journal of Physiology*, **164**, 318–329.

Sen, T. and Megaw, E.D. (1984). The effects of task variables and prolonged performance on saccadic eye movement parameters. In *Theoretical and Applied Aspects of Eye Movement Research*, edited by A.G. Gale and F. Johnson, (Amsterdam: North-Holland), pp. 103–111.

Simmerman, H. (1950). Visual fatigue. *American Journal of Optometry and Physiological Optics*, **27**, 554–561.

Simonelli, N.M. (1980). Polarized vernier optometer. *Behavior Research Methods and Instrumentation*, **12**, 293–296.

Starr, S.J., Thompson, C.R. and Shute, S.J. (1982). Effects of video display terminals on telephone operators. *Human Factors*, **24**, 699–711.

Stellman, J.M., Klitzman, S., Gordon, G.C. and Snow, B.R. (1987). Work environment and the well-being of clerical and VDT workers. *Journal of Occupational Behaviour*, **8**, 95–114.

Stern, J.A. and Skelly, J.J. (1984). The eye blink and workload considerations. In *Proceedings of the Human Factors Society 28th Annual Meeting*, edited by M.A. Alluisi, S. de Groot and E.A. Alluisi, (Santa Monica: Human Factors Society), pp. 942–944.

Stone, P.T., Clarke, A.M. and Slater, A.I. (1980). The effect of task contrast on visual performance and visual fatigue at a constant illuminance. *Lighting Research and Technology*, **12**, 144–159.

Tinker, M.A. (1939). The effect of illumination intensities upon speed of perception and upon fatigue in reading. *Journal of Educational Psychology*, **30**, 561–571.

Weber, A., Fussler, C., O'Hanlon, J.F., Gierer, R. and Grandjean, E. (1980). Psychophysiological effects of repetitive tasks. *Ergonomics*, **23**, 1033–1046.

Weber, R.A. (1950). Ocular fatigue. *Archives of Ophthalmology*, **43**, 2.

Weston, H.C. (1953). *The Relation Between Illumination and Visual Performance*. MRC Industrial Health Research Board Report (London: HMSO).

Wilkins, A. (1986). Why are some things unpleasant to look at? In *Contemporary Ergonomics 1986*, edited by D.J. Oborne, (London: Taylor and Francis), pp. 259–263.

Wilkinson, I.M.S., Kime, R. and Purnell, M. (1974). Alcohol and human eye movement. *Brain*, **97**, 785–792.

Wilkinson, R.T. and Robinshaw, H.M. (1987). Proof-reading: VDU and paper text compared for speed, accuracy and fatigue. *Behaviour and Information Technology*, **6**, 125–133.

Woo, G.C., Strong, G., Irving, E. and Ing, B. (1987). Are there subtle changes in vision after use of VDTs? In *Work with Display Units 86*, edited by B. Knave and P.G. Wideback, (Amsterdam: North-Holland), pp. 490–503.

Young, L.R. and Sheena, D. (1975). Survey of eye movement recording methods. *Behavior Research Methods and Instrumentation*, **7**, 397–429.

Zwahlen, G.T., Hartmann, A.L. and Rangarajulu, S.L. (1984). *Video Display Work with a Hard Copy—Screen and a Split Screen Data Presentation*. Final Report, Department of Industrial and Systems Engineering, Ohio University.

Part VI

Analysis and evaluation of work systems

One of the most obvious changes in ergonomics practice during the past two decades or so has been the expansion of its area of application. Most in the field would now accept that the 'system' in the human-machine system embraces more than the interface controls and displays, the workplace and environment. We must be interested in the wider context of work, the psychosocial environment and the impact of organisational factors such as production technology, organisation structure and finance. A very good example of this is the field of human-computer interaction, and specifically the introduction and use of Visual Display Terminals (VDTs or VDUs). In the context of workers' health and comfort, as much is written about good job design, work organisation, support and training and technology implementation as about the purely physical factors of lighting, seating, keyboards etc.

This part of the book then, and Part VII following, takes a fairly broad view of work systems. The first two chapters by Kirwan (28) and Brown (29) concentrate upon assessments which can be made in order to evaluate the 'quality' or success of a work system—performance measurement—and also as methods to identify areas for system improvement through redesign. Along with recognition of the vital role of the human-computer interface in determining the success of computer systems, a major boost to the perceived importance of ergonomics has been the Three Mile Island incident in 1979, and subsequent concern for nuclear power plant safety. This concern has continued through the occurrence of a number of well publicised disasters or near disasters involving complex systems in power plants, chemical processes, transport systems and so on. Fundamental to many of these incidents and to safety in such systems is the potential for human error. Therefore there is great interest in developing methods for the analysis and measurement, and subsequent enhancement, of human reliability. Kirwan discusses this as a component of a ten part generic methodology, and provides very much the perspective of an active practitioner.

There is a strong link between the consideration of human reliability and the use of accident reporting and analysis techniques. In Chapter 29 Brown shows how theories of human error may well be very productive in terms of ergonomics measures for safety improvements. He distinguishes human error theories from theories of accident causation; in very much a conceptual review, he shows how vital it is to understand such theories in order to best develop an accident reporting system and, most importantly, use its data in accident reduction. Many of the points he makes and the cautions he gives about reporting and analysis could be read in the context of event observation generally (see Chapters 2 and 3).

If we are to take a true systems approach in occupational ergonomics, and if we are to have any meaningful impact in our work, then we must understand much about the organisation which is the site of our investigations. We must be able to assess, for instance, how the way an organisation is structured, its management philosophy, it's willingness to change etc., may influence behaviour or other outcomes we are measuring, or may affect any developments we initiate. Shipley in Chapter 30 provides a view on such issues which will be food for thought for any ergonomics practitioner, and within it presents a powerful argument about *how* ergonomists should tackle investigations into work, whatever the *what* of the particular study's focus.

In Chapter 1 it was noted that ergonomics can have aims which relate to organisational as well as individual well-being, although the linkage between these was stressed. Whilst many of us would, in an ideal world, put individuals' interests as our major priority, we must work in the real world of industry at least as far as occupational ergonomics is concerned. Here, Simpson and Mason argue (in Chapter 31), an economic case generally must be made before an organisation will fund, or even play host to, an ergonomics investigation. This consideration can be extended; a powerful motivation for ergonomics input in consumer product design for instance is the economic threat of strict product liability provisions. Thus, we do require some means of showing the economic returns on our efforts, and Simpson and Mason present some relevant techniques.

Chapter 28

Human reliability assessment

Barry Kirwan

Introduction

Recent accidents such as at Chernobyl and Bhopal have demonstrated
unequivocally the importance of considering human error in high risk systems
(see USSR State Committee, 1986; Bellamy, 1986). For any existing plant,
or new one being designed, it is important to try to assess the likelihood of
such accidents and prevent them from occurring. This requires the assessment
of the impact of human errors on system safety and, if warranted, the
specification of ways to reduce human error impact and/or frequency. These
are the major goals of human reliability assessment and the primary
domain for application and development of human reliability assessment
methodologies has been high technology high risk plant.

Ironically, *human reliability assessment* (HRA) owes its current status to the
very accidents it would wish to prevent. Whereas research and development
in this field has been carried out since the early 1960s, the drive for
development of sound and practicable methodologies received a hitherto
unparalleled boost following the accident at Three Mile Island (USNRC,
1980). This accident in particular, which rocked the nuclear power world's
foundations and beliefs that such accidents simply couldn't happen, brought
home the realization that human error was of fundamental importance, and
that the survival of the nuclear power industry (and other similar industries)
would depend on the ability to prevent such accidents from recurring.
Accidents since Three Mile Island, such as at Bhopal, the *Challenger* disaster
(Rogers *et al.*, 1986) and Chernobyl, have entirely reinforced this view;
accidents such as the Zeebrugge ferry disaster have similarly demonstrated
the importance of human error in low-technology systems (Reason, 1988a).
Human reliability assessment clearly has an important role to play, and this
role is likely to extend to many industries, wherever human errors can
propagate within systems to lead to unacceptable events.

HRA is a hybrid area, arising out of the disciplines of engineering and
reliability on the one hand, and psychology and ergonomics on the other.

The former require human error probabilities to fit neatly into the logical mathematical framework of *probabilistic safety analysis* (PSA), and the latter urge more detailed and theoretically valid modelling of the complexity of the human operator (Wagenaar, 1986). PSA is the quantitative statement defining the expected frequencies of accidents, and hence it determines whether or not a plant's risk compares favourably or otherwise against pre-defined risk criteria (Green, 1983). Thus HRA must be incorporated into PSA if risk is to be properly estimated.

Very recently, significant attempts have been made to properly integrate HRA into PSA (Bellamy *et al.*, 1986), and to validate HRA approaches (Embrey and Kirwan, 1983; Comer *et al.*, 1984; Kirwan, 1988). It is likely that within a few years more cogent and valid methodologies will appear, and be adopted and prescribed by the bodies empowered to regulate the various industries at risk from human error. Until such a time it is only possible to describe the current general approach of HRA and the most prominent and promising of the current methods, and this is the intention of this chapter.

A generic approach or framework for HRA is presented. Although this framework has 10 steps within it (as will be described), there is a core of three goals, namely:

1. *Human error identification.* What can go wrong?
2. *Human error quantification.* How often will a human error occur?
3. *Human error reduction.* How can human error be prevented from occurring or its impact on the system reduced?

Most research has focused on (2), the development of human reliability quantification techniques, although logically (1) is at least (if not more) important, since unless all significant errors are identified a HRA will underestimate the impact of human error. Whilst a good deal of research has been carried out in the field of human error identification and particularly error classification, few practical techniques have been developed for use in risk assessments. It is likely that future research and development will focus on the development of such techniques.

Human error reduction has recently become more important since there is an obvious need for this capability once HRA is being applied in earnest, as there are bound to be identified some unacceptably probable human errors, which must in some way be reduced in frequency. This is where the role of the ergonomist in HRA is clearest and most useful, since the ergonomist can usually specify a number of ways to improve the reliability of operators' performance.

This chapter concentrates primarily on human error quantification, as this has been most heavily researched, but also discusses some of the most recent human error identification techniques, and elaborates on the human error reduction approaches currently available. It does not go into some of the more complex mathematical issues underlying the integration of HRA into

PSA (see, for example, Apostolakis *et al.*, 1987; or Park, 1987), as these would require a chapter in themselves, and are not required for an appreciation of how HRA works and is applied. Furthermore, as quickly becomes apparent to the practitioner, human reliability analysis is far from being a precise science (and nor is probabilistic safety assessment itself; Nicks, 1981), but is a useful means of identifying and prioritizing plant safety vulnerabilities to human error, and thereby reducing the frequency of accidents.

The following section details the generic HRA methodology. Following this, there is a discussion of future areas of investigation and development needs within the field of HRA.

Human reliability assessment: a generic methodology

Once it is decided that a human error or human reliability problem requires analysis, a means of systematically solving this problem is required. The 'problem' may range from possible errors during a nuclear power plant emergency, to a desire for improved performance in an offshore maintenance task. Whatever the objective of the assessment, a systematic methodology of HRA will help ensure that the problem itself is dealt with reliably, minimizing biases or errors distorting the analysis.

Each of the 10 steps of the methodology defined below, from initial problem definition to final documentation of the results, are briefly explained and further references are given on available guidelines and current research in these areas.

The generic human reliability assessment methodology encompasses the following:

1. *Problem definition.* To precisely define the problem and its setting in terms of the system goals and the overall forms of human-caused deviations from those goals.

2. *Task analysis.* To define explicitly the data, equipment, behaviour, plans, and interfaces used by the operators to achieve system objectives, and to identify factors affecting human performance within these tasks.

3. *Human error analysis.* To identify all significant human errors affecting performance of the system, and ways in which human errors can be recovered.

4. *Representation.* To model the human errors and recovery paths in a logical manner such that their impact on the system can be quantitatively determined. This usually necessitates integrating human errors with hardware failures in a fault or event tree.

5. *Screening.* To define the level of detail and effort with which the quantification will be conducted, by defining all significant human errors and interactions, and ruling out insignificant errors which can be effectively ignored by the study.

6. *Quantification.* To quantify human error probabilities and human error recovery probabilities, in order to define the likelihood of success in achieving the system goals.

7. *Impact assessment.* To determine the significance of human reliability with respect to the achievement of the system goals, to decide whether improvements in human reliability are required, and if so what are the primary errors and factors negatively affecting system reliability.

8. *Error reduction.* To identify error reduction mechanisms, means of supporting error recovery likelihood, and ways of improving human performance in achieving system goals, so that an acceptable level of system performance can be achieved.

9. *Quality assurance.* To ensure that the enhanced system satisfactorily meets system performance criteria, and will continue to do so in the future.

10. *Documentation.* To detail all information necessary to allow the assessment to be understandable, auditable, and reproducible.

Figure 28.1 shows the relationship between these various steps, as is discussed in detail later.

(1) Problem definition

There are two ways of defining the problem, dependent upon whether it is being considered as a problem in its own right, or as an integral part of a larger risk assessment. When a human reliability 'problem' has been identified (e.g. a desire to improve safety, or to assess risk or productivity), and defined in its system context (e.g. to assess the risks of offshore platform evacuation by lifeboat in severe weather), discussions should occur with system design and plant engineers, and if possible with operational and managerial personnel. Discussions around the problem will help define more precisely the scope of the project, and the range of conditions and scenarios which must be considered in order satisfactorily to resolve the problem, and/or fulfil the goals of a general safety assessment of a plant.

It is useful at this stage to define the system goals, at various levels, for which operator actions are required. This will in turn define higher level goals towards which the operators are aiming (e.g. maintain reactor core cooling, achieve highest output). It is also important to gain some understanding of these high level goals, and to determine in particular where and how the goal of safety fits in. Ideally there will be clear criteria via which the operators know when to 'drop' their production goals in favour of safety goals. If such criteria do not exist, then production goals may compromise safety goals and it is possible that the roots of a human reliability problem are already inherent in the system. Therefore, with existing systems it is well worth investigating this 'safety culture' aspect of the plant, as it can influence human reliability predictions dramatically, and is an important aspect of the problem

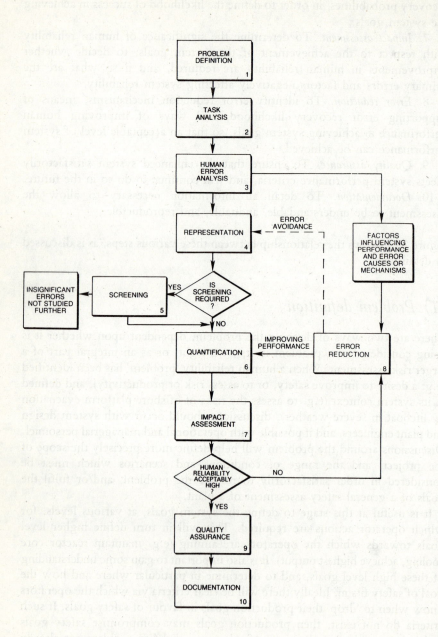

Figure 28.1. Steps in a human reliability assessment.

definition. If there are no clear criteria, it must not be assumed that operators will necessarily make the right decision in the heat of the moment.

If the HRA is being carried out as part of an overall risk assessment, the human reliability analyst will probably be given a set of scenarios which have been chosen for the risk analysis, and asked to consider human contributions to risk within these scenarios. In this case there are five types of human–system interaction which the analyst should consider with respect to an incident scenario (Spurgin *et al.*, 1987):

(a) maintenance/testing errors affecting safety system availability (latent errors),

(b) operator errors intiating the incident,

(c) recovery actions by which operators can terminate the incident,

(d) errors (e.g. misdiagnosis) by which operators can prolong or even aggravate the incident, and

(e) actions by which operators can restore initially unavailable equipment and systems.

Consideration of these types of interactions, and discussions with the system risk analysts at the problem definition stage will enhance the smooth integration of the human reliability analysis into the system risk analysis. It may also identify new important scenarios which the system analysts had not initially considered. This is important because the problem may otherwise be defined in too limited a scope. As an example, Kirwan's (1987) assessment of an emergency offshore depressurization system originally was only intended to look at operational failures during emergency scenarios and not maintenance aspects. However, when investigated further, a highly significant maintenance error was identified which had potentially dramatic effects on the whole platform; this error was probably of more significance than the entire set of operational failures put together.

At the end of the problem definition stage the problem to be addressed should be explicitly defined in its system context. A list of scenarios to be addressed and, within each scenario, a list of overall tasks required to achieve system and safety goals should also have been identified. This sets the scene for the task analysis phase. Figure 28.2 shows a brief example of the results of the problem definition phase.

(2) Task analysis

The object of task analysis is to provide a complete and comprehensive description of the tasks that have to be performed by the operator(s) to achieve the system goals. There are many forms of task analysis (Drury,

Problem:	To effect emergency shutdown (ESD) of a chemical plant during a loss of power scenario.
Problem Setting:	A computer-controlled, operator-supervised plant suffers a sudden loss of main power. The VDU display system wil also fail and so the operator, backed up by the supervisor, must initiate ESD manually using hardwired controls in the Central Control Room. However, due to valve failures on plant, these actions are only partially successful, and so the operator must send out another operator onto plant to determine which ESD valves have not closed. The CCR operator, via engineering drawings, can then determine which manual valves must be closed on plant. The outside operator must then go to close these valves, completing this action successfully within 2 hours from the onset of the scenario.
System Goals:	The overall system goals are safe shutdown of all feeds to the plant within 2 hours of loss of power. In this scenario there are no production goals once the event occurs since safety is clearly under threat. Prior to the event, the operator is concerned with achieving steady feed throughout via monitoring the top two levels of a VDU display hierarchy, and notifying the supervisor of any alarms higher than level 2. The outside operator (on plant) will have various duties associated with maintenance tasks.
Overall Human Error Considerations:	No operator initiating events were identified, and maintenance errors were not relevant except that identifying and moving the local manual valves could prove difficult. Recovery actions involve identifying the appropriate valves to close and closing them. Errors of failing to realise that ESD has not been 100% effective, and of mis-identifying the valves, appear most likely. Loss of power is so evident that misdiagnosis or failure to diagnose is not considered to be credible.

Figure 28.2. Example of problem definition.

1983), such as sequential task analysis which looks at operator actions as they occur in chronological order; hierarchical task analysis which considers tasks in terms of the hierarchy of goals the operator is trying to achieve (Shepherd, 1986); and tabular format decision task analysis (Pew *et al.*, 1987) which concentrates on cognitive decision-making aspects as a function of the information available, and operators' knowledge, expectations and beliefs about the situation. The latter type of analysis is useful for analysing operator diagnosis scenarios (e.g. for nuclear power plant emergencies). Other forms of analysis are discussed in terms of general use by Stammers *et al.*, elsewhere in this book.

If a detailed HRA is being carried out, then a task analysis is essential since it provides a detailed description of the operators' tasks from which it will be possible to identify errors in the next phase. The basic methods of deriving information for the task analysis are: observation; structured and unstructured interviews with operators, maintenance personnel, supervisors, managers and system designers; analysis of procedures; incident analyses; structured walkthroughs of procedures (where the operator talks the analyst through the procedure); and examination of system documentation such as engineering/process flow diagrams. In practice it is important not to rely on procedures/operating instructions as the sole source for defining the task, since often actual operating practices differ somewhat from the formal written documents.

For a proceduralized task in which the operator (or supervisor, or maintenance person is using familiar skills or written/remembered rules, a hierarchical task analysis is probably most appropriate. An example is shown in Figure 28.3, part of a task analysis for the task of starting up a plant, and considering the subtask 'warm up furnace'. Sequential information is represented in this form of hierarchical task analysis, via the numbers on the boxes at each level which determine the order in which the tasks should occur. Plan 4 in the diagram shows a more complex plan the operator must use while monitoring temperature and pressure (Shepherd, 1986).

Figure 28.4 illustrates another form of task analysis using a tabular format showing the type of information which can be recorded by the analyst during the task analysis. This type of task analysis may be more useful when the operator is in either a highly dynamic situation and/or one in which problem solving or diagnostic behaviour is required. The format focuses on the events as they occur in time, since the order in which events occur partly determines behaviour. However, the analyst must also be aware of the hierarchy of goals the operators are trying to achieve, which may shift during an emergency situation. In detailed analyses it may be necessary to utilize both hierarchical and tabular task analysis formats. Wherever possible the task analysis should be verified, if only by asking operations personnel to review it. Having defined the operators' tasks in detail, it is then appropriate to determine what can go wrong, in terms of what human errors can occur.

(3) Human error analysis

Human error analysis is arguably the most critical part of a human reliability analysis, since if a significant error is omitted at this stage then it will not appear subsequently in the analysis and hence the results may seriously underestimate the effects of human error on the system.

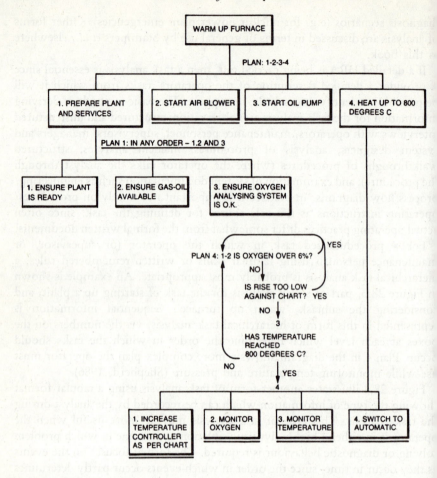

Figure 28.3. Example of hierarchical task analysis after Shepherd (1986).

The simplest approach is to consider the following possible 'external error modes' (Swain and Guttman, 1983) at each step in the procedure defined in the task analysis:

Error of omission   –   act omitted (not carried out)
Error of commission –   act carried out inadequately
                      –   act carried out in wrong sequence
                      –   act carried out too early/late
                      –   error of quality (too little/too much)
Extraneous error   –   wrong (unrequired) act performed

This approach is rudimentary but nevertheless can identify a high proportion of the potential human errors which can occur, as long as the assessor has a ood knowledge of the task and a good task description of the operator–system interactions.

Another method for human error analysis is embedded within the systematic *Human Error Reduction and Prediction Approach* (SHERPA: see Embrey, 1986a). This human error analysis method consists of a computerized question–answer routine which identifies likely errors for each step in the task analysis. The error modes identified are based on the 'skill rule and knowledge' model (Rasmussen *et al.*, 1981), and *Generic Error Modelling System* (GEMS: Reason and Embrey, 1986). Table 28.1 shows the psychological error mechanisms underlying the SHERPA system. An example of the tabular output from such an analysis is shown in Figure 28.5 (Kirwan and Rea, 1986). This 'human error analysis table' has similarities to certain reliability engineering approaches to identifying the failure modes of hardware components. One particularly useful aspect of this approach is the determination of whether errors can be recovered immediately, at a later stage in the task, or not at all, information useful if error reduction is required later in the analysis. This particular tabular approach also attempts to link error reduction measures to the causes of the human error, on the grounds that treating the 'root causes' of the errors will probably be the most effective way to reduce error frequency.

Another computerized system is the *Potential Human Error Cause Analysis* (PHECA) system (Whalley, 1988). Figure 28.6 shows the error causes and mechanisms inherent in the model, and Figure 28.7 shows the major performance shaping factors which interact with the error causes. This approach has also borrowed from the reliability world in the form of the well established HAZOP (Hazard and Operability Study) technique (Kletz, 1984), as all errors identified in PHECA can occur in only the following 'external error mode' forms:

Not Done	Repeated
Less than	Sooner than
More than	Later than
As well as	Mis-ordered
Other than	Part of

Human error identification technique development has, as noted above, generally received less attention than human error quantification. However, a general pattern is already emerging in terms of the functions required of any human error identification technique. Useful techniques must address the following:

External error mode identification. All human errors can be categorized into these descriptions. They are the level of description which is put into the probabilistic safety assessment, e.g. operator fails to respond to alarm ('error of omission', or 'not done', etc.).

T	Opr	System Status	Info Available	Operator Expectation	Procedure (written, memorised)	Decision/ Communications Act	Equipment/ location	Feedback	Secondary duties; Distractions; Penalties	Comments
0:30 mins	SS OO CRO	ESD; Cell 23 full of gas; mixture beyond explosion point at present; platform at muster status.	Gas cloud; Loud roaring noise;	Looking for leak in pipe union, flange, seal etc. Check near the gas detectors which were alarming.	1) Locate source and isolate if possible. 2) Maintain personal safety (use breathing apparatus) 3) Prevent ignition of gas	Ops search for leak in compressor module using sound and visual cues, as well as the gas detectors. Communicate to CCR.	Cell 23 gas detectors	Noise and smell of gas tactile cue if flesh exposed to gas jet path; if cold enough may see white gas plume from leak.	Maintain personal safety and avoid causing ignition. Extra delays if depressurise	Search will be more difficult in breathing apparatus. Deluge would make search safer, though less likely to succeed.
0:40	OFM CCRO SS	As at 0:30	Panel indications and from operators now outside of cell 23.	Gas leak now confirmed. Ignition possible.	Minimise chances of ignition.	CCR operator reviews vessel pressures on VDU system; gives OIM status report and confirms significant gas leak occurrence. CCR opts to consider whether to depressurise			Many enquiries may block communication channels, and must be responded to by CCR operators.	Consider merits of removing compressor liquid – How long would the vessels withstand fire if ignited? Results of hazard analyses should be immediately transmitted to offshore

| 0:42 | CCRO | As at 0:30 | Deluge Available (from light on console) | Ignition still likely | Tells outside operators to clear cell 23. Muster points broadcast on public address system. Deluge activated. | the liquid in the vessels which can only be removed if gas pressure remains in the compressors. However, no remote controls exist and it is considered too hazardous to try the operation. Decide therefore to master personnel at other end of the platform until delute cools down compressors and pressure drops. | | emergency centre to update their knowledge and enhance decision-making. |

Figure 28.4. Example of tabular task analysis.

TASK 51: TERMINATE SUPPLY AND ISOLATE TANKER. (SEQUENCE OF REMOTELY – OPERATED VALVE OPERATIONS)

TASK STEP	ERROR TYPE	RECOVERY STEP	PSYCHOLOGICAL MECHANISM	CAUSES, CONSEQUENCES AND COMMENTS	RECOMMENDATIONS		
					PROCEDURES	TRAINING	EQUIPMENT
51.1	ACTION TOO LATE	NO RECOVERY	PLACE LOSING ERROR	OVERFILL OF TANKER RESULTING IN DANGEROUS CIRCUMSTANCE.	OPERATOR ESTIMATES TIME/RECORDS AMOUNT LOADED	EXPLAIN CONSEQUENCES OF OVERFILLING	FIT ALARM-TIMING/VOLUME/TANKER LEVEL
51.2.1	ACTION OMITTED	5.2.4	SLIP OF MEMORY	FEEDBACK WHEN ATTEMPTING TO CLOSE CLOSED VALVE. OTHERWISE ALARM WHEN LIQUID VENTED TO VENT LINE			MIMIC OF VALVE CONFIGURATION
51.2.2	ACTION TOO EARLY	5.2.2	PLACE LOSING ERROR	ALARM WHEN LIQUID DRAINS TO VENT LINES	SPECIFY TIME FOR ACTIONS	OPERATOR TO COUNT TO DETERMINE TIME	
	ACTION OMITTED	5.2.2	SLIP OF MEMORY	AS ABOVE AND POSSIBLE OVER-PRESSURE OF TANKER (SEE STEP 5 1.2.3)			MIMIC OF VALVE CONFIGURATION
51.2.3	ACTION TOO EARLY	NO RECOVERY	PLACE LOSING ERROR	IF VALVE CLOSED BEFORE TANKER SUPPLY VALVE OVERPRESSURE OF TANKER WILL OCCUR		STRESS IMPORTANCE OF SEQUENCE AND EXPLAIN CONSEQUENCES	INTERLOCK ON TANKER VENT VALVE
	ACTION OMITTED	5.2.6	SLIP OF MEMORY	AUTOMATIC CLOSURE ON LOSS OF INSTRUMENT AIR.			MIMIC OF VALVE CONFIGURATION
51.2.4	ACTION OMITTED	5.2.2	SLIP OF MEMORY	AUDIO FEEDBACK WHEN VENT LINE OPENED.		EXPLAIN MEANING OF AUDIO FEEDBACK	MIMIC OF VALVE CONFIGURATION
51.3	ACTION OMITTED	NO RECOVERY	SLIP OF MEMORY	LATENT ERROR.	ADD CHECK ON FINAL VALVE POSITIONS BEFORE PROCEEDING TO NEXT STEP		MIMIC OF VALVE CONFIGURATION

Figure 28.5. Example extract of human error analysis (Kirwan and Rea, 1986).

Table 28.1. SHERPA: classification of psychological mechanisms (see Reason and Embrey, 1986)

1. *Failure to consider special circumstances.* A task is similar to other tasks but special circumstances prevail which are ignored, and the task is carried out inappropriately
2. *Short cut invoked.* A wrong intention is formed based on familiar cues which activate a short cut or inappropriate rule
3. *Stereotype takeover.* Owing to a strong habit, actions are diverted along some familiar but unintended pathway
4. *Need for information not prompted.* Failure of external or internal cues to prompt need to search for information
5. *Misinterpretation.* Response is based on wrong apprehension of information such as misreading of text or an instrument, or misunderstanding of a verbal message
6. *Assumption.* Response is inappropriately based on information supplied by the operator (by recall, guesses, etc.) which does not correspond with information available from outside
7. *Forget isolated act.* Operator forgets to perform an isolated item, act or function, i.e. an act or function which is not cued by the functional context, or which does not have an immediate effect upon the task sequence. Alternatively it may be an item which is not an integrated part of a memorized structure
8. *Mistake among alternatives.* A wrong intention causes the wrong object to be selected and acted on, or the object presents alternative modes of operation and the wrong one is chosen
9. *Place losing error.* The current position in the action sequence is misidentified as being later than the actual position
10. *Other slip of memory* (as can be identified by the analyst).
11. *Motor variability.* Lack of manual precision, too big/small force applied, inappropriate timing (including deviations from 'good craftmanship')
12. *Topographic or spatial orientation inadequate.* In spite of the operator's correct intention and correct recall of identification marks, tagging, etc., he unwittingly performs a task/act in the wrong place or on the wrong object. This occurs because of following an immediate sense of locality where this is not applicable or not updated, perhaps due to surviving imprints of old habits, etc.

Error cause or mechanism identification. These descriptions define, in psychologically and/or ergonomically meaningful terms, how the error actually occurred (e.g. the above 'failure to respond' may be due to a 'misinterpretation' of the signal, or due to 'reduced capabilities', etc.).

Identification of performance shaping factors (PSF). Factors which affect performance can obviously be usefully considered during the human error identification phase, although frequently they are not identified until the quantification phase. PHECA is currently the only technique which links PSF to error causes, which can be a source of useful information if error reduction is required at a later stage.

The distinction between the external error mode and the error cause or mechanism is particularly important if error reduction is required. Knowing the error mechanism, effective error reduction mechanisms will be more readily specified. It is likely that future human error identification techniques will follow this basic pattern as PHECA, and to a lesser extent SHERPA already do, making the errors identified more psychologically/ergonomically meaningful, and more helpful in reducing accident potential.

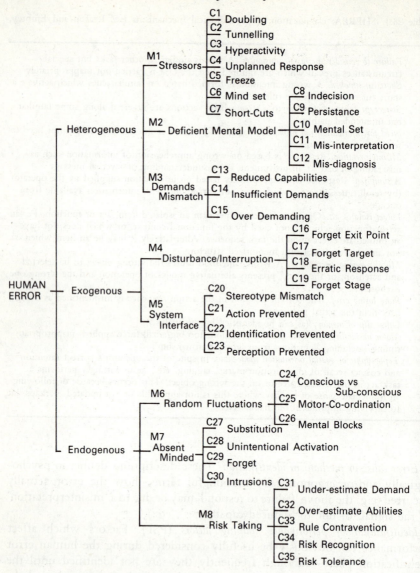

Figure 28.6. *Error causes grouped by error mechanisms (Whalley, 1988).*

There are very few reviews of human error identification techniques and approaches, although many paths have been explored (described in Kirwan, 1986). The one area in which all error identification methods have difficulty is in cognitive decision-making/diagnostic tasks, e.g. when an operator is trying to diagnose a complex set of symptoms in an emergency situation. Recent history, e.g. the Three Mile Island and the Davis-Besse (USNRC, 1985) incidents, has shown that one of the most significant potential human

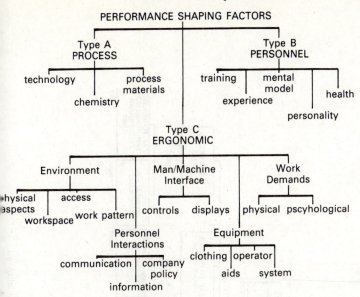

PERFORMANCE SHAPING FACTORS

Type A PROCESS
- technology
- chemistry
- process materials

Type B PERSONNEL
- training
- experience
- mental model
- personality
- health

Type C ERGONOMIC

Environment
- physical aspects
- access
- workspace
- work pattern

Man/Machine Interface
- controls
- displays

Work Demands
- physical
- pscyhological

Personnel Interactions
- communication
- company policy
- information

Equipment
- clothing
- operator aids
- system

Figure 28.7. Major sections of the performance shaping factors classification structure (Whalley, 1988).

errors is misdiagnosis during such an emergency, yet it is difficult to predict the form that a misdiagnosis may take. This form is important since a misdiagnosis may not only lead to an unsafe state or an accident, but may actually lead to a worse state of affairs than might have occurred had the operators done nothing. One method which has attempted to determine the nature of potential misdiagnoses is the confusion matrix approach (e.g. Potash *et al.*, 1981). This is basically a matrix of different possible scenarios, rated on similarity of symptoms (and hence confusibility) by operators and system dynamics experts. Other research has recently concentrated on expert–system based simulations of the operator (Woods *et al.*, 1987). However, usable tools are still not available, and so this is a primary target area for future research.

The identification of error recovery paths is largely carried out using the judgement of the analyst. The task analysis should also highlight points in the sequence at which discovery of an error will be possible, e.g. via indications (especially the occurrence of alarms) and checks or interventions by other personnel.

Analysts' judgement is a worthwhile resource to utilize in error identification. Many risk analysis practitioners build up experience of identifying errors in safety studies, either from their operational experience or from involvement in many different safety analyses. Whilst this is perhaps an 'art' rather than a science and as such is less accessible to the novice than a more formal method, its value must not be overlooked. Due to the specificity of every new safety analysis carried out, the human reliability practitioner 'standing

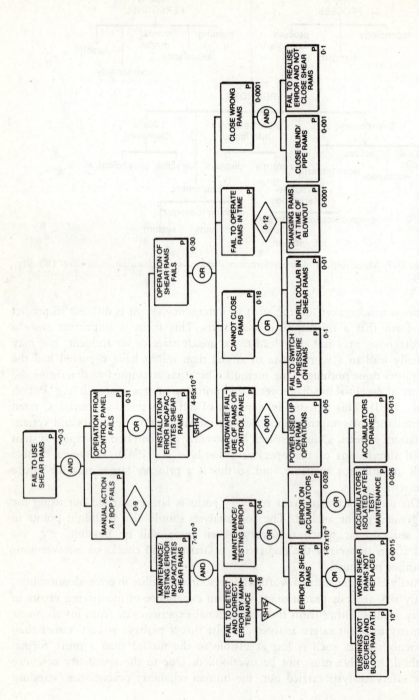

Figure 28.8. Offshore drilling blowout fault tree sub-tree: fail to use shear rams to prevent blowout.

on the outside' may often be more disadvantaged in terms of error identification than the hardware reliability analyst who knows the system details intimately. It is therefore worthwhile adopting a hybrid team approach to error identification as well as the use of formal systematic methods. The adage of 'two heads is better than one' is especially true in the area of human error identification.

The next step following identification of these error forms is to represent them in some logical format so that their effects on the system goals can be evaluated.

(4) Representation

A fault tree is a typical way of representing a set of human errors and their effects on the system goals. A fault tree is a logical structure which defines what events (human errors, hardware/software faults, environmental events) must occur in order for an undesirable event (e.g. an accident) to occur (Henley and Kumamoto, 1981). The undesirable event or outcome, usually placed at the top of the 'tree' and hence called the top event, may for example be 'failure to launch a lifeboat successfully at first attempt', or 'failure to achieve recirculation of primary coolant', etc. The tree is constructed primarily by using two types of 'gate' by which events at one level can proceed to the next level up until finally they reach the top event. The first type of gate is an 'OR' gate, and the event above this gate occurs if *any* one of the events joined below it by this gate occur. An event above an 'AND' gate only occurs if *all* the events joined below it by this gate occur. An example is shown in Figure 28.8.

A fault tree can be used to represent a simple or complex pattern of system failure paths, and may comprise human errors alone, or a mixture of human, hardware, and/or environmental events, depending upon the scenario. Once structured, the events (including human error probabilities) must be quantified to determine the overall top event frequency (or probability), and the relative contributions of each error to this undesirable event.

Another type of 'tree' is the operator action tree (OAT: see Figure 28.9). The OAT proceeds from an initiating event, usually placed at the left hand side of the tree (e.g. loss of power causes ESD demand), to consider a set of sequential events each of which may or may not occur, causing the tree to branch (usually) in a binary fashion at each event 'node'. The events and branching continue until an end state is reached for each path, which is either success in terms of achieving system safety, or else failure in terms of lost production, plant or equipment damage, injury, or fatality. OATs are especially useful when considering dynamic situations, and in general are preferable to fault trees when human performance is dependent upon previous actions/events in the scenario sequence (Hall *et al*, 1982).

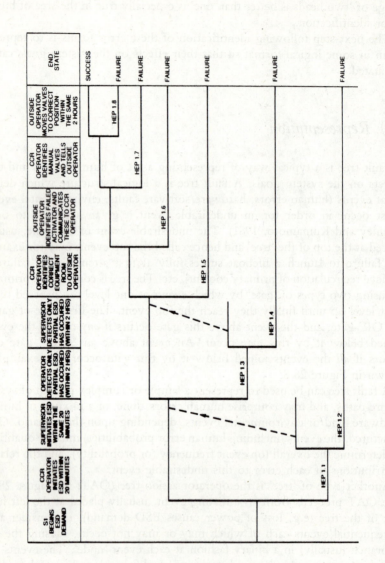

Figure 28.9. Operator action tree for ESD failure scenario (Kirwan, 1988).

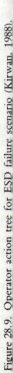

A recent variant on the OAT is to represent the emergency actions of operating personnel in the form of an event tree which uses three branches representing successful operation, failure to respond, and a mistake or misdiagnosis (USNRC, 1984). This type of OAT is able to represent important potential misdiagnoses that may have been identified in the human error analysis stage. If significant misdiagnoses are identified, the main event tree may branch off to another sub-event tree to consider the alternative consequences that may occur as a result of the mistake.

The above are formal methods used for representing moderately complex patterns and sequences of failures. Whilst the examples given here are simple, in practice such trees can become quite complex, with many 'nodes' and events, and in the case of event trees, a large number of possible final outcomes. It is something of an art to develop such trees so that the human errors are adequately and accurately represented, without letting the trees become too complex and unwieldy.

If in a particular assessment the number of errors is small, and their effect on the system goals is very simple, such representation may be unnecessary. Furthermore, some analyses may stop at this point if their objective was merely qualitative in nature, i.e. simply to identify human error modes without quantifying their probability or consequential effect on the system. In all other cases however, quantification of the human error probabilities will be necessary. Due to resource limitations screening may be applied to limit the amount of quantification required. If screening is not utilized, then quantification is the next step.

(5) Screening

A screening analysis identifies where the major effort in the quantification analysis should be applied. There may for example be particular tasks which are theoretically related to the system goals being investigated, but which in fact make little contribution to risk if they fail (due to diverse reliable back-up systems which adequately compensate for human error, or to the trivial nature of the tasks themselves. It is efficient to expend little effort on such tasks and instead focus on those in which human reliability is critical. The identification of those errors which can be effectively ignored by the rest of the study is the purpose of a screening analysis.

The *systematic human action reliability procedure* (SHARP) methodology defines three methods of screening logically structured human errors (see Spurgin *et al.*, 1987). The first method 'screens out' those human errors which can only affect the system goals if they occur in conjunction with an extremely unlikely hardware failure or environmental event. The second method involves allocating each human error a probability of 1·0, and examining the effects of the various errors on the system goals. Those that have a negligible effect even with a probability of unity are not considered

Table 28.2. Generic human error probabilities

Category	Failure probability
Simple, frequently performed task, minimal stress	10^{-4}
More complex task, less time available, some care necessary	10^{-3}
Complex, unfamiliar task, with little feedback and some distractions	10^{-2}
Highly complex task, considerable stress, little performance time	10^{-1}
Extreme stress, rarely performed task	10^{0}

further. The third method assigns broad probabilities to the human errors based on a simple categorization (such as the one shown in Table 28.2). This method works in the same way as the previous method but is a 'finer-grained' analytic method.

With most screening methods (particularly the third method above) there is a danger of ruling out of the study important errors and interactions, balanced against a need to reduce to a manageable level the complexity of, and resources required for, the analysis. As a general rule when applying any screening technique at any level in the study—if in doubt, leave the human error in the fault/event tree. The next step following screening, or following error analysis if screening was not applied, is quantification of the human errors in the fault or event trees.

(6) Quantification

Human reliability quantification techniques all quantify the human error probability (HEP), which is the metric of human reliability assessment. The HEP is defined as

$$\text{HEP} = \frac{\text{number of errors occurred}}{\text{number of opportunities for error to occur}}.$$

Thus, if when buying a cup of coffee from a vending machine on average one time in a hundred tea is accidentally purchased, the HEP is taken as 0·01 (it is somewhat educational to try and identify HEPs in everyday life with a value of less than once in a thousand opportunities, or even as low as once in ten thousand).

In an ideal world there would be many studies and experiments in which HEPs were recorded. In reality there are few such recorded data. The ideal source of human error 'data' would be from industrial studies of performance and accidents, but at least three reasons can be deduced for the lack of such data:

(a) difficulties in estimating the number of opportunities for error in realistically complex tasks (the so-called denominator problem),

(b) confidentiality and unwillingness to publish data on poor performance;

(c) lack of awareness of why it would be useful to collect data in the first place (and hence lack of financial incentive for such data collection).

There are other potential reasons (see Williams, 1983, for some of these) but the net result is a scarcity of HEP data. Other sources are simulator data (e.g. exercises using high-fidelity simulators), and data derived from experimental laboratory-based studies, reported in the human performance literature. Two problems exist with respect to simulator studies, the first being that such simulators are used almost exclusively for training purposes (and in the USA nuclear power industry for operator re-certification), and hence personnel on the simulator are highly motivated and frequently know what is on the training curriculum (i.e. they know which scenarios to expect). Secondly, it is not clear how realistic it is facing an emergency in a simulator compared with the real thing (the 'cognitive fidelity' issue). (See Meister elsewhere in this book for a more complete review of simulation.)

The human performance literature has a similar problem in that studies in this vein are usually highly controlled, often looking at one or two independent variables (unlike industry where many PSF vary and interact), and using reasonably motivated subjects for a short period of time. Generalizing from such studies to complex industrial multi-personnel situations is not easy, and is often a questionable exercise.

Overall, therefore, there is a 'data problem'. Furthermore, even if there is a sound datum for, e.g. a chemical plant operator failing to respond to an alarm in scenario X, how can this be generalized to scenario Y, or even to scenario X on a different chemical plant, with a different operating regime, or to a nuclear power plant? In other words, what defines the 'generalizability' of data? Such difficult and as yet unresolved issues as these have led to the development of non-data-dependent approaches, namely to the use of expert opinion. This is by no means necessarily a bad thing, and expert opinion has been used successfully in other areas (e.g. Murphy and Winkler, 1974; or Ludke *et al.*, 1977), and is in any case used at least occasionally in probabilistic safety assessments (Nicks, 1981) where similar problems often exist.

The human error quantification techniques described here all contain an element of expert judgement, even though some in particular give the appearance (not necessarily intended) of being based on 'hard' well-founded empirical data.

In a recent review by Kirwan *et al.* (1988), eight human reliability quantification techniques were qualitatively assessed. These were:

Absolute Probability Judgement (APJ) (Seaver and Stillwell, 1983).

Paried Comparisons (PC) (Hunns and Daniels, 1980),

TESEO (Bello and Columbari, 1980),

Technique for Human Error Rate Prediction (THERP) (Swain and Guttman, 1983),

Human Error Assessment and Reduction Technique (HEART) (Williams, 1986),
Influence Diagrams Approach (IDA) (Phillips *et al.*, 1983),
Success Likelihood Index Method (SLIM) (Embrey *et al.*, 1984), and
Human Cognitive Reliability Model (HCR) (Spurgin *et al.*, 1987).

Four of the techniques (APJ, PC, IDA, and SLIM) use a group of expert
judges to evaluate HEPs. APJ and PC largely leave the judgemental task to
the experts, with some help from the analyst or 'facilitator' who may point
out inconsistent judgements or biases in the judgement-making process.
SLIM and IDA also use expert judges, but the judges are asked to consider
what factors affect performance, and from the assessment of these factors
and modelling of their influence on performance, they then determine the
human error probability. They are assisted by the analyst in creating a
quantitative causal model of the influence of these factors on the HEP.
Typical performance shaping factors (PSF) utilized are stress, quality of
interface design, degree of training and adequacy of procedures.

TESEO, THERP and HEART, in contrast, either include a data base of
HEPs or specify procedures for generating numerical HEPs directly. These
techniques require only one analyst, rather than a group of experts. The data
on which they rely are a mixture of field experience and judgement in the
case of THERP, and a mixture of judgement together with data from
ergonomics and psychological performance literature in the case of HEART
and, to a lesser extent, TESEO.

Lastly the HCR model (also called the time reliability correlation approach)
attempts to quantify diagnostic/cognitive errors as a function of time elapsed
since the onset of the incident, and assumes that the likelihood of successful
diagnosis and, consequently action, increases as the time available increases.
This approach is currently a mixture of judgement and simulator data.

Within the scope of this chapter it is not possible to review all of these
techniques. Instead therefore, three are reviewed, namely SLIM, HEART
and THERP. The first two are probably of most interest to the ergonomist,
and the latter exemplifies the reliability engineering oriented approach, and
is also the technique which has been most widely used to date (for a review
of the others, see Kirwan *et al.*, (1988) or the indicated source references).

Success likelihood index method (SLIM)

SLIM can best be explained by means of an example human reliability
assessment, in this case an operator decoupling a filling hose from a chemical
road tanker. The operator may forget to close a valve upstream of the filling
hose, which could lead to undesirable consequences, particularly for the
operator. The human error of interest is 'failure to close V0602 prior to
decoupling filling hose'. In this case the decoupling operation is simple and
discrete, and hence failure occurs catastrophically rather than in a staged
fashion.

PSF identification

The 'expert panel' would typically be comprised of, for example, two operators with 10 years experience, one human factors analyst, and a reliability analyst familiar with the system who also has some operational experience.

The panel is initially asked to identify a set of *performance shaping factors* (PSFs), which are any factors relating to the individual(s), environment, or task, which affect performance positively or negatively. The expert panel could be asked to nominate the most important or significant PSFs for the scenario under investigation. In this example it is assumed the panel identify the following major PSFs as affecting human performance in this situation: training, procedures, feedback, perceived risk, and time pressure.

PSF rating

The panel are then asked to consider other human errors possible in this scenario (e.g. mis-setting or ignoring an alarm), and for each one, to decide to what extent each PSF is optimal or sub-optimal for that task in the situation being assessed. The 'rating' of whether a task is optimal or sub-optimal for a particular PSF is made on a scale of 1 to 9, in this case with 9 as optimal. For the three human errors under analysis, the ratings obtained are as follows:

| | Performance Shaping Factors | | | | |
Errors	Training	Procedures	Feedback	Perceived risk	Time
V0204 Open	6	5	2	9	6
Alarm mis-set	5	3	2	7	4
Alarm ignored	4	5	7	7	2

PSF weighting

If each factor was equally important, one might simply add each row of ratings and conclude that the error with the lowest rating sum (alarm mis-set) was the most likely error. However, this expert panel, as with most panels, does not feel the PSF are all equal. In this particular case (and with this particular panel of experts), the panel feels that perceived risk and feedback are most important, and are in fact twice as important as training and procedures, which are in turn one and a half times as important as time. (As it is a routine operation, time is not perceived by the panel to be particularly important.) Weightings for the PSF can be obtained directly from these considered opinions, as follows, normalized to sum to unity:

Perceived Risk	0·30
Feedback	0·30
Training	0·15
Procedures	0·15
Time	0·10
	1·00

SLIM, and the decision analysis technique it is based upon, called *simple multi-attribute rating technique* (Edwards, 1977) propose simply that preference can be derived as a function of the sum of the weightings multiplied by their ratings for each item (human error). SLIM does this and calls the resultant preference index a *success likelihood index* (SLI). This is illustrated using a table of weightings (W) × ratings (R): (SLI = WR) (see Table 28.3).

Table 28.3. SLI calculation

Weighting	PSF		V0204	Alarm mis-set	Alarm ignored
0·30	Feedback	(0·3 × 2) =	0·6	0·6	2·1
0·30	Perc Risk	etc.	2·7	2·1	2·1
0·15	Training		0·9	0·75	0·6
0·15	Procedures		0·75	0·45	0·75
0·10	Time		0·60	0·40	0·2
	SLI (Total)		5·55	4·30	5·75

In this case, the lowest SLI is 4·3, suggesting that 'alarm mis-set' is still the most likely error. However, due to the weightings used, the likelihood ordering of the other two errors have now been reversed (close inspection of the figures reveals that this is because feedback is held to be important, and there is ample feedback for 'alarm ignored' but not for 'V0204 open'). Clearly at this point, a designer would realize that increased feedback about the position of V0204 to the operator might be desirable.

However, the SLIs are not yet probabilities. Rather, they are indications of the relative likelihoods of the different errors. Thus the SLIs show the ordering of likelihood of the different errors, but do not yet define the absolute probability values. In order to transform the SLIs into HEPs, it is necessary to 'calibrate' the SLI values. (Note: the paired comparisons technique also requires this calibration using the same basic formula). Two earlier studies by Pontecorvo (1965) and Hunns (1982) have derived such a calibration relationship, both suggesting a logarithmic relationship of the form:

$$\mathrm{Log}_{10}\ (\mathrm{HEP}) = a\ \mathrm{SLI} + b.$$

If two tasks for which the HEPs are known are included in the task/error set which are being quantified, then the parameters of the equation can be derived via simultaneous equations, and the other (unknown) HEPs can be quantified. If in the above example, two more tasks (*A* and *B*) were assessed which had HEPs of 0·5 and 10^{-4} respectively, and were given SLIs of 4·00 and 6·00, respectively, then the equation derived would be:

Log (HEP) = 1·85 SLI + 7·1.

The HEPs would then be: V0204 = 0·0007; alarm mis-set = 0·14; alarm ignored = 0·003.

This is the body of the rationale underlying SLIM, but in practice SLIM is more complex and is computerized to facilitate its ease of use and to prevent bias, often found in the elicitation of expert opinions. The computerized version, known as SLIM-MAUD (SLIM using *M*ulti *A*ttribute *U*tility *D*ecompositon: Embrey *et al.*, 1984), due to the mathematics in the software which is present partly to avoid such bias, will produce slightly different values (HEPs) than the hand calculated method used above. In particular the simple summary of weightings and ratings is refined in several ways according to the more detailed mathematical requirements of multi-attribute utility theory. However, the above is the general rationale of SLIM, and enables the reader to understand more easily how SLIM works.

Human error assessment and reduction technique (HEART)

This technique is of particular interest to ergonomists as it is based on the human performance literature. It has been designed by its author as a relatively quick method for HRA, to be simple to use and easily understood. Its fundamental premise is that in reliability and risk equations one is interested in ergonomics factors which have a large effect on performance, e.g. causing a decrement in performance by a factor of three or more. Thus, whilst there are many well-studied ergonomics factors and consequent guidelines (e.g. lighting recommendations), many of these factors actually have (in reliability terms) a negligible effect on operator performance. HEART therefore concentrates on those factors which have a significant effect.

This point is important because in part it underlies something of a communications gap between engineers and ergonomists. Engineers and designers designing a plant cannot spend unlimited funds on the optimization of ergonomics aspects, and often ask how important (in quantitative terms) an ergonomics recommendation is. Frequently the ergonomist is unable to answer this question which to the engineer is fundamental. HEART in particular, and some of the other techniques (e.g. SLIM), allow the human reliability analyst to answer this question quantitatively.

The first part of the HEART assessment process is to refine the task in terms of its generic proposed nominal human unreliability, as shown in Table 28.4. Thus the task is first assigned a nominal human error probability by classifying it according to whether it is a complex task, a routine task, and

Table 28.4. Generic classifications (HEART, after Williams, 1986).

Generic task	Proposed nominal human unreliability (5th–95th percentile bounds)
(A) Totally unfamiliar, performed at speed with no real idea of likely consequences	0·55 (0·35–0·97)
(B) Shift or restore system to a new or original state on a single attempt without supervision or procedures	0·26 (0·14–0·42)
(C) Complex task requiring high level of comprehension and skill	0·16 (0·12–0·28)
(D) Fairly simple task performed rapidly or given scant attention	0·09 (0·06–0·13)
(E) Routine, highly-practised, rapid task involving relatively low level of skill	0·02 (0·007–0·045)
(F) Restore or shift a system to original or new state following procedures, with some checking	0·003 (0·0008–0·007)
(G) Completely familiar, well-designed, highly practised, routine task occurring several times per hour, performed to highest possible standards by highly-motivated, highly-trained and experienced person, totally aware of implications of failure, with time to correct potential error, but without the benefit of significant job aids	0·0004 (0·00008–0·009)
(H) Respond correctly to system command even when there is an augmented or automated supervisory system providing accurate interpretation of system stage	0·00002 (0·000006–0·0009)

so on. The next stage is to identify error producing conditions (EPCs) which are evident in the scenario and would negatively influence human performance. A table of the major EPCs in HEART is shown in Table 28.5.

Example

As a hypothetical example of how HEART is used to quantify a human error probability for a task, taken from Williams (1988), we assume that a safety, reliability, or operations engineer wishes to assess the nominal likelihood of an operative's failing to isolate a plant bypass route following strict procedures. The scenario necessitates a fairly inexperienced operator applying an opposite technique to that which he normally uses to carry out isolations and involves a piece of plant, the inherent major hazards of which he is only dimly aware. It is assumed that the man could be in the seventh hour of his shift, that there is talk of the plant's imminent closure, that his work may be checked and that the local management of the company is desperately trying to keep the plant operational despite the real need for maintenance because of its fear that partial shutdown could quickly lead to total permanent shutdown.

Table 28.5. HEARTS EPCs (Williams, 1986).

Error-producing condition	Maximum predicted nominal amount by which unreliability might change going from 'good' conditions to 'bad'
1. Unfamiliarity with a situation which is potentially important but which only occurs infrequently or which is novel	× 17
2. A shortage of time available for error detection and correction	× 11
3. A low signal-to-noise ratio	× 10
4. A means of suppressing or overriding information or features which is too easily accessible	× 9
5. No means of conveying spatial and functional information to operators in a form which they can readily assimilate	× 8
6. A mismatch between an operator's model of the world and that imagined by a designer	× 8
7. No obvious means of reversing an unintended action.	× 8
8. A channel capacity overload, particularly one caused by simultaneous presentation of non-redundant information	× 6
9. A need to unlearn a technique and apply one which requires the application of an opposing philosophy	× 6
10. The need to transfer specific knowledge from task to task without loss	× 5·5
11. Ambiguity in the required performance standards	× 5
12. A mismatch between perceived and real risk	× 4
13. Poor, ambiguous or ill-matched system feedback	× 4
14. No clear direct and timely confirmation of an intended action from the portion of the system over which control is to be exerted	× 4
15. Operator inexperience (e.g. a newly-qualified tradesman, but not an 'expert')	× 3
16. An impoverished quality of information conveyed by procedures and person/person interaction	× 3
17. Little or no independent checking or testing of output	× 3
18.* A conflict between immediate and long-term objectives	× 2.5
19. No diversity of information input for veracity checks	× 2.5
20. A mismatch between the educational achievement level of an individual and the requirements of the task	× 2
21. An incentive to use other more dangerous procedures	× 2
22. Little opportunity to exercise mind and body outside the immediate confines of a job	× 1.8
23. Unreliable instrumentation (enough that it is noticed)	× 1.6
24. A need for absolute judgements which are beyond the capabilities or experience of an operator	× 1.6
25. Unclear allocation of function and responsibility	× 1.6
26. No obvious way to keep track of progress during an activity	× 1.4

*18–26. These conditions are presented simply because they are frequently mentioned in the human factors literature as being of some importance in human reliability assessment. To a human factors engineer, who is sometimes concerned about performance differences of as little as 3%, all these factors are important, but to engineers who are usually concerned with differences of more than 300%, they are not very significant. The factors are identified so that engineers can decide whether or not to take account of them after the initial screening.

Using a simplified HEART, the safety reliability and operational engineer's assessment could look something like this:

Type of Task			Nominal Human Unreliability
F			*0·003*
Error-producing Conditions			
Factor	*Total HEART Affect*	*Engineer's Assessed Proportion of Affect* (from 0 to 1)	*Assessed Affect*
Inexperience	× 3	0·4	(3−1) × 0·4 + 1 = 1·8
Opposite Technique	× 6	1·0	(6−1) × 1·0 + 1 = 6·0
Risk Misperception	× 4	0·8	(4−1) × 0·8 + 1 = 3·4
Conflict of Objectives	× 2·5	0·8	(2·5−1) × 0·8 + 1 = 2·2
Low Morale	× 1·2	0·6	(1·2−1) × 0·6 + 1 = 1·12

Assessed nominal likelihood of failure
$$0·003 \times 1·8 \times 6·0 \times 3·4 \times 2·2 \times 1·12 = 0·27$$

Time-on-shift effects would be ignored as there is no indication of monotony.

Similar calculations may be performed if desired for the predicted 5th and 95th percentile bounds, which in this case would be 0·07 – 0·58. As a total probability of failure can never exceed 1·00, if the multiplication of factors takes the value about 1·00 the probability of failure has to be assumed to be 1·00 and no more.

The relative contribution made by each of the error-producing conditions to the amount of unreliability modification is as follows:

	% contribution made to unreliability modification
Technique unlearning	41
Misperception of risk	24
Conflict of objectives	15
Inexperience	12
Low morale	8

Thus an HEP of 0·27 (just over one in four) is calculated, which is a very high predicted error probability, and unlikely to be acceptable. In this case technique unlearning is the major contributory factor to this poor performance, and so clearly then either some form of retraining, or else redesign to make

Table 28.6. A subset of HEART remedial measures (Williams, 1986).

1.	Technique unlearning (× 6)	The greatest possible care should be exercised when new techniques are being considered to achieve the same outcome — they should not involve adoption of opposing philosophies
2.	Misperception of risk (× 4)	It must not be assumed that a user's perception of risk is the same as the actual level — if necessary a check should be made to ascertain where any mismatch might exist and what its extent is
3.	Objectives conflict (× 2.5)	Objectives should be tested by management for mutual compatibility, and where potential conflicts are identified these should either be resolved to make them harmonious or made prominent so that a comprehensive management control programme can be created to reconcile such conflicts as they arise, in a rational fashion
4.	Inexperience (× 3)	Personnel criteria should contain specified experience parameters thought relevant to the task — chances must not be taken for the sake of expediency
5.	Low morale (× 1·2)	Apart from the more obvious ways of attempting to secure high morale, by way of financial reward for example, other methods involving participation, trust and mutual respect, often hold out at least as much promise — building up morale is a painstaking process, which involves a little luck and great sensitivity — employees must be given reason to believe

the isolation procedures consistent across plant, must be considered. However HEART goes further than other techniques in error reduction, as for each EPC it gives corresponding suggested error reduction approaches, e.g. for the task above, the remedial measures in Table 28.6 would be proposed.

Thus HEART offers a quick and simple human reliability calculation method which also gives the user (engineer or ergonomist) suggestions on error reduction.

Technique for human error rate prediction (THERP)

THERP is in itself a total methodology for assessing human reliability. The quantification part of THERP comprises the following:

(a) a data base of human errors which can be influenced by the assessor to reflect the impact of PSFs on the scenario;

(b) a dependency model which calculates the degree of dependence between two operator actions (e.g. if an operator fails to detect an alarm, then failure to carry out appropriate corrective actions reliably cannot be treated independent from this failure);

(c) an event tree modelling approach to combine HEPs for steps in a task into an overall task HEP;

(d) the assessment of error recovery paths.

The basic THERP approach is shown in Figure 28.10, from Bell (1984).

Table 28.7. Task analysis — initiation of flow via stand-by train (Webley and Ackroyd; in Kirwan et al., 1988)

Task identifier*	Task of interest	Error identifier	Human errors	Human error probabilities (range of median values given in USSR state committe, 1986)	
A	Identify loss of flow via duty train	A_1	Fail to identify loss of flow	$10^{-4} - 0.25$	Alarm response model
B	Start correct procedure	B_1	Fail to start procedure	$10^{-3} - 10^{-2}$	Procedure used
				$10^{-2} - 5 \times 10^{-2}$	Procedure not used
C	Roving operator opens correct valve	C_1	Error of omission — verbal order	10^{-3}	
			Error of commission — select incorrect valve	$10^{-3} - 10^{-2}$	
D	1st operator starts stand-by pump via remote control	D_1	Error of omission — written procedures available	$10^{-3} - 10^{-2}$	Procedure used
				$10^{-2} - 5 \times 10^{-2}$	Procedure not used
		D_2	Error of commission — select incorrect control	$10^{-3} - 10^{-2}$	
E	Supervisor checks 1st operator	E_1	Error of omission — written procedures available	$10^{-3} - 10^{-2}$	Procedure used
				$10^{-2} - 5 \times 10^{-2}$	Procedure not used
		E_2	Error of commission — select incorrect valve	$10^{-3} - 10^{-2}$	
G	Shift Manager checks activation	G_1	Fail to initiate checking function	$10^{-3} - 10^{-2}$	
			Fail to identify errors	$5 \times 10^{-2} - 0.2$	

*Task F no errors identified for this step.

Phase 1: Familiarisation

Plant Visit

Review Information From
System Analysts

Phase 2: Qualitative Assessment

Talk- or
Walk-Through

Task Analysis

Develop HRA Event Trees

Phase 3: Quantitative Assessment

Assign Nominal HEPs

Estimate the Relative
Effects of Performance
Shaping Factors

Assess Dependence

Determine Success and
Failure Probabilities

Determine the Effects
of Recovery Factors

Phase 4: Incorporation

Perform a Sensitivity
Analysis, if Warranted

Supply Information to
System Analysts

Figure 28.10. Outline of a THERP procedure for HRA (adapted from Bell, 1984).

Probably because of its data base and similarities with reliability engineering approaches, THERP has been used more than any other technique in industry applications. This small section can only cover the rudiments of the technique, and for more information the reader is referred to Swain and Guttman (1983), Bell (1984) and Kirwan *et al.* (1988).

An example of the type of basic event data given in THERP is shown in Table 28.7.

A human reliability analysis event tree (HRAET, as shown in Figure 28.11) can be used to represent the operator's performance. Alternatively, as shown in Figure 28.12, an operator action event tree can be used (Whittingham,

1988). In each case the event tree represents the sequence of events and considers possible failures at each branch in the tree (omission, commission, and so on). These errors are quantified and error recovery paths are then added to the tree where appropriate. Figure 28.13 shows the PSFs which can be used in a THERP study, although often in a THERP study only one or at most a few of these are utilized quantitatively (e.g. stress).

It is important to model recovery, particularly when investigating highly proceduralized sequences, since often an operator will be prompted by a later step in the procedures to recover from an earlier error in a previous step. For example, if an operator omits a step to turn the power on, and then a second step is attended to which involves checking certain power-supplied instruments, then the operator will rapidly recover the first error. If recoveries in such highly proceduralized situations (for which THERP is typically used)

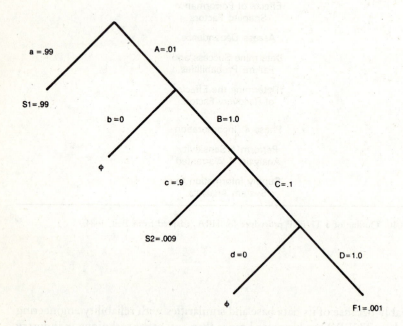

A = FAILURE TO SET UP TEST EQUIPMENT PROPERLY
B = FAILURE TO DETECT MISCALIBRATION FOR FIRST SETPOINT
C = FAILURE TO DETECT MISCALIBRATION FOR SECOND SETPOINT
D = FAILURE TO DETECT MISCALIBRATION FOR THIRD SETPOINT
φ = NULL PATH

Figure 28.11. HRA event tree of hypothetical calibration task (adapted from Swain and Guttmann, 1983).

are not identified, then human error may be overestimated. THERP is in fact the only technique which emphasizes error recovery in this way. Error recovery paths are shown on the operator action tree in Figure 28.12 as dashed lines, and in this particular tree these are largely recoveries of one person's error by another person.

THERP also models dependency between human errors. If, for example, an operator is in a high stress situation trying to carry out a procedure quickly, or is demotivated or fatigued, etc., then a whole section of procedures and recovery steps within these procedures may be carried out wrongly. This is an example of a 'human dependent failure'. An important example of dependency modelling concerns one operator checking another operator. It is unlikely that the action by operator 1 and the check by operator 2 will be completely independent, since the first operator may assume that the second will detect any faults, and the second may assume the first did the job properly. THERP is one of the few techniques which actually quantitatively models this dependency between actions/errors. It uses a simple five-level model of dependency from zero dependence (i.e. total independence) through low, medium and high dependence levels to complete dependence. The effects of these levels of dependence are mathematically calculated in the HRAET/OAT.

As a more general point, human dependent failures can be extremely important in any human reliability assessment, and care should be taken to identify these where possible (e.g. if an operator might become incapacitated, or a misdiagnosis might occur, or if production pressures might totally overrule safety considerations).

Advantages and disadvantages of SLIM, HEART, and THERP

SLIM-MAUD is a highly structured approach to the use of expert opinion, and often has high credibility with the experts taking part. The computerized version enables the user to investigate how improvements, in particular PSFs (e.g. interface design) can affect the HEPs, allowing the cost-effectiveness of error reduction strategies to be investigated (see impact assessment). However, SLIM relies on 'experts' who may be difficult to find and verify as experts, and on calibration data (at least two known HEPs). SLIM is an exhaustive technique and hence can use up a relatively large amount of personnel resources and time.

HEART is one of the quickest techniques available, and does not require experts, the EPCs instead being based on extensive analysis of the human performance literature. It also offers means of determining error reduction strategies and investigating the cost-effectiveness of such strategies. However it does not consider possible interactions between the various EPCs, and the nominal HEPs have not yet been validated. Many of the EPCs can only be assessed for an existing plant, and not for a plant being proposed or developed.

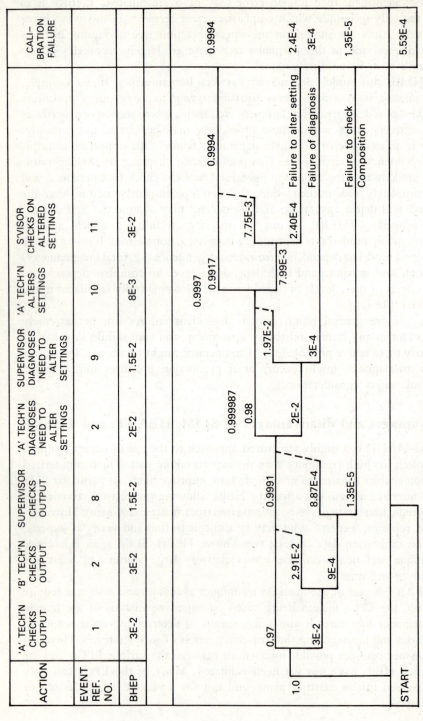

Figure 28.12. Supervisory check on altered settings (Whittingham, 1988).

EXTERNAL PSFs			STRESSOR PSFs	INTERNAL PSFs
SITUATIONAL CHARACTERISTICS	TASK AND EQUIPMENT CHARACTERISTICS:		PSYCHOLOGICAL STRESSORS:	ORGANISMIC FACTORS:
THOSE PSFs GENERAL TO ONE OR MORE JOBS IN A WORK SITUATION	THOSE PSFs SPECIFIC TO TASKS IN A JOB		PSFs WHICH DIRECTLY AFFECT MENTAL STRESS	CHARACTERISTICS OF PEOPLE RESULTING FROM INTERNAL AND EXTERNAL INFLUENCES
ARCHITECTURAL FEATURES	PERCEPTUAL REQUIREMENTS		SUDDENNESS OF ONSET	PREVIOUS TRAINING/ EXPERIENCE
QUALITY OF ENVIRONMENT: TEMPERATURE HUMIDITY. AIR QUALITY, AND RADIATION	MOTOR REQUIREMENTS (SPEED, STRENGTH, PRECISION)		DURATION OF STRESS TASK SPEED TASK LOAD	STATE OF CURRENT PRACTICE OR SKILL
LIGHTING	CONTROL-DISPLAY RELATIONSHIPS		HIGH JEOPARDY RISK THREATS (OF FAILURE LOSS OF JOB)	PERSONALITY AND INTELLIGENCE VARIABLES
NOISE AND VIBRATION DEGREE OF GENERAL CLEANLINESS	ANTICIPATORY REQUIREMENTS INTERPRETATION		MONOTONOUS, DEGRADING OR MEANINGLESS WORK	MOTIVATION AND ATTITUDES EMOTIONAL STATE
WORK HOURS/WORK BREAKS SHIFT ROTATION	DECISION-MAKING COMPLEXITY (INFORMATION LOAD)		LONG, UNEVENTFUL VIGILANCE PERIODS	STRESS (MENTAL OR BODILY TENSION)
AVAILABILITY/ADEQUACY OF SPECIAL EQUIPMENT, TOOLS, AND SUPPLIES	NARROWNESS OF TASK FREQUENCY AND REPETITIVENESS		CONFLICTS OF MOTIVES ABOUT JOB PERFORMANCE REINFORCEMENT ABSENT OR NEGATIVE	KNOWLEDGE OF REQUIRED PERFORMANCE STAN- DARDS
MANNING PARAMETERS	TASK CRITICALITY		SENSORY DEPRIVATION	SEX DIFFERENCES
ORGANISATIONAL STRUCTURE (eg AUTHORITY, RESPONSIBILITY, COMMUNICATION CHANNELS)	LONG AND SHORT-TERM MEMORY CALCULATIONAL REQUIREMENTS		DISTRACTIONS (NOISE, GLARE, MOVEMENT FLICKER, COLOR) INCONSISTENT CUEING	PHYSICAL CONDITION ATTITUDES BASED ON INFLUENCE OF FAMILY AND OTHER OUTSIDE PERSONS OR AGENCIES
ACTIONS BY SUPERVISORS CO - WORKERS, UNION REPRESENTATIVES, AND REGULATORY PERSONNEL REWARDS, RECOGNITION, BENEFITS	FEEDBACK (KNOWLEDGE OF RESULTS) DYNAMIC VS. STEP-BY-STEP ACTIVITIES TEAM STRUCTURE AND COMMUNICATION MAN-MACHINE INTERFACE		PHYSIOLOGICAL STRESSORS: PSFs WHICH DIRECTLY AFFECT PHYSICAL STRESS	GROUP IDENTIFICATIONS
JOB AND TASK INSTRUCTIONS:	FACTORS: DESIGN OF PRIME EQUIPMENT, TEST EQUIPMENT,		DURATION OF STRESS FATIGUE PAIN OR DISCOMFORT HUNGER OR THIRST	
SINGLE MOST IMPORTANT TOOL FOR MOST TASKS	MANUFACTURING EQUIPMENT, JOB AIDS, TOOLS FIXTURES		TEMPERATURE EXTREMES RADIATION G-FORCE EXTREMES	
PROCEDURE REQUIRED (WRITTEN OR NOT WRITTEN) WRITTEN OR ORAL COMMUNICATIONS			ATMOSPHERIC PRESSURE EXTREMES OXYGEN INSUFFICIENCY VIBRATION	
CAUTIONS AND WARNINGS WORK METHODS PLANT POLICIES (SHOP PRACTICES)			MOVEMENT CONSTRICTION LACK OF PHYSICAL EXERCISE DISRUPTION OF CIRCADIAN RHYTHM	

Figure 28.13. Comprehensive list of PSFs for THERP analysis (adapted from Swain and Guttmann, 1983).

742   *Evaluation of work systems*

Table 28.8. Summary of evaluation of techniques (from Kirwan et al., 1988)

	APJ	PC	TESEO	THERP	HEART	IDA	SLIM	HCR (Hannaman et al., 1984)
Accuracy	Moderate	Moderate	(Low)†	Moderate	Moderate	(Low)	Moderate	(Low)
Validity	Moderate/high	Moderate	Low	Moderate	Moderate	Moderate	Moderate	Low
Usefulness	Moderate/high	Low/moderate	Moderate/high	Moderate	High	Moderate/high	High	Low/moderate
Effective use of resources*	Moderate	Low/moderate	High	Low/moderate	High	Low/moderate	Low/moderate	Moderate
Acceptability	Moderate	Moderate/high	Low	High	(Moderate)	(Moderate)	Moderate/high	Low/moderate
Maturity	High	Moderate	Low	High	Low/moderate	Low/moderate	Moderate/high	Low

*A rating of high on this criterion means the technique is favourable with respect to effective use of resource (i.e. resource requirements are low), and vice versa.
†Ratings in parentheses were based on the subjective opinions of the authors and contributors to the document, since insufficient empirical evidence was available to justify a rating from applications alone.
Note: This table was constructed prior to the experimental study assessing accuracy (Kirwan, 1988).

THERP is similar in appearance to conventional reliability assessment approaches, yet it can bring PSFs into the equation, explicitly models human errors and error recovery, and considers dependency between errors. However, THERP also requires a good deal of resources, albeit in terms of a single assessor rather than a panel of experts, and often two different assessors may carry out THERP in rather different ways. There is no substantiation of the THERP data base, and detailed assessments, due to the fine level of description of the human errors in THERP analysis, can become quite complex.

The above are some of the basic pros and cons of the three techniques; for a more formal qualitative analysis of them, see Kirwan *et al.* (1988).

Selection of an appropriate quantification technique

Recent reviews by Kirwan *et al.* (1988) and Kirwan (1988) have attempted to aid the process of selecting which quantification techniques to use in a particular application. The first review qualitatively analysed the eight techniques mentioned earlier, drawing conclusions from the published literature. It assessed the techniques against a set of criteria, namely:

Accuracy: numerical accuracy (in comparison with known HEPs); consistency between experts and assessors.

Validity: use of ergonomics factors/PSF to aid quantification; theoretical basis in ergonomics, psychology; empirical validity; validity as perceived by assessors, experts, etc.; comparative validity (comparing results of one technique with results of another for the same scenario).

Usefulness: qualitative usefulness in determining error reduction mechanisms; sensitivity analysis capability, allowing assessment of effects on HEPs of error reduction mechanisms.

Effective use of resources: equipment and personnel requirements; data requirements, e.g. SLIM and PC require calibration data (at least two known HEPs); training requirements of assessors and/or experts.

Acceptability: to regulatory bodies; to the scientific community; to assessors; auditability of the quantitative assessment.

Maturity: current maturity; development potential.

The eight techniques are assessed against these criteria in Table 28.8, and a second selection guideline table based on perceived user requirements is presented in Table 28.9. With these two tables it is possible for the user to decide which technique or set of techniques to utilize.

A second, experimental review (Kirwan, 1988) looked particularly at the accuracy of five of the techniques (APJ, PC, SLIM, HEART, THERP), comparing their predictions against some known data points (e.g. see Figure 28.14), and comparing their predictions for a set of realistic PSA scenarios.

Table 28.9. Selection matrix (from Kirwan *et al.*, 1988)

	APJ	PC	TESEO	THERP	HEART	IDA	SLIM	HCR
Is the technique applicable for:								
Simple and proceduralized tasks?	Y	Y	Y	Y	Y	Y	Y	N
Knowledge-based, abnormal tasks?	Y	Y	Y	N	Y	Y	Y	Y
Misdiagnosis which makes a situation worse?	Y	Y	N	N	N	Y	Y	N
Are qualitative recommendations possible?	Y	N	Y	Y	Y	Y	Y	Y
Is sensitivity analysis possible?	N	N	Y	Y	Y	Y	Y	Y
Does the technique have:								
Requirements for calibration data?	N	Y	N	N	N	N	Y	N
Requirements for experts (judges)?	Y	Y	N	N	N	Y	Y	N

The results showed APJ and THERP to be accurate techniques with HEART exhibiting some degree of accuracy. The results for SLIM and PC were less clear since whilst they did not perform well in this experiment, this could have been due to the resources limitations of the experiment rather than the technique. It is likely that further validation studies will be carried out in the future, to narrow the most acceptable techniques down to two or three.

(7) Impact assessment

Once human error probabilities have been quantified, the system risk, or reliability, can be calculated and compared to an acceptable level to see if improvement is necessary. If so, it will first be necessary to determine whether human error was a major contributor to inadequate system performance. This involves carrying out an analysis of the importance of each 'event' in the fault tree, to see which events (quantitatively) most affect the predicted top event (accident) frequency. Many fault tree computer analysis packages will automatically determine the most important events.

If human errors do not significantly affect the predicted frequency of the event, then reduction of risk will be best effected by hardware/software improvements to the system. Otherwise, and particularly if human error 'dominates' the undesired event frequency, error reduction mechanisms should be investigated. Identification of the major human errors will enable the most effective error reduction mechanisms to be specified in the next phase. The positive effects of these mechanisms can be calculated and factored

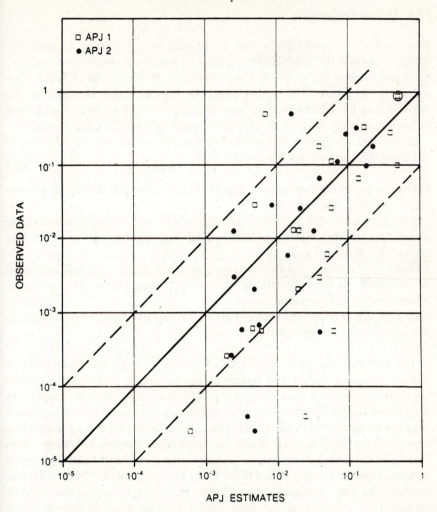

Figure 28.14. APJ estimates of the independent groups vs. observed data.

back into the quantitative analysis until system performance reaches an acceptable level. Similar calculations and 'sensitivity analysis' can also be carried out for event-tree-based assessments.

If human error cannot be reduced to an acceptable level, even with additional hardware recommendations, then significant redesign of the system and/or its operation will be required. Usually however, an effective combination of human and hardware modifications can be found to achieve an acceptable level of risk.

(8) Error reduction

Error reduction mechanisms will be unnecessary if human reliability is adequate, or not the most effective means of achieving system performance, or if not within the scope of the assessment. If on the other hand error reduction is required then either particular human errors identified in the analysis may be reduced in impact or frequency, or else a more general error reduction strategy may be developed to improve overall task performance.

In the case of specific identified critical errors, there are several ways of reducing their impact on the system:

Prevention by hardware or software changes: Use of interlock devices to prevent error; automate the task, etc.
Increase system tolerance: make the system hardware and software more flexible or self-correcting to allow a greater variability in operator inputs which will achieve the intended goal.
Enhance error recovery: enhance detection and correction of errors by means of increased feedback, checking procedures, supervision and automatic monitoring of performance.
Error reduction at source: reduction of errors by improved procedures, training, and interface or equipment design.

The first two measures require collaboration between the ergonomist or human reliability analyst and system design personnel, and may well prove expensive. Improved error recovery probabilities are often the simplest to implement, but may not reduce error likelihoods by a sufficient amount, and may not be feasible with all critical errors. Error reduction at source therefore may well be the primary means of improving human reliability. This may require consultation with a human factors specialist, although some quantification techniques (e.g. HEART) do prescribe specific error reduction mechanisms for identified errors. Error reduction measures can also be identified by considering the error causes and the PSFs identified in the human error analysis phase (e.g. via PHECA or a comparable system). Such measures should be effective as they are aimed at the root causes of the human error.

A second additional analysis of the results will be possible only with quantification methods which use a structured performance shaping factor (PSF) approach (e.g. SLIM, IDA, HEART, THERP, TESEO). With these approaches it is possible to determine the contributions of individual PSFs to human error goals (see the earlier HEART worked example). For example, the most significant PSF in a particular scenario may be 'quality of procedures' and, therefore, error reduction measures aimed at improving the quality of procedures will be most effective at reducing error likelihood. The potential reduction however, can actually be calculated since with these methods error probability is a direct function of the PSFs. Thus it may be that improving the quality of procedures for a particular error makes the error ten times less likely in a particular assessment, which may render the error probability 'acceptable'. If however, the error probability is still not low enough, then clearly other additional PSFs should be investigated. Furthermore, if for example quality of procedures is the most important PSF for a number of human errors, this then suggests that a single global error reduction strategy to generally enhance performance can be specified. This type of investigation of the results will enable the cost-effectiveness of potential error reduction strategies to be assessed.

If an overall improvement in performance is required, then the principal PSFs identified in the task analysis, human error analysis, and quantification phases should be addressed. It may be, for example, that training was continually cited as a prominent factor affecting performance in a particular process plant emergency scenario, and in fact was rated the most important factor when quantification took place. Clearly, focusing on training in this particular scenario should have an overall positive effect on performance, and may be the best single way of reducing the impact of human error on the system goals. Some case studies of PSAs, including the derivation of error reduction strategies, are given in Kirwan *et al.* (1988).

Where possible, the positive effects of the error reduction strategy adopted should be factored back into the quantitative analysis and it should be checked that the HEPs and overall system risk calculated become acceptable. This requires not only the precise operational definition of each aspect of the error reduction strategy, but also a method of ensuring that the strategy is properly implemented and maintained throughout the remaining life of the plant. This is part of the quality assurance phase.

(9) Quality assurance

If the assessment has generated certain error reduction mechanisms the implementation and effectiveness of these should be ensured. For example, if as a result of an analysis equipment design recommendations are accepted, the successful implementation of the design changes should be monitored, and if possible the effects on performance verified at a later stage.

A continual performance monitoring system represents a powerful quality assurance system. A risk assessment generally predicts reliability for many years ahead, and much can change in this time: a gradual degradation in performance standards; an increasing maintenance loading; the loss of personnel involved in the design; impromptu changes during commissioning and start-up of the plant; and increasing 'retrofit' changes to the plant. Due to such factors the plant risk level may slowly and almost imperceptibly rise above the initial predicted and acceptable level to one which is unacceptable. Long-term performance monitoring (i.e. recording human errors, near-misses, incidents, and accidents) will detect this. A performance monitoring system can identify the point in time at which the assumptions and results of human reliability analysis are no longer applicable, and signify the need for a further analysis to justify the acceptability of the risk of the plant. This type of quality assurance system can avoid the gradual erosion of safety barriers (the Bhopal syndrome).

(10) Documentation

It is important for the auditability and justifiability of the results of the study that it is fully documented. This will enable personnel unconnected with the assessment to understand how it was carried out, including all assumptions and judgements made during the study, and allow the study to be independently examined, updated, or even reproduced if necessary. It will also provide a potentially useful data base from which to monitor future progress of the system's performance in comparison to the predictions made in the analysis. It is also entirely possible that at some stage in the future an accident investigation may take recourse to reviewing the predictions of an earlier safety assessment, to assess why the PSA failed to predict the occurrence of the accident which has now occurred. This perhaps would at least allow HRA to learn from its own mistakes, should they occur.

Future directions in HRA

Three areas are noted below in which human unreliability can affect system risk, but which are not currently adequately addressed by HRA techniques. These are therefore predicted to be the most likely future directions for human reliability developments.

Low technology risk

The HRA field has mainly concerned itself with the high risk, high technology industry sector, including nuclear power plants and chemical plants. There exist however, a large number of other lower technology sectors, e.g. mining, which often incur a high risk via a large number of 'small' accidents (say one or two fatalities), rather than (high risk technology) industries where the high risk is caused by a very small probability of an accident with many and serious consequences. This is clearly an area where applied human reliability should be able to help reduce risk. Some workers have already begun to use HRA approaches in such areas (Collier and Graves, 1986).

Cognitive errors and misdiagnosis

If the operating crew in a plant misdiagnose a situation, they may be slow to realize they have made a 'mistake' or incorrect diagnosis. They may reinterpret system feedback within their mental view of what they think the problem is, rather than what it actually is (also called 'mind set'). This is in fact what occurred during the Three Mile Island accident in 1979, and can result in the human operators actually defeating safety systems and making matters worse than if they did nothing at all. Also there is evidence to suggest that there is a low probability of operators recovering such mistakes once made (Woods, 1984).

The 'modelling' and analysis of cognitive errors is a difficult problem area, and one that has not yet been resolved (Woods and Roth, 1986). Some work has been carried out on the psychology underlying cognitive errors in an attempt to predict the forms cognitive errors are likely to take (Reason and Embrey, 1986). The Influence Modelling and Assessment System (IMAS) can be used to elicit mental maps of operators' perceptions of causal relationships in nuclear power plant operation (see Embrey, 1986a). These maps could be used to consider what misdiagnoses may occur. Also, in the Artificial Intelligence field, an attempt is being made to simulate the cognitive operator with an expert system, and estimate what errors will occur (Woods *et al.*, 1987). These and other developments may eventually solve the difficult problem of cognitive errors.

Management, organizational and sociotechnical contributions to risk

Three of the most recent and salient accidents emphasizing the impact of human error are the Bhopal (1984), Chernobyl (1986), and Challenger Space Shuttle (1986) disasters. Each of these contained a significant human error contribution, without which the disasters would not have occurred. However, it is arguable that current HRA techniques would have difficulty in predicting them, because the fundamental types of error leading to the accidents were neither procedural nor diagnostic in the conventional sense (Watson and Oakes, 1988). Rather, all three involved a management and organizational error component. The Bhopal plant was possibly subject to economic pressures, resulting in management decisions which may have been more in favour of production than safety (Bellamy, 1986). The Challenger Space Shuttle disaster also appeared to contain a significant element of management-type decision error, also influenced by economic considerations (Rogers *et al.*, 1986). Lastly, the Chernobyl disaster (USSR State Committee, 1986) contained such a bizarre series of events affected by management that, had an analysis predicted this scenario prior to the accident, it possibly would have been treated with derision. Nevertheless, even such accidents as Chernobyl can be rationalized after the event, and indeed similar sequences can and have occurred (Reason, 1988b).

Errors may also occur when dealing with small groups, at a 'lower' management and organizational level and caused by individual, organizational, and sociotechnical factors (e.g. social pressures, personality conflicts; Bellamy, 1983). These factors and the errors they cause are rarely assessed in HRAs, yet clearly can influence risks; this may happen via risk-taking: decisions to ignore procedures; failure to communicate information due to personality conflicts, etc. Techniques could therefore be developed in the future which identify, quantify and specify how to reduce these types of error.

It is clear that HRA must eventually consider not only the operators in the control room, but also the management running the plant, since the latter can have a profound effect on safety.

In broad terms HRA is becoming an integral part of risk assessments and its usage is steadily increasing. In the future it would seem likely that this will continue and the 'science' of HRA will become an accepted tool for use in many areas, not only those dominated by the need to assess high risk systems.

Summary

This chapter has overviewed the field of human reliability assessment as a tool for reducing the risk of large scale accidents, although it has other uses (e.g. in improving productivity). A broad generic methodology has been outlined, detailing the various steps in HRA and its interactions with PSA. The more prominent methodologies and issues in HRA have been outlined. From the discussions in this chapter it is apparent that HRA is itself not yet a mature science, and has some way to go in the development of sound and theoretically and empirically valid methodologies. Nevertheless, a great deal can still be achieved with the existing approaches, and current research is progressing possibly faster than at any other time in HRA's history. The usefulness and veracity of HRA will, however, ultimately be proven one way or the other by the incidence of human-error induced large scale accidents in future years.

References

Apostolakis, G.B., Beer, V.M. and Mosleh, A. (1987). A critique of recent models for human error rate assessment. Paper presented at the *International POST-SMIRT 9th Seminar on Accident Sequence Modelling*, Munich, Federal Republic of Germany. Reprinted in *Reliability Engineering and System Safety* (1988), **22**, 201–217.

Bell, B.J. (1984). Human reliability analysis for probabilistic risk assessment. *Proceedings of the 1984 International Conference on Occupational Ergonomics*, 35–40.

Bellamy, L.J. (1983). Neglected individual, social and organisational factors in human reliability assessment. *Proceedingts of the 4th National Reliability Conference*, Reliability 85, NEC, Birmingham.

Bellamy, L.J. (1986). The Safety Management Factor: An analysis of Human Aspects of the Bhopal Disaster. Paper presented at the *1986 Safety and Reliability Symposium*, Southport.

Bellamy, L.J., Kirwan, B. and Cox, R.A. (1986). Incorporating human reliability into probabilistic risk assessment. Paper presented at *5th International Symposium in Loss Prevention and Safety Promotion in the Process Industries*, Societe de Chimie Industriale.

Bello, G.C. and Columbari, V. (1980). The human factors in risk analysis of process plants: the control room operator model, TESEO. *Reliability Engineering*, **1**, 3–14.

Collier, S.G. and Graves, R.J. (1986). Improving human reliability: practical ergonomics for design engineers. In *Proceedings of the 9th Advances in Reliability Technology Symposium*, University of Bradford.

Comer, M.K., Seaver, D.A., Stillwell, W.G. and Gaddy, C.D. (1984). *Generating Human Reliability Estimates Using Expert Judgement*. USNRC Report Nureg/CR-3688 (Washington, DC: USNRC).

Drury, C.G. (1983). Task analysis methods in industry. *Applied Ergonomics*, **14**, 19–28.

Edwards, W. (1977). How to use multi-attribute utility measurement for social decision-making. *IEEE Transactions on Systems, Man, and Cybernetics, SMC-7-5*.

Embrey, D.E. (1986a). Approaches to aiding and training operator's diagnosis in abnormal situations. *Chemistry and Industry*, **7**, 454–459.

Embrey, D.E. (1986b). SHERPA: a systematic human error reduction and prediction approach. Paper presented at the *International Topical Meeting on Advances in Human Factors in Nuclear Power Systems*, Knoxville, Tennessee.

Embrey, D.E. (1987). Error Analysis in Proceduralised Tasks. Material presented in the *Human Reliability Analysis Course*, SRD, UKAEA, Culcheth, Cheshire.

Embrey, D.E. and Kirwan, B. (1983). A comparative evaluation study of three subjective human reliability qualification techniques. In *Proceedings of the Ergonomics Society's Conference 1983*, edited by K. Coombes, (London: Taylor and Francis), pp. 137–141.

Embrey, D.E. Humphreys, P., Rosa, E.A., Kirwan, B. and Rea, K. (1984). *SLIM-MAUD: an Approach to Assessing Human Error Probabilities Using Structured Expert Judgement*. USNRC Report Nureg/CR-3518 (Washington, DC: USNRC).

Green, A.C. (1983). *Safety Systems Reliability* (Chichester: John Wiley).

Hall, R., Fragola, J. and Wreathall, J. (1982). *Post Event Human Decision Errors: Operator Action Tree/Time Reliability Correlation*. USNRC Report Nureg/CR-3010 (Washington, DC: USNRC).

Henley, E.J. and Kumamoto, H. (1981). *Reliability Engineering and Risk Assessment* (New Jersey: Prentice-Hall).

Hunns, D.M. (1982). The method of paired comparisons. In *High Risk Safety Technology*, edited by A.E. Green (Chichester: John Wiley).

Hunns, D.M. and Daniels, B.K. (1980). The method of paired comparisons. In *Proceedings of the 6th Symposium on Advances in Reliability Technology*, Report NCSR R23 and R24. NCSR, UKAEA, Culcheth, Cheshire.

Kirwan, B. (1986). *Techniques to Aid Human Error Identification in Human Reliability Assessment, Human Reliability Course Manual*. (London: IBC Technical Services Limited).

Kirwan, B. (1987). Human reliability analysis of offshore emergency blowdown system. *Applied Ergonomics*, **18**, 23–34.

Kirwan, B. (1988). A comparative evaluation of five human reliability assessment techniques. In *Human Factors and Decision Making*, edited by B.A. Sayers (Oxford: Elsevier), pp. 87–104.

Kirwan, B. and Rea, K. (1986). Assessing the human contribution to risk in hazardous materials handling operations. Paper presented at the *First International Conference in Risk Assessment of Chemicals and Nuclear Materials*, Surrey University.

Kirwan, B., Embrey, D.E. and Rea, K. (1988). *The Human Reliability Assessor's Guide*. Report RTS 88/95Q. NCSR, UKAEA, Culcheth, Cheshire.

Kletz, T. (1984). *HAZOP and HAZAN—Notes on the Identification and Assessment of Hazards* (Rugby: Institute of Chemical Engineers).

Ludke, R.L., Strauss, F.F. and Gustafson, D.H. (1977). Comparison of five methods for estimating subjective probability distributions. *Organisational Behaviour and Human Performance*, **19**, 162–179.

Murphy, A.H. and Winkler, R.L. (1974). Credible interval temperature forecasting: some experimental results. *Monthly Weather Review*, **102**, 784–794.

Nicks, R. (1981). Probabilistic approach to problems in reactor safety risk assessment. Paper presented at *Convegno Internacionale sin Foundmenti della Probabilita e della statistica*, Luino, Italy.

Park, K.S. (1987). *Human Reliability* (Oxford: Elsevier).

Pew, R.W., Miller, D.C. and Feehrer, C.G. (1987). *Evaluation of Proposed Control Room Improvements Through Analysis of Critical Operator Decisions* (Palo Alto, CA: Electric Power Research Institute).

Phillips, L.D., Humphreys, P. and Embrey, D.E. (1983). *A Socio-technical Approach to Assessing Human Reliability*. Technical Report 83-4. London School of Economics, Decision Analysis Unit.

Pontecorvo, A.B. (1965). A method of predicting human reliability. *Annals of Reliability and Maintenance*, **4**, 337–342.

Potash, L., Stewart, M., Dietz, P.E., Lewis, C.M. and Dougherty, G. (1981). Experience in integrating the operator contribution in the PRA of actions operating plants. In *Proceedings of the ANS/ENS Topical Meeting on PRA*, New York, American Nuclear Society, IL.

Rasmussen, J., Pederson, O.M., Carnino, A., Griffon, M., Mancini, C. and Gagnolet, P. (1981). *Classification System for Reporting Events Involving Human Malfunctions*. Report Riso-M-2240, DK-4000 (Roskilde, Denmark: Riso National Laboratories).

Reason, J. (1988a). Errors and violations: The lessons of Chernobyl. Paper presented at the *IEEE Conference on Human Factors in Nuclear Power*, Monterey, CA.

Reason, J. (1988b). Human fallibility. *Consortium of Local Authorities Proof of Evidence in the UK Hinckley 'C' Power Station Enquiry*, COLA 22.

Reason, J.T. and Embrey, D.G. (1986). *Human Factors Principles Relevant to the Modelling of Human Errors in Abnormal Conditions of Nuclear and Major Hazardous Installations*. Report for the European Atomic Energy Community (Lancs: Human Reliability Associates Ltd).

Rogers, W.P. *et al.* (1986). *Report of the Presidential Commission on the Space Shuttle Challenger Accident*, June 6.

Seaver, D.A. and Stillwell, W.G. (1983). *Procedures for Using Expert Judgement to Estimate Human Error Probabilities in Nuclear Power Plant Operations*. Nureg/CR-2743. (Washington DC: USNRC).

Shepherd, A. (1986). Issues in the training of process operators. *International Journal of Industrial Ergonomics*, **1**, 49–64.

Spurgin, A.J., Lydell, B.O., Hannaman, G.W. and Lukic, Y. (1987). Human reliability assessment—a systematic approach. In *Reliability 87*, NEC, Birmingham.

Swain, A.D. and Guttmann, H.E. (1983). *A Handbook of Human Reliability Analysis with Emphasis on Nuclear Power Plant Applications*. Nureg/CR-1278 (Washington, DC: USNRC).

USNRC (1980). *Three Mile Island: A Report to the Commissioners and to the Public (The Rogovin Report)*, USNRC Report Nureg/CR-1250-V. (Washington, DC: USNRC).

USNRC (1984). *Seabrook Station Probabilistic Safety Analysis. Section 10: Human Actions Analysis*. (Washington, DC: USNRC).

USNRC (1985). *Loss of Main and Auxiliary Feedwater at the Davis-Besse Plant on June 9, 1985*. Nureg-1154. (Springfield, VA: National Technical Information Service).

USSR State Committee (1986) *The Utilisation of Atomic Energy: The Accident at the Chernobyl Nuclear Power Plant and its Consequences*. Information compiled for the IAEA Experts' meeting, 19–25 August, Vienna (Part 1).

Wagenaar, W. (1986). *The Cause of Impossible Accidents*. Speech delivered on the occasion of the 6th Dujiker Lecture, 18 March, Netherlands Institute of Psychologists, University of Amsterdam.

Watson, I. and Oakes, F. (1988). *Human Reliability Factors in Technology Management*. SRD Report, UKAEA, Culcheth, Cheshire.

Whalley, S.P. (1988). Minimising the cause of human error. In *10th Advances in Reliability Technology Symposium*, edited by G.P. Libberton (London: Elsevier).

Whittingham, R.B. (1988). The design of operating procedures to meet targets for probabilistic risk criteria using HRA methodology. Paper presented at *IEEE Conference on Human Factors in Nuclear Power*, Monterey, CA. pp. 303–310.

Williams, J.C. (1983). Validation of human reliability assessment techniques. Paper presented at the *4th National Reliability Conference*, NEC, Birmingham,

Williams, J.C. (1986). HEART—a proposed method for assessing and reducing human error. In *Proceedings of the 9th Advances in Reliability Technology Symposium*, University of Bradford.

Williams, J.C. (1988). A data-based method for assessing and reducing human error to improve operational performance. Paper presented at the *IEEE Conference on Human Factors in Nuclear Power*, Monterey, CA, pp. 436–450.

Woods, D.D. (1984). Some results on Operator Performance in Emergency Events. In *Proceedings of the Institute of Chemical Engineers Symposium on Ergonomics Problems in Process Operations*, Symposium Series No. 90

(Rugby: Institute of Chemical Engineers), pp. 21–32.

Woods, D.D. and Roth, E.M. (1986). *Models of Cognitive Behaviour in Nuclear Power Plant Personnel: A Feasibility Study*, Nureg/CR-4532 (Washington, DC: USNRC).

Woods, D.D., Roth, G. and Pople, H. (1987). *An Artificial Intelligence Based Cognitive Model for Human Performance Assessment.* Nureg/CR-4862 (Washington, DC: USNRC).

Chapter 29

Accident reporting and analysis

Ivan D. Brown

Introduction

The nature and frequency of malfunctioning have important implications for the safety, productivity and efficiency of all human work and technological systems. This importance is recognized by the variety of accident reporting procedures which has developed as a means of work system evaluation. Indeed, in certain instances such procedures are a statutory requirement of system operation.

This widespread reporting of accidents is clearly in the interests of both society and the individual worker, given satisfactory interpretation of the data and translation of relevant findings into viable accident countermeasures. For a variety of reasons, including interindividual differences, intra-individual variability, fatigue and stress, it is virtually impossible to design and operate a perfectly safe human–technological system. Accident reporting thus represents the only practical way of evaluating system safety under real operating conditions and of identifying factors which may be contributing to accident causation.

The ubiquitous nature of accident reporting procedures should not, however, be allowed to obscure the essential differences in the way they are used to evaluate different areas of human activity. Good accident reporting systems are purpose designed. The information collected and the way it is sought will reflect an organization's need for that information and its expectancies about the various causes of malfunctioning. As Hale and Hale (1972) point out:

> The task of driving a car safely is very different from most industrial tasks in that it involves different skills, different motivation and a different degree of interaction with other people. Hence factors which could be expected to influence road accidents would not be expected to affect industrial accidents and vice versa.

Such differences between socio-technological systems may appear to be self-evident, yet all too often it is assumed that 'an accident is an accident is an accident'. This narrow view of accident causation has sometimes inhibited progress in accident prevention, at the level of both theory and practice. Hale and Hale (1972) go so far as to 'regard the influence that road accident research has had on industrial accident theories and conclusions as unfortunate in many instances'. This does not mean that it is impossible to develop a general model of human error and accident causation, but it may reasonably be concluded that no single, useful, accident reporting system will have universal applicability across the entire range of human activity.

In practice, the design of accident reporting systems will, in many instances, be compromised by the resources available within an organization to collect data, analyse and interpret them, and translate any important findings into viable accident countermeasures. Such compromises may be presented as an inescapable fact of working life. However, they must never be allowed to deflect attention from the fact that accident reporting is a means to an end, not an end in itself. Ease of data collection is of little importance if it compromises validity and renders data uninterpretable. It may appear more acceptable to compromise on the type of accident data collected in order to maximize reliability, but even this could have serious implications for the analysis of the data in question and its application to countermeasure design.

The rest of this chapter will discuss these and other aspects of accident reporting in more detail and consider ways in which it may be used more effectively to evaluate and improve human work and technological systems.

Definition of an 'accident'

Defining the term 'accident' would appear to be a prerequisite for the design and use of any accident reporting procedure. However, this is often not done explicitly, but is implied by the criteria used for categorizing those events which are to be reported as 'accidents'. Even where an explicit definition is offered, it may simply describe the subset of behavioural *outcomes* which the reporting procedure can record with an acceptable level of confidence, rather than describing the nature of system *malfunctioning* itself. For example, the British road accident recording procedure (Stats 19), like many other such procedures, defines an 'accident' as:

> One involving personal injury occurring on the public highway (including footpaths) in which a road vehicle is involved and which becomes known to the police within 30 days of its occurrence. The vehicle need not be moving and it need not be in collision with anything. (Department of Transport, 1987, p. 6)

Thus behavioural outcomes which result only in vehicle or other property damage are excluded from this 'accident' reporting system, largely to maximize the reliability of the available data and ease the burden of collecting and analysing it. However, such restricted definitions can have serious implications for the analysis and interpretation of accident data, as will be discussed later.

In addition to these effects of expediency, definitions of an 'accident' are biased by the specific interests of professional groups working in the fields of accident causation, prevention and treatment. The vast majority of accidents are probably multicausal, yet engineers and physicists will be interested mainly in technological malfunctioning and its rectification; behavioural scientists will be interested mainly in causes of human error and its prevention or reduction; physicians will be interested in injury patterns and ways of preventing or treating injury. Thus Shannon and Manning (1980), for example, classify accidents in terms of uncontrolled transfers of energy that result in injury. Farmer and Chambers (1926) asserted that, 'from a psychological point of view an accident is merely a failure to act correctly in a given situation'. In their terms, an 'accident' is simply equated with erroneous behaviour. By contrast, medical usage has equated 'accident' with 'injury', which may confuse the interpretation of road accident statistics, for example, where one 'accident' can give rise to several 'casualties'.

Because accident research and safety efforts cross a number of disciplinary boundaries, there is a need for an agreed definition of the term 'accident', which minimizes such confusion and maximizes communication among the many professional groups contributing to research and practice in this field. Unfortunately, lay language and dictionary definitions tend to equate 'accident' with 'chance' or 'luck', which appears to deny any attempt to understand or deal with the causes of accidents. For example, dictionary definitions include: 'arising from unknown or remote causes'; 'unforeseen, unplanned, or unpredictable event'; and 'any fortuitous or non-essential property, fact, or circumstance'. 'Suddenness' is often seen to be an essential characteristic of accidents, although it is clear that many accidents result from relatively prolonged insidious developments of discrepancy between actual and perceived circumstances. Other definitions come closer to capturing the essential features which accident reporting procedures aim to identify, such as: 'combination of causes producing an unfortunate result'; or 'injurious consequences'; and, in particular, 'lack of intention'. However, none of these definitions aids the design and use of accident reporting procedures. For this reason it has been argued, e.g. by the police, in relation to road traffic 'accidents', that such behavioural consequences are not random, chance events, but are usually predictable, or at least explainable. Therefore, they suggest, the term 'incident' should replace 'accident'. Whilst this alternative recognizes the small part played by chance or luck in accident *causation*, it does little to produce an informative, useful definition of the process which accident reporting procedures aim to record.

It seems preferable to base a definition of the term 'accident' on the fact that most human work involves one or more persons, using 'equipment' of greater or lesser complexity, and an 'environment' (physical or social) within which the work is performed. Each of these three 'main factors' may be solely responsible for malfunctioning of the work system. Alternatively, malfunctioning may be attributable to any of the three first-order interactions or the single second-order interaction between the three main system factors. This ergonomics approach (e.g. see Edwards, 1981), recognizes the multicausal nature of 'accidents', thus providing an opportunity to identify the behavioural antecedents of accidental consequences whilst retaining flexibility in the design and implementation of accident countermeasures. That is to say, an accident which is attributable to first- or second-order interactions between people, equipment and environmental factors may be remediable by treating any one or more of the relevant interface characteristics.

Using this approach, Brown (1976) defined an 'accident' as 'the unplanned outcome of inappropriate behaviour'. He defended this definition by pointing out that:

(a) It draws a clear distinction between antecedent *behaviour* and the *consequence* of that behaviour. By contrast with Farmer and Chambers' (1926) definition, only the *outcome* of the 'erroneous' behaviour is termed 'accidental'.

(b) This accidental outcome is characterized by its *unplanned* nature, rather than by the unpredictability which lay language and certain dictionary definitions have associated with accidents. Many accidental outcomes may in fact be quite predictable in probabilistic terms, even by the actor in question, but assigned a negligible probability of occurrence and then ignored for all practical purposes.

(c) The antecedent behaviour is termed *inappropriate* because it is mismatched to the actual demands of the task or the environment. (Although it may be perceived as appropriate where display faults misrepresent actual task demands.) It may be intentional, but unfortunately unwise behaviour, for example, when a car driver brakes sharply on an icy road. It may be intentional but misdirected or misplaced, for example, when a pilot attempts to land on the wrong runway. Alternatively, it may be unintentional, for example, when the captain of a cross-channel ferry puts to sea unaware that its bow loading doors are still open.

(d) The association between 'accidents' and 'chance', favoured by certain dictionary definitions, has been avoided because it has often blurred the distinction drawn above between antecedent behaviour and accidental outcome. Since identical inappropriate behaviour will not inevitably result in an accident each time it is repeated, it might be claimed that there is a chance relationship between such behaviour and the *occurrence* of accidents. However, there are clearly identifiable categories of accidents resulting from given combinations of inappropriate behaviour when performing specific tasks in known environmental conditions. The actual *nature* of the probable accidental outcome is therefore not simply a chance occurrence and it is the nature, rather than the frequency of occurrence of accidental consequences, which is more informative of the causes of inappropriate behaviours and which it is therefore important to capture in any accident reporting system.

This definition of an 'accident' can now be set within the general theoretical context of accident causation and human error, before considering in more detail the desirable characteristics of an accident reporting procedure and how its data analysis may be translated into effective accident countermeasures.

Theories of accident causation

A critical review of the main theories of accident causation has been provided by Hale and Hale (1972). These theories will be discussed here only from the standpoint of their possible implications for the recording and analysis of accident data.

The 'Pure Chance' theory

This theory holds that everyone exposed to the same objective risk has an equal liability to accidents, which are thus entirely chance determined. It is virtually impossible to test this theory, empirically, because of the difficulty of finding sufficiently large samples of working populations among which objective risk is actually equated. The theory also appears to beg the question of what constitutes 'objective risk'. Is it simply the risk associated with a particular task performed under given conditions? Or is it the risk associated with a particular level of operator skill when performing the task under given conditions? Clearly, novice and experienced workers would not be expected to have an equal liability to accidents, therefore the theory appears only to discount other individual differences which might contribute to accident causation, such as age, sex, intelligence, personality, temperament and motivation. But does it also discount, say, effects of impairment? As a null hypothesis, the 'pure chance' explanation may be a way of directing attention to accidents as an index of risk associated with particular tasks, working situations, or environmental conditions. But it would be a very uninformative accident reporting system which concentrated on these sources of causation to the exclusion of individual differences in accident liability.

The 'Biased Liability' theory

This theory holds that an individual's accident involvement either increases or decreases their liability to subsequent involvement. This appears eminently reasonable, since involvement in an accident may well increase apprehension, with its tendency to impair performance, when the circumstances surrounding that accident are perceived to recur. Alternatively, accident involvement may produce a tendency to avoid its attendant circumstances in future, or encourage the victim to improve the skills and knowledge required to perform more safely under those conditions. This theory may serve to explain the unreliability of accident data in predicting an individual's liability to future involvement. However, the theory is unhelpful in specifying the duration of biased liability, or the extent to which this bias will generalize to accidental circumstances similar to and different from those of the initial involvement. It therefore provides little guidance on the type of data which should be collected in any accident reporting system.

The 'Unequal Initial Liability' theory

This theory is probably better known as the theory of 'accident proneness'. For a conceptual review of the theory, see McKenna (1983). There are two versions of the basic theory. One holds that certain individuals are prone to accidents because of their innate personal characteristics. Thus their accident liability is assumed essentially to be a stable feature of their performance, irrespective of task, working conditions, time, or other non-personal factors. The other version, which owes much to the work of Cresswell and Froggatt (1963), holds that accident proneness is a variable factor, being associated with 'critical events' in the life of an individual rather than with situational risks. Both versions are inherently attractive, because they fit the common experience that some people have more accidents than others and that this pattern tends to vary with, for example, age and experience. The fallacies in this line of theoretical argument have been exposed by McKenna (1983), who suggests that it would be advantageous to abandon the concept of 'proneness' and employ a new term such as 'differential accident liability'. This would certainly permit a more flexible approach to the interpretation of accident data, by diverting attention away from the search for 'scapegoats' and towards the identification of contributory situational factors.

The 'Stress' theory

This theory holds that accidents happen when a task, environmental, or individual stressor reduces the capacity of an individual to meet task demands, or when the demands of a task increase beyond the normal capacity of an individual to meet them. Example of the former type of stressor would be fatigue, illness, or environmental heat, cold, noise, and so on. Examples of the latter would be increased informational load or work-rate requirements, or demands falling outside an operator's repertoire of skills. Whilst limited in scope, this theory does at least permit accident causation to be explained in terms of intra- and interindividual differences and also in terms of job, task and situational factors. (See Cox's chapter in this book for a review of theories of stress.)

The 'Arousal/Alertness' theory

This has been developed from research on physiological activation and human behaviour (e.g. see Duffy, 1962). The central hypothesis here is that a relationship exists between an individual's level of arousal/alertness and their performance on any task, the efficiency of which rises to a peak as arousal increases, but then declines as arousal becomes inappropriately high. This became known as the 'inverted-U hypothesis'. It predicts that accidents are more likely to occur both when arousal is low (e.g. when the person is underloaded, bored or drowsy) and also when arousal is high (e.g. when the person is anxious, or excessively motivated). This arousal theory is sometimes confused with stress theory because overarousal, like stress, tends to degrade performance. However, the concepts should be kept distinct because effects of stress are, by definition, harmful, since they represent a reduction of coping ability. Effects of arousal may, or may not be, harmful, depending upon the optimal level of arousal for performance of the task in hand. Thus the theories provide different guidance on the interpretation of accident data and on the design of accident countermeasures. In particular, it is important to avoid conceptualizing 'stress' as a factor which can be employed as a countermeasure against accidents attributed to underarousal of workers.

'Psychoanalytic' theories

These attribute accidents to subconscious processes with self-punitive aims, initiated by feelings of guilt, anxiety or motivational conflict. Although it is difficult to incorporate such theories into the design of accident reporting systems, clearly they have some guidance to offer when data analysis points to an individual's unexpectedly high level of differential accident liability.

'Epidemiological/Ergonomics/Situational' theories

These appear to have so much in common that, for all practical purposes, they are indistinguishable. Epidemiology, as its name implies, developed as an approach to the study of epidemics. It holds that causation is essentially a conjunction of a 'host' (the victim of the epidemic), an 'agent' (which transmits the disease) and an 'environment' within which 'host' and 'agent' interact. This seems perfectly analogous to the 'ergonomics', or 'situational' approach to explanations of accidents, in which a person (the 'host') interacts with a tool, or technological system (the 'agent') in a working environment (either physical or social). Both the medical and the ergonomics theories hold that it is the conjunction of all three factors which causes the problem (accident or epidemic) and that the solution can only be found by altering this conjunction. Furthermore, both approaches maintain a flexible view of accident/epidemic reduction, in that countermeasures are seen to result from manipulations of the person (host) and/or the technology (agent) and/or the environment.

This theoretical view of accident causation thus offers the greatest potential for improving safety, by directing attention to possible malfunctioning among all three components in human–technological systems and by keeping open all options for improving their independent and interacting performance characteristics.

The 'Domino' theory

This theory has been developed in both industrial and road safety fields in order to explain the sequential, multicausal, nature of accident causation. Hale and Hale (1972) list five stages in this sequential process, attributed to Heinrich (1950). These are:

1. Ancestry and social environment.
2. Individual fault.
3. Unsafe act and/or mechanical hazard.
4. Accident.
5. Injury.

These 'stages', as postulated, tend to confound the sequential with the situational factors contributing to accident production. They encompass the epidemiological/ergonomics view of accident causation, in that they highlight the involvement of people and objects within an environment, but they give more weight to the temporal nature of the accident process. Thus the value of this theory for accident reporting and analysis is that, apart from directing attention to the fact that accidents seldom have a single cause, it alerts us to the importance of antecedent behaviour and events. In particular, it draws attention to the possibility that certain antecedent behaviours, representing attempts to recover from initial errors, may actually be counterproductive in terms of accident avoidance, rather than simply 'inappropriate', as described in the earlier definition of an accident.

Theories of human error

Accident theories originated in attempts to describe statistical distributions of accidents using 'curve-fitting' exercises which often made unjustified assumptions about the homogeneity and stability of operator characteristics among the working populations in question. The specific nature of the accidents in question and the antecedent behaviour of victims and other involved persons were given little or no consideration. This lack of a conceptual basis for accident theorizing led to years of academic controversy and little of practical value for safety workers. By contrast, approaches to accident reduction via theorizing about human error, whilst not entirely uncontroversial, seem likely to be highly productive of ergonomics measures for improving safety; especially where the approach to explanation of error is made via theory of normal, error-free behaviour.

Reason (1988) has recently provided a good, succinct, overview of such current models of human performance and error. His account of these models

will briefly be discussed here only in relation to the guidance they provide
for accident reporting and its analysis.

Shiffrin and Schneider's (1977) theory

This theory of controlled and automatic information processing developed
from laboratory studies of search, detection, memory and retrieval. It
distinguishes two types of non-sensory memory: a long-term store, the
function of which is to retain information permanently but passively, and a
short-term store which has a dual function; (a) to provide an activated subset
of relevant information from the long-term store and (b) to provide a
'workspace' for conscious decision making and control processes.

This theory highlights the explanatory value of information obtained by
introspection among accident-involved individuals, preferably shortly after
the event. Given access to such information, the theory directs attention to
two aspects of human functioning which are important for accident prevention
and/or reduction:

1. Identification of operators' general knowledge of the task in question
and the particular subset of that knowledge which was perceived as specifically
relevant to the task during the period prior to the accident. Information of
this kind has obvious practical applications to operator training and the design
of informational displays.

2. Identification of those components of task performance which were
under the operator's conscious control during the run-up to the accident and
those which were sufficiently automated that they could be run off
subconsciously. Again, this type of information could provide guidance on
the need for attention to training and/or display design as an approach to
accident reduction and prevention. It could also highlight certain aspects of
inefficient work organization which are imposing an unnecessarily high load
on conscious control processes, thus resulting in accidents via stress or
distraction.

Broadbent's (1984) 'Maltese Cross' model of memory

This model was developed from studies of interference among concurrent
cognitive activities. It distinguishes 'representation', in the form of persisting
memory records, from 'processes' of translations between these memory
records. Representations, in the form of sensory store, abstract working
memory, motor output store and long-term associative store, communicate
via a central processing system.

As with Shiffrin and Schneider's theory, the 'representation' component
of this model highlights the need to obtain information on accident-involved
individuals' task knowledge and the intentions associated with their antecedent
behaviour. But, in addition, the 'process' component of the model emphasizes
the importance of the sequential nature of operator behaviour in providing

an understanding of the error(s) which contributed to the accident. Thus the practical applications of this theory to accident reduction and prevention will be via operator training and work design.

Norman and Shallice's (1980) 'attention to action' model

This is one of several models which have developed from observation of, usually inconsequential, cognitive failures, often in domestic or clinical contexts (e.g. see Reason, 1979; Norman, 1981). The control structure envisaged in this model consists of horizontal 'threads' which govern habitual activities, not under continuous conscious control, and vertical 'threads' representing the attentional processes which govern behaviour in novel or emergency conditions. The model also incorporates motivational factors, which are assumed to act via the vertical 'threads', with longer time constants than the attentional effects.

Thus this model, like the previous two discussed, emphasizes the need for accident reports to include information on the relative extent to which antecedent behaviour was under the operator's conscious control, rather than being automated. It also emphasizes the importance of knowing whether automated antecedent behaviour may have been rendered 'inappropriate' by the persistent activity of some specialized processing structure (or 'schema') beyond the point at which it was no longer required to be active by the task in hand. The inclusion of motivational factors in the model highlights the need for accident reports to include information on intentions of and reasons for antecedent behaviour, so that the 'why', as well as the 'what' and 'how' aspects of accidents can be examined, understood, and dealt with by remedial measures.

Rasmussen's (1982) 'skills, rules and knowledge' model

This model of behaviour was developed to account for the rather more serious errors and accidents that can occur in complex, industrial, process control operations, particularly the emergency situations that result from breakdown of hazardous nuclear and chemical processes. This three-level model of performance and error developed from investigations of trouble-shooting behaviour using verbal protocols. The information these provided on covert behaviour enabled the model to describe two types of error: 'slips' (unintended actions) and 'mistakes' (inappropriate plans or intentions). Skill-based behaviour is governed by stored patterns of preprogrammed instructions. Rule-based behaviour enables familiar situations to be dealt with by learned production rules (e.g. if 'this' state occurs then implement 'that' remedial action). Knowledge-based behaviour is used to deal with completely novel situations to which no actions have been pre-planned. Skill-based errors will be associated with inappropriate behavioural combinations of space, time and effort. Rule-based errors will be associated either with a misperception of situational demands, or the incorrect recall of appropriate procedures. Knowledge-based errors will result from limitations in operator resources, or from incomplete or incorrect knowledge.

This model again emphasizes the need to obtain detailed information on antecedent behaviour from accident-involved personnel, if the precise nature of contributions from human error are to be understood. The model also associates particular categories of error with different levels of skill among trained personnel and thus emphasizes the benefits that a well-designed accident reporting system can have for identifying training, or retraining, needs as part of any required countermeasure programme. More subtly, the model identifies the need to examine possible sources of interference between these different levels of skill-, rule- and knowledge-based behaviour, as potential contributors to accidents. For example, did the concurrent requirement to deal with a novel situation disrupt a skilled action sequence, or lead to the implementation of an inappropriate rule, thus provoking an accident? The model therefore has implications for the understanding of work design as a possible contributory factor in accidents, as well as for the place of training in accident reduction and prevention.

Baars' (1983) 'global workspace' model

This model was developed from the view of human cognition as a parallel distributed processing system, in which specialized processors cover all aspects of mental functioning without the need for central executive *control*, although co-ordinated by a central information *exchange*. This takes place within a 'working memory' to which the specialized processors compete for access as a function of their current level of activation. Once within this 'global workspace' they can recruit and control other processors. 'Consciousness' is identified with the current content of this workspace.

The model thus has similarities with Shiffrin and Schneider's model, in that it identifies a subset of cognitive operations on which attention is concentrated at any one time. It, too, therefore emphasizes the need to understand the 'how' and 'why' of behaviour antecedent to accidents, with its attendant implications for introspection and verbal reporting by accident-involved personnel as a method of revealing the relevant conscious contents of the operator's 'workspace'. Again, it identifies the possibility of reducing or preventing accidents by highlighting specific needs for training and/or work design.

Card *et al.*'s (1983) 'model human processor'

This theory is an attempt to construct a limited set of 'working approximations' to a broad range of cognitive activity. It has similarities with Broadbent's 'Maltese Cross' model, in that it consists of two parts: representations in memory, and principles of operation. These are represented at each of three interacting subsystem levels; perceptual, motor and cognitive. These three systems can operate in series or in parallel. Certain basic principles of operation are specified in the model; such as 'recognize-act', 'discrimination', 'uncertainty' and 'rationality'. Implications for the understanding of human error follow from the model's incorporation of constraints on goal attainment imposed by task structure, information input, limited resources and incomplete knowledge. These have obvious applications to the design of accident-reporting systems and implications for accident countermeasures based on (re)training and task design.

Anderson's (1983) ACT theory of cognitive architecture

This is a development of the 'Adaptive Control of Thought' model. It distinguishes three memory systems; working, declarative and production. Working memory interacts with five processes; encoding, performance, storage, retrieval and execution. Knowledge can be represented as: temporal strings, spatial images, or abstract propositions. Productions by the system can be matched in five ways in order to resolve conflicts: degree of match; production strength; data refractoriness; specificity; and goal dominance. 'Activation' controls the rate of information processing for production and is a function of environmental stimulation, production execution, and goal influences in working memory.

Immediate applications of this highly structured model to taxonomies of human error used in accident reporting systems are probably more apparent than real. However the model does, once again, emphasize the need to distinguish between representation and process in the knowledge that guided the antecedent behaviour of accident-involved individuals. It also highlights the need to examine and understands goals and reasons for such behaviour (the 'why' of accident causation), as well as the 'what' and 'how' of accident production.

Implications for accident reporting

From the foregoing discussion it is possible to identify the following implications for the reporting of accident occurrence.

Purposeful reporting

There is little point in spending time and effort on the collection of comprehensive data with little idea of the use to which its analysis will be directed, or in the absence of resources required to implement any necessary accident countermeasures. The latter will essentially be of two types:

1. *Primary safety measures.* The reduction or prevention of accident occurrence.

2. *Secondary safety measures.* The prevention, or reduction in severity, of injury associated with accidents that do occur.

The principal aim of any accident reporting system will be to highlight the need for primary safety improvements and identify the particular types of countermeasure which seem likely to be most efficient in preventing or reducing accident occurrence. The emphasis here will be on the collection of comprehensive data relevant to the antecedent behaviour of accident-involved personnel and its relation to concurrent task demands. This procedure has been carried to a high level of efficiency in, for example, civil aircraft operations, where the 'black box' flight recorder stores valuable information on technological functioning and ongoing activity at various behavioural interfaces for a prescribed period prior to the occurrence of any accident. It should also be possible to report antecedent behaviour comprehensively in accidents occurring within, say, the process industries, where tasks and operations are highly structured and sequences of operations are logged. It will be somewhat more difficult, even in the process industries, to report and reconstruct antecedent behaviour patterns and discrepancies at congruences between different tasks, e.g. process operation and equipment maintenance. Furthermore, it may be extremely difficult to obtain accurate behavioural reports from industries where work patterns are relatively unstructured, such as on building construction sites, or in small engineering workshops. Nevertheless, if inappropriate behaviour is to be avoided and accidents reduced or prevented, then the detailed behavioural antecedents of accidents must be captured in as much detail as possible, even in these difficult situations.

The purposeful needs of secondary safety are often thought to be met by reporting the nature, severity and causes of accidental injury, so that wounding agents can be removed, resited, or redesigned, and/or so that their potential to inflict injury can be minimized by the introduction of protective clothing or equipment. However, it will be clear that an understanding of inappropriate antecedent behaviour patterns can also contribute substantially to secondary safety. It will achieve this by identifying undesirable and avoidable associations between specific operator activities, transfer of energy, and injury accident occurrence. This knowledge can then be used to design and introduce safety devices, such as interlock systems and machine guards, which prevent such injurious associations from occurring. The type of behavioural data of importance here will not consist simply of physical movement patterns; although these will obviously contribute to the design of certain secondary safety measures, such as guard rails and protective clothing. Even where the purpose is secondary safety, advantage will be gained from reports on covert activity, such as intentions, reasons and judgement, since these too are susceptible to change via secondary safety measures.

Adequate reporting

As mentioned earlier, 'accident' reporting is often a misnomer since many systems report only that subset of accidents which result in injury. Such types of truncated accident reporting may appear to meet the principal needs of secondary safety and they are usually justified by their concentration on the 'more important' accidents and on 'cleaner' data. These claims have a certain face validity; however, they ignore the potential value of 'control' data on the behavioural antecedents of damage-only accidents, or near accidents which resulted in neither injury nor damage. Injury, damage and 'near misses' may all result from apparently identical human–technology interactions. The clue to the different consequences may be discernible only by detailed comparisons of the antecedent behaviours in question. If damage-only accidents and near accidents are not recorded, such comparisons are impossible and accident investigators can only speculate on the reasons why particular behaviour patterns sometimes result in accidental injury and sometimes do not. A concentration on injury accidents also severely limits the total number of data available for safety work and makes it difficult, if not impossible, to estimate the objective risk associated with specific behaviour patterns. A useful accident reporting system will therefore be one which collects data on *all* accidents, to the same level of detail, regardless of their consequences.

Factual reporting

Accident reports are frequently completed with the sole aim of attributing blame for human error. The result is usually a highly subjective, excessively brief account of system failure, which identifies a scapegoat but provides little or no information on the need for specific accident countermeasures in the work situation in question and certainly adds nothing to the general pool of knowledge on human error and accident causation. In order to meet even the first of these objectives, the 'first line' of any accident reporting system should avoid subjectivity and particularly the apportioning of 'blame' and, instead, concentrate on the factual reporting of task demands and operator behaviour in the relevant period prior to and surrounding the accident. As Rasmussen (1987) has pointed out: '. . . instead of focusing on human errors, data should be collected to represent situations of man–task mismatch and characterised accordingly'. In other words, a comprehensive factual description of technological system demands and human antecedent behaviour must be attempted initially, before there is any attempt to classify technical faults or human errors. These faults and/or errors should be identifiable from study of any obvious mismatches between task demand and operator response. But their nature and their 'causes' will become clear, if that stage is reached at all, only by further detailed analysis of the mismatches and the context in which they occurred.

Task-specific reporting

It follows from the above considerations that accident reporting systems which aggregate data across different tasks and thus profess to represent organizational safety via global statistics will contribute little to the understanding of human error, or to the design and development of accident countermeasures. They will merely index organizational safety, in a far from meaningful sense, since risk will largely be unquantifiable, and they may provide management with a crude measure of the cost–benefits of accident prevention and reduction. If accident data are to be more meaningful than this and contribute usefully to safety improvements, they must be recorded in a task-specific term and not aggregated across dissimilar tasks until an initial data analysis has been completed.

Verbal reporting

An important aspect of task-specific reporting is that it retains and enhances the value of introspective verbal reports provided by accident-involved personnel (see Patrick, 1987, for an overview of methodological issues associated with verbal reporting; see also Bainbridge's chapter and Sinclair's chapters in this book). Verbal reports will often be the only method of reconstructing the behavioural sequences antecedent to an accident and supporting these reconstructions with information on participants' covert reasons, intentions, perceptions, decisions, judgments, etc. Within tasks, such reconstructed sequences may be aggregated or compared in order to highlight any mismatches between system demands and operator behaviour which may have contributed to accident causation.

Analysis of accident data

The first general point to be stressed is that any accident analysis should aim principally to identify the potential for improving system safety, rather than simply identifying the 'causes' of past events. The latter objective may be considered essential to meeting the former, but this is true in principle rather than in practice, because of differences in the definition of human error as a 'cause' of accidents.

Where tasks are highly structured and operator behaviour can be clearly specified in a predetermined manner, error may be defined as any departure from operating instructions. In such work situations, mismatches between operator behaviour and task demands may be relatively easy to identify (depending upon the method used for concurrent automatic storing of antecedent behaviour in accidents and the truthfulness of subsequent verbal reporting). This will enable 'causal' or 'contributory' factors in operator performance to be identified with apparent ease and aid the production of global statistics on an organization's accident patterns. But it may be unhelpful in identifying acceptable options for accident countermeasures, because it will bias attention towards the elimination of any diversity in operator performance. Such diversity will be associated with differences between the various goals within system operation and it may not constitute 'erroneous' behaviour which needs to be eliminated if the system is to function safely.

A more productive view of human error, from the standpoint of safety improvement, is that it represents 'the effect of human variability in an unfriendly environment' (e.g. see Rasmussen, 1987; also Kjellén, 1984a,b). On this view, accident analyses should focus not simply on the frequency of discrepancy between prescribed task performance and actual operator behaviour, it should examine the precise nature and behavioural range representing such discrepancies. Instead of reflecting error to be eliminated in the name of accident prevention, antecedent behaviour analysed in this manner should suggest remedial measures which will result in a more error-tolerant system, capable of accepting safely a much wider range of behavioural diversity.

A second, related, general point is that accident analyses should include not only an examination of the range of antecedent behaviours actually exhibited by accident involved personnel, they should include consideration of alternative behaviours which would have met the task and system goals in question, but which were not exhibited (e.g. see Leplat's, 1987, description of 'fault tree analysis'). The results of such broader analyses are then capable of being utilized beyond the immediate aims of causal attribution, apportioning of 'blame' and error-tolerant system redesign; they may be suggestive of ways in which 'error-recovery' procedures can be developed and taught to operators, or incorporated into system software for automatic implementation. In other words, the analysis of accident data should aim not simply to make technological systems less 'unkind'; it should actively seek to identify any potential for improving operator support and overall user-friendliness of the system.

A third and final general point to make represents a criticism of the heavy reliance on operator biographical data by many accident reporting systems. This seems based largely on convenience in data collection, rather than usefulness in understanding accident causation or relevance to the design and implementation of acceptable remedial measures. Certainly it is simple and usually convenient to collect factual information on sex, age, education, socio-economic class, etc. However, these variables usually contribute less than informatively to data analysis and the identification of countermeasures. Sex and age information may be statutory requirements of certain accident data collection systems and these variables have some importance in certain work situations (e.g. the identification of radiation risks for pregnant women, or the possibility of 'thrill-seeking' as a risk factor among young male drivers). However, biographical data may be misleading in accident analyses, because many of them are confounded with experience factors. In addition, when biographical factors appear to be playing a significant role in the accident process, they usually point to selection and training as appropriate remedial measures. The latter may be acceptable, if potentially expensive, but considerable further analysis of the accident data is usually required in order to identify the nature and extent of any training requirements. The former may be unacceptable, in social if not in supply-and-demand terms, if selection conflicts with 'equal opportunities' legislation.

Where one aim of analysis is the classification of accident consequences, in order to identify methods for treatment and prevention of injury, it will be convenient to employ international injury coding systems (e.g. see Baker, 1982; Langley, 1982; Somers, 1983a,b). This will simplify the aggregation of data and their comparison across national or international organizations. But injury prevention will, primarily, require analysis of operator activities associated with injury production.

The considerations and arguments advanced in foregoing sections have highlighted the need to analyse antecedent behaviour in some detail, if the contributory factors in accidents are to be revealed and attendant options for remedial measures presented. How is this to proceed, given that the 'first line' of accident reporting has produced an adequate description of the accidental circumstances and consequences, the operator's antecedent behaviour and the related sequence of task demands?

The prime concern will be with the analysis of 'human error', in its broadest sense and specifically in relation to relevant task demands. What form is this analysis of error to take? Given that one main aim is to identify mismatches between task demands and operator behaviour, it would seem appropriate to base the investigation of human error on a detailed hierarchical task analysis of the work in question (see Patrick *et al.*, 1986; and the review of task analysis in this book from Stammers *et al.*). Such an analysis will break down a job into a hierarchical set of tasks, subtasks and 'plans' for describing when and in which sequence subtasks are performed. Task objectives form the basis of description. The structure of antecedent behaviour, composed from verbal reports and other more objective sources of data, can then be contrasted with this task analysis in order to reveal mismatches. However, it should be remembered that a task analysis may often represent the system designer's, or manager's view of operational requirements (although not if carried out correctly and fully, see Stammers *et al.* in this book). Mismatches will thus reflect 'errors' in the form of departures from these predetermined requirements, but they may not represent 'mistakes' in the sense of divergence from system goals. With this reservation, task analysis appears to provide an appropriate framework for the investigation of behaviour antecedent to accidents: it should readily permit the location and identification of 'error' in the task sequence, in a form which allows either operator behaviour or system demands to be modified and/or error–recovery mechanisms introduced, in order to eliminate future errors of that kind or increase system tolerance to them.

Using task analysis as the structure for investigation of error will, naturally, tend to produce a classification of antecedent behaviour based largely on met and unmet system demands. By contrast, it may be argued (see Rasmussen *et al.*, 1987) that human error must be classified in terms of *human* characteristics if the results of such analyses are to be applied to new as well as to existing systems. This seems eminently reasonable, given that human behaviour is the common element in any system requiring manual control or monitoring, and provided that any system for classifying human error can be identified with the system demands to which it referred.

A number of systems, or frameworks, have been developed for the classification of human error, based on the theories of error briefly described earlier. For example, Reason (1987a) presents a Generic Error Modelling System (GEMS) for locating common human error forms. It identifies three basic error types: skill-based slips, rule-based mistakes, and knowledge-based mistakes. Reason (1987b) also presents a classification based on the identification of eight 'primary error groupings' and eight 'information-processing domains', the interactions of which define five 'basic error tendencies' (see Table 29.1). The model also allows 'predictable error forms' to be associated with 'situational factors' known to promote that form of occurrence. It therefore appears to meet the joint criteria of classifying human error, relatively simply, in terms of normal human characteristics, without

Table 29.1. Matrix for classifying primary error groupings (reproduced from Reason (1987a) with permission of John Wiley & Sons Ltd)

	Ecological constraints	Change enhancement	Resource limitations	Scheme properties	Strategies heuristics
Sensory registration	False sensations* X	X			
Input selection			Attentional failures* X	X	O
Volatile memory			Memory lapses* X	X	O
Long-term memory			O	Inaccurate recall* X	X
Recognition processes	O	O	Misperceptions* X	X	X
Judgemental processes		X	Errors of judgement* X	X	X
Inferential processes			X	Reasoning errors* X	X
Action control			X	Unintended words/actions* X	O

X, Primary node; O, secondary node; *, primary error groupings.

losing sight of system demands and goals. As Reason (1987a) himself points out:

> To be of value in either a theoretical or a practical context, a classificatory framework must both simplify the available error data and acknowledge the multiplicity of possible causal interactions. Most importantly, however, it must recognise that predictable error and correct performance are two sides of the same coin, and hence demand common explanatory principles.

Rasmussen (1982) has presented a taxonomy of human error which appears less well-founded in the cognitive theories reviewed earlier, but which develops clearly from the concept of accidents as the consequence of 'inappropriate behaviour'. It describes features of the human–technological system mismatch in terms of inappropriate task performance, that is, omissions of activity in procedural sequences, action involving wrong components, reversals in an action sequence, inappropriate timing of action, and so on. The complete multifaceted taxonomy is illustrated in Figure 29.1, where Rasmussen uses the term 'external mode of human malfunction' to describe inappropriate task performance, in order to avoid the term 'human error', with its connotations of 'guilt' and 'blame'.

In order to characterize the covert mental activities involved in the mismatches examined by this taxonomy, Rasmussen's approach requires an analysis of the relevant decision processes in order to reveal the underlying cognitive errors (or 'internal mode of malfunction'). Given sufficient objective, factual information on the accidental circumstances and reliable verbal reports, an experienced accident investigator might pursue the analysis of cognitive error using Rasmussen's (1976) 'step-ladder' model of decision making (see Figure 29.2).

Kjellén (1987) presents an approach to accident analysis which has accident *control* as its principal objective. It is based on the concept of accident causation as the deviation of system variables from their norm. Kjellén discusses a range of taxonomies that may be used to classify such deviations and relates them to the underlying theories, or models from which they derive. Although this approach appears to have similarities with the concept of 'inappropriate behaviour' as the central feature of accident causation, it is more mechanistic and may not be well supported by the data available from industrial accident reporting systems.

Summary and conclusions

An accident is defined as: 'the unplanned outcome of inappropriate behaviour'. Accident reporting is seen essentially to require the provision of data from which the 'inappropriateness' of behaviour antecedent to accidents can be examined in fine detail. Factual information on technological system

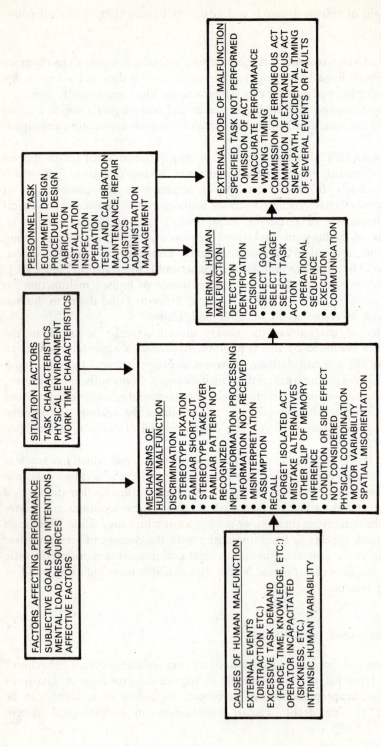

Figure 29.1. Multifacet taxonomy for description and analysis of events involving human malfunction (reproduced from Rasmussen (1982) with permission of Elsevier Science Publishers).

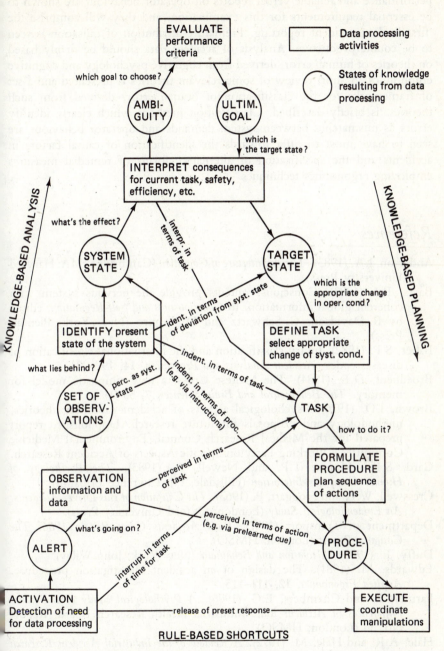

Figure 29.2. The 'step-ladder' model of decision making. Rectangles represent data processing activities, circles are states of knowledge from data processing. Reproduced from Rasmussen (1976) with permission of Plenum Press.

performance and reliable verbal reports on operator behaviour are shown to be essential requirements for this examination, and they will comprise the 'first line' of accident reporting. Premature attribution of causation is seen to be counterproductive. Analysis of accident data should be firmly based on theories of human error, derived from cognitive psychology and cognitive ergonomics. A short review of some relevant theories is presented and a set of frameworks for the classification of human error, derived from such theories, is briefly described. Classification methods which clearly identify errors as mismatches between system demands and operator behaviour are seen to have most to offer towards the identification of causal factors in accidents and the specification of alternative forms of remedial measures employing ergonomics techniques.

References

Anderson, J.A. (1983). *The Architecture of Cognition* (Cambridge, MA: Harvard University Press).

Baars, B.J. (1983). Conscious contents provide the nervous system with coherent global information. In *Consciousness and Self-Regulation*, edited by R. Davidson, G. Schwartz and D. Shapiro (New York: Plenum Press).

Baker, S.P. (1982). Injury classification and the international classification of diseases codes. *Accident Analysis and Prevention*, **14**, 199–201.

Broadbent, D.E. (1984). The Maltese Cross: a new simplistic model for memory. *The Behavioural and Brain Sciences*, **7**, 55–94.

Brown, I.D. (1976). Psychological aspects of accident causation: theories, methodology and proposals for future research. Unpublished report prepared for the Medical Research Council, Environmental Medicine Committee's Working Party on Specific Aspects of Accident Research.

Card, S.K., Moran, T.P. and Newell, A. (1983). *The Psychology of Human–Computer Interaction* (Hillsdale, NJ: Lawrence Erlbaum).

Cresswell, W.L. and Froggatt, P. (1963). *The Causation of Bus Driver Accidents: An Epidemiological Study* (London: Oxford University Press).

Department of Transport (1987). *Road Accidents Great Britain 1986: The Casualty Report* (London: HMSO).

Duffy, E. (1962). *Activation and Behaviour* (New York: John Wiley).

Edwards, M. (1981). The design of an accident investigation procedure. *Applied Ergonomics*, **12**, 111–115.

Farmer, E. and Chambers, E.G. (1926). *A Psychological Study of Individual Differences in Accident Rate*. Industrial Health Research Board, Report No. 30 (London: HMSO).

Hale, A.R. and Hale, M. (1972). *A Review of the Industrial Accident Research Literature* (London: HMSO).

Heinrich, H.W. (1950). *Industrial Accident Prevention*, 3rd edition (New York: McGraw-Hill).

Kjellén, U. (1984a). The deviation concept in occupational accident control—

I. Definition and classification. *Accident Analysis and Prevention*, **16**, 289–306.

Kjellén, U. (1984b). The deviation concept in occupational accident control—II. Data collection and assessment of significance. *Accident Analysis and Prevention*, **16**, 307–323.

Kjellén, U. (1987). Deviations and the feedback control of accidents. In *New Technology and Human Error*, edited by J. Rasmussen, K. Duncan and J. Leplat (Chichester: John Wiley), pp. 143–156.

Langley, J. (1982). The international classification of diseases codes for describing injuries and the circumstances surrounding injuries: a critical comment and suggestions for improvement. *Accident Analysis and Prevention*, **14**, 195–197.

Leplat, J. (1987). Accidents and injury production: methods of analysis. In *New Technology and Human Error*, edited by J. Rasmussen, K. Duncan and J. Leplat (Chichester: John Wiley), pp. 133–142.

McKenna, F.P. (1983). Accident proneness: a conceptual analysis. *Accident Analysis and Prevention*, **15**, 65–71.

Norman, D.A. (1981). Categorization of action slips. *Psychological Review*, **88**, 1–15.

Norman, D.A. and Shallice, T. (1980). *Attention to Action: Willed and Automated Control of Behavior*. Centre for Human Information Processing, Report 99 (La Jolla, CA: University of California).

Patrick, J. (1987). Methodological issues. In *New Technology and Human Error*, edited by J. Rasmussen, K. Duncan and J. Leplat (Chichester: John Wiley), pp. 327–336.

Patrick, J., Spurgeon, P. and Shepherd, A. (1986). *A Guide to Task Analysis: Applications of Hierarchical Methods* (Birmingham: Occupational Services).

Rasmussen, J. (1976). Outlines of a hybrid model of the process plant operator. In *Monitoring Behaviour and Supervisory Control*, edited by T.B. Sheridan and G. Johannsen (New York: Plenum Press), pp. 371–384.

Rasmussen, J. (1982). Human errors: a taxonomy for describing human malfunction in industrial installations. *Journal of Occupational Accidents*, **4**, 311–335.

Rasmussen, J. (1987). The definition of human error and taxonomy for technical system design. In *New Technology and Human Error*, edited by J. Rasmussen, K. Duncan and J. Leplat (Chichester: John Wiley), pp. 23–30.

Rasmussen, J., Duncan, K. and Leplat, J. (1987) (Eds). *New Technology and Human Error* (Chichester: John Wiley).

Reason, J.T. (1979). Actions not as planned: the price of automation. In *Aspects of Consciousness*, Volume 1, *Psychological Issues*, edited by G. Underwood and R. Stevens (London: John Wiley).

Reason, J.T. (1987a). Generic error-modelling system (GEMS): a cognitive framework for locating common human error forms. In *New Technology and Human Error*, edited by J. Rasmussen, K. Duncan and J. Leplat (Chichester: John Wiley), pp. 63–83.

Reason, J.T. (1987b). A framework for classifying errors. In *New Technology and Human Error*, edited by J. Rasmussen, K. Duncan and J. Leplat (Chichester: John Wiley), pp. 5–14.

Reason, J.T. (1988). Framework models of human performance and error: a consumer guide. In *Tasks, Errors and Mental Models*, edited by L.P. Goodstein, H.B. Andersen and S.E. Olsen (London: Taylor and Francis), pp. 35–49.

Shannon, H. and Manning, D. (1980). The use of a model to record and store data on industrial accidents resulting in injury. *Journal of Occupational Accidents*, **3**, 57–65.

Shiffrin, R.M. and Schneider, W. (1977). Controlled and automatic human information processing: II. Perceptual learning, automatic attending and a general theory. *Psychological Review*, **84**, 155–171.

Somers, R.L. (1983a). The probability of death score: an improvement of the injury severity score. *Accident Analysis and Prevention*, **15**, 247–257.

Somers, R.L. (1983b). The probability of death score: a measure of injury severity for use in planning and evaluating accident prevention. *Accident Analysis and Prevention*, **15**, 259–266.

Chapter 30

The analysis of organizations as a conceptual tool for ergonomics practitioners

Pat Shipley

Introduction

It is proposed in this chapter that the analysis of organizations is a useful conceptual tool for practitioners of ergonomic science. Whether acting in the role of practising ergonomist in industry, or researcher in the laboratory, the ergonomist works as part of an organization of some kind. Even the free-lance consultant encounters organizations which affect the practice of consultancy; government bodies that regulate its practice and organizations that consume its products and services. Organization is the endemic phenomenon of modern life in the developed world. To have some insight into how an organization functions, and why it functions in the way it does, is to appreciate how it influences you, the ergonomist, or aspirant practitioner, as one of its constituent parts. In turn you will learn how you can, and do, influence the organization as an agent of change, or of the status quo.

There is a philosophy underlying the chapter, that social relations are mutually constitutive, which is to say that both of the two theoretically extreme positions is avoided, of complete determinism on the one hand or pure voluntarism on the other. Ergonomists are not omnipotent, but if our powers of action and influence within organizations are quite definitely limited we are never, in principle, mindless and powerless dupes or dopes wholly incapable of changing anything. In the last resort we are invariably free, at least in our democratic society, to leave the organization altogether. In the course of the following argument, therefore, the ergonomist's power base as an organizational member will be examined. Also, it is taken as a given that the ergonomist's intentions include the promotion in an ethically acceptable way of ergonomic principles and practices in the production of goods and services, and in the quality characteristics of the goods and services themselves. It is also taken as given that those intentions include the promotion of their own and others' job satisfaction and well-being.

The concept of 'organization' is considered briefly, and a particular view of the development of organization studies, including the classical functionalist way of theorizing about organizations, is sketched. Ergonomic science is positioned in this framework. A stark comparison of the different world-views and paradigms broadly associated with the natural and social sciences is made, and their significance for understanding organizational life considered. The implications of these differences for the effectiveness of ergonomics interventions in organizations will also be considered.

The collaborative mode of intervention is mentioned as a possible way out of dilemmas posed by traditional ergonomic practice, and this approach will be elaborated elsewhere in the chapter on participative ergonomics by Shipley. The argument concludes with notes on assessing organizational culture, tempered with caveats about the unintended consequences of research and intervention practices, caveats which also point forward to other issues discussed in this book; participation and ethics.

Originating, we are told, in World War II in the UK, and located then in the research squadrons of the defence services, ergonomics science has since considerably broadened in practice beyond the science of aircraft seats and the design of the knobs and dials of military equipment to embrace activities and applications falling roughly within the general ideology of the 'quality of life', the 'quality of working life' especially. The prioritizing of safe and healthy practices at the workplace within resources limitations, the design of work activities to include an individual's complete job, even the job of a whole work team, as well as the traditional concern for the ergonomic design of engineered products, could now form part of this wider brief. Service functions such as communication networks and information providers, at railway stations, airports and town halls, for example, may in theory be subject to ergonomic scrutiny and design. These terms of reference cover the needs of consumers outside the workplace as well as of worker producers within. Such a wide and open frame of reference presupposes that, to be better effective ergonomists need, *inter alia*, an organizational survival kit, and the capacity to win others round to their point of view.

The study of organizations

There are those who would argue that the study of organizations as an 'organizational science' is an independent and discrete discipline. A contrary view is that the subject matter has no valid claim to special status but should be viewed as a branch of sociology, or social theory and philosophy, as applied to organizations. With the exception of some notable sociological studies, as an object of formal study organizations have come only quite recently under the spotlight. Given their relative youth, and (some would argue) given the social nature of their subject matter, the attainment of the goal of theoretical coherence is far removed from organization studies. In view of this, to propound a 'one best way' of analysing and modelling organizations would be folly. Any theory of organization is partial, at best. This partiality applies too, to the methodology of organizational analysis.

In the Department of Occupational Psychology at Birkbeck College teaching has come to reflect the need to respond to the dictates of organizational problems in practice. A problem-oriented approach forced confrontation of the complexities of those problems in a more 'eclectic' and interdisciplinary way than that fostered by the more solutions-oriented or prescription-based undisciplinary approach (see Shipley, 1982, p. 173 ff.). Concrete examples include course options on 'work and safety', in response to the UK 1974 Health and Safety legislation; 'intervention theory and practice' to meet the needs of growing numbers of independent consultants; and modules on 'industrial relations', 'conflict in organizations', and 'organizational development' and 'organizational change'. The 'quality of working life' banner is represented by a Tavistock*-inspired course in 'work design', and by a course on 'stress at work'.

Students of organizations have identified and described key organizational dimensions of structure, function, goals, decisions and environment, and more recently, values, power, conflict and culture. The definition of 'organization' has attracted debate. To Buchanan and Huczynski (1985) organizations are distinguishable from other social arrangements because their leaders are pre-occupied with a need for control. The reasons for organizing are of equal interest. Given that the human race is a remarkably adaptable species our ability to organize into effective working groups, where the needs and wishes of individuals are in principle subordinated to group goals, can be seen as a crucial strategy in this adaptability. In erstwhile imperial quests resources would have been located and efficiently utilized to explore and colonize successfully; similarly to build and manage public institutions, to conquer space, to run effective mountain rescue teams, to develop and market a product, the 'bottom line' is control, for non-profit and profit-making organizations alike. To enrol in an organization is to trade some of our freedom as individuals in exchange for benefits bestowed by its membership.

The allocation and distribution of resources and their accountability require skilled and committed management. Control and resource utilization is commonly achieved through the deployment of classic organizational tools of hierarchical structuring and the division of labour, enshrined in the typical organizational chart with the chief executive at the top. Established hierarchical organizations such as the Church of Rome have grown to be big, powerful and complex. More recent examples are giant multinational profit-making companies. Institutions have proliferated and some appear to have lost sight

---

* The Tavistock Institute of Human Relations in London.

of the goals they were originally set up to attain. The classic organizational structure, bureaucracy, projects a depersonalized image to the lowly servant toiling within its depths, or to the person in the street who relies on its services. To the consumer it may seem to be a law unto itself.

The place of ergonomic science and the classical view of organizations

Ergonomics can be seen as squarely within the 'managerialist' tradition of organizational theorizing. In practice ergonomics has usually operated in the role of management technology, except where consumer interests have been directly promoted by ergonomists on behalf of those consumers. The managerialist view is governed by a particular approach and a number of assumptions. Management practice is nothing if not pragmatic; its bias is in favour of what seems to work out well in practice. Much of actual management is intuitive rather than grounded in explicit theory. Management science, on the other hand, presupposes that it is helpful to know something in advance of trying to change it; a kind of 'cause–effect' knowledge, in particular about those things that inhibit the attainment of management's goals, such as restriction of output, wastage, absenteeism and so on, which it is hoped will be more effectively controlled as a result. Management practitioners vary in how far they regard management science as valuable, either in itself or as an attractive way of selling ideas and changes to the workforce.

A popular line of theorizing about organizations adopted by management scientists, often unconsciously, is the classical functional one. The functional view highlights formalities and structures and classical principles of organizational operation loosely defined as 'rationalist', rather than the social, informal, cultural and value-orientation side of organizations. For the functionalists, and the classicists, the organization is theoretically governed by rational principles; the rule of law, order, fact and logic, in the interests of efficiency. It is expected that management has an inalienable right to manage at all times in all conditions. The view of Burrell and Morgan (1979) is that functionalists stress the regulative and control features of organization: order and consensus, integration, and the status quo. It follows that questions of conflict, disharmony, imbalance and change will not be emphasized by functionalists, or are regarded as aberrant if they are discussed by them.

Early management and organization theorists of the classical school included Frederick Winslow Taylor of '*Scientific Management*' fame, Henry Fayol and Mary Parker Follett. They chose to specialize in one area that was extracted from the rich and seminal work of Max Weber of the turn of the century. Weber saw bureaucracies as a rational solution to the problems of efficiency posed by social arrangements and institutions which permitted non-rational influences, such as personal favours and feelings, to interfere with goal attainment. The classicists promoted simple formal principles which produced the dominant hierarchical organizational form as most of us know it today and which embodies the 'logic of efficiency'. The unforseen negative and 'pathological' consequences of this over-used prescription, where classically designed organizations have generated practices defeating the institution's original objectives, has since led to some attempts to dismantle common-mode structures and to the substitution of alternatives, such as matrix organizations.

Challenges to classic rationalism have come from quarters other than those primarily concerned with efficiency problems. Less well known among these challenges is a feminist one. In feminist psychoanalysis, powerful and heavily structured organizations are seen to be projections of the male ego. Rationalist values are dismissed as male values, as are the values of achievement, competitiveness and efficiency as priorities. Underlying rationalism, it is argued, is a deeper value—that of the active male principle of environmental mastery, in which man presumably copes with his own insecurities about life and nature by over-exploiting these for his own purposes. The feminist counterpoint to this asserts the value of living more in harmony with one's surroundings whilst prioritizing caring for others in authentic relationships, rather than relating to others in an exclusively instrumental way. Lest the reader should think the author, as female, has an axe to grind, no evaluation of either position is being offered. The feminist challenge is merely put forward out of interest, not as prescription nor as panacea, but as different.

The natural and social sciences contrasted

Ergonomics shares the epistemological values of the natural sciences. This epistemology is in the rationalist tradition, and a central feature of it is rigorous scientific method as a powerful form of enquiry. Through the vigorous application of this method nature has yielded up some of her secrets to man the inquisitor. In this tradition complex phenomena are reduced to analysable and controllable fragments by scientific experts (see Shipley and Harrison, 1974). This reductionism has its parallel in the division of labour principle and work specialization carried to extremes in the Taylorized factory and office. The rationalist western intellectual tradition acknowledges more readily the logical, cognitive, and goal-oriented sides of human functioning than it does the affective, intuitive and trans-rational. A-social man (sic!) is the ergonomist's chief focus as labourer or consumer. Most ergonomists have indeed been male, traditionally taught to develop and apply their knowledge through the medium of scientific expertise.

Weber was a social theorist, not a management consultant. His was a wide frame of reference sweeping outside and beyond the narrow confines of particular organizations, though these forms of social arrangement are microcosms of the society in which they are embedded. Statistical procedures invented by the State to serve its own control function were copied by corporations within that State. The functionalist bias is toward so-called hard, objective data. Functionalists, if Burrell and Morgan (1979) are to be believed, have a preference for explaining social functioning by appealing to the presumed underlying cohesion and unity of organizations and societies.

Illuminating metaphors are those which depict harmony and smooth functioning, such as machine metaphors, and organicist metaphors in which constituent parts of systems are taken to be dedicated to the viability of the system as a whole. 'Man–machine systems' and 'socio-technical systems' are the figures of speech of ergonomics. The epistemology, concepts and language of the natural sciences are consonant with these metaphors.

If the functionalist paradigm can be carved out of Weberian scripts, so also can the interpretive paradigm. In 'verstehen' an understanding of social activity is sought through the analysis of subjective experience as recounted by the social actors themselves. The natural sciences' predilection is to focus on behaviour; for the social scientist (or more precisely, the student of society) accounts of personal experience are of greater interest. The epistemology of the social sciences, its concepts, languages, methods and assumptions, belongs to a wholly different paradigm from that of the natural sciences, of which the interpretive, 'verstehen' tradition is a good example. A radically different way of understanding organizations is presented, and ergonomics science may benefit from studying this alternative tradition along with the natural sciences tradition. Its subject matter is people and their relationships; its metaphors are not those borrowed from biology and engineers. Dimensions of organizational structure and function, beloved of classicists, lose some of their salience and others come to the fore, such as the anthropological notion of culture. Discussion and dialogue is about issues normally suppressed in the classical view; about relationships between people, about co-operation and conflict, power, values, mythology and beliefs. The rational mind of the individual worker, seemingly operating in social isolation, dissolves as a figment in the minds of its inventors, scientists reared in the tradition of methodological individualism. Individual minds, like holograms, mirror the minds of those around them, just as organizations contain and institutionalize the beliefs of the wider society of which they form a part. It follows that an understanding of the organization as a whole can be achieved through an understanding of a single member, an organizational 'gate-keeper' perhaps.

Hard objective data and fundamental truths waiting to be discovered elusively slip the grasp of the organizational scientist. The data are the shifting sands of shared and negotiated meaning rather than measurable, logically-generated information. Buildings, documents and furniture physically exist, it is true, but the meaning they embody is symbolic, ambiguous and open to interpretation. No wholly automated system completely manned by robots as yet exists to my knowledge. It could be argued that in a society which has mastered the natural elements the important dimension of the environment is now the social, and this social environment is in our heads and hearts.

About 20 years ago the classical myth of how managers spent their time was exploded as a result of empirical studies by Rosemary Stewart in this country and Henry Mintzberg in the States (see Mintzberg, 1973; Stewart, 1979). Face to face talk, telephoning, frequent absences, and crisis management were much more common than the supposed rational and reflective management activities of planning, co-ordinating and organizing. In reality, managers seem little able to exercise control over their own work activities, reacting often to other people's behaviours rather than to their own priorities. That senior management rationally uses carefully accumulated information provided by a formal information system was found to be folklore.

Furthermore, managers actually much prefer to operate informally and verbally, and need the skills to do so; skills which are typically ignored in the textbooks. They need skills for making unprogrammed decisions, coping with ambiguity, managing conflicts and developing informal interpersonal information networks. The 'symbolic interactionist' school of the social sciences dwells on human relationships, and their symbolic expression, primarily through language as a symbolic medium. For the American social interactionist, George Herbert Mead (1863–1931), the concept of selfhood is generated through continuous social interaction, in which meaning and reality is mutually constituted by both parties to the interaction.

Culture and organization

Greenfield sites apart, every organization has its history, and culture is a reflection of that history. It would be easy to dismiss the concept of culture applied to organizations as a fashionable fad, or as a ragbag into which we dump all the unexplained and poorly understood loose ends of organizational life. But that would be throwing away a set of ideas which convey the realities of the organization that have stood the test of time. Culture is shared meaning. Currently, many social theorists and management scientists increasingly view culture as central to understanding control and resistance to change in organizations and in society.

Schein (1985) suggests that there are three levels of culture: *artefacts* such as buildings, documents and policies; *values* such as what people agree should be the case, e.g. that safety and welfare should take precedence over profit and efficiency; and *basic assumptions*, the often unquestioned guesses and hunches about how things work and how problems should be dealt with. Artefacts are surface symbols, while assumptions and values lie at a deeper level.

The British management theorist, Charles Handy, offers us an organization typology in which organizational types are differentiated as distinct cultures.

The structural features of organizations are embodiments of the culture (see Handy, 1985). We are introduced by Handy to four organizational cultures: power, role, task and person. The metaphor for the power culture is the spider's web with power at the centre. The role culture is the bureaucratic norm, and the metaphor is the Greek Temple with structural columns holding up the pediment or strategic unit. In the task culture the prime value is to get the job done and organizational form will be used as a means to this end. Its metaphor is the matrix. The remaining stereotype is the person culture of the alliance of consultants or craftsmen sharing common facilities and working as partners in mutual consent. The metaphor of the person culture is of a cluster or galaxy of stars. (If you work in an ergonomics laboratory, consultancy or other group ask yourself, 'Does my organization look like a cluster of stars or a Greek Temple?')

When an ergonomist enters an organization, absorption into its culture is hard to resist. The older the institution the stronger and more entrenched the culture and consequently its resistance to change. We follow the dictates of cultures often intuitively and unconsciously. The ergonomist's background, a disciplinary and educational background which is usually that of the methodological individualism of the natural scientist, may represent an alternative culture that clashes with the host culture. The process of socialization or social conditioning begins once the portals of the organization have been passed through. Handy (1985) defines socialization as a process which is designed to encourage the individual to adapt to the organization's values and customs. This includes the accepted norms of behaviour (how the individual is expected to behave), and the accepted mode of organizational operation ('how things get done around here'). Contact with people outside the organization, a professional reference group like the Ergonomics Society, for example, can be a solace and a support to the single ergonomics practitioner coping with organizational reality. The chances are that the ergonomist will blend quite well into the host culture where classical organizational principles are in operation on the surface. Ergonomics science has, after all, perpetrated its own myths about rational man and rational practice. The clashes then might arise at the level of the informal subculture, if they arise at all.

As a consultant the ergonomist is probably first introduced to the organization by a client from a subculture within that organization; someone, say from production, marketing, design or occupational health. To be effective, ergonomists may have to work across subcultural boundaries, as practitioners who apply a comparatively holistic perspective to their task, although they will be located physically in a single unit within a particular subculture. A worse fate may be a base in a corporate department at HQ, where trying to work with antipathetic people in the field from that base proves difficult.

These occupational or functional subcultures and identities cut across and through the organization, just as ethnic and gender identities do so, serving as a basis for stereotyping. We hear of 'hard-nosed engineers', 'software types', and so on. Each functional area has its own vested interests, and competing cultures have communication problems which frustrate attempts at co-ordination. Each is socialized and rewarded differently. Identities are shared with reference groups outside the organization, such as professional bodies. The production man is rewarded by getting the product out, the research scientist by producing data for the next scientific paper.

Culture enables an organization and individuals to cope with uncertainty, about the future, the present, the meaning and purpose of life and so on. Because the culture and the organization have survived this long there will be widespread dependency on it; it is consolidated by powerful executives and reinforced by the unquestioning habitual practice of countless peons who are 'just following the rules'. It conveys a comfortable feeling of security, and outsiders who appear to be rocking the boat may soon find themselves rapidly ejected. Violent and powerful change forced from outside however, such as from a take-over or merger, may reveal the falseness of that security, and under such circumstances tensions held just below the surface can erupt into conflict. Many a nominal take-over and merger has failed to work out in practice, because of the failure of the two cultures in the forced marriage to hit it off. Organizational change can be extremely painful under such conditions. Where a powerful status quo is resistant, to technological change, for example, the ergonomist expecting a rational response to a rational suggestion may be the only one to suffer when that suggestion is rebuffed.

There are two sides to an organization: its task side and its relationships side. Changes intended to improve or change the task requirements may disrupt relationships. Human beings have needs from peers for support and friendship. Some form of informal dominance hierarchy may exist too. This informal side to organizations, not visible in the organizational chart, office layout and job specifications, can be equally powerful as a block to change or as a facilitative channel. A new technology may be seen to threaten a way of working and of relating which has built up around an existing technology. A good example of this is the study in the 1950s by the Tavistock researchers of the short-wall form of mining in North East England (Trist and Bamforth, 1951). A common set of beliefs, languages and practices evolve to enable a group to adapt and survive and cannot be demolished overnight.

Power in organizations

A bland symbolic interactionist view of people negotiating meaning, and sharing a definition of social reality, is consistent to some critics with a unitarist model of organizational life. Organization members are united and grouped under a common banner in this account; they are rather like a well-conducted orchestra. For some theorists, this view omits an important variable; it largely overlooks the role of power in organizations. People may be unequal parties to a negotiating process. Indeed, they may have no bargaining rights whatsoever. The pluralist view on the other hand emphasizes the diverse vested interests and potential conflicts being contained and managed within organizational boundaries; the organization is a loose coalition or set of coalitions pursuing different interests. The introduction of information technology will be supported by a coalition, or alliance, of like-minded people whose status and power is bound up with the promotion of such technology. Dominant coalitions change as conditions change. A powerful technology one day may become obsolete the next. For pluralists the existence of power is a fact of organizational life; manipulation and conflict is as commonplace as co-operation. Conflicts are resolved through the manipulation of power. The 'political' facts of organizational life present problems of access for ergonomics consultants, either access to people or information, whether operating from outside or inside the organization.

The social commentator, Alan Fox, contrasts the pluralist with the radical position (Fox, 1985). For Fox even the pluralist view is too benign. Agreements not entered into freely, but which are the outcome of coercive or manipulative power rather than of equitable bargaining between parties of roughly comparable strength, cannot be morally binding. The unitarist/ pluralist views both unwittingly legitimize much current amoral and immoral practice, whereas the radical perspective draws out the power inequalities more sharply. The trades unions, for example, do not challenge management on fundamental social issues about the public implications of the corporation's goals, the hierarchical structure of the organization, the massive inequalities in pay differentials, and so on. The reasons why this challenge is not taken up include the industrial indoctrination of people through the mass media and training and educational programmes which are closely controlled by the power elite. Socially-dominant ideas and rhetoric are simply taken for granted by the masses. All management strategies designed to secure compliance and commitment are dismissed as manipulative in Fox's radical critique. They rob people of dignity and self-respect. Even taken for granted routines and practices may be responsible for the recreation of injustices. Unlike mindless indoctrination and socialization, commitment entails conscious choice.

Organizations are stratified into power positions and ergonomist new-comers learn their position in the organizational geology. Almost certainly, as scientific experts, their position power will be quite strictly limited, relegated to an administrative and executive support role. The scientist is often a member of service management, a back-up to front-line management. This position power is supplemented by expert power, depending on the attitude other organization members hold towards experts. Handy (1985) describes the various forms of individual power that may be available to the organization member. Coercive power usually springs to people's minds whereas reward power, legitimate power, expert and resource power, and charismatic power perhaps do not do so readily. Expert power can be derived from valued contributions made on the job; it also derives from the status of science and the professions in society at large. The professions, like the sciences, have their pecking order, and ergonomics science is far from being near the top.

Because of the restraints on the ergonomist's position power, and because the expert power of traditional ergonomics is largely limited by training and indoctrination in the natural sciences, the organizational power base of ergonomics needs to be extended to improve the chances for effective ergonomic intervention in organizations. In theory there are means of doing this through a better understanding of organizational culture and function, and an awareness of the power that trans-rational factors possess to resist or facilitate changes. Hanging on to outmoded models, such as machine and systems models, retards this development. The acquisition of social and interpersonal practical skills to enhance information assimilation and persuasive powers, in the way that successful managers themselves have done, as a supplement to rational skills, is another valuable addition (see Shipley and Harrison, 1974). Consorting with top management, as a management agent, is often not enough. In the organizational underground information flows up from the bottom and across peer groups, and is often blocked at the interface with management. Informal social activity is spontaneous and can be emotionally-charged, even if at their higher levels management would prefer such communication to be overt, and that people's feelings were left at home and not brought into the workplace. To get safety standards improved usually requires more than a change to explicit company policy, although that helps. The value of it has to permeate through to the grassroots. People have to believe in it and want to change.

There may be no shortcut to the laborious groundwork involved in identifying and developing the appropriate coalitions and alliances. This groundwork may have to be done at all levels from the level of company policy, through group levels, and at the level of the individual opinion leader. Even senior management is restricted in how far it can get things changed, hence the notorious use of subversive and secretive tactics at the top. Sufficient numbers of the right people have to recognize they have an ergonomics problem to accept an ergonomics intervention. Having recognized the problem is not enough, because there has to be a willingness to accept the change required, as opposed to the alternative of continuing to live with the problem. Then the practical implementation of the agreed change has to be achieved.

As in the master–slave paradox power is a relational process; power has to be 'accepted' by both parties in the relationship. The balance of power can alter with the conditions. The most powerful will be those seen to have the key to the organization's viability; those who safeguard the status quo in a stable period as in the public or professional bureaucracy, and those who have the vision to lead it through turbulence in an unstable period. Unlike a machine bureaucracy, an adhocracy is a flexible organization and its power distribution is dispersed to enable it to produce rapid responses as the changing situation requires (see Mintzberg, 1979). The unwieldy bureaucracy, however, is cumbersome in the face of rapidly shifting demands. Role structure and rule-following behaviour are meant to ensure continuity, reliability and predictability; a strong culture which can act as a curb on divergent and aberrant behaviour. It does not cope well in emergencies when non-programmed decisions are called for.

The contingency view of organizations presupposes that, like adaptable species in nature, organizations take on the shape most suitable to their habitat and conditions. Some would be tempted to believe, no doubt, that certain ossified institutional forms are rather like organizational dinosaurs in the contemporary world where change rather than stability appears to be the reality, as pressure to compete and to innovate spirals upward. But, however well planned in advance these changes are by ergonomists and their collaborators, the consequences are rarely wholly predictable, although a valuable ergonomic contribution can be made through systematic and disciplined enquiry to reduce some of that ambiguity. It seems that options and choices may be open to us all the while we have a future, but complete power and control is never one of those options. To think otherwise is to be less effective in one's interventions than might have been possible.

The unintended consequences of the attempts by ergonomists at applying expert power single-mindedly in organizations is a case in point (Shipley and Harrison, 1974). The collaborative or participative mode of intervention, where power is equalized between consultant and client, is one possible way forward for ergonomic practice as an alternative to the expert mode.

Learning about what goes on in organizations: beyond dualisms

The ergonomist is well-versed in the principles of the scientific method, and these have their place in the testing of hypotheses in controlled settings, such as that of the laboratory, or its near-equivalent 'in the field', the industrial simulation. The physical dimensions of the environment and the cognitive aspects of people's minds lend themselves better to analysis by such principles than do many social and organizational variables, where social context is of great importance.

Some theorists may argue that the organization has a life of its own, independently of its constituent human parts. This is a reification of the organization, and the implications of this reification for the measurement of organizational dimensions remains to be worked out. The ontological status of organizations is a controversial issue and a contrasting view is that organizations are no more than the people who make them up. The Gestalt idea that the organization is more than the sum of its parts does not appear to be inconsistent with either view, but it does have specific implications for organization assessment and data collection. Organizational members do not work at their jobs in a social vacuum. Work is done in constant interaction with other people, and even in the most closely-prescribed work tasks there is always room for a little discretion, cutting corners or filling in gaps, supplementing or bending the rules, as opposed to 'working to rule'. Without the exercise of discretion by proactive, creative members organizations would grind to a halt. Because the future is always open, because it is never wholly predictable, then this must be so. Collecting data in organizations, therefore, must capture the processes and products of those interactions. Collecting the contents of individuals' minds and observing routine behaviours involves learning about the organizational culture and its body of shared knowledges and beliefs, in a way that transcends the dualism inherent in the semantic opposition of 'individual versus environment'.

A generalist or nomothetic view is that it is possible to discover universals about organizations and therefore to generalize across organizations. The alternative position is that it is never possible to generalize. The midway position is that organizations do share similarities but at a very general level— the preoccupation of management with control, the development of a set of operating rules and so on—and that the forms these generalities take, such as culture, vary in particular ways within each organization. In Birkbeck College courses students are encouraged to analyse their employing organiz- ations using general conceptual frameworks or crude maps. Concepts are tested out in practice through personal work experience, and these personal experiences and concepts are shared and evaluated in class.

The students' models of organizational functioning inevitably influence how they go about finding out what is going on in an organization. Behaviour can be monitored and documents and other records can be scrutinized, but these do not tell us anything about why people have or have not taken various actions. Social accountability theory (see Shotter, 1984) seeks to explain people's behaviour as illuminated by their own accounts of the situation. However, attributional error alerts us to the biases in our adjudication of our own or another's behaviour; the observer is prone to attribute causality to the actor, the actor is more likely to draw attention to constraining situations that influenced what was done or not done. To attribute accidents so frequently to human error is a case in point.

To scapegoat a relatively powerless person may be a good way for management to get out of a sticky situation. Whereas crises and emergencies may be reacted to negatively, and covered up, they could be opportunities for a courageous management to better understand, re-evaluate and revise their organization's culture. Attribution theory warns us to expect attributional bias in evaluating other people's actions even if our intentions are just and fair, and that to solicit the actor's account is to solicit a more balanced picture. After all, in the courtroom those charged are given an opportunity for self-defence when they are called to account.

When organizational members act out their roles there is tacit knowledge or agreement about how things should be done and what situations mean; a kind of 'negotiated order'. The environment is 'internalized', taken inside one's thoughts and worked on by the exercise of imagination. These mental models will be more or less accurate versions of organizational reality, approximations which guide individual behaviour and actions. To learn about these models, to get inside people's heads, we can talk to people, and we can explore with them the meanings of particular situations, how an individual's views compare with others, and how far behaviour is related to intentions or how far it appears to be outside individual control. This is the mutual exploration of organizational members' views, but largely on their terms. Interviews with individuals and group discussions can be recorded systematically and then analysed *post hoc* using analysis protocols, such as content analysis of interviews and discourse, or dialogue analysis of interactions. Pauses and silences can be analysed as well as verbal content, and the 'music' or tone of the utterances as well as the semantics. Analyses by different observers can be compared for their degree of consensus and treated statistically. The analyst may wish to use some underlying perspective, such as psychodynamic theory, in the analysis. Contradictions can occur between what is said and what is done, or between two or more verbal statements. Emotionally neutral language used to describe emotionally laden behaviour may, for example, lend clues about the organization's culture.

Ethnography is the name for a body of anthropological techniques, whose object of study is culture. Culture can reside in physical embodiments, such as buildings, or in the fluidities of behaviours and languages. Oral language is a primary transmitter of all cultures, whereas written, formal language, such as the academic and scientific, and the bureaucratic, is not common to all. Literacy is a central feature of rationalist culture.

Language is also a vehicle for getting things done. Organizational vocabularies express communal values, reinforcing and legitimizing the status quo. They express shared and collective beliefs about reality. The organizational ideology conveys a set of beliefs about what the organization is supposed to be about; how to get things done. What is *not* said can be as illuminating as what *is* said. 'Morality' may generally be a taboo word; 'stress' may be regular verbal currency in non-macho organizations; members may be enjoined not to bring their feelings into the workplace. Symbols such as 'old school' ties, and rituals such as annual factory outings and retirement parties, serve their own cultural purposes—all these cultural processes can provoke emotion and action and a sense of shared identity.

Language can be examined through dialogue and documentation. The analyst may or may not be part of the dialogue. There is no one best way of finding things out; no rigorous methodology in uncontrollable naturalistic settings. The change agent who ignores culture prevalent in the organization is not going to be effective always. Sometimes culture may be ignored with impunity, sometimes change can be managed around it. But sometimes attempts may need to be made to change the culture. Anyway, an effective choice of strategy depends on a good prior sense of what is acceptable, and what is changeable.

Often the 'old hands' can be the least aware of the basic assumptions underlying the culture. To ask them is to attract a mixture of fact, fantasy and propaganda. But the 'native view' remains a good source of information to the wary, especially when supported with evidence from other sources, or even if it is contradicted by this other evidence. Contradictions are at least interesting, and sometimes illuminating. In the studies of management time a variety of methods have been used, including interviews, questionnaires, diary methods, and behavioural observations of meetings and telephone calls, which can produce differing records of time spent.

How we go about collecting this information has an important bearing on the quality of the data; the agent has to draw on more skill than the mere manipulation of numbers and binary logic. Perhaps we need organizational anthropologists in all our fieldwork teams to get behind the scenes. A preferred method of analysis for anthropologists is 'participant observation', where the culture temporarily absorbs and transforms its stranger researcher. But if the ergonomist is having access problems, is seen to be potentially disruptive, a threat even, then this cultural embrace is not readily forthcoming. However, the alternative status, non-participant observation, can raise its own brand of ethical problems, as discussed by Drury in his chapter on direct observation techniques.

The quality of such data, as for any source of data, must be considered carefully. How trustworthy, how valid are the data? It would be too easy to dismiss intrinsically qualitative data out of hand as too subjective. To do so may be to throw away the valuable. In my experience much qualitative data has been collected in an invalid way and this has contributed to the negative reputation acquired by this category of data. To use the jargon, the 'process' as well as the content is important. The conditions under which such data collection takes place affect the quality of those data in an important way (see Shipley, 1987). Under collaborative conditions the chances of acquiring valid data are much greater. This is further elaborated in the chapter on participative ergonomics by Shipley.

Intentions and effects: ethical dilemmas in ergonomics

How far ergonomists are prepared to offer their services depends on who is paying and how much is offered, and on the ergonomists' own value orientations. It may not be clear whose interests are being met, your own or your client's, or both. Sometimes it may not be clear who your client is, whether it really is the user of the equipment, for example, or the supplier, the workforce's representatives or the firm's management. To practise according to our professional ethical code, it is encumbent upon us to be constantly on our guard. Yet our training as scientists may not have included these ethical and value considerations. If so, we could be acting unconsciously in a complicit and collusive manner which may shock us if it were explicitly pointed out to us (see Shipley, 1982).

We would have to be open to such criticism, although the chances are that over the years we have defended ourselves well against challenges to our practices and the effects such practices have. In other words, we may block our minds and ears. To hide behind science as an objective value-neutral enterprise may be a common defence and be responsible for our own resistance to change.

A father figure of industrial psychology claimed that the business of the applied psychologist was wholly instrumental, to supply the means for the fulfilment of another's (usually the business owner's) aims. The job of the new discipline was to 'produce most completely the influences on human minds which are desired in the interest of business' (Munsterberg, 1913, p. 24). He earlier stated: 'But no technical science can decide within its limits whether the end itself is really a desirable one' (Munsterberg, 1913, p. 17 ff.). In his day such beliefs were commonplace but I doubt whether any contemporary psychologist would get away so easily with such a bold prescription. In a special issue of the *BPS Occupational Psychology Newsletter* dedicated to the debate about values and ethics, a Birkbeck colleague introduced an added slant to the debate (Hollway, 1986). For her, good intentions are simply insufficient, and applied scientists should reclaim responsibility as human beings for the effects their practices have.

One way of doing this is by enlarging our own awareness to take account of social factors, such as cultural blocks and power dynamics at work in those organizations in which ergonomics functions. Another way is to extend our power base, as discussed previously, to increase the prospects of the desired effects following from our good intentions. The possibilities afforded by the collaborative approach for dealing with ethical dilemmas will be discussed in the chapter on participative ergonomics by Shipley.

Consultants, I suspect, do not often collude knowingly, but is that sufficient to absolve us of responsibility? A comfortable presumption is that there is no underlying conflict of interests. But there is now no excuse for holding onto obsolete models and presumptions. A body of literature on social and organizational studies, and a relevant set of practical skills for learning about organizations, many of them quite everyday, exist, from which the practitioners of ergonomics science and their clients stand to benefit.

Acknowledgment

The generous help of my Birkbeck Colleague, John Coopey, previously with Esso management, is acknowledged in the preparation of this chapter.

References

Buchanan, D.A. and Huczynski, A.A. (1985). *Organizational Behaviour: An Introductory Text* (London: Prentice-Hall).

Burrell, G. and Morgan, G. (1979). *Sociological Paradigms and Organizational Analysis* (Aldershot: Gower).

Fayol, H. (1916). *General and industrial management* (London: Pitman). (Translated into English by C. Storrs, 1949.)

Follett, M.P. (1918). *The new state.* (London: Longman).

Fox, A. (1985). *Man Mismanagement,* 2nd edition (London: Hutchinson).

Handy, C.B. (1985). *Understanding Organisations,* 3rd edition (Harmondsworth: Penguin).

Hollway, W. (1986). Effects not intentions: the ethical criterion for occupational psychology. *Occupational Psychology Newsletter, Special Issue on Values and Ethics in Occupational Psychology,* **23**, pp. 5–8.

Mintzberg, H. (1973). *The Nature of Managerial Work* (New York: Harper and Row).

Mintzberg, H. (1979). *The Structuring of Organisations* (London: Prentice-Hall).

Munsterberg, H. (1913). *Psychology and Industrial Efficiency* (Boston: Houghton Mifflin).

Schein, E. (1985). *Organisational Culture and Leadership* (London: Jossey-Bass).

Shipley, P. (1982). Psychology and Work: The growth of a discipline. In *Psychology in Practice,* edited by S. Canter and D. Canter (Chichester: John Wiley), pp. 165–176.

Shipley, P. (1987). The methodology of applied ergonomics: validity and value. In *New Methods in Applied Ergonomics,* edited by J.R. Wilson, E.N. Corlett and I. Manenica (London: Taylor and Francis).

Shipley, P. and Harrison, R.G. (1974). The Ergonomics Practitioner as Change Agent. Unpublished paper to Ergonomics Society Annual Conference, St. John's College, Cambridge (Available from Birkbeck College).

Shotter, J. (1984). *Social Accountability and Selfhood* (Oxford: Blackwell).

Stewart, R. (1979). *The Reality of Management* (London: Pan Books).

Taylor, F.W. (1911). *Principles of scientific management* (New York: Harper & Row).

Trist, E.L. and Bamforth, K.W. (1951). Some social and psychological consequences of the longwall method of coal-getting. *Human Relations,* **4**, 3–38.

Weber, M. (1947). The theory of social and economic organisation. (New York: Free Press).

Edited readings

There are some British (Penguin) paperbacks which are edited readings, manageable in size and good prices. Readers may like to have copies of the following:

Pugh, D.S. (Ed.) (1984). *Organisation Theory,* 2nd edition (Harmondsworth: Penguin).

A mixture of classic articles by famous people such as F.W. Taylor and Max Weber through to modern theorists like Fred Fiedler and Henry Mintzberg. Organizational level of analysis bias.

Warr, P. (Ed.) (1987). *Psychology at Work*, 3rd edition (Harmondsworth: Penguin).

A successful seller and until recently virtually the only text available for use in occupational psychology courses which was not written by Americans. Bias is less towards individual level of analysis and the classical theory prominent in the earlier editions. The third edition covers an even wider range of topics moving from emphasis on individuals, through groups, to the study of organizations.

Annotated bibliography

Beetham, D. (1987). *Bureaucracy* (Milton Keynes: Open University Press).

An excellent review of theory on bureaucracy, eclectic, and concluding with a critique which incorporates a personal perspective emphasizing the democratic objection to bureaucracy . . . a specialist text.

Buchanan, D.A. and Huczynski, A.A. (1985). *Organisational Behaviour* (London: Prentice-Hall).

A clearly written, good value, comprehensive paperback by University-based theorists with consultancy experience. Not noticeably extremist and controversial—more an unbiased review but does not go deeply into important areas. A worthwhile first level course text which is somewhat like (but not much) 'distance learning' in style: i.e. it might just have come out of the Open University. Solid and basic.

Burrell, G. and Morgan, G. (1979). *Sociological Paradigms and Organisational Analysis* (Aldershot: Gower).

Another good-value paperback by University theorists: well-written, and a stimulating but more controversial critique and historical analysis of organizational sociology, organized around four paradigms—i.e. functionalist, interpretive, humanist and structuralist. More appropriate for sociologists than practitioners perhaps. Should be guided reading and informed by practical experience.

Fox, A. (1985). *Man Mismanagement*, 2nd edition, (London: Hutchinson).

A modest-sized well-written, forceful and stimulating paperback by this Ruskin fellow deservedly running into a 2nd edition (first published in 1974) to accommodate Britain's much altered industrial relations climate under 'Thatcherism'. Out of the eight chapters, three are devoted to the subject of participation. Fox's position is a radical one and his treatise is rhetorical in flavour. It may be too polemical for some people's taste, however. Every industrial relations specialist should be familiar with it, even if in disagreement with it.

Handy, C.B. (1985). *Understanding Organisations*, 3rd edition (Harmondsworth: Penguin).

A popular good-value 'business school' introductory paperback in the managerialist mould clearly written by a popular energetic management theorist with substantial industrial experience. Of very 'handy' proportions, too! Uncritical and relatively superficial but not a bad start to the field. Need more than this for a solid course textbook, however.

Katz, D. and Kahn, R.L. (1978). *The Social Psychology of Organisations*, 2nd edition (New York: John Wiley).

An American text, unrivalled in its field for a long time, until the stranglehold over the field by Americans was broken by some British competitors in recent years. Replete with 'evidence' and carefully put together as an argument with the theoretical stamp on it of the Institute for Social Research, Michigan. Solid. A good price.

Chapter 31

Economic analysis in ergonomics

Geoff Simpson and Steve Mason

Introduction

Several authors (e.g. Alexander, 1985; Galloway, 1985; Schneider, 1985; Simpson, 1985a, 1988) have argued strongly that there is an increasing need to emphasize economic arguments in the promotion of ergonomic research and in the justification for ergonomic change. As Alexander (1985) states: 'A manager may allow ergonomics to be tried, but without bottom-line improvements (or other equally convincing measures of effectiveness) the use of ergonomics will soon diminish and then disappear'. A fundamental problem within this context is that health and safety research is still seen in many organizations as largely altruistic with no significant, or even tangible, returns on investment. While ergonomics continues to be considered as primarily a health and safety discipline it will, inevitably, face the same problem.

Unfortunately, ergonomics, like its collaborators within health and safety, has tended to shy away from economic argument. In part this is due to genuine and justifiable reservations on the assumptions necessary to cost issues in health and safety. However, it is also in part due to a lack of realization of the extent to which equally 'debatable' assumptions are made in the 'legitimate' accountancy field and, in part, through a lack of awareness of the techniques and data which can be used to build an economic justification.

The purpose of this chapter, therefore, is to identify the kind of data which can be used by ergonomists to build an economic case and some of the procedures and calculations which can be used to present the case. By way of conclusion, a number of recent studies which have used economic analysis of ergonomics are quoted to show that not only is the objective advocated desirable, but it is also achievable.

The basic requirements for an economic analysis of ergonomics

Traditionally ergonomics has been justified on the basis of health and safety with occasional, though usually vague (and somewhat embarrassed), references to production improvements. The first step necessary in developing an economic base for ergonomics is the resurrection of production improvements as a legitimate objective in ergonomics. The second is the acceptance that health and safety issues can be legitimately considered as loss prevention topics. Most industrial organizations have, in recent years, undergone some form of rationalization in pursuit of a 'leaner and fitter' operation. This has almost inevitably involved staff reductions and often the breakdown of skill boundaries to promote an increasingly multi-skilled work-force. While the 'leaner and fitter' organization is undoubtedly more productive and profitable under normal circumstances, it is also less resistant to disruptions in its reduced work-force. If disruptions occur as a result of, for example, 'organizationally self-inflicted' sickness such as repetitive strain injury, then clearly it is in the organization's financial interest to remove that drain on resource—in other words to engage in a loss prevention exercise. Similarly the increasing use of human reliability approaches in ergonomics can be used in conjunction with the costing models used in engineering reliability studies. The third requirement, having defined some of the costing approaches relevant to ergonomics, is to identify the kind of data which can be used in costing calculations. Finally, the fourth stage is familiarization with some of the calculations and procedures appropriate to the analysis and promotion of ergonomic change in economic terms.

Developing economic arguments for ergonomic change

Whether or not ergonomists feel comfortable with the idea, industries are profit-making centres and every job has been created as a necessary part of a larger profit-making machine. An ergonomist, like any other person in the industry, will therefore be expected to help the organization achieve its goals. For example, if ergonomists conduct a study of a group of workers who are standing at a bench all day assembling small components, it is highly probable that they will conclude that the workstations should be redesigned so that the work-force can be seated. If they then approach the works manager and argue that money should be spent on modifying benches and buying seats to improve working postures, the chances are very slim that the changes will be authorized. The manager may say that there is no apparent problem, after all people have been working like that for as long as can be remembered with no complaints, that people are not paid to be comfortable, and that anyway the budget is not available.

The reality of the situation is that the ergonomists and managers are talking different languages. The management's remit is centred around profit-making, product quality and meeting production time-scales. They are likely to see the benefits of ergonomics change as, at best, peripheral to their objectives. The ergonomists then get frustrated by the management's lack of immediate enthusiasm for their ideas and wander off wondering why the engineers/management cannot understand them. The simple fact is that if ergonomists want to communicate effectively with people in industry, then it is up to them to try and learn industry's language. This does not mean that the ergonomist's role has to concentrate exclusively on production-related issues. Getting recommendations implemented which improve the health and safety of the work-force will be much easier if the ergonomist talks the industry's language. To return to the previous example, the ergonomist discovered that the workstudy department had previously assessed the job and when working out the appropriate 'relaxation allowance' had given 3% to cover them standing all day. They then approached the manager and said 'Do you realise that you are paying that group of workers an extra 3% to stand up?'. The manager then told the ergonomist to change the workstations to allow operators to work seated. The end result was exactly what the ergonomist wanted but the route was novel and extremely effective.

Some of the many factors which can be used to develop such economic arguments are now discussed.

Available data

The data that an ergonomist may find useful are usually easy to obtain; they are probably being collected simply to run the business. The following six functions in a company are potential sources of information, issues or criteria that ergonomists are likely to find useful, either to provide an economic justification for new studies, or to show the benefits achieved as a result of ergonomic intervention. The first three functions relate to those data and issues which are normally considered to be 'close' to the ergonomics remit. The second group represent functions, issues and approaches which are perhaps less frequently considered by many ergonomists. (See chapter 4 by Drury on archival data.)

Personnel departments

(1) Absenteeism records.
(2) Turn over rates.
(3) Training costs.
(4) Compensation costs, e.g. for injury.

Safety departments

(1) Accident black-spots (whether by site, job or equipment used can be particularly revealing).

(2) Jobs needing special safety precautions.
(3) Jobs needing unusual safety equipment.

These problems all cause delays either directly or through additional procedures needing to be followed to maintain safety standards.

Medical units

(1) The nature, severity and length of absence from injury.
(2) The nature and length of absence for health problems.
(3) Type and frequency of minor injuries dealt with at the medical centre.
(4) The type and frequency of symptoms reported (e.g. headaches, eyestrain).

Time off the job can easily be costed in terms of wage charges plus overheads, cost of temporary cover and lost production.

Work study/method study departments

1. *Workstudy 'relaxation allowance' payments.* It is normal for workstudy engineers to agree to pay the work-force extra to 'compensate' for: standing, heat, physical effort, noise levels, thermal environment, and poor lighting.

These are usually termed relaxation allowances and are meant to compensate for reduced performances caused by the presence of the various influences. The accuracy of these allowances is debatable in many instances. However, since it is implicit that the management and work-force accept them, they are very useful to the ergonomist in building a proposal to improve all aspects of working conditions.

2. *High variation of individual performance.* The performance of people on incentive schemes will vary. However where this variation is larger than normal, this is likely to reveal elements of a task which demand extremes of physical effort or skills. An ergonomist could therefore be directed at reducing the needs for such excessive abilities through redesign.

3. *Inspection reject rates.* The costs of reject components are often much higher than generally appreciated. Of course some rejects will be passed and some acceptable parts rejected, and where this occurs an ergonomist could do well to study the actual inspection procedures. Where the inspection is generally accurate, high reject rates could be caused by a large number of factors many of which are within the remit of the ergonomist.

4. *High inspection costs.* A high investment in inspection will generally be justified where there are high costs and warranty claims associated with component failure. Ergonomics can make a considerable impact in controlling or maintaining the performance levels of inspectors and hence can be viewed as a loss prevention aid.

Plant engineers

1. *Excessive downtime*. Although it is traditionally considered that downtime is purely a function of poor engineering (and hence not concerned with ergonomics), recent studies suggest that operating mistakes, through poor ergonomic features, lead to some equipment breakdowns (see, for example, Williams, 1982). Likewise if routine or repair maintenance is not performed strictly in line with the procedures laid down in the manufacturer's handbooks, then the likelihood of subsequent breakdowns is increased. For example, if a fitter is under pressure to repair a machine quickly, or if the importance of the particular task is not perceived, then a hose union may not be cleaned sufficiently before disconnection. The subsequent ingress of even very small amounts of dirt can block filters very quickly and result in a rapid failure of an expensive hydraulic pump.

2. *Equipment difficult to maintain*. The costs of maintaining equipment can be up to 30% of the total operating costs of an organization and yet designers (and ergonomists) often neglect simple design solutions which can overcome or minimize basic faults. It has been estimated, for example, that repair times could be reduced by 30% in many industrial tasks by improving access alone (Seminara and Parsons, 1982).

3. *Excessive scrap wastage*. Many industries will not routinely collect this information but the costs of operating errors which result in scrapping a component are surprisingly high, especially for those which have already undergone many machining operations.

Industrial relations departments

Problems of industrial relations could occur for a variety of reasons which may be considered as outside the scope of ergonomics. (See chapter 1 for a different, wider view of ergonomics—Eds.) Nevertheless, poor working conditions may have been a contributor. The apparent 'cause' of the difficulty is often a symptom of a wider problem. For example, complaints of VDU operators' eye strain through poor lighting, although often the case, may also arise from other sources, e.g. organizational changes. Solving industrial relations problems is difficult enough without diversions through irrelevant issues. The ergonomist often has information which can reduce the chances of the debate centring on symptoms rather than causes; solving industrial relations problems is difficult enough without being side tracked into negotiating on the wrong issues.

Of course all these measures can be supplemented by data from an ergonomics survey. For example, near-miss accident data, attitudes to risk-taking and attitudes relating to job satisfaction can all combine to help construct valuable economic arguments for making ergonomic changes which can improve both the operating performances and the health and safety standards. However, moving immediately to a proposal to institute an ergonomics study to define/refine problems involves a rather strange argument in financial terms. In essence this states: 'Give me some money so that I can justify you giving me some more'! If at all possible, it is best to be armed with some information, from sources of the type described above, even if the study proposed is only an exploratory one.

Techniques and data which are useful for cost-justifying proposals

Making investment return predictions for ergonomics

The production of predictive costing figures is essentially part of the marketing exercise of presenting the proposal itself. Professionals in the marketing field emphasize the importance of presenting a product in the most favourable light without actually telling lies. The latter point is crucial, for if you exceed what is credible the proposal is likely to be lost and it is possible that all future proposals will be viewed very sceptically. Inevitably given a shortage of some data, assumptions have to be made. It is therefore important to specify the assumptions made and always to be conservative in terms of predicted achievements. It is better to offer little and deliver more, than to offer a great deal and risk disappointing people by delivering less than they expected. Anyone who has ever attempted a predictive cost–benefit analysis is fully aware of the dangers. For example, you predict a saving of £10 000, but in reality provide only £8 000; you can almost guarantee the reaction which will not be (as you had hoped), 'Thanks, that £8000 is very useful', but rather 'What happened to the other £2000'!

High levels of accuracy are therefore not actually crucial. In fact it is probably advantageous if the calculations are seen to be only as reasonable estimates, as any subsequent debate with specialists will almost certainly improve the calculations. For example, if the manager does not agree with your approximation of, say, overheads on salary cost and proposes a different figure, if this is then incorporated there can be little further argument since it is after all the manager's own figure. Some examples of economic cost arguments are developed below for both production-related issues and for health and safety.

Productivity

Of the several accountancy approaches to predicting costs, two which are of particular value to ergonomics are shown below. One is to calculate the contribution of poor ergonomics to the total cost of ownership of a machine or group of machines using the methods of life-cost accounting (e.g. the accumulated costs throughout the life cycle of the machine). This framework can also be useful in 'loss prevention' arguments. The second approach is the prediction of increased revenue arising from the contribution of ergonomics to improved productivity.

An example of each is given in relation to a proposal to examine the ergonomics of a coal winning machine (shearer). Although the figures quoted are out of date, this is of little importance in the current context as the objective is simply to show the approach.

'Life-cost' calculation

An estimate of the total life-cost to the organization must be derived covering all of that type of machine in the industry. This is approximated by the equation:

$$\text{Total life cost} = n\,(x + Lt)$$

where, L = the life expectancy of the machine, n = the number of machines in use, x = the capital cost per unit, t = the operating cost of a machine per annum.

For shearers, the figures are: L = 8 years; n = 540; x = £250 000, t = £500 000.

Using this equation, the total life cost of shearers in UK mining is £2.3 billion.

PREDICTED TIME SAVINGS THROUGH IMPROVED SHEARER DESIGN

Predetermined time and motion systems (e.g. MTM 1; Maynard *et al.*, 1948) are very powerful tools for the ergonomist. Where these cannot be easily applied, even simple conservative estimates of improvements on a breakdown of the various operating costs may be all that are required to argue for change.

Ideally the performance implications of particular ergonomic limitations should be known, in this way the overall job implications can be built up from the component task limitations. Unfortunately, given the complexity of industrial jobs and the specific context in which they are carried out, this ideal approach is only possible retrospectively using current data. For predicting the influences of ergonomic design deficiencies on performance, a number of avenues can be used:

1. If possible an assessment of the machine (or sample of machines) should be made.

2. If relevant knowledge is not available and pilot studies are not possible, task synthesis techniques (e.g. Annett *et al.*, 1971; Maynard *et al.*, 1948) should be considered.

3. In addition, some aspects of the ergonomics literature can be used either by careful analogy with the machine under consideration, or by the use of task-oriented human error/reliability data (e.g. Swain and Guttman, 1983).

This information on task-related ergonomic limitations can then be related to the operational cycle of the equipment involved and *conservative* estimates derived of the performance improvement likely to arise. In the shearer calculation, elements of (1), (2) and (3) were all used to derive the figures

Table 31.1. Potential saving from improved ergonomics of coal winning machines

Shift breakdown	Average duration (min)	Percent saving through application of ergonomics of coal winning machines	Potential saved time (min)
Men travelling	89	0	0
Preparation and meals	26	1% through better design, reducing preparation	0.26
Machine running	107	2% resulting from improved performance	2.14
Ancillary time	29	1% through better design	0.29
Operational time	63	0	0
Lost time			
Electrical	12	5% through diagnostics	0.6
Mechanical	31	10% through access & handling improvements	3.1
Mining/geological	77	0	0
		Total min/shift	6.39

contained in Table 31.1. The example in Table 31.1 shows that if fully implemented, ergonomic improvements in shearer design are likely to save 6.4 min each shift for a typical shearer design. These machines are available for production for 330 min each shift and therefore the current ergonomic limitations lose 2% of the total potential operating time.

It should be noted that these estimates include a potential saving through improving the maintainability features of shearers. Maintenance aspects of machinery are often overlooked, however, as maintenance costs are typically 30% of the costs of ownership, substantial savings can usually be found simply through recommending improved access to those components requiring frequent attention during routine maintenance operations.

Life-cost accounting is being used increasingly in industry as an aid to purchasing decisions. For example, faced with a choice between two machines (A and B) both of which seem adequate for their purpose, but without any additional information most would normally choose the cheapest (say, A). However a life-cost analysis may show, for example, that spares are more expensive for A, and/or that routine maintenance takes longer. When the cost of these issues are projected over the life expectancy of the machine, the capital cost differential may become irrelevant with B in fact proving to be cheaper over the long-term. While this may seem a long way from ergonomics, there are in fact many opportunities to incorporate ergonomic issues into life-cost analyses. Say, for example, that A is noisier than B. In this case it can be argued that the life-cost equation should include the cost of noise reduction, screening or the provision of hearing defenders. Similarly it could also include the cost of routine audiometry, changes to warning signals, and even estimates of the possible cost of litigation for hearing loss. Even if these fail to negate the capital cost differential which favours purchase of the noisier machine, they will have sensitized the management to the financial benefit of an ergonomic investment in hearing protection.

Increased revenue calculation

The time penalty of poor ergonomics of shearers over a 1 year period = the number of machine shifts per week × number of working weeks × the cost of poor ergonomics per shift. This calculation shows the cost to be 32 431 h/year. An average high technology coal face produces 3 tonnes/min at a saleable price of £40/tonne, and cuts coal for an average of 107 min per shift. The revenue per shift is therefore £12 840 or £7200 per machine hour.

Assuming all the lost hours per year from poor ergonomics (32 431) could be converted into extra machine running time over all the shearers, then:

increased revenue = 32 431 × 7200
= £234 million for the 540 units,
i.e. £0.4 million per machine.

The cost or ergonomic limitations per annum in this instance is equivalent to:

$$\frac{234 \times 10^6}{2.3 \times 10^9} \times 100 = 10\% \text{ of the total life-cost of shearers in the UK.}$$

The increased revenue figures may look spectacular, however the additional production may not have a market, at least at the existing prices, and the cost of the retrofit design changes needed may be prohibitive.

This problem can be reduced or eliminated by providing designers with all the necessary ergonomic criteria in a form that can be used at the concept and drawing board stages of new designs. This may seem ambitious, however this is the approach adopted by the British mining industry. Ergonomists there have produced separate design handbooks for a number of different types of mining machines including both their operational and maintenance requirements. By producing separate design handbooks for different types of machines, the ergonomists were able to develop much more specific guidelines than would have been possible if a single handbook was provided for all mining machines (Simpson and Mason, 1983). Once produced, the same data can be used and re-used by the designers in all the supplier companies and hence appropriately derived and presented ergonomics information can in fact be used to influence all machines of a particular type. Thus the long-term benefits can be very considerable indeed.

One step short of the increased revenue calculation (which avoids some of the reservations such as, can the extra product be sold?) is to examine machine/system availability. Most managers would listen to any proposal to reduce downtime and reasonable, easy to use estimates have been proposed to calculate availability. For example:

$$\text{Availability} = \frac{\text{MTBF}}{\text{MTBF} + \text{MTTR} + \text{MTPM}}$$

where, MTBF = mean time between failure; MTTR = mean time to repair; and MTPM = mean time for preventative maintenance.

As several authors (e.g. Seminara and Parsons, 1982; Ferguson *et al.*, 1985) have emphasized the role of ergonomics in the design of machines for ease of maintenance, it is possible to see immediately two aspects of the equation (MTTR and MTPM) where an economic case for ergonomics can be made.

Health and safety

The costing of health and safety issues is often avoided because of the inability to 'cost a life' given the natural ethical and/or moral reservations surrounding such an exercise. However, if one accepts that no organization actively wants to kill its staff and thus work simply on the costs to the organization, then the moral and ethical reservations on costing health and safety disappear, especially if by doing so ergonomists increase the probability of obtaining funding to promote improvements in health and safety.

There are, of course, many cost implications to an organization arising from health and safety issues; however the 'core costs' are as follows:

1. Costs incurred by disruptions in manning from non-work related sickness absence, e.g. influenza.

2. Costs incurred by disruptions in manning from work-related, chronic health issues, e.g. back pain, respiratory disease, dermatitis.

3. Costs incurred by disruption in manning from lost time injuries.

4. Costs incurred from compensation payments against injuries, e.g. loss of limbs.

5. Costs incurred from compensation payments against health effects, e.g. back pain, hearing loss, tenosynovitis.

6. Direct cost in lost production arising from accidents, i.e. time lost in accident recovery, investigations.

7. Social costs incurred in sickness benefit payments.

Although generalized costs are usually relatively easy to obtain, it is much more difficult to find information on particular subgroups of the work-force, e.g. a particular job category. However, with some effort a reasonable approximation is often possible.

Using the shearer example, the potential health and safety costs could be developed as follows:

– Define the relevant costs from (1) to (7) above for the industry.

– Reduce these costs proportionately to the numbers of men associated with shearer driving.

– Multiply these costs by the life expectancy of the machines.

Note: The costs in (7) above can be ignored for this group as they are effectively off-set by payment in lieu of wages. The costs in (1) are irrelevant. The costs in (5) may be influenced but a significant improvement is unlikely. Ergonomic improvements should reduce the costs in (2) and (3) but the amount is difficult to estimate. The national costs of (4) and (6) are known and will be influenced by improvements in the ergonomics although the total cost is not recoverable.

For convenience, appropriate at this level, assume that the ergonomic savings in (2) and (3) are equivalent to the costs in (4) and (6) which cannot be related to ergonomics. The relevant national health and safety costs can therefore be approximated to the costs in categories (4) and (6) which, in this case, allowing for inflation since the last published data = £8 million (Collinson, 1980). Assuming health and safety problems are equivalent across all production job categories, then:

$$\text{cost per underground worker} = \frac{£8 \times 10^6 \times (1 - s/u)}{u} = £50 \text{ p.a.}$$

where, s = surface work-force; and u = underground workforce.

Shearers are one or two man-operated, working two or three shifts per day and therefore we can assume four operators for each of the shearers. The maintainers of the shearers will also benefit from improved design and therefore using the industry's comparison of production and maintenance man-shifts, it is reasonable to assume that the same number of maintenance men are also involved. Health and safety associated purely with the ergonomics of shearer design = cost per man (£50) × number of men (8) × number of shearers (540) = £216 000 p.a.

Clearly, such a calculation can be made more accurate by knowing, for example, exactly the number of shearer drivers, by using a ratio which avoids the assumption of equivalent risk over all job categories, by a closer approximation to the actual costs in (2), (3), (4) and (6). However, the principle involved is the same and the degree of accuracy obtained will depend on the availability of information and the argument to be developed. For example, if it is simply to show that real costs are involved, a superficial approximation as above may suffice, whereas to predict the rate of return on a redesign investment would require a more careful analysis.

Examples of economic analysis in ergonomic studies

Previous sections have suggested the type of performance, health and safety issues which can be used to develop an economic case to support ergonomic studies. This section provides examples of a number of studies which have included economic analysis. Taken as a whole they cover each of the three main areas of ergonomic activity, showing clearly that ergonomics can, and often does, contribute not only to the health and safety of the work-force, but also to the financial health of the organization.

Problem of heat stress in the steel industry

Feinstein and Crawley (1968) described a study of the design and siting of a slab shear pulpit in a steelworks which proved to be of considerable economic benefit. The job was to remotely-operate a shear blade to remove the end defects in steel slabs prior to rolling. The defect of most importance was a 'pipe', a hollow indentation, which occurred at each end of the slab. The 'pipe' was an inevitable consequence of the cooling process and was therefore entirely predictable. Unfortunately however, the depth of the 'pipe' was not. The problem was exacerbated by the surface temperatures of the steel which created a considerable radiant heat problem. In order to avoid the heat, the pulpit had been sited over 10 m away from the shear blades. This distance, together with line of sight problems (which meant that to check the front cut, the slab had to be reversed beyond the pulpit), forced the operator into one of two equally disadvantageous practices. In order to get an accurate cut, the operator had to loop the slab several times to and from the blades, thus creating a bottleneck. Alternatively, if he wished to avoid bottlenecks, he had to deliberately overcut which, of course, incurred the cost of 'wasting' good steel.

The benefits of a pulpit closer to the blades were of course obvious. However there remained the problem of how to protect the operator from the radiant heat. The ergonomists suggested that the pulpit should be glazed with gold laminate glass, which was known to significantly reduce heat flux, and that a number of other improvements should also be included at the same time. When the redesign was costed the management considered it to be far too expensive given that no estimates of the return on the cost had been presented. The authors went back to their laboratory and carried out a series of studies to approximate the performance improvements which could be expected from the changes proposed. When these were presented to the management, it was decided to build the new pulpit. It was also decided that the actual benefit should be examined. The plant immediately instigated a study to measure the delays caused by bottlenecks at the shears and checked all the off-cuts to establish the amount of good steel being recycled due to overcutting. The same measurements were continued for a year after the pulpit had been installed. This study identified a saving of slightly over £120 000 in the first year. The new pulpit had cost £10 000. This represented a payback time on the capital invested of less than one month!

Back pain

This is in many senses the classic example of an ergonomic/health issue which can be treated as a loss prevention exercise. Details of the overall cost of the problem have been presented elsewhere (see, for example, Simpson, 1985b) so one, particularly graphic, statistic will suffice in this context. Manstead (1984) presented data which showed that in the UK during 1982 more than six times as many man-days were lost due to back pain than the total lost due to industrial disputes! Moreover, it should be remembered that back pain is by no means restricted to the 'heavy' jobs or those involving manual handling. Lloyd *et al.* (1986), for example, have shown in a study which compared miners and office workers, that there was no significant difference in the incidence of back pain in the two groups under the age of 45 years. There can be no doubt that back pain is both widely prevalent and expensive to industry, however can it be shown that ergonomic intervention will reduce those costs to the benefit of both the individuals and their employers? Teniswood (1982) describes a study of back pain in an Australian mining company. The study covered both surface and underground operations and examined manual handling, workspace and vibration in the drivers' cabs of mobile plant and the introduction of a new training programme. The lost time back injuries were halved in two years. Although the author himself places no financial value on this achievement, a subsequent paper (Anon, 1983) states that the company's operations were $A157 000 per annum more efficient.

A paper by colleagues of the authors (Chan *et al.*, 1987) describes an interesting use of the cost implications of back pain to promote ergonomic change. They had been asked to advise on the workstations for a new engine assembly line which, at that stage, was still on the drawing board. An initial examination of the plans for the new line and studies of a similar existing line suggested that four tasks on the line were likely to need improvement in ergonomics terms. Simulations of these tasks in the new configuration showed that the workplace design was likely to make the target cycle-times set for the operations unachievable.

One of the major changes on the proposed new line was a significant reduction in the number of repair loops on the line in comparison with previous practice. The combination of less repair loops and failure to meet cycle times on the tasks studied suggested that bottlenecks would be inevitable. However there was also considerable concern that redesigning the four workplaces could incur unacceptable delays on the progress of the whole development which was on an exceptionally tight schedule.

It seemed possible that the ergonomic arguments would be lost. It was then decided to use the fact that on two of the tasks, back pain was likely to be a major problem in conjunction with the discovery that musculoskeletal problems (in particular, back pain) was the largest cause of sickness absence among the company's assembly line workers, to attempt an economic argument. Using information from the simulations and data from the company's medical service, it was possible to predict that the absenteeism from back pain on the new line (70 operators) would be of the order of 50 man-weeks/year. Covering this absence within the proposed manning levels was almost impossible and 'carrying' spare manning was also considered to be unacceptable by the company in terms of break-even costs. The predicted financial implications created a new basis for the ergonomic argument.

When the ergonomic changes were proposed on the basis of yielding improvements in cycle time of between 6 and 16% (dependent on task) and a reduction of approximately 20% in musculoskeletal absence for those tasks, they were accepted.

Although no information is yet available to show whether these predictions were achieved, this example shows how important an economic argument is, even when the potential of ergonomics was recognized by the company, as evidenced by the fact that they called in the advice at a relatively early stage.

Workplace design

The first study discussed in this section is another example of how a loss prevention argument can ensure that ergonomic considerations are taken seriously. Work carried out by the British mining industry on prototype mining machinery (Mason *et al.*, 1980) had shown that there were often serious ergonomic limitations in terms of both the control layout and the sightlines on a wide range of mobile plant. Unfortunately as new prototypes are hardly an everyday occurrence and have to be studied in surface simulations, it was difficult to obtain any feel for the real implications of such shortcomings. Moreover it became apparent that a fully developed working prototype was too late in the process to suggest major changes on ergonomic issues, especially as the penalties which would arise from ignoring the ergonomic improvements were unknown. A subsequent study (Chan *et al.*, 1985) was able to examine the operational implications of poor layout and restricted sightlines (as well as other ergonomic issues) in some detail for a particular class of mining equipment—underground development machines.

These machines are used to drive underground roadways. They have the facility to remove strata, either by using rotating picks to cut it down or by drilling for shot-blasting. They are also able to collect the debris and load it onto a conveyor system for removal, either to pack the roadway sides to improve stability, or to take it to the surface. Such machines tend to be rather large, yet by the nature of mining, they are expected to operate in relatively confined spaces. This creates particular problems in terms of both controls and sightlines. The position of the driver and the bulk of the machine often restrict vision. Additionally all the machines studied were tracked vehicles which, given their large size and the confined space, made them difficult to position accurately.

The fact that these reservations were not simply a failure to meet some form of academic/ergonomic ideal was shown by the field studies in that it was standard practice in 80% of the machines studied for an additional man from the development team to act as a 'spotter' for the driver. The spotter positioned himself at a point which gave unrestricted vision and signalled instructions to the driver using hand and caplamp signals. The use of the spotter immediately identified that there were economic implications behind the ergonomists' concerns. What was intended as a one man operation was in reality closer to a 1.5 man operation. The study was also able, using a variety of techniques (including some from work study), to estimate the contribution of poor sightlines and control layout to the overall cycle time for the operation. The studies suggested that the ergonomic limitations were adding approximately 5% to the cycle time. This, used with a knowledge of the cost of drivage operations, together with estimates of the salary costs incurred from the 'unnecessary' function of the spotter (multiplied across all similar machines in use in the industry) revealed a total cost of between £8 million and £18 million, depending on which figures were used for drivage and wage costs.

As a result of highlighting the loss prevention argument the initial reluctance to 'impose' ergonomic considerations on the designers, because of lack of tangible benefits, was reduced considerably. For example, all suppliers of such machinery are being issued with ergonomics design manuals which were also produced during the study. Achieving this end point would have been almost impossible without the economic argument.

The second example in this section concerns a recent Norwegian study (Spilling et al., 1986) of the influence of workplace design on musculoskeletal problems, and is probably the most thorough example published so far showing the economic analysis of ergonomic change. The plant studied was primarily concerned with the assembly and wiring of telephone switching panels, and employed a predominantly female work-force. The workstations involved considerable muscular loading and awkward postures. This was reflected in high sickness absence records for the plant.

During 1975 the authors carried out an extensive ergonomic redesign. Particular emphasis was placed on the need to give each operator greater flexibility allowing, for example, for both seated and standing operation. Several other changes were also made, including improved seating, improved tools and major changes to both lighting and ventilation. A number of the factors mentioned earlier in this chapter, e.g. sickness absence and labour turnover, which could be expressed in financial terms were then monitored throughout the period to 1983.

Prior to the improvements, musculoskeletal sickness absence was running at 5.3% of the production time available; in the period from 1975–82 it had dropped to 3.1%, a difference which was significant. While this is obviously a major improvement, a closer look at the period around the change shows an even greater effect. Although the average prior to 1975 was 5·3% as stated, the trend was rising steeply—the figures for 1973 and 1974 being 6.8% and 10%, respectively. Between 1979 and 1982 however, the absence rate was almost static at just below 3%. The analysis of labour turnover was even more marked. Prior to 1975 turnover was running at about 30%, whereas in the period after the change (to 1982) it had been reduced to an average of slightly over 7.5%. Obviously the labour turnover could have been influenced by many factors other than the ergonomic improvements. However, in interviews with the staff the improved working conditions were the most frequently mentioned issue. These reductions in labour turnover also created additional financial benefits beyond the immediate production improvements, for example, there were attendant savings in terms of both training and recruitment costs, which at a turnover of 30% were considerable.

The authors report a long and detailed financial analysis including indirect savings such as the training cost reduction. All calculations were normalized and fully amortized (over a 12 year life). The savings totalled approximately 3.25 million NKr on an investment, covering both the study and the cost of implementation, of approximately 0.5 million NKr.

While there are a number of other studies in the literature which include some form of economic analysis of ergonomic intervention (see for example, Simpson, 1988), the above are sufficient in the present context to show that economics can be a powerful tool for the ergonomist. Unfortunately, however, despite these examples economic analysis remains rarely used in ergonomics, despite the fact that many studies collect the type of data which would make it extremely easy. A study reported by Ong (1984) is a good example. The study covered VDT operations in an airline computer centre in Singapore. Improvements were made to the workplace, the lighting and the work patterns and extremely detailed records taken covering reported fatigue and performance. Reported muscle fatigue dropped considerably, in some areas by a half, and visual fatigue reduced by about 30%. Performance in terms of keystrokes/hour improved by almost 25% simultaneously with an error rate reduction from around 1% to 0.1%. Obviously these results are impressive as they stand but with relatively little effort, they could have been turned into monetary figures, which would have had even more impact. Moreover, a financial statement is, in effect, context free—it is, if the pun can be forgiven, the 'universal currency' of management discussion.

Conclusions

Arguably, if ergonomics is as important to industry as ergonomists believe, then the extent to which it is not routinely used must suggest some failing in the way it is presented to management. It would seem reasonable to presume that as most managers already have plenty of problems, they are unlikely to be enthusiastic when someone turns up telling them they have more, especially if they are of the previously unheard-of ergonomic variety! Given that they already know they have safety problems, health problems, productivity problems and cash flow problems amongst others, it can hardly be surprising if they 'turn a deaf ear' to someone telling them they also have ergonomics problems. Moreover, a manager's job is to solve problems not to go looking for them.

The acceptance of these simple points immediately identifies a new approach to the promotion of ergonomics. First, there are no such things as ergonomic problems—there are however a wide range of problems including safety problems, occupational health problems, productivity problems and labour turnover problems in which ergonomics knowledge may be of assistance. Second, ergonomics must be seen increasingly as the provider of solutions rather than as the identifier of problems. Once it is accepted that the role of ergonomics is to aid in the solution of 'accepted industrial problem areas', then the development of economic arguments to support and justify the expenditure on ergonomics becomes much easier as the majority of relevant data are already being collected somewhere in the organization. The presentation of ergonomics as relevant to accepted problem areas supported with an economic bias immediately gives the ergonomists the advantage of talking the same language as the manager they are trying to convince. The increasing use of economic benefit in support of ergonomics change will in no way undermine the basic objective of improving health and safety. In many circumstances, it may in fact be the only way to ensure that the improvements are implemented.

This chapter has shown the kind of data which can be used to develop a cost–benefit approach to ergonomics and some of the procedures available to utilize such data. It has also shown, from the literature, that such analyses are in fact feasible and that they often have a major influence on the acceptance of ergonomic proposals. All that remains is for the use of economics in ergonomics to become the rule rather than the rare exception. This will hopefully lead to more publications on the economic analysis of ergonomic benefits, which will in turn enable more ergonomists to present their proposals and results in a form of immediate interest to the industries who fund them.

References

Alexander, D.C. (1985). Making ergonomics pay: adopting a business approach. *Industrial Engineering*, July, 32–39.

Annett, J., Duncan, K.D., Stammers, R.B. and Gray, M.J. (1971). *Task Analysis*. Training Information Paper No. 6. (London: HMSO).

Anon (1983). The human factor. *Mining Magazine (MIMAG)*, December, 15–18.

Chan, W.L., Pethick, A.J. and Graves, R.J. (1987). Ergonomic implementation in the design of an engine assembly line. In *Contemporary Ergonomics*, edited by E.D. Megaw. (London: Taylor and Francis) pp. 140–145.

Chan, W.L., Pethick, A.J., Collier, S.G., Mason, S., Graveling, R.A., Rushworth, A.M. and Simpson, G.C. (1985). *Ergonomic Principles in the Design of Underground Development Machines*. Final Report on CEC Contract 7247/12/007. (Edinburgh: Institute of Occupational Medicine) (IOM Report TM/85/11).

Collinson, J.L. (1980). Safety—the cost of accidents and their prevention. *The Mining Engineer*, January, 561–571.

Feinstein, J. and Crawley, J.E. (1968). *The Ergonomic Design of a Slab Shear Pulpit at Colvilles Limited*. Report No. BISRA OR/HF/8/68. (London: British Steel Corporation).

Ferguson, C.A., Mason, S., Collier, S.G., Golding, D., Graveling, R.A., Morris, L.A., Pethick, A.J. and Simpson, G.C. (1985). Final Report on CEC Contract 7247/12/008. (Edinburgh: Institute of Occupational Medicine) (IOM Report TM/85/12).

Gallaway, G.R. (1985). Marketing ergonomics: influencing developers, managers and customers. In *Ergonomics International 85*, edited by I.D. Brown, R. Goldsmith, K. Coombes and M. Sinclair. (London: Taylor and Francis) pp. 991–993.

Lloyd, M.H., Gould, S.R. and Soutar, C.A. (1986). Epidemiologic study of backpain in miners and office workers. *Spine*, **11**, 136.

Mason, S., Simpson, G.C., Chan, W.L., Graves, R.J., Mabey, M.H., Rhodes, R.C. and Leamon, T.B. (1980). *Investigation of Face-end Equipment and Resultant Effects on Work Organisation*. Final Report on CEC Contract 6245-12/8/47. (Edinburgh: Institute of Occupational Medicine) (IOM Report TM/80/11).

Manstead, S.K. (1984). The work of the Backpain Association: past, present and future. In *Occupational Aspects of Back Disorders*, edited by J. Brothwood. (London: Society of Occupational Medicine).

Maynard, H.B., Stegmarten, G.J. and Schwarb, J.L. (1948). *Methods—Time Measurement*. (New York: McGraw-Hill).

Ong, C.N. (1984). VDT work place design and physical fatigue: a case study in Singapore. In *Ergonomics and Health in Modern Offices*, edited by E. Grandjean. (London: Taylor and Francis) pp. 484–494.

Schneider, M.F. (1985). Ergonomics and economics. *Office Ergonomics*, May/June, p. 8, 12, 30.

Seminara, J.L. and Parsons, S.O. (1982). Nuclear power plant availability. *Applied Ergonomics*, **13**, 177–189.

Simpson, G.C. (1985a). Some requirements for improving the industrial utility of ergonomics. In *Ergonomics International 85*, edited by I.D. Brown, R. Goldsmith, K. Coombes and M. Sinclair. (London: Taylor and Francis) pp. 334–336.

Simpson, G.C. (1985b). Cost benefits of ergonomic action. In *Ergonomics in the ECSC industries 1980–1984*. Community Ergonomics Action Report 5, Series 3, Volume I. (Luxembourg: European Coal and Steel Community).

Simpson, G.C. (1988). The economic justification of ergonomics. *International Journal of Industrial Ergonomics*, **2**, 157–163.

Simpson, G.C. and Mason, S. (1983). Design aids for designers: an effective role for ergonomics. *Applied Ergonomics*, **14**, 117–183.

Spilling, S., Eitrheim, J. and Aaras, A. (1986). Cost benefit analysis of work environment investment at STK telephone plant at Kongsvinger. In *The Ergonomics of Working Postures*, edited by E.N. Corlett, J. Wilson, and I. Manenica. (London: Taylor and Francis) pp. 380–397.

Swain, A.D. and Guttman, H.E. (1983). *Handbook of Human Reliability Analysis with Emphasis on Nuclear Power Plant Applications*. Report

NUREG/CR-1278 (Washington DC: US Nuclear Regulatory Commission).

Teniswood, C. (1982). Back injury prevention in metaliferous mining operations. In *Ergonomics and Occupational Health*, edited by P. Rawlings. (Melbourne: Ergonomics Society of Australia and New Zealand).

Williams, J.C. (1982). Cost effectiveness of human factors recommendations in relationship to equipment design. *Proceedings of the Institution of Chemical Engineers Symposium*, Series No. 76. (London: Pergamon Press).

Part VII

Introduction and implementation of systems

How often, in our professional or personal lives, have we had what we believed to be an excellent idea, scheme or proposal turned down, and how often has that happened not because of lack of intrinsic merits or due to flaws in the plan but because of the way we introduced it to our colleagues or friends? Likewise the success or failure of many new systems, or at least the degree of their successful utilisation, will be determined by how they are implemented. Eason, in chapter 33, looks at this issue generally which is relevant to all the other content of this book. First however, in chapter 32, Shipley discusses what is usually held to be a major plank in any change initiative, including new systems introduction; participation. In so doing she treats participation as something much more fundamental than a mechanism for ergonomics change; she discusses it as a basic philosophy in an organisation, distinguishing participation ideology from participation methodology. She also though provides examples of ergonomics in a participative strategy, and of participation-based initiatives within ergonomics.

The type of implementation strategy adopted, as identified by Eason, will determine the degree of participation that is possible. What he sees as concepts of value—the user 'champion', user representation, and full user involvement—will be more feasible and relevant in a strategy of incremental implementation than in a 'Big Bang' type of scenario. Whilst there are some downsides—for instance the time and other resources required, participant coping problems and so on—Eason generally argues for a user-orientation in systems implementation, which includes development and introduction. In doing so he provides apt commentary for the concerns of the whole of this book. Concern for, involvement of, and information gathering from or with the people who work in offices, factories, transport, services etc., must underpin ergonomics methods and techniques.

Chapter 32

Participation ideology and methodology in ergonomics practice

Pat Shipley

Participation as ideology

The promotion of 'participation' in the workplace is a controversial subject. In the first place, we are dealing with an abstract construct which means different things to different people. Does it mean, for example, genuine power sharing; that is the reform and modification of all formal organizational policies, practices, structures and procedures to enable all organizational members down to the grassroots, to have a positive central involvement in all decisions taken in an organisation? Or does it mean something far less radical when, for example, the work force is merely informed or minimally involved if a change is planned in the organization which could have a bearing on the work they do? 'Involvement' implies many different possibilities from genuine sharing in any decisions that led to a proposed change, through token but ineffective representation and consultation, down to mere warning that some situation has to be adjusted to which has been decided somewhere else.

In the second place, given that the aims of and the form that the involvement should take are quite clear, it may not be at all certain how best to attain that involvement. Thus the goal, ideology or philosophy of participation needs to be distinguished from participative methodology; i.e. participation as a principle and good as an end itself in contrast with participation as an instrument or tool for easing in change of some kind.

Caught up in controversy the advocates of the participation movement have to confront their own values and assumptions, how far for example these are compatible with the wishes of the different interest groups working in industry. They must also address the severe restrictions that apply to their own competence and expertise in helping organizations to change, in a participative management direction, say; restrictions which may originate in themselves as well as be imposed from outside. The promotion of participation and worker involvement, therefore, may originate from idealistic or pragmatic impulses.

There is an established 'Quality of Working Life' movement in North America. The Ontario Ministry of Labour in Canada has its own Quality of Working Life (QWL) Centre which publishes its regular news journal, *'QWL Focus'*. Volume 6, issue 1 (March, 1988) of this journal is dedicated to 'New Technology and Organizational Choice'. In that issue an author from the Rideau (Ontario) Regional QWL Centre (Best, 1988) saw fit to clarify the difference between QWL and 'excellence', the latter being a fashionable buzz word in the business world which has recently travelled to the UK from across the Atlantic. To quote him:

> The concept of 'excellence' is based on values which are largely, if not solely, management's values: productivity, growth and commitment to the company (management). Quality of Working Life is based on a belief in democracy—in sharing of responsibilities and authority between workers and managers and in the basic dignity of people. . . Proponents of 'excellence' have discovered that a happy worker is a better worker. To this end, managers may devise a number of incentives or 'goodies' to help workers feel better about their workplace or more involved in what managers are deciding. However, these giveaways stop well short of recognizing the worker as an equal partner in the operation of the workplace. (p. 59)

This is participative management which grows out of enlightened self interest.

My colleague at Birkbeck College, Jean Hartley, reminds me of the usefulness of the distinction between direct and indirect participation, in her paper on industrial relations (Hartley, 1984). Direct participation involves employees in decision making and control at the level of their jobs. Indirect participation is involvement through channels of representation at various levels, plant or company. This could stretch as far as worker directors on company boards. Hartley maintains that both forms should be seen as complementing rather than conflicting with each other. This theme will be explored a little more later on in the chapter in the context of relevant legislation. Suffice to leave it here at the moment, pointing out that no explicit legislation exists in the UK or America, in contrast with Scandinavia and the European Continent, to promote worker participation and democracy as a principle, an end in itself.

Some time ago another colleague in the Department of Occupational Psychology at Birkbeck College ran a postgraduate course module on 'Democracy and Organizations' and 10 weeks later neither he nor his students felt much clearer about what democracy meant and how to define it. It would be safer, wouldn't it, to leave the problem to someone else, or to confine the debate to State politics and to deny any relevance to the workplace? We would certainly get away with such a comfortable position as ergonomics practitioners in, for instance, present day British industry; not so in Sweden however, where work people are required by law to participate and to collaborate.

The sections that follow include a summary of the European legislative position on worker participation and the state of our knowledge through research on participation and its applications. They also elaborate the action research perspective introduced in this book's chapter on organizational analysis as a possible approach for both the technical and ethical enrichment of ergonomic practice. Some attempt is made finally to look into the future, but proactively, and not as passive recipients of 'come what may'.

Participation as methodology: A particular perspective

There is a large and separate body of literature on decision making. Much of it is apolitical in that power (the capacity to control and influence others) as a potential influence on decision making, is often unappreciated. The literature covers a 'history' of studies falling broadly into two parts. The first part, roughly pre-1970, was dominated by laboratory studies of individuals, pairs or small groups, doing artificial short-term tasks in contrived settings, and the dominant model was the rational model. Decision making was taken to be the rational choice of the best alternative among a number of carefully weighed-up possibilities. In the second part, a number of studies explored real-life decisions under complex conditions, many of them in organizational settings where power factors can be influential. One organizational theorist, Pfeffer (1981), is among those who have criticized rational choice theory for failing to take into account the diversity of interests contained within typical organizations. In hierarchically structured organizations specialization and task hierarchy reduce power for the many at the base of the hierarchical pyramid; power or lack of it is structured into the organization.

The management prerogative, the right to manage at all times, and to make all the important decisions, is usually unquestioned, taken for granted and assumed to be rational. This shared ideology, so the argument goes, leads to habitual practices which legitimate the ideology. People can be trusted to follow instructions, after socialization into the norms of the workplace. Socialization, therefore, transforms power into authority. 'Technical rationality', of procedures, rules, norms and techniques, is legitimated and authenticated as rational and reasonable. In contrast, with 'substantive rationality' the values underlying a practice are fully understood and espoused, and practice is based on genuine commitment to that rationality.

The ideology of functional or technical rationality, prioritizes the goal of 'efficiency', presumably with the implicit de-prioritizing of other values, if they are seen to be potentially in conflict with 'efficiency'.

Limited imagination and the narrow horizons of some powerful careerist managers, whose careers incidentally are usually outlived by the organization's more durable life span, may be reasons for so-called 'cost-effective' or efficiency decisions being made on short-term criteria; their efficiency over the longer term may be found to be more apparent than real. Many political models of organizational function propose or imply that decision makers make decisions according to their own vested interests and preferences quite regardless of the welfare of the whole organization or of wider social issues.

Those with or without formal power in organizations may therefore be interested in promoting the project of participation for different reasons. They may be committed to the granting and securing of participation as a moral or civic right; a commitment which may be externally imposed by law and may be at best half-hearted or partial, or it may be a genuine personal commitment to the ideal whatever the law dictates. An indirect form of commitment, an alternative to direct value commitment, is an instrumental commitment associated with the practical harnessing of participation as a means of furthering some goal other than individual or group rights. This other goal may or may not be intrinsically good or bad.

My own view is that the present revival of interest in the United States in worker involvement which was unprovoked by legislation is due to a widespread belief among North American management that employee involvement and participation enhances worker motivation and satisfaction, which in turn enhances efficiency through work commitment. (See for example the paper by Mohrman and her associates in the 1986 international review of mainly American work on industrial/organizational psychology.) Reasons for this might include: antecedent conditions of high labour turnover and absenteeism during a period of full employment, and emergent norms about social and power relations in organizations, which sharpened the ever present problem for management of motivating low status workers. Traditional means of solving such problems, such as improved financial incentives, closer supervision and greater subdivisions of work, have been found wanting, and participation and involvement may be the latest fad therapy or prescription for dealing with them. However, that is far from being the whole picture. Increased pressure from international competition has played its part.

In the special issue of the '*QWL Focus*', Warrian (1988), previously Research Director of the United Steelworkers Union of America, claims that the unions share a cynical view of participative management; that QWL is exploited for management's productivity reasons, leading to greater management control, 'speed up' and indoctrination. This is participation without genuine shared control.

Legislation about participation

In a paper which compared UK law with European and Swedish law within the context of health, safety and welfare at work (Shipley, 1987), two broadly different industrial relations models with their different background histories, philosophies and practices have been contrasted. One model is the model of institutionalized participation and external regulation, the Continental model, for which West Germany and the Netherlands are particularly known. After the second Great War and the Yalta agreement which resulted in the division of Germany, democracy became an important concern for the State authorities; Germans were expected by the allies to learn to be more democratic (in order to forstall a further outbreak of fascism). The application of democratic principles in the workplace, through means of worker participation for instance, was to be a logical and major step in that direction.

The proximity of institutionalized Communism in the Eastern bloc countries also played its part in the growth of Continental participation, as a response to the challenge of perceived growing and dangerous demands from the political left, as well as a reaction to the authoritarian right. I described the alternative model in that paper as, at one time, characteristically British; that of informal collective bargaining, focused locally at plant level rather than industry-wide. This model is variously referred to as 'voluntaristic', even 'laissez-faire', 'adversarial', or 'muddling through'. Many members from either side, employer and labour, view the collective bargaining process as indispensable and a widespread and important form of participation in industry.

In the institutionalized model, as in representative political democracy, participation is not a direct experience for the rank and file, who participate indirectly through their representatives and on whose commitment, values, competence and power they are dependent. Decisions are made on their behalf by these representatives. As this kind of representation formalized into union representation the bargaining process itself became institutionalized, and some of the attributes of institutionalization, such as behavioural rituals, came to characterize the process. In a state democracy indirect participation may be the only practicable method but problems of genuine democractic representation of individual and grass-roots opinion still remain. The majority of individual workers under both schemes, it could be said, remain divested in practice of the enriching experience afforded by substantial direct personal involvement in important decision making. The net outcome of this process can sometimes be stress, inefficiency and/or dissatisfaction at a number of levels.

'Joint consultation', which became popular in the UK for some time after the second Great War, in contrast to the traditional 'bargaining' process emphasized co-operation and consent around unifying issues such as the organization's viability. Joint consultation was meant to come about through regular meetings of management-employee committees. Popularity for this kind of participation eventually waned among unions and wage earners whose opinions were solicited but without concessions made by authority. No change in power structures occurred because management continued to make the important decisions and retained control.

The Continental model of participation was institutionalized through the establishment of Works Councils and the appointment of worker directors on company 'supervisory' boards. These structures have since been the subject of Continental research. With its entry to membership of the Common Market Britain fell under the influence of European legislation, culminating in the Bullock Report on industrial democracy (Bullock, 1977). Both sides of industry at first took a lukewarm view of the ideas on worker involvement that it embodied, for different reasons. For unions it could have been viewed as a way of increasing worker protection and power; for management it projected an attractive picture of industrial relations harmony. But 'Bullock' was thrown out because of the influence of powerful interests. Traditionalist unions opposed it on doctrinal grounds, while some employer spokesmen feared for the loss of management prerogatives. Subsequent attempts to revive the idea of industrial democracy in Britain have also failed, and unions have since lost considerable ground under recent changes to Union Law in the UK.

Both sides of British industry have also glanced with interest in the direction of Scandinavia, particularly since the enactment of the Swedish Co-determination Law on the joint regulation of working life, and the Norwegian and Swedish Work Environment Laws in the period 1977–78. The Work Environment Laws from both these Scandinavian countries encourage industry to adopt and implement QWL principles, promoting worker involvement and development, and minimizing stress at work. The intentions behind the legislation in the case of Sweden appear to some to be frankly political in the spread of social democracy and decision making power through all levels of society, although the restraining hand of central bureaucracy has turned out in practice to contradict this dispersion of power to some extent.

However Swedish law is explicit in requiring employers and employees to collaborate in various ways, including the provision of opportunities in the design of work, to enable workers to influence their local work situations themselves. The Co-determination Act is a legal shrine for a principle, that of collaboration between labour and employer, which was already a relatively long-established norm in Sweden. Swedish labour already had considerable power compared with Britain and elsewhere, although Swedish labour critics of the Act claim that the balance of power still rests firmly in the hands of the employer.

There have been recent signs of cracks showing in the famous Swedish practice and ideology of harmony and consensus, although this remarkable country has managed to hold onto her prosperity during the recent recession along with relatively low unemployment and inflation levels. If Sweden joins the EEC she may use her political will and experience to move other EEC member states more in the direction of worker participation. Her economic revival came about under a Social Democratic administration with its strong labour sympathies, which was returned to power in 1981 after a shortish interruption to its long reign of some 40 years. Little of Swedish industry is publically owned but there is a large public budget available for work force retraining in this country which is committed to technological innovation.

Other Scandinavian countries, Norway and Denmark, share cultures of notional self-reliance and autonomy. Norwegian industrial and economic growth was achieved through the continuous accommodation of pre-existing tradition and culture. The statutory presence of worker representatives on company boards occurred as far back as 1948 in Norway, but these were boards of companies either wholly or partly government-owned. The Oslo Work Research Unit pioneered worker participation in schools, industry and shipping, as a result of co-operation with the UK Tavistock Institute, and Norway's QWL movement spread to Sweden whose most well-known application is the Volvo system of job design. Experiments with Tavistock-inspired 'autonomous work groups' were more possible in such cultures than in the Britain where they were first conceived. In 1987 the Norwegian government announced a new wave of QWL promotion.

The QWL movement did not really begin in Britain until after the prosperous 1960s and into the 1970s during a Labour administration, as worker absenteeism and unemployment rates began to creep up. It was then that the Work Research Unit was set up, steered by a tripartite group from the TUC (Trades Union Congress), CBI [Confederation of British Industry (Employers' Federation)], and State, to promote job satisfaction, and later, participation, in British industry. Promoting formal worker democracy, as such, is not its business; worker participation is, particularly during the introduction of new technology.

The Japanese model of labour management consensual decision making, quality circles, and single union agreements, locates the emphasis squarely on informal and non-legal methods of involvement. Leaders decide after a process of informal consultation. Participation Japanese-style appears more instrumental than idealistic in its goals; i.e. participation without real worker control again.

Recent research on participation in complex decision-making

For clarity's sake, a simple distinction is made here between research focused predominantly on management groups and recent research, mainly Continental, assessing the actual involvement in company decision making by workers or their representatives within the framework of industrial democracy. Also it would be helpful here to remind the reader of the distinction made earlier between participation entered into through commitment to an ideal, that of the moral right of people to be involved in decisions affecting their welfare and their life and work opportunities, and participation as a methodology for helping managers to deal with complexity.

Management-centred research is about methodological participation, and a classic study by Vroom and Yetton (1973), is often cited to illustrate this. This study is included in the book of readings edited by Pugh (1984) (see Vroom, 1974). The industrial or State norm is that of a top management, or Cabinet, making logical decisions aided by a support crew of specialists, and with everyone in broad agreement about goals and methods for attaining them. American scholars appear to shy away from a global concept of 'participation'. In their analysis of several complex decisions made by managers in real-life organizations Vroom and Yetton (1973) identified decisions as: autocratic, consultative, or democratic. The underlying theory was contingency theory, and their research supported the theory in that they found greater variance between decision styles across problems rather than across managers. Decision rules are implemented, it was argued, to enhance decision quality, and acceptability by subordinates of such decisions. A branching model for good decision making practice was drawn up, taking into account in turn: how far there was a quality requirement for rationality; how far sufficient information was available; the importance of subordinate acceptance; and how far conflict among subordinates might be generated by the boss's solution. In a genuine decision sharing situation single individuals or ruling elites do not impose their own wishes.

In Europe, a multidisciplinary team of researchers has analysed complex decision making at senior organizational levels across different cultures and countries (see Heller and Wilpert, 1981). Shared decision making was found to be more likely if there was 'intellectual closeness' between manager and privileged subordinate (between those of similar educational background), i.e. when expertise gaps were narrow. [Presumably the boss' position power can be counterbalanced, or even exceeded, by the subordinate's expertise, the expert power balance can be more equal even if the positions are unequal, or even shifted temporarily in the subordinate's direction.] A number of organizational decisions in several organizations were scrutinized, and overall a high level of participation was found at management levels.

Ergonomists typically work in multidisciplinary teams, and the paper by Clegg and Wall (1984) is a reminder that what they refer to as 'lateral participation' can be as important for organizational and task efficiency as 'vertical participation', and is indeed much less controversial. These authors point out that the participation literature is focused almost exclusively on the latter, and "how to transcend the boundaries between the 'managers' and 'managed'",(p. 429) to the neglect of the "integration of traditionally separate functional hierarchies." (p. 430) They see this as especially problematic in a contemporary era of rapid and continuous change when organizations have to develop quick-acting and transient organic structures, such as temporary project teams, under conditions of chronic uncertainty in order to remain viable. (See also chapter 30 on organizational analysis in this book.)

Clegg and Wall (1984) report case studies from the Sheffield (UK) Applied Psychology Unit demonstrating communication problems arising across functional divisions. They warn us that "specialisms will attract and recruit experts in their own areas, people with specific professional training and standards who may well have quite dissimilar perspectives as well as a language of their own" (p. 436), and that "the very factors which promote the need and opportunity for participation, at the same time encourage organisations to develop in ways which undermine its efficacy". (p. 437) "Put bluntly, the non-expert should not 'interfere'". (p. 438)

American research is largely descriptive and confined to individual organizations, whereas a growing European tradition in decision making research is international and more interested in structural influences on decision making, such as the distribution of power within organizations. This is what one would expect within a European context of compulsory worker participation and co-determination. In their study of 200 decision processes in seven organizations covering the Netherlands, the UK and Yugoslavia, the DIO International Research Team (1983) also appeal to contingency theory as an explanation for what was found in their investigations. Quality of decision made was prioritized less than was commitment to the solution and the early involvement of all parties to the process.

Markedly asymmetrical power relationships between employer and non-management employee continue to exist however despite the formal introduction of worker democracy. A Dutch team based at Amsterdam is reputed for its description and analysis of complex organizational decision making both amongst management, and at worker representative levels. They found that worker power can be neutralized by the deployment by management of certain blocking techniques, tantamount to psychological hostage-taking. This is made easier by labour's comparative shortfall in expertise and advance information. Management typically possesses maximum available technical know-how and advance information.

The problem is amplified when the egos of the worker representatives who occupy a minority or marginal position on decision making boards get in the way of their opportunities for participation. Rather than look foolish they may just keep quiet! The Swedish State recognized this problem and provided substantial funds for training and advice to be bought in by labour to help to redress this power and skill imbalance.

Structures on their own are rarely enough, as working women in Britain learnt when the limitation of sex discrimination and equal opportunities legislation were to unfold in practice. A useful and comprehensive review paper by Amsterdam researchers on decision making research is that by Koopman *et al.* (1984). The deficiencies of early industrial engineering and bad industrial practice left a legacy of ills; labour welfare and well-being problems, and company efficiency problems. Formal worker participation as a theoretical solution to these problems has had only limited effect. Much more needs to be done in honing the tools available for promoting participation goals.

Action research and participative methodology in ergonomics

The chapter in this volume on organizational analysis touched on ideas about collaborative methodology as a means additional to the traditional expert mode of doing ergonomics research, and of putting that research into effect.

In a paper which made the case, probably for the first time, for the ergonomist practitioner in the role of change agent, we suggested (Shipley and Harrison, 1974) that ergonomics research and practice may be managed traditionally in a manner that is modelled on the classic military and bureaucratic organizational hierarchy, and that this could generate some unintended consequences associated with traditionally 'rigorous' research. We have seen little sign that ergonomics practice has changed much since these early observations.

This issue has re-emerged in other recent papers; see, for example, the published proceedings of the 1987 Ergonomics Society annual conference (Megaw, 1987), in which authors writing in a subsection of the proceedings devoted to ergonomics and trades unions, address the problem of non-participative methods used by ergonomists. One paper (Thorne and Russell, 1987) advocates consultation by ergonomists of worker representatives and a commitment to their involvement, to secure better co-operation. Another paper (Winterton, 1987) holds out the promise of a greater role for ergonomists in the generation of an alternative design philosophy, if they were to use the power base and institutional channel that the shop floor union provides. Finally, a paper by Forrester (1987) is focused on the role of the ergonomics practitioner as expert which is contrasted with that of the 'resource person' who is seen to be more attractive.

In their paper outlining alternative models for introducing new information technology into organizations, Blackler and Brown (1986) criticize traditional ergonomics and Tavistock socio-technical theory and contrast them with their scenario for more participative systems design. The former they argue are examples of a misleading and oversimplified linear and rational 'task and technology approach' to introducing new systems, in comparison with the more organic and cyclical 'organization and end-user approach' of participative method. Although they readily acknowledge that ergonomics has had some success in persuading designers to make their software and hardware more 'user-friendly' they believe there is considerable room for improvement; that ergonomics' boundaries are too limited and its political position too weak to have much real effect. Ergonomics apparently fails to take into account the powerful social and structural factors which influence strategic decision making.

Illuminated by ideas from Perrow (1983) on American human factors engineers, Blacker and Brown (1986) suggest that traditional ergonomists may unwittingly contribute to mechanistic social system (where machines are installed to prop up existing social structures, such as authoritarian political arrangements), because of their predominantly narrow physiological perspectives and individualistic levels of analysis. These authors' views are not entirely theoretical. They are both empirical researchers and their paper was inspired also by their recent empirical study assessing current British industrial practices in the utilization of microelectronics. They found the standards of implementation to be generally poor.

In an earlier paper, Shipley and Harrison (1974) attempted to analyse why methods based on the politics of expertise can fail in a culture like our own. It was proposed, for example, that the expert too readily assumes solutions to the client's problem, recommending prescriptions which the system either rejects or may implement only in amended form, and that a psychological reason for this rigid expert behaviour is a love of status and the other symbols of expertise, which may bring a warm glow to the ego. The problem of limited ergonomics influence may however have other causes, such as limited imagination and competence. (See the chapter in this book on organisational analysis which analyses the power basis of the ergonomist as a potential organizational agent for change.)

Changes to ergonomics training to include the study of the theory and practice of client–consultant relationships in a collaborative, power sharing way were also proposed in Shipley and Harrison (1974). This participative approach is often referred to as 'interventionist research', or 'action research' and a useful text by a British exponent is that by Clark (1972). Shipley and Harrison (1974) also advocated an expansion of ergonomics training to incorporate knowledge of organizations, the diagnosis of complex systems, and the management of change to enchance competence. As an example of resistance to ergonomics recommendations, we pointed to the management tactic of searching for ways to avoid or cast doubt on the consultant's findings; questioning the sampling and statistical analysis, for example.

It may be easier to indulge in scape-goating and denial rather than deal with the rational improvements called for. A further difficulty is that it takes time to achieve results and few of us, consultants or management, are willing these days to find that time. The diagnosis and problem solving phases in action research must proceed at a pace the client feels happy with. A continuous and unhurried organizational educational process may be required which is designed to give ergonomics gradually away to the client system. Then the consultant has to retain a hold in the organization to be able to help that process along. Retaining this hold can be difficult.

If it is seen to be of value, then practical problems must be the main barrier to the implementation of action research in ergonomics, coupled with perhaps a too naive view taken by some theorists of the personal implications for practitioners as members of organizations in trying to be participative. The ideology of participation should perhaps be tempered by a careful consideration of the practical realities of organizational life. The more we can appeal to the productivity benefits of methodological participation the more will management want to use such an approach. We appear to suffer however from a dearth of evidence pointing to such benefits.

An extreme view of the more traditional approach to ergonomics is that the expert is reluctant to give away control, the client is relegated to the status of object, and the detached observer stance creates psychological distance and a failure to generate trust and client commitment.

I recently came across an ergonomics project using an action research approach carried out by a team from an ergonomics laboratory in Paris. In their recent presentation to a symposium in Zadar (Yugoslavia), Teiger and Laville (1987) described how the methods of ergonomics can be used by non-ergonomists, in their report of shiftworking changes involving 2000 workers in a chemical plant. The company management gave the 'solution' to the ergonomics team initially along with the problem; a 'scientific' solution using biological criteria. Instead, the ergonomists were instrumental in setting up a task force which included some of the shift operators, with themselves acting as facilitators. The result was an outcome that was acted upon; a revised practice based on a number of compromises between ergonomic, social, economic, technical and organizational criteria. The final choice was made by the operators themselves, and the company retained the ergonomics team for other projects. The ergonomists here lent their research skills and knowledge expertise as members of a team operating on lateral participative lines in which the potential users, as the experts on the job, had considerable say and influence about how the project was managed.

Participation and the future

Ergonomists are sitting on an important area; new information technology can be introduced into organizations for good or for ill; ergonomists, along with equipment engineers, designers and systems analysts are channels for the introduction of such technology. (See Eason's chapter in this book.) The technology can be used to promote efficiency, and stress may or may not be a side effect depending on how the technology is introduced into the organization and what adaptations to the new system are required by users over the longer term. (Stress, and its assessment, is discussed elsewhere in this book by Cox.)

A recent book by the Director of the UK Policy Studies Institute (Daniel, 1987) summarized surveys made by the Institute, of technical change and industrial relations. Unfortunately the general conclusion is an overwhelmingly negative one, as far as promoting participation is concerned. "Even major changes have been introduced with surprisingly little consultation", we are told. (But it was also argued that the support from workers and their representatives for technical change was already available on site and consultation was not therefore needed.)

Sell (1986) who works at the UK Work Research Unit, advocates the use of the 'action research' principles of good client–consultancy practice, outlined above, as a means to the humane as well as efficient use of technology. Worker involvement imposes stress on some management who also need help in the management of their own stress; time is the biggest price to be paid by all management in being participative. The most successful programmes, he tells us (Sell, 1986), occur when there has been joint responsibility of union and management for seeing them through, and where the involvements are positively rewarded; not necessarily by pay rewards.

Furthermore if participation policy and procedure itself is not introduced participatively, if it is done in a manipulative way, then that project will ultimately fail also. Just so, if management succeeds in getting some other controversial project through 'participatively' the longer term damage to the organizational climate, to the level of trust, may be the price paid for it. The Ergonomics Society in Britain expects its registered practitioners to practise in accordance with its code of ethical professional practice. This code enjoins us to safeguard the interests of our clients, and the participants in our research programmes, particularly their safety and welfare interests. It enjoins us not to allow our standards of practice to be compromised by politics. Ethical practice is not the central subject of this chapter on participation, nor indeed of that on organizational analysis. But the role, values, skills, imagination and power base of the ergonomics practitioner as a proactive agent for change are.

What kind of change are ergonomists involved in, and whose agents are they, and why are they? All organizations have to be managed, even worker co-operatives. They are not easily managed effectively and fairly. How can ergonomists extend their powers to help the good intentions of management decisions and practices to be translated into good effects? And what can they do to protect themselves and others in their care should management expect them to collude in practices which are inconsistent with their personal values and their professional code of practice? My observation is that the ergonomist is first of all a person, a fellow or sister human being, and the expert role comes second to that.

Acknowledgments

My thanks are extended to John Coopey and Jean Hartley at Birkbeck College, and to Reg Sell and Frank Blackler, for their valuable suggestions and comments on a draft of this paper.

References

Best, G. (1988). Understanding QWL. *QWL Focus*, **6**, 59–60.
Blackler, F. and Brown, C. (1986). Alternative models to guide the design and introduction of the new information technologies into work organisations. *Journal of Occupational Psychology*, **59**, 287–313.
Bullock (Lord Bullock) (1977). *Report of the Committee of Inquiry on Industrial Democracy*. CMND 6706 (London: HMSO).
Clark, P. (1972). *Action Research and Organisational Change* (London: Harper and Row).
Clegg, C.W. and Wall, T.D. (1984) The lateral dimension to employee participation. *Journal of Management Studies*, **21**, 430–442.
Daniel, W.W. (1987). *Workplace Industrial Relations and Technical Change* (London: Policy Studies Institute and Francis Pinter Publishers).

D10 International Research Team (1983). A contingency model of participative decision-making: an analysis of 56 decisions in three Dutch organisations. *Journal of Occupational Psychology*, **56**, 1–18.

Forrester, K. (1987). Restructuring the relationship: the ergonomist as a resource person. In *Contemporary Ergonomics 1987*, edited by E.D. Megaw (London: Taylor and Francis), pp. 365–371.

Hartley, J. (1984). Industrial relations psychology. In *Social Psychology and Organizational Behaviour*, edited by M. Gruneberg and T. Wall (Chichester: John Wiley), pp. 149–183.

Heller, F.A. and Wilpert, B.W. (1981). *Competence and Power in Managerial Decision-Making* (Chichester: John Wiley).

Koopman, P.L., Broekhuysen, J.W. and Meijn, O.M. (1984). Complex decision-making at the organisational level. In *Handbook of Work and Organisational Psychology*, edited by P.J.D. Drenth, H. Thierry, P.J. Willems and C.J. de Wolff (Chichester: John Wiley), pp. 831–854.

Megaw, E.D. (Ed.) (1987). *Contemporary Ergonomics 1987*, Proceedings of the Ergonomics Society's 1987 Annual Conference, Swansea, Wales (London: Taylor and Francis).

Mohrman, S.A., Ledford, G.E., Lawler, E.E. and Mohrman, A.M. (1986). Quality of worklife and employee involvement. In *International Review of Industrial and Organisational Psychology 1986*, edited by C.L. Cooper and I.T. Robertson (Chichester: John Wiley), pp. 189–216.

Perrow, C. (1983). The organisational context of human factors engineering. *Administrative Science Quarterly*, **28**, 521–524.

Pfeffer, J., (1981). *Power in Organisations* (Boston: Pitman).

Pugh, D.S. (1984). *Organisation Theory* (Harmondworth: Penguin).

Sell, R.G. (1986). The politics of workplace participation. *Personnel Management*, **18**, 34–37.

Shipley, P. (1987). The management of psychosocial risk factors in the working environment: UK law compared. *Work and Stress*, **1**, 43–48.

Shipley, P. and Harrison, R.G. (1974). The ergonomics practitioner as change agent. Unpublished paper to the Ergonomics Society Annual Conference, St. Johns College, Cambridge, (available from P. Shipley at Birkbeck College, Malet Street, London WCIE 7HX).

Teiger, C. and Laville, A. (1987). How ergonomic methods can be used by non-ergonomists. Paper to the *International Occupational Ergonomics Symposium: Applied Methods in Ergonomics*, Zadar, Yugoslavia.

Thorne, C. and Russell, D. (1987). On the till: worker's perceptions of health and safety in supermarket checkout design. In *Contemporary Ergonomics 1987*, edited by E.D. Megaw (London: Taylor and Francis), pp. 352–356.

Vroom, V. (1984). A normative model of managerial decision-making. In *Organisation Theory*, edited by D.S. Pugh (Harmondworth: Penguin), pp. 256–276.

Vroom, V.H. and Yetton, P.W. (1973). *Leadership and Decision-Making*. (Pittsburg: University of Pittsburgh Press).

Warrian, P. (1988). Changing Structures, Changing Strategies. *QWL Focus*, **6**, 11–15.

Winterton, J. (1987). New Technology and re-designing work. In *Contemporary Ergonomics 1987*, edited by E.D. Megaw (London: Taylor and Francis), pp. 365–371.

Annotated readings

Ottaway, R.N. (Ed.) (1979). *Change Agents at Work* (London: Associated Business Press).

This small volume consists of contributions from eight 'change agents' or practitioners; as trainers and as consultants. Their problems and achievements in changing organizations in actual practice is the volume's focus, not organizational change in abstract. A number of the contributors are women. Mary Weir of the Manchester Business School, for example, shares with us her experiences and reflections of an action research project commissioned by the Work Research Unit, which involved her in improving work life in an American factory based in Scotland, the goal of which was the encouragement of "human development according to a set of criteria which would be defined by the people themselves". (p. 86) What follows is an account of her largely successful intervention to build a "self-maintaining learning process" (p. 91) in that organization.

This was done through assuming the role of learner herself as well as educator and facilitator, and by helping promote appropriate structures, such as work groups and a work improvement committee; formal but not rigid entities capable of working independently after her exit, continuing the process of organizational learning and change. Traditional measures, such as absentee rates and turnover levels, she reminds us, help us very little, and the generation of baseline data is often important but by methods more valid than the traditional unilateral questionnaire method. Talking to people, once you have established trust and credibility, is far more effective, i.e. the 'tried and true' method of everyday exchange in folk communities.

The gains were lasting: "everyone cared enough to want to make it work and not let the Project fade through apathy or disillusion", (p. 99); the costs were "far more time and energy than the casual observer would imagine". (p. 101) Both action researcher and collaborative participants have to be ready to give real commitment to it.

Argyris, C., Putnam, R and McLain Smith, D. (1985). *Action Science: Concepts, Methods, and Skills for Research and Intervention* (San Francisco: Jossey-Bass).

In a personal communication to me Frank Blackler felt that a clearer idea should be conveyed of how ergonomists, with their 'finely developed set of skills' might begin to serve their clients in a more participative way. He reminded me of this book, in which change agents are termed 'action scientists'. The book is valuable because it gives a good analysis of the grooved thinking into which we professionals are so readily prone to fall, but, more importantly, it gives substantial practical guidelines.

To quote Frank again: "people have real difficulty in being creative in their thinking". . . and. . . "Blocks to the imagination may. . . be as significant as power plays". . . I agree with this, and I think Argyris and colleagues do a good job in suggesting ways of unblocking professional minds through their 'maps of social action' and normative suggestions ('rules') for helping clients to change by unblocking their minds too, instead of raising their psychological defences.

A concrete example is the 'Easing-in Script': "a description of the sequence of expectancies and moves hypothesised to be in an actor's head when he produces that sequence." (p. 251) These practitioner tools can be used descriptively or prescriptively. The authors suggest that their script for combining enquiry with advocacy is normative because it puts forward an alternative set of suggestions or expectancies for the client to consider and accept or reject. Note the word 'script' (actor's *talk*), and the participative values underlying action science.

Their ideas are well-grounded in psychological theory and empirical findings. The action scientist is teacher, counsellor and facilitator within this normative theory of practice. A caveat though, is that this is an American text; the organizational power realities will be different from Britain and other countries (the context in which change agents work); so also will be the problems and opportunities for access because of the differences between cultures.

Chapter 33

New systems implementation

Ken D. Eason

Introduction

Technical change is a pervasive feature of organizational life. Enterprises devote a considerable proportion of their resources to the planning, purchasing and development of new technical systems to help them become more efficient and effective. At the end of each technical system development process there has to be an implementation phase when the ability of the new system is tested with real tasks and when the employees of the organization adapt to the changes in their working lives occasioned by the new system.

Implementation of a new technical system is a demanding time because different issues have to be addressed concurrently:

1. *Installing and testing the technical system.* It is one thing to have tested the system under experimental conditions, it is quite another to subject it to the full range and demands of everyday working conditions. There will therefore be many technical problems to overcome before the technical system operates as planned.

2. *Local workplace design.* If the system is to be used by many employees it will have to be installed in the workplaces of these employees. This may involve the creation of new workplaces as in a new factory, the redesign of existing workplaces as when the tills are modified in a bank to accommodate a terminal, or accommodating new equipment on an existing desk.

3. *Training and support.* People with new technical tools will need training to make good use of them. They may also need training in new ways of working. Training before implementation may have to be supplemented by support after implementation since it may take a considerable period of time to develop all of the skills necessary to cope effectively in the new situation.

4. *Organizational change.* Technical change of a substantive nature almost certainly goes hand in hand with organizational change which may require major adaptation from the employees in the organization.

5. *Acceptance of change.* All of the foregoing changes constitute a new and uncertain future for the employees in the organization and an important task in implementation is to develop strategies which will foster positive attitudes towards the change process by those to be affected by it.

It would be difficult to address all of these issues simultaneously under ideal conditions, but an additional factor often adds substantially to the problem. Most technical systems are introduced into organizations that are ongoing concerns and which have to maintain the volume and integrity of their activities throughout the process of change, such as serving customers and making products.

Given the range of issues to be addressed, it is perhaps not surprising that implementation of new systems is often problematic. It is quite common for systems eventually never to be implemented, or for the implementation process to become a very painful and protracted period. The reasons may be many. If, for the purposes of analysis, we disregard the possibility that the technical system proves ineffective or inappropriate (a major cause of problems in fact) we can examine the problems of implementation itself.

The problem that has been widely recognized since the days of the Luddites is resistance to change; people acting against technical change in an attempt to preserve their jobs, their livelihoods and the ways of working that they have known. Change inevitably brings uncertainty and unfamiliarity and may well bring undesirable consequences, so it is hardly surprising that people who consider themselves to be losers in the change process mobilize their resources to protect their interests. Resistance to change is often presented as a blind, irrational attempt to hold back progress, but it can be a well developed strategy with clear and defendable goals. In reviewing the implementation of information technology, for example, Keen (1981) identifies a number of 'counter-implementation strategies' which user groups may employ to ensure the outcome is in their best interests. These include 'keeping the project complex', 'minimizing the implementer's legitimacy' and 'exploiting their lack of inside knowledge' which are all ways of deflecting or limiting the efforts of those attempting to introduce the system.

Technical systems design is usually portrayed as a rational, objective process but, at implementation, the presence of resistance to change may reveal that the different 'stakeholders' (groups of people affected by the proposed system) have different goals and the process has also to be seen as one in which conflicts have to be resolved or negotiated.

Even when resistance to change is not a major inhibitor of change, the human and organizational systems may take a considerable period of time to adjust to the new circumstances. Change is frequently followed by an 'initial dip' in performance and work output, and it may be weeks or months before the improved levels for which the change was introduced are achieved in practice. In this period individual users may be developing and consolidating the skills they need for the new form of working and new organizational structures and procedures may be evolving. It may be possible to switch from one technical system to another overnight but the human and social systems, depending as they do upon human motivation and learning, are better described as changing by evolution.

Implementation is therefore a complex socio-technical change process in which technical developments are accompanied by a wide range of human changes. The most common reason for failure in implementation is that the process is treated as a technical problem, and the human and organizational issues go unrecognized and untreated. In this chapter we will examine strategies and tactics for implementation which focus upon the process as socio-technical change. In the next section we will consider overall strategies to cope with the different circumstances of implementation with particular reference to the roles of potential users. Subsequently the function of user involvement is examined as the principal mechanism by which the human issues of implementation can be managed.

Implementation strategies

Figure 33.1 identifies six major strategies used for technical system implementation and is derived from Eason (1988). The strategies are loosely arranged on a dimension from the most revolutionary to the most evolutionary. At one extreme we have the major start up of a complete new system overnight and, at the other, the steady introduction of technical facilities over an extended period of time.

The aim is to identify the main principles of each strategy and the circumstances in which it might be used. We will then subsequently consider

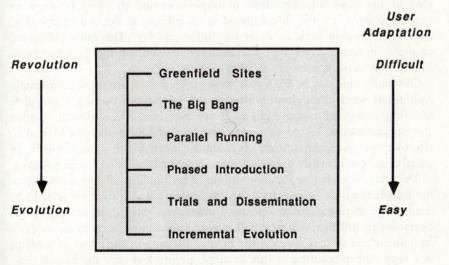

Figure 33.1. Implementation strategies.

the implications for the management of the human and organizational aspects of implementation.

The 'greenfield site' implementation.

This is when the new system is introduced in an entirely new situation to any previous work system, i.e. a new factory is opened on a different site, or a new shop is opened in a town. Technically the system can be developed without too many constraints associated with existing forms of technology. Organizationally there is also the opportunity to break with tradition and to explore new working practices. The problems for implementation are often that planning has to be done without the help of the people who will operate the system because they have yet to be recruited, and there may be extra 'teething troubles' because the staff are initially strangers to one another. This kind of change process can be greatly facilitated by the recruitment of key staff sometime before the implementation date, so that they can play a major part in detailed planning. Systematic attention also has to be given to team building exercises because new staff will be entering new roles with no pre-existing knowledge of one another to help them. Team building exercises provide people with the opportunity to explore the nature of their role, its relationships with other roles and to get to know the people who will occupy these roles. If these explorations can be made before the new system is implemented, there will be a strong basis for co-operation and team action when it is demanded under operational conditions.

The big bang

One of the most difficult kinds of implementation to effect is when an existing system is being discontinued in its entirety at the end of one day, and a new system replaces it on the following day. The most publicized example in recent years was the overnight switch of the London Stock Market to electronic trading in 1986.

Obviously this can be high risk strategy because it attempts to maintain continuous work throughout whilst, at the same time, making a complete technical change in a single phase. Every part of the new system, human and organizational, has to be functioning well if this strategy is to lead to effective operation immediately. It is almost inevitable that there will be an initial dip in performance before smooth, integrated operation can be achieved.

The chances of failure can be minimized by holding off-line trials before the launch not only to test the technical system but also to allow staff to be trained and to rehearse new operating procedures. However, scaling up the operation to full line working will inevitably throw greater strain on both equipment and staff. Provision for manual back-up in the event of mishaps is a very wise precaution in this strategy. Before and after the launch date, there will be a need for extra resources. The need for specialist resources such as technical staff and trainers is expected, but what often comes as a surprise is the need for additional user resources. Since it is often the intention to run the new system with fewer staff, management may be surprised and resistant to the idea that extra staff are needed during the transition period. If normal work is to be sustained, staff are to be trained and be involved in off-line trials, overtime or additional resources will be required if the stress of overload is not to be added to the inevitable stress of change. Some large organizations, who accept that change is the norm rather than the exception, maintain a team of staff who can go into user departments and take over routine duties whilst the local staff prepare for change.

Given the problem of integrating the many faceted aspects of a complex process in a single implementation phase, there need to be good reasons for adopting this approach. One such reason is that the work system needs all the facets working together to be effective; a phased approach cannot be adopted. If, for example, all user sites need to be given simultaneous, real-time access to a new data base, a 'big bang' strategy will be needed. Another reason is that the strategy ensures the change has a high profile and there can be no loss of momentum in the change process which is a characteristic of some of the other strategies. Finally, the opportunity of a complete change means that organizational changes can be made as well as technical change.

Parallel running

One popular way of minimizing the risks to the ongoing workload is to introduce the new system alongside the old one, and to run them in parallel until everybody is confident that the new system will be effective. When the maintenance of the quantity and quality of existing work is of the highest priority, this is the most popular strategy.

Whilst this strategy provides an insurance policy, it is not without costs. The common way of operating is to perform the work by the old system and then to repeat it with the new system. This will inevitably involve extra work and will constitute an additional strain on staff unless extra resources are provided. The process of using the new system for no real purpose can also produce negative attitudes amongst users. If the system keeps failing, they may not develop confidence in it and may cling to existing procedures. If it works well and saves a lot of time and energy, it can be very frustrating to continue using the old system. In the freightforwarding case study described by Shackel et al. (1989), the new system provided a way of capturing data once and using it in many ways whereas the old system required users to repeat the data capture for every use. After a short period of parallel running the staff pleaded to be allowed to use the new system for real tasks.

If parallel running is used as a strategy, it needs an agreed programme of tests so that everybody can see the progress that is being made towards the switch to the new system and can participate in the decision making.

Another problem with parallel running is how to make organizational changes. It is difficult enough for staff to operate two technical systems simultaneously; it is still more difficult for them to operate in two organizational structures simultaneously. There is a tendency to keep the existing organizational structures and procedures during parallel running which may inhibit the search for structures which can take best advantage of the new technical system.

Phased introduction

The problems of making massive changes can be eased by phasing in the changes over a period of time. There are two ways in which large scale changes can be subdivided to facilitate phased introduction. First, the functionality of the technical system can be introduced in phases so that the basic task processes can be supported in the early phases and subsequently facilities can be added which support, for example, the management and development tasks. Secondly, it may be possible to introduce the system in different parts of the organization at different times. A combination of these two approaches can also be used.

There are several advantages to introducing functionality in stages. It means that users do not have to master a very sophisticated system all at once but can engage in a progressive learning process as the system is gradually implemented. It also means that system implementation can start before the whole system has been designed. Indeed it may be possible for early user experiences with the system to shape the development of the facilities that are to be introduced later. This point will be elaborated further in the section on evolutionary design. A final advantage is that an explicit policy of phased introduction is in fact a recognition of an almost universal experience; even when a 'big bang' strategy is adopted, the system is never finished and revisions, elaborations and refinements are continually being added.

An important consequence of adopting this strategy is that decisions have to be made about what parts of the system to deliver first. There will be a tendency to deliver the foundation parts of the technical system in the first phase. If it is an information technology system, for example, the data bases may be designed together with the data capture procedures and only a few of the major output facilities may be provided. The danger of this approach is that the first experience the users have of these systems is that they demand input whilst giving very little of value in return. Another possibility is that all the useful outputs go to one group of users whilst the input load falls on others. Neither of these strategies is likely to promote a good user reaction. A good planning rule is that phasing should ensure that each group of users receives a service they value early in implementation to provide a positive experience of the system upon which later phases can build.

The alternative to introducing functionality in phases is to introduce the complete system in one site after the other. This is a popular strategy in organizations which have similar functional units spread over a geographical area, for example, a chain of shops or branches. In this strategy an implementation team can be developed which becomes very skilled at handling the implementation issues because they encounter them many times over. There is, however, the danger that they will forget how unfamiliar these issues are to the staff next on the list to make the change and they may expect implementation to be progressively faster with each new site they visit. The trainer for whom everything is 'obvious' and 'common sense' often serves only to increase the would-be users' sense of failure and frustration.

Trials and dissemination

This strategy explicitly recognizes that there will be 'teething troubles' when a new system is introduced, by holding a major trial before embarking upon full scale implementation. Unlike prototype testing which is often used to revise the specification for a system, the trial is usually undertaken with the technical system it is planned to implement. The purposes of the trial may be many; to test whether the technical system can do the job, whether it is robust and reliable and so on, to establish the training and support that is necessary, to examine the organization changes that are needed. To learn such lessons it is necessary to make the trial as realistic as possible. Ideally, a representative user community will use the new system for real work and thoroughly evaluate the implications, although off-line trials may be necessary when the maintenance of work standards is critical. (Christie and Gardiner look at ways of undertaking full scale trials of systems in their chapter on evaluating human computer interfaces.)

This strategy provides a valuable opportunity for the user organization to thoroughly prepare itself for implementation before it is in the throes of full scale change. However, many organizations fail to make good use of the opportunities that the trial presents. They may, for example, choose as a trial site one in which the staff are positive in their attitudes and which is relatively isolated from other sites so that the change can be made without knock-on effects. These factors may make it relatively easy to implement the change but it should come as no surprise that it is much more difficult to make widespread changes subsequently. A related problem is the 'Hawthorne effect' (Rothlisberger and Dickson, 1939). This is the situation where change proves very effective in terms of productivity not because of the change itself but because the staff are 'in the spotlight' and respond to the extra attention they are receiving. This factor is often regarded as a confounding variable in experimental designs which compare the results of a change group with a control group where no change is introduced. In implementation it is a positive force and displays the motivational effects of user participation. The danger in a trial and dissemination strategy is that the staff in the trial get a lot of attention compared with the staff in the later dissemination. This factor, allied to the tendency of the implementation team to expect each successive implementation to go faster, can turn a very positive reaction in the trial into a very negative reaction to the wider dissemination.

Another typical failing is not to learn the lessons of the trials. Too often thorough evaluation is not undertaken and, as a result, detailed faults in the technical system are not identified and corrected, training regimes are not tailored to the needs of the users, and organizational changes are not planned. The trial simply becomes the first implementation and makes this strategy indistinguishable from a phased introduction across sites.

Incremental implementation

The logical alternative to a revolutionary change is gradual evolution. The advantage of such an approach is that users are never confronted by major change but have only to cope with small incremental steps. Ideally the learning they achieve from early steps can help them to select the next stages of the evolution. If this is achieved, the strategy may truly be called user led or user centred.

The growing sophistication and flexibility of technology is making the incremental implementation of systems an increasingly practical proposition. It is already a good characterization of the method of technical change used to implement technical tools for managers and professionals. When the user community is discretionary and powerful and their requirements are varied, not to say idiosyncratic, the service has to be tailored to individuals and incremental implementation may be the only option that is acceptable to users.

It is useful to distinguish between *ad hoc* and planned evolution. *Ad hoc* evolution is simply introducing technical change bit by bit with no overall objective to guide development. At its worst, it allows systems to 'grow like Topsy' so that the overall structure has incompatible parts, early stages are rendered redundant and later developments are forced into ineffective routes because of early decisions. To avoid this situation requires the identification of broad requirements and the planning of an overall system architecture and organizational structure which can accommodate many ways of meeting these requirements. Within such a structure, it should then be possible for individual users to follow an evolutionary route adding incremental parts to their own services to suit their requirements and their rate of learning.

A user's view of implementation strategies

The six strategies listed in Figure 33.1 form a range from the most revolutionary to the most evolutionary. Viewed from the perspective of the potential user, some are more daunting to cope with than others. The 'greenfield' and 'big bang' strategies are clean and comprehensive but mean the user has a lot of changes to cope with in a very short time. If the user has to cope with normal business at the same time, this kind of change can become a major strain even when the users are positively in favour of the change. 'Parallel running' and 'phased implementation' are safer and slower ways of introducing major changes in circumstances where normal work must be maintained. 'Trials' and 'incremental implementation' provide early opportunities for users to experience the technical change and can enable them to use this experience to guide the later stages of development. The more evolutionary forms of implementation therefore give users more opportunity to cope and to contribute to the form of development.

User involvement in system implementation

Whatever the strategy that is followed, a crucial factor affecting the success of implementation is the degree to which potential users are involved in the process. At worst, the users are confronted by a completely determined system which may have negative effects on their working lives and which they are expected to master very quickly whilst maintaining their normal work output. In these circumstances, which are not uncommon, even the most well designed system will run into implementation difficulties.

To avoid this situation, mechanisms for user involvement are necessary. Evolutionary developments give more opportunity for these mechanisms but they are possible and necessary in all implementation strategies. Figure 33.2 identifies the major mechanisms by which user involvement can be effected.

The first requirement is to establish a temporary organization structure which can carry the organization through the change. This structure will charge individuals with the variety of responsibilities that are necessary to implement the change. In many situations, these roles will be taken by professionals in the change process, i.e. technical designers, trainers and ergonomists. It is important, however, that many of the roles are taken by potential users.

It is, of course, necessary for users to participate in the design process from the outset to ensure, for example, that the specification meets the requirements of the users. In a large scale development, it is likely that only a few users can actively participate in the design phases. When it comes to implementation, however, everybody will be affected and therefore a mechanism is needed to involve everybody. Shipley looks at the ideology and methodology of participation as a general issue in this book. In a number of implementation strategies, we have found three concepts of value in treating temporary organizational structures.

The user 'champion'

It is important to involve a senior user manager in the implementation process, if possible vesting responsibility for implementation in the manager.

Temporary Organizational Structures for Managing the Change Process	
Technical Roles	User Roles
Technical System Implementors	Senior User Champions
Trainers	User Representatives as Local User Support
Ergonomists	Local User Decision Making

Figure 33.2. Temporary design structures for system implementation.

This will clearly identify the process as user driven and will announce the commitment of the user organization to the development. It will also recognize that, whilst there are technical issues to consider, the primary problems in implementation are human and organizational adaptation.

User representatives as local implementors

We will assume that, from the beginning of the system development, a small number of users have been involved in the process to represent user interests. If these people have been chosen to represent the major user groups they can play a significant role during implementation. By this time they are the users who are most knowledgeable about the system plans. They are in an excellent position therefore to help plan local implementation; the start-up procedures, training, workplace changes etc. They may also come to play a significant role subsequent to implementation as the principal sources of support to users learning to cope with the new system. The combination of user experience and system knowledge can mean user representatives are much more able to understand and resolve the problems experienced by users than are technical experts.

Full user involvement in local implementations

If resistance to change is to be avoided it is necessary to involve all potential users in the change process, not merely a selected few. In a large implementation it is difficult to involve everybody in the strategic decisions but there are many local decisions in which everybody can participate; the timing and sequence of implementation, the extent of parallel running, allocations of duties to work roles are examples of such decisions. Working under the direction of local management, user representatives can play an important role in organizing working parties of local users to plan and undertake these activities. It is important to note that involvement of this kind gives people considerable influence over the decisions that affect them personally, and it is this kind of influence which most successfully combats feelings of external threat. For example, Wall and Lischeron (1977) demonstrated that whilst people liked to feel their interests were represented in strategic decision making, they wanted to be personally involved in the decisions that affected their daily work.

User oriented implementation topics

The topics that the temporary organizational structure must address during implementation relate to both technical design and organizational change and some of the most important topics are listed in Figure 33.3. The most

Technical Design	Organizational Change
Minimum Critical Specification	Training and Support
Customisation	Job Design
Participative Work Place Design	Team Building
Trials, Evaluations and Evolution	

Figure 33.3. Topics in user-oriented implementation.

important feature of the technical design is that the earlier stages of development should have left considerable freedom during implementation for people to shape the system to meet local requirements. Involving people in a design process is not sufficient in itself to overcome resistance to change. It is important that their participation leads to decisions about significant issues and that can only be achieved if the technical system leaves options. In listing the principles of socio-technical design, Cherns (1976) cites 'minimum critical specification' as the vehicle which provides for discretion during implementation and beyond. This is a requirement laid on designers to specify only that which has to be specified and to leave open as much as possible for later decision making. Effective implementation can depend heavily upon the flexibility and adaptability built into the technical design so that it can be 'personalized' or 'customized' to local needs at the point of implementation. One encouraging feature of the increasing sophistication of the computer technology which underpins most new systems, is the degree of flexibility and adaptability that can be provided. Whereas a few years ago, for example, a user might be restricted to a standard printout provided for a multitude of users, a system might now include a report generator, a tool which permits users to define their own reports. The provision of such tools within systems means that users, during implementation, may well find they have major areas of discretion and therefore important decision making tasks to undertake.

Another area which requires detailed local design constitutes one of the traditional areas of ergonomics. The integration of a new technical system within a workplace will involve issues of workstation layout, furniture selection and environmental design. In most large scale applications with a diverse user population there will be many different workplaces which require attention. There are, of course, a wide variety of standards, guidelines, data tables, techniques and tools available to support workplace design; see, for example, Grandjean (1986). For many ergonomists the principal contribution of the discipline to implementation is workplace and environmental design and it can be a major contribution to the subsequent health and efficiency of the work-force. (Many chapters in this book are indeed devoted to such contributions.) It is important to note, however, that if the practitioner approaches this role as the expert who designs all the workplaces for the implementation, the effect may be the same as the technical specialist who completely designs the technical service for the would-be users. The work-force may react negatively because, however well intentioned the specialist may be, decisions are being made *for* the users and not *by* them. Gower and Eason (1989) have provided a case study of office automation in which participative workplace design was employed. Secretaries were helped to select furniture, plan office layout and so on, with the ergonomists providing advice on relevant standards and guidelines whilst the secretaries provided information on local requirements, customs and culture.

Finally, with respect to technical design, in the implementation strategies that recognize the need for evolution there will be a need to conduct realistic trials. Users in trials need to be treated, not as subjects in an experiment, but as participants in the creation of their future work environment. This means that users not only get to use the future system but plan the trial, collect evidence about its value and its deficiences, and work with technical specialists on the future development of the system.

In addition to the decisions that relate to the technical design, there are in any socio-technical change process a wide range of human and organizational changes to be considered. Figure 33.3 also lists some of these topics. Whereas for technical design there are likely to be specialists available who will see it as their responsibility to be concerned with technical aspects of implementation, there are not likely to be specialists in organizational change available and these responsibilities will fall upon local management and work-force. One consequence of this is that these are topics which frequently are largely unrecognized and are dealt with in an *ad hoc* manner, as and when it becomes evident they must be given attention. The exception is training because it is widely recognized that potential users need to be able to understand and operate the new technical systems that are being provided. Too often, however, the training is limited to technical knowledge when much of the adaptation that is required relates to changing role responsibilities and relationships. There is also a tendency to attempt 'one shot' training before implementation in a single intensive burst, when many users can cope much better with a series of short training sessions spread over a period of time, combined with access to a range of 'point of need' support facilities (e.g. manuals, prompt cards and liaison staff).

Job design issues most clearly need attention in 'greenfield' and 'big bang' implementation strategies, when full scale change is being attempted in a single phase. There are a number of examples in the literature of procedures for job design in these circumstances, for example, Aberg (1981) in manufacturing industry, and Mumford (1983) in office automation. The most important requirement is that the technical system should have the flexibility to support a range of job design options so that participating users can choose a structure to which they can commit themselves and which will be effective. Eason and Sell (1981), for example, cite a case in which an order entry system was sufficiently flexible to permit branch offices around the country to allocate tasks to jobs in quite different ways and still make effective use of the technical system. One difficulty user groups often experience when planning for technical system implementation is that they cannot foresee the implications for job structures. Running a live trial offers a very effective way of revealing the implications, and it can be used to experiment with different job structures and test the flexibility of the technical system.

Job design issues take a different form in parallel running and evolutionary forms of implementation. Often there is no plan to change job structures but, as the technical system begins to be exploited, people begin to expand their job horizons and to encroach upon the work of others. The shop floor employees may, for example, begin to encroach on the role of the foreman or the inspector whilst the manager, by doing some of the typing, may encroach on a secretary's role. These factors can cause a gradual erosion of the existing job structures, often with poor relations developing between role holders. In the office automation case study described by Gower and Eason (1989) an evolutionary implementation strategy was followed. Regular monitoring audits were conducted to check for job design 'drift' in order to bring it to the attention of senior staff. By this mechanism emerging *ad hoc* practices were assessed and those job changes that were beneficial were formally recognized and implemented.

It will be apparent that, where major organizational changes are taking place, people will find themselves entering into new work relations with their colleagues. In the 'greenfield' situation they may be entirely new colleagues, whereas in the other strategies it may be the same people in different roles. In each case there is a need to help people appreciate not just the new technical system but also the new social system that is in operation. Team building in which people explore the new responsibilities and the relationships as well as getting to know how their encumbents may interpret them, is a very useful technique for preparing for implementation. It is particularly necessary in 'greenfield' and 'big bang' implementations because people will be expected to slot straight into the new job structure. Again the concept of running a trial can be used as a way of building a team understanding ahead of the time when it has to become an operational reality.

Conclusions

Implementation procedures can make or break a technical innovation by the way potential users are introduced to the innovation and the extent to which they have an opportunity to shape the new system to meet their requirements. Different implementation strategies require different treatments. The evolutionary patterns offer more time for user learning and reflection and therefore give more opportunity to influence the subsequent course of events. The problem in these cases is often to sustain the impetus of the implementation because it will take time and may become fragmented. In the strategies that have major implementation points, much more needs to be done before implementation if the users are not to be overwhelmed. In these cases there must be user involvement during planning and realistic trials are needed to minimize the shock of the implementation.

The need to involve users in planning and decision making for implementation raises an interesting question about the appropriate role for the ergonomist in this process. There are many areas in which the ergonomist could contribute directly by making 'design' decisions, for example in workstation layout, environmental design, and manuals and training. It is essential, however, to ensure that the ultimate users play their part in these processes. There is also a much more challenging role the ergonomist can play. As one of the few disciplines that can take a socio-technical view of implementation the ergonomist could assist in the establishment of an implementation strategy which facilitates organizational change and human learning as well as technical change. The ergonomist can also assist in the creation of the temporary organizational institutions that are necessary to manage the transition from old to new. Within this structure the ergonomist should be able to identify the most effective way to make direct contributions. In short, where the users have not faced this kind of change before and the technologists see technical issues as the major priority, many of the issues addressed in this chapter will not be effectively handled unless someone with a socio-technical perspective is able to facilitate the construction of a process of implementation in which the users can play a full part.

References

Aberg, U. (1981). Techniques in redesigning routine work. In *Stress, Work Design and Productivity*, edited by E.N. Corlett and J. Richardson (Chichester: John Wiley), pp. 157–164.

Cherns, A.B. (1976). The principles of socio-technical design. *Human Relations*, **28**, 783–792.

Eason, K.D. (1988). *Information Technology and Organizational Change* (London: Taylor and Francis).

Eason, K.D. and Sell, R.J. (1981). Case studies in job design for information processing tasks. In *Stress, Work Design and Productivity*, edited by E.N. Corlett and J. Richardson (Chichester: John Wiley), pp. 195–208.

Gower, J.C. and Eason, K.D. (1989). The introduction of information technology in a city firm. In *The Application of Information Technology*, edited by S.D.P. Harker and K.D. Eason (London: Taylor and Francis).

Grandjean, E. (1986). *Ergonomics in Computerized Offices* (London: Taylor and Francis).

Keen, P. (1981). Information Systems and Organizational Change, *Communications of the ACM*, **24.1**, 24–33.

Mumford, E. (1983). *Designing Secretaries* (Manchester: Business School).

Rothlisberger, F.J. and Dickson, W.J. (1939). *Management and the Worker* (Cambridge, MA: Harvard University Press).

Shackel, B., Eason K.D. and Pomfrett S.M. (1989). Organizational prototyping—a case study in matching the complete system to the organization. In *The Applications of Information Technology*, edited by S.P.D. Harker and K.D. Eason (London: Taylor and Francis).

Wall, T.D. and Lischeron, J.A. (1977). *Worker Participation* (London: McGraw-Hill).

Chapter 34

Methods in context

E. Nigel Corlett

The method and the meaning

Which of us has not been aware, at some time in our professional lives, of a solution chasing a problem? The temptation to see problems from the standpoint only of certain techniques is one of which we must always remain aware. The measurements we make constrain our understanding of the problem and where we have a bias towards certain methods it is almost inevitable that we shall have limited our understanding of the problem.

Since the above argument suggests that we have to understand the problem before we choose the methods it leaves us in something of a quandary. If we understand the problem we don't always need to do the study, but if we don't then how do we choose a method? The quandary is more a literary device than reality, however, since the argument just emphasises that experimentation for problem solution is an iterative as well as an intellectual exercise. We choose methods according to our understanding but, as the results then give us additional information, we are in a position to re-assess the utility of our methods and re-work the investigation in other ways. We try, from which we learn, and then we try again, benefitting each time from a critical analysis of our results and their meanings.

Of course, as we gain experience we have a greater ability to see the likely shape of a problem, and become more adept at choosing methods which are appropriate for its investigation. We know the problem field better and we have a better understanding of the likely effects arising from the situations we are studying. However, with increased familiarity we can become biased. We use favourite methods or use favoured models of the situation, without sufficient initial thought.

People who move into ergonomics from another discipline can also fall into this trap at the start, seeing a problem in limited terms. If their work is presented only to a specialist group with similar backgrounds, the bias is reinforced and leads to a body of knowledge purporting to cover ergonomics issues but in reality providing a sub-optimal coverage of the problems. Terms can become misused and distorted, such as 'optimal' being equated with 'minimum', 'user friendly' representing childishly chatty dialogue, or 'efficient' being applied to an increase in short term output but not related to the human consequences of that output.

Contributing to knowledge

These are all very normal behaviours for people. We are not suggesting any evil intent or stupidity, but if we are aware that they are likely to occur, we must counteract their effects. Their likelihood gives good reason for the sorts of activities in which academics indulge but which commercial people sometimes feel are not productive. These are attendance at conferences, the widening of interests beyond the confines of one's own studies to see what others do, and the writing and publishing of papers.

Conferences present an opportunity, which we do not always take, to hear about a wider spectrum of work than just our own interests, as well as to hear of work in our own area done by people who have taken a different approach. Different approaches may lead to different conclusions, which is when we can gain increased insights into our own work.

To take an interest in other areas introduces us to different techniques, or methods used in different ways. Results, or their interpretation, in one area can illuminate our own work in another. Not least, studying people can be different from studying inanimate matter as the interactions within a person can be relatively unknown or only inferred. Thus considering other areas of work can help us to appreciate this complexity and, hence, to interpret better our own investigations.

The third leg in our support structure for better investigation is publication. Many people outside the academic world see this as an indulgence, the gathering of a list of publications for promotion purposes. Whilst this does happen it will be recognized that good publications demonstrate a person's contribution to the subject and, at the same time, their competence in the discipline. The reader will note the qualification 'good' in the previous sentence! All of us rely upon published contributions; we quote them not to demonstrate our own erudition, or even to convince others that we are right, but to outline the context of our own work and the foundations on which it is built. If we take our subject seriously, it is necessary that we describe to others the work we do, so that they may benefit from our experiences, use our knowledge and avoid repeating our mistakes. No matter how much the commercial world may financially reward an ergonomist, professional quality is established by the evidence of results publicly judged by our peers, as well as by the contributions we make to improving the capacity of our discipline to achieve its purposes. In particular, the commercial world which benefits from the discipline has a major responsibility to support its development, by allowing its professionals to take part in professional activities. This responsibility cannot be shirked by (sometimes spurious) claims of commercial confidentiality or by insistence on the need for a close attention to maximising its own profitability.

Presentation and purposes

It could be argued with a good deal of justification that a book on methods should have had a section on presentation of results. Although the editors had some sympathy with this there is obviously a limit. A cogent argument for including presentation is in the area of interpretations. Statistical or other forms of analyses are only the first step, for it is what we make of the analyses which counts.

What we analyse depends to a great extent on what we have measured but many workers will have had the experience of the gradual illumination and understanding which comes from poring over the data; how new arrangements of the data demonstrate a better understanding of their meaning and how different ways of presenting the findings clarify the relationships. The recommendation to decide on the analysis and significance levels *before* starting the measurements is sound, particularly as the experimental design quite often defines the analysis method, and it should be decided beforehand if the method adopted is relevant to the effects being studied. However, this should not prevent workers from turning over their results in their minds and wondering, for example, why the dimensions they have measured are as they are, or whether there are yet other factors which will contribute to an understanding.

When an eminent scientist, many years ago, broadcast a talk on British radio on 'the fraud of the scientific paper', he pointed out that the standard sequence in a paper, of literature review, experimental design, experiment, analysis and conclusions, was not how science tended to be pursued. More usually, a question arises in the worker's mind, which is turned over and speculated upon until a means for trying it out is developed. It is then tried out, to 'see if it works', and then tested again and again, interspersed with thought and discussion, to understand its mechanisms and its boundaries. Good experimental and analytical controls are exercised throughout, but only at the writing-up stage is the conventional, logical sequence exposed which implies that the whole activity was the natural outcome from some previous stage.

The whole process, from idea to presentation, is one where a worker's curiosity must be alert. 'Why have these things happened?', 'What happens if. . .?', 'Is what has happened caused by what I think it is?' are the questions which should always be in a worker's mind. Although much work is inevitably narrowly focused, this should not necessarily restrain speculation about effects. For example, studying the likely problems occurring in routine assembly of electronic circuits on a 20 s cycle should also embrace the likely effects on the worker of the focused attention and suppression of internal or external mental activity needed to achieve consistent error-free production. Carpal tunnel syndrome and local muscular fatigue are not the only responses to this kind of repetitive work.

Cost-benefits?

The question of 'benefit' is a further matter for concern, since a narrow view of what it means can inhibit or distort the investigation of ergonomics problems. Quite often, in reported cost-benefit analysis studies, the benefits given are limited to the additional surplus on the direct activities accruing from the changes. Although this is undeniably a benefit, it is a limited interpretation of the word even for its everyday use. This is not to say that benefits should not be expressed in monetary terms, but that some effort should be made to express, on a common metric which is usually money, the many other gains from the ergonomics changes. As yet, our abilities in this respect are limited but a recent review (Corlett, 1988) has indicated that there are methods which can be pressed into service and developed in this area.

Developments in cost-benefit analysis are important because they will allow us to give a truer measure of the value of our work. 'Making workpeople comfortable', when expressed like that, can seem to have low priority in many studies; nice to do but hardly essential. To a great extent this is because we can put numbers on some things related to performance, but not on the effects of discomfort; we cannot quantify its costs. So it is not visible in the decision equation.

From a community health viewpoint, to take but one perspective, there is extensive evidence of the outcomes of situations which are recognised in their early stages through the presence of discomfort. Because these consequences take a long time to develop, or are paid for by others than the employer, are not good reasons for seeing them as less important than performance in the short term. Their visibility, and consequent recognition and incorporation in the criteria for designing investigations will arise in great part from a quantification of the costs of their existence. A consequence of this development would be a change in the methodologies for many studies, which would incorporate techniques for cost-benefit analysis as a normal component in evaluation.

What are we working for?

By looking at the wider effects of the interactions between people and their environment we are inevitably brought face to face with some other questions. These are not new questions; many people face them in the course of their professional lives, but they are quite fundamental to our ergonomics activities, to the interpretation of our results and to how we propose they should be implemented.

If our objectives include, for example, improving the effectiveness of human performance, do our measures of effectiveness stop at, say, speed and errors? Can we accept such a limited definition in every case? For whom are we working? Some employers would be happy if we accepted this limited definition, and in some cases of course it would be quite sufficient; but if there is anything in the claims by ergonomists to be a profession, then there are evidently many cases where a limited perspective on performance is quite insufficient.

To return to an earlier point, this is where the wider contacts and interactions within professional activities are important. The broadening of understanding about the whole person which comes from such widened contacts, together with discussions with one's peers, help to maintain a human-centred view of ergonomics work which maintains ethical boundaries appropriate for a profession which has people as its central interest. If misused, our profession has the means, within it, for the ruthless exploitation of others and unfortunately there are always a few people in the world who would wish to use this knowledge in such a way. It is, of course, the duty of every professional to block such a process and to see that there are true benefits for the subjects of our studies.

These are ethical matters, much influenced by the individual's personal beliefs and values. It has been found in many of the human sciences that it is not sufficient to leave these matters unspecified and rely on the assumption that all people will gravitate to the same point of view in caring for others. Professional associations, universities and others have drawn up codes of conduct to be observed when their members practise their profession. There are also codes of ethics for observance by society members in their professional activities which involve human subjects.

As examples of these, the code of practice presented to applicants to go on the Professional Register of the Ergonomics Society is given (Figure 34.1), as well as the British Psychological Society code of professional ethics for research with human subjects (Figure 34.2). Where readers have no sources of their own, we commend these to them for guidance in their decisions.

---

1.   In pursuit of their profession, those on the Professional Register of the Ergonomics Society shall at all times value integrity, impartiality and respect for evidence, and shall sustain the highest ethical standards.

2.   Within their obligations under the law, they shall hold the interest, safety and welfare of those in receipt of their services or affected by those services to be paramount at all times. Those carrying out reasearch shall safeguard the interests of participants, and ensure that their work is in keeping with the highest standards of scientific integrity.

3.   They shall endeavour to maintain and develop their professional competence, and to recognise and work within its limits, striving always to identify and overcome factors restricting this competence.

4.   They shall not lay claim, directly or indirectly, to have competence in any area of ergonomics in which they are not competent, nor to have characteristics or capabilities which they do not possess.

5.   They shall take all reasonable steps to ensure that their qualifications, capabilities or views are not misrepresented by others, and to correct any such misrepresentations of which they become aware.

6.   If requested to provide services outside their personal competence, or if they consider that services of such nature are appropriate, they shall give every reasonable assistance towards obtaining such services from those qualified to provide them.

7.   They shall take all reasonable steps to ensure that those working under their supervision act in concordance with this Code of Conduct.

8.   They shall refrain from making misleading, exaggerated or unjustified claims for the effectiveness of their methods, and they shall not advertise services in a way likely to encourage unrealistic expectations about the effectiveness and results of those services.

9.   They shall take all reasonable steps to preserve the confidentiality of information acquired through their professional practice or research, and to protect the privacy of individuals or organisations about whom information is collected or held. Subject to the requirements of the law they shall prevent the identity of individuals or organisations being revealed without their expressed permission. When working in a team or with collaborators, they should inform recipients of services or participants in research of the extent to which personally identifiable information may be shared between colleagues, and will ensure as far as lies within their powers that those with whom they are working will respect the confidentiality of the information.

10.   With the exception of recordings of public behaviour, they shall only make sound or visual recordings of recipients of services or participants in research with the expressed agreement of the individuals or their representatives both to the recording being made and to the subsequent conditions of access to the recordings.

11.   They shall conduct themselves in their professional activities in ways which do not damage the interests of the recipients of their services or participants in their research and which do not undermine public confidence in their ability to perform their professional duties.

12.   They shall neither solicit nor accept from those receiving their services any significant financial or material benefit beyond that which has been contractually agreed, nor shall they accept any benefits from more than one source for the same work without the consent of all the parties concerned.

13.   They shall not allow their professional responsibilites or standards of practice to be diminished by considerations of religion, sex, race, age, nationality, class, politics or extraneous factors.

14.   Where they become aware of professional misconduct by a professional colleague that is not resolved by discussion with the colleague concerned, they shall take steps to bring that misconduct to the attention of the General Secretary of the Society, doing so without malice.

Figure 34.1. Ergonomics Society current Code of Conduct for those admitted to the Professional Register.

Psychologists are committed to increasing the understanding that people have of their own and others' behaviour in the belief that this understanding ameliorates the human condition and enhances human dignity. These ethical values must characterize not only applications of psychological knowledge but also the means of obtaining knowledge. Performing an investigation with human subjects may occasionally require an ethical decision concerning the balance between the interests of the subject and the humane or scientific value of the research.

Psychologists require an atmosphere of free inquiry and communication without misrepresentation of their knowledge and methods by others. Psychologists must match this freedom with ethical concern, competence, objectivity and the non-wasteful use of material resources and human resources. Psychologists have an obligation to prevent misuse through personal influence, public statement and professional sanction. Psychologists can and should promote the public understanding of psychological knowledge in such a way as to prevent its misuse or render misuse ineffective.

The psychologist has a general obligation to make the results of his research available to other psychologists, to related scientists, and to allied professions. No psychologist should seek to restrict the availability or publication of his own or colleagues' research without seeking the opinion of experienced and disinterested colleagues. Until such publication has permitted the verification of results and the evaluation of their apparent implications by the scientific community, psychologists have an obligation to resist the premature citation of results in wider discussions on policy, and especially their premature use in policy formulation. This general principle does not prevent a psychologist from undertaking explicitly confidential research on restricted topics (e.g. for commercial development or national security) where that research does not violate these principles.

The following set of ethical principles is issued by the British Psychological Society in the belief that a detailed list of prescribed and proscribed procedures would be impractical. It is the Society's belief that the degree of awareness and responsibility that follows from adherence to this general set of principles will serve to raise standards in psychological science and will safeguard the welfare of human subjects who contribute to it. While it would be appropriate to use this set of principles as an indication of the level of awareness that a psychologist should display, the psychologist's compliance with these principles can only be determined by those of his peers who are experienced with the problems which the principles encompass. Accordingly, the principles should not be used as a substitute for a considered judgement in which a case is examined on its merits in all aspects. The principles place reliance upon the opinion of the psychological community as an extension of the individual investigator's ability to anticipate the ethical issues raised and to assess the extent to which any consequences for the subject may be serious. The opinion of colleagues should also assist the investigator in determining whether the research is justified scientifically or pragmatically.

Scientific justification involves the assessment of both the conceptual importance of the potential results and their usefulness to mankind. Pragmatic justification involves assessing, for example, the likely effects of participants' guesses about the objectives of the research upon public attitudes to psychological inquiry in general and upon local voluntary participation in particular.

1. Whenever possible the investigator should inform the subjects of the objectives, and, eventually, the results of the investigation. Where this is not possible the investigator incurs an obligation to indicate to the subject the general nature of the knowledge achieved by such research and its potential value to people, and to outline the general values accepted by psychologists as listed in the introduction to these principles. The investigator's name, status and employer or affiliation should be declared.

2. In all circumstances the investigator must consider the ethical implications and the psychological consequences for his subjects of the research he is carrying out. The investigator must actively consider, by proper consultation, whether local cultural variations, special personality factors in the subjects or variations in his procedure from procedures reported previously may introduce unexpected problems for the subject.

3. An investigator should seek the opinion of experienced and disinterested colleagues whenever his research requires or is likely to involve:

(a) Deception concerning the purpose of the investigation or the subject's role in it.
(b) Deception concerning the basis of subject selection.

Figure 34.2. British Psychological Society's 'Ethical Principles for Research with Human Subjects' 1978.

(c) Psychological or physiological stress.

(d) Encroachment upon privacy.

Geographical and institutional isolation of the investigating psychologist increases rather than decreases the need to seek colleagues' opinions.

4. Deception of subjects, or withholding of relevant information from them, should only occur when the investigator is satisfied that the aims and objects of his research or the welfare of his subjects cannot be achieved by other means. Where deception has been necessary, revelation should normally follow participation as a matter of course. Where the subject's behaviour makes it appear that revelation could be stressful or, when to reveal the objectives or the basis of subject selection would be distressing, the extent and timing of such revelation should be influenced by consideration for the subject's psychological welfare. Where deception has been substantial, the subject should be offered the option of withholding his data, in accordance with the principle of participation by informed consent.

5. In proportion to the risks of stress or encroachment upon privacy the investigator incurs an obligation to emphasize to the subject at the outset his volunteer status and his right to withdraw, irrespective of whether or not payment or other inducement is offered, and to describe precisely the demands of the investigation.

   Wherever a situation turns out to be more stressful for an indivudual subject than anticipated by the investigator or than might be reasonably expected by the subject from his introduction, the investigator has an obligation to stop the investigation and consult an experienced and disinterested colleague before proceeding.

6. In proportion to the risks under 3 (a)—(d) and to the personal nature of the information involved the investigator incurs an obligation to treat data as confidential and to conceal identities when reporting results.

7. Studies on non-volunteers, based upon observation or upon records (whether or not explicitly confidential) must respect the privacy and psychological well-being of the subjects.

8. Investigators have the responsibility to maintain the highest standards of safety in procedure, equipment and premises.

9. Where research involves infants and young children as subjects, consent should be obtained from parents or from those *in loco parentis,* according to the foregoing principles. In the case of children of appropriate age, the informed consent of subjects themselves should also be obtained in advance. In research involving children caution should be exercised when discussing results of research with parents, teachers or others *in loco parentis* since evaluative statements may carry unintended weight.

10. If a subject solicits advice concerning educational, personality or behavioural problems, extreme caution should be exercised and if the problem is serious the appropriate source of professional advice should be recommended.

11. It is the investigator's responsibility to ensure that research executed by associates, employees or students conforms in detail to the ethical decision taken in the light of the foregoing principles.

12. A psychologist who believes that another psychologist or related investigator may be conducting research not in accordance with the foregoing principles has the obligation to encourage the investigator to re-evaluate the research in their light, if necessary consulting a responsible senior colleague as a source of further opinion or influence.

Figure 34.2. Continued

Working groups should formalise assessments in this area, to be sure that their working attitudes never drift into a perception of their work as dealing with people as objects.

Of course, it is not always simple to see the ethical problems. In a discussion dealing with moral questions which arise during the design and implementation of computer systems, Pullinger (1989) points to the difficulties in this respect of a programmer concerned with providing only one part of a complex body of software. The increased specialization reduces the opportunities for understanding and reflection on the functioning of the total system, and hence the opportunities for individuals to assess their contributions in terms of their own moral outlook.

Pullinger raises a number of points which are important for these workers developing computer interfaces, but he points out that the exercise of ethical judgement requires a wide understanding of the technological limits, the system's purposes and people's social and personal needs. It also requires a position to be taken on the forms of society which are acceptable, for example on the question of security and confidentiality of information. He concludes that computer specialists are not in a position to exercise such judgements, primarily because their information base is deficient in much of the requisite knowledge. A group better able to consider this area are those who have moved into the Human–Computer Interface area from ergonomics, occupational psychology or the like.

This recognition, that good work cannot contribute to a good society unless it is informed by more than technological competence, is one which we stress here. The presentation of information from an investigation which is clear as to its meaning and its limitations is as important as Pullinger's point that the designer of an expert system has a duty to make it clear that the system has limits, which will vary with the user's expertise, and that to transfer judgement from the individual to the system is not appropriate.

Conclusion

It will be evident that this book has but scratched the surface. It could have been more detailed on many methods which are mentioned, there could have been other areas introduced which are important for ergonomists, and there is much room for expansion in the ways that methods may be chosen, used and their results interpreted. What the Editors hope will be one result of this text is that there will be an increased awareness of the need to select methods in relation to the model of the system as seen by the investigator, and the careful selection of a minimum number of methods sufficient for the investigator's purposes. Of course, experimentation is expensive and if more data of relevance can be gathered without damaging a study, then they should be gathered. However we trust that the user of the book will be able to avoid a touching faith in the value of 'high tech' equipment rather than an intelligent use of the minimum necessary technology. We also hope that, no matter what the readers' opinion of the book may be, they will not throw it at the problem! The relevant analogy for experimental work is the precision of the scalpel rather than the universal attack of the shotgun. The latter is often combined with a blind faith in statistical packages to solve the problems of thoughtless data collection. If our text has contributed to the efficiency of experimentation, the clarity of understanding the results and not least a recognition of the important contributions which ergonomics investigations make to the well-being of individuals and society, we will be well content.

References

Corlett, E.N. (1988). Cost Benefit Analysis. *International Reviews of Ergonomics*, **2**, 85–104

Pullinger, D.J. (1989). Moral judgement in designing better systems. *Interacting with Computers*, **1**, 93–104.

Contributors

Jane Astley is a consultant with Technica Ltd., London. She has worked in industry as an ergonomist and as an Aston-based Ph.D. student was part of the human factors group at the Department of Trade & Industry Warren Spring Laboratory. There she developed task analysis methods for display design in process control environments. Her current interests are in the ergonomics of complex system design and human reliability. She is registered with the Ergonomics Society as a practitioner.

Lisanne Bainbridge is Reader in Psychology at University College London. Most of her research has been on industrial process operation, on the cognitive mechanisms of planning activities and using knowledge, including their implications for mental workload and advanced interface design. Much of this work has been based on analysis of verbal protocols. In 1976 she was awarded the Bartlett Medal of the Ergonomics Society.

Maurice Bonney is Professor of Production Management and Head of the Department of Production Engineering & Production Management at the University of Nottingham. His research areas have centred around the development of computer aids in the field of robot simulation, computer aided ergonomics design, computer aided work study, line balancing and computer aided production management. He is a director of BYG Systems Ltd and SAMMIE CAD Ltd.

Rosemary Bonney graduated in Ergonomics from Loughborough University. She then undertook Ph.D. research at Nottingham University, specialising in the effects of vibration and other factors on spinal load. She is now employed as an ergonomist at the Vermont Rehabilitation Engineering Centre, Burlington, Vermont, USA.

Ivan Brown is Assistant Director of the Medical Research Council's Applied Psychology Unit at Cambridge and Extra-mural Professor of Traffic Science at the University of Groningen, in the Netherlands. He has researched accident causation and prevention for some thirty years, specializing in behavioural aspects of road safety.

Mark A. Bullimore received his B.Sc. in Ophthalmic Optics (Optometry) in 1983, and his Ph.D. in 1987, both from Aston University, Birmingham.

He is presently a Research Scientist at the School of Optometry, U.C. Berkeley. He has published papers in the areas of ocular accommodation, myopia, visual performance in ocular disease, and visual optics.

Mike Burton is Lecturer in Psychology at the University of Nottingham, where he also received his Ph.D. His principal research interests are Knowledge Acquisition and the Modelling of Human Cognitive Processes. He has published extensively in both these areas.

Michael Carey is a consultant with RM Consultants Ltd. Warrington. He was previously a contract research officer at Aston University. At Aston he carried out research on interface design in industrial process control systems for the Warren Spring Laboratory and investigated user modelling techniques in command and control contexts on a Ministry of Defence sponsored project. He is registered with the Ergonomics Society as a practitioner.

Dr **Keith Case** is a Senior Lecturer in Computer Aided Engineering in the Department of Manufacturing Engineering at Loughborough University of Technology. His current research interests include the study of user interfaces to CAD systems and their effect upon learning, the development of learning aids for CAD, off-line programming of robots by CAD and configuration tools for modular machines and highly integrated CAD/CAM systems. He is a director of SAMMIE CAD Ltd.

Dr. **Bruce Christie** lectures in information technology at the City of London Polytechnic and is active as an I.T.consultant through HeptaCon Ltd. He is a Member of the Institute of Management Consultants, a Chartered Psychologist and a Fellow of the British Psychological Society. He has a number of publications and patents in I.T. and human factors.

Professor **Nigel Corlett** has recently retired as Head of the Department of Production Engineering and Production Management at Nottingham University. He is Scientific Adviser to the University's Institute for Occupational Ergonomics. Prior to this he was Professor of Industrial Ergonomics at the Department of Engineering Production at Birmingham University, joining that Department in 1957. His previous posts included factory management and head of Design and Development for a large domestic equipment company. He has a D.Sc. from London University and is a Fellow of the Ergonomics Society, the Human Factors Society and the Fellowship of Engineering.

Tom Cox is Professor of Organizational Psychology and Director of the Centre for Organizational Health in the Department of Psychology at Nottingham University. He is a chartered occupational psychologist and a Fellow of the British Psychological Society and the Royal Society of Health. Tom is senior partner in the consultancy, Maxwell & Cox Associates and Managing Editor of the international journal *Work & Stress*. He has published extensively in relation to occupational health and stress.

Colin G. Drury is Professor of Industrial Engineering at State University of New York at Buffalo. He is also Executive Director of The Center for Industrial Effectiveness, an organization which works with industry to define competitive needs, and address those needs with University expertize. He is a Fellow of the Ergonomics Society and the Human Factors Society, and was awarded the Bartlett medal in 1981. He volunteers too readily to write book chapters.

Professor **Ken. D. Eason** is Head of the Department of Human Sciences at Loughborough University of Technology and Co-director of the HUSAT Research Centre. Over a 20-year period he has researched the impact of computer systems upon their users. His doctoral studies investigated the impact of systems upon managers. Latterly, he has researched the system design process in order to identify methods by which the characteristics of users and their organisations can be better recognised by design teams. Professor Eason has published two books '*Managing computer impact*' and '*Information technology and organisational change*'.

Jane Fulton is currently Director of Human Factors at IDTWO, an Industrial Design Company in San Francisco, California. She came to IDTWO from the Institute for Consumer Ergonomics, Loughborough, where she worked as a Senior Research Officer for 9 years. She has degrees in Psychology and Architecture.

Dr. **Margaret Gardiner** is a Director of HeptaCon Ltd., an information technology and human factors consultancy serving clients in the public and private sectors internationally. She is a Chartered Psychologist and an Associate Fellow of the British Psychological Society. Her human factors experience includes several R&D projects under ESPRIT, human factors training for engineers, and product development work.

Peter Hancock received his B. Ed. and M.Sc. degrees from Loughborough University where his work concerned modelling of physiological systems. He received his Ph.D. from the University of Illinois where his research focused upon time perception. He has followed broad interests in human factors with particular concern for the energetic aspects of operator performance in association with complex systems. He is the editor of *Human Factors Psychology*, and co-editor of *Human Mental Workload* with his colleague Najmedin Meshkati. Hancock's most recent text is a co-edited book with Mark Chignell entitled *Intelligent Interfaces: Theory, Research and Design*. He has recently left his position as an Assistant Professor at the University of Southern California to take up an Associate Professorship at the University of Minnesota in Minneapolis.

Dr. **James Hartley** is Reader and Head of the Department of Psychology at the University of Keele. He has published widely in the field of text design, educational technology, and university teaching.

Christine Haslegrave is a lecturer in Occupational Ergonomics at the University of Nottingham, currently engaged on research into the biomechanical demands of manual handling tasks as well as working with the Institute for Occupational Ergonomics on health and safety problems in industry. She was Head of the Ergonomics Section at the Motor Industry Research Association for several years, with interests in vehicle safety and ergonomic legislative testing.

Peter Howarth received a B.Sc. in Ophthalmic Optics (Optometry) from The City University, London; an M.Sc. in Ergonomics from Loughborough University; and a Ph.D. in Physiological Optics from the University of California at Berkeley. He is presently a Research Scientist at the Lawrence Berkeley Lighting Laboratory, School of Optometry, U.C. Berkeley, where he is conducting human factors research in lighting. As well as visual ergonomics, his vision research interests include the pupillary system, and the chromatic aberration of the eye.

Åsa Kilbom is Professor of Work Physiology at the Swedish Institute of Occupational Health in Stockholm, and head of its Applied Work Physiology Division. After a degree in medicine at the Karolinska Institute, she trained as a specialist in clinical physiology. Her research has been on physical training, physiological effects of static exercise and cardiovascular demands in occupational work. During the last 5–6 years her main focus has been occupational musculoskeletal disorders, especially their multifactorial causation and the quantitative relationship between occupational exposure and severity of symptoms. She is a member of the editorial board of several international scientific journals and has contributed in numerous scientific publications and textbooks.

Barry Kirwan studied psychology (B.Sc.) and ergonomics (M.Sc.), specialising in human reliability. In various consultancies he has worked on offshore, marine, chemical, nuclear power and service sector projects involving human factors and human reliability assessment. He is currently running a large human factors and reliability programme at British Nuclear Fuels Limited, and is developing and validating a prototype system for comprehensive human reliability assessment.

Steve Mason obtained a first degree in Engineering from the University of Leicester, followed by a Masters degree in Work Design and Ergonomics from the University of Birmingham. On completing his Masters degree he worked for a number of years as an ergonomist in the Methods Study Dept. of the General Electric Company. He has spent the last 13 years working on mining ergonomics and is currently Deputy Head of the Ergonomics Branch, British Coal Technical Department.

Ian McClelland first trained as a Mechanical Engineer and then moved to Loughborough University in 1970 where he completed the post-graduate course in Ergonomics. In 1972 he joined the Institute for Consumer Ergonomics where he undertook research and consultancy work for a wide range of clients within industry, commerce and government. In 1986 he joined Corporate Industrial Design, Philips as Manager of the Applied Ergonomics group. Interests include the integration of usability engineering principles into user-interface design, simulation tools for user-interface design development, and product evaluation methods.

Ted Megaw graduated from Cambridge University in 1964 with a degree in Experimental Psychology. He subsequently obtained an M.Sc. in Work Design and Ergonomics from Birmingham University where he then stayed to complete a Ph.D. in 1970 on manual control. He continued research into eye movement control, visual search, industrial inspection, visual fatigue and ergonomics databases. Since 1974 he has been a lecturer in Ergonomics at Birmingham University where he is also now Director of the Ergonomics Information Analysis Centre.

Najmedin Meshkati received a B.Sc. in Industrial Engineering, a B.A. in Political Science and Economics, a M.Sc. in Engineering Management and a Ph.D. in Industrial and Systems Engineering, the last from the University of Southern California, in 1983. His dissertation on mental workload measurement won the 1983 Phi Beta Kappa Alumni Award for Research and Innovation. His research interests include mental workload assessment, micro- and macroergonomics of complex, large-scale technological systems, human factors of technology transfer, and ergonomics of developing countries. He is co-editor of *Human Mental Workload* with his colleague Peter A. Hancock. Presently, he is an Assistant Professor of Human Factors at the Institute of Safety and Systems Management and Lecturer at the Department of Industrial and Systems Engineering at the University of Southern California. He is a recipient of the Presidential Young Investigator Award (PYI) from the National Science Foundation in 1989.

Dr. **David Meister** is presently a Senior Scientist at the Naval Ocean Systems Center, San Diego, California. He is the author of 7 textbooks on various aspects of human factors, including system development, testing and methodology. He is a former President of the Human Factors Society and the 1984 winner of the Franklin V. Taylor award for outstanding contributions to engineering psychology, given by the American Psychological Association.

Ken Parsons is a senior lecturer in ergonomics within the Department of Human Sciences at Loughborough University. He has a B.Sc. (Hons) in ergonomics from Loughborough, a postgraduate certificate in education from Cambridge University and a Ph.D. from Southampton University. He is chairman of the British Standards Institute committee concerned with ergonomics of the thermal environment and acts as principal U.K. expert on C.E.N. and I.S.O. committees. He is Director of the

Human Modelling Group within the Department of Human Sciences at Loughborough.

Dr. **Stephen Pheasant** lectures in anatomy and ergonomics at the Royal Free Hospital School of Medicine, London. He is also an honorary consultant at the Robens Institute Industrial and Environmental Safety Centre at Surrey University. His books include *Bodyspace, Anthropometrics: an Introduction* and *Ergonomics – Standards and Guidelines for Designers.*

Dr. **Mark Porter** is a Senior Lecturer in Ergonomics in the Department of Human Sciences at Loughborough University of Technology. He is also the Head of the Vehicle Ergonomics Group in this department and a director of SAMMIE CAD Ltd.

Mansour Rahimi has a B.Sc. and an M.Sc. in Industrial Engineering and a Ph.D. in Industrial Engineering and Operations Research (with specialization in Human Factors Engineering) from Virginia Polytechnic Institute and State University. He is an Associate Professor of Safety Science and has been teaching in different areas of system safety and ergonomics. His current research interests are in human reliability, risks associated with computerized technology, and rehabilitation engineering. He is the Chair of Technical Programs for the *Second International Conference on Human Aspects/Ergonomics of Advanced Manufacturing and Hybrid Automated Systems.*

Nigel Shadbolt is Lecturer in Psychology at the University of Nottingham. After receiving his Ph.D. in Artificial Intelligence from Edinburgh University, he moved to Nottingham in 1984 when he founded the A.I. group. He has published and researched extensively in the areas of knowledge acquisition, expert systems and planning.

Pat Shipley is Reader in Occupational Psychology at Birkbeck College, London and Director of the Stress Research and Control Centre there. She had several years industrial management experience before taking up an academic career. She is a Fellow of the British Psychological Society and of the Ergonomics Society.

Geoff Simpson obtained a first degree in Occupational Psychology from the University of Wales, followed by a Masters degree in Ergonomics from the University of London. He began work with the Human Factors Section of British Steel, followed by a period lecturing in the Psychology Dept. at the University of Nottingham. Since leaving Nottingham, he has spent the last 13 years working primarily on the ergonomics of mining operations and is currently Head of Ergonomics Branch, British Coal Technical Department. He is a Chartered Psychologist, a Fellow of the British Psychological Society, a Fellow of the Ergonomics Society and a member of the Institution of Occupational Safety and Health.

Murray Sinclair became an Ergonomist in 1966, graduating with an M.Sc. in Ergonomics from Loughborough University. He worked in industry for a few years, before joining Loughborough University in 1970 as a Lecturer in Ergonomics. His research work has been in industrial applications in Ergonomics, and has recently included major collaborative projects such as the Alvey *Design to Product* large-scale demonstrator, and *CADCAM in the automotive industry* in RACB, a commission of the European Community programme in advanced communications. Recently he has joined the HUSAT Research Centre as a Principal Consultant.

Rob Stammers is a Lecturer in Applied Psychology at Aston University. His current research interests are in task analysis, human–computer interaction and training. He has carried out research projects for the Department of Trade & Industry, the Ministry of Defence, Alvey, the Central Electricity Generating Board and the Economics and Social Research Council. He is a Fellow of the Ergonomics Society and the British Psychological Society and a Chartered Psychologist. In 1989 he won the Otto Edholm Award of the Ergonomics Society. He is Director of User Technology Ltd., Aston Science Park.

Moira Tracy studied for a physics degree in Liège, followed by some teaching and a year at the Antwerp University sleep-laboratory, where she was involved with a *Spacelab* project. She has since been working in biomechanics at Nottingham University, obtaining her Ph.D. in 1988.

John Wilson trained first in Engineering and Management and received his Ph.D. in Work Design and Ergonomics from Birmingham University. He has held academic teaching appointments at Birmingham University, Nottingham University and University of California, Berkeley. His research interests include human factors in manufacturing and development of methods. He is Scientific Editor of *Applied Ergonomics*, and author of a number of ergonomics books.

Author Index

Subject Index

Page numbers in bold denote the first page of the chapter on that particular subject. Where subjects are referred to on a number of consecutive pages, only the first one is indicated.

876